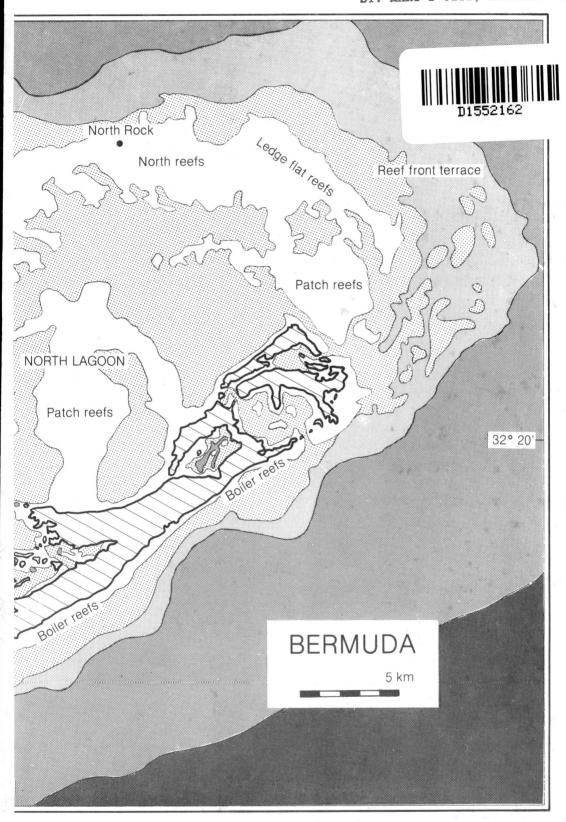

# MARINE FAUNA
# AND FLORA OF BERMUDA

# MARINE FAUNA AND FLORA OF BERMUDA

A Systematic Guide to the Identification of Marine Organisms

*Edited by*
**WOLFGANG STERRER**

*Bermuda Biological Station
St. George's, Bermuda*

*in cooperation with*
**Christiane Schoepfer-Sterrer**
*and 63 text contributors*

A Wiley-Interscience Publication
**JOHN WILEY & SONS**
**New York** · **Chichester** · **Brisbane** · **Toronto** · **Singapore**

Copyright © 1986 by John Wiley & Sons, Inc.

Illustrations © 1986 by Wolfgang Sterrer.

All rights reserved. Published simultaneously in Canada.

Reproduction or translation of any part of this work beyond that permitted by Section 107 or 108 of the 1976 United States Copyright Act without the permission of the copyright owner is unlawful. Requests for permission or further information should be addressed to the Permissions Department, John Wiley & Sons, Inc.

*Library of Congress Cataloging in Publication Data:*

Sterrer, Wolfgang.
  Marine fauna and flora of Bermuda.

  "A Wiley-Interscience publication."
  Bibliography: p.
  Includes index.
  1. Marine fauna—Bermuda—Identification. 2. Marine flora—Bermuda—Identification. I. Schoepfer-Sterrer, Christiane. II. Title.

QH110.S74   1985       574.92′463       85-10600
ISBN 0-471-82336-8

Printed in the United States of America

10 9 8 7 6 5 4 3 2 1

*To*
RUPERT J. M. RIEDL,
who has shaped this book as much as he has shaped our lives

# CONTRIBUTORS

CHARLES ARNESON, Scripps Institution of Oceanography, LaJolla, California, United States

ILSE BARTSCH, University of Hamburg, Hamburg, West Germany

BODO VON BODUNGEN, Institute of Marine Science, Kiel, West Germany

TORLEIV BRATTEGARD, Biological Station, Espegrend, Norway

JAMES BURNETT-HERKES, Department of Agriculture and Fisheries, Paget, Bermuda

STEPHEN CAIRNS, National Museum of Natural History, Smithsonian Institution, Washington, D.C., United States

DALE CALDER, Royal Ontario Museum, Toronto, Ontario, Canada

RICHARD CASTENHOLZ, Department of Biology, University of Oregon, Eugene, Oregon, United States

A. RALPH CAVALIERE, Department of Biology, Gettysburg College, Gettysburg, Pennsylvania, United States

A. FENNER CHACE, JR., National Museum of Natural History, Smithsonian Institution, Washington, D.C., United States

KERRY B. CLARK, Florida Institute of Technology, Melbourne, Florida, United States

ROGER CRESSEY, National Museum of Natural History, Smithsonian Institution, Washington, D.C., United States

ROGER J. CUFFEY, Department of Geosciences, Pennsylvania State University, University Park, Pennsylvania, United States

CHRISTIAN C. EMIG, Marine Station D'Endoume, Marseille, France

# CONTRIBUTORS

CHRISTER ERSÉUS, Department of Zoology, University of Göteborg, Göteborg, Sweden

SHIRLEY FONDA, Department of Geosciences, Pennsylvania State University, University Park, Pennsylvania, United States

JANE A. FRICK (deceased), Alpine, New Jersey, United States

STEPHEN GARDINER, Department of Zoology, University of North Carolina, Chapel Hill, North Carolina, United States

OLAV GIERE, Zoological Institute and Museum, University of Hamburg, Hamburg, West Germany

MASON HALE, JR., National Museum of Natural History, Smithsonian Institution, Washington, D.C., United States

MICHAEL G. HARASEWYCH, National Museum of Natural History, Smithsonian Institution, Washington, D.C., United States

COOS DEN HARTOG, Museum of Natural History, Leiden, Netherlands

EIKE HARTWIG, Zoological Institute and Museum, University of Hamburg, Hamburg, West Germany

CHRISTOPH HEMLEBEN, Geological Institute, University of Tübingen, Tübingen, West Germany

ROBERT P. HIGGINS, National Museum of Natural History, Smithsonian Institution, Washington, D.C., United States

IDWAL W. HUGHES, Department of Agriculture and Fisheries, Paget, Bermuda

LEONARD C. IRELAND, Marine Biological Laboratory, Woods Hole, Massachusetts, United States

RUSSELL H. JENSEN, Delaware Museum of Natural History, Greenville, Delaware, United States

H. WILLIAM JOHANSEN, Department of Biology, Clark University, Worcester, Massachusetts, United States

SAMUEL E. JOHNSON, Nature Conservancy, Portland, Oregon, United States

MEREDITH L. JONES, National Museum of Natural History, Smithsonian Institution, Washington, D.C., United States

ERNST KIRSTEUER, The American Museum of Natural History, New York, New York, United States

ERIKA KOHLMEYER (deceased), Institute of Marine Science, Morehead City, North Carolina, United States

JAN KOHLMEYER, Institute of Marine Science, Morehead City, North Carolina, United States

LOUIS KORNICKER, National Museum of Natural History, Smithsonian Institution, Washington, D.C., United States

REINHARDT M. KRISTENSEN, Institute of Cell Biology and Anatomy, University of Copenhagen, Copenhagen, Denmark

PIERRE LASSERRE, Biological Station of Roscoff, Roscoff, France

SUSAN M. LESTER, Department of Biology, Clemson University, Clemson, South Carolina, United States

ROSALIE F. MADDOCKS, Geology Department, University of Houston, Houston, Texas, United States

RAYMOND B. MANNING, National Museum of Natural History, Smithsonian Institution, Washington, D.C., United States

EDWARD A. MANUEL, Department of Agriculture and Fisheries, Paget, Bermuda

JOHN C. MARKHAM, Arch Cape Marine Laboratory, Arch Cape, Oregon, United States

JOHN J. MCDERMOTT, Department of Biology, Franklin and Marshall College, Lancaster, Pennsylvania, United States

PATSY A. MCLAUGHLIN, School of Marine

and Atmospheric Science, University of Miami, Miami, Florida, United States

CLAUDE MONNIOT, National Museum of Natural History, Paris, France

FRANÇOISE MONNIOT, National Museum of Natural History, Paris, France

BYRON F. MORRIS, Outer Continental Shelf Office, Bureau of Land Management, Anchorage, Alaska, United States

LOUIS MOWBRAY (deceased), Bermuda Aquarium, Museum and Zoo, Flatts, Bermuda

CLARISSE ODERBRECHT, Department of Oceanography, Rio Grande, Brazil

JÖRG OTT, Institute of Zoology, University of Vienna, Vienna, Austria

DAVID L. PAWSON, National Museum of Natural History, Smithsonian Institution, Washington, D.C., United States

MARY E. PETERSEN, The University Zoological Museum, Copenhagen, Denmark

JOHN F. PILGER, Department of Biology, Agnes Scott College, Decatur, Georgia, United States

THOMAS G. RAND, Bermuda Aquarium, Museum and Zoo, Flatts, Bermuda

MARY E. RICE, Smithsonian Marine Station at Link Port, Fort Pierce, Florida, United States

REINHARD M. RIEGER, Institute of Zoology, University of Innsbruck, Innsbruck, Austria

CLYDE F. E. ROPER, National Museum of Natural History, Smithsonian Institution, Washington, D.C., United States

KLAUS RÜTZLER, National Museum of Natural History, Smithsonian Institution, Washington, D.C., United States

LUITFRIED VON SALVINI PLAWEN, Institute of Zoology, University of Vienna, Vienna, Austria

GEORGE A. SCHULTZ, Hampton, New Jersey, United States

REINHART SCHUSTER, Institute of Zoology, University of Graz, Graz, Austria

ALAN J. SOUTHWARD, The Laboratory, Plymouth, England, United Kingdom

EVE C. SOUTHWARD, The Laboratory, Plymouth, England, United Kingdom

MICHAEL SPINDLER, Alfred Wegener Institute for Polar Research, Bremerhaven, West Germany

DON C. STEINKER, Department of Geology, Bowling Green State University, Bowling Green, Ohio, United States

WOLFGANG STERRER, Bermuda Biological Station, St. George's, Bermuda

DAVID WINGATE, Bermuda Aquarium, Museum and Zoo, Flatts, Bermuda

# FOREWORD

The appearance of this book marks the first publication of a detailed, well-illustrated, American field guide to an extensive marine flora and fauna. It is a significant event for both the laymen and scientists to whom it is directed. Over 1500 species, most of which because of technical problems and scattered literature have been difficult to identify, will be made far more accessible. A large number of these species have never before been included in a general guide. Since the marine life of Bermuda broadly overlaps that of the West Indies and that of the southeastern United States, the guide will have use well beyond the Island of Bermuda. The impressive roster of specialists who have contributed sections and the care with which the format and illustrations have been organized and put together give this work an enviable authority and distinction.

ROBERT D. BARNES

*Professor of Biology*
*Gettysburg College*

# PREFACE

> "Là, tout n'est qu'ordre et beauté . . ."
> (Charles Baudelaire, *Les Fleurs du Mal*)

To reveal the beauty and order in marine life by introducing the reader to the organisms that inhabit the ocean around Bermuda is the purpose of this book. It has been designed as a guide for all those—scientists, teachers, students and other nature lovers—who make it their business or pleasure to explore the unfamiliar shapes, colors and motions that await them beneath the surface of the sea.

More than 1,500 species are described and illustrated. Selected to represent the majority of species a serious explorer is ever likely to encounter in the Bermuda area, they also cover the entire range of marine organisms—plants and animals, large or small, from the intertidal to the deep sea—in a systematic manner, beginning with single-celled blue-green bacteria and including the giant whales.

The natural system, i.e., the hierarchy that best reflects the evolutionary history and relatedness of organisms, provides the structure for the book and its 114 chapters. Definitions for 1,200 categories above the genus level, from family through phylum, make up the framework for an orderly representation of the diversity of marine life. Rather than being an expedient means of assigning empty names to living creatures, species identification thus becomes a learning process that retraces both evolution and our understanding of it.

The ocean around Bermuda lends itself well to such a comprehensive task: of the

56 major groups (phyla) into which living organisms can be divided, 45 occur in the ocean, and of these only 4 have not yet been recorded in local waters. Because Bermuda's marine flora and fauna represent a significant proportion of that found in the Caribbean, and because of the systematic way in which information is presented, including more than 800 references, the book should be useful as an identification guide and an introduction to systematic marine biology throughout the tropical West Atlantic.

The language is necessarily that of the scientist who has described and evaluated each organism and its parts. Yet care has been taken to explain, in the chapters or the glossary, all but the most general terms. The text therefore tries to strike a balance between meeting the specialist's demands for completeness and accuracy, and the needs of the naturalist for simplicity.

The 2,870 illustrations should satisfy both. Black-and-white drawings were prepared from a combination of live and preserved specimens, color slides, and published material, preferably in that order; color plates were selected from more than 5,000 photographs. All illustrations emphasize the appearance of the living organism.

Ten years of research, writing and editing, and the patient cooperation from more than 100 scientists, artists, volunteers and donors went into the making of this book. I am grateful to all of them for their help and faith. In addition to the chapter authors, the following colleagues contributed unpublished information, collected or identified specimens, or commented on parts of the manuscript: Angeles Alvarino, Frederick M. Bayer, John Berrill, Douglas Biggs, Brian P. Boden, Thomas E. Bowman, Robert Bullock, John H. Cardellina II, Clayton B. Cook, Arthur Guest, Thomas M. Iliffe, Sue Jickells, Jean Kenyon, Thomas D. Sleeter, Robin South, Diane Stoecker, William Randolph Taylor and Bernhard Werner. Jack Lightbourn's many years of collecting from R/V North Star have greatly enriched our knowledge of organisms inhabiting the flanks of the Bermuda seamount.

The work of 12 artists was largely coordinated by Christiane Schoepfer-Sterrer, who also executed the majority (57%) of the black-and-white drawings; the rest are by Mary White (21%), Mary Redfern (10%), Molly Kelly Ryan (7%), and Christine Phillips, Emmy Hoffmann, G. Kristine Jensen, Elizabeth Stehli, Dorset Trapnell, Eldon Trimingham, Tina Hutchings and David Faroo. G. Kristine Jensen took most of the 212 color photographs and assembled them into plates; the remaining pictures are by Klaus Ruetzler (3), Oxford Scientific Films (1) and myself.

The final manuscript was patiently wordprocessed by Zina Francis, with Jill Cadwallader and Margaret Emmott lending a hand at various stages. Ursula Ison checked and edited the references; Susan M. Lester, Margaret Carroll and Rebecca Day assisted in proofreading.

Mary Conway and her colleagues at John Wiley & Sons are particularly appreciated for providing prompt and expert assistance when needed but otherwise letting the project take its own course and pace.

Financial support was given by the Robert Sterling Clark Foundation, Mrs. Claire B. Jonklaas, and a number of donors who contributed toward the production of color plates: The Bank of Bermuda Ltd.; The Bank of Butterfield, Ltd.; L. P. Gutteridge Ltd.; Appleby, Spurling & Kempe, Ltd.; Shell Company of Bermuda; Electronic Communications—The K&J Group; Mr. William de V. Frith; Sir James and Lady Pearman; and the Trustees of the Bermuda Biological Station, in particular Mr. and Mrs. Gerard M. Ives, Mr. and Mrs. William T. Kemble, Dr. Rudolph F. Nunnemacher and Dr. W. Redwood

Wright. The Bermuda Biological Station's fellowship program and grants from the National Science Foundation enabled many authors to do the fieldwork for their contributions.

I finally want to thank Associate Director Anthony H. Knap and my fellow workers as well as President W. Redwood Wright and the trustees of my home institution for their unfaltering confidence that my frequent disappearances from the office to the laboratory would eventually result in something tangible and, I hope, useful.

WOLFGANG STERRER

*St. George's, Bermuda*
*Summer 1984*

# CONTENTS

| | |
|---|---|
| LIST OF BLACK-AND-WHITE PLATES | xxi |
| LIST OF COLOR PLATES | xxix |
| INTRODUCTION  *W. Sterrer* | 1 |
| LIVING ORGANISMS  *(W. Sterrer)*\* | 7 |
| **KINGDOM MONERA** (Bacteria and blue-green algae)  *(W. Sterrer)* | 9 |
| **Phylum Cyanobacteria ( = Cyanophyta) (Blue-green algae)**  R. W. Castenholz | 9 |
| **KINGDOM FUNGI**  *(J. Kohlmeyer & E. Kohlmeyer)* | 17 |
| **Phylum Eumycota (True fungi)**  *J. Kohlmeyer & E. Kohlmeyer* | 17 |
| Appendix: Lichenes (Lichens)  *M. E. Hale* | 24 |

\*Authors of Section Introductions are in parentheses.

## KINGDOM PLANTAE (Plants)  *(W. Sterrer)*  26

**Phylum Chlorophyta (Green algae)**  A. R. Cavaliere  27
**Phylum Phaeophyta (Brown algae)**  A. R. Cavaliere  41
**Phylum Rhodophyta (Red algae)**  A. R. Cavaliere & H. W. Johansen  50
**Phylum Chrysophyta**  *(B. v. Bodungen & C. Oderbrecht)*  66
   Class Bacillariophyceae (=Diatomeae) (Diatoms)  B. v. Bodungen & C. Oderbrecht  67
   Class Chrysophyceae (Golden-brown algae)  B. v. Bodungen & C. Oderbrecht  70
   Class Xanthophyceae (Yellow-green algae)  B. v. Bodungen & C. Oderbrecht  73
**Phylum Euglenophyta**  B. v. Bodungen & C. Oderbrecht  74
**Phylum Dinophyta (=Dinoflagellata, Pyrrhophyta) (Dinoflagellates)**  B. v. Bodungen & C. Oderbrecht  75
**Phylum Anthophyta (=Angiospermae) (Flowering plants)**  E. A. Manuel  79

## KINGDOM ANIMALIA (Animals)  *(W. Sterrer)*  92

### Subkingdom Protozoa (Single-celled animals)  *(W. Sterrer)*  92

**Phylum Sarcodina**  M. Spindler, C. Hemleben & D. Steinker  93
**Phylum Sporozoa**  M. Spindler  103
**Phylum Ciliophora (Ciliates)**  E. Hartwig  104

### Subkingdom Metazoa (Many-celled animals)  *(W. Sterrer)*  110

**Phylum Porifera (Sponges)**  K. Rützler  111
**Phylum Cnidaria**  *(W. Sterrer)*  127
   Class Hydrozoa  *(D. R. Calder)*  127
     Hydroida (Hydroid polyps)  D. R. Calder  128
     Hydromedusae  D. R. Calder  140
     Order Siphonophora  D. R. Calder  149
   Class Scyphozoa (Jellyfish)  W. Sterrer  155
   Class Anthozoa (Corals, anemones)  S. Cairns, C. den Hartog & C. Arneson  159
**Phylum Ctenophora (Comb-jellies)**  W. Sterrer  194

### BRANCH BILATERIA (Bilaterally symmetric animals)  *(W. Sterrer)*  197

**Phylum Platyhelminthes (Flatworms)**  *(W. Sterrer)*  197
   Class Turbellaria (Free-living flatworms)  R. M. Rieger & W. Sterrer  198
   Class Trematoda (Flukes)  W. Sterrer  203
   Class Cestoda (Tapeworms)  W. Sterrer  205
**Phylum Mesozoa**  W. Sterrer  206
**Phylum Nemertina (=Rhynchocoela) (Ribbon worms)**  E. Kirsteuer  208
**Phylum Gnathostomulida (Jaw worms)**  W. Sterrer  211
**Phylum Gastrotricha**  W. Sterrer  213

| | |
|---|---|
| **Phylum Nematoda (Roundworms)**  *J. Ott* | 216 |
| **Phylum Rotifera (Wheel animalcules)**  *W. Sterrer* | 219 |
| **Phylum Kinorhyncha**  *R. P. Higgins* | 220 |
| **Phylum Priapulida**  *W. Sterrer* | 222 |
| **Phylum Acanthocephala (Spiny-headed worms)**  *W. Sterrer* | 223 |
| **Phylum Sipuncula (Peanut worms)**  *M. E. Rice* | 224 |
| **Phylum Echiura (Spoon worms)**  *J. F. Pilger* | 228 |
| **Phylum Pogonophora (Beard Worms)**  *E. C. Southward* | 230 |
| **Phylum Annelida (Segmented worms)**  *(M. L. Jones)* | 232 |
|     Class Polychaeta (Bristle worms)  *M. L. Jones, S. L. Gardiner, M. E. Petersen & W. Sterrer* | 232 |
|     Class Oligochaeta  *O. Giere, C. Erséus & P. Lasserre* | 258 |
|     Class Hirudinea (Leeches)  *T. G. Rand* | 263 |
| **Phylum Tardigrada (Water bears)**  *R. M. Kristensen & W. Sterrer* | 265 |
| **Phylum Arthropoda**  *(W. Sterrer)* | 268 |
|     **Subphylum Chelicerata**  *(W. Sterrer)* | 268 |
|     Class Arachnida (Spiders, mites, scorpions, etc.)  *(W. Sterrer)* | 269 |
|         Order Pseudoscorpiones (=Chelonethida) (False scorpions)  *R. Schuster* | 269 |
|         Order Acari (Mites and ticks)  *R. Schuster & I. Bartsch* | 270 |
|     Class Pycnogonida (=Pantopoda) (Sea spiders)  *J. Markham* | 275 |
|     **Subphylum Crustacea**  *(W. Sterrer)* | 277 |
|     Class Branchiopoda  *B. F. Morris* | 278 |
|     Class Ostracoda (Mussel shrimps)  *R. F. Maddocks & L. S. Kornicker* | 280 |
|     Class Copepoda  *B. F. Morris & R. Cressey* | 288 |
|     Class Cirripedia (Barnacles)  *A. J. Southward* | 299 |
|     Class Malacostraca (Higher crustaceans)  *(W. Sterrer)* | 305 |
|         Order Leptostraca  *G. A. Schultz* | 305 |
|         Order Stomatopoda (Mantis shrimps)  *R. B. Manning* | 306 |
|         Order Euphausiacea (Krill)  *W. Sterrer* | 310 |
|         Order Decapoda (Shrimps, lobsters and crabs)  *F. A. Chace, Jr., J. J. McDermott, P. A. McLaughlin & R. B. Manning* | 312 |
|         Order Mysidacea (Opossum shrimps)  *T. Brattegard* | 358 |
|         Order Mictacea  *W. Sterrer* | 361 |
|         Order Cumacea  *J. Markham & W. Sterrer* | 362 |
|         Order Tanaidacea  *J. Markham* | 364 |
|         Order Isopoda (Pill bugs, wharf lice, fish lice)  *G. A. Schultz* | 366 |
|         Order Amphipoda  *S. E. Johnson* | 372 |
|     **Subphylum Uniramia**  *(W. Sterrer)* | 381 |
|     Class Chilopoda (Centipedes)  *R. Schuster* | 382 |
|     Class Insecta (Insects)  *I. W. Hughes & R. Schuster* | 383 |
| **Phylum Mollusca (Mollusks)**  *(W. Sterrer)* | 392 |
|     Class Solenogastres (Worm mollusks)  *L. v. Salvini Plawen* | 393 |
|     Class Polyplacophora (Chitons)  *R. H. Jensen & M. G. Harasewych* | 394 |
|     Class Gastropoda (Snails, limpets and slugs)  *R. H. Jensen & K. Clark* | 397 |

Class Scaphopoda (Tusk shells)   *R. H. Jensen & M. G. Harasewych*   459
Class Bivalvia (=Pelecypoda, Lamellibranchia) (Clams, mussels, shipworms, etc.)   *R. H. Jensen & M. G. Harasewych*   460
Class Cephalopoda (Cuttlefishes, squids and octopuses)   *C. F. E. Roper*   492

**SUPERPHYLUM LOPHOPHORATA (=Tentaculata)**   *(W. Sterrer)*   500

**Phylum Bryozoa (Sea mats, moss animals)**   *R. J. Cuffey & S. Fonda*

**Phylum Phoronida (Horseshoe worms)**   *C. C. Emig*   516

**Phylum Brachiopoda (Lamp shells)**   *C. C. Emig*   518

**Phylum Chaetognatha (Arrow worms)**   *W. Sterrer*   519

**Phylum Echinodermata**   *(D. L. Pawson)*   522
Class Asteroidea (Starfish, sea stars)   *D. L. Pawson*   523
Class Ophiuroidea (Brittle stars)   *D. L. Pawson*   527
Class Echinoidea (Sea urchins and sand dollars)   *D. L. Pawson*   531
Class Holothuroidea (Sea cucumbers)   *D. L. Pawson*   537

**Phylum Hemichordata**   *(W. Sterrer)*   541
Class Enteropneusta (Acorn worms)   *W. Sterrer*   541
Class Pterobranchia   *W. Sterrer & S. M. Lester*   544

**Phylum Chordata**   *(W. Sterrer)*   545
**Subphylum Urochordata (=Tunicata)**   *(W. Sterrer)*   545
Class Ascidiacea (Sea squirts)   *F. Monniot & C. Monniot*   545
Class Thaliacea (Salps, pyrosomids and doliolids)   *W. Sterrer*   558
Class Appendicularia (=Larvacea)   *W. Sterrer*   561
**Subphylum Cephalochordata (=Acrania) (Lancelets)**   *M. Spindler*   562
**Subphylum Vertebrata (=Euchordata)**   *(W. Sterrer)*   565
Class Chondrichthyes (Cartilaginous fishes)   *J. Burnett-Herkes*   566
Class Osteichthyes (Bony fishes)   *J. Burnett-Herkes*   571
Class Reptilia   *(W. Sterrer)*   651
Order Testudines (=Chelonia) (Turtles)   *J. A. Frick & L. C. Ireland*   651
Class Aves (Birds)   *D. Wingate*   656
Class Mammalia   *(W. Sterrer)*   668
Order Cetacea (Whales, porpoises and dolphins)   *L. Mowbray & W. Sterrer*   668

# REFERENCES   672

# GLOSSARY   695

# TAXONOMIC INDEX   702

# BLACK-AND-WHITE PLATES

| | | |
|---|---|---|
| 1 | CYANOBACTERIA (Blue-green algae) | 14 |
| 2 | EUMYCOTA (Fungi) | 21 |
| 3 | LICHENES (Lichens) | 25 |
| 4 | ULVALES, CLADOPHORALES (Green algae 1) | 31 |
| 5 | CODIACEAE, UDOTEACEAE (Green algae 2) | 33 |
| 6 | CAULERPACEAE, BRYOPSIDACEAE, DERBESIACEAE (Green algae 3) | 36 |
| 7 | SIPHONOCLADALES (Green algae 4) | 38 |
| 8 | DASYCLADACEAE (Green algae 5) | 40 |
| 9 | ECTOCARPALES, DICTYOTALES 1 (Brown algae 1) | 44 |
| 10 | DICTYOTALES 2, SPOROCHNALES, SCYTOSIPHONALES (Brown algae 2) | 47 |
| 11 | FUCALES (Brown algae 3) | 49 |
| 12 | BANGIALES, NEMALIALES (Red algae 1) | 54 |
| 13 | CRYPTONEMIALES (Red algae 2) | 57 |
| 14 | GIGARTINALES, RHODYMENIALES (Red algae 3) | 60 |
| 15 | CERAMIACEAE, DELESSERIACEAE, DASYACEAE (Red algae 4) | 63 |
| 16 | RHODOMELACEAE (Red algae 5) | 65 |
| 17 | BACILLARIOPHYCEAE (Diatoms) | 69 |

| | | |
|---|---|---|
| 18 | CHRYSOPHYCEAE (Golden-brown algae), XANTHOPHYCEAE (Yellow-green algae), EUGLENOPHYTA | 72 |
| 19 | DINOPHYTA (Dinoflagellates) | 78 |
| 20 | CHENOPODIALES—GERANIALES (Flowering plants 1) | 82 |
| 21 | MYRTALES, SOLANALES, CAMPANULALES (Flowering plants 2) | 85 |
| 22 | NAJADALES, BUTOMALES (Sea grasses) | 89 |
| 23 | GRAMINALES (Beach grasses) | 90 |
| 24 | LOBOSIA, FILOSIA (Amoebas, etc.) | 96 |
| 25 | FORAMINIFERIDA (Forams) 1 | 97 |
| 26 | FORAMINIFERIDA (Forams) 2 | 100 |
| 27 | ACANTHARIA, RADIOLARIA; SPOROZOA (Radiolarians, etc.) | 102 |
| 28 | CILIOPHORA (Ciliates) | 107 |
| 29 | PORIFERA: Development | 114 |
| 30 | KERATOSA (Sponges 1) | 116 |
| 31 | HAPLOSCLERIDA, POECILOSCLERIDA (Sponges 2) | 119 |
| 32 | HALICHONDRIIDA, AXINELLIDA (Sponges 3) | 122 |
| 33 | HADROMERIDA (Sponges 4) | 123 |
| 34 | ASTROPHORIDA, SPIROPHORIDA, LITHISTIDA (Sponges 5) | 125 |
| 35 | CALCAREA (Sponges 6) | 127 |
| 36 | CNIDARIA and CTENOPHORA: Development | 131 |
| 37 | ATHECATA (Hydroid polyps 1) | 133 |
| 38 | HALECIIDAE, LAFOEIDAE, CAMPANULARIIDAE (Hydroid polyps 2) | 135 |
| 39 | SERTULARIIDAE, PLUMULARIIDAE (Hydroid polyps 3) | 138 |
| 40 | ANTHOMEDUSAE (Hydromedusae 1) | 142 |
| 41 | LEPTOMEDUSAE, LIMNOMEDUSAE (Hydromedusae 2) | 145 |
| 42 | ACTINULIDA, TRACHYMEDUSAE, NARCOMEDUSAE (Hydromedusae 3) | 147 |
| 43 | CYSTONECTAE, PHYSONECTAE, CALYCOPHORAE 1 (Siphonophores 1) | 151 |
| 44 | DIPHYIDAE, ABYLIDAE (Siphonophores 2) | 153 |
| 45 | SCYPHOZOA (Jellyfish) | 157 |
| 46 | ALCYONACEA, GORGONACEA 1 (Soft corals 1) | 166 |
| 47 | *Eunicea, Plexaurella* (Soft corals 2) | 168 |
| 48 | *Muricea* (Soft corals 3) | 170 |
| 49 | GORGONIIDAE, ELLISELLIDAE (Soft corals 4), PENNATULACEA (Sea pens) | 171 |
| 50 | BOLOCEROIDARIA, ACONTIARIA (Sea anemones 1) | 174 |
| 51 | ENDOMYARIA (Sea anemones 2) | 177 |
| 52 | ASTROCOENIINA (Stony corals 1) | 180 |
| 53 | FUNGIINA (Stony corals 2) | 181 |

# BLACK-AND-WHITE PLATES

| | | |
|---|---|---|
| 54 | FAVIIDAE (Stony corals 3) | 183 |
| 55 | RHIZANGIIDAE—MUSSIDAE (Stony corals 4) | 185 |
| 56 | CARYOPHYLLINA, DENDROPHYLLINA (Stony corals 5) | 187 |
| 57 | CORALLIMORPHARIA (False corals), ZOANTHIDEA (Sea mats) | 189 |
| 58 | CERIANTHARIA (Tube anemones), ANTIPATHARIA (Black corals) | 193 |
| 59 | CTENOPHORA (Comb jellies) | 196 |
| 60 | PLATYHELMINTHES—POGONOPHORA (Lower worms): Development | 199 |
| 61 | CATENULIDA—RHABDOCOELA (Flatworms 1) | 201 |
| 62 | POLYCLADIDA (Flatworms 2) | 202 |
| 63 | TREMATODA (Flukes), CESTODA (Tapeworms) | 205 |
| 64 | MESOZOA | 207 |
| 65 | NEMERTINA (Ribbon worms) | 210 |
| 66 | GNATHOSTOMULIDA (Jaw worms) | 213 |
| 67 | GASTROTRICHA (Gastrotrichs) | 215 |
| 68 | NEMATODA (Roundworms) | 218 |
| 69 | ROTIFERA (Wheel animalcules), KINORHYNCHA (Kinorhynchs), PRIAPULIDA (Priapulids) | 220 |
| 70 | ACANTHOCEPHALA (Spiny-headed worms), SIPUNCULA (Peanut worms), ECHIURA (Spoon worms) | 227 |
| 71 | POGONOPHORA (Beard worms) | 231 |
| 72 | POLYCHAETA (Bristle worms): Development | 235 |
| 73 | POLYNOIDAE—HESIONIDAE (Bristle worms 1) | 237 |
| 74 | SYLLIDAE (Bristle worms 2) | 239 |
| 75 | NEREIDIDAE, GLYCERIDAE (Bristle worms 3) | 241 |
| 76 | ONUPHIDAE—DORVILLEIDAE (Bristle worms 4) | 243 |
| 77 | ORBINIIDAE—SPIONIDAE (Bristle worms 5) | 245 |
| 78 | MAGELONIDAE—COSSURIDAE (Bristle worms 6) | 248 |
| 79 | OPHELIIDAE—OWENIIDAE (Bristle worms 7) | 251 |
| 80 | SABELLARIIDAE—TRICHOBRANCHIDAE (Bristle worms 8) | 253 |
| 81 | SABELLIDAE, SERPULIDAE (Bristle worms 9) | 256 |
| 82 | ARCHIANNELIDA (Bristle worms 10) | 258 |
| 83 | OLIGOCHAETA (Oligochetes), HIRUDINEA (Leeches) | 261 |
| 84 | TARDIGRADA (Water bears) | 267 |
| 85 | PSEUDOSCORPIONES (False scorpions), ACARI (Mites) | 273 |
| 86 | PYCNOGONIDA (Sea spiders): Development | 276 |
| 87 | PYCNOGONIDA (Sea spiders) | 277 |
| 88 | "ENTOMOSTRACA" (Lower crustaceans): Development | 279 |
| 89 | CLADOCERA (Water fleas) | 282 |
| 90 | MYODOCOPIDA, HALOCYPRIDA, PLATYCOPIDA (Mussel shrimps 1) | 283 |
| 91 | PODOCOPIDA (Mussel shrimps 2) | 287 |

| # | Title | Page |
|---|---|---|
| 92 | CALANOIDA (Copepods 1) | 291 |
| 93 | CYCLOPOIDA (Copepods 2) | 293 |
| 94 | HARPACTICOIDA (Copepods 3) | 296 |
| 95 | SIPHONOSTOMATOIDA, NOTODELPHYOIDA (Parasitic copepods) | 298 |
| 96 | CIRRIPEDIA (Barnacles) | 303 |
| 97 | STOMATOPODA (Mantis shrimps): Development | 307 |
| 98 | LEPTOSTRACA, STOMATOPODA (Mantis shrimps) | 309 |
| 99 | EUPHAUSIACEA (Krill): Development | 311 |
| 100 | EUPHAUSIACEA (Krill) | 312 |
| 101 | DECAPODA (Decapods): Schematic | 314 |
| 102 | DECAPODA (Decapods): Development | 316 |
| 103 | PENAEIDEA (Whip shrimps) | 318 |
| 104 | PASIPHAEIDAE—GNATHOPHYLLIDAE (Shrimps) | 321 |
| 105 | ALPHEIDAE (Snapping shrimps) | 324 |
| 106 | HIPPOLYTIDAE, PROCESSIDAE (Shrimps) | 327 |
| 107 | STENOPODIDEA (Coral shrimps), ASTACIDEA (Lobsters) | 329 |
| 108 | PALINURIDAE (Spiny lobsters), SYNAXIDAE | 331 |
| 109 | SCYLLARIDAE (Locust lobsters) | 332 |
| 110 | THALASSINOIDEA (Mole shrimps) | 334 |
| 111 | COENOBITOIDEA, PAGUROIDEA (Hermit crabs) | 337 |
| 112 | GALATHEOIDEA (Porcelain crabs), HIPPOIDEA (Sand fleas) | 339 |
| 113 | GYMNOPLEURA (Frog crabs), DROMIACEA (Sponge crabs), OXYSTOMATA (Box crabs) | 341 |
| 114 | PORTUNIDAE (Swimming crabs) | 344 |
| 115 | XANTHIDAE (Mud crabs) | 347 |
| 116 | GRAPSIDAE (Shore crabs) | 351 |
| 117 | GECARCINIDAE (Land crabs), OCYPODIDAE (Ghost crabs) | 353 |
| 118 | OXYRHYNCHA (Spider crabs) | 357 |
| 119 | MYSIDACEA (Opossum shrimps) | 360 |
| 120 | MICTACEA (Mictaceans), CUMACEA (Cumaceans) | 362 |
| 121 | TANAIDACEA (Tanaids) | 365 |
| 122 | ANTHURIDEA, FLABELLIFERA, VALVIFERA (Isopods 1) | 369 |
| 123 | ASELLOTA, EPICARIDEA, ONISCOIDEA (Isopods 2) | 371 |
| 124 | GAMMARIDEA 1 (Amphipods 1) | 375 |
| 125 | GAMMARIDEA 2—INGOLFIELLIDEA (Amphipods 2) | 379 |
| 126 | CHILOPODA (Centipedes), APTERYGOTA (Wingless insects) | 383 |
| 127 | INSECTA (Insects): Development | 385 |
| 128 | PTERYGOTA (Winged insects) | 389 |
| 129 | MOLLUSCA (Mollusks): Development | 393 |
| 130 | SOLENOGASTRES (Worm mollusks), POLYPLACOPHORA (Chitons) | 394 |

| | | |
|---|---|---|
| 131 | MOLLUSCA (Mollusks): Schematic | 400 |
| 132 | GASTROPODA (Snails): Egg cases | 401 |
| 133 | PLEUROTOMARIACEA (Slit shells), FISSURELLACEA (Keyhole limpets), PATELLACEA (Limpets) | 403 |
| 134 | TROCHACEA (Top and star shells), NERITACEA (Nerites) | 405 |
| 135 | LITTORINACEA (Periwinkles), RISSOACEA (Risso shells) | 408 |
| 136 | CERITHIACEA (Worm and horn shells) | 411 |
| 137 | EPITONIACEA (Wentle traps, a.o.), EULIMACEA | 416 |
| 138 | HIPPONICACEA, CALYPTRAEACEA (Slipper and carrier shells) | 418 |
| 139 | STROMBACEA (Conchs) | 419 |
| 140 | TRIVIACEA, CYPRAEACEA (Cowries, a.o.) | 421 |
| 141 | HETEROPODA (Heteropods) | 423 |
| 142 | NATICACEA (Moon snails), TONNACEA (Helmets, tuns) | 425 |
| 143 | CYMATIACEA (Tritons) | 427 |
| 144 | MURICACEA (Rock shells, a.o.) | 430 |
| 145 | BUCCINACEA (Dove shells, whelks, a.o.) | 433 |
| 146 | VOLUTACEA (Olives), MITRACEA (Miters), CONACEA (Cones) | 435 |
| 147 | ACTEONACEA—PYRAMIDELLACEA | 439 |
| 148 | THECOSOMATA (Sea butterflies), RHODOPACEA, ONCHIDIACEA | 441 |
| 149 | GYMNOSOMATA | 444 |
| 150 | ANASPIDEA (Sea hares), NOTASPIDEA | 445 |
| 151 | ASCOGLOSSA (Sea slugs) | 449 |
| 152 | NUDIBRANCHIA (Nudibranchs) | 453 |
| 153 | PULMONATA (Pulmonate snails) | 456 |
| 154 | SCAPHOPODA (Tusk shells) | 460 |
| 155 | NUCULACEA, ARCOIDA (Arks), MYTILOIDA (Mussels) | 465 |
| 156 | PTERIACEA, PECTINACEA (Scallops) | 469 |
| 157 | ANOMIACEA, OSTREACEA (Oysters), LIMACEA | 471 |
| 158 | LUCINACEA (Lucines), CHAMACEA (Jewel boxes) | 474 |
| 159 | GALEOMMATACEA—MACTRACEA (Cockles, a.o.) | 477 |
| 160 | TELLINACEA (Tellins) | 480 |
| 161 | ARCTICACEA, VENERACEA (Venus clams, a.o.) | 485 |
| 162 | MYOIDA (Shipworms, a.o.), ANOMALODESMATA | 488 |
| 163 | SEPIOIDEA, TEUTHOIDEA (Squids) | 495 |
| 164 | VAMPYROMORPHA, OCTOPODA (Octopuses) | 498 |
| 165 | LOPHOPHORATA: Development | 502 |
| 166 | ENTOPROCTA (Entoprocts) | 503 |
| 167 | CTENOSTOMIDA (Ectoprocts 1) | 504 |
| 168 | ANASCINA, CRIBRIMORPHINA (Ectoprocts 2) | 507 |
| 169 | ASCOPHORINA (Ectoprocts 3) | 511 |
| 170 | CYCLOSTOMIDA (Ectoprocts 4) | 515 |

| | | |
|---|---|---|
| 171 | PHORONIDA (Horseshoe worms), BRACHIOPODA (Lamp shells) | 517 |
| 172 | CHAETOGNATHA (Arrowworms) | 521 |
| 173 | ECHINODERMATA (Echinoderms): Development | 524 |
| 174 | ASTEROIDEA (Starfish) | 525 |
| 175 | OPHIUROIDEA (Brittle stars) | 529 |
| 176 | CIDAROIDA—ECHINOIDA (Regular sea urchins) | 534 |
| 177 | HOLECTYPOIDA—SPATANGOIDA (Irregular sea urchins) | 536 |
| 178 | HOLOTHUROIDEA (Sea cucumbers) | 539 |
| 179 | HEMICHORDATA: Development | 542 |
| 180 | HEMICHORDATA (Hemichordates) | 543 |
| 181 | UROCHORDATA (Tunicates): Development | 547 |
| 182 | POLYCLINIDAE, DIDEMNIDAE (Sea squirts 1) | 549 |
| 183 | POLYCITORIDAE (Sea squirts 2) | 551 |
| 184 | PEROPHORIDAE, ASCIDIIDAE (Sea squirts 3) | 553 |
| 185 | STYELIDAE (Sea squirts 4) | 555 |
| 186 | PYURIDAE (Sea squirts 5) | 557 |
| 187 | THALIACEA (Salps) | 560 |
| 188 | APPENDICULARIA (Larvaceans) | 563 |
| 189 | ACRANIA (Lancelets): Development | 564 |
| 190 | ACRANIA (Lancelets) | 565 |
| 191 | CHONDRICHTHYES (Cartilaginous fishes): Schematic | 567 |
| 192 | CHONDRICHTHYES (Cartilaginous fishes) | 569 |
| 193 | OSTEICHTHYES (Bony fishes): Schematic | 573 |
| 194 | OSTEICHTHYES (Bony fishes): Development | 576 |
| 195 | ELOPIFORMES, CLUPEIFORMES (Anchovies, a.o.) | 578 |
| 196 | ANGUILLIDAE (Eels), CONGRIDAE (Conger eels), OPHICHTHIDAE (Sand eels) | 580 |
| 197 | MURAENIDAE (Morays) | 583 |
| 198 | MYCTOPHIFORMES, SALMONIFORMES (Lantern fishes, a.o.) | 585 |
| 199 | LOPHIIFORMES, GADIFORMES (Frogfishes, a.o.) | 588 |
| 200 | ATHERINIFORMES (Flying fishes, a.o.) | 591 |
| 201 | BERYCIFORMES, GASTEROSTEIFORMES (Sea horses, a.o.) | 593 |
| 202 | SERRANIDAE 1 (Coney, a.o.) | 596 |
| 203 | SERRANIDAE 2 (Groupers) | 599 |
| 204 | GRAMMISTIDAE—MALACANTHIDAE (Soapfishes, a.o.) | 601 |
| 205 | ECHENEIDAE (Shark suckers), CARANGIDAE (Jacks), CORYPHAENIDAE (Dolphinfishes) | 605 |
| 206 | LUTJANIDAE (Snappers), GERREIDAE (Mojarras) | 607 |
| 207 | HAEMULIDAE (Grunts) | 611 |
| 208 | SPARIDAE (Porgies)—KYPHOSIDAE (Sea chubs) | 613 |
| 209 | CHAETODONTIDAE (Butterfly fishes) | 615 |

| | | |
|---|---|---|
| 210 | POMACANTHIDAE (Angelfishes) | 617 |
| 211 | POMACENTRIDAE (Damselfishes) | 619 |
| 212 | LABRIDAE (Wrasses) | 623 |
| 213 | *Scarus* (Parrot fishes 1) | 626 |
| 214 | *Sparisoma* (Parrot fishes 2) | 628 |
| 215 | MUGILIDAE (Mullets), SPHYRAENIDAE (Barracudas) | 630 |
| 216 | CLINIDAE (Blennies), BLENNIIDAE (Combtooth blennies), GOBIIDAE (Gobies) | 633 |
| 217 | ACANTHURIDAE (Surgeonfishes) | 635 |
| 218 | SCOMBRIDAE (Tunas), GEMPYLIDAE (Oilfishes), ISTIOPHORIDAE (Billfishes) | 637 |
| 219 | NOMEIDAE (Man-of-war fishes), SCORPAENIDAE (Scorpion fishes), BOTHIDAE (Lefteye flounders) | 640 |
| 220 | BALISTIDAE 1 (Triggerfishes) | 643 |
| 221 | BALISTIDAE 2 (Filefishes) | 645 |
| 222 | OSTRACIIDAE (Trunkfishes) | 647 |
| 223 | TETRAODONTIDAE (Puffers), DIODONTIDAE (Porcupine fishes), MOLIDAE (Sunfishes) | 649 |
| 224 | TESTUDINES (Turtles), AVES (Birds): Development | 653 |
| 225 | TESTUDINES (Turtles) | 655 |
| 226 | PODICIPEDIFORMES—FALCONIFORMES (Seabirds 1) | 661 |
| 227 | CHARADRIIFORMES, CORACIIFORMES (Seabirds 2) | 665 |
| 228 | CETACEA (Dolphins and whales) | 670 |

# COLOR PLATES

Color plates appear between pages 352 and 353.

| | Title | Sponsored by |
|---|---|---|
| 1 | Chlorophyta (Green algae) | *Bank of Bermuda Ltd.* |
| 2 | Phaeophyta (Brown algae) and Rhodophyta (Red algae) | *L. P. Gutteridge Ltd.* |
| 3 | Porifera 1 (Sponges) | *Trustees of the Bermuda Biological Station* |
| 4 | Porifera 2 (Sponges) | *Shell Company of Bermuda* |
| 5 | Anthozoa 1 (Octocorals) | *Bank of Butterfield Ltd.* |
| 6 | Anthozoa 2 (Sea anemones and stony corals) | *William de V. Frith* |
| 7 | Anthozoa 3 (False corals, sea mats, tube anemones) | *Mr. & Mrs. Gerard M. Ives* |
| 8 | Polychaeta (Bristle worms) | *Trustees of the Bermuda Biological Station* |
| 9 | Decapoda 1 (Shrimps) | *Sir James and Lady Pearman* |
| 10 | Decapoda 2 (Spiny lobsters, mole shrimps, hermit crabs, a.o.) | *Dr. W. Redwood Wright* |
| 11 | Decapoda 3 (Crabs) | *Mr. & Mrs. William T. Kemble* |
| 12 | Gastropoda 1 (Sea hares and other sea slugs) | *Bank of Bermuda Ltd.* |
| 13 | Gastropoda 2 (Sea slugs) | *Dr. Rudolph F. Nunnemacher* |

COLOR PLATES

| Title | Sponsored by |
|---|---|
| 14 Echinodermata (Starfish, brittle stars, sea urchins and sea cucumbers) | *Appleby, Spurling & Kempe Ltd.* |
| 15 Ascidiacea 1 (Sea squirts) | *Electronic Communications Ltd.—The K&J Group* |
| 16 Ascidiacea 2 (Sea squirts) | *Bank of Butterfield Ltd.* |

# MARINE FAUNA
# AND FLORA OF BERMUDA

# INTRODUCTION

At VERRILL's (1901a) time about 900 *species* of marine invertebrates were known from Bermuda, and 200 species of fishes. COLLINS & HERVEY (1917) list 342 species of marine algae, and if we add the few marine vertebrates other than fishes (about 25) and the few flowering plants associated with the marine environment (about 25) we arrive at a total of no more than 1,500 species.

Since then—not least thanks to the efforts that went into the preparation of this book—the total number of species known from Bermuda waters and shores has tripled, yet this may still only represent half of the number of species that actually occur on and in the vicinity of the Bermuda seamount.

A judicious *selection* had to be made, therefore, if we wanted to present the entire range of such diversity in a one-volume field guide. The following criteria were used:

1. Large, conspicuous (in color or behavior) and abundant species were preferred to small, inconspicuous and rare ones.
2. Nearshore and surface-dwelling species were selected over those occurring offshore or in the deep sea.
3. Parasitic species are given as examples rather than as a representative sample.
4. Species that were only recorded once or whose record is doubtful were generally excluded.
5. Plankton species, in spite of their mostly small size and offshore occurrence, have been given relatively more space because they are easily collected and often used in course work.
6. However, species that did not satisfy these criteria but are the only Bermudian representatives of major taxa (orders, classes, phyla) have generally been included.

The *presentation* follows the natural system of organisms, i.e., a hierarchic arrangement that reflects degrees of relationships that organisms acquired in the course of evolution. The overall system used in the book is semi-conservative (see LIVING ORGANISMS, p. 7) and is the editor's choice, whereas the systematic arrangement within individual chapters follows largely the preferences of chapter contributors. Chapters deal with taxa on a level no higher than phylum (e.g., Eumycota) and no lower than order (e.g., Decapoda); they are grouped into sections that diagnose and summarize superposed taxa. In the hierarchic presentation, certain main levels are used consistently throughout (italicized in the following tabulation), and intermediate levels are inserted as needed. They are abbreviated as below when used in the taxonomic part of a chapter.

*Kingdom*
  Subkingdom
    Branch
      Superphylum
        *Phylum* ( = Division of botanists)
          Subphylum (S.PH.)
            *Class* (CL.)
              Subclass (S.CL.)
                Superorder (SUP.O.)
                  *Order* (O.)
                    Suborder (S.O.)
                      Infraorder (I.O.)
                        Section (SECT.)
                      Cohort
                        Superfamily (SUP.F.)
                        *Family* (F.)
                          Subfamily (S.F.)
                            *Genus*
                              *Species*
                                Subspecies
                                  Form (f.)

## Section Introductions

characterize taxonomic categories that embrace several chapters (e.g., Kingdom Animalia or Subphylum Crustacea). They give, in a first paragraph, a brief systematic diagnosis within the next-higher category (e.g., Crustacea are "Mostly aquatic ARTHROPODA...").

A second paragraph typically reviews major features and organs of taxonomic importance, and introduces scientific terms applicable to the entire group.

A third paragraph gives a breakdown of the next-lower taxa in the group, separating those that are not represented in Bermuda from those that are, with chapters and page numbers. Where needed, taxonomic categories not used in this book are mentioned, to aid cross-referencing with other scientific literature, and references are given to recent monographs or critical literature.

## Chapters

consist of an *introduction*, which familiarizes the reader with general characteristics of the group, and a *taxonomic part* designed for species identification. The order in which data are presented is identical in each chapter, and any lack of information in a specific area is usually indicated at the appropriate place.

## The Introduction

contains the parts CHARACTERISTICS, OCCURRENCE, IDENTIFICATION, BIOLOGY, DEVELOPMENT and REFERENCES, each composed of several paragraphs.

CHARACTERISTICS: *A brief systematic diagnosis of the taxon within the NEXT-HIGHER CATEGORY* (e.g., Pisces are *"Aquatic VERTEBRATA with..."*; Vertebrata are *"Variously shaped CHORDATA..."* etc.). General features likely to be of value in the field follow, such as body size range, color and patterns, consistency and movement.

A separate paragraph discusses the systematic position of the taxon (if doubtful

# INTRODUCTION

or controversial), and its representation in Bermuda and in the book. It concludes with the total number of species known worldwide, followed by the number of species known from Bermuda and the number selected thereof for the book.

OCCURRENCE provides information on characteristic environments and substrates, depth range and abundance.

A second paragraph lists methods and devices for collecting and extracting specimens. The more commonly used procedures and instruments are described in the Glossary at the end of the book. Because of the small scale and fragility of Bermuda's nearshore marine biotopes, all collecting of living organisms should be kept to the absolute minimum necessary for research and teaching, and conservation rules and legislation strictly adhered to (for details on current policies see brochures issued by the Division of Fisheries and by the Bermuda Biological Station). Although all observation may be intrusive to some extent, live observation is preferable to collecting and preserving specimens; rocks should be turned back to their original position and organisms taken back to where they were found; in short, the motto of the U.S. National Speleological Society to "Take nothing but pictures, leave nothing but footprints, kill nothing but time" provides an admirable guideline for exploring any vulnerable natural system, living or inanimate.

A third paragraph may discuss the importance of the taxon to humans.

IDENTIFICATION first establishes the basis on which species determination is made (e.g., external features, dissection), and then lists technical aids (e.g., low-power microscope) needed.

The main section introduces those characteristics and explains those technical terms that are used for identification in the taxonomic part of the chapter. If necessary, an explanation of the sizes given in the plates follows here; unless stated otherwise sizes given are the largest measurement (length or width, but usually exclusive of appendages) of live, mature specimens.

A final paragraph provides methods and formulas used for anesthetizing, killing, fixing and preserving specimens for scientific purposes, with reference to the Glossary where necessary.

BIOLOGY gives data on sexuality, reproduction, method of fertilization, number of progeny, growth, life-span, behavior, feeding, predators, parasites and other biological features peculiar to the taxon.

DEVELOPMENT discusses primarily larval stages and alternation of generations; these are pictured on separate plates. Where applicable, methods for collecting larvae are given here.

REFERENCES are kept to a minimum, with preference given to recent ones because these generally contain older ones in their bibliographies. A first paragraph lists some general, comprehensive introductions to the taxon, with emphasis on the North Atlantic geographic area wherever possible.

The second paragraph summarizes major studies on the taxon in Bermuda, or states the absence thereof. Some additional references may be given in the taxonomic part. All references are listed in full at the end of the book.

## The Taxonomic Part

consists of short paragraphs that diagnose taxa down to species. Within these paragraphs, information is presented in the following sequence and typographic pattern.

**GROUP (=SYNONYM) (Common Name)** Diagnosis, as a rule within the framework of the NEXT-HIGHER LEVEL, followed by general remarks about biology, ecology, or systematic subgroupings. (An abbreviated sentence in parentheses, usually at the end of order, superfamily or family diagnosis, indicates how many species of this group are known—but not necessarily identified—from Bermuda, e.g., "20 spp. from Bda.")

**Genus, species,** rarely **subspecies** or form (f.). Unidentified species are listed as sp. and species whose identity is doubtful as cf. (i.e., confer, meaning "compare with"). The name of the first describer (author) of the species follows; it stands in parentheses if the author described the species under a name different from the valid one given here. Botanical practice usually provides both the original describer and the author who assigned the valid name. Authors are sometimes abbreviated; L. stands for Carl Linné, the founder of binomial nomenclature. Common synonyms are given in parentheses, preceded by the = sign, and followed, where necessary, by a reference to the paper in which this synonym was used. Common English names are given wherever available, as well as local Bermudian names, the latter preceded by the letter B:

Diagnoses of genus and species. Wherever there is only one species listed in a given genus, the genus diagnosis is usually not separated from the species diagnosis; if there are two or more, the genus diagnosis is given under the first species listed as "Genus with . . ." and the species diagnosis, separated by a hyphen, as "Species with . . ." In all following species, the genus name is abbreviated, and the genus diagnosis is referred to as "Genus as above." Sizes are in metric units, and refer to typical mature specimens, or upper limits. The second part of this paragraph contains data on occurrence, abundance, biology and reproduction; with the roman numerals I-XII denoting months of sexual maturity, spawning or other seasonal events. Note that neither geographic ranges nor specific localities in Bermuda are given; the former because it would have gone beyond the scope of this book, the latter because of the danger of over-collecting once a particular locality has been singled out. (Where applicable this paragraph ends with a reference to the Color Plate number, and the number of the figure thereon.)

## Black-and-White Plates

are positioned as close to the corresponding text as possible, and individual species are identified in the figures; plates are therefore not referred to in the text except at the beginning of the text section to which they belong. Black-and-white plates are numbered (from 1 to 228) and named. There are three types of black & white plates: (1) TAXA, (2) DEVELOPMENT and (3) SCHEMATIC.

1. Plates of TAXA are identified by the smallest common taxonomic denominator of the species on the plate (e.g., Ciliphora or Syllidae). If there are more than 3 groups represented, the range is given by a dash (e.g., Grammistidae—Malacanthidae). Most plates also carry a common name (e.g., Sea spiders) for easier reference. The order in which species are illustrated on a plate, as read from the upper left to the lower right, follows largely that of the text. Sizes of figures are approximately relative within a given plate, i.e., the largest species in nature also appears largest on the plate; however, small species are drawn relatively larger both within a plate and throughout the book. Unless stated otherwise in the text (see end of the paragraph on IDENTIFICATION) sizes indicated refer to the largest dimension (length or width) of typical, mature, live specimens, usually excluding appendages. All sizes are in metric units and abbreviated as follows:

meter, m (1 m = 100 cm = 39.37 inches)

centimeter, cm (1 cm = 10 mm = 0.3937 inch)

millimeter, mm (1 mm = 1000 μm = 0.03937 inch)

micrometer, μm

2. DEVELOPMENT plates appear near the corresponding paragraphs and show typical developmental stages (eggs, larval forms, metamorphosis) of species (or higher taxa) from Bermuda. The same applies to relative sizes as in plates of taxa.

3. SCHEMATIC plates are provided for those few groups (Decapoda, Mollusca, Chondrichthyes, Osteichthyes) that have a voluminous or detailed terminology. They should be used for general orientation, together with the paragraphs on IDENTIFICATION and the taxonomic sections. (Note that the organisms shown are chimaeras made up of often incompatible taxonomic features!)

**Color Plates**

are numbered (1-16) and named according to the chapter(s) they belong to (e.g., Polychaeta); in addition they carry common, descriptive names (e.g., Bristle worms). Color illustrations are individually referred to at the end of species diagnoses (e.g., Color Plate 3.6). Note that color plates of some taxa also contain species of other taxa (as substrates or associated organisms); these are identified and referred to in the same way as the rest.

**The Process of Species Identification**

is a difficult undertaking even for the specialist, but fraught with pitfalls for the generalist. Before attempting to assign a name to an organism, the user of this book should keep the following in mind:

1. Many species display an often astonishingly wide variability of characters such as size, shape, color and other features— each organism is an individual first, and only the high degree of similarity in structure and function characterizes individuals as belonging to the same species. To distinguish intraspecific (i.e., within the species) variability from the consistent (though sometimes inconspicuous) differences that separate one species from another is an unending challenge on which the final (though not necessarily correct) verdict has to be left to the taxonomic expert.

2. The species selected for this book constitute only about a third of those recorded from Bermuda, and an even smaller fraction of the total number that can be expected to occur here. In many instances a species listed may be very similar to others not listed or not yet recorded. (The number of species recorded from Bermuda, as stated at the end of the diagnoses for orders, superfamilies or families, when compared with those actually described in the book, can be used—with caution!—as a crude indicator for the reliability of a species identification.) Furthermore, Bermuda's geographic isolation and location at the northern extreme of the subtropical climate zone make it likely that its species composition fluctuates naturally at fairly short periods, so that the appearance of previously unrecorded species has to be expected, just as the disappearance, natural or human-induced, of established species is relatively common.

Most people prefer pictures to keys. The easiest route to identification is first to find the closest fit between a specimen and an illustration, and then to verify and refine this approximation by checking the written diagnosis. The presentation of species and their characteristics within the natural system aims at introducing the reader to the hierarchic order inherent in a naturally evolved system, an experience that is not conveyed by dichotomous keys in which characters are used on the basis of their conspicuousness or exclusiveness rather than their systematic rank.

Once a tentative species identification has been made, it needs to be checked by going both upward in the system (i.e., by ensuring that *all* (!) of the diagnoses of the genus, family, order, etc., apply) and downward (i.e., that *all* (!) of the species characters apply). If only *one* character (which often means *one* word in a diagnosis!) disagrees, the identification has to be discarded as invalid, and the process repeated. Failure to arrive at a satisfactory identification means that the species in question is not among those represented in the book, and the reader is referred to taxonomic monographs and original descriptions given under REFERENCES.

**General References**

Because chapter references are restricted in scope and number, a few general books and treatises that may help the reader in acquiring a broader background in the field and gaining further access to specialized literature are listed here.

The most recent synopsis and classification of living organisms to the level of families is by PARKER (1982); a much briefer survey on the level of phyla is by MARGULIS & SCHWARTZ (1982), and references useful in identifying marine animals have been compiled by SIMS (1980). Marine macro-algae are diagnosed by TAYLOR (1960). The vast field of invertebrate zoology has been summarized by BARNES (1980); more detailed accounts of most phyla are by HYMAN (1940-1967) and KAESTNER (1979-1980). GIESE & PEARSE (1974-1979) provide exhaustive data on the reproduction of marine invertebrates, and COSTELLO et al. (1957) suggest methods for obtaining and handling marine eggs and embryos. A good introduction to collecting and preserving invertebrates is by LINCOLN & SHEALS (1979), and laboratory techniques such as culturing, fixation, and microanatomy can be found in MAHONEY (1973). HALSTEAD (1978) gives a complete treatment of poisonous and venomous marine animals of the world.

Addison E. VERRILL must be credited with the first comprehensive survey of Bermuda's marine fauna (and many other aspects of the island's natural history), which he published in several dozen papers between 1898 and 1923. Most of these are referred to in the respective chapters, although their nomenclature is often out-of-date. Knowledge of Bermuda's algae was first reviewed and enlarged by COLLINS & HERVEY (1917), and more recently updated by TAYLOR & BERNATOWICZ (1969). Much additional information on Bermuda's marine organisms can be found in the Bermuda Biological Station's Contributions, which have been issued since 1903 and in which series this book bears the number 1000.

# LIVING ORGANISMS

CHARACTERISTICS: *Complex systems of largely carbon-based molecules organized as cells, and capable of metabolism, self-replication, development and evolution by taking elements and energy from their non-living environment and/or other organisms.*

In the most primitive of living organisms, bacteria and blue-green algae, the genetic material of the cell nucleus is not separated from the cell plasm by a membrane nor is it in the form of stainable, large chromosomes; membrane-bounded chloroplasts and mitochondria are lacking. This condition is called prokaryotic. All other organisms have chromosomes, a nuclear membrane and complex cell organelles, the eukaryotic condition. Prokaryotes divide by simple fission whereas eukaryotes undergo a complex nuclear division sequence called mitosis. All groups except the majority of Fungi possess flagella as a means of cellular locomotion; these are simple and rotate at the base of their attachment in prokaryotes, but have a complex structure of usually 9+2 microtubules and beat along their full length in eukaryotes. Other features commonly used to separate kingdoms of organisms are the number of cells (uni- or multicellular), and whether nutrition is autotrophic, i.e., by photo- or chemosynthesis of organic matter from inorganic ingredients (as in plants, blue-green algae, and many bacteria), or heterotrophic, i.e., by absorption (fungi) or ingestion (animals) of organic matter from the surroundings. Viruses, because of their non-cellular nature and inability to perform any life functions outside their hosts, are now generally considered to be specialized parasitic cell fragments rather than complete organisms.

The traditional concept of two kingdoms, plants and animals, has gradually been modified to reflect new discoveries of organisms and changing concepts of phylogenetic relationships. Many biologists now favor a 5-kingdom system (Monera, Protista, Fungi, Plantae, Animalia), first proposed by WHITTAKER (1969) and recently illustrated by MARGULIS &

SCHWARTZ (1982). Although this approach resolves the artificiality of older systems that assign unicellular organisms to either plants or animals (when some in fact share characteristics of both), it creates the rather unwieldy and heterogeneous kingdom Protista—a level of organization rather than a monophyletic taxon. For this book we have adopted a semi-conservative 4-kingdom system that acknowledges Monera (p. 9) and Fungi (p. 17) as well-circumscribed kingdoms while retaining the traditional assignment of unicellular organisms to either Plantae (p. 26) or Animalia (p. 92). (The recently established Archaebacteria (WOESE 1981) are not considered here.) The most comprehensive, up-to-date synopsis of living organisms is by PARKER (1982).

Of an estimated 3-10 million species of organisms now alive, only about 1 million have been scientifically described. Bermuda's marine environment may well contain 10,000 species of which about 4,500 are known to science; 1,526 have been selected for this book.

# KINGDOM MONERA

## Bacteria and blue-green algae

CHARACTERISTICS: *Motile or non-motile prokaryotic ORGANISMS with heterotrophic or autotrophic nutrition, the latter by photo- or chemosynthesis.*

Major distinguishing characteristics within the kingdom are the structure of the cell wall, type of motility and mode of nutrition.

Of 3 recognized phyla (or superphyla), Bacteria are not treated here because of the great difficulties in classification and identification; and the newly described Prochlorophyta (LEWIN 1981), symbionts of Ascidiacea, have not been reported from Bermuda. The third phylum, Cyanobacteria (formerly called Cyanophyta), is well represented.

## Phylum Cyanobacteria
(=Cyanophyta)(Blue-green algae)

CHARACTERISTICS: *Uni- or multicellular, aerobic, photosynthetic MONERA with pigments in single thylakoids (not in pairs or stacks) which are not enclosed by membranes so as to form plastids. The predominating pigments are phycocyanin (blue) and chlorophyll a (green); in addition there are carotenoids and other phycobiliproteins, often including phycoerythrin.* The width of individual cells varies from less than 1 μm to 60 μm; cushion-shaped colonies may measure 10 cm across. The color ranges from a deep blue-green to deep brown or nearly black to violet red or red and (less commonly) yellow to green, depending on the proportion of pigments present. Many filamentous species are able to move by gliding slowly over surfaces.

There is no consensus regarding the systematics of the phylum. The system of GOLUBIC (1976) is followed for categories above the rank of genus. The species of blue-green algae as listed here are "form-species"; i.e., they match descriptions and names based on simple characteristics such as size and shape. What constitutes a "true species" in these non-sexually reproducing prokaryotic organisms is unknown, but a new and more soundly based system is being developed (RIPPKA et al. 1979). Of 2 classes, the generally inconspicuous Coccogoneae, although present in Bermuda, are not considered here. Of a total of about 1,500 species of blue-greens, about 65 are reported or known from marine collections in Bermuda; 13 of the more common or conspicuous species are included here. Endolithic species are not considered. Blue-greens have not been studied extensively in Bermuda, and it is likely that many more species will be discovered.

OCCURRENCE: In a wide range of aquatic or humid habitats, planktonic or benthic. Blue-greens in Bermuda are known mostly from summer collections. Little seasonal information is available. Attached blue-greens are widely distributed on rocky intertidal surfaces and similar surfaces below low tide, but never deeper than the penetration of effective light for photosynthesis and growth (<100 m). They also occur on sandy and other unstable surfaces, and on the roots or rhizophores of mangroves. On rocks, blue-greens are generally inconspicuous, but in protected mangrove inlets or seawater ponds large filamentous masses of a few species may become the dominant vegetation. Planktonic blue-green algae are restricted to a single genus (*Trichodesmium*) that occurs mainly offshore as small brownish red to yellowish clusters or fascicles of filaments just visible to the unaided eye.

Collecting from rock or sand surfaces is done mainly by hand, forceps or knife. Some filamentous aggregates or cushions may be a few to several centimeters in diameter or thickness and may be plucked. Delicate ephemeral dark coatings on sand can only be collected together with the sand in jars or plastic bags. Leathery, tightly attached patches may have to be removed with a small knife or chisel, preferably with a small fragment of the rock at their bases. Planktonic blue-greens may be collected together with other plankton by standard nets.

IDENTIFICATION: Except for a few species, examination for identification purposes requires a compound microscope. Small portions of the sample should be teased out and spread on a slide with pointed forceps and needles in a drop of seawater, and covered with a cover glass. The color of single cells or filaments can best be seen with bright field illumination, but cell structure, cross-walls, sheaths and internal granulation are more obvious using a phase contrast optical system. When filaments or cells are about 1 μm or less in size the color of individuals becomes difficult to judge.

The organisms consist of single cells, cells in aggregates or colonies, single rows or filaments of cells, or multiple series of cells, with or without branching. The microscopic organization of blue-greens and its terminology are as follows:

1. Cell: The basic unit, which may be spherical, rod-shaped (straight or curved) or a blunt-ended cylinder if 1 of many in a row (trichome).
2. Colony: A mass of cells that have divided in more than 1 plane and have remained adhering to one another.
3. Trichome: A row of cells that has resulted from the division of cells generally in 1 plane only, thus creating a thread.
4. Sheath: When present, a firm or soft gelatinous excretion from cells that forms a casing around trichomes or capsules around single cells or colonies. May be stained with ruthenium red, or seen with dilute India ink in living material.
5. Filament: When used correctly, a trichome together with its sheath.
6. Gliding: The only form of movement in blue-greens, a slow sliding of the trichome or cell along a firm material (e.g., glass, agar, mud, other trichomes). The movement is steady in one direction for a few minutes or longer, then commonly the direction is reversed. Movement may continue, however, in a single direction towards a source of brighter light (phototaxis). Rates are 1-35 mm/hr.
7. Hormogonium: A row of cells or short section of trichome that breaks free of the remainder of the trichome and by its own movement migrates away. Cells of a hormogonium are often different in shape and size from the parent trichome.
8. Heterocyst: A cell in a trichome of cells that appears different because of a paler, less granular

nature, a thicker wall, and usually larger size. Small wall thickenings called nodules often appear at 1 or both poles of the cell. These cells under a microscope are usually brighter and glisten. There is usually no more than 1 heterocyst for every 10 vegetative cells and often fewer. Many types of blue-greens never produce heterocysts. In some, heterocysts appear only as the end cell of a trichome.

9. Akinete (resting spore): Cell larger than the vegetative cells of a trichome, but not present in many types. It has a thicker wall and granular interior, and is often yellow or orange in contrast to the greener or redder vegetative cells. Often adjacent to a heterocyst, but in some species almost all vegetative cells develop into akinetes at the end of favorable growth conditions. Akinetes do not occur in most species found in Bermuda.
10. Necridial cell: A cell that dies in some filamentous types, leaving a collapsed or flattened wall case. These occur at sometimes regular intervals in some species, forming points of trichome separation or breakage.
11. Gas vacuoles: Bright refractile areas irregular in shape within the cells of some blue-greens. They often appear red because of the refraction of light around their edges. They are composed of smaller gas vesicles that contain the common atmospheric or cellular gas mixtures but are impermeable to water. They usually occur only in buoyant planktonic types, such as *Trichodesmium*.
12. True branch: The result of a division of an occasional cell of a trichome in a plane at a 90° angle from the normal plane of cell division, thus initiating a side trichome.
13. False branch: The result of a portion of a trichome erupting from the normal line of cells in a sheath and initiating a new side trichome without changing the plane of cell division. These eruptions or false branches often occur next to a heterocyst.

Samples composed of or containing blue-greens are best examined soon after collection, while still alive. Collections will keep for some days if quite dilute in seawater (i.e., about 1 part sample to 20 parts water). These should be kept cool (not cold) and in darkness or in dim light. For long-term preservation introduce about 4-5% of a concentrated solution of formaldehyde (ca. 37% formaldehyde) into the seawater with the samples, then store in the dark to keep color longer. Additions of a pinch of sodium borate (borax) to each jar will keep the solution alkaline and a few drops of glycerin will keep the specimens moist. Most blue-greens may also be preserved by simply drying on paper or muslin cloth and depositing in envelopes. They can then be rewetted for miroscopic examination. Most of the robust and tougher forms can also be frozen and stored for many years without change in appearance. However, after thawing, the material should be examined rapidly before decomposition (cell lysis) occurs.

BIOLOGY AND REPRODUCTION: Reproduction is by cell division which is a binary fission. A parent cell usually cleaves into 2 usually identical daughter cells that may separate or adhere. Filamentous forms may simply elongate and new cross-walls (septa) grow from the wall to the center. A few colonial forms may reproduce by the larger cells cleaving internally into many small cells (endospores or baeocytes), which are then released as daughter cells.

Growth occurs when cells divide and the daughter cells elongate or enlarge to parental size. In aggregates or colonies and in many filamentous forms, any or all cells may be capable of division. In others that have apico-basal polarity, only certain basal area cells may be capable of repeated divisions, but the other cells may continue to grow by elongation. There often is no life-span as such, because most cells are capable of dividing indefinitely. No sexual process (conjugation) is known for the blue-greens, but some of the genetic material (DNA) can be transferred by an as yet unknown mechanism from one type to another.

Dispersal mechanisms may involve migrating hormogonial stages or the sinking of akinetes in planktonic forms, but usually involve merely the carrying by water of single cells, aggregates or filaments that may have become free of their attachments. The gliding motility of various blue-greens may be preferentially toward, or away from, a higher light intensity, or concentration of particular substances (e.g., oxygen, sulfide). In some species, migrating behavior involves the disappearance and reappearance of filamentous blue-greens on the surface of marine sands with changing light conditions.

Blue-green algae photosynthesize in the same manner as green plants. Although all blue-greens contain chlorophyll *a* many are reddish in coloration owing to the predominance of phycoerythrin. Some of these species produce this red pigment only under the very dim light of shaded habitats or under the greenish light of many deeper submarine situations, a phenomenon called complementary chromatic adaptation.

REFERENCES: A comprehensive discussion of the biology and problems of classification of blue-greens may be found in STANIER & COHEN-BAZIRE (1977); recent books are by CARR & WHITTON (1973, 1982) and FOGG et al. (1973). Only one comprehensive book is available in English for the purpose of identification (DESIKACHARY 1959). However, a good summary of classification to the generic level, with many illustrations, is available in French (BOURRELLY 1970). The most used standard work for identification is in German (GEITLER 1932; reprinted 1971). One book (in French) covers marine species (FRÉMY 1934; reprinted 1972); another, more readily available book is by HUMM & WICKS (1980) but the taxonomic system of F. Drouet is used. This system does not stand up to critical evaluation and is not used here (see GOLUBIC 1976).

For Bermuda specifically, very little has been published except for accounts by COLLINS & HERVEY (1917) and BRITTON (1918). GEBELEIN (1969), SHARP (1969) and GOLUBIC & FOCKE (1978) discuss the layered mats and "biscuits."

**Plate 1**

**CL. HORMOGONEAE:** Multicellular, filamentous Cyanobacteria.

**O. OSCILLATORIALES:** Hormogoneae in which the trichomes are with or without sheaths; no heterocysts, akinetes or branching. Cells divide in 1 plane only, and all retain ability to divide; trichomes without apico-basal differentiation. No specialized cells except sometimes necridial cells that allow breakage of the trichomes; end cells may be of different shape. Motility of trichomes common but not of whole filament when prominent sheath is present. Hormogonia common, often gliding out of sheath. (Over 30 spp. from Bda; single family: Oscillatoriaceae.)

***Trichodesmium thiebautii*** Gomont (= *Oscillatoria thiebautii*): Genus with planktonic trichomes, reddish brown to yellowish, with gas vacuoles; occurs in fusiform fascicles or radial clusters of many trichomes; trichomes glide in contact with one another.—Species: Trichome width usually 6-10 μm, cells 1-2× longer than broad, granular with gas vacuoles; trichomes arranged in radially oriented clusters of up to about 3 mm in diameter. Common in upper 50 m in summer offshore, less commonly inshore; a major component of the phytoplankton in the Bermuda region.

***Oscillatoria miniata*** (Zanard.) Hauck ex Gomont: Genus with unbranched trichomes, no specialized cells except occasional necridia; tip cells may be narrower, tapered, or otherwise different from remaining cells; no apparent or persistent sheath; capable of gliding motility at most times.—Species (Bermuda variety): Trichomes 12-16 μm wide, cells much shorter than wide; trichome constricted and granular at crosswalls, slightly tapered over last few cells, tip cell rounded; prominent longitudinal "stripes" in cells, probably groups of pigment-bearing membranes; gliding motility. Trichomes a deep red color forming (in summer) a loose, felt-like covering (with luxuriant tufts) on shaded rhizophores of mangroves below low tide mark.

***O. borneti*** (Zukal) Forti: Genus as above.—Species: "Giant" variety in Bermuda; trichomes 40-56 μm wide, pale pink or salmon colored to reddish; cells much shorter than broad and with distinct reticulate pattern of internal polygonal sap vacuoles, separated by protoplasm, giving the appearance of compartments; no taper to trichome, end cell rounded; trichomes delicate and easily broken; gliding motility. Occurs in quiet seawater ponds below low tide mark (0-2 m) on soft sediments and attached to green algae, particularly *Caulerpa*.

***Phormidium penicillatum*** Gomont: Genus with trichomes similar to *Oscillatoria*, but with individual sheaths and much gel-like material excreted, which binds the filaments together into a common mass.—Species: Trichomes about 6 μm wide, cells about 1.5× longer than broad, non-granular, with prominent cross-walls; individual sheaths colorless but distinct. Forms large (3-10 cm broad), rather soft gel-like balls (mottled yellow-brown) attached by paler "stalk" to rocks below low tide mark to a few meters' depth; common and conspicuous in summer on rocks even in areas of fast current movement.

***P. hendersonii*** Howe: Genus as above.—Species: Trichomes 1.5-2.0 μm wide (usually 1.7-1.8 μm), cells 2-3× longer than broad, non-granular, yellow-brown in color; prominent cross-walls; sheaths thin and colorless. Forms brownish cushions or "biscuits" (1-5 cm broad) of entangled filaments (with interior accretion strata); cushions are generally 2-3× broader than high, spongy in texture; microscopically, the live filaments may be seen together with many empty sheaths. On intertidal rocks in protected bays that are generally bordered with mangroves, mainly in summer.

***P. corium*** Gomont: Genus as above.—Species: Trichomes 4-5 μm wide, cells somewhat longer than broad, non-granular, olive-brown in color; trichome is slightly constricted at cross-walls, which are distinct; sheath, when present, distinct but colorless. This species, together with an even more abundant filamentous blue-green (trichome 0.8-1.0 μm wide), forms large subtidal cushions or "biscuits," usually 1-8 cm broad and irregular in shape. *P. corium* is the same blue-green alga referred to in SHARP (1969) and GEBELEIN (1969) as *Oscillatoria submembranacea*. The cushions form mainly at the sandy bases of *Thalassia* or *Syringodium* where moderate currents occur. The cushions are brown to greenish white and contain large amounts of sand bound into a spongy but still friable structure. Most of the pigmented cells are near the surface of the structure, whereas the empty colorless sheaths remain inside still binding the sand. The "biscuits" occur mainly in summer.

***Lyngbya confervoides*** Agardh: Genus with trichomes each with a distinct sheath and not embedded in a common mucilaginous material.—Species: Trichomes about 14-24 μm wide, cells 3-8× shorter than broad, granular, olive-brown to purple;

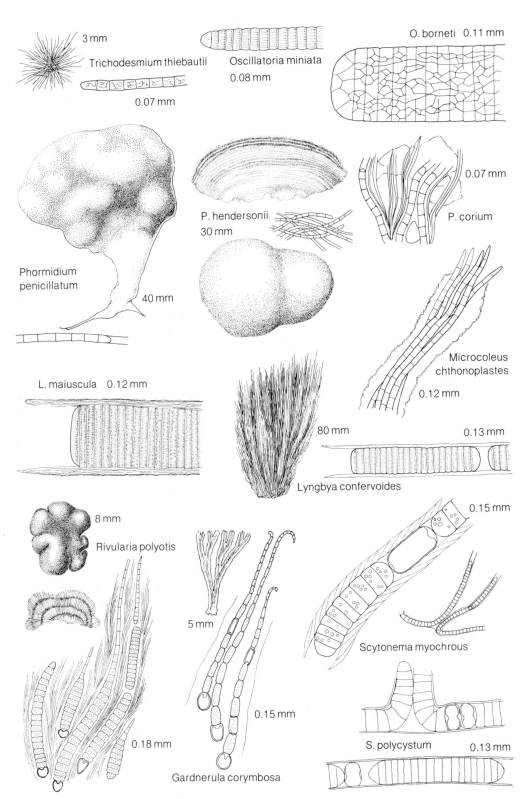

**1 CYANOBACTERIA (Blue-green algae)**

sheath colorless, 2-3 μm thick or more, often laminated. Forms dark (almost black) tufts or waving pendants attached to rocks between tide marks and below; very common over many parts of the coast; also tangled "bushes" of filaments in protected bays and seawater ponds.

***L. maiuscula*** Harvey ex Hooker: Genus as above.—Species: Trichome 30-45 μm wide or more, cells 6-15× shorter than broad; sheath colorless, 3-8 μm thick and often laminated; cells finely granular and dark yellow-brown under microscope. Forms tough hairy masses (brown to blackish) with filaments commonly over 20 cm long; often attached to mangrove rhizophores or rocks in quiet seawater ponds or inlets.

*Microcoleus chthonoplastes* Thuret: Trichome 3-4 μm wide, cross-walls indistinct, cell length 1-2× breadth, non-granular, end cell tapered to a tip; many blue-green to green colored trichomes twisted in a common colorless sheath. Forms greenish, felt-like mat on rocks and on consolidated sediment in intertidal areas protected from heavy wave action.

O. **NOSTOCALES:** Hormogoneae in which trichomes are with or without sheaths; heterocysts present under most circumstances; akinetes present under special conditions in some species; false branches occur in some species, usually next to 1 or more heterocysts.

F. **RIVULARIACEAE:** Nostocales having trichomes with attenuated apex and apico-basal differentiation; heterocyst ordinarily present at the wider (basal) end of the trichome; also intercalary heterocysts in some species; cell divisions in restricted "meristematic" region of trichome nearer base; akinete present next to basal heterocyst at times in some species; distinctive motile hormogonia may be formed. (Over 20 spp. from Bda.)

***Rivularia polyotis*** (Agardh) Bornet & Flahault: Genus with tapered trichome with basal heterocyst; packed in confluent sheaths in radial arrangement forming gel-like to leathery thallus (cushion or nodule); akinetes not present.—Species: Mature trichome in basal region 8-10 μm wide, cells shorter than broad except towards tapering apex, constricted at cross-walls, yellow-brown in color; sheath colorless to yellowish; trichomes arranged more or less radially and parallel, forming concentric strata. Hormogonium consists of separated row of short cells, germinates with immediate apico-basal differentiation. Dark brown to blackish leathery cushions (1-10 mm broad or more), circular to irregular in shape, on high intertidal rocks over most of rocky coastline; tough and best removed with knife.

***Gardnerula corymbosa*** (Harvey) De Toni (=*Polythrix corymbosa*): Genus and only species with tapered trichome similar to *Rivularia*, but many grow upright in a common sheath that "branches" dichotomously to form a tufted thallus similar to a candelabrum, usually a few millimeters high and sometimes over 1 cm; trichomes 6-8 μm wide at base, cells somewhat longer than broad but variable, trichome tapering into straight or curved tip. Green, sheath colorless to yellow (when older); heterocysts basal and intercalary; multitrichome branches 100-200 μm thick. Common over much of the coast on intertidal rocks, forming greenish turf-like patches, often protruding through sand; tough and best removed with knife.

F. **SCYTONEMATACEAE:** Nostocales in which the trichome is not tapered; sheath present, heterocysts present (sometimes rare); false branches occur singly or as twins; cell divisions usually restricted to intercalary region but near distal end of trichome or trichome "branch"; motile hormogonia formed; akinetes very rare.

***Scytonema myochrous*** Agardh: Genus with most of false branches as twins.—Species: Trichome 16-24 µm wide, cells shorter to slightly longer than broad, usually constricted at the cross-walls; greenish, granular or not; heterocysts yellowish, squat to elongate; sheath 4 µm thick, yellow and laminated when old. Forms moss-like turf (dark green to brown or blackish) from mid to high tide mark on rocks over many parts of the coast, wave-protected and exposed; filaments may be over 1 cm in length, usually less.

***S. polycystum*** Bornet & Flahault: Genus as above.—Species: Trichome 12-18 µm wide, cells quadrate to 2-3× shorter than broad, not constricted at cross-walls, non-granular, red; sheath colorless and non-laminated, 1-2 µm thick; heterocysts frequent, same width as vegetative cells but longer; twin false branches but infrequent. Forms tangled bushes of red filaments often several centimeters long, attached to rocks or mangrove rhizophores below low tide mark in quiet protected waters; easily mistaken for a red alga until examined microscopically.

R. W. CASTENHOLZ

# KINGDOM FUNGI

CHARACTERISTICS: *One- to many-celled eukaryotic ORGANISMS largely without flagella. Cell division by mitosis and division of cytoplasm; nutrition heterotrophic (never photosynthetic), mostly by absorption.*

The thallus may be unicellular or filamentous, septate or non-septate, usually non-motile in Eumycota, plasmodial in Myxomycota; motile stages occur in some classes. Cell walls of non-plasmodial forms are usually composed of chitin; they rarely contain cellulose. Reproduction is sexual or asexual, homo- or heterothallic; typical reproductive stages are spores. Chromatophores are absent, but pigments may occur. Fungi are saprobes or parasites, or symbionts of algae with which they form lichens.

The kingdom as defined by WHITTAKER (1969) has 2 phyla (divisions), one of which (Myxomycota, or Mycetozoa) has recently (OLIVE 1975) been assigned to the kingdom Protista. Myxomycota have not been reported from Bermuda. The other phylum, Eumycota, is represented, as are Lichenes (included here as an appendix, p. 24).

**Phylum Eumycota** (True fungi)

CHARACTERISTICS: *Mycelial or sometimes unicellular FUNGI without amoeboid or plasmodial phase.* Fruiting bodies of marine fungi are generally microscopic and rarely exceed 4 mm, in contrast to many terrestrial fungi, which are well known as macroscopic, fleshy mushrooms.

The taxonomic arrangement of many groups of Fungi is presently in a state of flux. Ascomycotina and Basidiomycotina are probably natural monophyletic groups, whereas the majority of the Deuteromycotina are asexual states (anamorphs) of Ascomycotina. The perfect states (teleomorphs) of most imperfect fungi are unknown. The total number of known fungi is about 45,000 but an estimated number of at least 250,000 species may exist. Of 5 subphyla (subdivisions), Mastigomycotina and Zygomycotina have not been reported from Bermuda. The following species have been recorded from marine habitats: 150 filamentous Ascomycetes, 4 Basidiomycetes and 55 Deuteromycetes, plus about 140 yeasts which can be assigned to the three subdivisions. Of 22 species recorded from Bermuda, 11 are included here.

OCCURRENCE: Fungi known from the marine and estuarine environment are saprobes or parasites of algae and sea grasses, or they occur saprobically on cellulosic substrates, such as wood or marsh plants. Lower fungi are found in marine animals and may cause diseases, e.g., in fish; others are commensals that live in the intestine of marine arthropods without damaging their hosts. Some Ascomycetes and Deuteromycetes form submarine, lichen-like associations with algae. Marine fungi occur wherever organic material is available in oceans and estuaries. They have been mostly found in intertidal habitats, but have also been retrieved from the deep sea to 5,315 m.

Fungi are collected with the substrates, such as rotting wood of pilings, roots and submerged branches of mangroves and other shoreline trees, algae, sea grasses or marsh plants. Some Ascomycetes and Deuteromycetes grow in organic debris washed up on beaches, and fruiting bodies may be attached to grains of sand. Ascospores and conidia of such arenicolous fungi accumulate in foam along the shore and can be caught by collecting sea foam with a skimming ladle.

IDENTIFICATION: Because of their small size, marine fungi have to be identified under the microscope, preferably with phase optics. Squash mounts of Ascomycetes, Basidiomycetes and Deuteromycetes in distilled water suffice to obtain ascospores, basidiospores and conidia, respectively. These reproductive organs usually show the characters needed for identification.

Higher fungi consist of thin filaments (hyphae) which, together, form a mycelium. Hyphae are usually septate, simple or branched, colorless (hyaline) or pigmented. Fruiting bodies of Ascomycetes (ascocarps) are mostly dark and contain asci (sing. ascus), sac-like cells that usually harbor 8 ascospores. Asci are thin- or thick-walled, and dissolve (deliquesce) to release ascospores or eject them forcibly. Two-walled (bitunicate) asci of the Loculoascomycetes have a thick outer wall (ectoascus), which ruptures at the apex to release the thin, expanding inner wall (endoascus), which in turn ejects the ascospores through its apex. Ascospores of marine fungi are often covered by gelatinous sheaths or provided with characteristic fiber-, spine- or thorn-like appendages. The interior of immature ascocarps of marine Pyrenomycetes is filled with a sterile cellular tissue (pseudoparenchyma) that forms chains of thin-walled, inflated cells (catenophyses) or dissolves when asci and ascospores are fully developed. Immature ascocarps of Loculoascomycetes are filled with filaments (pseudoparaphyses) which are attached with their bases and tips to the inner wall of the fruiting body. Asci usually develop at the base of the ascocarp centrum and grow upward between catenophyses or pseudoparaphyses. Fruiting bodies of Basidiomycetes (basidiocarps) form spores (basidiospores) on the outside of a large cell, the basidium. Hyphae of Basidiomycetes are characterized by buckle-like outgrowths (clamp connections) that connect the 2 cells resulting from a cell division. Reproductive organs of Deuteromycetes are conidia that are formed on the mycelium, on special cells (conidiophores) or inside fruiting bodies (pycnidia).

Higher marine fungi can be preserved by drying them together with the substrates (wood, leaves, algae), they can be frozen, or they can be preserved in 75% ethanol or 5% seawater-formalin. The best method for preservation is to mount fungi in glycerin in permanent slides by using a double-cover-glass method (KOHLMEYER & KOHLMEYER 1972). Saprobic fungi, especially wood-inhabiting species, can be cultured on artificial media in the laboratory.

BIOLOGY: Fungi are heterotrophic organisms with parasitic, saprobic or symbiotic modes of life. Obligate parasites are rare among the higher marine fungi. Some Ascomycotina induce the formation of berry-like galls in living Phaeophyta, e.g., in floating *Sargassum*. Others cause discolorations in the host plants, namely, black blotches or light-colored areas. Certain species of Ascomycetes cause a wild growth of hairs in the algal hosts. Infectious fungal diseases in algae are usually rare and have never reached epidemic proportions.

Symbiotic associations between higher fungi and marine algae can be compared with lichens. There are three types of such associations in marine habitats: (1) primitive submarine lichens, which are loose symbioses in which the algal or fungal partner, or both, are able to occur in a free-living state; (2) "true" lichens, forming obligate morphological-physiological units in which the fungal partner determines the habit of the association; and (3) mycophycobioses, i.e., obligate symbiotic associations between a systemic marine fungus and a marine macroalga in which the habit of the alga dominates. Primitive lichens usually occur as epiphytes among microscopic algae on the surface of macroalgae. (True lichens live above the highwater mark, e.g., on branches of mangrove trees, but are often exposed to salt spray.) Mycophycobioses border on parasitism, but both partners probably depend on each other and are in most cases not found separately in nature.

Saprobic higher fungi are involved in the decomposition of many different kinds of organic material in the marine environment and play a role similar to that of fungi in terrestrial habitats, e.g., in forest litter. Some degrade alginates, and others animal products (keratin of worm tubes, chitin of Hydrozoa and tunicin of tunicates), but the majority of the Ascomycetes, Basidiomycetes and Deuteromycetes attack cellulosic plant substrates (driftwood; wood of shoreline fortifications; untreated fishing vessels, especially in the tropics; submerged leaves, seedlings, trunks, roots and branches of mangroves; leaves and rhizomes of sea grasses, e.g., *Thalassia*). Wood-inhabiting fungi cause a type of destruction known as "soft rot" and, together with bacteria, seem to favor the attack of wood by shipworms (Teredinidae) and gribbles (*Limnoria*). Sand-inhabiting (arenicolous) Ascomycetes grow on organic substrates in sandy beaches and form their fruiting bodies on grains of sand or on calcareous shells of marine animals. Higher fungi found in the deep sea between 631 and 5,315 m differ morphologically from species living in littoral and sublittoral zones. Indigenous deep-sea fungi appear to grow slowly and need more than 2 years for the production of fruiting bodies.

Many species of higher marine fungi are physiologically adaptable and can endure extreme changes of salinity. The geographic distribution of marine fungi appears to be mainly controlled by temperature, and there are species typical of the tropics and of temperate zones. Some ubiquitous species occur in all oceans, regardless of the water temperatures. The Bermuda marine fungi are typical representatives of tropical and subtropical waters. Availability of oxygen is a limiting factor for the occurrence of the higher marine fungi, because low contents of dissolved oxygen in the water seem to prevent their growth.

REPRODUCTION: The development of Ascomycetes starts with the germination of an ascospore. The germ tube is a uni- or multinucleate hypha that continues to grow and develops a mycelium of assimilative hyphae. The mycelium may produce secondary structures, such as aerial hyphae, chlamydospores (thick-walled, non-deciduous asexual spores), conidia and sexual fructifications. Sexual reproduction starts with the copulation of male and female cells (often gametangia). The behavior of sexual nuclei differs in the various groups of Ascomycetes. After plasmogamy (plasma fusion) the nuclei pair without fusing and divide simultaneously. Ascogenous (ascus-producing) hyphae containing nuclear pairs (dikaryons) are formed. The upper dikaryotic cells of the ascogenous hyphae transform into asci in which nuclear fusion (karyogamy), meiosis and finally mitosis occur. Ascospores develop endogenously through free cell formation, the process by which the 8 nuclei are cut off by walls inside the immature ascus.

In Basidiomycetes karyogamy and meiosis are followed by exogenous formation of usually 4 basidiospores on basidia.

Deuteromycetes lack sexual generations and reproduce vegetatively by conidia. Most Deuteromycetes are imperfect states of Ascomycetes, and increasing numbers of connections between recognized representatives of the 2 groups are discovered. However, it can be assumed that most of Deuteromycetes have lost the ability to form sexual states.

REFERENCES: General treatises on marine fungi are by JOHNSON & SPARROW (1961), HUGHES (1975), and KOHLMEYER & KOHLMEYER (1979). Illustrations of 90 species are found in KOHLMEYER & KOHLMEYER (1964-1969).

The only paper on marine fungi of Bermuda is by KOHLMEYER & KOHLMEYER (1977); reports on species from the Sargasso Sea and drifting *Sargassum* are by KOHLMEYER (1971, 1972) and ULKEN (1979). A recent informal report on Bermuda's marine species including parasites is by MILLER (1984).

## Plate 2

**S.PH.** **ASCOMYCOTINA (Sac fungi):** Mostly microscopic Eumycota with branching, always septate hyphae. Characterized by the ascus, which produces typically 8 ascospores endogenously. Asci originate in fruiting bodies (ascocarps) that are absent in the yeasts. Asexual reproduction by conidia may occur besides sexual reproduction. Ascospores are non-motile, 1- to multi-celled. (15 spp. from Bda.)

**CL.** **PYRENOMYCETES:** Ascomycotina with a thin, unitunicate ascus membrane that often dissolves at or before maturity of the ascospores. Ascocarps ostiolate (with opening), flask-shaped; interior with pseudoparenchyma or basally attached paraphyses. (Of 3 orders with marine representatives, only Sphaeriales recorded from Bermuda.)

*Lindra thalassiae* Orpurt, Meyers, Boral & Simms: Genus with immersed or partly exposed, semiglobose or ellipsoidal, carbonaceous or coriaceous (leathery), papillate or epapillate ascocarps. Asci early deliquescing, cylindrical to clavate, thin-walled. Pseudoparenchyma filling venter of immature ascocarps, at maturity deliquescing without forming catenophyses. Ascospores filiform, multiseptate, hyaline, with inflated tips, but without appendages or mucus-filled apical chambers.—Species with (220-) 230-390 μm long ascospores that taper towards the apices and are 1-1.5 μm in diameter at the tips. They are 14-26-septate and mostly curved (S-, U- or alpha-shaped). Ascocarps frequently immersed in dead or senescent leaves of *Thalassia*, but also found in air vesicles of *Sargassum*, causing a softening of the algal tissue ("raisin disease"). The ascospores are found in foam

# EUMYCOTA

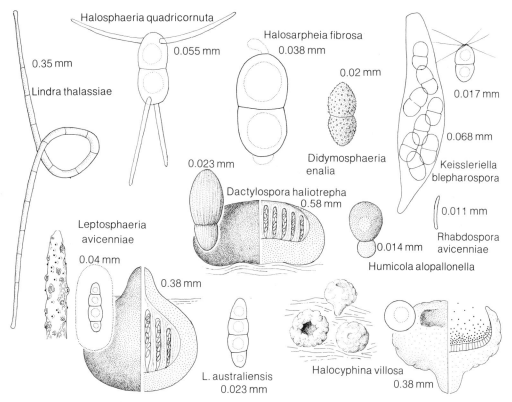

**2 EUMYCOTA (Fungi)**

along the shore near beds of *Thalassia*.

***Halosphaeria quadricornuta*** Cribb & Cribb: Genus with immersed or partly exposed globose to ellipsoidal, coriaceous or membranaceous, rarely subcarbonaceous, papillate ascocarps. Asci early deliquescing, clavate or subfusiform pedunculate, thin-walled throughout. Pseudoparenchyma filling centrum of immature ascocarps, at maturity deliquescing or forming catenophyses. Ascospores ellipsoidal, rarely rhomboid or cylindrical, 1-septate, hyaline, with apical or also lateral appendages.— Species with 20-35 μm long ascospores that bear at each end 2 subterminal, cylindrical, attenuated, stiff appendages, 20-37 μm. Catenophyses absent. In submerged wood, e.g., dead branches of *Conocarpus*. Most frequently found in tropical or subtropical waters, on wood or bark of a variety of substrates, especially mangroves, but also on drifting coconuts. The ascocarps occur often in or under calcareous linings of empty shipworm tubes.

***Halosarpheia fibrosa*** J. Kohlmeyer & E. Kohlmeyer: Genus with immersed or partly immersed, obpyriform to subglobose, coriaceous, papillate ascocarps. Asci persistent, clavate, pedunculate, thick-walled below the apex, thin-walled at the base, at maturity breaking off at the base from the ascogenous tissue. Pseudoparenchyma filling centrum of immature ascocarps, eventually forming chains of cells (catenophyses). Ascospores ellipsoidal, 1-septate, hyaline, with a cap-like appendage at each apex. After release into the water appendages become soft, scoop-like, and transform into coils of long fibers.—Species with 32-44 μm long ascospores. Frequently in submerged wood or bark of roots and branches of mangroves (*Avicennia, Conocarpus, Rhizophora*) and other intertidal wood.

CL. **LOCULOASCOMYCETES:** Ascomycotina with a thick, bitunicate ascus membrane that does not dissolve before maturity of the ascospores. Ascospores are forcibly ejected after rupture of the outer ascus wall and expansion of the inner wall. Ascocarps often enclosed in a stroma. (Of 3 orders with marine representatives found in Bermuda, only Dothideales are not treated here.)

O. **PLEOSPORALES:** Loculoascomycetes with ostiolate, flask-shaped ascocarps.

*Didymosphaeria enalia* Kohlmeyer: Genus with immersed or erumpent, globose or flask-shaped, often clypeate, papillate ascocarps. Asci 4- or 8-spored, thick-walled, cylindrical or clavate, pedunculate. Pseudoparaphyses wide or filiform. Ascospores elongate, 1-septate, brown.—Species with 16.5-23 μm long, ellipsoidal, dark brown, verrucose (warty) to verruculose ascospores. In intertidal wood, often in dead roots of mangroves.

*Keissleriella blepharospora* J. Kohlmeyer & E. Kohlmeyer: Genus with immersed, rarely erumpent, globose or semiglobose, often clypeate, rarely papillate ascocarps. Asci 4- to 8-spored, often thin-walled with thickened apex, cylindrical, clavate or saccate. Pseudoparaphyses thin, attached at base and apex. Ascospores 1- or multiseptate, hyaline, often with gelatinous sheath or apical appendages.—Species with 12-21 μm long, ellipsoidal, thin-walled, appendaged ascospores; 4-7 radiating terminal setae, about 13 μm long at the apex. Common in bark of seedlings, roots, submerged branches and bases of trunks of *Rhizophora*.

*Leptosphaeria avicenniae* J. Kohlmeyer & E. Kohlmeyer: Genus with immersed, mostly broadly conical or flattened, papillate ascocarps. Asci 8-spored, thick-walled, clavate to subcylindrical, short pedunculate, in some species with a refractive apical apparatus. Pseudoparaphyses numerous, filamentous. Ascospores with 3 or more transverse septa, yellowish brown to subhyaline, ellipsoidal, fusiform or cylindrical, rarely worm-like.—Species with 17.5-25 μm long, ellipsoidal, 3-septate, hyaline ascospores, enclosed in a 2.5-5 μm thick gelatinous sheath. Asci cylindrical, without apical apparatuses. In pneumatophores, bark and wood of submerged branches and tree trunks of *Avicennia*.

*L. australiensis* (Cribb & Cribb) G. C. Hughes: Genus as above.—Species with 19-27 μm long, ellipsoidal, fusiform or clavate-fusiform, 3-septate, hyaline ascospores without sheath or appendages. Asci clavate-fusiform, relatively thin-walled, with an apical apparatus. Common in bark and wood of submerged dead parts of shoreline trees and in other intertidal wood; also in prop roots and rooted seedlings of *Rhizophora*.

O. **HYSTERIALES:** Loculoascomycetes with open cup- or saucer-shaped ascocarps (apothecia) in which asci are uncovered at maturity.

*Dactylospora haliotrepha* (J. Kohlm. & E. Kohlm.) Hafellner: Genus with superficial, sessile, flat or convex, fleshy-leathery, brown to black ascocarps that are rooted in the substrate. Pseudoparaphyses clavate, surpassing the asci. Asci 8-spored, clavate, pedunculate, apically thick-walled; the ectoascus forms a gelatinous sheath that reacts blue with iodine. Ascospores ellipsoidal or obovoid, 1-septate in the lower 1/3, brown.—Species with (15-) 18-27.5 (-31.5) μm long ascospores with delicate, forked, longitudinal striations. Ascocarps 0.36-1 mm in diameter. On intertidal wood and bark especially of shoreline trees.

S.PH. **DEUTEROMYCOTINA** (=FUNGI IMPERFECTI) (Imperfect fungi): Microscopic Eumycota with branching, septate hyphae. Characterized by the absence of a sexual state. Reproducing asexually by non-motile conidia, which are often imperfect states of Ascomycetes. (6 spp. from Bda.)

CL. **HYPHOMYCETES:** Deuteromycotina with conidia formed on hyphae or superficial conidiophores, not in or on fruiting bodies.

*Humicola alopallonella* Meyers & Moore: Genus with 1-celled, globose, rarely obovoid or pyriform, smooth, light to dark brown conidia.—Species with obpyriform, ovoidal or subglobose, 1- or 2-celled, fuscous (dusky) conidia, 10-22.5 (−37.5) μm long, (8.5-) 10-18 μm in diameter; when 2-celled, apical cell large and dark, basal cell small and light brown. On intertidal wood, e.g., submerged parts of mangroves, often on the calcareous lining of empty shipworm tubes.

CL. **COELOMYCETES:** Deuteromycotina with conidia formed inside a fruiting body.

*Rhabdospora avicenniae* J. Kohlmeyer & E. Kohlmeyer: Genus with subglobose or depressed, immersed or erumpent, ostiolate pycnidia. Conidia bacilliform or filiform, hyaline.—Species with 9-12.5 μm long, 1-celled, straight or curved conidia. In bark of pneumatophores and trunks of *Avicennia*.

S.PH. **BASIDIOMYCOTINA:** Eumycota with macroscopic or rarely microscopic fruiting bodies (basidiocarps). Characterized by the basidium, which produces typically 4 non-motile basidiospores exogenously. (1 sp. from Bda.)

CL. **HYMENOMYCETES:** Basidiomycotina with well-developed

basidiocarp, opening before basidiospores are mature, or having completely exposed basidia; basidiospores forcibly ejected (ballistospores).

O. **APHYLLOPHORALES:** Hymenomycetes with flat, sessile, stalked, clavate or cup-shaped basidiocarps.

*Halocyphina villosa* J. Kohlmeyer & E. Kohlmeyer: Monotypic genus with funnel- or cup-shaped, pedunculate, whitish or yellowish, hairy basidiocarps. Basidia 4-spored, clavate or cylindrical with a narrow base, produced on the inner side of the fruiting body. Basidiospores subglobose, 1-celled, smooth, hyaline.—Species with basidiospores 8-10.5 μm diameter. In wood, often in submerged parts of mangroves or other shoreline trees; mostly protected in cracks or depressions at the sea-air interface.

J. KOHLMEYER & E. KOHLMEYER

## Appendix: Lichenes (Lichens)

CHARACTERISTICS: *Symbiotic ORGANISMS consisting of a fungal component and a unicellular or filamentous alga, the fungus making up the bulk of the plant body (thallus) and forming sexual reproductive structures.* Three major growth forms: crustose, foliose (leafy) and fruticose (bushy). Thallus 1-10 cm or more in diameter, growing on tree bark or rock.

Classified under Class Ascomycetes of Fungi or maintained as a separate class Lichenes. Of approximately 15,000 species with about 100 in Bermuda, only a few are associated with the marine environment; 3 are included here.

OCCURRENCE: On trees and shrubs or rocks in the littoral zone and rocks in the intertidal and spray zone. No tropical species grow permanently submerged.

Corticolous (bark-dwelling) species can be removed with a knife and air dried, later mounted on $3 \times 5$ cards. Saxicolous (rock-dwelling) species are collected with hammer and chisel and also glued to cards to minimize damage during later handling.

IDENTIFICATION: The great range in characters and sizes used to identify lichens demands various techniques. Crustose and foliose species with apothecia (disc-shaped fruiting bodies) are sectioned with a razor blade; the sections are mounted on slides in water and examined under a compound microscope for spore details that separate genera. The presence of a rim around the disc (lecanorine structure) is also important. Foliose and fruticose species are also distinguished by morphological characters seen under a low-power dissecting microscope, noting especially thallus size, growth form, color and presence of diaspores (organs for vegetative reproduction such as powder-like soredia or finger-like, erect isidia). Finally, chemical color tests or even chromatographic analysis are necessary in many cases for species identification.

BIOLOGY: Virtually nothing is known of the life cycle of lichens and their means of reproduction. Most crustose species are presumed to propagate with microscopic, wind-borne spores, the foliose species with vegetative diaspores.

DEVELOPMENT: Most species are perennial and grow during the rainy season. Ra-

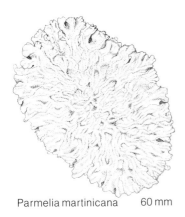
Parmelia martinicana 60 mm

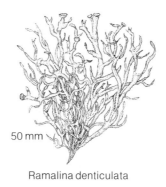
Ramalina denticulata 50 mm

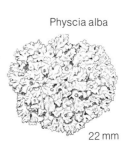
Physcia alba 22 mm

## 3 LICHENES (Lichens)

dial growth rates are estimated at 2-5 mm/yr, life-span at 10-25 yr.

REFERENCES: For general information see AHMADJIAN & HALE (1974) and HALE (1979).
Some foliose species occurring in Bermuda are treated in HALE (1971). The only comprehensive study of Bermuda lichens is by RIDDLE (1916).

## Plate 3

F. **PARMELIACEAE:** Thallus foliose, lobes fairly broad; apothecia lecanorine, the spores simple, colorless.

*Parmelia martinicana* Nyl.: Thallus pale tannish gray, 3-9 cm broad, adnate (hugging) on bark. Upper surface coarsely isidiate, the isidia becoming irregularly thickened. Lower surface black, moderately rhizinate (root-like). Apothecia lacking. Common on trees near the shore.

F. **RAMALINACEAE:** Thallus fruticose, yellow, branches flattened; apothecia terminal or lateral; spores 2-celled, colorless.

*Ramalina denticulata* (Eschw.) Nyl.: Thallus tufted, fruticose, 1-3 cm long, strongly flattened. Apothecia subterminal. Common on shrubs and trees in the littoral zone.

F. **PHYSCIACEAE:** Thallus foliose, whitish, lobes narrow; apothecia lecanorine; spores 2-celled, brown.

*Physcia alba* (Fée) Müll. Arg: Thallus closely adnate on bark, 1-2 cm broad. Upper surface smooth, lacking soredia; lower surface white, corticate, sparsely rhizinate. Apothecia common, 0.5-1 mm in diameter. Common on trees and shrubs near the shore.

M. E. HALE

# KINGDOM PLANTAE

## (Plants)

CHARACTERISTICS: *Mostly non-motile eukaryotic ORGANISMS usually with autotrophic nutrition by photosynthesis.*

As defined here, plants range from single-celled, microscopic plankton organisms to large forms with either comparatively little tissue differentiation (such as green, brown and red algae), or the vascular plants with distinct tissue and organ development, the latter making up most of the familiar terrestrial vegetation. Plant cells are usually enclosed in a rigid cell wall made of cellulose, and contain photosynthetic pigments (e.g., chlorophyll) that enable them, in the presence of sunlight, to synthesize organic molecules from carbon dioxide, water and inorganic nutrient salts. Only some have lost this ability and live as saprophytes or parasites. Reproduction is asexual by spores or by fragmentation of the vegetative body; or sexual by the fusion of two reproductive cells (gametes) that may be similar (isogamous), dissimilar (anisogamous), or differentiated as a smaller, motile "sperm" and a larger, non-motile "egg" (oogamous).

Phyla (=divisions) are separated according to the presence of various pigments, storage products, complexity of the plant body, production of an embryo following fertilization, production of seeds and development of a vascular (conducting) system for the movement of materials throughout the organism.

The kingdom as defined here comprises 15 living phyla of which 8 (mosses, horsetails, ferns, conifers, etc.) do not have species that live in or near the ocean and are therefore not included. Phyla represented are Chlorophyta (p. 27), Phaeophyta (p. 41), Rhodophyta (p. 50), Chrysophyta (p. 66), Euglenophyta (p. 74), Dinophyta (p. 75) and Anthophyta (p. 79), the latter with a few marine grasses and a larger number of species that are regularly encountered in shore environments. Anthophyta, in older systems, are called Angiospermae within the now abandoned taxon Spermatophyta.

## Phylum Chlorophyta (Green algae)

CHARACTERISTICS: *Unicellular, colonial or multicellular PLANTAE, often as coenocytes (without cell walls), which are simple or form complex leaf- or bush-like structures; cells uni- or multinucleate. Chloroplasts variable in shape and number, containing pigments similar to those found in vascular plants; chlorophylls a, b, and various carotenoids (carotenes and xanthophylls) most common. Food reserve generally starches stored in pyrenoids; cell wall composed mainly of cellulose, and often encrusted with calcium carbonate.* The size of Bermuda's species ranges from microscopic to 50 cm, the color from all shades of green to blackish, yellow or almost white. Though some are of a soft, spongy or jelly-like consistency, others are papery, leathery or calcareous.

Of 3 classes, 2 (Charophyceae and Prasinophyceae) are not represented in Bermuda. The third, Chlorophyceae, presently comprises 15 orders of which 5—representing the more conspicuous, multicellular forms—are listed here. Of some 800 marine species, about 100 have been reported from Bermuda; 41 are included in this account.

OCCURRENCE: Although the great majority of species are found in the plankton and benthos of fresh water, marine forms are an important and diverse component of most shallow (to 200 m) bottoms. The intertidal and immediate subtidal zones harbor the most luxuriant growth as well as the greatest number of species. Many species are found adjacent in the same habitat, whereas others form distinct bands according to depth, or grow on each other (epiphytic) or on shells or exoskeletons of marine animals (epizoic). Some unicellular forms are endophytic (in other algae), others endozoic (e.g., in opisthobranch mollusks). Another type of symbiosis is that with fungi (as lichens, p. 24). Endolithic species, finally, are embedded in the surface of rocks, or form bands in the periphery of coral skeletons.

Collect with mask and snorkel in shallow water, or by hand at low tide. In deeper water, rakes, dredges or SCUBA are commonly employed.

IDENTIFICATION: A microscope is necessary for the identification of all unicellular green algae and most of the simple filamentous forms, and is usually essential for the accurate, specific identification of most macroscopic forms. Whenever possible, algae should be studied live (they will do well, even for extended periods, if kept uncrowded in running seawater, but will spoil quickly in standing water). Endolithic forms are separated from the substrate by dissolving small pieces of coral in a 5% solution of glacial acetic acid, then fixed in Holm's fixative (10 g iodine, 20 g potassium iodide, 20 g glacial acetic acid, 200 ml distilled water) before being spread on a slide for microscopic examination (LUKAS 1974).

Filamentous forms may be either regularly septated and unicellular, or multinucleate and coenocytic (without cell walls). The former filaments are evenly compartmentalized, with each compartment (cell) provided with one or several nuclei (smmi-coenocytic, e.g., *Rhizoclonium*), whereas coenocytic filaments are generally free of septations, or if such structures occur they are rare and usually associated with the delimiting of reproductive organs (e.g., *Codium*). In many species the filaments do not remain monosiphonous (individual filaments) but become polysiphonous (fused together) to form thin leaves (foliose) or bush-like shapes that superficially resemble simple vascular plants. In *Codium, Caulerpa, Udotea* and others a mass of intertwining, anastomosing, coenocytic filaments are fused together to form the characteristic plant body (thallus). In septate species, divisions of the nucleus in each of the filaments occur primarily in a single plane; the resulting filaments are unbranched (uniseriate). If nuclear division proceeds along a second plane; branching occurs. In *Ulva*, divisions of the nucleus occur regularly along 2 planes, with divisions in a 3rd plane resulting in a thallus that is sheet-like and 2 cells thick (distromatic). Cell walls may contain mucilage (a glue-like substance). Other criteria used in species identification are the structure of the reproductive organs (gametangia), the flagellation of zoospores and/or gametes and the deposition of calcium carbonate in the cell wall.

Preserve in 3-4% formalin-seawater, or press as herbarium specimens.

BIOLOGY: Little is known regarding the periodicity or seasonality of most algae in Bermuda; it is quite clear, however, that local and seasonal variations exist. It is not uncommon to locate a particular species in the same habitat for years, and then find it to be absent for no obvious reason. Many algae are annual plants that germinate from asexual spores or zygotes. In several cases, the large vegetative plant, which represents only 1 phase of the life cycle, may die back at the end of its growing season; all that remains of the alga is a holdfast or stolon that regenerates new fronds when conditions are favorable again. A major factor determining the presence and condition of algae in Bermuda is the grazing by various herbivorous feeders, particularly sea urchins and fishes. Distinct cropping of algal stands is evident several times during the year.

REPRODUCTION: Both sexual and asexual reproduction are common. Asexual reproduction in filamentous forms commonly occurs by fragmentation of the thallus owing to water movement or other forces. The production of flagellated spores (zoospores) in sporangia is a widespread and important method of asexual reproduction among the filamentous and multicellular species. Small to large numbers of zoospores are produced by nuclear division (mitosis), released from sporangia and disseminated into the water. If attachment occurs at a favorable site, the spores lose their flagella, germinate and grow into adults of a genetic makeup identical to the parent plant. Sexual reproduction may be isogamous (both gametes morphologically identical), anisogamous (gametes morphologically dissimilar with the "female" usually being larger) or oogamous (a motile "male" gamete and a larger, non-motile egg). Adult plants may be homothallic (male and female gametes are produced on the same thallus) or heterothallic (male and female gametes are produced on separate plants). Three sexual life cycles are known in green algae, haplontic, diplontic and diplohaplontic patterns. In the haplontic pattern (mostly unicells), meiosis occurs immediately after fertilization (postzygotic), resulting in haploid mature plants, a diploid stage being limited to the zygote. The haploid plant carries both asexual and sexual reproductive organs; hence, there is no alternations of generations. In the diplontic pattern (e.g., *Codium*), the mature plant is diploid; reproductive cells (gametangia) undergo reduction division producing haploid gametes; fertilization results in a diploid zygote that divides by mitosis into the mature plant. In the diplohaplontic pattern (e.g., *Ulva*), the normal sexual cycle involves the alternation of a haploid with a diploid generation. The diploid phase (sporophyte) produces asexual, haploid zoospores by meiosis, the resulting plant (gametophyte) is haploid and produces gametes by mitosis. The zygote formed by fertilization is diploid and grows into the diploid generation. With few exceptions, the gametophyte and sporophyte generations are similar in appearance (isomorphic).

REFERENCES. For thorough, general treatments see DAWES (1981), BOLD & WYNNE (1978), ROUND (1973) or PRESCOTT (1968); biology, physiology and ecology are covered in LOBBAN & WYNNE (1981), and a comprehensive treatment of physiology and biochemistry is available in STEWART (1974).

A taxonomic account of the green algae of the West Indies is found in CHAPMAN (1961). COLLINS & HERVEY (1917) and, more recently, TAYLOR & BERNATOWICZ (1969) treat the Bermuda species; seasonal aspects of the Bermuda algal flora are considered by BERNATOWICZ (1952), and

LUKAS (1974) deals with endolithic algae. The most extensive coverage of the algae of Bermuda is by TAYLOR (1960).

**Plate 4**

O. **ULVALES:** Filamentous to foliaceous Chlorophyceae; filamentous forms solid, tubular or ribbon-like; foliaceous forms 1 cell (monostromatic) or 2 cells thick (distromatic). Most plants erect from a basal holdfast; cells having a single, parietal chloroplast with pyrenoids and a single nucleus. Asexual reproduction by fragmentation and zoospore formation; sexual reproduction by formation of iso- or anisogametes; sexual cycle haplontic or diplohaplontic.

F. **ULVACEAE:** Filamentous, tubular or broad, sheet-shaped Ulvales; 1 or 2 cells thick; attached or free floating. (9 spp. of *Enteromorpha*, 1 sp. of *Monostroma* and 2 spp. of *Ulva* from Bda.)

*Enteromorpha flexuosa* (Wulf.) J. Ag.: Genus capillary or broadly tubular; filaments rounded in cross section, simple or branched, at times constricted along length.—Species generally unbranched or only rarely so at base; simple, intestiform above, tapered to a semisolid base; to 15 cm long, 5-10 mm at its widest point, occasionally flattened and constricted along its length. Found floating or attached to various substrates in the intertidal zone.

*E. lingulata* J. Ag.: Genus as above.—Species simple or branched near the base; branches few to many, the younger initiating lower than the older ones; tubular above, seldom over 2 mm diameter; mostly smooth, not constricted, to 10 cm long. Attached, often abundant and dense near low tide level or in temporary pools in the intertidal zone.

*Monostroma oxyspermum* (Kütz.) Doty: Genus monostromatic; tubular at first, usually expanding into a broad, flat, thin blade; attached to the substrate by a small basal disc.—Species in tufts, light green, pale, soft; to 10 cm high, superficially resembling *Ulva*, but much thinner and more delicate. Attached to rocks or other solid objects in the intertidal zone or below; often in quiet, calm tide pools and on mangrove roots. Common during the summer.

*Ulva fasciata* Delile (Sea lettuce): Genus foliaceous and firmly attached to the substrate; adult plant distromatic, with orbicular, lobed or elongate blades.—Species typically divided into long, lanceolate segments, 10-150 cm. Intertidal or near-subtidal belt, on rocks, in water of moderate to strong action; when growing under severe wave action, the alga is compact and dwarfed.

*U. lactuca* L.: Genus as above.—Species orbicular or irregular in outline; not divided into linear segments but more or less broad and sheet-like. Attached to rocks at the intertidal zone; attains best growth in shallow, quiet pools. (Color Plate 1.11.)

O. **CLADOPHORALES:** Filamentous, simple or branched, uniseriate Chlorophyceae; cells multinucleate, with cell divisions occurring independently of nuclear divisions; semicoenocytic; walls thick, with or without mucilage. Reproduction asexually by fragmentation or by biflagellated zoospores; sexual reproduction isogamous; diplohaplontic, isomorphic.

F. **CLADOPHORACEAE:** Cladophorales forming basally attached filaments that are simple or branched; multinucleate cells having reticulated chloroplasts. (7 spp. of *Chaetomorpha*, over 20 spp. of *Cladophora* and 5 spp. of *Rhizoclonium* from Bda.)

***Chaetomorpha linum*** (Müll.) Kütz.: Genus with cylindrical, basally attached, unbranched filaments; cells multinucleate, firm, heavy-walled, barrel-shaped.—Species light green, with entangled or twisted, loose, stiff filaments; cells swollen, cylindrical, firm. Common in protected pools and ponds, often associated with mangrove roots during the warmer months.

***C. media*** (C. Ag.) Kütz.: Genus as above.—Species dark green; to 10 cm high, attached; filaments erect, stout, may taper toward basal attachment. Abundant on intertidal rocks exposed to moderate surf; found throughout the year, but more frequently during the warmer months.

***Cladophora fascicularis*** (Mert.) Kütz.: Genus filamentous, branched, commonly bushy, attached by rhizoids, prostrate or erect; sometimes forming free-floating masses; growth apical or intercalary; cells multinucleate, chloroplasts reticulate.—Species 20-50 cm high, freely branching; branchlets fasciculate (parallel); apical cells tapering; a relatively massive alga. In the intertidal zone on rock in protected situations, or forming conspicuous green, free-floating strands; found throughout the year.

***C. fuliginosa*** Kütz.: Genus as above.—Species in mats or tufts, 2-8 cm high; bright green when living, blackening upon drying or pressing; main filaments erect or recurved above; lower branches dichotomous, alternate or unilateral, coarse throughout; cells cylindrical, terminal ones many times longer than broad. Abundant throughout the year; found intertidally or on rocks exposed to surf or moderate wave action.

***C. prolifera*** (Roth) Kütz.: Genus as above.—Species tufted, commonly occurring as a ball or entangled filaments (to 5 cm or more in diameter). Filaments coarse and stiff, dark green, becoming dark olive or black upon drying; cells 4-6× longer than broad, apical ones with blunt tips. On rocks and reefs below low tide mark; common in shallow inshore waters where it may form thick unattached mats.

***Rhizoclonium hookeri*** Kütz.: Genus with attached filaments that are dark and spreading; branching irregular.—Species filamentous, stiff, entangled, branched, 40-80 μm in diameter; cells thick-walled. Common on rocks at the intertidal zone. Often found entangled with other small brown and red algae on *Rhizophora* roots.

O. **CAULERPALES:** Coenocytic and siphonous Chlorophyceae; septa occur sparingly, usually only to delimit reproductive structures. Plants single, uniaxial; or more complex multiaxial, branching forms that develop into flat, broadly cylindrical, segmented, feather- or bush-like structures.

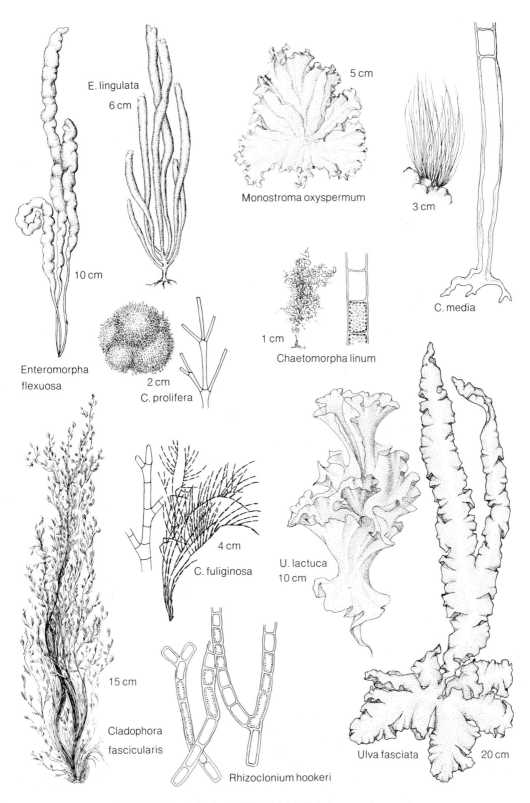

**4 ULVALES, CLADOPHORALES (Green algae 1)**

## Plate 5

**F. CODIACEAE:** Soft, spongy, attached and erect Caulerpales with generally dichotomous rope-like branches composed of intertwined or anastomosing coenocytic filaments. Sexual reproduction by biflagellated anisogametes produced in simple gametangia; life cycle diplontic. (5 spp. of *Codium* and 3 spp. of *Avrainvillea* from Bda.)

***Codium decorticatum*** (Woodw.) Howe: Genus usually erect and dichotomously branched; texture soft, spongy, uncalcified; rope-like branches composed of interwoven, coenocytic filaments, each of which terminates in an expanded, swollen tip (utricle).—Species large, erect (10-15 cm or more); typically flattened at the dichotomies of branches, which reach 10 mm in diameter; utricles 200-500 μm in diameter. In shallow and sublittoral waters; attached to bottom or on rocks and old coral heads. Common year round, abundant in spring and early summer. (Color Plate 1.13.)

***C. intertextum*** Coll. & Herv.: Genus as above.—Species gregarious, usually in colonies, prostrate, sparingly and not dichotomously branched; branches slender and delicate, to 5 mm in diameter; utricles 70-150 μm in diameter. In shallow waters during spring and summer, often abundant; not common during colder months.

***C. taylori*** Silva: Genus as above.—Species small, rarely to 10 cm high; branches dense, cylindrical and flattened below the dichotomies; branching irregular and not often evenly dichotomous; utricles 125-325 μm in diameter, spindle-shaped or cylindrical. In shallow waters, on old coral heads and rocky bottoms; grows year round, but not common.

***Avrainvillea nigricans*** Dec.: Genus usually with a stalk and a terminal, flattened, expanded blade (flabellum); both composed of interwoven, coenocytic filaments to 75 μm in diameter. Uncalcified, attached to substrate with large, bulbous mass of rhizoids.—Species 10-20 cm high, blackish green or olive-green; sessile, but more commonly with a terete or slightly flattened stalk to 5 cm long; flabellum leathery, soft or spongy. Usually in quiet, muddy waters, ponds or coves; often completely or partially covered with sediment. Common in winter and summer. (Color Plate 1.9.)

**F. UDOTEACEAE:** Multiaxial, macroscopic Caulerpales; several representatives deposit calcium carbonate. Sexual reproduction, where known, by anisogametes produced in compound, multicellular gametangia. (3 spp. of *Udotea*, 4 spp. of *Halimeda* and 3 spp. of *Penicillus* from Bda.)

***Udotea cyathiformis*** Dec.: Genus with thallus composed of a single stalk terminating or gradually expanding into a flat, fan-like or funnel-shaped blade (flabellum); entire structure made up of dichotomously branched, cocnocytic filaments; usually erect and well anchored into the sediment.—Species to 15 cm high; flabellum funnel-shaped, to 5 cm in diameter; walls of funnel firm but occasionally splitting, composed of closely packed, parallel filaments. Below low tide zone, growing and usually submerged in sand or finer sediment; found throughout the year, but not common.

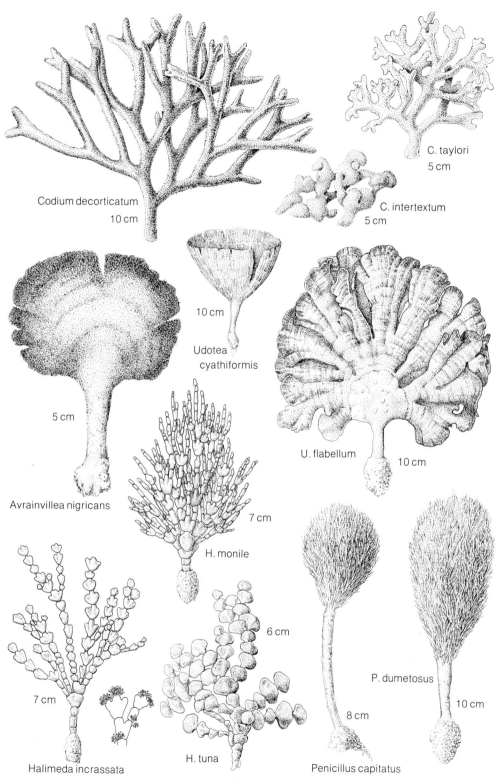

**5 CODIACEAE, UDOTEACEAE (Green algae 2)**

***U. flabellum*** (Ell. & Sol.) Lamour.: Genus as above.—Species 5-10 cm (to 15 cm) high; flabellum flat, broadly fan-shaped; the single to multiple blade is smooth, firm and may become ribbed and circumferentially zoned with calcium carbonate. Widely distributed throughout the year; common on shallow, sandy to muddy bottoms. Often found among sea grasses with *Penicillus*.

***Halimeda incrassata*** (Ell.) Lamour.: Genus having articulated segments of calcified discs or cylinders alternating with flexible, noncalcified joints, erect and anchored in substrate by fibrous holdfast. Microscopic examination of the calcified surfaces reveals a polygonal pattern formed by the expanded, peripheral utricles.—Species 10-25 cm high, with a distinct stalk and branches composed of segments 5-10 mm in diameter; segments flat, irregularly crenate (round-toothed), triangular or lobed; lower segments cylindrical to ovate, upper ones primarily trilobed and distinctly ribbed. In shallow, quiet waters; usually well anchored in sandy or loose calcareous sediments; also among sea grasses and well below low tide zone. Frequently found in spring and summer. (Color Plate 1.5.)

***H. monile*** (Ell. & Sol.) Lamour.: Genus as above.—Species 10-25 cm high: lower segments flattened or terete, tri-lobed at junction of new branches; upper segments distinctly cylindrical or terete, 1-2 mm in diameter, 3-8 mm long. Frequently found throughout the year in muddy waters or shallow lagoons. (Color Plate 1.4.)

***H. tuna*** (Ell. & Sol.) Lamour.: Genus as above.—Species 5-10 cm high, branching in a single plane; lower segments slightly flattened or terete, upper ones flat, disc-shaped, cuneate (wedge-shaped), reniform (kidney-shaped), thin, 10-15 mm in diameter. Common on sandy bottoms as well as on rocks and reefs; seldom in large quantities. Found throughout the year, but more frequently in spring.

***Penicillus capitatus*** Lamck. (Merman's shaving brush): Genus consisting of a single, terete or slightly flattened stipe terminating in a cluster or tuft of loose, tangled filaments; stalk anchored to substrate with a mass of rhizoids; the head or terminal filaments are dichotomously branched and become calcified with age.—Species 7-10 cm (to 15 cm), with a terminal, well-defined head of slender filaments usually under 0.5 mm in diameter and all arising from a more or less common point near the apex of the stipe. Throughout the year in shallow water; usually found growing in sand or mud, frequently with sea grasses.

***P. dumetosus*** (Lamour.) Blainv.: Genus as above.—Species 10 cm high or less (occasionally to 15 cm); stalk irregular and somewhat flattened; head tapers gradually down the stalk. Filaments to 0.8 mm in diameter; may be calcified at base, but usually not at ends. In well protected locations, on sandy bottoms; found throughout the year with *Halimeda* and *Udotea*. (Color Plate 1.10.)

### Plate 6

F. **CAULERPACEAE:** Caulerpales typically composed of 3 more or less distinct parts; long, rope-like or stolon-like, prostrate strands along the surface of the substrate; rhizoidal extensions projecting from these strands, downward onto or into the substrate; and erect, photosynthetic fronds of variable structure extending upward into the water. Reproduction by fragmentation, or dissemination and fusion of gametes. (8 spp. of *Caulerpa*, the only genus in this family, from Bda.)

***Caulerpa verticillata*** J. Ag.: Genus with rhizoids all generally similar in structure, but with a wide variation in shape and structure of the erect portion. The coenocytic filaments making up the plant have internal, rod-, tube- or tongue-like projections (trabeculae) extending into their lumina.—Species 5-10 cm high, delicate, tufted; erect branches usually forked or irregularly branched, lower portion naked; whorls of branchlets at intervals along the uppermost portion of the thallus; the erect branches are naked at the internodes of whorls. Frequently located in quiet water on rocks, walls and other submerged objects; not uncommon on mangrove roots or loose sediment; collected year round. (Color Plate 1.7.)

***C. prolifera*** (Forssk.) Lamour.: Genus as above.—Species upright, flat; lanceolate or linear blades arising from the stolon; 5-10 cm high; blades 5-8 mm wide, attached to the stolon by erect, slender stalks. Seldom found in abundance; it occasionally forms colonies; frequent in shallow water, on rocks or sandy bottoms.

***C. mexicana*** (Sond.) J. Ag.: Genus as above.—Species to 10 cm high (seldom over 15 cm); foliose, erect branches at intervals along the stolon are stalked, flat and pinnate; pinnae flat, more or less evenly spaced along the main stolon, 2-10 mm long. In protected places in shallow or deep waters; occasionally on rocky outcrops; not uncommon on mangrove roots.

***C. sertularioides*** (Gmel.) Howe: Genus as above.—Species 10-15 cm high; feather-like, erect branches numerous, stalked, flat, 1-2 cm wide, simple or sparingly branched; pinnules close, needle-shaped, slightly upcurved, to 1 cm long. Intertidal on rocks or sandy substrates; generally colonial; dense populations growing in shaded areas of moderate wave action. (Color Plates 1.8, 14.8.)

***C. cupressoides*** (West) C. Ag.: Genus as above.—Species 15 cm high (occasionally 20-25 cm); variable in shape but always erect, with distinct dichotomous branches bearing few to many ranks of short branchlets which may be cylindrical, conical or flattened; tips of branchlets always pointed. In shallow, quiet to wave-exposed water throughout the year; anchored to substrate by stolons and rhizoids often covered with fine sediment.

***C. racemosa*** (Forssk.) J. Ag.: Genus as above.—Species extremely variable in size and structure; erect, foliar branches simple or sparingly branched; branchlets positioned around each stalk or branch are spherical, subspherical, clavate or cylindrical. Short forms (3-5 cm) resembling grapes; in many varieties the main branches are elongate, 8-15 cm, and bear the branchlets along their entire length. One of the most frequently encountered green algae; it forms clumps on rocky shores, seawalls and occasionally sandy bottoms; it tolerates extreme wave action, but is

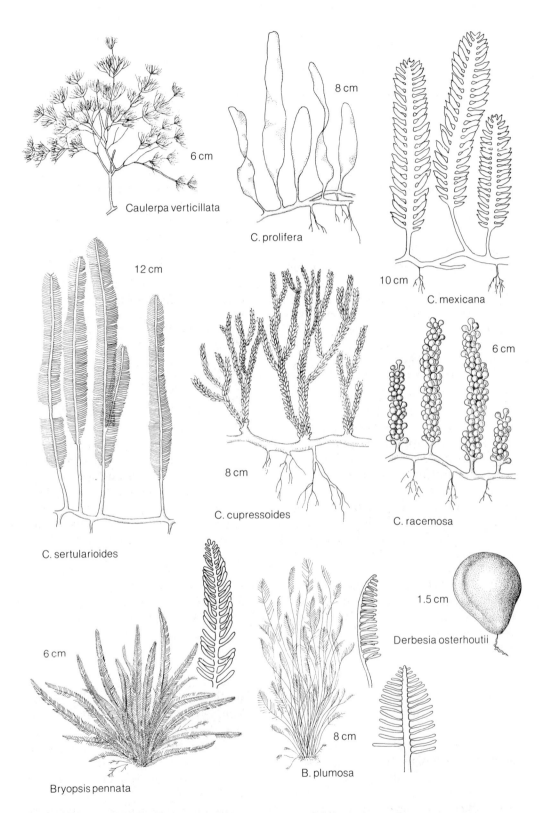

**6 CAULERPACEAE, BRYOPSIDACEAE, DERBESIACEAE (Green algae 3)**

usually dwarfed in such habitats; best growth in spring and winter.

F. **BRYOPSIDACEAE:** Caulerpales with erect thallus arising from an extended rhizoidal system. Uniaxial; siphon simple, not aggregate or fused; feather-like branches produced pinnately or radially from the main axes. Sexual reproduction by anisogametes; mature plants diploid. (4 spp. of *Bryopsis* from Bda.)

***Bryopsis pennata*** Lamour.: Genus bushy, tufted and erect; several main axes or branches arising from a rhizoidal system; feather-like branchlets forming pyramidal fronds.—Species to 5 cm; forming 1 or 2 rows of branchlets from the base to the tip of the main axes. Usually below low tide level attached to rocks, seawalls or wooden objects that have been submerged for some time; occurs year round.

***B. plumosa*** (Huds.) C. Ag.: Genus as above.—Species to 10 cm; always producing fronds with 2 rows of branchlets; branchlets formed primarily on upper branches and concentrated toward the tips; the more or less triangular shape of the fronds is composed of branchlets which are sharply constricted at their point of attachment to the main axis. Frequent in quiet waters of bays or saltwater ponds; appears to be indiscriminately attached to several submerged substrates such as rocks, walls and objects of wood.

F. **DERBESIACEAE:** Caulerpales having a vesicular gametophyte stage alternating with a filamentous sporophyte; the gametophyte, which is more easily recognized, is a coenocytic, spherical vesicle, solitary or clustered. (3 spp. of *Derbesia* from Bda.)

***Derbesia osterhoutii*** (Blinks & Blinks) Page (=*Halicystis osterhoutii*): Genus solitary, gregarious or in clusters; spherical, subspherical or oval spheres attached to substrate by a short stalk or peg.—Species dark green, usually under 1 cm in diameter, subspherical to pyriform; colonial; this phase, which represents the *Halicystis*-phase, is gametophytic. On red encrusting algae; on rocks and old coral heads; not abundant, but common.

### Plate 7

O. **SIPHONOCLADALES:** Coenocytic Chlorophyceae occasionally with septations and cellular compartments. Some produce erect segments, others are vesicular, sac-like or flattened. A unique characteristic of this order is the manner in which septations are formed within the thallus (segregative cell division). Cytoplasm along the periphery of the coenocyte concentrates and cleaves; the cleavage becomes isolated by a septum and the segregated portion elongates or enlarges. The macroscopic plant is diploid, and flagellated gametes are produced by meiosis.

F. **SIPHONOCLADACEAE:** Siphonocladales producing loosely branched or irregular tufts of distinctly segmented, branched filaments to 0.5 mm in diameter. (2 spp. of *Cladophoropsis* from Bda.)

***Cladophoropsis membranacea*** (C. Ag.) Børg.: Genus a loose mass of branched, uniseriate filaments; dense and tufted; no main axis.—Species

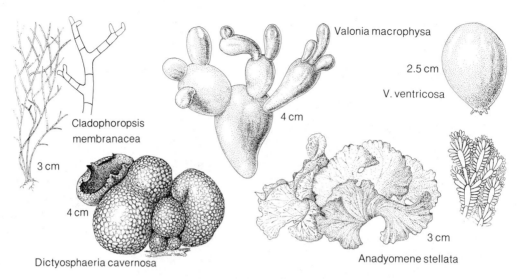

## 7 SIPHONOCLADALES (Green algae 4)

with regular septations of the stiff, broad filaments (0.1-0.3 mm in diameter); pale olive-green. A common and abundant alga of the intertidal zone; forms tufts and mats in ponds and on rocks and sea walls; frequently attached to *Rhizophora* and *Avicennia* roots; reported in winter and spring; not uncommon year round.

F. **VALONIACEAE:** Siphonocladales with solitary or clustered vesicles (0.5-4 cm in diameter); single-celled, coenocytic, but occasionally with cells within the larger membrane resulting in parenchymatous tissue; algal spheres, which are more or less translucent, are attached to the substrate by rhizoids. (1 sp. of *Dictyosphaeria*, and 4 spp. of *Valonia* from Bda.)

***Dictyosphaeria cavernosa*** (Forssk.) Børg.: Genus with sessile, subspherical or lobed vesicles; algae hollow or solid and forming pseudoparenchymatous tissue as a result of cell division.—Species 3-5 cm in diameter (to 8-10 cm), spherical or irregularly lobed; outer wall consisting of angular cells readily distinguishable with the aid of a lens. Not widely distributed and seldom found in good shape; often torn and collapsed; usually associated with old coral heads, rocks and reefs.

***Valonia macrophysa*** Kütz.: Genus clavate, spherical, sausage-shaped or subcylindrical; plants may be solitary or clustered, prostrate or erect.—Species generally forming large masses of vesicles that are variable in shape and sometimes slightly branched; cells to 1.5 cm in diameter, lower ones associated with rhizoids. Commonly found on rocks, old coral heads, under stones in shallow water and rock jetties; year round. (Color Plate 1.3.)

***V. ventricosa*** J. Ag.: Genus as above.—Species solitary, seldom growing together; spherical or subspherical vesicle attached to substrate with rhizoids; translucent, to 4 cm in diameter. A widely distributed, easily identified alga; found throughout the year on submerged seawalls, rock jetties, old coral heads, stones and in crevices; frequently associated with smaller red algae, and commonly covered with encrusting algae.

F. **ANADYOMENACEAE:** Siphonocladales with sheet-like or foliose, monostromatic thallus composed of branching coenocytic filaments that anastomose into an erect, flat sheet or blade. (1 sp. of *Anadyomene* from Bda.)

*Anadyomene stellata* (Wulf.) C. Ag.: Genus foliose, inconspicuously stalked and attached to the substrate with rhizoids; a series of microscopic, branching filaments giving a rib-like appearance to the fan-shaped or lobed blade.—Species composed of one or more erect, crisp blades, usually 2-8 cm high and equally as broad; fan-like ribs or vein-like cells branch throughout the thallus. Often mistaken for *Ulva*, this species is very common on rocks in both exposed and sheltered areas; frequent in tide pools along exposed shores; under favorable conditions it will cover a large rocky area forming a crisp, green felt. Found throughout the year, abundant in spring and summer. (Color Plate 1.6.)

O. **DASYCLADALES:** Radially symmetrical Chlorophyceae composed of an axis with variously arranged whorls of lateral filaments, some sterile and some fertile; the thallus remains uninucleate until reproduction. Sexual reproduction is initiated by the formation of gametangia in lateral filaments; the single large nucleus, usually situated basally, undergoes several divisions; secondary nuclei migrate into the gametangia, and uninucleate cysts develop as a result of cytoplasmic cleavage; meiotic divisions of the cysts result in a number of biflagellated gametes that are liberated and fuse, forming a diploid zygote which germinates into the vegetative plant. (Only 1 family.)

### Plate 8

F. **DASYCLADACEAE:** Dasycladales displaying simple to complex structure based on a main axis with radially symmetrical parts; arrangements are spindle-shaped, finger-like or subcylindrical; stalked circular discs or saucers; or delicate feathery plumes and branched structures terminating in fine, penicillate filaments. Except for *Batophora*, algae in this family may be partially or totally encrusted with calcium carbonate. (2 spp. of *Acetabularia*, and 1 sp. each of *Batophora*, *Cymopolia*, *Dasycladus* and *Neomeris* from Bda.)

*Neomeris annulata* Dickie: Genus solitary or gregarious, cigar-shaped, finger-shaped or cylindrical; the calcified, whitish structure is composed of a central axis bearing whorls of closely packed, compound branchlets; the main axis is terminated with a tuft of bright green, branching monosiphonous filaments.—Species cylindrical or slightly spindle-shaped; 1.5-3 cm high, 1-3 mm in diameter; often gregarious. Main axis appears solid, but when viewed microscopically, lateral branchlets, some of which bear rows of spherical gametangia, become evident. Widespread and common; usually found with *Cymopolia* or in similar habitats and on similar substrates; frequents sunny, warm water and often found just below the low tide zone; collected year round. (Color Plate 1.2.)

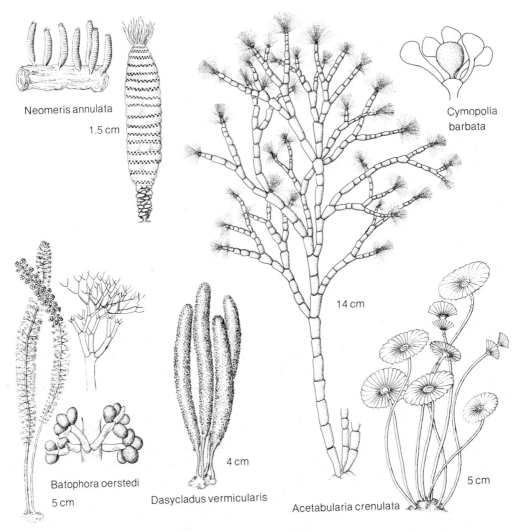

**8 DASYCLADACEAE (Green algae 5)**

***Batophora oerstedi*** J. Ag.: Genus with basally attached, long erect axis bearing whorls of repeatedly dichotomous or trichotomous branches that are monosiphonous, and terminate in fine filaments that are both fertile and vegetative; not calcified.—Species to 10 cm high, delicate, green, soft; main axis usually naked below but provided with whorls of branchlets above which are forked 4-6x forming a soft cylinder 3-5 mm in diameter; small, spherical sporangia 0.5-0.8 mm in diameter are borne among the forked branches. Found throughout the year in pools and lagoons of quiet waters, as well as in mangrove swamps; grows on rocks, shells and occasionally on mangrove roots.

***Dasycladus vermicularis*** (Scop.) Kras.: Genus with compact, closely arranged whorls of branches around the main axes; olive-green, soft and usually not calcified.—Species to 6 cm high, 2-6 cm in diameter; 2 or more axes per plant; lateral filaments repeatedly dichotomously or trichotomously branched, closely arranged on the main axis and usually produced along its entire length. Sporangia solitary and produced on the inner side of branchlets. In quiet waters attached to rocks and shells; widespread, seldom found, but occurring year round.

*Cymopolia barbata* (L.) Lamour.: Genus branched and heavily calcified; segmentations of main branches rounded and cylindrical; ends of branches provided with tufts of bright green, dichotomously and trichotomously branched filaments.—Species typically 5-15 cm high, branches cylindrical, white, segmented and flexible at nodes. Segments usually narrower at the basal portion; lower segments 2-3 mm in diameter, upper segments thinner and shorter. Sporangia spherical and solitary, borne at ends of coenocytic filaments making up the major cylinders. Most commonly found growing on rocks just below the low tide zone; frequent in warmer, quiet waters throughout the year. (Color Plate 1.12.)

*Acetabularia crenulata* Lamour. (Mermaid's wineglass): Genus composed of a slender, usually calcified stipe that terminates in a concave or flattened disc.—Species to 8 cm tall; discs 5-8 mm in diameter, single or layered one upon the other, composed of a whorl of radiating segments or filaments originating from the stipe; sometimes with an additional whorl of smaller radiating segments above and below the main disc; stipe and/or discs may be partially or completely calcified. Frequently overlooked but nonetheless widely distributed; most commonly found in quiet waters attached to stones, rocks, shells and other submerged objects; occasionally collected from *Rhizophora* roots and as an epiphyte on larger green algae; found year round. (Color Plate 1.1.)

A. R. CAVALIERE

## Phylum Phaeophyta (Brown algae)

CHARACTERISTICS: *Multicellular PLANTAE with uniseriate, unbranched and branched filaments, tufts of filaments or multiaxial branching systems. Most produce parenchymatous thalli that are large, macroscopic, leathery, rubbery or tough; several with root-like, stem-like and leaf-like structures; all but a few are attached. Cell wall composed of an inner cellulose and an outer gelatinous layer, the latter being made up of pectic material and alginates. Cells generally uninucleate, containing chlorophylls a and c, beta-carotene, and several accessory pigments, the most important being fucoxanthin. Reserves stored as sucrose, glycerol or more complex soluble carbohydrates, laminarin (starch) or mannitol (alcohol). Chloroplasts 1 to many per cell, disc-shaped, simple, lobed, branched or reticulate.* The size of Bermuda's species ranges from microscopic to 80 cm, the color from brown or yellow-brown to olive-brown, dark brown or almost black.

Only a single class, Phaeophyceae, is presently recognized, and is almost exclusively marine. Somewhat under 1,500 species are known. Thirty genera encompassing over 60 species have been reported from Bermuda. The present account includes 22 species.

OCCURRENCE: Brown algae are the most conspicuous seaweeds along rocky coasts in both northern and southern hemispheres. Not as numerous or large in tropical waters as in more northern or southern latitudes; frequently occur attached firmly to rocks; intertidal zone or subtidally. In Bermuda, *Dictyota* is commonly attached whereas *Sargassum,* although occasionally attached, constitutes the bulk of the plant biomass floating in the Sargasso Sea.

Collect with mask and snorkel in shallow water, or by hand at low tide. In deeper water, rakes, dredges or SCUBA are employed.

IDENTIFICATION: Phaeophyta in Bermuda are usually macroscopic (with the exception of the Ectocarpales) and distinct enough to be separated into genera without microscopy; identification beyond generic level requires some knowledge regarding methods of growth and the structure of reproductive organs.

Cell division and growth of brown algae occur at the apex of the plant body, between the apex and base of filaments (intercalary), or at the base of a filament (trichothallic); in the larger, parenchymatous forms, division may occur in any cell of the thallus, or be restricted to meristematic (growth) regions. Motile gametes and spores are usually oval or pyriform, biflagellated, heterokontic (of unequal lengths), and inserted laterally, one of the tinsel type and the other of the whiplash type. Plurilocular and unilocular organs are stalked or sessile, terminal or intercalary, single or grouped in a pustule (sorus); under 200 μm in length and 20-50 μm in diameter. In Fucaceae, conceptacles containing the male and/or female sexual reproductive cells are located in receptacles that may be larger, swollen ends of branches, or specialized, terminal or axillary branchlets formed along the thallus.

Preserve in a 3-4% formalin-seawater solution, or press as herbarium specimens. Larger brown algae are often treated in a solution of FAG (formaldehyde 10%; ethyl alcohol 25%; glycerin 5%; distilled water 60%). After allowing the alga to remain submerged in the solution for at least 24 hr, drain for several hours and store in a plastic bag. The specimen will remain soft and pliable indefinitely if not allowed to dry out. The solution is reusable and algae may be submerged again when necessary.

BIOLOGY AND REPRODUCTION: Asexual by fragmentation or laterally biflagellated zoospores; sexual reproduction isogamous, anisogamous or oogamous. With the exception of the Fucales, the brown algae undergo an alternation of sporophyte with gametophyte generation (diplohaplontic); generations may be isomorphic (morphologically similar) or heteromorphic (morphologically dissimilar); in most cases, the gametophyte is haploid and produces gametes; the sporophyte is diploid and produces both diploid and haploid spores, the former growing into sporophytes, the latter into gametophytes. In the Fucales, the plants are diploid, gametes being produced by meiosis; the diploid zygote that results from oogamous fertilization grows directly into the parent generation (diplontic). A number of reproductive structures are formed; unilocular (1-celled) and plurilocular (multicellular) organs (sporangia or gametangia) produce zoospores or gametes, the former by meiosis, the latter generally by mitosis; in oogamous reproduction, both eggs (oogonia) and sperm (antherozoids) are produced in oval or pear-shaped cavities (conceptacles) embedded within the thallus; conceptacles are usually confined to specialized branches of the thallus (receptacles). Male and female gametes may be formed on the same plant (homothallic) or on separate plants (heterothallic).

REFERENCES: A general treatment is found in DAWES (1981), BOLD & WYNNE (1978), ROUND (1973) or PRESCOTT (1968). For a more thorough study of physiology and taxonomy see LOBBAN & WYNNE (1981), IRVINE & PRICE (1978) and STEWART (1974).

Floristic studies of brown algae of the West Indies are covered by CHAPMAN (1961), and of Bermuda by COLLINS & HERVEY (1917), TAYLOR & BERNATOWICZ (1969), BERNATOWICZ (1952) and TAYLOR (1960). Data on the floating *Sargassum* community have been summarized by MORRIS & MOGELBERG (1973) and BUTLER et al. (1983).

**Plate 9**

O. **ECTOCARPALES:** Small, often microscopic Phaeophyceae composed of uniseriate, branched filaments that produce a partially prostrate and partially erect, loose or tufted thallus; growth apical, intercalary or trichothallic. Reproduction by alternation of generally isomorphic generations; asexual spores flagellated; gametes anisogamous or isogamous.

F. **ECTOCARPACEAE:** Simple, mostly uniseriate Ectocarpales to 10 cm in length. Flagellated zoospores in unilocular sporangia; gametes in plurilocular gametangia. (11 spp., 3 of *Ectocarpus*, from Bda.)

***Ectocarpus siliculosus*** (Dillw.) Lyngb. (=*E. confervoides* (Roth) Le Jolis): Genus a tuft or mass of branched, uniseriate filaments, prostrate or erect, tapered toward apex and basally attached by rhizoids; cells containing few, ribbon-shaped or branched chloroplasts; both plurilocular and unilocular reproductive organs produced.—Species yellow-brown to dark brown, slimy; variable in size, normally to 10 cm long; with irregular, tapering branches. Sporangia 20-30 μm in diameter, to 50 μm long; gametangia slightly thicker and longer. Ubiquitous; epiphytic, epizoic; on all substrates. Usually in quiet, warm waters of ponds, coves and caves; intertidal; year round.

O. **DICTYOTALES:** Flat and membranous Phaeophyceae; thallus 1 to several cells thick; growth by a single apical cell or a margin of apical cells; diplohaplontic; isomorphic; asexual spores nonmotile; sexual reproduction oogamous; gametophytes heterothallic. (Only 1 family.)

F. **DICTYOTACEAE:** Flat, ribbon-like, strap-like or fan-like Dictyotales; usually dichotomously branched. (8 spp. of *Dictyota*; 4 spp. of *Padina*, 3 spp. of *Dictyopteris* and 1 sp. each of *Lobophora*, *Stypopodium* and 2 other genera reported from Bda.)

***Dictyota dichotoma*** (Huds.) Lamour.: Genus yellow-brown to dark brown, dichotomously branched, flat, ribbon-like; erect and attached by rhizoids; thallus usually 3 cell layers thick, often iridescent; hairs arise in tufts, scattered over the thallus. Growth by means of a single apical cell.—Species with true dichotomous branching; usually 10-20 cm high; strap-shaped, occasionally twisted branches 2-5 mm wide; tips usually blunt. On rocks, old coral heads, epiphytic on other larger algae, or occasionally on mangrove roots; intertidal or below; sparse during summer.

***D. dentata*** Lamour.: Genus as above.—Species 10-20 cm high; bushy, erect; main axes well defined; terminal branches alternate; 1-2 mm long, spur-like branchlets at intervals along main branches; tips of branchlets rounded or acute. Frequent in moderately exposed habitats; epiphytic and epizoic; common throughout the year; from intertidal to subtidal zone.

***D. ciliolata*** Kütz.: Genus as above.—Species with fine, dentate projections along the margins of subdichotomous or alternate branches; dark brown; to 15 cm high, branches to 7 mm wide,

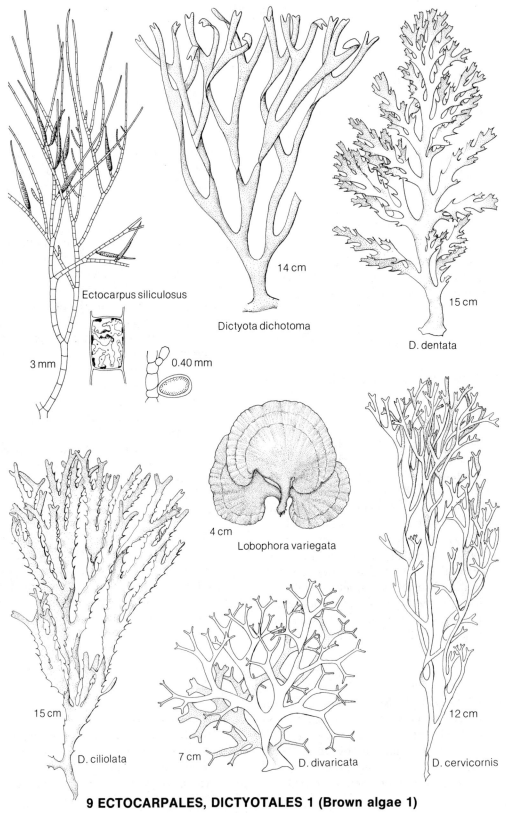

9 ECTOCARPALES, DICTYOTALES 1 (Brown algae 1)

frequently spirally twisted; *D. ciliolata* var. *bermudensis* Taylor is slightly smaller, with fewer and less prominent dentations. On rocks or other algae; intertidal zone or below; widespread; grows year round, abundant in winter and spring.

***D. divaricata*** Lamour.: Genus as above.—Species generally under 10 cm high; branching regularly dichotomous and widely angled; branches broad in lower portions of thallus (to 3 mm wide) and becoming abruptly narrow (0.1-0.2 mm) in the upper portions. On rocks, shells, mangrove roots; in quiet, shallow waters, but in moderately exposed places as well; common year round, sparse in late summer. (Color Plate 2.2.)

***D. cervicornis*** Kütz.: Genus as above.—Species with asymmetric to subdichotomous or cervicorn (1-sided) branching; 10-15 cm high, branches slender, 1-3 mm wide, tips acute. In moderately exposed areas; intertidal to subtidal; on rocks, plants, shells; often dredged.

***Lobophora variegata*** (Lamour.) Wom. ( = *Pocockiella variegata* (Lamour.) Papenf.): Genus prostrate or erect; flat, deltoid or orbicular; simple at first, becoming lobed with age; with marginal rows of apical cells.—Species light brown to tan; fan-shaped, kidney-shaped or orbicular blades flat, to 6 cm wide, 1-3 mm thick; erect or decumbent, attached to substrate by rhizoids. Common in intertidal zone; on rocks, old coral heads, shells; in moderate to extreme exposure; year round; sparse in late summer. (Color Plate 2.2.)

**Plate 10**

***Stypopodium zonale*** (Lamour.) Papenf.: Genus erect, flat, leafy; 10-30 cm high; attached by substantial, rhizoidal holdfast; older plants split into segments.—Species characteristically blue-green iridescent under water; blades broad at first, becoming split into smaller, strap-shaped or funnel-shaped segments when older, transversely banded by rows of hairs; blackening upon drying or pressing. Widespread in subtidal zone and areas of moderate wave action; best growth in autumn and winter; not common. (Color Plate 2.1.)

***Dictyopteris justii*** Lamour.: Genus large and coarse, 10-40 cm high, erect; wide dichotomous branches strap-shaped, with a conspicuous midrib.—Species dark brown, occasionally iridescent; blades several cells thick, 1-5 cm wide, flat, usually irregularly serrate, with a prominent midrib extending from the base of the thallus up through the dichotomously branched blades. On exposed shores or below the low tide zone. Abundant in late spring and during the winter; sparse or absent in summer.

***Padina vickersiae*** Hoyt: Genus with single or multiple fan-shaped blades, splitting when larger or older; margin generally involute or rolled in; blades 2 to several cells thick; zoned with moderate to heavy rows of microscopic hairs; calcium carbonate commonly over 1 or both surfaces of blade.—Species producing clusters of fan-shaped blades that are entire, split or segmented; calcification on upper surface, light to none below; blades 2-8 cell layers thick; gametophytes heterothallic. Common on rocks below the low tide zone; often partially or totally covered with fine sediment; year round; abundant during warmer months. (Color Plate 2.5.)

*P. sanctae-crucis* Børg.: Genus as above.—Species usually 5-10 cm high; blades stalked and attached to substrate by rhizoids, 4-6 cm wide, split, involute, zonate, and calcified only on the upper surface; antheridia and oogonia homothallic. In moderately exposed areas or below the low tide zone; in ponds, coves; usually attached to rocks or old coral heads. Found throughout the year; most luxuriant growth during summer.

O. **SPOROCHNALES:** A small group of Phaeophyceae that undergo a heteromorphic alternation of generations; the gametophyte is small, filamentous and microscopic; the sporophyte develops a macroscopic thallus that characteristically produces tufts of hairs at the tips of all branches; growth trichothallic.

F. **SPOROCHNACEAE:** Sporochnales with firm branches terminating in tufts of hairs. (3 spp., 2 of *Sporochnus*, from Bda.)

*Sporochnus bolleanus* Mont.: Genus dark brown, bushy, erect, with a rhizoidal holdfast; main axes distinct; many lateral branches bearing fertile area terminate in tufts of filaments.—Species wiry, large, generally 20-30 cm high, with 1 or more filiform axes supporting many branches; branches provided with short branchlets which, when young, tipped with tufts of brown filaments 5-8 mm long. Forms colonies on rocky shores below low tide level; often washed up as drift.

O. **SCYTOSIPHONALES:** Variously shaped Phaeophyceae: subspherical, lobed, hollow or flat, stalked, tapered or tubular. Macroscopic plant produces only plurilocular sporangia or gametangia; life cycle heteromorphic. (A single family, Scytosiphonaceae, in Bda.)

F. **SCYTOSIPHONACEAE:** Scytosiphonales producing parenchymatous thalli. (1 sp. each of *Colpomenia*, *Hydroclathrus*, *Scytosiphon* and 2 other genera from Bda.)

*Colpomenia sinuosa* (Roth) Derb. & Sol.: Genus with balloon-shaped, convoluted, spherical to lobed, hollow thalli. (1 sp. each of *Colpomenia*, *Hydroclathrus*, *Scytosiphon* and 2 other genera from Bda.)
surface.—Species sessile, irregularly shaped, hollow, 3-8 cm in diameter; plurilocular gametangia located among tufts of hairs over thallus. On rocks in the intertidal zone; occasionally forms band of yellow-brown growth at lower intertidal zone; year round; abundant in spring.

*Hydroclathrus clathratus* (Bory) Howe: Genus usually flat, net-like, clathrate (resembling lattice work), attached; hollow and spherical when young, flattened with age.—Species spherical and hollow or flattened and perforated; 1 to several cm in diameter. On rocks in the intertidal zone; not common; most frequent in spring.

*Scytosiphon lomentaria* (Lyngb.) C. Ag.: Genus with brown, unbranched, tubular thallus; having constrictions along its length; commonly gregarious filaments are attached to the substrate by a single, basal holdfast.—Species dark brown or olive-brown, usually 10-20 cm high; tubular fronds 5-10 mm in diameter; irregularly constricted;

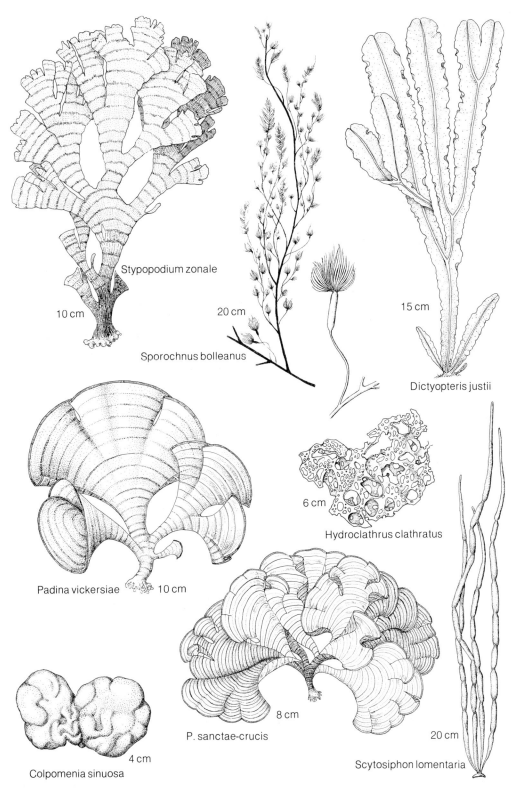

**10 DICTYOTALES 2, SPOROCHNALES, SCYTOSIPHONALES (Brown algae 2)**

plurilocular sporangia extended in sori over surface of thallus. Frequently on stones, rocks, old coral heads; in shallow water of moderately protected areas; most commonly seen in winter and early spring.

## Plate 11

O. **FUCALES:** Radially, bilaterally or dichotomously branched Phaeophyceae that may be flat or terete; commonly produce leaf-like blades and spherical or subspherical air vesicles (pneumatocysts). Life cycle diplontic; meiosis precedes gamete production; the diploid zygotes resulting from the fusion of biflagellated anterozoids and oogonia develop directly into the diploid plant; vegetative reproduction by fragmentation; asexual zoospores do not occur.

F. **SARGASSACEAE:** Fucales of radial organization; axes much branched; terminal branchlets bearing foliar, leaf-like or turbinate organs; a single egg produced in each oogonium. Because of the morphological variations that occur within and among species, and the suspected hybridization among sexually reproducing forms of *Sargassum*, species separation is often difficult if not impossible. (Over a dozen spp. of *Sargassum* and 2 spp. of *Turbinaria* from Bda.)

***Sargassum bermudense*** Grun. (Sea holly): Genus yellow-brown, olive-brown or dark brown; free-floating, or attached with large, simple or lobed holdfast; cylindrical main axis or axes support flattened, leaf-like, serrated or dentate branches; some leaf-like branches provided with cryptostomata (sunken cavities having hair-like cells). Spherical or subspherical vesicles develop on separate stalks in the axils of branches; fertile branchlets bear elongate, finger-like or cylindrical reproductive receptacles.—Species normally 10-30 cm long; generally attached, bushy, with normally a single main axis that is coarse, rough or spiny. Leaves thick, 1-3 cm long, 1-5 mm wide, rigid, crowded toward the apex, irregularly serrate, apices acute, attachment to petiole abrupt. Cryptostomata scattered or in rows along prominent midrib; vesicles numerous, spherical; receptacles alternately branched. In sheltered areas of moderate wave action; attached to rocks; intertidal habitat or lower into subtidal zone. Crowded and dwarfed in areas of heavy wave action; year round. (Color Plate 2.4.)

***S. filipendula*** C. Ag.: Genus as above.—Species 20-80 cm long; attached or free-floating; usually one or few long main axes, sparingly branched. Leaves thin, linear, serrate, prominent midrib, 4-15 cm long, usually under 8 mm wide, often 20-30× longer than broad. Cryptostomata scattered; vesicles spherical to subspherical; receptacles long, slender, on single or divided axillary branch. On rocks and old coral heads below low tide zone; not common.

***S. natans*** (L.) J. Meyen: Genus as above.—Species 20-50 cm long; free-floating; widely branched; no main or dominant axis. Leaves linear or narrow, 2-10 cm long, 2-6 mm wide, teeth aculeate (spine-like). Cryptostomata and midrib not apparent; vesicles spherical, with spines 2-5 mm long; receptacles unknown; reproduction asexual by fragmentation. The most

# PHAEOPHYTA

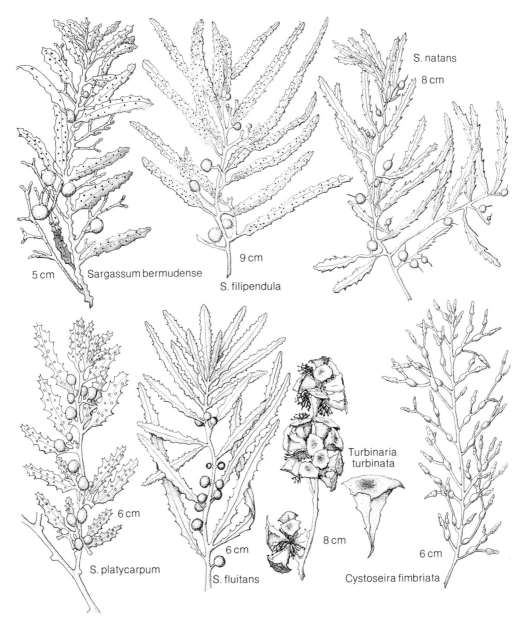

**11 FUCALES (Brown algae 3)**

abundant species occurring in waters or washed ashore; the dominant weed of the Gulf Stream. Usually with *S. fluitans* but more dominant, and differentiated from it by the presence of needle-like or spine-like projections on vesicles; common throughout the year.

***S. platycarpum*** Mont.: Genus as above.—Species small, usually 10-20 cm long; normally attached; with short, stout main axis sparingly branched. Leaves lanceolate, acute, crowded, coarsely and deeply serrated, 2-6 cm long, 2-6 mm wide, tapering into petiole. Cryptostomata in a single row on each side of a prominent midrib; vesicles oval, on short stalks; receptacles with dentate wings. Intertidal on rocks or rock outcrops; not common; occurs in summer and autumn.

***S. fluitans*** Børg.: Genus as above.—Species 20-50 cm long; free-floating; widely branched; no dominant axis. Leaves firm, thick, lanceolate, serrations triangular and broad at base, asymmetric, 2-6 cm long, 3-8 mm wide. Cryptostomata absent; vesicles oval to subspherical; receptacles unknown; reproduction asexual by fragmentation. In drift; known only from pelagic state; often with *S. natans*, but not as abundant, and separated from it by the lack of spines on vesicles; year round; frequent in spring and summer. (Color Plate 2.6.)

***Turbinaria turbinata*** (L.) Kuntze: Genus to 15 cm high or more, dark brown; attached; holdfast slender, lobed, or branched; thallus erect, branched; with numerous trumpet-shaped or inverted pyramid-shaped (turbinate) leaves.—Species 10-15 cm high; main axes long, with short lateral branches. Leaves turbinate, to 1 cm long, the top rounded, triangular or squared. Receptacles branched, dichotomous and densely packed between the leaves. Washed ashore with *Sargassum*; attached to intertidal rocks at or below low tide.

F. **CYSTOSEIRACEAE:** Much branched, radial Fucales; receptacles formed on regular branches; each axis terminated in a single apical cell; 1 egg formed in each oogonium. (1 sp. of *Cystoseira* from Bda.)

***Cystoseira fimbriata*** (Desf.) Bory: Genus slender, bushy, 1 to several main axes, to 35 cm long; stems smooth or warty, branching; vesicles single or catenulate (chained), formed by the swelling of branches.—Species bushy above; lower lateral and younger branches flattened and leaf-like, main and older ones terete; 2-3 × pinnate; vesicles intercalary, catenulate, serrate, and spindle-shaped. Occurs as drift; not common.

A. R. CAVALIERE

## Phylum Rhodophyta (Red algae)

CHARACTERISTICS: *Unicellular or multicellular PLANTAE of diverse morphology: endolithic, minute filamentous epiphytes, branching filaments or foliose. Without motile structures or flagellated cells. Chloroplasts containing chlorophylls a and d, various carotenoids as well as the accessory photosynthetic pigments phycocyanin and phycoerythrin (phycobilins). Food reserve is a complex polysaccharide, floridian starch; sugars and glucosides serve as storage products. Cell wall consists of an inner cellulose component and an outer mucilage or pectic layer. Several species deposit calcium carbonate, either from certain regions of the plant (articulated forms) or from the whole thallus (encrusting forms).* Size ranges from microscopic to several decimeters. Among the larger red algae 2 more or less distinct groups are recognizable: the more conspicuous, fleshy forms (filamentous, foliose or bushy), and the hard, often brittle encrusting corallines. Although shades of red, from almost black to pink to almost white, predominate, shades of yellow, olive, green or greenish brown are common as well.

Slightly under 4,000 species of marine red algae are known, and placed in a single class, Rhodophyceae, which is divided into 2 subclasses, Bangiophycidae and Florideophycidae. Of over 200 species representing 6 orders reported from Bermuda, 44 species are included here.

OCCURRENCE: In all latitudes, but their numbers and conspicuousness markedly increase in temperate and tropical regions where they outnumber both green and brown algae. Although many intertidal species are known, most appear to be subtidal. All reds appear to originate as attached forms, but because of wave action and grazing, they are often found floating. Larger brown and green algae often serve as hosts for epiphytic red algae. Species of *Ceramium*, *Bostrychia*, *Polysiphonia* and others are often epiphytes as well as being attached to non-living substrates. Epiphytic red algae usually do not penetrate their hosts but are superficially attached. In many cases a single host may harbor a dozen different species of epiphytes. It is not uncommon, for example, to find *Ceramium byssoideum* attached to *Caloglossa leprieurii*, which is in turn attached to *Codium decorticatum*. Encrusting forms (Corallinaceae) grow on every type of substrate below the surface of the water. Because of their deposition of calcium carbonate, these algae contribute substantially to the maintenance and stability of reef structure by filling in and cementing spaces and crevices between old coral heads.

In Bermudian waters several factors influence the abundance of red algae during the year: day length, relative light conditions, temperature, and periodic or continual grazing by herbivorous vertebrates and invertebrates, especially starfish and sea urchins.

Collect by hand at low tide, by snorkeling or SCUBA, or with rakes and dredges in deeper water.

IDENTIFICATION: Usually distinguished from green or brown algae by color which alone, however, is not a reliable characteristic. For positive identification, some of the tissue may be boiled in water; the red, phycobilin pigments are water-soluble and leach out. All Rhodophyta turn dull green or grass green as a result of this treatment. Calcareous algae may be recognized by their firm, crystalline nature or when examined under a dissecting microscope. Several drops of dilute hydrochloric acid will release carbon dioxide bubbles when added to the tissue. The taxonomic limits of species in several genera of Corallinaceae are unclear, and positive determinations may require the study of decalcified and sectioned plants.

All corallines consist of delicate filaments jointed together by cell walls impregnated with calcium carbonate; crusts of these filaments comprise 3 tissues: a hypothallium, in which the filaments are close to the substrate and approximately parallel to it; a perithallium, in which the filaments are above the hypothallium and approximately perpendicular to the substrate; and an epithallium, a superficial layer of cells, which consists of the terminal cells of many filaments. In articulated branches, a central core of filaments (medulla) is surrounded by a pigmented tissue (cortex).

Press as herbarium specimens, or store in a 3-4% formalin-seawater solution. Calcified or encrusted algae, which are 3-dimensional and/or fragile, may be soaked for several hours in formalin-seawater, allowed to dry and stored or mounted in boxes. Joints of articulated algae will remain flexible if soaked in formalin-seawater containing 25% glycerin and allowed to air dry.

BIOLOGY: Red algae are widely distributed bathymetrically as well as geographically. The depth to which algae grow is determined by the availability of light penetrating the water and their possession of specific photosynthetic pigments. Red algae in high intertidal zones absorb the full light spectrum, especially red wavelengths, which enhance the production of phycocyanin. Under these conditions the predominant color is green, olive or dark brown. At greater depths the absence of red light enhances the production of phycoerythrin, which results in a rosy red or pink color.

REPRODUCTION: Sexual reproduction in the subclass Florideophycidae is typically oogamous, and composed of 3 distinct phases: haploid gametophytes, a diploid carposporophyte, and a diploid tetrasporophyte. The gametophytes and tetrasporophyte are free-living; the carposporophyte, which arises through sexual reproduction on the female gametophyte, remains attached and dependent on that plant. In the typical life cycle, the male gametes (spermatia) are released from spermatangia on the male gametophyte and fertilize the egg contained in the carpogonium on the female gametophyte; as a result of fertilization, the zygote nucleus begins to divide or initiates other divisions that result in the development of the diploid carposporophyte. Occasionally, the carposporophyte may be surrounded by a covering of haploid tissue (pericarp) derived from the gametophyte; the entire structure is termed a cystocarp. The carposporophyte bears gonimoblast filaments that generate carpospores from carposporangia; upon release and growth, carpospores grow into the second diploid phase, the tetrasporophyte, which matures and produces tetrasporangia; within each tetrasporangium, 4 haploid spores are produced as a result of meiosis; haploid tetraspores are disseminated, germinate and develop into male and female gametophytes. Members of the Rhodophyta may exhibit isomorphic or heteromorphic alternation of generations (diplohaplontic) in which the gametophytes and tetrasporophytes are either similar or dissimilar, respectively.

The orders and families of the subclass Florideophycidae are often separated on the basis of reproductive structures and processes in the sexual reproductive cycles. Several types of asexual spores are known: monospores are individually produced in thin-walled sporangia; tetraspores are formed in groups of 4, and multiple divisions result in polyspores. The male sexual reproductive organ (spermatangium) may develop from a simple, vegetative branch of the thallus or be produced on specialized branches (trichoblasts). The female sexual reproductive organ (carpogonium) may also develop from a simple, vegetative cell or be produced on a specialized filament (carpogonial branch). The typical flask-shaped or pear-shaped carpogonium contains the egg in the basal portion and an extended trichogyne at the apex of the structure. Following fertilization, the carposporophyte may develop directly from the carpospore (diploid, sexual spore) in the carpogonium, or the fertilized nucleus may migrate into one or more auxiliary cells from which the carposporophyte develops.

In the Corallinaceae, reproductive cells are produced within roofed chambers (conceptacles) that are located on the surface of crusts or segments; tetrasporangia contain large red spores in linear tetrads; bisporangia are sometimes formed. Several tetrasporangia form within conceptacles that open by a single or many pores. Sexual conceptacles are always uniporate, producing large numbers of spermatia or 2-celled carpogonial filaments. Following fertilization, cells on the floor of a female conceptacle fuse to form a more or less flat fusion cell. Short, specialized filaments grow from the margins or upper surface of a fusion cell forming large, red carposporangia, each of which contains a single carpospore.

Asexual reproduction may occur by the formation of spores in sporangia, mono-, bi- or polyspores in most, or by fragmentation.

REFERENCES: For a general treatment see BOLD & WYNNE (1978), ROUND (1973) or PRESCOTT (1968); classification, physiology and ecology are covered in KYLIN (1956), DIXON (1973), IRVINE & PRICE (1978) and LOBBAN & WYNNE (1981).

Red algae of coastal North Carolina are reported by KAPRAUN (1980) and of the West Indies by CHAPMAN (1961); Bermuda species were originally treated by COLLINS & HERVEY (1917) and more recently by TAYLOR (1960), BERNATOWICZ (1952) and TAYLOR & BERNATOWICZ (1969).

S.CL. **BANGIOPHYCIDAE:** Unicellular, filamentous or membranous Rhodophyta; cells uninucleate; pit connections (cytoplasmic strands between cells) almost always absent; asexual reproduction by monospores; sexual reproductive structures simple. (4 orders known, 1 reported from Bda.)

### Plate 12

O. **BANGIALES:** Filamentous or membranous Bangiophycidae. (1 family from Bda.)

F. **BANGIACEAE:** Filamentous Bangiales; asexual reproduction by the formation of spores produced by internal divisions of the protoplasm (monospores); life cycle involves alternation between the filamentous gametophyte and a shell-boring filamentous "conchocelis" phase, the latter producing sporangia where meiosis probably occurs. (About 16 spp. from Bda.)

*Bangia atropurpurea* (Roth) C. Ag. (=*B. fuscopurpurea* (Dillw.) Lyngb.): Genus uniseriate at base, becoming biseriate and multiseriate toward upper portion; filiform, unbranched; attached by basal rhizoids or free-floating; filaments terete, sometimes irregularly constricted, enveloped in a gelatinous sheath.—Species may be attached; soft, slimy, to 50 cm long; pale yellow, yellow-green, brown or purplish. Produces blooms in ponds and temporary pools. Often intertidal; most common in winter and spring.

S.CL. **FLORIDEOPHYCIDAE:** Multicellular Rhodophyta variable in form; both uni- and multiaxial; cells uni- or multinucleate; growth predominantly apical; pit connections prominent; sexual reproductive organs more complex than in Bangiophycidae. (5 of the 6 orders known reported from Bda.)

O. **NEMALIALES:** Filamentous, creeping or erect Florideophycidae; minute, uniseriate, uninucleate, epiphytic forms, to some that are larger and more complex, uniaxial, multiaxial, calcified and corticated (having a covering of cells); soft or cartilaginous; no auxiliary cells formed; possessing a wide variety of life history types. (6 families, approximately 35 spp. from Bda.)

F. **HELMINTHOCLADIACEAE:** Erect, coarsely branched Nemaliales; sometimes partially calcified; cystocarps immersed among filaments; lacking a pericarp.

*Liagora ceranoides* Lamour.: Genus soft, bushy, usually terete; may be slightly or heavily calcified; branches irregular or dichotomous, reddish or purplish, white if calcified.—Species compact, moderately calcified, white; soft, tufts to 8 cm in diameter; branching irregular or dichotomous. Attached in shallow, exposed and sheltered

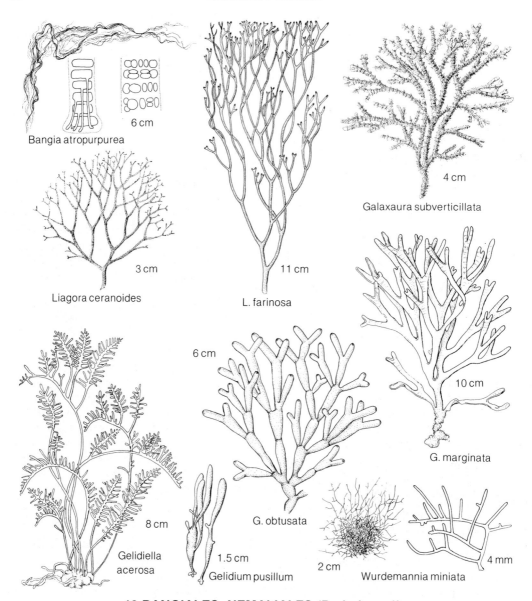

**12 BANGIALES, NEMALIALES (Red algae 1)**

water; common during winter and spring.

***L. farinosa*** Lamour.: Genus as above.—Species dichotomously branched, loose, tangled; calcified or partially so; pinkish or reddish; tufts to 12 cm in diameter. Commonly attached; in shallow water of sheltered locations; intertidal zone; abundant in winter and spring. (Color Plate 2.8.)

F. **CHAETANGIACEAE:** Erect, multiaxial, bushy, soft and gelatinous, or cartilaginous Nemaliales; sometimes slightly or moderately calcified; a pericarp develops around the carposporophyte, the spores discharging through a pore from a conceptacle-like structure.

***Galaxaura subverticillata*** Kjellm.: Genus firm to wiry in texture; dichotomously branched and bushy; branches usually truncate or depressed at ends; terete or flattened, occasionally jointed.—Species to 7 cm high; branches irregular or dichotomous; characteristically covered with assimilative filaments (hair-like cells) resulting in a rough texture; red to purplish; various degrees of calcification. Attached in shallow waters and on old coral heads; found throughout the year, most common in winter and spring.

***G. obtusata*** (Ell. & Sol.) Lamour.: Genus as above.—Species to 10 cm high; dichotomous or irregularly branched; branches terete, stiff, regularly segmented; pink to reddish depending on degree of calcification; usually stiff in texture; tips of branches truncate or depressed. Extremely common throughout the year, attached in sheltered areas, in the intertidal and subtidal zones; frequent on old coral heads. (Color Plate 2.9.)

***G. marginata*** (Ell. & Sol.) Lamour.: Genus as above.—Species to 15 cm high, smooth; non-segmented, flattened, frequently transversely banded; terminated with a brush of deciduous hair-like cells. Intertidal to below the low tide mark; in exposed or protected areas; reported year round, infrequent in fall.

F. **GELIDIACEAE:** Firm, cartilaginous, tough, wiry, branched Nemaliales; uniaxial or multiaxial; growth apical; tetrasporangial divisions cruciate (cross-shaped) and embedded in the cortex of localized areas or branchlets of the tetrasporophyte.

***Gelidiella acerosa*** (Forssk.) Feldm. & Hamel: Genus to 15 cm high, bushy, slender, branches terete; basal portions attached by rhizoidal branches.—Species with several long axes having many short determinate branchlets 2-6 mm long, radially or bilaterally attached; feather-like; greenish yellow to purple. Attached to rocks and old coral heads in the intertidal zone, and to 3 m below low tide mark; year round.

***Gelidium pusillum*** (Stackh.) Le Jol.: Genus with main axes erect; lateral branches cylindrical, or flattened and firm.—Species small, solitary or forming a dense mat; erect blades to 1.5 cm high, cylindrical below, flattened above; sparsely pinnate; olive-green to red or purple. Common and widespread; forming turf on rocks, pebbles and shells; in shallow water and slightly below the intertidal zone; throughout the year.

F. **WURDEMANNIACEAE:** Gregarious Nemaliales with slender, wiry branches; multiaxial; tetrasporangial divisions zonate (transversely parallel).

***Wurdemannia miniata*** (Drap.) Feldm. & Hamel: Genus bushy, small, gregarious; wiry, with multiaxial growth; forming dense mats; attached to substrate by frequent rhizoids.—Species forming dull, red to pink clumps, or often bleached on substrate; under 3 cm high; branches entangled below, becoming loose and sparingly branched above; terminal branchlets erect and attenuated. Abundant in exposed and sheltered areas; common on rocks in the intertidal zone; on mangrove roots; year round.

**Plate 13**

O. **CRYPTONEMIALES:** Filamentous or soft, fleshy, uni- or multiaxial, or pseudoparenchymatous Florideophycidae; some cartilaginous, crusty or calcified. Asexual reproduction by tetraspores; carposporophyte originates from an auxiliary cell; life cycle diplohaplontic, diplontic, isomorphic or heteromorphic. (4 families, approximately 30 spp. from Bda.)

F. **CORALLINACEAE:** Calcified, pink or bleached Cryptonemiales forming thin crusts, thick, knobby crusts, nodules or delicate segmented branches; segments separated from one another by uncalcified joints. Reproductive cells occur within conceptacles usually opening by a single pore.

*Amphiroa fragilissima* (L.) Lamour.: Genus branched and jointed from attached holdfast; sometimes detached and living entangled among other algae; segments cylindrical; frequently considerably longer than broad; branching irregular or repeatedly dichotomous; medulla composed of small cells alternating with long ones in tiers; conceptacles on the surfaces of segments.—Species usually entangled masses of branches, often with other algae; branching regularly and widely dichotomous at junctions of segments; segments swollen at each end; 2-4 mm long, 150-600 μm broad; conceptacles prominent when present, 300-400 μm in diameter. Abundant in quiet water; attached to old corals, shells, or epiphytic; abundant year round. (Color Plates 2.3, 3.8.)

*A. rigida* Lamour.: Genus as above.—Species growing as small clumps of branches up to 1 cm across; dichotomies regular, wide, diverging; branchlets at intervals between forks; segments cylindrical, not swollen at ends; lower segments 1-1.5 mm thick, upper segments thinner. Restricted distribution inshore; attached, often epiphytic on *Lithophyllum*; intertidal zone and below. Year round, more frequent in winter.

*Fosliella farinosa* (Lamour.) Howe: Genus forming thin crusts; only one to a few cells thick; on other algae and sea grasses. Trichocytes (enlarged hair-bearing cells) present among the smaller cells of the crust.—Species forming delicate whitish crusts; monostromatic in vegetative parts; cells quadrangular, filaments radiating outward from a central area on surface of host; trichocytes common, paler than vegetative cells; tetrasporangial and cystocarpic conceptacles protruding markedly, 150-250 μm in diameter; ♂ conceptacles smaller. On marine grasses, especially *Thalassia*; in shallow bays; common throughout the year.

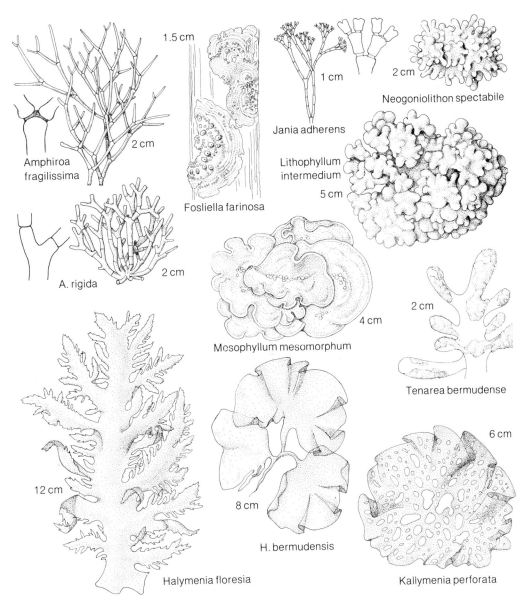

## 13 CRYPTONEMIALES (Red algae 2)

***Jania adherens*** Lamour.: Genus erect and jointed from holdfasts attached to hard surfaces or algae; segments cylindrical, branching dichotomous; joints consisting of a single zone of cells; single conceptacles with an apical pore form in the swollen tip of each fertile segment from which 2 branches usually arise.—Species less than 2 cm high, segments cylindrical except where forked into 2 branches; segments 60-200 μm broad, less than 1 mm long; tetrasporangial conceptacles 200-300 μm in diameter. Growing on algae or on hard surfaces; often intermingled with other algae. Uncommon; most often along exposed shores with and on fleshy algae.

***Neogoniolithon spectabile*** (Fosl.) Setch. & Mason: Genus forming crustose growths on rocks and corals, tightly adherent or as massive knobs; usually beset with protuberances; parts of plants often broken free, on the sea bottom. Trichocysts present as isolated large cells or vertical series of large cells; conceptacles all uniporate.—Species with crowded, irregularly branched, anastomosing protuberances, 1-3 mm in diameter; forming attached cushions that may break off to become free-living marl. Common in subtidal reef areas; the type locality is Bermuda.

***Lithophyllum intermedium*** (Fosl.) Fosl.: Genus crustose, on rocks or shells, sometimes forming free-lying nodules, with protuberances of various sizes and shapes. Tissues made of many layers of small rectangular cells; all conceptacles open by 1 pore.—Species with protuberances variable in shape, size and spacing; low and broadly rounded, 1-5 mm in diameter or more. Attached to corals, rocks or broken loose; in relatively calm areas.

***Mesophyllum mesomorphum*** (Fosl.) Adey: Genus crusty, thin, brittle, especially at the edges; surface smooth near the margins where small ridges paralleling the edge may be present; older surfaces somewhat irregular and often bearing crowded convex conceptacles. Perithallium relatively thin; hypothallium thick and cells in decumbent, arched tiers; roofs of tetrasporangial conceptacles having numerous tiny pores; sexual conceptacles opening by single pores.—Species thin, fragile, 300-500 μm thick, leafy, overlapping lobes loosely attached and easily separating from substrate; tetrasporangial conceptacles crowded in older crusts; roofs protruding as much as 400 μm in diameter. Growing loosely over rocks, algae or debris. Common in subtidal zones; abundant in aquarium tanks; the type locality is Bermuda.

***Tenarea bermudense*** (Fosl. & Howe) Adey: Genus forming tightly or loosely attached crusts on other algae or rocks; cells longer than wide, arranged in tiers with elongated axes perpendicular to the substrate; conceptacles open by 1 pore; the center of each tetrasporangial conceptacle contains a prominent tuft of sterile filaments; sporangia restricted to periphery of the chamber.—Species a crust 1-2 mm thick. On rocks, pebbles or other algae.

F. **CRYPTONEMIACEAE:** Foliaceous, compressed or terete Cryptonemiales; alternately, radially or bilaterally branched; multiaxial; tetrasporangia tetrahedral; auxiliary cells produced on special accessory branches of the gametophyte.

***Halymenia floresia*** (Clem.) C. Ag.: Genus attached, flat, foliaceous or bushy; soft and gelatinous, fleshy; variously lobed and/or branched; branches flattened or terete.—Species profusely branched; to 30 cm high or more, soft and gelatinous; main axis or axes 2 cm wide and 2 mm thick, flattened, having pinnate, slender, terete branchlets; red, purplish or olive-green. In sheltered places, attached to rocks or clefts; most abundant in winter. (Color Plate 2.13.)

*H. bermudensis* Coll. & Howe: Genus as above.—Species with small basal stipes that are simple or branched to 1 cm long; several large foliaceous fronds arise from the stipes; fronds firm to gelatinous, ovate, obovate or cordate (heart-shaped), simple or lobed; margins plane or ruffled, entire or coarsely dentate; fronds to 30 cm wide, 1 mm thick; red, purplish or olive-green. In shallow water, under rock ledges or clefts; usually in sheltered places; smaller specimens on mangrove roots; throughout the year.

F. **KALLYMENIACEAE:** Foliaceous Cryptonemiales with blades or multiaxial branches, erect, soft; 3-celled carpogonial branch arising from a large, lobed supporting cell.

*Kallymenia perforata* J. Ag.: Genus forming expanded, flattened blades, simple or broadly lobed, with or without a stipe at base.—Species to 20 cm or more in diameter, soft, gelatinous; fronds lobed, orbicular or reniform; regularly or irregularly perforated with holes 1-3 cm in diameter; some fronds imperforate; pink or pale red. In deep water; attached to rocks or in crevices; in fall and winter.

**Plate 14**

O. **GIGARTINALES:** Membranous, leaf-like, or filiform to branching Florideophycidae; may be calcified, crustose or corticated; auxiliary cell delimited from a simple, vegetative filament. (The largest order, with over 25 families known; 6 families, approximately 20 spp. from Bda.)

F. **GRACILARIACEAE:** Gigartinales with slender to coarse branches that are terete to flattened or strap-like; firm and cartilaginous; medullary cells pseudoparenchymatous; tetrasporangia cruciately divided; carposporophyte develops toward exterior of gametophyte, the cystocarp contained in pericarp with a pore.

*Gracilaria debilis* (Forssk.) Børg.: Genus bushy; branches flattened or terete, fleshy or cartilaginous, regularly or irregularly branched.—Species to 20 cm high, cartilaginous, one or more main axes with branches forming in a single plane and usually recurved toward axis; straw-yellow to pink when fresh, drier specimens becoming purple to dark. Not common; in shallow water, usually attached; spring and summer.

F. **SOLIERIACEAE:** Gigartinales having compressed, flattened blades, or terete branches; multiaxial; tetrasporangia immersed in the cortex, zonately divided; cystocarps with a pore.

*Eucheuma isiforme* (C. Ag.) J. Ag.: Genus bushy, radially or bilaterally

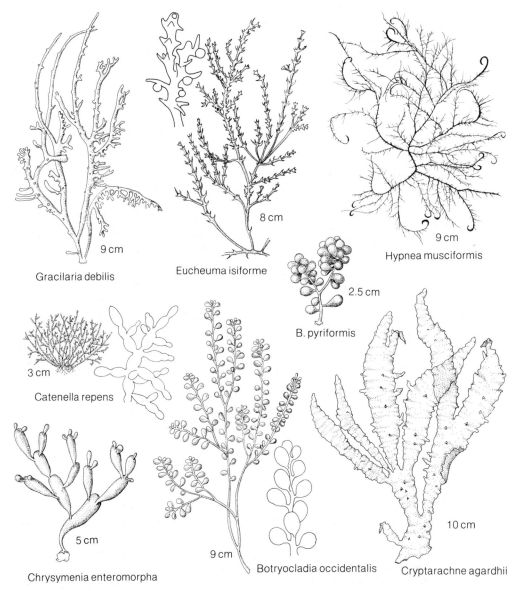

**14 GIGARTINALES, RHODYMENIALES (Red algae 3)**

branched; branches terete or slightly flattened, smooth or with nodules or spines.—Species variable in height, 5-50 cm; firm to cartilaginous; branches sparse to moderate, having whorled spines or nodules; usually 1 or a few main holdfasts; pale yellow, straw, brownish or reddish. Attached at or below the intertidal zone; in shallow and sheltered areas; not abundant.

F. **RHABDONIACEAE:** Flattened or cylindrical Gigartinales; branches radial or bilateral, constricted, oval; tetrasporangia zonately divided.

*Catenella repens* (Lightf.) Batt.: Genus prostrate; articulate branches spindle-shaped, oval or less commonly terete.—Species creeping or prostrate, to 3 cm high, consisting of oval or spindle-shaped segments measuring approximately 2-5 mm in length; branches of segments dichotomous or trichotomous; rhizoids produced at joints; red, purple, olive-brown or brown. On rocks and old corals; common on mangrove roots; often mixed with *Bostrychia, Caloglossa, Cladophora*, etc.; common throughout the year; abundant in winter.

F. **HYPNEACEAE:** Bushy Gigartinales; branches terete, lateral; cortex pseudoparenchymatous; with short, spine-like lateral branches; tetrasporangia localized in branchlets, zonately divided.

*Hypnea musciformis* (Wulf.) Lamour.: Genus bushy, spreading; branches terete, slender, occasionally with spiny or spur branchlets.—Species to 50 cm high, main branches to 2 mm in diameter, with many spur branchlets to 10 mm long; upper branches bare and typically curved or with a swollen hook; red, dull brown or purple. Not abundant; near low tide mark or intertidal zone; attached.

O. **RHODYMENIALES:** Florideophycidae with multiaxial growth, simple or bushy; soft to cartilaginuus; sometimes hollow; tetrasporangia tetrahedrally divided; 3-4-celled carpogonial branches produced; auxiliary cells that give rise to carposporophyte are delimited along cells supporting carpogonial branches. (2 families, approximately 10 spp. from Bda.)

F. **RHODYMENIACEAE:** Solid or hollow, foliaceous or bushy Rhodymeniales having a main axis; occasionally dichotomously branched.

*Chrysymenia enteromorpha* Harv.: Genus with terete branches that are hollow and may be slightly flattened; alternately branched.—Species with thallus divided by constrictions; branching segments cylindrical, hollow; plant to 25 cm high. Intertidal species or below; attached to rocks, coral, or old shells.

*Botryocladia occidentalis* (Børg.) Kylin: Genus loosely branched; main axes slender, terete, cartilaginous, bearing oblong, oval or spherical bladder-like branchlets.—Species to 25 cm high, axes frequently branched, with small, numerous, ovoid to subspherical pedicellate bladders 2-5 mm long; pale rose, purple or brownish. In the intertidal zone or deeper; on rocks or old coral heads, never abundant; usually sparse.

*B. pyriformis* (Børg.) Kylin: Genus as above.—Species to 2 cm high, branches few; bladders pyriform, to 9 mm long, usually 2-6 per plant; rose-red, purple or brownish. On intertidal rocks and rock crevices; not abundant.

*Cryptarachne agardhii* (Harv.) Kylin: Genus foliaceous, lobed or cleft; with several secondary blades.—Species with blades to 20 cm long, with a short stipe; basal part of plant broad; numerous distal branches to 3 cm wide; rose-red. Rare, attached to old coral heads.

O. **CERAMIALES:** Filamentous and branched Florideophycidae; uniaxial, polysiphonous; may be corticated; auxiliary cell formed after fertilization. (4 families, approximately 200 spp. (!) from Bda.)

## Plate 15

F. **CERAMIACEAE:** Ceramiales with main, uniseriate axes, corticated at nodes; bushy, branches terete; tetrasporangia tetrahedrally divided; carposporophyte exposed, without a pericarp.

***Ceramium byssoideum*** Harv.: Genus erect, sparse or matted, dichotomously branched; central axial filaments relatively large-celled with bands of smaller, corticated cells at the nodes.—Species simple, microscopic, dichotomously or seldom alternately branched, to 90 mm in diameter; nodal bands composed of several rows of cells; tips of branches often pincer-like or forceps-like; red, purplish or dark. Growing on larger algae, roots of mangroves, sea grasses, shells, rocks, etc; very common but inconspicuous because of size; occurs year round. (Color Plate 2.11.)

***Centroceras clavulatum*** (C. Ag.) Mont.: Genus bushy, filamentous, dichotomously or laterally branched; internodes corticated; nodes spinulose.—Species 10-20 cm high, stiff, matted or entangled; filaments 50-200 μm in diameter; internodes short above, longer below; corticated throughout; 2-celled spines at each node prominent in apical branches. Common as short tufts, entangled with other algae; ubiquitous. Frequent in winter, less during summer.

***Spyridia aculeata*** (Schimp.) Kütz.: Genus bushy, erect; branches alternate, corticated; ultimate branchlets filamentous, uniseriate, nodal.—Species to 25 cm high; major branches dense, corticated; terminal branchlets to 1 mm long, tips ending with a spine; rose-red, olive-brown or dull with age. Common on rocks and coral heads; in subtidal and intertidal zone; fall and winter.

***Wrangelia penicillata*** C. Ag.: Genus bushy, filamentous; central axis producing 2-ranked branches in a single plane; ultimate branches tufted and verticillate; hairy.—Species with wide and irregularly alternate branches; main axes to 15 cm high; branches extending to 20 cm wide; axes corticated by longitudinal filaments; branchlets terminating in attenuated cells or single spine-like cells; red to purplish. On old coral heads and rocks, smaller specimens epiphytic; usually below intertidal zone; common in winter.

F. **DELESSERIACEAE:** Foliaceous Ceramiales; blades constructed of lateral filaments, with or without midrib; branches, if present, flattened; tetrasporangia tetrahedrally divided; carposporophyte enclosed in a pericarp.

***Caloglossa leprieurii*** (Mont.) J. Ag.: Genus prostrate, spreading

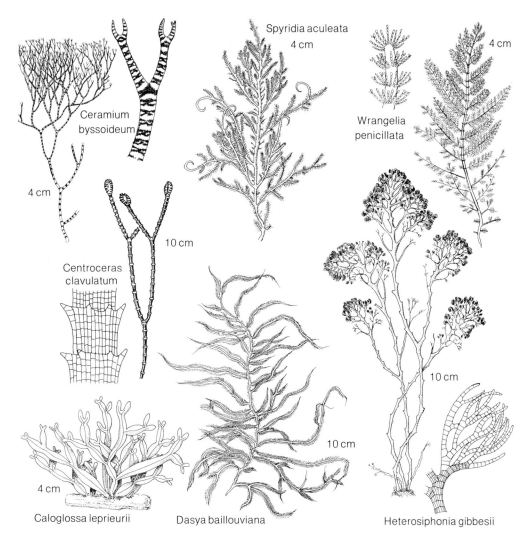

**15 CERAMIACEAE, DELESSERIACEAE, DASYACEAE (Red algae 4)**

or erect; blades flat, dichotomously forked; prominent midrib conspicuous.—Species flattened throughout, 4-5 cm across; branching segments lanceolate, linear, 2 mm broad, to 6 mm long, bearing a midrib; rose, reddish or purple. Commonly with members of *Bostrychia* on mangrove roots as an epiphyte, or attached to rocks, old coral heads and shells; common throughout the year.

F. **DASYACEAE:** Slender Ceramiales with cylindrical axis covered with fine, monosiphonous filaments; development sympodial; tetrasporangia tetrahedrally divided, borne in specialized branches; cystocarp enclosed in pericarp with a pore.

***Dasya baillouviana*** (Gmel.) Mont.: Genus erect, bushy, soft; main branches heavily corticated, terete, covered with fine, filiform hairs; branchlets crowded, whorled, poly- to monosiphonous.—Species to 50 cm high, delicate, freely branched; main branches covered with slender, filiform branchlets (ramelli) to 8 mm long; pink, red or dark red. Common throughout the year; in shallow, sheltered areas; epiphytic, on shells, rocks, and old coral heads. (Color Plate 2.7.)

***Heterosiphonia gibbesii*** (Harv.) Falk.: Genus erect, main branches simple, lacking extensive cortication, sometimes flattened; ultimate branchlets (ramelli) alternately branched and monosiphonous.—Species to 20 cm high, main axis corticated, sparsely forked; branches more or less denuded at base, becoming branched above with densely placed ramelli; rosy red, pink or purplish brown. Common during winter in sheltered areas; on rocks, old coral heads; may be epiphytic on larger algae.

## Plate 16

F. **RHODOMELACEAE:** Bushy Ceramiales; main axes moderate or sparingly branched, occasionally delicate, terete, polysiphonous; tetrasporangia cruciately divided; cystocarp enclosed by a pericarp with a stalk.

***Polysiphonia denudata*** (Dillw.) Kütz.: Genus erect, bushy, dichotomously branched; all branches polysiphonous, terete; siphons usually aligned parallel to the longitudinal axis of the plant, or spirally arranged; delicate tricho- blasts present in some.—Species 10 cm (to 25 cm) high, main axis pink to red or purple, consisting of approximately 6 pericentral cells; lower branches denuded, upper branches having many soft, pliable, reddish to brown branchlets. On mangrove roots; as an epiphyte on a number of larger algae; not uncommon on rocks, shells and crevices of old coral heads.

***Digenia simplex*** (Wulf.) C. Ag.: Genus with thick, widely dichotomous branching; axes cartilaginous, fleshy and provided with closely radiating branchlets.—Species 10 cm (to 25 cm) high, cartilaginous or wiry in texture; irregularly or dichotomously branched, covered with slender, stiff branchlets to 5 mm long; branches may be denuded below; red, dull red to brown. Throughout the year in intertidal or subtidal locations; on rocks, old coral heads, shells; often a host for epiphytic algae.

***Bostrychia montagnei*** Harv.: Genus rhizoidal, stoloniferous, polysiphonous; branching bilateral with terminal branches often incurved; ramelli at tips of branches often monosiphonous.—Species to 8 cm high, alternately pinnate, terminal branchlets incurved; tufted, coarse, dull red, purple or black. On mangrove roots with *Caloglossa*; common throughout the year. (Color Plate 2.12.)

***Herposiphonia secunda*** (C. Ag.) Ambronn: Genus small, composed of creeping rhizomes and erect, compressed or terete branches; polysiphonous, usually unbranched at base, becoming

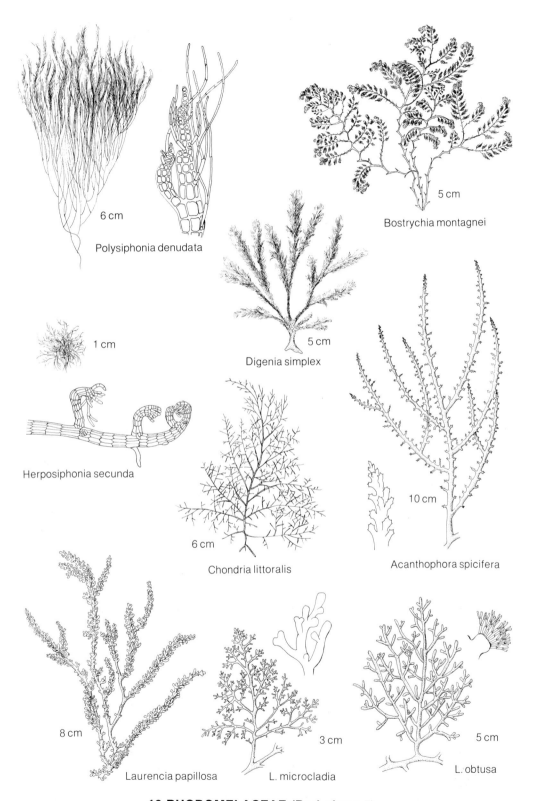

## 16 RHODOMELACEAE (Red algae 5)

sparingly or moderately branched at apex.—Species to 2 cm high; closely attached to substrate by rhizoidal cells; branching at apex sparse, alternate; mat-like; dark red, red-brown or purplish. On rocks in the intertidal zone; component of algal felt; epiphytic on larger algae, especially *Codium* and *Caulerpa*; found in fall, winter and spring.

***Chondria littoralis*** Harv.: Genus erect, bushy; alternate, terete branching; terminal branches and branchlets usually constricted at the bases, ultimate branchlets thin, spindle or club-shaped and tipped with clusters of trichoblasts.— Species with older branches denuded at base; ultimate branchlets spindle-shaped, constricted at base and tipped with a tuft of trichoblasts; reddish or purple. Attached below low tide zone; rare.

***Acanthophora spicifera*** (Vahl) Børg.: Genus moderate to large, sparsely branched, bushy, cartilaginous or fleshy; branches alternate, terete; upper branches with short, spiny, spur branchlets spirally arranged.—Species to 25 cm high, loosely branched, with short spur branchlets having spirally placed spines; main axis lacking spines; pale rose to reddish, or purple; occasionally pale yellow or brown. Common in shallow water; on rocks, old coral heads; usually in protected areas; frequent throughout the year.

***Laurencia papillosa*** (Forssk.) Grev.: Genus bushy and erect, fleshy or cartilaginous; branches alternate or irregular, terete; terminal branchlets clavate with a slightly constricted base; tip of clavate branchlets usually darker in color, occasionally rosy red.— Species usually under 5 cm high, densely packed; cartilaginous, branches to 2 mm diameter; branches covered with short, clavate branchlets; olive or dark green throughout. In exposed areas of the littoral zone; may appear dwarfed and simple when exposed to wave action; common in winter and spring. (Color Plate 3.8.)

***L. microcladia*** Kütz.: Genus as above.—Species to 10 cm high, wiry, main axis to 2 mm diameter; branching narrowly pyramidal; greenish throughout main branches, terminal branchlets rose to pinkish. Common; attached to rocks in the intertidal zone; throughout the year.

***L. obtusa*** (Huds.) Lamour.: Genus as above.—Species to 25 cm high, bushy, may be matted, sparingly to moderately branched, with short, 1 mm truncate branchlets having sunken apices; olive-green to yellow throughout, rose to pink branchlets. One of the most commonly encountered algae in Bermuda. Widespread throughout the year; in sheltered locations; on a number of substrates. (Color Plate 2.10.)

A. R. CAVALIERE & H. W. JOHANSEN

## Phylum Chrysophyta

CHARACTERISTICS: *Unicellular or colonial, yellow-green to golden-brown PLANTAE; cells have 2, 1 or no flagella. Cell walls when present containing cellulose, silica, calcium carbonate, mucilaginous substances or chitin. Pigments concentrated in defined chromatophores containing chlorophyll a and c and various accessory pigments (carotenoids and xanthophylls). Oil and leucosin are the customary food storage; sta ch is not present. Reproduction asexual or sexual.*

All 3 classes are well represented in Bermuda: Bacillariophyceae (p. 67), Chrysophyceae (p. 70) and Xanthophyceae (p. 73).

## Class Bacillariophyceae ( = Diatomeae) (Diatoms)

CHARACTERISTICS: *Unicellular CHRYSOPHYTA without a flagellum; cells enclosed in a frustule impregnated with silica, and consisting of 2 halves overlapping each other. Chloroplasts lobed, or made up of many discoid bodies.* From 0.015 to 2 mm in size, diatoms are of cylindrical, elliptical, polygonal or lancet shape, some forming long chain-like colonies. Some planktonic forms can regulate their buoyancy, and many benthic species are able to crawl slowly over the substrate.

Both orders are represented in Bermuda (most systems do not recognize families). Of about 4,000 marine species, 165 have been identified from Bermuda, of which 10 are included here.

OCCURRENCE: Freshwater and marine; in plankton and benthos (also as epiphytes). Though the majority are free-living, some are symbionts or parasites. They are most abundant in colder latitudes.

Collect plankton species with a small-mesh (10 μm) plankton net or—for very small forms—use water bottles in combination with settlement chambers. Benthic species are best collected with SCUBA, scooping the uppermost sediment layer into a jar or scraping rocks with a knife. Feces of holothurians are usually very enriched with frustules of benthic species.

IDENTIFICATION: A microscope with good magnification (> 100×) and preferably phase contrast is necessary to identify species, live or preserved.

The diatom frustule (external skeleton) consists of 2 parts (valves) the larger of which (epitheca) overlaps the smaller (hypotheca), much as a lid covers a Petri dish. The valves are connected in the girdle region either by a ring-like band (pleura) or by small appendices fringing the inner part of the girdle. Intercalary bands may occur between the valve and the connecting band. The frustules may be ornamented with patterns of pores and raised areas, and may carry spines or long processes (setae). Centric diatoms have radial symmetry, pennate diatoms bilateral symmetry. For orientation of the frustule it is useful to distinguish between 3 axes: the apical axis passing horizontally between the valves; the pervalvar axis connecting the surfaces of the 2 opposing valves; and the transapical axis lying at right angles to the 2 other axes. In pennate diatoms the valve surface may have a slit (raphe) along the apical axis, which may be provided with polar nodules and bordered by regular lateral structures. The cytoplasm is in contact with the medium via the raphe. Other pennate diatoms may have a pseudoraphe, an unstructured area along the apical axis flanked by regular lateral markings (costae, areolae).

Fixation and removal of organic matter are required for reliable species identification. Specimens can be put in a muffle oven at 550°C, or carefully heated with potassium permanganate and concentrated hydrochloric acid. After rinsing in distilled water store in a mixture of distilled water and ethanol. Most planktonic diatoms can be fixed in 4% formaldehyde or Lugol's solution, for later identification.

BIOLOGY: Most diatoms are autotrophic. Those benthic forms that have a raphe are capable of gliding over the substrate; others are attached. Some species live as symbionts or parasites in Foraminifera, corals and algae; others serve as substrate to dinoflagellates, fungi and ciliates. In Bermuda waters, plankton diatoms reach their peak of abundance in winter and spring, but occur throughout the year in considerable numbers.

REPRODUCTION: Reproduction is by cell division, which separates the epitheca from the hypotheca, both becoming epithecae of the daughter cells. Because this results in progressive reduction of cell size, cells eventually form auxospores (sexually or asexually), which then develop into cells of original size. In some species this cell size reduction is circumvented by an adjustment of the frustule during cell division. During sexual reproduction (oogamy) mostly non-siliceous spermatogonia are formed.

REFERENCES: For a general introduction see BOLD & WYNNE (1978) and SOURNIA (1978).
HULBURT et al. (1960) give a list of species of the Sargasso Sea. CLARKE (1934) and BODUNGEN et al. (1982) list species from inshore waters. The benthic species of Bermuda have yet to be studied.

## Plate 17

SECT. **CENTRICAE:** Cells cylindrical or disc-shaped. Valves always radially symmetrical, commonly provided with processes. Very often forming chains.

*Leptocylindrus danicus* (Cleve): Cells long, cylindrical, linked into chains by entire valve surface. Cells weakly siliceous without visible sculpturing. Numerous small chromatophores. Diameter 6-11 μm, pervalvar axis 30-60 μm. In inshore and offshore plankton; common.

*Guinardia flaccida* (Castracane) H. Peragallo: Cells cylindrical, single, or united to chains by entire valve surface. Valve slightly concave, with intercalary bands and an irregular tooth at the margin. Chromatophores numerous, more or less lobed. Diameter 30-50 μm, pervalvar axis 75-150 μm. Most abundant in offshore plankton.

*Rhizosolenia shrubsoleii* Cleve: Cells single or in short chains. Valves conical, strongly eccentric, forming a continuous line with the girdle margin on 1 side, but oblique on the other side. Valves end in a short hollow spine, with small wings extending to about 1/3 of the spine's length. Intercalary bands forming a scale-like pattern. Chromatophores small, numerous. Diameter 15 μm, pervalvar axis 300-500 μm. In inshore plankton; very common.

*Chaetoceros glaudazii* Mangin: Cells united to straight chains; valves elliptical bearing a small median process which fuses with that of the adjoining cell. Setae long, straight, extending from the central area of the valve surface. Setae on the lower valve of the terminal cell with short spines. Chromatophores small, rounded, numerous, extending into setae. Diameter 15-20 μm, pervalvar axis 30-35 μm. Mainly offshore, but sometimes the dominant species of inshore phytoplankton.

*Hemiaulus hauckii* Grunow: Cells long, straight or twisted, often forming chains. Cells with long, strongly pointed processes. Cell wall weakly siliceous, often without visible sculpturing. Chromatophores small, numerous. Diameter 15-18 μm, pervalvar axis 28-35 μm. Most common in offshore plankton.

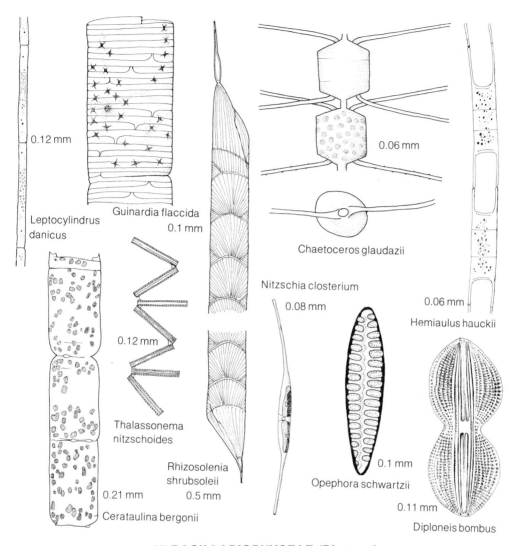

**17 BACILLARIOPHYCEAE (Diatoms)**

***Cerataulina bergonii*** H. Peragallo: Cells cylindrical; valves slightly convex with 2 short, stout processes at the margin that terminate in short spines protruding into the adjoining cell. Chains often twisted. Chromatophores small, numerous. Diameter 25-52 μm, pervalvar axis 70-120 μm. In inshore and offshore plankton; abundant.

**SECT. PENNATAE:** Valves bilaterally symmetrical. Raphe or pseudoraphe present.

***Thalassonema nitzschoides*** Grunow: Cells linear-narrow in girdle view; valves with parallel sides and bluntly rounded ends. Cells united into star-shaped or zigzag chains. Chromatophores numerous small granules. Length 30-90 μm, width 2-5 μm. In inshore plankton, very common.

***Nitzschia closterium*** W. Smith: Spindle-shaped single cell with flexible hair-like ends. Length 50-80 μm, width 3-5 μm. Ubiquitous; most common in the littoral zone but also frequently found in plankton.

***Opephora schwartzii*** (Grunow) Petit: Frustules rectangular in girdle view. Valves linear, elongate with rounded ends. Coarse areolae, 3-4 in 10 μm, at right angle to the pseudoraphe. Valve length 60-100 μm, width 10-12 μm. In the littoral and sublittoral zones, attached to sand grains, not motile; common.

***Diploneis bombus*** (Ehrenb.) Cleve: Cells solitary. Valves deeply constricted, sections often unequal in size. Valve surface with transverse and some longitudinal costae. Two chromatophores along the girdle. Length 80-110 μm, width 22-35 μm. A common benthic diatom able to move by means of its raphe.

B. v. Bodungen & C. Oderbrecht

## Class Chrysophyceae (Golden-brown algae)

CHARACTERISTICS: *Mostly unicellular and biflagellated (less commonly coenocytic and nonmotile) CHRYSOPHYTA. Cell wall often contains calcareous or siliceous intercalations. Cell mostly with 1 or 2 chromatophores; the main pigments are chlorophyll a, fucoxanthin and xanthophyll. Reserve substances are the polysaccharide chrysolaminarin and oil. Asexual reproduction more common than sexual reproduction. Minute (4-90 μm), many-shaped; some move by means of flagella.*

The number of species is estimated at 1,000; 28 have been reported from Bermuda, of which 8 are included here.

OCCURRENCE: All are plankton organisms with the greatest diversity in fresh water. Calcareous forms (Coccolithophorida) are most common in warm oceans, whereas the (strictly marine) silicoflagellates prefer colder waters and were particularly abundant in the Tertiary. Both groups contribute significantly to the formation of ocean floor sediments. In Bermuda's inshore waters silicoflagellates can seasonally dominate the phytoplankton.

Collect live specimens by filtering or centrifuging water samples that have been treated with buffered formaldehyde.

IDENTIFICATION: Examine live or fixed specimens under a high-power microscope, preferably with phase contrast and polarized light. For very small species, scanning or transmission electron microscopy techniques may be necessary. Skeletal structures become clearer after incineration at 450°C.

Species with flagellate organization can be sedentary and develop a cellulose case (lorica). Others may form colonies in which pear-shaped cells are linked together at their posterior ends, and are covered with siliceous scales arranged like pan tiles. The protoplast of the coccolithophorids has a cover in or on which small calcareous plates (coccoliths) are located. The cell cover may calcify, in which case the coccoliths are firmly connected to the cover. Besides the coccoliths, sculptured scales of organic material occur in the cell cover. The coccoliths vary widely in size (1-40 μm) and shape: round, oval, rhombic, or club-like, solid or pierced. The silicoflagellates have a siliceous skeleton inside the cell, consisting of a polygonal basal ring with spines and mostly an apical, arched frame. This frame may be composed of small transverse, bow-shaped bars or form a concentric, elevated ring (apical ring). Amoeboid Chrysophyta are naked and may have pseudopodia. Other species may be embedded in a gelatin or surrounded by a cell wall; both types lack flagella. Cells can be connected to branched or unbranched threads or grow together to parenchymatic tissues.

Lugol's solution or 4% buffered formaldehyde is best used for fixation and storage. Coccolithophorids must be stored in a buffered solution to prevent the calcareous parts from dissolving.

BIOLOGY: Both the motile and sedentary Chrysophyceae are mostly autotrophs. Some of the motile species are mixotrophs; despite their 1 or 2 chromatophores they can take in organic substances. The colorless species generally are heterotrophs. In the non-sedentary groups locomotion results from the flagella or the pseudopodia. To survive unfavorable conditions cysts are formed inside the protoplast, which are mostly impregnated with silicate.

DEVELOPMENT: The most common mode of reproduction is asexual bipartition or formation of zoospores or aplanospores. Sexual reproduction is isogamous; 2 vegetative cells, acting as gametes, combine into a zygote. The protoplast of the coccolithophorids divides into 2 naked monades with flagella. These develop into a calcareous flagellate or into a sedentary form of coccal or trichal organization.

REFERENCES: For general information see SOURNIA (1978) and BOLD & WYNNE (1978).
HULBERT et al. (1960) give a list of coccolithophorids from the Sargasso Sea, and BODUNGEN et al. (1982) report on species from inshore waters.

## Plate 18

O. **CHRYSOMONADALES:** Solitary or colonial, unicellular; with 1 or more flagella; naked or with a lorica.

S.O. **COCCOLITHINEAE:** Unicellular, solitary, mainly marine Chrysomonadales; cell wall often calcified, with coccoliths, 2 yellow-brown chromatophores, and 2 flagella of nearly equal length. (26 spp. from Bda.)

F. **SYRACOSPHAERACEAE:** Coccolithineae with unpierced coccoliths which are disc-, cup-, hump-, or beaker-shaped.

*Scyphosphaera apsteinii* (Lohmann): Cell cover spherical with small, dimorph coccoliths; large beaker-shaped coccoliths grouped around the largest periphery, serving as suspension apparatus. Diameter of cell cover 10-25 µm, size of the beaker coccoliths 30-50 µm, the small dimorph coccoliths are 8-10 µm long. In offshore plankton, rarely inshore; mostly found in deeper samples.

F. **DEUTSCHLANDIACEAE:** Coccolithineae with unpierced coccoliths that are leaf- or band-shaped.

*Calciosolenia murrayi* (Gran): Long cylindrical cells, pointed at both ends, with 2 calcareous spines at one end. Band-like coccoliths surround the protoplast in 2 helical layers. No space between the coccoliths. Length 50-80 µm, width 4-6 µm. In- and offshore.

*Anoplosolenia brasiliensis* (Deflandre): Cell in the middle bulging, towards the ends narrowing, both ends slightly bent to the same side. No processes. Long, band-shaped coccoliths cover the protoplast in 2 layers giving the impression of little rhombic plates. Flagella never observed. Length 70-110 µm, width 4-7 µm. Frequently found in inshore and offshore plankton.

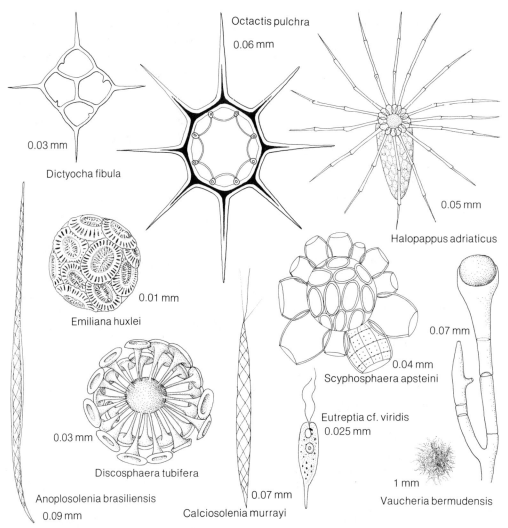

**18 CHRYSOPHYCEAE (Golden-brown algae), XANTHOPHYCEAE (Yellow-green algae), EUGLENOPHYTA**

F. **HALOPAPPACEAE:** Coccolithineae with protoplasts surrounded by a homogeneous, more or less calcified cover with long calcareous spines.

*Halopappus adriaticus* (Schiller): Cell cover conical, with convex sides; posterior end pointed, anterior end with a large orifice in the cover surrounded by 15-17 3-jointed calcareous spines. Cell cover without coccoliths; 2 large yellow chromatophores. Length 15-20 μm, width 10 μm, spines 30-40 μm long. In offshore and inshore plankton near the surface; occasional.

F. **COCCOLITHACEAE:** Coccolithineae with pierced coccoliths that may consist of a short tube with a small disc at both ends or a basal disc with multiform processes.

*Emiliana huxlei* (Lohmann; Hay & Mohler): Spherical cell with elliptic coccoliths, built of 2 small discs (2 µm in diameter) connected by a short tube. With 1-2 yellow-green chromatophores. Diameter of the cell 5-12 µm. Very common in the offshore plankton during winter.

*Discosphaera tubifera* (Murray & Blackmann): Cell cover spherical to oval; coccoliths with calyx-shaped processes, the stalk-shaped part solid. At 2 opposite locations on the cell cover the processes are 2× as long as the others. With 2 chromatophores. Diameter of the cell 10 µm, processes about 10 µm long. In in- and offshore plankton; abundant.

S.O. **SILICOFLAGELLINEAE:** Exclusively marine, unicellular, solitary Chrysomonadales with siliceous endoskeleton; cell body naked, with 1 flagellum and numerous small yellow-brown chromatophores. (2 spp. from Bda.)

F. **DICTYOCHACEAE:** Silicoflagellineae with an endoskeleton consisting of a basal ring with a pyramid-shaped top or an apical ring. The siliceous bars of the frame are hollow.

*Dictyocha fibula* (Ehrenberg): Basal ring quadratic, rectangular, or rhombic, with 4 radial spines of which 1 opposing pair is longer than the other. Skeleton hat-shaped. Cell 10-40 µm in diameter. Seasonally forming mass blooms below 18 m in the inshore waters.

*Octactis pulchra* (Schiller): Octagonal massive basal ring; apical ring also octagonal, nearly the same diameter as the basal ring, but very delicate. One opposing pair of the 8 radial spines longer than the others. Flagellum never noticed. Diameter of basal ring 25-35 µm. Seasonal forming mass blooms below 18 m in the inshore waters.

B. v. BODUNGEN & C. ODERBRECHT

**Class Xanthophyceae** (Yellow-green algae)

CHARACTERISTICS: *Unicellular or coenocytic, green to yellow-green CHRYSOPHYTA; motile, planktonic or sedentary. Cell wall lacking, or consisting of 2 parts overlapping in the middle of the cell. Cell wall built of pectin or cellulose, and may be impregnated with silicate. Pigments are chlorophyll a, beta-carotenoids, and 3 xanthophylls. Reserve substances are glucose (chrysolaminarin) and oil. Vegetative reproduction is common; some species have sexual reproduction. Cell size ranges from a few micrometers to about 600 µm.*

Of about 800 described species only a few are marine; 5 sedentary species have been reported from Bermuda, of which 1 is included here.

OCCURRENCE: Mostly in fresh water, planktonic or benthic. The sedentary species live on moist soil and sheltered, shallow bottoms. They can produce dense felts and play an important role in stabilizing the sediment in *Salicornia* turf. In Bermuda, the sedentary species grow on sheltered, sandy or muddy shores around the low tide level and under mangroves.

Collecting is best done by scraping a thin layer of sediment on which the algae grow into a jar filled with seawater.

IDENTIFICATION: Live or fixed, with a microscope or stereoscope. Species identification is difficult because many resemble those of the phylum Chlorophyta; treatment with iodine-potassium, however, stains the starch-containing Chlorophyta violet.

Xanthophyceae occur in a variety of forms. The monadal type is unicellular, lacks a cell wall but possesses 2 flagella of unequal length (one of which can be reduced). The rhizopodial type lacks flagella and a cell wall; cells are sometimes connected by plasma threads to form complicated nets. The capsal type consists of a protoplast without a cell wall or flagella, embedded in gelatin. The most common, coccal type has a solid cell wall that may be sculptured with silicate intercalations, but lacks flagella. In the trichal type cells are united to form simple or branched threads; the cell wall consists of H-shaped, interlocking parts. The siphonal algae of this class are balloon-shaped to filamentous, branched plants, without partitions between the numerous nuclei (coenocytic) or differentiation between rhizoidal and assimilative parts. The filaments develop oogonia and antheridia for sexual reproduction, carrying the oocyte and the spermatozoids, respectively.

A 7% alcohol or 4% formaldehyde solution can be used for storage and identification. Lugol's solution may be used for the monadal forms.

BIOLOGY: Most species are photosynthetic, but mixotrophic species are also known. Some are phagotrophs, taking in bacteria, diatoms, and other small particles with their pseudopodia. Numerous species are epiphytes on other algae. Monadal species and the reproductive cells (zoospores) are able to swim by means of their flagella; the rhizopodial forms perform an amoeboid locomotion.

DEVELOPMENT: Vegetative reproduction takes place via biflagellate zoospores or non-motile aplanospores; those of the unicellular species already attain full growth to adult cells in the parental cell. Bipartition of the cells is also very common. Sexual reproduction is only known as oogamy in *Vaucheria*. The mature oogonia contain 1 oocyte. Numerous biflagellate spermatozoids each containing a complete set of cell organelles are released from the antheridia. After fertilization the zygote encases itself in a thick wall, and a new alga develops after a resting period.

REFERENCES: For general information see BOLD & WYNNE (1978).

Bermuda's *Vaucheria* species have been described by TAYLOR & BERNATOWICZ (1952a, b).

## Plate 18

O.  **HETEROSIPHONALES** (=BOTRYDIALES): Siphonal, filamentous or balloon-shaped Xanthophyceae, without partitions between the numerous nuclei. Motile cells with 1 nucleus only during reproduction stages. (2 genera, each representing a family.)

F.  **VAUCHERIACEAE:** Irregularly or dichotomously branched, filamentous Heterosiphonales; central, large vacuole surrounded by a thin plasma layer clinging to the cell wall. Numerous small chromatophores, nuclei and oil droplets in the plasma layer.

*Vaucheria bermudensis* (Taylor & Bernatowicz): Irregularly branched filaments without constrictions, 20-50 μm in diameter. Antheridia terminal, more or less cylindrical; oogonia stalked. Forms very diffuse, light green colonies on sheltered sandy bottoms.

B. v. BODUNGEN & C. ODERBRECHT

## Phylum Euglenophyta

CHARACTERISTICS: *Mostly unicellular PLANTAE usually with 2 flagella at the apical end; cell with helical symmetry. Most are photosynthetic, but some lack chloroplasts and are heterotrophic. Pigments are chlorophyll a and b, and beta-carotene and derivates; the main re-*

*serve substance is paramylon, not starch. Vegetative reproduction is common, sexual reproduction apparently absent.* Cell size ranges from 10 to 600 μm; many are green, some colorless. Marine forms swim rapidly by flagellar action.

The taxonomy is artificial, partly because of the "animal" characteristics of some species. Of about 800 species, all in the single class Euglenophyceae, 1 species is given here as an example.

OCCURRENCE: Predominantly in fresh water; some are marine. They prefer water with a high content of organic matter where they may produce blooms. Some species may form a green, slimy coating on mud and sand bars in estuaries. Fine mesh plankton nets and/or water bottles in combination with settlement chambers are used for collecting. Sedentary forms may be scraped off their substrate.

IDENTIFICATION: Live or preserved, with a high-power microscope with phase contrast.

The cell body is more or less helically distorted, mostly with a superimposed bilateral symmetry. It is covered by a periplast (or also a pellicula), which may be thin and flexible; or composed of rigid, strong ribs sometimes with pearl-like knobs between them. The number of strips and the surface sculpture of the periplast are important taxonomic criteria. In the anterior part of the cell a large pulsating vacuole adjoins a bottle-like invagination of the anterior end of the cell body. In this so-called ampoule mostly 2 flagella are inserted. The flagella can be of equal length or one can be so short that it does not protrude from the ampoule. The paramylon grains, the main assimilation product, have a roundish or longish, ring-like shape.

Osmium tetroxide, 2-4% formaldehyde solution, or Lugol's solution can be used as fixatives and for storage.

BIOLOGY: Most species are autotrophs; however, under poor light conditions they can live as optional heterotrophs. Pigment-containing species have an eyespot (stigma), colored red by carotenoids, at the ampoule. The stigma serves as shade for a receptor which controls the motions of the flagella and hence locomotion with regard to light (phototaxis). Under unfavorable conditions resting cells are formed, the flagella are shed and the cells are covered with a gelatinous layer (palmella stage).

DEVELOPMENT: Vegetative reproduction often takes place during the palmella stage, but also during the motile stage. After mitosis and division of ampoule and stigma, the cell divides longitudinally so that the daughter cells are of the same length, but only half as wide as the parental cell. Sexual reproduction is not known.

REFERENCES: For general information see LEEDALE (1967).
Bermuda's species have not been studied.

**Plate 18**

F. **EUTREPTIACEA:** Euglenophyta of diverse body shape, provided with 2 long flagella, both of which are used for swimming.

*Eutreptia* cf. *viridis* (Perty): Cell green, spindle-like, elliptic or oval; posterior end tapered. With 15-20 chromatophores, ringlike paramylon grains, and 2 flagella of equal length. Length of the cell 50-65 μm, breadth 5-15 μm; flagella 40-65 μm. Usually found in eutrophic waters; rare.

B. v. BODUNGEN & C. ODERBRECHT

## Phylum Dinophyta (=Dinoflagellata, Pyrrhophyta) (Dinoflagellates)

CHARACTERISTICS: *Mostly unicellular PLANTAE, usually with 2 flagella 1 of which rests in a groove encircling the cell. Cell enclosed by a thin cell wall or cellulose plates sometimes encrusted with silica. Nucleus large, in form of persistent chromosomes; outer plasma with green, yellow or brownish plastids. Pigments are chlorophyll a and peridinin; reserve substances are starch and oil. Reproduction commonly by cell division, rarely sexual by fusion.* Cell size ranges from 5 to 2,000 μm; most are 20-150 μm. Many species are bioluminescent. Locomotion is typically a combination of rotation and forward movement by means of the 2 flagella.

Of more than 1,200 known species in 2 classes, 75 species have been identified from Bermuda, of which 11 are included here.

OCCURRENCE: The vast majority of species are members of the marine plankton; only a few occur in fresh water. Their main abundance is in warm and temperate oceans where they sometimes form mass blooms ("red tides"). Some are symbionts or parasites.

Red tides may be toxic and result in mass die-offs of fishes and other organisms; some species, however, accumulate the toxin and pass it along to predators, among them many commerically valuable fish and shellfish species. Consumption of these by humans may cause severe, sometimes lethal poisoning (ciguatera and paralytic seafood poisoning). Except for isolated, insufficiently documented cases such poisoning is not known to occur in Bermuda.

Planktonic species can be collected with a small-mesh net (10 μm) and/or water bottles combined with settlement chambers.

IDENTIFICATION: Live or preserved, with a high-power microscope provided with phase contrast.

The vegetative cell is covered by several layers of membranes; cells may be unarmored (naked) or armored with cellulose plates (thecate). The armored cells are distinguished by the formation of their theca and number of thecal plates (2-100). Most species have a longitudinally oriented groove (sulcus) and a transverse groove (cingulum), each of which contains 1 of the flagella. The cingulum separates the anterior and the posterior halves of the cell (called epicone and hypocone for the naked, epitheca and hypotheca for the thecate dinoflagellates, respectively). Epi- and hypotheca often have wing- or horn-like appendices extending out from the cingulum and/or sulcus. These membranous structures (lists) may be supported by ribs.

Osmium tetroxide, 2-5% glutaraldehyde, 2-4% formaldehyde, or Lugol's solution can be used as fixatives and for storage. Diamine blue 3B is commonly used for staining the theca. For examination of the thecal plates the theca may be dissociated with sodium hypochlorite (commercial bleach).

BIOLOGY: Most are photo-autotrophic, but hetero- and phagotrophic forms as well as parasites are also known. Vertical migration with and without diurnal rhythms occurs. Dinoflagellates are very abundant as symbionts (zooxanthellae) in a wide variety of marine organisms including Sarcodina, Porifera, Coelenterata (esp. the reef-building corals), Turbellaria and Mollusca. The degree of benefit that hosts derive from these relationships may vary from one association to another.

REPRODUCTION: Cell division, which may involve a longitudinal or transverse bipartitioning of the parental cell, is the common mode of reproduction. Armored dinoflagellates split the parental cell and theca into 2 portions, each of which synthesizes the missing portion; or the parental theca may be shed, necessitating the formation of a new theca. Sexual reproduction can be triggered by nutrient deficiency but also occurs spontaneously. Small male gametes, similar to the vegetative cells, attach themselves to and are resorbed by the female cells; the resulting zygote remains motile or forms a resting cyst. Meiosis occurs in the zygote, which is the only diploid phase. The persistence of stainable, condensed chromosomes beyond the division cycle and other peculiarites of their nuclear organization may place them somewhat apart from most other eukaryotic organisms.

REFERENCES: For general information see BOLD & WYNNE (1978) and SOURNIA (1978); a systematic synopsis is by LOEBLICH (1982). Free-living, unarmored species can be found in KOFOID & SWEZY (1921). There is a vast literature on zooxanthellae; recent summaries of various aspects are by TAYLOR (1974), TRENCH (1981), COOK (1983), ANDERSON (1983a, b) and LEE (1980).
HULBURT et al. (1960) recorded species from the Sargasso Sea, and some of Bermuda's inshore species are listed in BODUNGEN et al. (1982). TRENCH (1974) and SCHOENBERG & TRENCH (1980a, b, c) studied the zooxanthellae of coelenterates occurring in Bermuda.

## Plate 19

CL. **DESMOPHYCEAE:** Naked or armored Dinophyta, nearly always laterally compressed. Flagella mostly inserting apically at the same location.

O. **PROROCENTRALES:** Armored Desmophyceae with theca consisting of 2 plates; laterally compressed. (7 spp. from Bda.)

*Prorocentrum gracile* (Schütt): Body narrow, lanceolate, 50 μm long; prominent tooth at the anterior end of the cell. Photosynthetic. Very common in the inshore plankton; an important food source for the benthos.

O. **DINOPHYSIALES:** Laterally compressed, armored Desmophyceae; cingulum close to the anterior end, epitheca very short. List well developed. Theca composed of up to 18 plates. (4 spp. from Bda.)

*Dinophysis caudata* var. *pedunculata* (Schmidt): Epitheca very small, cell strongly compressed. Hypotheca prolonged in pendulum-like process, shapes vary widely. Lists supported by ribs extend from cingulum and sulcus. In lateral view cell length 50-90 μm, cell width 40-50 μm. More common inshore than offshore.

CL. **DINOPHYCEAE:** Armored or naked Dinophyta with distinct longitudinal and transverse grooves in which the flagella are located.

O. **PERIDINIALES:** Armored Dinophyceae with varying cell shapes. Cingulum more or less median. Epitheca and hypotheca often extended into horns. Thecal plates may bear pits, knobs, spines or ridges. Epitheca always distinct. (46 spp. from Bda.)

*Gonyaulax polygramma* (Stein): Epitheca ends in a blunt horn. Cingulum descending by 1-1.5 of its width. Hypotheca round with 2 unequal spines. Plates with numerous pores and small spines. Photosynthetic. Diameter about 50 μm. A very common neritic and oceanic species of the subtropical plankton.

*Peridinium brochii* (Broch): Epitheca extends in a small short horn, hypotheca with 2 short horns. Cingulum median, hardly descending; sulcus seldom on the epitheca. Heterotroph; 80 μm long. In the plankton throughout the year.

*Oxytoxum tesselatum* (Stein): Pointed epitheca much smaller than

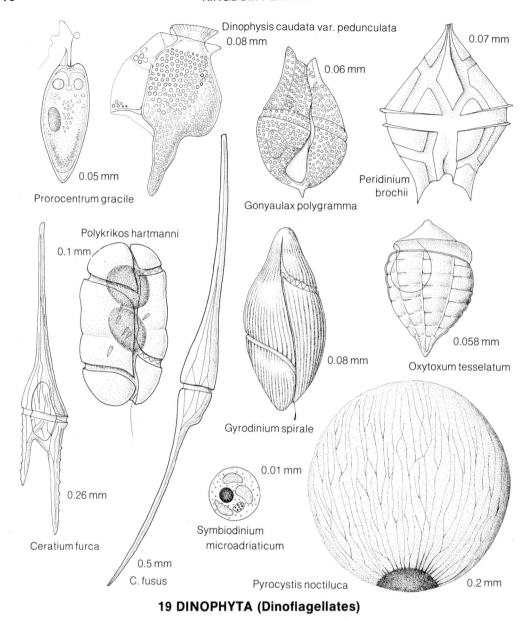

**19 DINOPHYTA (Dinoflagellates)**

hypotheca. Cingulum deep and broad, not descending. Five closely joined wedge-like plates leave a small opening at the end of the hypotheca, which is closed by a 6th plate carrying a pointed projection. Photosynthetic; 50-65 μm long, 25-30 μm wide. Sporadically occurring in the plankton.

***Ceratium furca*** (Ehrenberg): Genus biflagellate, photosynthetic, with prominent fusiform theca, with 1 elongate epithecal horn and 1-3 hypothecal horns.—Species not very elongate, 210-380 μm long; hypotheca with 2 unequal, parallel, strongly sculptured horns. Cingulum not descending. Very common in inshore plankton, but also quite abundant offshore.

***C. fusus*** (Ehrenberg): Genus as above.—Species very elongate, 300-500 μm long; hypotheca with 2 horns, 1 of which is reduced; the other is long and more curved than that of the epitheca. Very common in- and offshore.

***Pyrocystis noctiluca*** (Murray): Spherical or slightly elongated cells with a thick cellulose membrane. The protoplast (cell body) is only a thin coating at the cell wall, forming a much ramified plasma net connected to the nucleus. Photosynthetic and strongly bioluminescent; 500-1,000 μm in diameter. Regularly in the offshore plankton, but also common inshore, especially during the warmer months.

O. **GYMNODINIALES:** Naked Dinophyceae without plates; cell surrounded by a thin plasma skin. Plasma body with cingulum and sulcus and 2 flagella. Cell shape rotund to spindle-like, very variable even within a species. Cingulum mostly in a median position. (About 20 spp. from Bda.)

***Polykrikos hartmanni*** (Zimmermann): Colony of 2 or seldom 4 cells, rounded at both ends of the colony. Cingula staggered at their ends. Sulci grown together. With numerous small plastids; nematocysts also present, located along the sulcus. Length 80-120 μm, width 55-75 μm. Common, but never numerous in- and offshore.

***Gyrodinium spirale*** (Bergh): Large, slender, asymmetric epicone pointed, hypocone blunt. Entire body with prominent striation. Heterotroph. Length (60-100 μm) and width (30-60 μm) variable. Very common inshore.

***Symbiodinium microadriaticum*** (Freudenthal) (=*Gymnodinium microadriaticum*): Symbiotic zooxanthellae. Cells associated with hosts are spherical, lacking epicone, hypocone and flagella. Dividing cell pairs 7-10 μm in diameter; non-dividing cells are smaller. Zooxanthellae from some hosts possess an "accumulation body," a structure distinct from the pyrenoid and that apparently serves a storage function. Occasionally biflagellated motile cells with an epicone and smaller hypocone are produced. Endosymbiotic in cells of Foraminiferida, Radiolaria, some Porifera, many Actiniaria, and virtually all Octo- and most Hexacorallia in Bermuda; also in the mesoglea of *Cassiopea*.

B. v. BODUNGEN & C. ODERBRECHT

## Phylum Anthophyta (=Angiospermae) (Flowering plants)

CHARACTERISTICS: *Multicellular PLANTAE mostly with vessels (specialized cells that are involved in the transport of water and nutrients) in the stems; male and female gametes in highly specialized structures called flowers in which the ovules are enclosed by tissue, forming a carpel. Fertilization double, with endosperm development accompanying the development of the embryo; development of seeds enclosed within a fruit whose coat is derived from the carpel wall.* The size range is vast, from herbs of a few centimeters to trees 150 m high.

Of about 300,000 species (!) in two classes (Dicotyledoneae and Monocotyledoneae), only the permanently submerged sea grasses (4 spp.) and a selection of 18 species that are habitually or preferentially found on the shoreline (halophytes) are considered here.

OCCURRENCE: Flowering plants form the dominant vegetation cover in most terrestrial environments; some are aquatic. Most are mesophytes (growing under medium moisture conditions) but many are xerophytes (adapted to low moisture); some have become epiphytes (growing on other plants), and a few are parasites.

Freshly collected plants are pressed in a plant press made up of a pair of frames, corrugated cardboard, newsprint and 2 webbing straps. The frames, generally $30 \times 45$ cm, are made of slats of a light wood fastened together with two similar slats. A similar size piece of cardboard is placed on the lower frame and the plant specimen enclosed in a folded piece of newsprint, which is placed on the cardboard. A second piece of cardboard is placed on top of the newsprint and the process repeated, with the upper frame finally placed on the last sheet of cardboard. The 2 webbing straps are tightened around the frames (stand on the press as the straps are tightened) and secured. The press is placed in a dry room for 7 days. The dried pressed plants are then removed and mounted on cardboard (herbarium sheets).

IDENTIFICATION: Examine floral parts under a dissecting microscope for number of carpels, placement of ovary or ovaries (superior or inferior), and number and position of stamens. Other plant parts are usually examined macroscopically.

Flowering plants range in shape from trees (single trunk) to shrubs or herbs. They may be annuals, biennials or perennials and are composed of a root, stem and leaves. Leaves occur at the nodes on the stem and are usually differentiated into stalk and blade. The edges of the blade may be entire or variously toothed or lobed. Simple leaves have 1 main stalk and a blade whereas compound leaves have a main stalk with several leaflets. In pinnate leaves there is a main stalk with leaflets arranged on either side. Bipinnate leaves have the leaflets themselves divided in a pinnate manner. Leaves may be variously arranged on the stem: alternate (1 leaf at each node), opposite (2 leaves on either side of a node), or whorled (several leaves radiating from a node). Stipules are leafy appendages, often found paired, occurring at the base of the leaf stalk.

The sexually reproducing part is the flower, basically composed of 4 series of elements arranged in whorls around a central axis. The outer whorl is composed of sepals, usually green and collectively termed the calyx. Next are the petals, often brightly colored and collectively known as the corolla. The corolla and calyx, together termed the perianth, may be arranged radially symmetrically (actinomorphic) or bilaterally symmetrically (zygomorphic). Modified leaves, or bracts, are often found as a whorl below the flower and are collectively known as an involucre. The 2 inner series are the gamete-producing parts of the flower. The male reproductive organ (stamen) consists of a stalk (filament) bearing the anther in which pollen is produced. The female reproductive organ (pistil) consists of one or more carpels (ovule-bearing organs), which may be free or fused. At the lower end of the pistil is the ovary with 1 or more chambers (locules) containing the ovules; above this is the style, bearing the stigma at its tip. A compound ovary is formed of 2 or more fused carpels. The ovary can be superior (above the level at which the perianth and stamens are attached) or inferior (perianth and stamens are attached at its apex). The apex of the flower stem is known as the receptacle; if the petals, sepals and stamens are attached to it below, free from the ovary, the arrangement is termed hypogynous. A perigynous arrangement is where the perianth and stamens are joined to the rim of a cup-shaped hypanthium that surrounds the ovary but is not usually fused with it. In an epigynous arrangement the ovary is truly inferior with the perianth and stamens borne on top of the ovary. Flowers may be bisexual (both male and female reproductive organs present) or unisexual; and plants may be monoecious (male and female flowers are present on the same plant) or dioecious (borne on separate plants). A polygamous arrangement has separate male, female and bisexual flowers on the same plant.

REPRODUCTION AND DEVELOPMENT: The plant with its roots, stems, leaves and flowers is the sporophyte, because it consists of diploid cells and produces microspores (pollen grains) as well as megaspores (ovules). The gametophyte in the life history is represented by 2 structures, the microgametophyte (male gametophyte) and the megagametophyte (female gametophyte). The microgametophyte is represented by the pollen grain and the subsequent pollen tube, which forms after coming into contact with the stigma. The pollen tube contains 1 tube nucleus and 2 sperm nuclei. The megagametophyte is represented by the embryo sac consisting of 1 egg nucleus, 2 synergid nuclei, 2 polar nuclei and 3 antipodal nuclei. Pollen is produced in the anther sacs of the stamens and transferred to the stigmas of the pistil. The transfer of pollen from anther to stigma may be brought about by wind currents, raindrops or insects, depending on the plant species. The pollen tube grows down the style to the ovule. It penetrates the micropyle and enters the embryo sac into which it discharges its two sperm nuclei. One sperm fuses with the egg nucleus (fertilization), the other with the 2 polar nuclei to form the endosperm nucleus. The remaining 5 nuclei of the embryo sac usually disintegrate after the unions. The zygote (fertilized egg) undergoes cell division and grows into the embryo, and the endosperm nucleus and cell divide to form the endosperm (nutrient) tissues; together they form the seed. The ovary enlarges to form the fruit (seed-bearing organ). There are several types of fruit: nut (a dry, single seeded and non-opening (indehiscent) fruit with a woody outer layer); capsule (a dry fruit which normally splits open (dehisces) to release its seeds); berry (a fleshy, indehiscent fruit); and drupe (resembles a berry with 1 or more seeds, each seed surrounded by a stony layer). The seed upon reaching maturity is dispersed and, with suitable environmental conditions, germinates and grows into a new plant that in turn produces flowers. Depending on species, the life-span of a plant may be but a few weeks or several centuries.

REFERENCES: For a general introduction to the systematics of vascular plants see LAWRENCE (1951), of flowering plants HUTCHINSON (1973). Caribbean plants are covered by ADAMS (1972), and a monograph of terrestrial grasses is by HITCHCOCK (1971). PHILLIPS & McROY (1980) summarize sea grass biology, and DEN HARTOG (1959, 1964) describes several marine grasses. Many shore plants can be found in SILBERHORN (1982).

There is no specific work dealing with Bermuda's marine and coastal flowering plants, but most species have been listed by BRITTON (1918).

CL. **DICOTYLEDONEAE:** Herbaceous or woody Anthophyta. Vascular tissue of the stem arranged either in a hollow cylinder around a relatively small pith, or in bundles arranged in a single circle; leaves usually with net venation; floral parts in 4's or 5's or multiples of same; seed embryos with 2 cotyledons (seed leaves).

*Plate 20*

O. **CHENOPODIALES:** Mostly herbaceous Dicotyledoneae with leaves alternate or opposite; stipules absent or very small; flowers without petals; carpels numerous or solitary, free or joined; seeds without curved embryo around the endosperm.

F. **CHENOPODIACEAE** (Goosefoot family): Annual or perennial Chenopodiales, herbs or shrubs; stems sometimes jointed; leaves alternate, rarely opposite, simple or reduced to scales; stipules absent. Flowers small, bisexual or unisexual, polygamous or dioecious; calyx 3-5-lobed, usually persistent in fruit, or absent; stamens as many as calyx lobes and opposite them, hypogynous or inserted on a disc or on the calyx. Ovary superior or inferior, 1-locular; ovule solitary; styles 1-3. Fruit a nutlet.

# KINGDOM PLANTAE

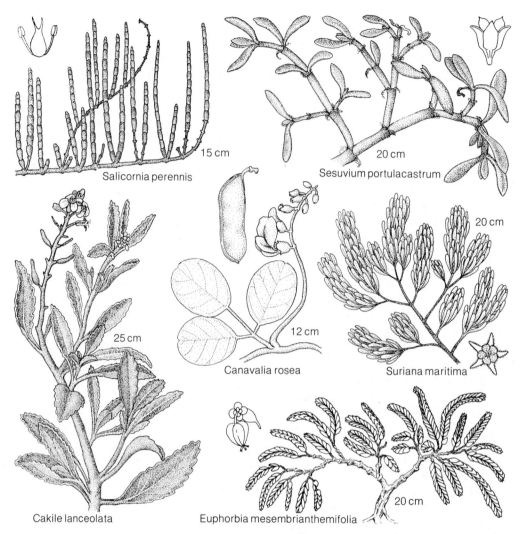

**20 CHENOPODIALES - GERANIALES (Flowering plants 1)**

*Salicornia perennis* Mill. (Woody glasswort): Erect-branched, perennial herb to 30 cm high, the main stem often trailing and woody. Stem fleshy, jointed, often reddish; leaves reduced to opposite scales joined in pairs; flower spikes 3-5 cm long; flowers green and inconspicuous; seed with hooked hairs. On coastal rocks and sands and among mangroves. Flowers in spring and summer.

F. **AIZOACEAE** (Carpetweed family): Herbaceous or low-shrubby, often fleshy Chenopodiales; leaves simple, alternate, opposite or whorled, fleshy or reduced to scales, with or without stipules. Flowers bisexual; calyx of 5-8 joined sepals; stamens few to many; ovary superior or inferior, 1 to several locular; style 1 or absent; stigmas as many as loculi. Fruit a capsule, enclosed by a persistent calyx.

*Sesuvium portulacastrum* L. (Seaside purslane): Perennial herb, branches trailing and rooting at nodes; stems and leaves succulent, often reddish; leaves 2.5-5 cm long. Flowers solitary; calyx deeply 5-lobed, green outside, pink to deep rose within; ovary superior, usually 3-locular; seeds black, nearly smooth. In marshes and coastal sands. Flowers from spring to autumn.

O. **RHOEADALES:** Usually herbaceous Dicotyledoneae with clustered, bisexual and actinomorphic flowers; petals and sepals separate; stamens hypogynous, ovary superior, compound, composed of 2 to many carpels.

F. **CRUCIFERAE** (Mustard family): Annual, biennial or perennial Rhoeadales; herbs, rarely subshrubs; forked or stellate unicellular hairs commonly present; leaves usually alternate, simple or pinnate, without stipules. Flowers bisexual, usually actinomorphic; perianth in 3 series, calyx of 4 free sepals in 2 whorls, corolla of 4 free petals, usually clawed, rarely absent; stamens 6 in 2 whorls, tetradynamous—the 2 outer shorter and opposite the lateral sepals, the 4 inner, longer and opposite the petals, usually free but filaments of each pair sometimes joined; ovary superior, usually 2-locular by a complete false partition; style simple or obsolete; stigmas mostly 2. Fruit usually opening by means of valves and referred to as a silique, if enlongated, or a silicle, if short. Fruit sometimes a 1- to few-seeded nut.

*Cakile lanceolata* (Willd.) O. E. Schulz (=*Raphanus lanceolatus* Willd.) (Southern sea rocket, Scurvy grass): Annual; stem weak, often ascending, 30-100 cm long; leaves entire or coarsely toothed, up to 8 cm long and 1.5 cm broad; petals pale purplish; pods 1.7-2 cm long, the upper joint longer than the lower. On beaches and coastal rocks. Flowers from spring to autumn.

O. **LEGUMINALES:** Tree-, shrub- or herb-shaped Dicotyledoneae; leaves simple to bipinnate; flowers actinomorphic or zygomorphic; petals free, or some partially united; stamens few to numerous, often diadelphous (1 free and the remainder with filaments fused together); carpel solitary, superior; fruit a pod.

F. **PAPILIONACEAE** (Pea family): Herb-, shrub-, vine- or tree-shaped Leguminales. Flowers mostly bisexual, zygomorphic; calyx 4 or 5 cleft, sometimes 2-lipped; corolla of 5 unequal, separate or nearly separate petals comprising 2 keel (lower) petals, 2 wing (lateral) petals, and 1 standard (upper) petal that surrounds the others in the bud; stamens 10, sometimes 9, rarely 5, monadelphous (filaments all fused together), or diadelphous or sometimes separate; pistil simple, superior, 1-locular; ovules 1 to many. Fruit a dehiscent or indehiscent pod.

*Canavalia rosea* (Sw.) DC. (Bay bean): Perennial, prostrate vine 1-8 m long; leaflets 3, leathery, 4-10 cm long; corolla pink or rose-purple; pod 10-12 cm long; seeds oblong, brown. On beaches and sand dunes. Flowers all year.

O. **GERANIALES:** Herb-, shrub- or tree-shaped Dicotyledoneae. Sepals usually overlapping one another;

petals present and often clawed, usually free; stamens as many as or twice as many as petals and then in 2 whorls; disc-glands often present; ovary superior, mostly compound; ovules pendulous; seeds mostly without endosperm.

F. **SIMAROUBACEAE** (Quassia family): Shrub- or tree-shaped Geraniales; leaves alternate, simple or more usually pinnately compound, usually without stipules; flowers bisexual or more usually unisexual and mostly dioecious, actinomorphic; perianth in 1 or 2 series, sepals 3-8 free or partly joined, petals 3-8 free or united, sometimes absent; stamens as many as petals or 2× as many, free, inserted on or at base of a disc; ovary superior, often raised on the disc, comprised of 2-5 (rarely 1) unilocular simple pistils or the pistils joined at the base, or by the styles, into a lobed 2-5 chambered ovary; ovules 1 or 2 in each chamber. Fruit a capsule, samara (dry indehiscent fruit with a flattened membrane or wing), or schizocarp (locules separate at maturity to form single-seeded units).

*Suriana maritima* L. (Tassel plant): Shrub, 1-2.5 m high; leaves clustered, alternate, hairy, 1.5-4 cm long; 5 sepals 6-10 mm long; 5 yellow petals, 6-10 mm long. Common on coastal rocks and beaches. Flowers in spring and summer.

F. **EUPHORBIACEAE** (Spurge family): Tree-, shrub- or herb-shaped Geraniales, often with milky sap, sometimes fleshy and cactus-like; leaves alternate, sometimes opposite or whorled, simple or compound, stipules present although sometimes reduced to spines or hairs. Flowers unisexual, monoecious or dioecious; perianth of 1 or 2 series or lacking, usually small and actinomorphic; sepals and petals usually free, 5-parted; staminate ($\male$) flowers with stamens usually as many or 2× as many as petals or reduced to 1; a non-functional ovary is often present; pistillate ($\female$) flowers with or without staminodes (sterile stamens, often modified); ovary superior, of 3 fused carpels, 3-locular, 1 or 2 ovules in each locule; styles 3, free or united; stigmas 3 or 6. Fruit usually a 3-lobed capsule, splitting, often elastically, into 3 1-seeded portions that open to shed seed.

*Euphorbia mesembrianthemifolia* Jacq. (=*E. buxifolia* Lam.; *Chamaesyce buxifolia* (Lam.) Small) (Coast spurge): Perennial, without hairs, stems erect or decumbent, 20-60 cm long, branching; leaves fleshy 0.8-1.2 cm long. Stipules conspicuous, about 1 mm long, white. Fruit 2 mm in diameter. On coastal rocks and beaches. Flowers throughout the year.

## Plate 21

O. **MYRTALES:** Herb-, shrub- or tree-like Dicotyledoneae; leaves mostly opposite and often gland-dotted; flowers actinomorphic or zygomorphic, often showy, or reduced to a stamen and pistil attached to the hypanthium or the ovary enclosed by it; stamens show tendency to group in bundles, sometimes of 2 types and then opening by terminal pores.

F. **COMBRETACEAE** (Combretum family): Tree- or shrub-shaped

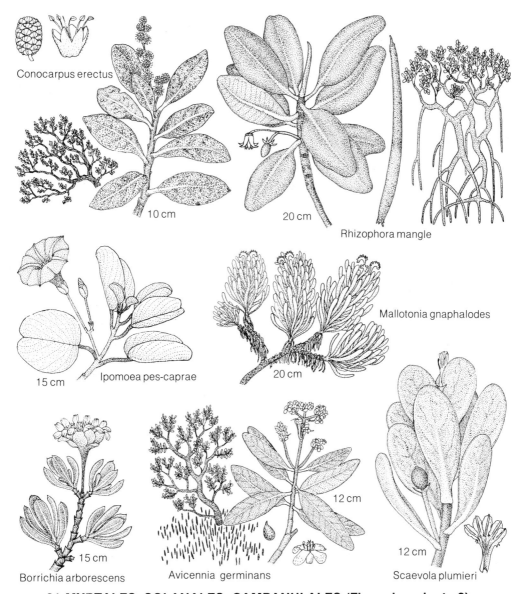

## 21 MYRTALES, SOLANALES, CAMPANULALES (Flowering plants 2)

Myrtales; leaves opposite, alternate or rarely whorled, without stipules. Flowers bisexual, rarely unisexual, actinomorphic or occasionally zygomorphic, often clustered in globular or elongated heads. Perianth typically of 2 series, the parts fusing to form a hypanthium that is joined to the ovary; calyx lobes 4-8, persistent, corolla lobes of same number or absent; stamens usually 2-5 or 2× as many as corolla lobes; pistil 1, the ovary inferior, unilocular. Fruit leathery and drupaceous, 1-seeded, often winged.

***Conocarpus erectus*** L. (Buttonwood): Shrub or small tree 3-5 m high, rarely larger. Leaves 2.5-5 cm long, softly leathery, usually without hairs, with glandular pits on the underside of the leaves near the midrib. Flower heads solitary, in clusters or more usually in spikes. Fruit a 2-winged drupe. On coastal rocks. Flowers in autumn and winter.

F. **RHIZOPHORACEAE** (Mangrove family): Shrub- or tree-shaped Myrtales with swollen nodes; leaves usually opposite, leathery, simple, with stipules falling soon after development, or leaves rarely alternate and without stipules. Flowers usually bisexual, actinomorphic, perigynous to epigynous; sepals 3-14, more or less joined; petals of same number as sepals, usually fleshy or leathery; stamens 2-4 times as many as sepals, inserted on the edge of a disc; ovary superior to inferior, of 2-12 fused carpels, 2-12 locular; ovules usually 2 in each locule. Fruit a berry or drupe.

*Rhizophora mangle* L. (Red mangrove): A shrub or tree to 10 m high with arching aerial roots that form impenetrable thickets; leaves 5-15 cm, leathery. Petals yellow, cleft at tip, often with brownish balsam-scented exudate; stamens 8; seed solitary, germinating in the persistent fruit, the elongating root sometimes reaching the ground before the fruit falls. Common along borders of saltwater lagoons and saline swamps. Flowers in summer and autumn.

O. **SOLANALES:** Mostly herbaceous or twining Dicotyledoneae, rarely shrubs or trees. Leaves alternate, stipules absent; corolla actinomorphic; stamens the same number as and alternate with the corolla lobes; ovary superior, 1-4 locular, ovules numerous to solitary.

F. **CONVOLVULACEAE** (Morning-glory family): Herbaceous or woody, often climbing Solanales; sap usually milky; leaves alternate, simple, without stipules. Flowers bisexual, often with paired bracts; sepals 5, usually free; corolla of joined petals, 5-lobed or entire; stamens 5, fused to the base of corolla tube and alternate with lobes; ovary superior, of 2 fused carpels each usually 2-locular. Fruit a globular or plump 2-6-seeded capsule.

*Ipomoea pes-caprae* (L.) R. Br. (Seaside morning glory): Perennial, succulent, without hairs; stems prostrate and creeping to 18 m; compound leaf of 3 leaflets each 6-10 cm long and as broad. Flowers showy; corolla funnel-shaped, purple, 4-5 cm long. Common on beaches and sand dunes. Flowers in summer and autumn.

F. **BORAGINACEAE** (Borage family): Herb-, shrub- or tree-shaped Solanales; leaves usually alternate, without stipules. Flowers bisexual, usually actinomorphic, mostly blue; flower spikes often curved like a scorpion's tail or coiled like a spring. Calyx 5-lobed, parted or cleft, usually remaining on fruit. Corolla of 5 fused petals; stamens as many as the corolla lobes and alternate with them, inserted on the tube or throat. Ovary superior, 2-locular, often becoming falsely 4-locular at maturity; style simple, entire or 2-cleft. Fruit mostly of 4 nutlets or a drupe.

*Mallotonia gnaphalodes* (L.) Britton (Seaside lavender): Succulent bushy shrub, 0.6-2 m high. Leaves silkily hairy, often appearing silvery-gray, 4-11 cm long. Corolla white. Fruit ovoid, of 2 nutlets. Frequent on beaches and coastal rocks. Flowers from spring to autumn.

F. **VERBENACEAE** (Vervain family): Herb-, shrub- or tree-shaped Solanales, often with quadrangular branchlets. Leaves opposite or whorled, rarely alternate, simple or compound, stipules absent. Calyx 4-5 lobed, cleft or 2-lipped; stamens 4, 2 long and 2 short, rarely only 2, inserted on the corolla tube and alternating with the lobes. Ovary superior, 2-4 locular; style terminal, stigma 1 or 2; ovules solitary or paired. Fruit berry-like or drupaceous or an aggregate of 2 or 4 nutlets.

*Avicennia germinans* (L.) L. (Black mangrove): Evergreen tree to 12 m high; developing numerous pneumatophores (aerial roots). Twigs quadrangular. Leaves leathery, opposite, entire, hairy when young but becoming smooth above, 3-8 cm long. Corolla white. Fruit an oblique capsule. Common along borders of saltwater lagoons and saline swamps. Flowers from spring to autumn.

O. **CAMPANULALES:** Annual or perennial Dicotyledoneae, herbs or shrubs. Calyx 5-lobed, often joined to the ovary. Corolla usually of 5 petals joined at least at the base; stamens as many as corolla lobes. Ovary inferior, often unilocular with a single ovule.

F. **GOODENIACEAE** (Goodenia family): Herbaceous or shrubby Campanulales. Leaves alternate or sometimes opposite. Flowers bisexual; calyx 5-(rarely 3-)toothed, tubular, joined to the ovary, sometimes obsolete. Corolla 5-lobed, split on one side; stamens 5. Ovary mostly inferior, 1-2 locular; stigma surrounded by a terminal outgrowth. Fruit drupaceous, berry-like or a capsule.

*Scaevola plumieri* (L.) Vahl (Beach lobelia): Fleshy shrub to 1.5 m high. Leaves 3-8 cm long. Flowers zygomorphic, corolla white or pinkish, woolly within, its tube split to the base on 1 side. Fruit an oval berry, black, juicy, 2-seeded. Occasionally on beaches. Flowers from spring to autumn.

F. **COMPOSITAE** (Dandelion family): Herbaceous, shrubby, rarely tree-shaped Campanulales, often with milky sap. Leaves usually alternate, opposite or rarely whorled. Flowers crowded onto a head on a common receptacle, surrounded by a whorl of one or more free or joined bracts. Flowers bisexual, monoecious or dioecious, the outer ones (ray florets) often ligulate (strap-shaped), the inner ones (disc florets) tubular; calyx tube united with the 1-locular ovary, the limb (pappus) crowning its summit in the form of bristles, awns, scales, teeth, or lacking; stamens 5, rarely 4, attached to the petals, filaments free, anthers joined in a tube about the style. Ovary inferior, 1-locular, 1-ovuled, style usually 2-cleft. Fruit an achene (small, dry, indehiscent, single-seeded).

*Borrichia arborescens* (L.) DC. (Seaside oxeye): Shrub to 3 m tall. Leaves opposite, 2.5-6 cm long, gray with fine hairs or bright green without hairs, sometimes with both gray and green leaves on the same plant. Flower heads to 2.5 cm across; ray flowers yellow, few and ♀; disc flowers yellow and bisexual. Common on coastal rocks. Flowers all year.

**CL. MONOCOTYLEDONEAE:** Mostly herbaceous Anthophyta; vascular tissue of the stem usually in scattered vascular bundles; leaves with parallel veins; floral parts in 3's or multiples of 3; seed embryos with a single cotyledon.

## Plate 22

**O. NAJADALES:** Submerged aquatic annual or perennial Monocotyledoneae; leaves alternate or opposite, sheathing. Flowers minute, unisexual, in leaf axils; perianth of small scales or absent; stamens 1-3; carpels free 1-9; ovule 1. Widely distributed in fresh or salt water.

**F. CYMODOCEACEAE:** Submerged marine Najadales, either herbaceous or woody. Leaves arranged in 2 vertical rows, linear or awl-shaped with a sheathing base. Flowers unisexual with ♂ and ♀ flowers on separate plants; ♀ flower of 2 free carpels with 2 or 3 stigmas and a single ovule in each carpel.

*Syringodium filiforme* Kütz. (=*Cymodocea manatorum* Aschers.) (Manatee grass): Leaf blades more or less circular in cross-section, 4-30 cm long, 0.8-1.8 mm in diameter, sheaths 1-5 cm long; fruit ellipsoid, flattened, beaked by a persistent style base; seed 2.5-3 mm. Abundant in some shallow inshore waters, often washed up on shores. Flowers in spring and summer.

*Halodule bermudensis* Den Hartog: Rhizome creeping, with 1 or 2 roots and a short, erect stem at each node; sheaths 1-2 cm long; leaf blade 2.5-5 cm long, narrowed near the base, midrib conspicuous; intramarginal veins inconspicuous, both ending in a strongly developed tooth. Flowers unknown. Shallow water.

**O. BUTOMALES:** Perennial, aquatic, herbaceous Monocotyledoneae; leaves arising from the stem or the base of the stem, alternate or whorled. Flowers showy to small and minute, bisexual or unisexual; perianth of 2 series, the outer green and sepal-like, the inner petal-like; stamens numerous or reduced to 3; ovules numerous, scattered on the walls of the carpel.

**F. HYDROCHARITACEAE:** Butomales with leaves along elongated stems. Flowers actinomorphic, within a pair of bracts or bifid spathe (bract); perianth free, 3 in each series; staminate (♂) flowers with 3 to many stamens; pistillate (♀) flowers having an inferior ovary of 2-15 united carpels; l-locular with numerous ovules. Fruit berry-like; seeds many, without endosperm.

*Thalassia testudinum* Konig & Sims (Turtle grass): Rhizomatous herb; sheaths brown; leaves green, linear, ribbon-like, blunt-tipped, 8-15 mm broad, 15-30 cm long. Flowers white, dioecious, solitary. In shallow, fully marine areas, in- and offshore.

*Halophila decipiens* Ostenf.: Slender-stemmed, much branched herb; leaves opposite, simple, differentiated into stem and blade, with conspicuous midrib. Spathes of 2

ANTHOPHYTA

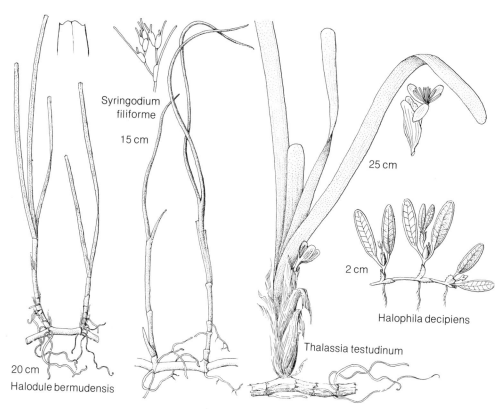

**22 NAJADALES, BUTOMALES (Sea grasses)**

free bracts containing 1 staminate and 1 pistillate flower; styles 3. In shallow lagoons and brackish ponds.

## Plate 23

O. **GRAMINALES (Grasses):** Annual or perennial herbaceous, rarely shrubby or tree-shaped Monocotyledoneae; stems erect, ascending or prostrate and creeping, cylindrical, rarely flattened, jointed, usually hollow in the internodes, closed at the nodes. Leaves solitary at the nodes, alternate and 2-rowed, consisting of a sheath surrounding the stem passing into the blade. At the junction of the sheath and blade is a membranous structure called the ligule. Flowers usually bisexual, sometimes unisexual, small, usually consisting of 3 stamens, a superior, 1-locular ovary with 2 styles, and 2 or 3 minute transparent or fleshy scales (lodicules) representing the perianth arranged between 2 bracts (lemma and palea), the whole forming a floret. Florets 1 to many, on a short axis (rachilla) and bearing at the base 2 empty bracts (upper and lower glumes), the floret and glume forming a spikelet. Spikelets may be variously arranged to form inflorescences such as spikes, racemes and panicles. A spike has stalkless spikelets arranged along a central axis whereas a raceme has stalked spikelets similarly arranged. A pani-

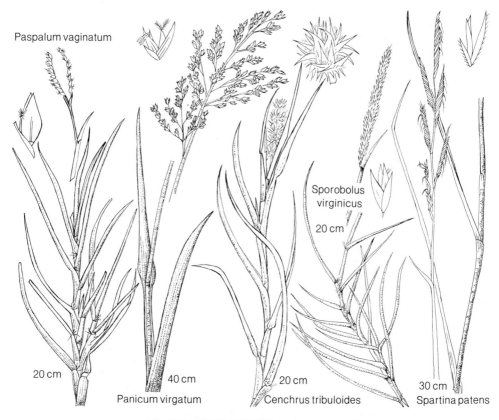

**23 GRAMINALES (Beach grasses)**

cle is a branched raceme, each branch bearing a raceme of spikelets. (Only family: Gramineae.)

***Paspalum vaginatum*** Sw. (Sheathed paspalum): Stems 20-60 cm tall from a stout rootstock. Leaf sheaths compressed, usually crowded and overlapping, at least at the base. Leaf blades to 15 cm long, 1.5-3 mm wide, without hairs or with a few hairs, margins rolled in when dry. Racemes terminal, 5-7 cm long, usually in pairs; spikelets 3-4 mm long in a single row. In brackish marshes. Flowers in summer and autumn.

***Panicum virgatum*** L. (Switch grass): Stems erect from a stout rootstock, 1-2 m tall; leaf blades 3-15 mm wide, to 30 cm long; lower and middle sheaths exceeding the internodes, the upper shorter; panicles terminal to 48 cm long, projecting beyond plant; spikelets ovoid, 2.8-6 mm long. On coastal rocks. Flowers in summer and autumn.

***Cenchrus tribuloides*** L. (Bur grass): Perennial, with flat blades 6-12.5 cm long, 0.4-0.8 mm wide and terminal flower spikes 2.5-6 cm long. Stems prostrate forming mats to 1 m wide. Spikelets 2 in an ovoid involucre consisting of 2 thick, hard valves that are exteriorly armed with stout spines. On sandy beaches and dunes. Flowers from spring to autumn.

***Sporobolus virginicus*** (L.) Kunth (Seashore rush grass): Perennial, flowering stems to 40 cm or more; rhizomes hard and scaly; leaf blades firm, sharply pointed, very variable in length and breadth, up to 15 cm long; spikelets 2 mm long; lower glume a little more than half as long as the spikelet, the upper as long as the spikelet. On beaches and mangrove borders. Flowers in summer and autumn.

***Spartina patens*** (Ait.) Muhl. (Salt grass): Stems 30-120 cm tall, erect, or decumbent at base. Leaves 15-75 cm long, 2-4 mm broad, appearing round owing to margins rolling in. Flower spikes 2-10, 2.5-5 cm long; spikelets 6-8 mm long; outer glumes pointed, hairy on the keel, the first usually rather less than half as long as the second. Rocky and sandy coasts. Flowers in summer and autumn.

E. A. MANUEL

# KINGDOM ANIMALIA

## Animals

CHARACTERISTICS: *Mostly motile eukaryotic ORGANISMS with exclusively heterotrophic nutrition, normally by ingestion.*

As defined here, animals range from the single-celled flagellates, a few micrometers in size, to the whales. Active movement and locomotion are characteristic of the majority of animals; some are sedentary or even permanently attached to the substrate but retain a free-moving form (e.g., a larva) during part of their life cycle. Asexual reproduction by budding is common in the lower phyla; sexual reproduction is by fusion of 2 nuclei (in Protozoa) or 2 reproductive cells (gametes; in Metazoa) usually followed by meiosis.

Major distinguishing characters within the kingdom are whether the organism consists of 1 or many cells (which defines subkingdoms), organization of the body into tissue layers and cavities, and type of symmetry (none, radial or bilateral).

Both subkingdoms are well represented in Bermuda: Protozoa, and Metazoa (p. 110). It should be noted that the Metazoa as defined here correspond to "Animalia" as defined in 5-kingdom systems (see Organisms, p. 7).

## SUBKINGDOM PROTOZOA (Single-celled animals)

CHARACTERISTICS: *Mostly minute ANIMALIA consisting of a single cell that may be simple or highly differentiated and contain 1 or several nuclei. Sexual reproduction by fusion of 2 entire individuals or their nuclei only, or by fusion of 2 reproductive cells (gametes).*

The cell can be differentiated into regions (organelles) which carry out specific functions; e.g., pseudopodia (temporary cytoplasmic protrusions) and undulipodia (=flagella and cilia; permanent, hair-like structures) for locomotion, digestive vacuoles and a cytostome (cell mouth) for feeding, contractile vacuoles for water balance.

With some 50,000 known species Protozoa are now considered a level of organization rather than a homogeneous phylogenetic entity. One of the 4 major "phyla," Mastigophora (= Flagellata), comprises organisms that possess chloroplasts and are autotrophic ("phytoflagellates"); they are treated here as plants (see Chrysophyta, p. 66, Euglenophyta, p. 74 and Dinophyta, p. 75). The heterotrophic Mastigophora ("zooflagellates") are not considered here because no marine, free-living representatives have been reported from Bermuda as yet. Three phyla are included: Sarcodina (= Rhizopoda, p. 93), Sporozoa (p. 103) and Ciliophora (= Ciliata, p. 104). A 5th phylum, "Cnidosporidia", which contains some parasites of fishes, is not considered. Because of close affinities, Mastigophora and Sarcodina are sometimes combined in the superphylum Sarcomastigophora. Sporozoa are now frequently considered a class within the phylum Apicomplexa.

## Phylum Sarcodina

CHARACTERISTICS: *PROTOZOA typically with pseudopodia; flagella, if present, restricted to development stages.* The majority are microscopic, many solitary Amoebida are as small as 15 μm (!); some planktonic colonies of Radiolaria, however, may reach 3 m (!) in length. Most are colorless, some red or white and others greenish-yellowish owing to symbiotic algae. Locomotion is a slow gliding by means of pseudopodia.

This vast phylum comprises some 5,000 recent species of Foraminiferida, 500 of Radiolaria, 100 of Acantharia and an additional 100 species in the smaller classes. Of about 200 species reported from waters in and around Bermuda, 24 are included here as examples.

OCCURRENCE: Predominantly marine and free-living; some Amoebida are parasitic. Radiolaria and Acantharia are pelagic in open ocean conditions as is a small minority of Foraminiferida. The great majority of all foraminiferan species is benthic, from the intertidal to the greatest known depths of the oceans, with the highest diversity at 100-300 m.

The benthic foraminiferans can be divided into 2 groups again, those moving freely on the substrate and those living attached to plants, other animals or sediment. Some species are even buried in the sediments of the sea floor. Amoebida have recently been found in great diversity in the surface film of the ocean, whereas others live on mud and plants.

Collecting methods are dependent on the habitat. Surface film organisms are taken by a neuston sled, or a screen that is laid on the surface of the water and subsequently rinsed to collect the adhering specimens. Planktonic organisms are sampled with plankton nets (mesh size of the net can vary from 10 to several hundred μm). Benthic animals can be obtained with dredges or grab samplers of different constructions or by washing them off stones and vegetation. To isolate Foraminiferida, sediment samples have to be sieved first to get rid of mud and clay particles. Heavy liquids (e.g., carbon tetrachloride) will help in separating the remaining sediment from the foraminiferans.

Foraminiferan tests are important constituents of the sediments in many marine depositional environments, especially in tropical carbonate provinces. Because of their abundance, and rapid evolution as documented in the test, Foraminiferida (and Radiolaria) provide useful clues for the dating of sediment layers (stratigraphy) and the interpretation of paleoenvironments.

IDENTIFICATION: For all but the largest species a compound microscope with phase contrast is essential.

Important distinguishing features for all Sarcodina are provided by the size and the shape of cells, nuclei and pseudopodia (temporary or semi-permanent projections of the cell plasm). Pseudopodia can be rounded (lobopodia), thread-like (filopodia) or finger-like (dactylopodia), or they may anastomose in a complex manner (reticulopodia). Although naked Sarcodina are distinguished on the basis of cell features alone, all others are identified primarily by characteristics of the test.

In Foraminiferida, important taxonomic features (in order of decreasing phylogenetic stability) include nature of the test wall, plan of growth, chambers, sutures, aperture and ornamentation. A few primitive species have a test that is organic in composition; others construct an agglutinated (arenaceous) test by cementing foreign particles gathered by the pseudopodia to the test surface. Most secrete a calcareous test (usually of calcite, rarely of aragonite). Calcareous tests may be perforate (with pores) or imperforate. The test may consist of 1 chamber (unilocular) or of many chambers (multilocular). Unilocular tests exhibit various shapes, including spherical, flask-shaped, tubular and planispiral. In multilocular tests, the first chamber, which is usually spherical or globular, is termed the proloculus. Subsequent chambers are added in a variety of plans of growth, such as milioline (with narrow elongate chambers), planispiral (coiled flat), trochospiral (snail-like), streptospiral (coiled like a ball of twine), triloculine (with 3 externally visible chambers), triserial (with chambers in 3 rows), biserial (2 rows) and uniserial (1 row). Successive chambers are separated from one another by septa, the positions of which are marked at the test surface by sutures; the latter may be bridged by retral processes. The axis around which chambers are coiled is termed the umbilicus (navel). The aperture (the main opening of the test) can be simple, sieve-like (cribrate) or provided with teeth (dentate). Various types of superficial test ornamentation may occur, including areolae (small areas within the larger surface), costae (ribs), carinae (keels), striations, knobs and spines.

In Acantharia, the cytoplasm is differentiated into 2 regions that are sharply delimited by a membranous, mucoid or chitinous structure, the central capsule. The outer (extracapsular) cytoplasm consists of an assimilative layer (matrix) and a vacuolated layer (calymma), whereas the inner (intracapsular) cytoplasm contains the nucleus and symbiotic algae (zooxanthellae). The acantharian skeleton, made of strontium sulfate, consists of 20 very regularly arranged radial spines that can be simply pointed, blunt, forked or bearing processes (apophyses). Spines can be circular, elliptic or square in cross section.

In Radiolaria, the cytoplasm is also divided by the central capsule into intra- and extracapsular regions. The central capsule is perforated by 1, 3 or many pores and can consist of a single or a double membrane. The radiolarian skeleton is siliceous and consists of spicules that can be simply needle-like, or more commonly in the form of tripods, rings, lattice shells or a combination of these. Shells often bear spines, horns or other structures. The shells can be very complex, consisting of 5 or more concentric lattice layers.

Preserve in 2% osmium tetroxide for histology. Foraminiferan tests should be rinsed in fresh water and stored dry. To obtain skeletal elements of other Sarcodina treat with concentrated sulfuric acid (to remove the cytoplasm), rinse with ethanol, then make a permanent mount, dry or with Canada balsam.

BIOLOGY: Life-span of Sarcodina varies from a few days to more than a year in some larger foraminiferans. In most groups the pseudopodia serve to trap and gather food particles, which may include organic detritus, bacteria, diatoms, flagellates and other microorganisms. Some of the planktonic foraminiferans are carnivorous. Their diet consists of copepods and amphipods. Predators of Sarcodina are larger detritus feeders and plankton feeders. Parasites include bacteria, other protozoans and nematodes. Some larger benthic and some planktonic Foraminiferida contain zooxanthellae (dinoflagellates) and zoochlorellae (chlamydomonans). Zooxanthellae are also found to be symbionts in Radiolaria and Acantharia.

REPRODUCTION: The mode of reproduction varies among the different groups. Asexual and sexual reproduction occur, and there may be an alternation of different modes of reproduction. Binary fission (mother cell divides into 2 daughter cells), multiple fission (mother cell divides into many daughter cells), and budding (mother cell releases 1 daughter cell) are ways of asexual reproduction. Sexual reproduction can be by gametogamy (gametes are formed by different individuals which eventually fuse), autogamy (gametes or gamete nuclei from the same individual fuse), or gamontogamy (sexual reproduction is initiated by the union of gamonts). In the Amoebida binary and multiple fission dominate, but in a few species sexual reproduction has been reported. An alternation of generations has been described in the Trichosida (*Trichosphaerium*), but remains to be confirmed. Although the complete life cycle is known for relatively few Foraminiferida, an alternation of sexually and asexually reproducing generations appears to be common. It is a heterophasic alternation of generations, one generation being haploid (gamont), the other diploid (agamont). Meiosis occurs before the agamont reproduces. In at least some of the more advanced species, this alternation of generations results in test dimorphism in which the agamont usully has a smaller proloculus (first chamber) and a larger test than the gamont. Asexual reproduction is by multiple fission. Sexual reproduction involves gametogamy (whereby either flagellated or amoeboidal gametes are produced), autogamy and gamontogamy. A few are apogamous, the sexual generation having been lost. The life cycle of Radiolaria and Acantharia is very incompletely known. Binary and multiple fission, however, have been observed. Sexual reproduction is known in some forms. The gametes can be isogametes (gametes of equal size) or anisogametes (gametes of different size).

REFERENCES: A recent book covering all Protozoa is by GRELL (1973). The most comprehensive up-to-date treatment of Sarcodina is by LOEBLICH & TAPPAN (1964; Foraminiferida, and smaller groups), CAMPBELL (1964; Acantharia and Radiolaria) and ANDERSON (1983a; Radiolaria). The 2 latter taxa are well represented in the classical study by HAECKEL (1887); see also TREGOUBOFF (1953). A good new summary of Foraminiferida is by BOLTOVSKOY & WRIGHT (1976). Marine Amoebida have recently been described by SAWYER (1975) and DAVIS et al. (1978). Trichosida are dealt with by SHEEHAN & BANNER (1973) and ANGELL (1975). The most comprehensive treatment of Gromiida is by ARNOLD (1972).

Many of Bermuda's Foraminiferida are included in CUSHMAN (1918-1931). More recent studies of planktonic foraminiferans are by BÉ (1959, 1960), BÉ & ANDERSON (1976), BÉ & HAMLIN (1967), BÉ et al. (1977), ANDERSON & BÉ (1976), SPINDLER et al. (1978, 1979) and HEMLEBEN et al. (1978). The only recent studies of benthic foraminiferans are by STEINKER (1980), STEINKER & BUTCHER (1981) and PESTANA (1983). Bermuda's Radiolaria have been studied by ANDERSON (1976a, b, c, 1978, 1980).

**CL.** **RHIZOPODEA:** Sarcodina with lobopodia, filopodia or reticulopodia.

### Plate 24

**S.CL.** **LOBOSIA:** Rhizopodea with typically lobose pseudopodia.

**O.** **AMOEBIDA:** Naked Lobosia without fixed body shape; locomotion by lobopodia. Typically uninucleate, but several species multinucleate. Majority free-living, many parasitic; marine and fresh water. Several marine species with polymorphic forms. (More than 10 spp. from surface waters of the Atlantic and the Caribbean.)

***Vexillifera pagei*** Sawyer: Locomotive form with slender pseudopods normally formed at the anterior margin; to 18 μm long. Floating form with 6-10 filose pseudopods normally much longer than the irregularly spherical body mass. Uninucleate, diameter of nucleus about 3 μm. Commonly found offshore.

**O.** **TRICHOSIDA:** Lobosia with dactylopodia and/or lobopodia, and a test of calcite spicules. Di-

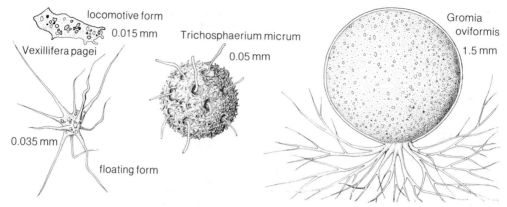

**24 LOBOSIA, FILOSIA (Amoebas, etc.)**

ameter up to 2 mm. Locomotion by means of lobopodia. Marine; known mainly from vegetation in littoral and shallow sublittoral environments. (1 sp. from Bda.)

***Trichosphaerium micrum*** Angell: Schizont usually hemispherical when attached to the substrate; to 50 μm. Calcite spicules form a sheath around the body, with small openings for the pseudopodia. Multinucleate; gamontic generation not known. Commonly found among detritus washed from shallow-water algae and from marine grass blades.

S.CL. **FILOSIA:** Rhizopodea with tapering and branching, rarely anastomosing filopodia.

O. **GROMIIDA:** Filosia with organic test, commonly with agglutinated particles; test with distinct aperture. Locomotion by means of filopodia. (1 sp. from Bda.)

***Gromia oviformis*** Dujardin: Probably with alternation of generations: the relatively large testate gamonts undergo gamontogamy to produce stellate young that apparently develop into inconspicuous nontestate amoeboid schizonts. The origin of the gamont remains obscure. Gamont spherical to ovate in shape and up to 3 mm in diameter; test colorless, transparent, and organic in composition. Protoplasm usually brown; pseudopodia protrude out of the test through the aperture. An oral apparatus consists of an elevated collar with septal bars extending into the apertural opening. The pseudopodia gather food particles, which are taken into the test of the gamont through the aperture. Commonly found among detritus on shallow, subtidal algae and grass blades.

S.CL. **GRANULORETICULOSIA:** Rhizopodea with delicate, finely granular reticulopodia that sometimes anastomose in a complex manner.

**Plate 25**

O. **FORAMINIFERIDA:** Granuloreticulosia with unilocular or

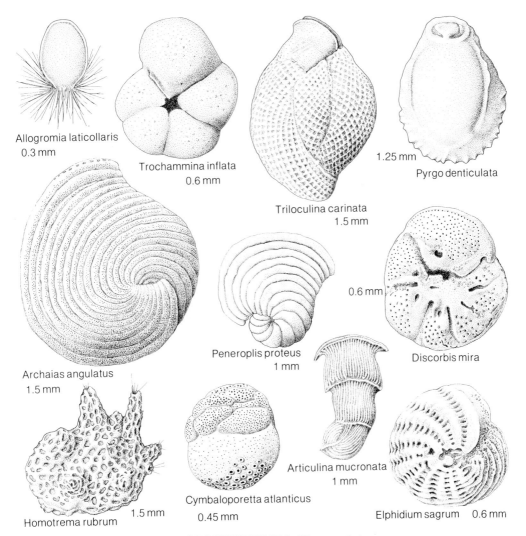

**25 FORAMINIFERIDA (Forams) 1**

multilocular test; heterophasic alternation of generations in at least some species. Almost exclusively marine. (Over 100 spp. known from Bda.)

**SUP.F. ALLOGROMIIDEA:** Foraminiferida with organic test, sometimes with agglutinated material on organic base.

*Allogromia laticollaris* Arnold: Test unilocular, ovoidal to spherical; simple aperture with internal extension (entosolenian tube); test transparent, cytoplasm orange. Life cycle consists of an alternation of generations; sexual reproduction is by autogamy, gametes are amoeboid. Adults range from 100 to 500 μm in diameter. Found among detritus on algae and grass blades in shallow waters.

**SUP.F. LITUOLIDEA:** Foraminiferida with agglutinate test; multilocular, with simple or labyrinthic septation.

***Trochammina inflata*** (Montagu): Wall finely arenaceous, smooth, light brown; test trochospiral, composed of 3 to 3 1/2 whorls, with 5 to 5 1/2 chambers in last whorl. Chambers inflated; dorsal sutures slightly curved, ventral sutures nearly straight, radiating out from umbilicus; aperture a slit at base of last chamber, extending nearly from periphery to umbilicus. Diameter to 0.65 mm. Common in shallow water on algae among mangroves.

**SUP.F. MILIOLIDEA:** Foraminiferida with calcareous test, imperforate, occasionally with surficial agglutinated material; typically planispiral or coiled in varying planes; aperture terminal, simple, dentate or cribrate.

***Triloculina carinata*** Orbigny: Adult test typically with 3 external chambers added at an angle of 120°; test longer than broad, surface reticulate, periphery typically carinate; sutures somewhat depressed; aperture an elongate slit at end of last chamber, with long narrow tooth. To 1.5 mm in length. Live individuals may be found on subtidal vegetation such as *Padina*, or among bottom sediments.

***Pyrgo denticulata*** (H. B. Brady): Test with 2 external chambers, elongate, biconvex; wall smooth; aboral end denticulate; aperture narrow and broad, with a prominent broad tooth extending along the width of the aperture. To 1.25 mm or more in length. Occurs in low numbers in shallow water; most numerous on reefs.

***Archaias angulatus*** (Fichtel & Moll): Test lenticular, planispiral and involute in early stages; later chambers becoming flaring and even annular. Early chambers simple, later chambers with interseptal pillars. Cytoplasm green owing to presence of zoochlorellae; aperture a series of pores on peripheral face. Diameter to 3 mm. A nearly ubiquitous species in shallow waters, commonly found living on vegetation such as *Thalassia*, *Padina* and calcareous Codiaceae. Test wall is thick and resistant to mechanical destruction, so that empty tests are important constituents of the shallow-water sediments.

***Peneroplis proteus*** Orbigny: Test planispiral, early portion thickened and tightly involute, later portion flaring; chambers typically undivided; sutures slightly depressed; wall smooth; apertures consist of pores along the periphery. To 1.5 mm or more in length. Widespread and common in shallow waters where it is frequently found living on marine grasses and algae; usually abundant on reefs.

***Articulina mucronata*** (Orbigny): Early portion of test triloculine, later chambers (usually 1 or 2) uncoiled; test compressed, surface with numerous costae; apertural end widely flaring and recurved; aperture elongate, with everted lip. To 1.25 mm in length. Common on nearshore mud flats; present on inner reefs.

**SUP.F. ROTALIIDEA:** Foraminiferida with calcareous, perforate test; septate; plan of growth trochospiral in ancestral forms, may later become planispiral, uncoiled or complex; wall composed of concentric laminae; aperture initially a basal slit, multiple or complex in some later forms. Includes planktonic as well as benthic forms.

*Discorbis mira* Cushman: Test trochospiral, plano-convex; periphery slightly lobulate and carinate; adult test composed of about 3 1/2 whorls, with 5-7 chambers in last whorl; wall coarsely perforate. Dorsal sutures flush with test surface, oblique, slightly curved; ventral sutures depressed and slightly curved; ventral chambers with umbilical flaps; umbilical knob typically present on ventral side; aperture a slit at base of last chamber, extending from umbilical region toward periphery. Early portion of test brown, later chambers commonly with clear walls. Diameter to 0.65 mm. Common on intertidal and subtidal algae, as well as marine grasses. Because of the sturdy test, it may be found among the bottom sediments.

*Cymbaloporetta atlanticus* (Cushman) (= *Tretomphalus atlanticus*): Test trochospiral in early stages, later chambers added in an alternating annular series, usually with 4 chambers in last whorl; dorsal test surface perforate, ventral surface imperforate. Dorsal sutures oblique, ventral sutures radial and deeply depressed; aperture a slit at base of last chamber. The final chamber of the gamontic test is a large, globular float chamber, perforated by coarse pores, and covering the ventral test surface. Dorsal test surface brown. Diameter to 0.5 mm. A widespread species. Because of the planktonic stage of the gamont during gametogenesis, it can be transported into a wide variety of habitats.

*Homotrema rubrum* (Lamarck): Test attached; early chambers trochospiral, later chambers added irregularly in concentric layers; adult test may be branched, globose, or encrusting. Outer surface with cribrate areolae surrounded by imperforate rims; color dark red. Diameter to several millimeters. Common as encrusting masses on reefs. The tests of dead individuals are abundant in sand, causing the pink hue of many beaches. (Color Plate 6.7.)

*Elphidium sagrum* (Orbigny): Test planispiral, biconvex, periphery rounded, margin not lobulate; about 15 chambers in last whorl. Retral processes short and broad, about 10-15; apertural face triangular, apertures consist of rounded openings at base of apertural face. To 0.5 mm in diameter. Common in inshore basins.

## Plate 26

*Hastigerina pelagica* (Orbigny): Test planispiral, chambers spherical to ovate; adult test with

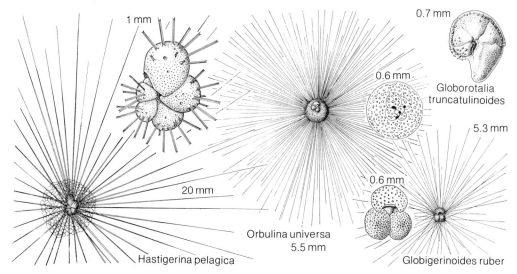

**26 FORAMINIFERIDA (Forams) 2**

about 6 chambers per whorl; wall coarsely perforated; sutures deeply depressed. Aperture a broad opening at the base of the last chamber; test with triradiate calcareous spines to 1 cm in length; color of cytoplasm brilliant to dark red. Test diameter to 1.2 mm without spines, with spines more than 2 cm. In living animals the test is surrounded by bubble-like cytoplasm. Often associated with large dinoflagellates (e.g., *Pyrocystis*). Common in surface waters of the open ocean, III-X.

***Orbulina universa*** Orbigny: Test trochospiral to streptospiral in early stages, last chamber globular, entirely enclosing the earlier ones; wall with perforations of varous sizes; no aperture in adult stage, in earlier stages with umbilical apertures; test with fine calcareous spines. Diameter of adult test to 0.9 mm, with spines more than 5 mm. Living specimens are surrounded by thousands of symbiotic algae during daytime. Common in surfaces waters of the open ocean, IV-IX.

***Globigerinoides ruber*** (Orbigny): Test trochospiral, chambers spherical to ovate; 5 chambers per whorl in juveniles, later 3 chambers per whorl; walls perforated; sutures deeply depressed. Aperture umbilical with 2 secondary apertures; test with long calcareous spines. Diameter of adult test to 0.6 mm, with spines more than 5 mm. Test in 2 variations of color: pink and white. Living specimens with symbiotic algae exhibiting a diurnal rhythm. Common in open-ocean surface waters, IV-VIII.

***Globorotalia truncatulinoides*** (Orbigny): Test trochospiral, angular-conical, with 5 or 6 chambers per whorl; wall perforated; test surrounded by thin, unperforated keel; sutures thickened and depressed. Aperture elongated from umbilicus to periphery, bordered by a lip; test without spines but with pustules. Diameter of test to 0.8 mm. Very common in open ocean-surface waters, XII-III.

## Plate 27

CL.  **ACTINOPODEA:** Sarcodina typically with axopodia or filopodia. Normally floating forms, some attached secondarily. Naked or with test of silica or strontium sulfate.

S.CL.  **ACANTHARIA:** Actinopodea with skeleton of strontium sulfate, consisting of 20 regularly arranged radial spines that emerge from the center along 5 circles (comparable to the equatorial, 2 tropical and 2 circumpolar circles of the terrestrial globe). Exclusively marine. (About 20 spp. from around Bda.)

F.  **ACANTHOLONCHIDAE:** Acantharia with spines of different size; 2 opposite equatorial spines in the longitudinal axis are much longer than the 18 others.

*Amphilonche elongata* Müller: The 2 equatorial spines are much thicker and longer than the 18 others, which are equal. Most of the cytoplasm is arranged in a spindle shape around the 2 equatorial spines. The intracapsular cytoplasm contains a large number of zooxanthellae. Total length 0.3-0.5 mm. Commonly found in the upper 100 m in open-ocean conditions.

F.  **ACANTHOSTAURIDAE:** Acantharia with spines of very different sizes, 4 equatorial ones much larger than others.

*Lithoptera tetraptera* Haeckel: The 4 equatorial spines with apophyses lying in the meridian plane of the spine. The apophyses consist of 2 transverse beams that are connected at equal distances by 4 rods parallel to the spine, which results in 4 square meshes in a single row. The 16 smaller spines are within apophyses. Cytoplasm with 4 lobes around the equatorial spines, zooxanthellae within intracapsular cytoplasm. Diameter to 0.4 mm. Often found in open-ocean waters within the upper 100 m.

S.CL.  **RADIOLARIA:** Actinopodea usually with a siliceous skeleton generated by the peripheral cytoplasm and enclosing the central capsule. Skeleton when present composed of needle-like spicules or, most commonly, of tripods, rings, lattice shells or a mixture of these. (About 100 spp. from around Bda.)

O.  **PORULOSIDA:** Radiolaria with pores distributed over entire surface of central capsule.

F.  **SPHAEROZOIDAE:** Exclusively colonial Porulosida with spicules of different types.

*Sphaerozoum punctatum* Müller: Skeleton consisting of spicules scattered around the central capsule in the calymma. Spicules composed of a middle rod with 3 needle-like shanks on each end. Shanks with many small spines.

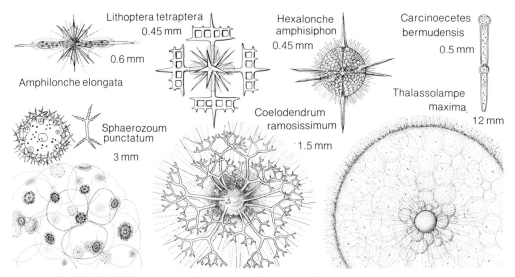

**27 ACANTHARIA, RADIOLARIA; SPOROZOA (Radiolarians, etc.)**

Individuals of the colony are embedded in the jelly-like calymma. Colony with zooxanthellae. Colony to a few millimeters, single individual about 0.1 mm. Very common in surface waters.

F. **THALASSICOLIDAE:** Porulosida without skeleton or spicules; not forming colonies.

*Thalassolampe maxima* Haeckel: Body spherical, colorless. Central capsule with a thick but transparent membrane. Extracapsular cytoplasm very thin, intracapsular cytoplasm with large round alveoles and the spherical nucleus in the center. With a diameter of 10-12 mm its central capsule is the largest of all known Radiolaria. Never associated with zooxanthellae. Diameter to 15 mm. Commonly found near the surface of the open ocean.

F. **CUBOSPHAERIDAE:** Porulosida with lattice shell single or concentrically multiple, with 6 main radial spines in 2 planes.

*Hexalonche amphisiphon* Haeckel: With 2 concentric lattice shells, one intracapsular, the other extracapsular. With 6 main radial spines of equal size and numerous smaller bristle-shaped radial by-spines covering the surface. Pores in the outer shell regularly hexagonal with very thin bars, prolonged on the outer as well as the inner surface into short conical tubes. Inner shell octahedral with irregular polygonal meshes. Diameter of the outer shell to 0.2 mm. Regularly found in the upper 100 m of the open ocean.

O. **OCULOSIDA:** Radiolaria with pores restricted to 1 pole or to tubular openings of central capsule.

F. **COELODENDRIDAE:** Oculosida with 2 thin-walled valves,

each with a conical process (galea) from which branched tubes originate.

***Coelodendrum ramosissimum*** Haeckel: From the galea hollow tubes arise that are dichotomously branched; terminal branches diverging at right angles, their end-knobs with 4 crossed short teeth. The central capsule is enclosed between the 2 valves; its 3 openings lie in the open frontal fissure between them. Often found in open-ocean waters within the upper 100 m.

M. SPINDLER, C. HEMLEBEN & D. STEINKER

## Phylum Sporozoa

CHARACTERISTICS: *PROTOZOA with spores as resting stage. Spores simple, without polar filaments and with one to many sporozoites. Undulipodia are absent (except for flagellated microgametes in some groups).* A few micrometers to over 10 mm in length (some gregarines); mostly whitish-opaque. Locomotion (if any) is a slow gliding.

Of some 3,900 known species, 4 have been described from Bermuda, all belonging to the order Gregarinida. Of the other order, Coccidia, no marine representatives have been reported from Bermuda as yet.

OCCURRENCE: Exclusively parasitic. Besides terrestrial and freshwater animals they occur in a wide range of marine organisms, from flatworms to chordates. Sporozoans are of importance to humans because they cause illnesses such as toxoplasmosis and malaria.
 Sporozoa can be obtained by dissecting infected organisms, particularly the hindguts of crabs of the genera *Pachygrapsus*, *Panopeus*, *Mithrax* and *Calappa*.

IDENTIFICATION: Because of the small size of sporozoans and their parasitic way of life histological methods and a good compound microscope are essential.
 Species identification is based primarily on size and shape of cell and spores.

BIOLOGY AND REPRODUCTION: Both sexual (gamogony) and asexual reproduction (sporogony) occur, exhibiting a haplohomophasic alternation of generations (both generations are haploid; only the zygote is diploid). In a number of sporozoans a further asexual multiple fission (schizogony) takes place. In Eugregarinida (to which *Carcinoecetes* belongs), the gamont is divided into a short, anterior protomerite and longer, posterior deutomerite. The protomerite can elongate to form an extension or epimerite. In several species the gamonts unite; one gamont attaches by its protomerite to the deutomerite of another, thus forming a syzygy. The anterior gamont is called primite, the posterior satellite. Food uptake in Sporozoa is through micropores. The food consists of cytoplasm, body fluids and dissolved food material of the host. Several sporozoans contain myonemes (contractile organelles) that enable them to perform gliding movements.

REFERENCES: For a general orientation see GRELL (1973) or KUDO (1966).
 Bermuda's Sporozoa have been investigated by BALL (1951).

## Plate 27

***Carcinoecetes bermudensis*** (Ball): Gamonts usually form syzygy. Size of a solitary gamont to 126 μm by 36 μm. Protomerite about

1/6 of total length of gamont. Rudiment of epimerite indistinctly delimited. Protomerite may take on hammer shape. Primite and satellite loosely fused. In the mid- and hindgut of *Pachygrapsus transversus*; more often found in ♂ than in ♀ crabs.

M. SPINDLER

## Phylum Ciliophora (Ciliates)

CHARACTERISTICS: *PROTOZOA with simple cilia or compound ciliary organelles in at least 1 stage of life cycle; subpellicular infraciliature universally present even when cilia absent. With 2 types of nuclei (except in a few homokaryotic forms).* Body size ranges from 0.01 mm to 4.5 mm. The cytoplasm is colorless, yellowish, or brownish-greenish. Errant forms swim steadily in a revolving, spiral motion, or jump, glide or stalk. On contact with an obstacle they often reverse the ciliary beat and "back up." Many sessile forms are capable of bending motions, and of quick contractions of the stalk.

Of 7 subclasses in 3 classes, 2 (the subclasses Vestibulifera and Hypostomata) are left out. Some 8,000 species of Ciliophora have been described; of 55 species known from Bermuda, 16 are included here as examples.

OCCURRENCE: In all humid environments, including the body cavities of other animals. Marine species are mainly benthic, to 4,000 m depth, but are most common in the littoral on algae, in mud, or interstitially between sand grains; some are planktonic (especially the suborder Tintinnina). The mainly sessile Peritricha and Suctoria are symphorionts (i.e., are attached to and carried by many hosts including vertebrates, invertebrates and even other ciliates). Many species are cosmopolitan.

Collect pelagic forms with a fine plankton net. For benthic species collect likely sediment and extract specimens with the climate deterioration, magnesium chloride or seawater ice methods. Sessile forms are obtained by carefully scanning likely carrier organisms (algae, Polychaeta, Acari, Crustacea, Mollusca and Bryozoa) under a dissecting microscope.

IDENTIFICATION: Always start with live specimens, in a drop of seawater, under a cover slip. Sensitive species can be kept in a hanging drop and observed with an inverted microscope.

To impede ciliary movement add methyl cellulose (commercially called Methocel or Protoslo, used as 2% aqueous solution: sketch a circle of Methocel on the slide, fill it with water containing organisms, cover the whole with a slip) or nickel sulfate solution (1 part of 1% solution in 5 parts seawater). The nuclear apparatus can be stained by adding small amounts of methyl green solution (without previous fixation; 1% aqueous methyl green in 0.5-1% glacial acetic acid) or Feulgen nucleal reaction (with previous fixation). Staining of cilia and cirri is by Noland's solution (a combined fixative and stain, especially used in Hypotrichida) or silver impregnation by the Chatton-Lwoff technique or Bodian's Protargol (protein silver) method.

The presence, distribution and structure of cilia, thread-like organelles used for locomotion and/or feeding, provide the most important criteria for identification. Cilia can be simple and distributed evenly over the body, or grouped into cirri, or fused into long rows (membranes) or short platelets (membranelles). In Suctoria only juvenile stages have cilia. Another surface differentiation of systematic consequence is the oral region (peristome), particularly the location of the cell mouth (cytostome), which may open directly or at the bottom of a buccal cavity and lead into the endoplasm by way of a variously strengthened, non-ciliated tube (cytopharynx). The cytoplasm is divided into an (internal) endoplasm and an (external) ectoplasm; the latter forms a membranous, elastic or rigid cover (pellicle). The ectoplasm contains contractile fibrils (myonemes) and explosive capsular organelles that are used in self-defense (trichocysts) or predation (toxicysts). The nuclear apparatus is located in the endoplasm. It is dual; i.e., it consists of (large) macronuclei and (small) micronuclei whose shape and numbers are important identification criteria. Contractile vacuoles, organelles used for osmoregulation, are common.

Fixation prior to permanent staining is best done with Schaudinn's fluid (2 parts of saturated aqueous mercuric chloride with 1 part 96% ethanol; before use, add 5 ml of glacial acetic acid to 95 ml of this mixture). Two other excellent fixatives are osmium tetroxide (use the vapor from a 2% aqueous solution) and Bouin's.

BIOLOGY: Nutrition is mostly holozoic; i.e., food particles or prey organisms are ingested whole. One group (deposit feeders) specializes in picking up food (e.g., diatoms) from the substrate, or hunting ciliates, flagellates or small metazoans. Members of a second group (suspension feeders) fold out their large undulating membrane through which they filter water propelled by the adoral membranelles. According to their preferred food the ciliates can be classified as bacteriophages, herbivores, carnivores or histophages (feeding on decaying tissue of larger animals). Locomotor responses to stimuli are either by trial and error (such as backing away from an obstacle and changing course) or by movement along a gradient of stimulus intensity. Endosymbionts (bacteria, zooxanthellae, zoochlorellae) are common. Many Ciliophora are symbionts, living attached (as symphorionts) on other organisms, or in the digestive tract, body cavities or organs of numerous invertebrate and vertebrate hosts; some species are parasites. Ciliates, in turn, are food for many small Metazoa.

REPRODUCTION: Binary fission, in which the mother cell typically splits transversely into 2 (sometimes more) daughter cells, is the only mode of reproduction. It can give rise to filial products of equal or unequal size; these can be temporarily held together in a chain, or occur as buds. Some Suctoria and other sedentary species produce ciliated larvae (swarmers) by budding, which disperse and then settle to grow into the adult organism. Conjugation is a widespread sexual phenomenon (though not necessarily tied to reproductive events); 2 individuals pair up to exchange genetic material derived from the micronucleus.

REFERENCES: For a general introduction consult CORLISS (1979); one of the most extensive taxonomic treatments of free-living Ciliophora is by KAHL (1930-1935). Marine interstitial species have been studied by DRAGESCO (1960), and a bibliography has been compiled by HARTWIG (1980a). BORROR (1973, 1980) deals with many aspects of free-living marine species. A useful new key based on stained specimens is by SMALL & LYNN (1985).

Bermuda's symbiotic ciliates have been dealt with by BIGGAR (1932) and LUCAS (1934, 1940), and free-living species by HARTWIG (1977, 1980b).

## Plate 28

**CL. KINETOFRAGMINOPHORA:** Ciliophora with simple cilia, never joined to form membranelles or cirri. Body ciliature ranging from a uniform covering to total absence. Cytostome often apical or near-apical, and never in a buccal cavity. Cytopharyngeal apparatus typically prominent.

**S.CL. GYMNOSTOMATA:** Kinetofragminophora with uniform somatic ciliation (holotrichous). Cytostome more or less devoid of cilia. Commonly with trichocysts.

**O. KARYORELICTIDA:** Gymnostomata with dual nuclear apparatus with diploid, non-dividing macronuclei. Oral area apical, or a ventral slit. Somatic toxicysts present. Elongate, fragile, highly thigmotactic species.

*Tracheloraphis incaudatus* (Kahl): Slim, with parallel corrugated sides visible at contraction; broadly rounded posteriorly. Cytoplasm filled with granules. Mouth terminal, funnel-shaped, without a longitudinal slit at its dorsal edge. Ciliature of 31 or 32 longitudinal rows. Dorsal glabrous stripe, characteristic of the genus, 4 or 5 cilia rows wide. The nuclear apparatus consists of 6 macronuclei and 2 micronuclei forming a single group. A typical inhabitant of sandy sediments.

*Geleia nigriceps* Kahl: Body slightly flattened, nearly parallel-sided; anterior end rounded. A frontal groove, above the cell mouth, is lacking. The 25-35 somatic ciliary rows reach the anterior end with undiminished width, but end posteriorly polarly. Body with brownish pigment; cytoplasm with variously distributed granules. Nuclear apparatus, in the center of the cell, consists of 2 macronuclei with an intercalary micronucleus. Algae are taken as food. A purely interstitial species.

O. **PROSTOMATIDA:** Gymnostomata with dual nuclear apparatus with polyploid macronuclei. Oral area apical or subapical, or in a shallow atrium; cytostome oval or round. No toxicysts in the immediate mouth area, but some species with somatic toxicysts.

*Helicoprorodon gigas* (Kahl): Worm-like and contractile. The cytopharyngeal apparatus possesses a skeletogenous armature ("trichites"). The somatic ciliature curves into more or less horizontal rows at the anterior end, a characteristic of the genus. Cell mouth apical. The nuclear apparatus is a chain of 15-36 macronucleus elements; micronuclei in great number. Contractile vacuole terminal. Carnivore. Widely distributed in sediments.

O. **HAPTORIDA:** Gymnostomata with dual nuclear apparatus with polyploid macronuclei. Cytostome terminal, oval or slit-like. Toxicysts concentrated in oral region. Thigmotactic or clavate "sensory" cilia near anterior end.

*Mesodinium pupula* Kahl: A medial constriction separates a pear-shaped anterior from a rounded posterior section. Several dark inclusions in the anterior part, colorless bodies in the posterior. The ciliature is reduced to 2 medial rings of membranelles. Nuclear apparatus of up to 2 macronuclei. Contractile vacuole terminal. Known only from interstitial sand biotopes.

S.CL. **SUCTORIA:** Kinetofragminophora without a mouth; instead with multiple, contractile, capitate "sucking" tentacles (often also with thin, prehensile tentacles). Adult stage rarely with cilia; always sessile, either directly or by means of a non-contractile stalk.

*Ephelota gemmipara* (Hertwig): With a stalk of variable length (to 1.5 mm), poorly ringed, and proximally broadened. Body rounded-trapezoidal, extremely variable. With both non-sucking prehensile and suctorial tentacles. Macronucleus irregular, branching. With 1 or more contractile vacuoles. On drifting *Sargassum*.

CL. **OLIGOHYMENOPHORA:** Ciliophora with compound ciliary

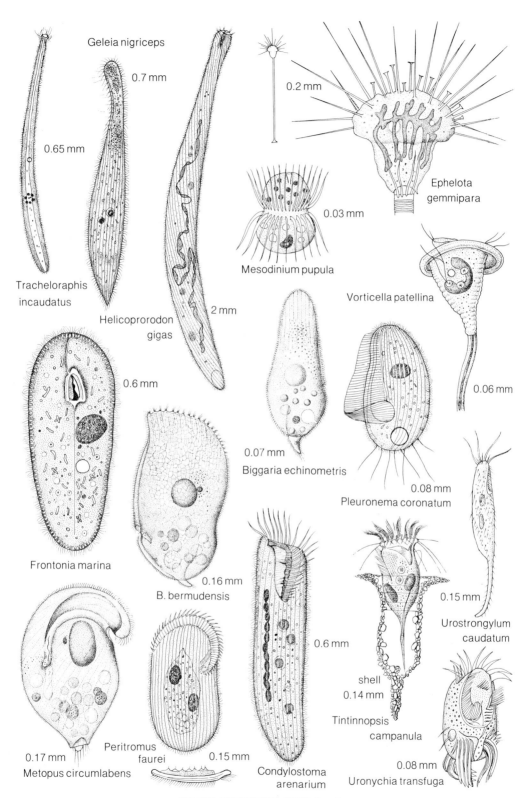

## 28 CILIOPHORA (Ciliates)

membranes and membranelles forming a well-defined (though sometimes inconspicuous) oral apparatus. Body ciliature never forming cirri. Cytostome usually ventral and near the anterior end of the body, always in a buccal cavity. Cytopharynx generally inconspicuous. Colony formation is common in some subclasses.

**S.CL. HYMENOSTOMATA:** Oligohymenophora often with uniform and heavy body ciliature. Oral structures when present usually not prominent.

**O. HYMENOSTOMATIDA:** Hymenostomata with buccal cavity containing membranelles. Oral area ventral, usually on anterior body half. Contractile vacuoles with permanent openings in the pellicle; cell anus present.

*Frontonia marina* Fabre-Domergue: Body slightly flattened. Somatic ciliature in numerous longitudinal meridians. Buccal cavity in the anterior 1/3. A clearly visible seam extending from behind the cell mouth to the posterior body end. Macronucleus central, with 3 adjacent micronuclei. Trichocysts ending at pellicle. Herbivorous. A eurytopic species, known from sandy and muddy bottoms.

**O. SCUTICOCILIATIDA:** Hymenostomata with buccal cavity often dominated on right side by large tripartite paroral membrane. Body uniformly to sparsely ciliated; distinctive caudal cilia common.

*Biggaria echinometris* (Biggar): Genus with posterior region terminating in a rudder-like style.—Species with anterior end symmetrically rounded, with few longitudinal striations. Oral apparatus posterior to the middle of the body. Protoplasm transparent but in anterior region densely granular. Rounded macronucleus near center of body accompanied by 1 micronucleus. Found in the intestinal tract of the sea urchin *Echinometra lucunter*.

*B. bermudensis* (Biggar): Genus as above.—Species with anterior 2/3 of body thin and leaf-like. Cilia arranged in rows, indicated by 35-40 longitudinal striations. Oral apparatus in the posterior 1/3 of cell. Protoplasm colorless, transparent, highly alveolar in structure. Single, rounded macronucleus, near middle of the body, with a single micronucleus. Found in the intestinal tract of the sea urchins *Lytechinus variegatus* and *Diadema antillarum*.

*Pleuronema coronatum* Kent: Body oval to ovoid. Peristome narrow anteriorly, circular posteriorly, reaching nearly up to posterior end of the body. At right margin of peristome a pocket-shaped membrane surrounds the cell mouth. In posterior 1/3 of body several elongate cilia. Macronucleus spherical. Contractile vacuole posterior. Feeds mainly on bacteria. In marine interstitial sand and limnetic biotopes.

**S.CL. PERITRICHA:** Oligohymenophora with body ciliature reduced to a subequatorial locomotor fringe. Oral apparatus predominantly at apical pole. Membranelles and an outer membrane of oral area arranged in circular wreaths that finally dip inwardly through the buccal overture into the buccal cavity. Many species stalked and sedentary, others mobile, with basal disc for attachment. Trichocysts absent. Binary fission longitudinal. Some species colonial.

O. **PERITRICHIDA:** Peritricha with the characters of the subclass. (Of 2 suborders, Mobilina are not included here.)

S.O. **SESSILINA:** Peritrichida with mainly sessile and sedentary forms. With rigid or contractile stalk or adhesive disc. Mostly symphoriants on bodies of animals from many marine and freshwater metazoan taxa.

*Vorticella patellina* Müller: Body funnel-shaped. Peristome pad broad and clearly vaulted. The invaginated area of the anterior body end ("vestibulum") extends only 1/3 of the body. Pellicle with fine transverse striations. Macronucleus horseshoe-shaped. Stalk contractile. On various algae and invertebrates.

CL. **POLYHYMENOPHORA:** Ciliophora with compound ciliature forming a well-defined adoral zone of numerous buccal membranelles often extending out onto the body surface. Body ciliature usually restricted to cirri. Cytostome generally anterior, in a buccal cavity.

S.CL. **SPIROTRICHA:** Polyhymenophora with the characters of the class.

O. **HETEROTRICHIDA:** Spirotricha with prominent buccal membranelles. Body ciliature commonly uniform and dense. Macronuclei oval, often beaded.

*Metopus circumlabens* (Biggar): Elongate-oval, rounded anteriorly and tapering posteriorly; highly flexible. Region anterior to oral groove hook-like. Peristome a deep furrow. Numerous food vacuoles posteriorly. Body cilia in oblique rows, with a tuft of much longer cilia at posterior end. Macronucleus oval, accompanied by a single micronucleus. In the intestinal tract of the sea urchins *Diadema antillarum* and *Echinometra lucunter*.

*Peritromus faurei* Kahl: Flattened, rounded at both ends. Dorsal hump low, irregular, with transparent papillae and spines. Body ciliature ventral, in longitudinal meridians. Adoral zone with about 45 membranelles. Two macronuclei. Food vacuoles contain diatoms and algae. On sandy bottoms.

*Condylostoma arenarium* Spiegel: Body variable in shape and length. Buccal cavity wide, with numerous membranelles on the left lip; right lip with large undulating membrane. Macronucleus moniliform with as many as 11 fragments. Food vacuoles contain algae. On sandy bottoms; very common.

O. **OLIGOTRICHIDA:** Spirotricha with conspicuous adoral zone of membranelles, often extending out of the buccal cavity and winding around the apical end of the body. Body ciliature reduced. Many species with a lorica, a loosely fitted shell of organic material or foreign matter.

***Tintinnopsis campanula*** (Ehrenberg): Shell typically cylindrical, with a widely flaring opening and a stout pedicel. Shell composed of irregular particles secreted by the organism itself. Opposite to the mouth is an area of cilia limited by an undulating membrane. Two macronuclei. Swims with oral end forward. A widespread member of the marine plankton.

O.  **HYPOTRICHIDA:** Spirotricha with a prominent adoral zone of membranelles near anterior end of body, and rows or groups of cirri on ventral surface. "Sensory bristles" in rows are common on dorsal surface. Body flattened dorsoventrally, peristomial field often broad.

***Urostrongylum caudatum*** Kahl: Spindle-shaped. Buccal region not markedly narrow, and relatively short (not over 1/4 body length) with several prominent frontal membranelles. Somatic ciliature reduced to 4 rows of cirri. Body cytoplasm with intense dark granulation, clearly distinguished from the light, long, non-ciliated, spine-like tail. The main food is diatoms. Widely distributed in sandy sediments.

***Uronychia transfuga*** (Müller): Adoral zone of membranelles with membrane-like portions along left edge of the buccal cavity. Undulating membrane well developed. Field of fronto-ventral cirri reduced to 3 cirri near anterior end of the adoral zone of membranelles. With 2 or 3 strong left lateral cirri, and 3 strong right caudal cirri inserted in a dorsal groove; these 2 groups of cirri used for jumping. Four strong transverse cirri present. Usually with 2 macronuclei. Feeds on algae and Protozoa. Common in sandy sediments.

E. Hartwig

## SUBKINGDOM METAZOA
(Many-celled animals)

CHARACTERISTICS: *Small to large ANIMALIA consisting of few to many cells (with usually 1 nucleus each) that are often variously specialized and may form tissues and organs. Sexual reproduction always by fusion of 2 gametes.*

The fertilization of the larger, usually non-motile egg by the smaller, motile sperm, both haploid (i.e., with a single set of chromosomes), results in a diploid zygote, which then divides by mitosis. The embryo, a hollow ball (blastula) at first, organizes itself (in the gastrula stage) into primary cell layers (ecto-, endo-, and mesoderm). These form the basis for the differentiation of the adult tissues (i.e., groups of cells similar in structure and function) and organs (tissues forming functional units). Metazoa are by far the most diverse major group of organisms, not only in their size range (from 0.1 mm long Gastrotricha to 30 m long whales!), but also in terms of shape, locomotion, behavior and ecological adaptations.

The subkingdom comprises 32 phyla, of which 27 have known marine representatives in Bermuda: Porifera (p. 111), Cnidaria (p. 127), Ctenophora (p. 194), Platyhelminthes (p. 197), Mesozoa (p. 206), Nemertina (p. 207), Gnathostomulida (p. 211), Gastrotricha (p. 213), Nematoda (p. 216), Rotifera (p. 219), Kinorhyncha (p. 220), Priapulida (p. 222), Acanthocephala (p. 223), Sipuncula (p. 224), Echiura (p. 228), Pogonophora (p. 230), Annelida (p. 232), Tardigrada (p. 265), Arthropoda (p. 268), Mollusca (p. 392), Bryozoa (p. 500), Phoronida (p. 516), Brachiopoda (p. 518), Chaetognatha (p. 519), Echinodermata (p. 522), Hemichordata (p. 541), and Chordata (p. 545). The 5 phyla not represented are the microscopic, primitive, marine Placozoa, Nematomorpha (nematode-like worms with few marine representatives), the recently described marine interstitial Loricifera, the terrestrial Onychophora and the parasitic Pentastomida. Recent phylogenetic analyses support the subdivision of Metazoa into Placozoa, Parazoa (=Porifera) and Eumetazoa (all other phyla), and the latter into Coelenterata (=Radiata, i.e., Cnidaria and Ctenophora) and Bilateria. Only the latter term is used for classification purposes here (p. 197).

## Phylum Porifera (Sponges)

CHARACTERISTICS: *Sedentary, aquatic, filter-feeding METAZOA bounded by a 1-cell layer of flat pinacocytes (pinacoderm) and containing flagellated choanocytes (choanoderm) that create a unidirectional water current through the body. Water enters numerous small ostia (pores) and leaves through larger oscula. Mesohyle (between pinacoderm and choanoderm) contains various mobile cells, collagen and, usually, a skeleton of spongin, mineral (silica or calcium carbonate) or both.* The size ranges from millimeters to more than 1 m diameter, commonly 0.1-10 l in volume. Consistency varies with nature and density of the skeleton from soft crumbly to stiff elastic and stony hard. Colors are often vivid (yellow, red, blue), particularly in dark locations; shades of green and brown are commonly caused by symbiotic algae. Sponges are crustose or massive, cushion-, fan-, tube-, tree- or cup-shaped. Attached to the substrate, they show little movement except contraction of the entire body or of the openings upon disturbance. They are most commonly confused with compound ascidians.

Of the 4 recent sponge classes, the reef-dwelling Sclerospongea have hitherto not been found in Bermuda, and Hexactinellida occur in the deep sea. Owing to many new data on embryology, histology and chemistry, the positions of higher taxa remain in a state of flux. Of about 5,000 species known, 70 (approximately, because of many uncertain identifications) are found in Bermuda; 49 are reported here.

OCCURRENCE: Predominantly marine (3 Demospongea families in fresh water, but none reported from Bermuda), on stable substrates. Calcarea are most common in very shallow water; Demospongea (95% of all recent species) occur in all depths and climatic zones but, except for small, crustose and endolithic forms, avoid high-energy environments. Substrates include rock (particularly in caves), dead coral, subtidal mangrove roots, sea grass rhizomes, algae, other sponges, shells of mollusks and crabs, and artificial structures (buoys, pilings). Soft bottoms in calm water are colonized by initial settlement on rubble fragments.

Collect by wading, snorkeling or SCUBA diving (turn over rocks, look inside caves); on deep level bottoms also by dredging. Cut with knife or chisel; include substrate where possible (with encrusting or excavating forms in particular). Wear gloves for some forms are irritating to the skin (by spicules, toxins or epizoic cnidarians). Commercially usable species have been reported from Bermuda but are now very rare or absent. Other forms are under investigation for their antimicrobial properties.

IDENTIFICATION: Some species can be identified from color, shape, consistency and surface structures. These characters should be noted from fresh specimens, together with possible presence of pigment exudate or color changes shortly after collecting. Color photographs for documentation are useful if immediate identification is not possible. Presence and type of skeleton should be determined from microscope preparations. Make 0.2-0.5 mm sections (razor blade) of dehydrated (alcohol or air dried) material; cut perpendicular and, subsequently, parallel to the surface and note whether a surface layer (ectosome) is detachable from the internal layer (choanosome); clear in xylene and mount in balsam under cover glass. Alcohol-hardened material can also be stained to make soft tissue components (e.g., spongin, choanocyte chambers) better visible. A convenient stain for this purpose is a saturated solution of basic fuchsin in 95% ethanol; rinse well with alcohol. Spicules are isolated by digesting small but representative tissue fragments in cold sodium hypochlorite (Clorox) or (siliceous spicules only) boiling nitric acid; examine under high-power optics. Permanent mounts of dry spicules in balsam require thorough rinsing with water and absolute alcohol. Let spicules settle in test tube after each change (minimum 1 hr in water, 0.5 hr in alcohol) to prevent loss of microscleres. Use standard histological techniques to determine shape, size and arrangement of choanocytes and choanocyte chambers for classification of some Keratosa and Calcarea.

Common sponge shapes are crustose, chambered (excavating), irregularly massive, spherical, tubular or cylindrical. Most colors are represented, but fade or change quickly after collecting (especially when exposed to air), even before preservation. Consistency can be mucous soft, compressible, elastic, stiff, cartilaginous or hard and brittle. Conspicuous surface structures include exhalant openings (oscula), protrusions (conuli, papillae), embedded sand grains and special spiculous reinforcements (cortex). The surface layer (ectosome) covers an internal layer (choanosome) that contains skeletal material and is traversed by canal systems. Three types of canal systems are recognized in sponges; all occur in Calcarea. Ascon is a simple tube (spongocoel) lined with choanocytes (choanoderm), e.g., in *Clathrina*. Sycon, named after the sponge where it occurs, has a folded choanoderm; the choanocytes line short tubes that radiate from a common atrium. Leucon occurs in many Calcarea (e.g., *Leucandra*) and in all Demospongea; the choanocytes are restricted to small chambers that are dispersed through the thick mesohyl; the choanocyte chambers are connected to each other and to the outside by a system of canals without flagellated cells.

Note reticulation and structure of spongin fibers, and mineral composition, size classes, position and type (shape) of spicules. Collagenous spongin can occur as patches connecting spicules or it can build up a substantial elastic framework. Skeleton structure can be reticulate (net-like), with ascending primary and connecting secondary fibers, dendritic (tree-like branching) or intermediate such as dendroreticulate or plumoreticulate (feather-like). Spongin fibers can be clear or cored by a pith, by sediments, or by spicules. Spongin fibrils (*Ircinia*) and spongin spicules (*Darwinella*) are structures not connected to the framework. Most sponge spicules are siliceous, except in Calcarea where they are of calcium carbonate (test with acid!). One distinguishes megascleres (commonly >50 µm) that are structurally important, and microscleres that occur unoriented in certain parts of the tissue. Megascleres can be arranged in radiate, reticulate or felted fashion, in strands, coring (fully embedded), or echinating (partly embedded in) spongin fibers. Megascleres can have 1 (monaxon), 3 (triaxon) or 4 (tetraxon) axes. Monaxon forms can be pointed at both ends (oxea); pointed at one end, rounded at the the other (style); rounded at both (strongyle); pointed at one, knobbed at the other (tylostyle); or knobbed at both ends (tylote). Cladotylotes are tylote at one end, anchor-like with recurved clads (rays) at the other. Spined spicules have the prefix acantho- (e.g., acanthostyle).

Monactines, diactines, triactines and tetractines are radiate spicules with 1-4 rays, respectively. Triaenes are tetractines with 1 ray (rhabd) commonly much larger than the other 3, which can point forward of (protriaen) or toward (anatriaen) the rhabd. Desmas are irregular complex branching and interlocking megascleres. Microscleres can be various forms of asters where rays originate from 1 point (euaster) or from an axis (streptaster). Euasters include those with long free tapered rays (oxyaster), with short ray and thick centrum (spheraster) and with coalescent rays and special surface ornamentation (sterraster). The most common kind of streptaster has a spiraled axis and is termed spiraster. A related type, the amphiaster, has spines radiating from both ends of the shaft. Microscleres can also be hairlike (raphid), C- or S-shaped (sigma; sigmaspire, if contorted) or bow-shaped (toxa). Another common group of microscleres are anchors (chelae) with either equal-sized (isochela) or unequal-sized (anisochela) recurved ends. Arcuate chelae have a bow-shaped shaft. Hexactinal spicules (e.g., hexaster) characteristic of the class Hexactinellida have 6 rays arranged in 3 perpendicular axes (triaxon).

Sizes given in plates refer to the largest dimension (height or width) of the specimen shown.

Fix in 10% formalin-seawater, neutralized and buffered by methenamine (20 g/l final solution), store in 70-80% ethyl alcohol (change twice). Dry large specimens for more convenient storage but fix representative portions in liquid. Use formalin or Bouin's fixation and Mallory's triple stain for routine histological examination. If necessary decalcify in 5% nitric acid (rinse well, neutralize in 5% aqueous sodium sulfate solution). For sectioning embed in polyester wax or epoxy resin, or in 12% gelatin for freezing and cryostat microtomy.

BIOLOGY: Most sponges are successive hermaphrodites, producing male and female gametes anywhere in the endosome. Male specimens emit sperm through their oscula into the water column. In viviparous species, which constitute the majority, females receive sperm through the inhalant water current and incubate fertilized eggs until they are expelled as free-swimming larvae. The few known oviparous species release numerous eggs enveloped in mucous sheets for outside fertilization. Several methods of asexual reproduction are common. Most species are able to regenerate from fragments. External budding is commonly observed (e.g., *Tethya* spp.), and gemmules, cell aggregates comparable to but less complex than those of freshwater sponges, are formed by some species (e.g., *Ulosa ruetzleri*, *Cliona lampa*). Life-span ranges from a few months to about 10 yr; some very large (0.8-2.5 m diameter) specimens of Demospongea observed in deep reef zones, outside the range of wave action, are estimated to be 50-100 years old. Sponges can be classified as unselective suspension feeders. They create a unidirectional water current by uncoordinated beating of choanocyte flagella, and filter bacterial and other cells and detrital organic particles under 50 μm in diameter, the maximum size of incurrent openings (ostia). Thirty to over 100 l water can be micro-filtered in 1 hr by a sponge of 1 l volume. Digestion takes place intracellularly (phagocytosis). Many tropical and subtropical shallow-water species harbor symbiotic bacteria (e.g., *Aplysina* spp., *Pseudoceratina crassa*) and unicellular algae (zoocyanellae, e.g., *Ircinia felix*, *Chondrilla nucula*; zooxanthellae, e.g., *Cliona caribbaea*) that are partly phagocytized, partly used as a source of dissolved nutrients. Sponges provide substrate or hiding places for many epibionts such as algae (*Jania*), Hydrozoa (*Sphaerocoryne*), Zoanthidea (*Parazoanthus*), Entoprocta (*Loxosomella*), Amphipoda (*Caprella*) and endobionts (many polychaetes and crustaceans). Some of these are parasites or predators because they regularly feed on the tissue of the host (e.g., Alpheidae). Predators not living in or on sponges are many fishes and some sea turtles. Some crabs (Dromiidae, Majidae) hold sponges on their backs for camouflage without harming them except for an occasional trimming. Several species are remarkable for their capacity of excavating limestone (Clionidae, in particular).

## Plate 29

DEVELOPMENT: Two principal types of larvae occur in Porifera; both have a ciliated anterior portion. The parenchy-

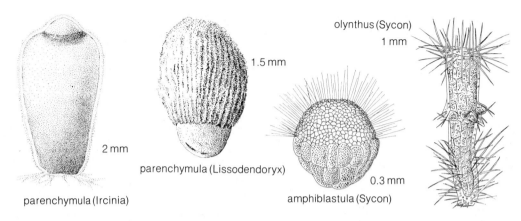

### 29 PORIFERA: Development

mula proper to most Demospongea is oval or pear-shaped, solid, 0.5-2 mm long; it is comparable to the cnidarian planula. The amphiblastula of Calcarea is spherical, smaller (50-300 μm), and has a central cavity. Hexactinellid larvae, as far as they are known, are of the parenchymula type but may not be able to swim actively. Parenchymula larvae, in contrast to amphiblastulae, can have a high degree of differentiation when released from the parent. Most sponge larvae swim in characteristic rotating or spiraling fashion close to the bottom. Settlement is preceded by a short creeping phase. Attachment by the anterior pole occurs usually after a few hours or a day. Then the larva flattens and develops a canal system and 1 osculum within 2 days. Higher Calcarea (e.g., Sycettida) go through a tubular asconoid development stage (olynthus) shortly after settlement.

Sponge larvae are best caught by stationary plankton nets installed near the bottom and facing a unidirectional current. Nearly mature larvae are often released by the parent sponge in the lab under adverse conditions (e.g., heating, oxygen depletion).

REFERENCES: Recent comprehensive presentations of the phylum are found in BRIEN et al. (1973) and BERGQUIST (1978). Techniques for systematic and ecological study are outlined in RÜTZLER (1978).

Bermuda sponges have been treated by DE LAUBENFELS (1950) and RÜTZLER (1974). A recent monograph on a related fauna from the Bahamas, including chapters on systematic procedures and an illustrated glossary, is presented by WIEDENMAYER (1977). Modern revisions of 3 orders of Caribbean demosponges are by SOEST (1978, 1980, in press).

CL. **DEMOSPONGEA:** Porifera with siliceous spicules or spongin fibers, commonly both, rarely neither present. Megascleres monaxon or tetraxon. Great variety of color, shape and size.

### Plate 30

O. **KERATOSA:** Demospongea lacking proper mineral spicules. Commonly elastic in life owing to spongin skeleton. Generally in shallow water. (About 20 spp. from Bda.)

F. **SPONGIIDAE:** Keratosa with reticulum of primary (ascending) and secondary (interconnecting, thinner) spongin fibers. Fibers without pith, some cored by moderate quantities of foreign inclusions (sediment grains). Small, spherical choanocyte chambers (<50 μm diameter).

*Ircinia felix* (Duch. & Mich.) ( = *I. fasciculata* sensu de Laub.): Genus with filamentous spongin threads filling the choanosome, which make the sponges extremely tough and difficult to tear; reticulated sand pattern on surface.—Typical form (f. *felix*) of species crustose, with raised oscula, or massive, with 1-2 mm conules on surface. Forma *fistularis* (Verrill) with simple erect hollow branches and terminal oscula; f. *acuta* (Duch. & Mich.) (non *I. strobilina*) conical or massive, with elevated oscula, with large (2-4 mm) conuli. All forms grayish to chestnut-brown, up to 25 cm. With a very distinctive odor when exposed to the air. On protected sediment and rock bottoms (f. *felix*, f. *fistularis*) and on the reefs (f. *felix*, f. *acuta*), 0.5-5 m; very common. (Color Plate 3.3.)

F. **APLYSINIDAE:** Keratosa with laminated fibers cored by a granular pith, without foreign inclusions. Small (<50 μm) ovoid choanocyte chambers.

*Aplysina fistularis* (Pallas)( = *Verongia fistularis*): Genus with regular hexagonal reticulum of amber fibers.—Species fleshy spongy, with small low conuli. Color deep yellow with brown or greenish (shallow water) tinge; turns dark purple to black in air (aerophobic). Typical form (f. *fistularis*) consists of clusters of smooth tubes; f. *ansa* has more irregular tubes with short or fairly long (15 cm) and sometimes branching digital processes rising from the rim of the cylinder. Maximum size 30 cm. Common in all reef and open-lagoon environments, 1-6 m; f. *fistularis* only in outer reefs, 30-40 m.

*Pseudoceratina crassa* (Hyatt) ( = *Ianthella ardis* de Laub.) (Blue bleeder): Dendritic knotty fibers, dark amber, to 0.5 mm thick. Surface with rounded conuli, otherwise smooth. Rubbery, firm consistency. Cushions or clusters of massive conical chimneys. Dull green, brown yellow or golden, aerophobic (turns bluish-purple-black when killed); stains fingers with purple exudate when handled. Size to 25 cm. Very common in shallow caves, on open rocks and reefs, 0.5-5 m. (Color Plate 3.1.)

F. **DYSIDEIDAE:** Keratosa with laminated primary (ascending) and secondary (interconnecting) spongin fibers, commonly packed with sand grains. Choanocyte chambers ovoid or sack-shaped and fairly large (>50 μm diameter).

*Dysidea etheria* de Laub.: Genus with primary fibers packed, secondaries cored by sand grains and broken (foreign) spicule fragments; choanosome also charged with foreign material. Unelastic, easily torn.—Species occurs as crusts or lumpy cushions, to 12 cm, with small (1 mm) conules. Two color varieties: brilliant clear blue and grayish blue; rarely transitions. Mainly on vertical hard substrates in inshore waters, 0.5-2 m. (Color Plate 3.4, 5.)

*D. janiae* (Duch. & Mich.) ( = *Desmacella janiae* Verr.; *D. fragilis* forma *algafera* de Laub.): Genus as above.—Species as aggregates of cylindrical lobes, to 6 cm, with apical oscula. Entire sponge permeated by the branching alga *Jania* sp. The alga replaces the spongin skeleton partly or entirely; it is dead (white) in the

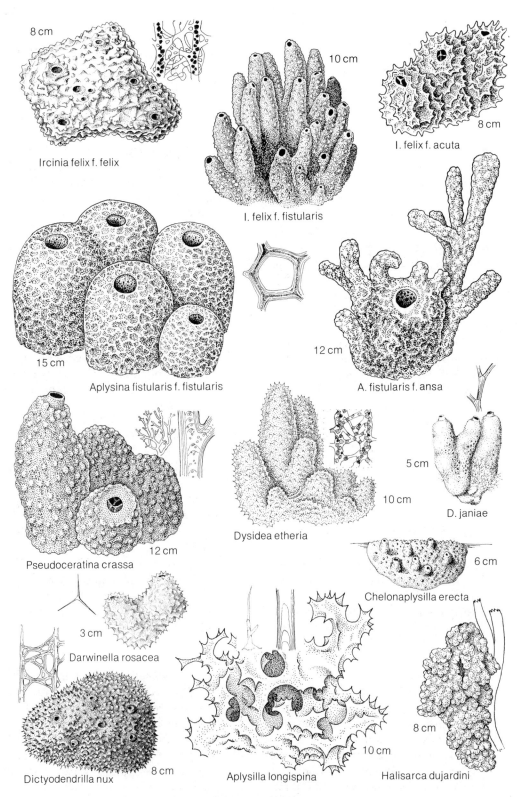

**30 KERATOSA (Sponges 1)**

deeper portions of the sponge, mostly live (pinkish to greenish purple) near the surface. Sponge color accordingly whitish, pinkish, light greenish purple. Common but inconspicuous among *Jania* turfs on patch reefs of inshore waters, 0.5-2 m. (Color Plate 3.6.)

F. **DICTYODENDRILLIDAE:** Keratosa with reticulate, distinctly pithed, laminated, and dark puplish colored fibers and delicate, cavernous tissue. Choanocyte chambers ovoid and large (>50 μm diameter).

*Dictyodendrilla nux* (de Laub.) (=*Dendrilla nux*): Primary and secondary fibers with only a few isolated foreign inclusions. Massive, to 15 cm, with finely conulose surface, cavernous choanosome; very soft and compressible. Dark grayish blue, appearing black in the field; bleeds clear blue pigment when handled. Moderately common on inshore hard bottoms, 1-2 m.

F. **APLYSILLIDAE:** Keratosa with simple or branched (dendritic), pithed and laminated ascending fibers and large (>50 μm) choanocyte chambers.

*Aplysilla longispina* George & Wilson (=*A. sulfurea* sensu de Laub.): Thickly incrusting, covering large areas, to 20 cm × 15 cm. Surface with numerous large (1-5 mm), slender, frequently compound conules. Soft cavernous tissue, stiff elastic fibers. Bright sulfur yellow color turns dark purple upon preservation. Abundant in shallow inshore caves, 1 m. (Color Plate 3.10.)

*Chelonaplysilla erecta* (Row): Genus with neat surface reticulation of sand grains.—Species encrusting, covering up to 20 cm × 20 cm, with small oscular chimneys. Consistency soft, easy to tear. Deep purplish black color. Common in inshore waters, fouling on buoys, pilings and similar structures, 1-5 m.

*Darwinella rosacea* Hechtel: Genus with spongin spicules in addition to the fiber skeleton. Thin rose-red conulose crusts, to 5 cm. In inshore waters, 1 m; uncommon. (Color Plate 7.11.)

F. **HALISARCIDAE:** Keratosa in which the spongin skeleton is absent. Large (>50 μm) sack-shaped choanocyte chambers.

*Halisarca dujardini* Johnston: Soft yellowish brown lumpy incrustations, 5 cm × 10 cm, on shells and seagrass blades. In protected bays, 1-2 m.

## Plate 31

O. **HAPLOSCLERIDA:** Demospongea with reticulate skeleton of simple spicules (oxeas, strongyles). Spicules single (unispicular) or in bundles (multispicular), connected by more or less spongin, or coring solid spongin fibers. Usually no microscleres, occasionally sigmas. (About 12 spp. from Bda.)

F. **HALICLONIDAE:** Haplosclerida with small (generally <150 μm long) oxeas of uniform length, lacking microscleres and specialized tangential ectosomal skeleton.

*Reniera hogarthi* (Hechtel) (=*Haliclona permollis* sensu de Laub.): Genus with isodictyal (equal-sided mesh) reticulation, some multispicular tracts (spongin restricted to spicule nodes).—Species encrusting or forming clusters of volcano-shaped or tubular elevations to 15 cm. Very soft, limp consistency. Color violet. Common on rubble pieces among sea grass in protected inshore bays, 1-3 m.

*Amphimedon viridis* Duch. & Mich. (=*Haliclona viridis*): Genus with ascending multispicular tracts cemented by spongin, with unoriented spicules in between.—Species appears as massive cushions with raised oscula, as groups of conical tubes, occasionally as solid branches. Maximum size 18 cm. Consistency fleshy, doughy, mucous when rubbed. Color dull green, occasional specimens purplish brown. Very common in shallow caves, on open rocks and on mud bottoms in inshore waters, 0.5-5 m. (Color Plate 3.9.)

*Haliclona molitba* de Laub.: Genus characterized by reticulation of spongin fibers cored by spicules.—Species amorphous, encrusting or digitate-ramose, to 10 cm, very soft, compressible and limp. Clear vivid violet color. Typically attached to seaweeds in inshore bays, 1-3 m. (Color Plate 3.8.)

*H. monticulosa* (Verr.) (=*Liosina monticulosa*): Genus as above.—Species encrusting or digitate, to 12 cm. Soft spongy, bright scarlet. On rocks or, characteristically, encrusting other sponges (e.g., *Ircinia felix*), in shallow inshore caves, 0.5-1 m. (Color Plate 3.2.)

F. **NIPHATIDAE:** Haplosclerida with irregular reticulation of stiff spongin fibers cored by robust oxeas. Dendritic or frazzled fiber ends protruding above the surface.

*Niphates erecta* Duch. & Mich. (=*Haliclona variabilis* sensu de Laub.): Cushions or single creeping or erect branches to 20 cm. Flush oscular rims ragged owing to protruding fiber ends. Stiff spongy, resilient consistency. Color bluish to pinkish lavender. Some specimens have sigmas for microscleres. Common on shallow (0.5-3 m) patch reefs. Most frequently found with *Parazoanthus parasiticus* covering surface. (Color Plates 3.12, 7.9.)

F. **CALLYSPONGIIDAE:** Haplosclerida with spongin fibers cored by spicules, and with special tangential surface reticulation of primary and secondary meshes.

*Callyspongia vaginalis* (Lam.): Clusters of large, thin-walled tubes, to 25 cm. Surface of most specimens covered by pronounced cone-shaped projections ("spines," conules). Spongy elastic, grayish green to lavender. Surface commonly colonized by *Parazoanthus parasiticus*. Most abundant on outer reefs and inshore patch reefs, 1-5 m.

F. **ADOCIIDAE:** Haplosclerida with spicules in isodictyal reticulation and ascending multispicular tracts. Spicules in a range of sizes. Tangential ectosomal spicule skeleton present.

*Adocia amphioxa* (de Laub.) (=*Strongylophora amphioxa*): Genus with oxeas, no microscleres.—Species as dull gray brittle cushions, to 10 cm, with elevated oscula. Moderately common on rocks in inshore waters, 1 m.

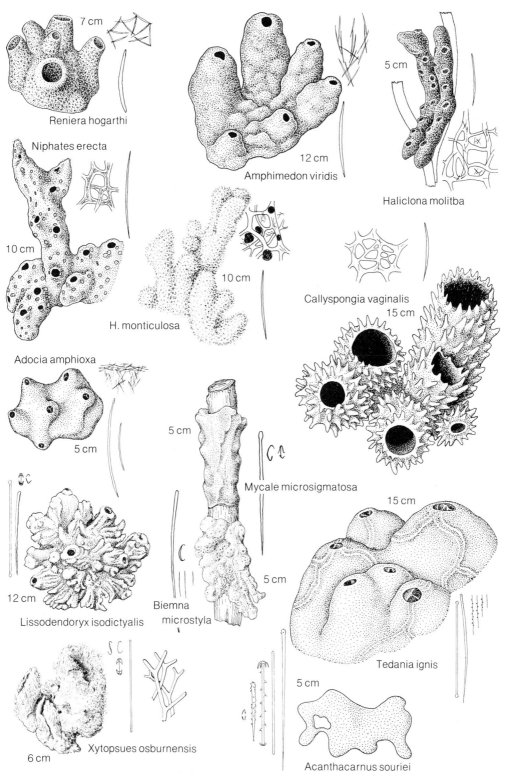

**31 HAPLOSCLERIDA, POECILOSCLERIDA (Sponges 2)**

O. **POECILOSCLERIDA:** Demospongea with skeleton composed of spicules and spongin fibers. Megascleres (monactine and diactine, frequently spiny) and microscleres (varied; chelae, sigmas, microxeas and toxa; no asters). Great variety in shape, structure and color. (About 14 spp. from Bda.)

F. **MYCALIDAE:** Poeciloslcerida with dendroreticulate spicule strands (styles or subtylostyles), interstitial anisocheles and sigmas.

*Mycale microsigmatosa* Arndt: Thin crusts of up to 8 cm × 15 cm area with meandering exhalant canals. Brick red over drab interior. Coating subtidal mangrove roots; inshore waters.

F. **BIEMNIDAE:** Poeciloslcerida with dendroreticulate spicule strands (styles), with sigmas and raphids.

*Biemna microstyla* de Laub.: Thin crusts (8 cm × 15 cm) with uneven, tuberculate surface. Dull yellow to yellow-orange. Common, coating subtidal mangrove roots in inshore waters.

F. **TEDANIIDAE:** Poeciloslcerida with dendroreticulate monactines (styles), ectosomal diactines (tylotes) and roughened oxeote microscleres.

*Tedania ignis* (Duch. & Mich.) (Fire sponge): Encrusting to massive, to 20 cm, with vents located on conical elevations; at some locations erect digitate. Cavernous, soft, easily torn. Bright red ectosome, brownish red inside. Occasionally specimens appear blackish at the surface owing to dense populations of epizoic *Loxosomella tedaniae*. Very common on mangrove roots, other organisms (e.g., other sponges, crabs), rock and mud bottoms of protected inshore water, from low tide level to 4 m. (Color Plates 3.11, 11.6.)

F. **MYXILLIDAE:** Poeciloslcerida with endosomal monactines (styles) in isodictyal reticulation, ectosomal diactines (tylotes), and isochelae and sigmas.

*Lissodendoryx isodictyalis* (Carter) (Garlic sponge): Massive amorphous to lobate, with meandering surface convolutions. Size to 20 cm. Spongy compressible. Golden yellow, frequently with tinges of bluish green. Tendency to incorporate large quantities of foreign matter, e.g. sand, rubble, algae, sea grass blades. Common on mangrove roots and sediment bottoms of inshore waters, 1-3 m. (Color Plate 3.7.)

F. **DESMACIDONIDAE:** Poeciloslcerida with plumoreticulate arrangement of diactine megascleres (strongyles), chelate and sigmoid microscleres.

*Xytopsues osburnensis* (George & Wilson): Genus packed with foreign materials.—Species rounded-massive, some oscular tubes, to 14 cm. Lumpy surface with meandering subsurface canals. Endosome between spicule strands permeated densely with strands of the red alga *Jania*. Soft, mucous, easily torn. Color brownish purple, mottled, with tinges of pink and green. Inshore waters, moderately common, 1 m.

F. **CLATHRIIDAE:** Poeciloslcerida with ascending tracts of monactines (styles, acanthostyles) that are echinated by accessory megascleres. Microscleres are isochelae and toxa.

***Acanthacarnus souriei*** Levi: Genus with echinating acanthostyles and characteristic cladotylotes (rose-stem spicules).—Species a thin, bright red-orange film, to 20 cm, coating and permeating substrate, possibly boring. Common on coral rock, patch reefs, 1-3 m.

### Plate 32

O. **HALICHONDRIIDA:** Demospongea with monaxonid megascleres (oxeas, styles, strongyles) and spongin, without microscleres. Endosomal spicules in confusion (crisscross) with ectosomal organization, or arranged in ascending tracts. (About 5 spp. from Bda.)

F. **HYMENIACIDONIDAE:** Halichondriida with ascending spicule strands.

***Ulosa ruetzleri*** Wiedenmayer (=*Dysidea crawshayi* sensu de Laub.): Genus with flat, ribbon-like dendroreticulate spongin fibers cored by spicules; skin-like conulose ectosome resembling keratose sponges; quantities of foreign materials throughout the choanosomes, including the fibers.—Species with spicules as long styles. Encrusting to massive, to 20 cm, very soft, compressible, limp; bright orange, with darker conspicuous gemmules. On rocks, other sponges (e.g., *Ircinia felix*, *Amphimedon viridis*), mangrove roots, in inshore waters and on patch reefs, 1-3 m. (Color Plate 4.12.)

***U. bermuda*** (de Laub.) (=*Fibulia bermuda*): Genus as above.—Species with spicules as strongyles. Massive, with oscula on rounded lobes or low tubes, to 12 cm. Very spongy, limp, conulose surface, chestnut to blackish brown. Fairly common, inshore waters, 1-3 m.

O. **AXINELLIDA:** Demospongea with great variety of monaxon megascleres, including acanthose forms and spongin. Microscleres absent, except in a few families. Typically with condensed axial, and plumose or plumoreticulate extra-axial skeleton. (3 spp. from Bda.)

F. **AXINELLIDAE:** Axinellida with styles, with or without oxeas, without microscleres.

***Homaxinella rudis*** (Verr.) (Red tree sponge): Genus ramose, with simple spiculation of styles. Erect, treelike branching, to 12 cm, with lumpy, hispid surface. Spongy, but firm elastic, slightly mucous. Clear red. Common on rock in shaded locations, shallow caves, inshore waters, 0.5-2 m. (Color Plate 4.2.)

***Pseudaxinella explicata*** (Wiedenmayer) (=*P. rosacea* sensu de Laub.): Genus massive, axial condensation replaced by ascending plumoreticulate spicule columns.—Species red orange crusts or cushions, to 10 cm, stiffly spongy; strong mucus production when handled. Common on rock in shallow caves, inshore waters, 0.5-2 m. (Color Plate 4.1.)

F. **EURYPONIDAE:** Encrusting Axinellida with principal megascleres perpendicular to the substrate, echinated by secondaries.

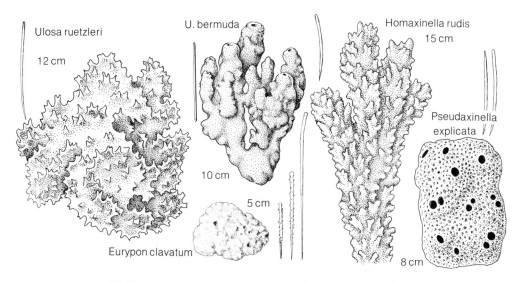

## 32 HALICHONDRIIDA, AXINELLIDA (Sponges 3)

***Eurypon clavatum*** (Bowerbank): Genus with erect tylostyles echinated by acanthostyles of 2 size classes.— Species as very thin, hispid, orange red encrustation of up to 18 cm. Moderately common, coating rocks; inshore waters, 1 m.

## Plate 33

O. **HADROMERIDA:** Demospongea with monactinal megascleres (tylostyles or subtylostyles, rarely styles) organized on a radial pattern. Spongin present, but never as fibers. Microscleres, if present, astrose (asters, spirasters) or oxeote (micro-oxeas). (About 12 spp. from Bda.)

F. **SPIRASTRELLIDAE:** Encrusting or massive Hadromerida with tylostyles and spirasters. The spirasters are mostly stout and form a substantial part of the skeleton.

***Spirastrella mollis*** Verr. (=*S. coccinea* sensu de Laub.): Encrusting, to 14 cm with meandering subsurface canals; yellow-orange, reddish orange to brownish red. Small specimens very common under rocks of reef environments, intertidal to 5 m. (Color Plate 4.14.)

F. **CLIONIDAE:** Papillate, encrusting or massive Hadromerida, with tylostyles and spirasters or amphiasters, some species with oxeas or raphids. Microscleres constitute only a small portion of the skeleton. All species excavate limestone, at least in the early stages of their life cycles.

***Spheciospongia othella*** de Laub.: Large, massive, to 25 cm, boring when young; with robust tylostyles and minute spirasters and amphiasters with compound spines. Grayish brown to black. Common in and on rocks of inshore waters, intertidal to 2 m. Sometimes associated with snapping shrimps (*Alpheus cylindricus*) and barnacles (*Membranobalanus declivis*).

***Cliona caribbaea*** Carter (non *C. caribboea* sensu de Laub.): Genus

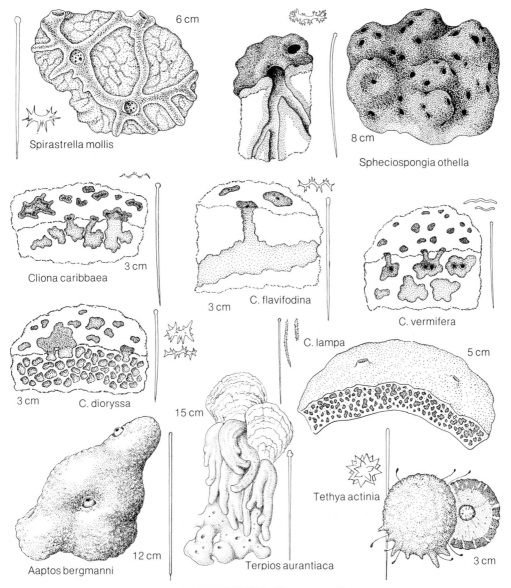

## 33 HADROMERIDA (Sponges 4)

mostly endolithic, with surface papillae; some stages encrust substrate. Boring throughout life cycle.— Species with tylostyles and 1 size class of thin wavy spirasters. Bores small chambers, with tendency of merging inhalant and exhalant papillae to form small (3-20 mm) crusts. Greenish, olive or brown, depending on density of symbiotic zooxanthellae. Very common in shells and rock, inner and outer reefs, 0.5-3 m. (Color Plate 4.11.)

***C. flavifodina*** Rützler: Genus as above.—Species with tylostyles and robust coarsely spined spirasters of 1 size class. Small (5 mm) discrete yellow-brown papillae, large (25 mm) ragged excavations filled with yellow tissue. Very common in inshore waters, open bays and outer reefs, 0.5-10 m.

***C. vermifera*** Hancock: Genus as above.—Species with tylostyles and smooth, undulated rods as microscleres. Vivid orange-red papillae (2 mm), large but discrete chambers (to 8 mm). Very common in rock, inshore waters and outer reefs, 1-5 m.

***C. dioryssa*** (de Laub.) (= *Spirastrella dioryssa*): Genus as above.—Species with tylostyles and 2 size classes of spirasters. Orange to yellow-orange papillae or crusts (confluent papillae, 16 mm), chambered excavations (8 mm). Very common in rock, coral and shell, inshore waters and outer reefs, 0.5-5 m. (Color Plate 4.10.)

***C. lampa*** de Laub.: Genus as above.—Species with tylostyles, spined micro-oxeas and spiny microrhabds. Encrusting and boring large (to several square meters) areas of substrate, brilliant red. Less common yellow variety with same habit and distribution. Bears gemmules. Kills coral and clam substrates (e.g., *Siderastrea*, *Chama*). Very common at shallow locations with strong water currents, inshore waters, 0.5-1 m. (Color Plate 4.8, 9.)

F. **SUBERITIDAE:** Hadromerida without microscleres.

***Aaptos bergmanni*** de Laub.: Genus with styles for spicules. Massive or subspherical, to 20 cm; firm consistency. Dark brown appearance, with rich yellow interior. Moderately common in shaded inshore habitats, 0.5-2 m.

***Terpios aurantiaca*** Duch. & Mich.: Genus characterized by tylostyles with 3-lobed heads. Irregularly massive, lobate, some specimens grape- or finger-shaped with processes fused sideways. To 20 cm. Color yellow, orange or greenish blue. Greenish-bluish tinges owing to symbiotic bacteria. Common on mangrove roots and rocks in enclosed waters, even under low salinity conditions, 0.5-2 m. (Color Plate 4.13.)

F. **TETHYIDAE:** Hadromerida with pronounced radiate structure, with strongyloxeas or styles for megascleres, asters for microscleres.

***Tethya actinia*** de Laub. (Tangerine sponge): Spherical, to 5 cm diameter. With lumpy surface, buds and attachment fibers protruding. Bright orange or green outside, dull orange inside. Very common inshore and on reefs, intertidal to 1 m. (Color Plate 4.3, 4.)

### Plate 34

O. **ASTROPHORIDA:** Demospongea with tetractine and oxeote megascleres in some radial arrangement, and astrose microscleres. Either or all spicule types can be lost. (5 spp. from Bda.)

F. **STELLETTIDAE:** Astrophorida with long-shafted triaenes, and with euasters.

***Myriastra crassispicula*** (Sollas): With one category of oxyasters. Drab, spherical, to 5 cm; attached to pieces of rubble. Common on offshore secondary hard bottoms, 80 m.

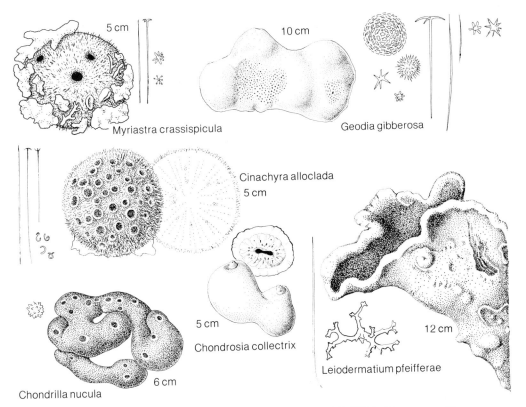

**34 ASTROPHORIDA, SPIROPHORIDA, LITHISTIDA (Sponges 5)**

F. **GEODIIDAE:** Astrophorida with long-shafted triaenes, and with sterrasters forming surface armor.

*Geodia gibberosa* Lam.: Encrusting to amorphous massive, 12 cm. Smooth, partly pitted surface, leathery tough consistency. Color white to dark gray, dependent on light exposure. Moderately common but inconspicuous under rocks or among seaweeds, frequently overgrown by other sponges; inshore waters and outer reefs, intertidal to 10 m.

F. **CHONDROSIIDAE:** Astrophorida with reduced spicule complement.

*Chondrilla nucula* Schmidt (Chicken liver sponge): Spherasters of 1 size class only. Thickly encrusting, to 15 cm, with slippery smooth surface, firm cartilaginous consistency. Brown, greenish brown (from cyanelles) to cream color, depending on light exposure. Very common on rocks, mangrove roots and fouling artificial structures, in caves as well as illuminated habitats, inshore waters and reefs, 0.5-5 m.

*Chondrosia collectrix* (Schmidt): Genus without spicules.—Species reniform or lobate, cartilaginous cushions, 5 cm, off-white, gray to black. Moderately common in caves and under rocks, inshore waters, 0.5-2 m.

O. **SPIROPHORIDA:** Spherical Demospongea with radial skeleton of triaenes and oxeas. Microscleres are contorted sigmas (sigmaspirae). (1 sp. from Bda.)

F. **TETILLIDAE:** Spirophoridae with characteristic inhalant and exhalant depressions (porocalyces).

*Cinachyra alloclada* Uliczka (=*C. cavernosa* sensu de Laub.): Hemispherical or subspherical, to 8 cm, with hispid surface caused by protruding spicules. Firm consistency, ability to contract strongly. Yellow color frequently obscured by trapped sediments. Common in caves and under rocks; inshore waters, 0.5-2 m. (Color Plate 4.5.)

O. **LITHISTIDA:** Demospongea with interlocked desmas forming a hard skeleton. (2 spp. from Bda.)

F. **LEIODERMATIIDAE:** Lithistida with oxeas.

*Leiodermatium pfeifferae* (Carter) (=*Azorica pfeifferae*): Thin erect folded plate, 13 cm, stony hard, ochreous white. Dredged twice, from 800 and 1,900 m.

### Plate 35

CL. **CALCAREA:** Porifera with mineral skeleton composed of calcium carbonate. No distinction between megascleres and microscleres.

O. **CLATHRINIDA:** Calcarea with simple tubular spongocoel lined by choanocytes (asconoid type); choanocyte nucleus in basal position (Only 1 family: Clathrinidae, with about 3 spp. from Bda.)

*Clathrina coriacea* (Montagu) (=*Leucosolenia canariensis* sensu de Laub.): Bright yellow cushions, to 8 cm, made up of trelliswork of ascon tubes with regular triactines packed in the wall. Moderately common in caves, under rocks and on mangrove roots, inshore waters and outer reefs, 0.5-3 m. (Color Plate 4.6.)

O. **LEUCETTIDA:** Calcarea with leuconoid construction; choanocytes restricted to chambers, with nucleus in basal position (Only family: Leucettidae, with 1 sp. from Bda.)

*Leucetta microraphis* (Haeckel) (=*L. floridana*) (Dead man's fingers): Irregular lobate to digitate, up to 50 cm, fragile with rough surface. Color white to shades of pink. Two size classes of triaxons. Very common in shallow caves inshore (1 m), also from secondary hard bottoms in 80 m. (Color Plate 4.15.)

O. **SYCETTIDA:** Calcarea with syconoid or leuconoid canal system (choanocytes lining tubes or chambers); choanocyte nucleus apical, connected to flagellum. (About 5 spp. from Bda.)

F. **SYCETTIDAE:** Sycettida with tubular choanocyte chambers radiating from the atrium.

*Sycon ciliatum* (Fabricius) (=*Scypha ciliata*): Vase-shaped, 12 mm, with spicule crown around osculum. Fragile, white. Varieties of monaxon, triaxon and tetraxonspicules present. Common in caves and under rocks, inshore and near shore, 0.5-1 m. (Color Plate 4.7.)

F. **GRANTIIDAE:** Sycettida with choanocyte chambers and cortical wall.

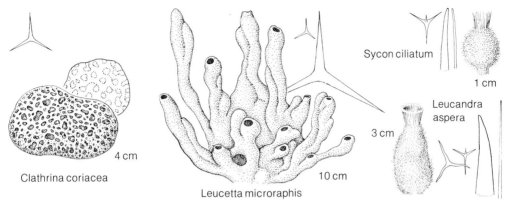

**35 CALCAREA (Sponges 6)**

*Leucandra aspera* (Schmidt) (=*Leuconia aspera*): White, sack-shaped, to 5 cm, tapering toward an oscular spicule crown. Brittle, with hispid surface. Thin and thick oxeas, triaxons and tetraxons. Moderately common in shallow inshore water, 0.5-2 m.

K. RÜTZLER

## Phylum Cnidaria

CHARACTERISTICS: *Polyp- or medusa-shaped METAZOA of generally radial symmetry and with nematocysts (stinging capsules); body wall of 2 cellular layers enclosing a (mostly) non-cellular membrane (mesoglea); the sole body cavity (gastrovascular cavity, or coelenteron) may be subdivided by radially arranged partitions (mesenteries). The larva is typically a planula.*

Distinguish between individuals and colonies, the usually sedentary polypoid stage and the usually free-swimming medusoid stage, and various skeletal materials and types. The mesoglea can be thin-membranous, or massive-gelatinous. Nematocysts, each produced by and contained in an epidermal cell (cnidocyte), are small egg- to spindle-shaped capsules that explode when stimulated, discharging thread-like tubes variously adapted to coil around, stick to or penetrate and poison a prey organism or aggressor. There are 3 basic and a number of subtypes of nematocysts whose occurrence has significance for cnidarian classification (but is of only limited use for the purposes of identification within the framework of this book and therefore not considered in any detail).

The presentation of the 3 classes Hydrozoa (p. 127), Scyphozoa (p. 155) and Anthozoa (p. 159) follows the conservative order; more recent arguments variously propose either Scyphozoa or Anthozoa as the most primitive Cnidaria. In some systems the Cubomedusae, here treated as an order of Scyphozoa, are given the rank of a class, Cubozoa.

## Class Hydrozoa

CHARACTERISTICS: *CNIDARIA occurring as either polypoid or medusoid stages or both, with tetramerous or polymerous radial symmetry (infrequently bilateral), non-cellular mesoglea, and gonads that are usually epidermal. A stomodeum and nematocyst-bearing structures in the gastrovascular cavity, such as septa and gastric cirri, are lacking. Medusae are typically craspedote (i.e., with a velum or shelf of tissue about the opening of the subumbrellar cavity).*

Hydrozoan systematics has been complicated by separate classification systems for polyps (hydroids) and hydromedusae; moreover, hydroid and medusa stages of a given species have often been known by different scientific names. Progress is being made toward uniting different stages of a species under a single name and in eliminating the dual classification. The system used here recognizes 7 orders in the class, all of which occur in Bermuda. The presentation as "Hydroida", "Hydromedusae" and Siphonophora is for convenience only. A group of genera (*Velella*, *Porpita*) often united as Chondrophora (either as a separate order or within Siphonophora) are here considered as Athecata within the hydroid polyps.

Nearly 3,000 species of hydrozoans are known worldwide, 160 of which have been recorded from Bermuda. Of these, 70 are included here.

## HYDROIDA (Hydroid polyps)

CHARACTERISTICS: *(Polypoid HYDROZOA; usually colonial, benthic and sessile; with or without a medusa stage in the life cycle).* From about 1 mm to 2 m or more in height; most reach a maximum size of a few centimeters. Although the living parts are typically soft and rather delicate, hydroids are commonly protected to some degree by a chitinous envelope (perisarc). Hydrocorals secrete a hard calcareous skeleton whose form superficially resembles that of the true corals. Hydranths may be variously colored owing to pigment in the gastrodermis, and the perisarc may be clear, golden or even black. Several species are brown owing to the presence of algal symbionts. Movements consist basically of hydranth and tentacle bending and contraction, and mouth opening. *Velella* and *Porpita*, regarded as colonies by some authors and as solitary polyps by others, float at the surface and are propelled by water currents and the wind. Several genera are known to have species that are luminescent.

About 2,000 species of hydroids are recognized worldwide. Of the 70 species identified to date from Bermuda, 32 are included here.

OCCURRENCE: Largely sessile and epifaunal, although representatives of the group are found in the plankton, the neuston, and even the meiofauna. Bathymetrically, they occur from the intertidal zone to the deep sea. In shallow waters around Bermuda, they are most diverse and abundant in areas swept by tidal currents or subjected to wave action. Some species display a marked substrate preference, whereas others occur on a large number of substrate types. Around Bermuda, hydroids commonly occur on rocks and rock rubble, algae, turtle grass, invertebrates including other hydroids, pilings, floats, buoys and wrecks. Little is known about seasonal cycles and reproductive periodicities of Bermuda hydroids.

Hydroids in shallow waters are best collected by snorkeling or SCUBA diving. Collections from deeper waters may be made using a dredge. Discovery of small species usually requires careful examination of a variety of substrates in the laboratory using a dissecting microscope or magnifying glass. *Velella*, *Porpita*, and species growing on pelagic *Sargassum* may be collected in neuston nets, by dipnetting or by beachcombing.

Some hydroids, most notably *Millepora alcicornis* and *Macrorhynchia philippina*, are capable of stinging humans.

IDENTIFICATION: Based principally upon external characters, species determination requires the use of a stereoscope or microscope. Gonophores must be present or medusae obtained for the identification of some species. Availability of live material is therefore often advantageous.

Hydroids are morphologically quite simple, consisting of a base or other means of attachment, a stem and a terminal hydranth (polyp). In solitary species the proximal portion of the stem may be variously adapted for attachment, or a pedal disc may be present. Colonial species attach to their substrate by a branching and sometimes anastomosing stolon system called a hydrorhiza, by a lamelliform holdfast or by a laminated crust. One, several or many stems may arise from a given hydrorhiza. The hydroid stem may consist of simple unbranched stalks (pedicels), each bearing a single terminal hydranth, or a branched or unbranched main stalk (hydrocaulus) supporting pedicels, ultimate lateral branches (hydrocladia) or both. Stems and branches usually are divided by nodes into internodes of varying length. Internodes may have internal septa. The hydrocaulus may be either monosiphonic (consisting of a single stolon) or polysiphonic (with several fused stolons). Bundles of hydrorhizal tubes band together in some species to form an upright stalk (rhizocaulus) similar in appearance to a polysiphonic hydrocaulus. The hydranth is divisible into an oral end (hypostome or proboscis) bearing a mouth, and a column with tentacles externally and a gastric region internally. An aboral muscular region (sphincter) is usually present in colonial forms. The arrangement and types of tentacles vary. They may occur in 1 or more whorls, or be scattered over the hydranth. A number of different types have been categorized, the more common of which include filiform (thread-shaped), capitate (with a terminal knob) and moniliform (with a series of knobs). Tentacles are greatly reduced or lacking in a few hydroids. The hydranths of a given colony are interconnected by a tube of living tissue called the coenosarc. The coenosarc is usually protected and supported by a non-living, chitinous sheath, the perisarc, which may be ringed or annulated. In *Millepora*, a calcareous skeleton (coenosteum), perforated by openings (larger gastropores and smaller dactylopores), is secreted by the coenosarc. Hydranths may be more or less protected by a chitinous hydrotheca. When present, the hydrotheca is one of the most useful characters for identification. It displays great diversity in form from one species to another, yet is relatively constant in shape within a given species. Hydrothecae may be sessile (without a stalk) or pedicellate (with a stalk), well developed or reduced, smooth or ribbed, and vary from radially to bilaterally symmetrical and from bell-shaped to tubular. Some hydrothecal structures to look for include the diaphragm (basal circular shelf), operculum (lid over the hydrothecal aperture), intrathecal ridges and intrathecal teeth (internal partitions or flanges), puncta (internal dot-like thickenings) and marginal teeth. Hydroids are usually polymorphic; included among the various types of polyps are gastrozooids (feeding polyps), gonozooids (generative polyps) and dactylozooids (protective polyps). The latter include nematophores, which have fixed or movable nematothecae. Reproductive individuals (gonophores) may arise from the hydrorhiza, stem, branches, pedicels, hydranths or a hollow stalk (blastostyle). In about 1/3 of all known hydroid species, gonophores develop into free medusae; they remain fixed in the remainder. Fixed gonophores vary in degree of reduction from medusa-like to highly atrophied, and in a few they are reduced to gonads in the body wall. In thecate hydroids, the gonophores are protected by a chitinous gonotheca. The gonophores of some Plumulariidae are protected by modified hydrocladia or hydrocladial appendages (phylactocarpia); these may be elaborated into basket-shaped structures called corbulae, as in *Aglaophenia*.

Hydroids may be anesthetized in chloral hydrate or magnesium sulfate before fixing and preservation; this is usually more important for athecate than thecate species. Fix in Bouin's or Zenker's fluid, or use 4% formalin. Preserve in 70% alcohol. Some specimens may require treatment with a clearing reagent. Hydroids are best examined as wet mounts.

BIOLOGY: Most hydroids are dioecious, with individuals or single colonies ordinarily producing medusae or fixed gonophores of 1 sex only. In species with fixed gonophores, the egg is typically fertilized in the female gonad and retained to an advanced stage of development. Eggs of medusae are brooded in a number of species but in most are shed into the water. The number and size of the eggs produced vary greatly from one species to another; eggs are usually larger in size and fewer in number if they are retained. Some hydroids grow rapidly and form large colonies, making them important as fouling organisms. Life spans are poorly known. Hydranths of some species (*Obelia*) undergo regular regression-replacement cycles, whereas others (*Tubularia*) evidently do not. Although colonies of certain species may live for only a few weeks or months, colonies of *Millepora* may live for several years. Hydroids have well-known powers of regeneration. Some, especially those in temperate and polar areas, undergo distinct seasonal cycles of activity and dormancy. All are carnivorous, subsisting on a variety of planktonic and benthic prey; only a few are parasitic. Predators include nudibranchs and pycnogonids; certain larval pycnogonids, as well as protozoans, are parasitic on hydroids.

**Plate 36**

DEVELOPMENT: Cleavage of the zygote typically leads to the formation of an elongate, ciliated planula larva that ordinarily attaches within a few hours or days, giving rise to a new hydroid. In some species, the planula develops into a tentaculate actinula larva or even into a polyp before being released from the gonophore. The distinctive larva of *Velella* is called a rataria. Colonies arise by growth and asexual budding from the metamorphosed larva. Medusa buds, sessile gonophores, frustules, buds and some resistant cysts also arise asexually.

Planulae may be taken occasionally in plankton tows, and can also be obtained from the fertilized eggs of hydrozoans maintained in the laboratory.

REFERENCES: Despite its obsolete nomenclature and systematics, the most comprehensive taxonomic reference currently available on hydroids of the western North Atlantic is that of FRASER (1944). Monographs by VAN GEMERDEN-HOOGEVEEN (1965) and VERVOORT (1968) are useful, not only because of the affinity of the local fauna with that of the Caribbean, but also because the species are discussed within the context of a modern classification system. Another useful reference is the monograph on the hydroids of southern Africa by MILLARD (1975).

Approximately a dozen papers containing taxonomic information on the hydroids of Bermuda have been published; the 2 most important are by CONGDON (1907) and BENNITT (1922). No modern survey is available yet.

**Plate 37**

O. **ATHECATA:** Hydrozoa with hydranths lacking a true hydrotheca; tentacle arrangement variable. Reproductive polyps, if present, lacking gonothecae; fixed gonophores or free medusae.

S.O. **CAPITATA:** Athecata with some tentacles capitate in the larva, the adult or both; gonophores usually on hydranth.

F. **TUBULARIIDAE:** Solitary or colonial Capitata; hydranths with 2 whorls of filiform tentacles in adult; hydrocaulus with firm perisarc; anchoring filaments seldom present. Fixed gonophores or free medusae. (1 sp. from Bda.)

*Ectopleura pacifica* Thornely: Colonies in tangled masses to 5 cm high; hydrocaulus unbranched or irregularly branched, perisarc smooth or wrinkled but not annulated; tentacles about 20 in each of 2 whorls, proximal ones larger and longer than distal. Medusa buds arising between 2 tentacular whorls; tentacle bulbs 4, tentacles 2, opposite. Occasional, inshore on floats and offshore on buoy chains.

F. **VELELLIDAE:** Highly specialized, floating Capitata with or without an upright sail. (2 spp. from Bda.)

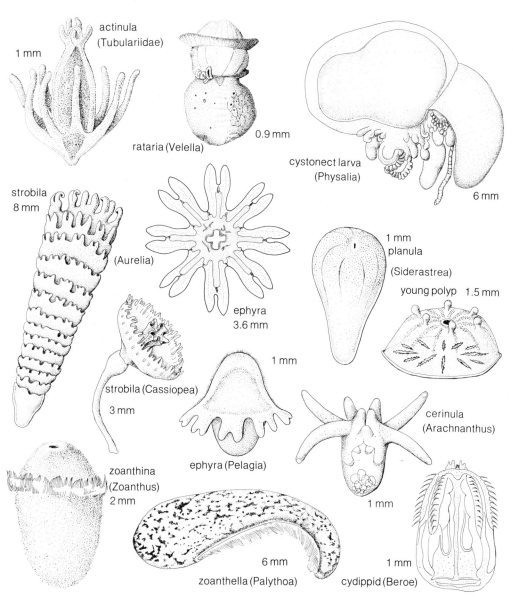

**36 CNIDARIA and CTENOPHORA: Development**

***Velella velella*** (L.) (By-the-wind sailor): Float oval, to 10 cm wide with an upright, triangular sail; specimens in mirror images. Margin soft, central region with concentric air chambers; undersurface with a central gastrozooid ("proboscis") surrounded by gonozooids ("blastostyles") and dactylozooids ("tentacles"). Medusae free, arising on gonozooids. A sporadic, oceanic species, occasionally beached in large numbers (together with its predators *Janthina* and *Glaucus*).

***Porpita porpita*** (L.): Float circular, to 3 cm wide, flat or slightly arched. Margin soft, central region with concentric air chambers, sail lacking; undersurface with a central gastrozooid surrounded by gonozooids and dactylozooids; tentacles with 3 longitudinal rows of nematocyst batteries. Medusa free, arising on gonozooids. A sporadic, oceanic species.

F. **HALOCORDYLIDAE:** Capitata with upright, branched colonies; perisarc firm, tubular. Hydranths with an oral whorl of capitate tentacles, an aboral whorl of well-developed filiform tentacles, and additional capitate tentacles between the two. Free medusae where known. (1 sp. from Bda.)

*Halocordyle disticha* (Goldfuss) (=*Pennaria tiarella*): Colonies large, to 15 cm or more high; hydrocaulus monosiphonic with dark perisarc; branches alternate; hydranths on annulated pedicels, club-shaped with a proximal whorl of filiform tentacles and 2-5 whorls of capitate tentacles. Medusa buds arising just distal to filiform tentacles. Common to abundant inshore in shallow water and offshore on buoy chains. Frequently used in experimental work.

F. **SPHAEROCORYNIDAE:** Capitata with colonies having unbranched pedicels or sparingly branched hydrocauli; perisarc firm. Hydranths pyriform with a conical proboscis; tentacles capitate, in a scattered whorl around bulbous hydranth base. Gonophores are free medusae. (1 sp. from Bda.)

*Sphaerocoryne bedoti* Pictet: Hydrocaulus embedded in sponge; perisarc thin, terminating below hydranth. Hydranths bulbous, hypostome extensible, tentacles capitate, scattered, often 25 or more. Medusa buds arising just distal to tentacles; tentacle bulbs 4, equally developed. On sponges in shallow inshore waters.

F. **CLADONEMATIDAE:** Capitata with colonies having short stems arising from a hydrorhiza; perisarc firm. Hydranths with an oral whorl of capitate tentacles and usually with an aboral whorl of vestigial filiform tentacles. Free medusae. (1 sp. from Bda.)

*Cladonema radiatum* Dujardin: Colonies minute. Hydranths elongate, arising singly from hydrorhiza on short pedicels; tentacles in 2 whorls, oral tentacles 4, capitate, aboral tentacles 4-8, filiform. Medusa buds arising just distal to filiform tentacles; tentacles of medusa branched. Fairly common on *Thalassia*.

F. **ZANCLEIDAE:** Capitata with hydranths having scattered tentacles; tentacles all capitate, all filiform, or of both types. Free medusae. (1 sp. from Bda.)

*Zanclea costata* Gegenbaur: Colonies minute. Hydranths cylindrical to oval, arising singly from hydrorhiza on short pedicels; tentacles short, scattered, all capitate, rather numerous. Medusa buds arising among or just below tentacles. Common to abundant on pelagic *Sargassum*.

F. **MILLEPORIDAE:** Capitata with encrusting calcareous skeleton having pores containing polyps; pores without styles (basal spines); tenta-

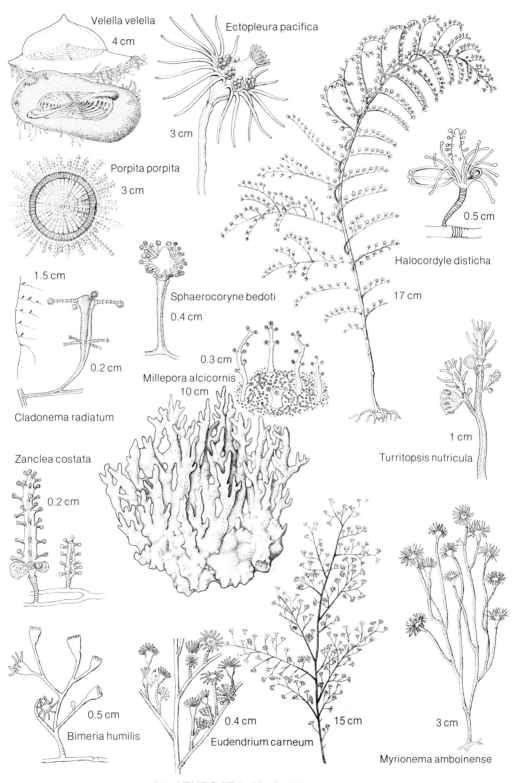

## 37 ATHECATA (Hydroid polyps 1)

cles capitate. Free medusae. (1 sp. from Bda.)

***Millepora alcicornis*** L. (Fire coral, sea ginger): Colonies massive, calcareous, forming horn-shaped, upright branches or plates; coenosteum perforated by larger gastropores containing gastrozooids, usually surrounded by smaller dactylopores containing dactylozooids. Gastrozooids short, stout, with an oral whorl of 4-6 capitate tentacles; dactylozooids long, slender, mouthless, tentacles capitate, scattered. Medusae free but degenerate, formed in special chambers (ampullae). Common to abundant in shallow water, on inner ledges and outer reefs. Color variable, usually yellowish-brown due to algal symbionts. A severe stinger! (Color Plate 8.11.)

S.O. **FILIFERA:** Athecata with all tentacles filiform.

F. **CLAVIDAE:** Filifera with elongate hydranths; hypostome conical; tentacles scattered. Fixed gonophores or free medusae. (3 spp. from Bda.)

***Turritopsis nutricula*** McCrady: Colonies small, usually 1 cm or less high; hydrocaulus branched or unbranched, arising from a tangled rhizocaulus; perisarc mostly smooth. Hydranths elongate with scattered filiform tentacles. Medusa buds arising on stems or pedicels below hydranths. Common inshore in shallow water and offshore on buoy chains.

F. **BOUGAINVILLIIDAE:** Filifera with hydranths of moderate length; hypostome conical; tentacles in one whorl. Fixed gonophores or free medusae. (4 spp. from Bda.)

***Bimeria humilis*** Allman: Colonies a few millimeters high, with single pedicels or short hydrocauli arising from a stolon; perisarc wrinkled or annulated. Hydranths oval, tentacles filiform, covered with perisarc basally. Gonophores fixed, arising on stems or pedicels, or from the stolon. Common inshore in shallow waters and offshore on wrecks and buoy chains.

F. **EUDENDRIIDAE:** Filifera with cup-shaped hydranths; hypostome flexible, club-or funnel-shaped; tentacles in one whorl. Gonophores fixed. (4 spp. from Bda.)

***Eudendrium carneum*** Clarke: Colonies large, to 10 cm or more high; hydrocaulus and main branches polysiphonic; branches and pedicels annulated basally. Hydranths with 25-30 tentacles. Gonophores fixed, on atrophied hydranths; ♂ gonophores 4-5-chambered, ♀ gonophores covered with thick perisarc, often zigzag on pedicel. Common to abundant, in shallow inshore waters. Color reddish when alive. All records of *E. ramosum* from Bermuda are probably based on specimens of this species.

***Myrionema amboinense*** Pictet (=*Eudendrium hargitti*): Colonies to 5 cm high; hydrocaulus monosiphonic, branches few, irregularly placed, annulated basally. Hydranths with 35-55 tentacles in 2 closely appressed whorls. Gonophores fixed, on entire or atrophied hydranths, ♂ gonophores 3-chambered, ♀ gonophores ovoid.

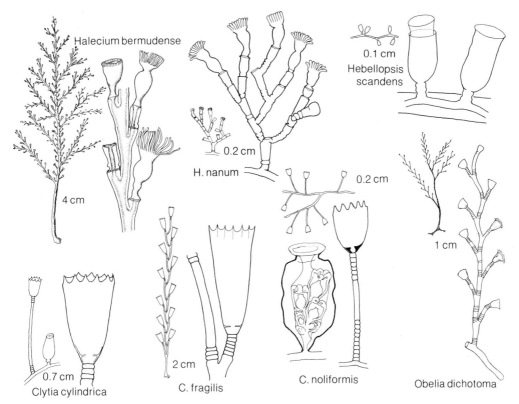

**38 HALECIIDAE, LAFOEIDAE, CAMPANULARIIDAE (Hydroid polyps 2)**

Common to abundant in shallow, quiet inshore waters. Color rusty-brown owing to algal symbionts.

O. **THECATA:** Hydrozoa with hydranths usually having a hydrotheca; tentacles in one whorl; reproductive polyps, if present, with gonothecae. Fixed gonophores or free medusae.

**Plate 38**

F. **HALECIIDAE:** Thecata with hydrothecae reduced, not fully accommodating contracted hydranth; margin entire, operculum absent, puncta present. Gonophores fixed. (7 spp. from Bda.)

*Halecium bermudense* Congdon: Genus with erect hydrocaulus; hydrothecae on hydrocaulus and branches. Nematophores lacking; gonophores sporosacs.— Species with colonies to 8 cm high; hydrocaulus polysiphonic basally; branches occasionally rebranched. Hydrothecae shallow, separated from internode by diaphragm, sometimes renovated, margin flaring but little, puncta present. Gonothecae of ♂ slender, truncate, tapering gradually to base, of ♀ obovate with large lateral aperture, aperture margin wavy. Common to abundant, especially inshore in shallow water.

***H. nanum*** Alder: Genus as above. —Species with small colonies, with single pedicels or short hydrocauli to 3 mm high. Hydrothecae occasionally renovated, shallow, basal diaphragm present, margin flaring but little, puncta present. Gonothecae of ♂ oval, of ♀ sac-shaped with terminal aperture. Common, especially on pelagic *Sargassum*.

F. **LAFOEIDAE:** Thecata with hydrothecae tubular or bell-shaped; margin entire, operculum lacking, diaphragm usually absent. Fixed gonophores or free medusae. (8 spp. from Bda.)

***Hebellopsis scandens*** (Bale): Colonies very small with short, unbranched pedicels. Hydrothecae cylindrical, sometimes curved but not contorted, margin flaring, occasionally renovated, diaphragm present. Gonothecae on stolon, club-shaped, with terminal aperture. Occasional, in shallow inshore and deeper offshore waters, on other hydroids.

F. **CAMPANULARIIDAE:** Thecata with hydrothecae more or less bell-shaped; margin entire or toothed, operculum lacking, diaphragm present. Fixed gonophores or free medusae. (8 spp. from Bda.)

***Clytia cylindrica*** L. Agassiz: Genus with stolonial or branching colony. Hydrotheca usually with toothed margin; diaphragm present. Free medusae with 4 tentacles at liberation.—Species with small colonies with unbranched pedicels. Hydrothecae cone-shaped, constricted at diaphragm, marginal teeth 9-12, triangular. Gonothecae cylindrical, terminal aperture large. Common to abundant, in shallow inshore waters.

***C. fragilis*** Congdon (=*Laomedea tottoni*): Genus as above.—Species with colonies to 2 cm high; hydrocaulus monosiphonic, zigzag. Hydrothecae deep, diaphragm thin, margin scalloped in cross section; teeth 10-16, acute. Gonothecae ovoid, truncate distally. Occasional, in shallow inshore waters.

***C. noliformis*** (McCrady): Genus as above.—Species forming small colonies with unbranched pedicels. Hydrothecae cup-shaped, perisarc often thickened basally, marginal teeth 10-12, rounded. Gonothecae on stolons, broadly oval, with terminal collar. Common to abundant on pelagic *Sargassum*.

***Obelia dichotoma*** (L.): Colonies to 2 cm high; hydrocaulus monosiphonic, zigzag. Hydrothecae cup-shaped, diaphragm thin, margin entire. Medusae with 16 or more tentacles at liberation; gonothecae oval, with terminal collar. Common inshore in shallow waters and offshore on buoy chains.

**Plate 39**

F. **SERTULARIIDAE:** Thecata with sessile or pedicellate, radially or bilaterally symmetrical hydrothecae, margin entire or toothed; operculum usually present, consisting of 1-4 components, diaphragm usually present. Hydranth when retracted with or without a pouch (abcauline caecum). Gonophores fixed. (13 spp. from Bda.)

***Thyroscyphus marginatus*** (Allman) (=*Campanularia marginata*): Genus with pedicellate hydrothecae; margin entire or with 4 teeth; operculum of 4 valves, persistent or shed early; diaphragm present; hydranth with an annular fold basally.—Species forming colonies to 20 cm or more in height; hydrocaulus monosiphonic, more or less regularly branched. Hydrothecae large, on short pedicels, tumbler-shaped; margin entire, with a ring-shaped thickening. Gonothecae cylindrical, with transverse lobes. Common, in shallow inshore and deeper waters offshore.

***T. intermedius*** Congdon: Genus as above.—Species forming very small colonies, with single pedicels or short hydrocauli arising from hydrorhiza. Hydrothecae fusiform, rugose, margin with 4 teeth, operculum of 4 flaps. Gonothecae not known. Common, on *Thalassia* and pelagic *Sargassum*.

***Dynamena disticha*** Bosc (=*Sertularia cornicina*): Genus with sessile hydrothecae; 3 unequal marginal teeth; operculum of 2 valves; abcauline caecum lacking.—Species forming colonies usually 1-3 cm high; hydrocaulus monosiphonic, unbranched. Hydrothecae cylindrical, curved outward distally, in opposite pairs, adnate frontally, separated in back, margin with 1 small median and 2 large lateral teeth. Gonothecae oval, rugose, usually on stolon. Common inshore in shallow waters on reefs and banks offshore, and on pelagic *Sargassum*. Largely epiphytic.

***D. quadridentata*** (Ellis & Solander): Genus as above.—Species forming colonies usually 1 cm or less in height; hydrocaulus monosiphonic, infrequently branched. Hydrothecae in 1-6 contiguous pairs between nodes, margin with 1 small median and 2 large, lateral teeth, intrathecal teeth present. Gonothecae on hydrocaulus, broadly oval, rugose. Common, in shallow inshore waters, and on pelagic *Sargassum*. Largely epiphytic.

***D. crisioides*** Lamouroux: Genus as above.—Species forming colonies to 10 cm high; hydrocaulus monosiphonic with alternate branches. Hydrothecae tubular, subopposite, deeply immersed; operculum of 3 flaps. Gonothecae on hydrocaulus and branches, oval, distal end with distinct neck. Abundant, intertidal to 1 m on hard substrates, inshore.

***Sertularella conica*** Allman: Genus with sessile hydrothecae; 4 equal marginal teeth; operculum of 4 valves, abcauline caecum present.—Species forming colonies to 2 cm or more in height; hydrocaulus monosiphonic, irregularly branched. Hydrothecae alternate, fusiform, rugose, marginal teeth sharp, intrathecal teeth well developed. Gonothecae oval or pear-shaped, rugose; orifice small. Occasional inshore, common offshore on banks and on shallow reefs.

***S. speciosa*** Congdon: Genus as above.—Species forming colonies to 20 cm high; hydrocaulus polysiphonic basally, branches alternate. Hydrothecae alternate, short, cylindrical, largely or entirely immersed, with thickened ridge just below margin; marginal teeth low. Gonothecae cylindrical, tapering

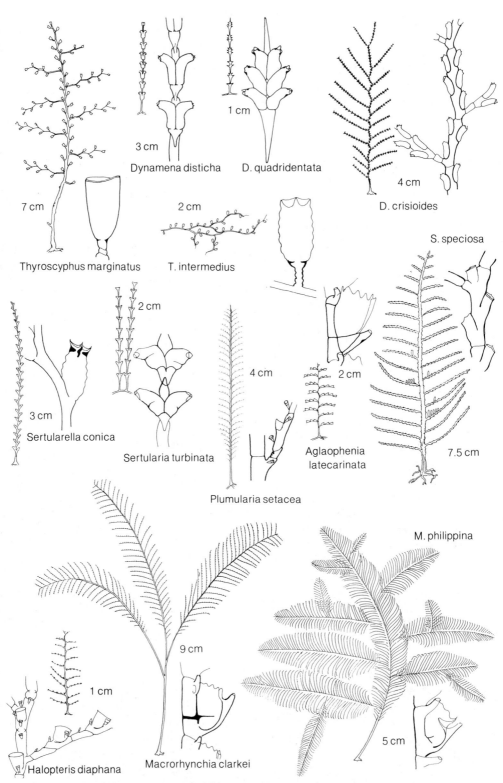

**39 SERTULARIIDAE, PLUMULARIIDAE (Hydroid polyps 3)**

basally, ridged longitudinally, distal end lobed. At cave entrances; rare.

***Sertularia turbinata*** (Lamouroux): Colonies to 2 cm high; hydrocaulus monosiphonic, unbranched. Hydrothecae in opposite pairs, close or adnate frontally, separated in back; outer wall with a median indentation marking location of internal ridge, margin with 1 small median and 2 large lateral teeth; operculum present. Gonothecae rugose, flattened, plano-convex in cross section on hydrocaulus. Common, inshore in shallow waters and offshore on banks and buoy chains.

F. **PLUMULARIIDAE:** Thecata with sessile, often bilaterally symmetrical hydrothecae; margin entire or toothed, operculum absent, nematothecae usually present. Gonophores usually fixed (may be liberated in *Macrorhynchia*). (16 spp. from Bda.)

***Halopteris diaphana*** (Heller): Colonies to 2 cm high; hydrocaulus monosiphonic, unbranched, zigzag, bearing hydrothecae and alternate hydrocladia; hydrocladia with alternating athecate and thecate internodes. Hydrothecae cup-shaped, distal 1/3 free, margin entire; nematothecae 2-chambered, trumpet-shaped. Gonothecae cornucopia-shaped, distal end truncate. Common inshore in shallow waters and on pelagic *Sargassum*.

***Plumularia setacea*** (L.): Colonies to 5 cm high; hydrocaulus monosiphonic, straight, usually unbranched; hydrocladia alternate with alternating athecate and thecate internodes. Hydrothecae small, margin entire; nematothecae 2-chambered, trumpet-shaped. Gonothecae elongate with long neck, aperture circular. Common offshore on fish trap lines.

***Aglaophenia latecarinata*** Allman: Colonies to 2 cm high; hydrocaulus monosiphonic, unbranched; hydrocladia alternate, internodes thecate with internal septa. Hydrothecae cone-shaped, margin with 9 teeth, intrathecal ridge distinct, straight; nematothecae 1-chambered, immovable, adnate to hydrotheca, distal end of median nematotheca free. Corbulae short, stout, with 7-10 pairs of ribs. Common to abundant on pelagic *Sargassum*.

***Macrorhynchia clarkei*** (Nutting): Genus with paired, immovable lateral nematothecae, gonothecae protected by phylactocarpia consisting of modified hydrocladia.—Species forming colonies to 35 cm high, alternately or dichotomously branched; hydrocaulus and branches polysiphonic basally; hydrocladia alternate, internodes thecate with internal septa. Hydrothecae cup-shaped, margin with 9 teeth, intrathecal ridge distinct, straight; nematothecae 1-chambered, adnate to hydrotheca, distal end of median nematotheca free. Gonothecae disc-shaped. Occasional inshore, common to abundant in deeper waters offshore.

***M. philippina*** (Kirchenpauer) (=*Lytocarpus philippinus*): Genus as above.—Species forming colonies to 20 cm high; hydrocaulus and branches polysiphonic basally, alternately or irregularly branched; hydrocladia alternate, internodes thecate with internal septa. Hydrothecae with distinct lip near margin; margin with low, rounded teeth; nematothecae 1-chambered, adnate to hydrotheca, free distal end of median nematotheca long. Gonothecae disc-shaped. Common inshore in shallow waters and on banks offshore. Mildly venomous.

D. R. CALDER

# HYDROMEDUSAE

CHARACTERISTICS: *(Medusoid HYDROZOA; usually solitary and planktonic; with or without a polyp stage in the life cycle).* Most hydromedusae are little more than a few millimeters in size when fully grown; others attain a diameter of several centimeters. They vary in consistency from soft and gelatinous to firm and gristle-like. Although the jelly is usually transparent, the manubrium, tentacles, tentacle bulbs, and gonads of living medusae may be delicately and variously colored. Unfortunately, these colors usually fade in preserved specimens. Certain deep-sea forms are heavily pigmented. Bioluminescence occurs in some species. Swimming is accomplished by rhythmic pulsations of the umbrella resulting from the opposing forces of muscular contraction and mesogleal elasticity. The velum is important in enhancing swimming velocity and in turning. A number of species are known to undergo vertical migrations daily.

Of some 700 known species, about 45 have been recorded from Bermuda; 25 of these are included here. Neritic medusae appear to be less well known in the Bermuda area than the oceanic species.

OCCURRENCE: Planktonic, in the neritic (nearshore) zone and in oceanic waters from the surface to the deep sea. Some neritic genera (*Cladonema, Staurocladia*) are more or less adapted to a benthic existence; a few (*Halammohydra*) are interstitial in sand. Oceanic species are categorized as epipelagic (from the surface to 200 m) or bathypelagic (200-3,500 m). Holoplanktonic hydromedusae (those that are planktonic throughout their life cycle) are normally far better represented in plankton collections from Bermuda than are meroplanktonic forms (those that are planktonic for only part of their life cycle). Marked seasonal changes in abundance have been noted for some species. Hydromedusae, along with several other zooplanktonic taxa, may be useful indicators of the movement of certain water masses.

Although they may be dipnetted and are occasionally taken in trawls, hydromedusae are usually collected with plankton nets. When collecting medusae, consideration must be given to net diameter, mesh size and towing speed. Hydromedusae are very delicate, and nets should be outfitted with a terminal bucket. To obtain specimens in best condition, collections should be left undisturbed in the plankton bucket for a few minutes after retrieval before preservative is added. Collections and observations may also be made using SCUBA.

IDENTIFICATION: Hydromedusae are identified largely on the basis of structures that are apparent without dissection. However, a microscope is necessary for the examination of some morphological features.

Hydromedusae display tetramerous or polymerous radial symmetry. The swimming bell (umbrella) has 2 surfaces, an outer, convex exumbrella and an inner, concave subumbrella. Observe the shape of the umbrella and the amount of jelly (mesoglea) present. The umbrella margin is entire in most species but is lobed in the Narcomedusae. The subumbrellar cavity is partially enclosed by an inward-projecting ectodermal shelf called the velum. A tube-like manubrium (mouth stalk) hangs from the center of the subumbrellar cavity. Occasionally the manubrium is located at the end of a cone-shaped projection, the gastric peduncle. The mouth, at the distal end of the manubrium, may be simple or bordered by clusters of nematocysts, lips, lobes, or oral tentacles. Note the number and arrangement of the radial canals extending from the manubrium to the periphery of the umbrella, where they usually meet a continuous ring canal circling the margin of the umbrella. In some genera (*Olindias, Geryonia, Liriope*), blind centripetal canals radiate centrally from the ring canal. Observe the shape of the gonads, and discern whether they occur on the radial canals, manubrium, or both. Study the structure, number and arrangement of the marginal tentacles. These tentacles are either solid or hollow, are typically very extensile, and are usually heavily armed with stinging organelles or nematocysts. They vary in shape from filiform (thread-shaped) or claviform (club-shaped) to capitate (with a terminal knob) or moniliform (with a series of knobs). Marginal tentacles are usually dilated basally, forming variously shaped tentacle bulbs. In some genera (*Bougainvillia*), several tentacles arise from each tentacle bulb. The tentacle bulbs of *Aequorea* and some others have openings to the exterior called excretory pores. Very small tentacle-like structures (cirri) occur on the margin of some medusae (*Lovenella*). Pigmented light receptors (ocelli) may be present on the tentacle bulbs. Other sense organs (statocysts) may occur on the umbrella margin. Ectodermal statocysts (marginal vesicles) develop in the velum as small pockets, and are categorized as open or closed depending upon whether or not their top is sealed off by velar tissue. One or more calcareous concretions (statoliths) occur in these vesicles. Endodermal statocysts (sensory clubs) arise from the ring canal and may be free or enclosed. Other club-like structures called cordyli are found on the margin of certain Leptomedusae. Some Narcomedusae have exumbrellar tracks of bristle-bearing ectodermal cells known as otoporpae.

For anatomical purposes, hydromedusae may be fixed and preserved in a 5% solution of formalin in seawater. If desired, specimens may first be anesthetized in a solution of magnesium chloride or chloral hydrate. Alcohol causes medusae to shrink and should be avoided unless statocysts or nematocysts are of particular interest. For histological purposes Bouin's or weak Fleming's solution may be used.

BIOLOGY: Most hydromedusae are dioecious, but microscopic examination of the gonad is usually necessary to differentiate the sexes. Eggs of most planktonic hydromedusae are fertilized after being shed into the water, although brooding occurs in some. Life expectancy varies; a few large medusae (*Aequorea*) may live for several months, whereas others (*Halocordyle*) survive for only a few hours. A typical lifespan would probably last about 2 months. Growth is therefore quite rapid. Medusae are carnivorous, often preying heavily upon copepods and other small crustaceans, larval fish and even other species of medusae. Prey is usually captured by the marginal tentacles and transferred to the mouth in a variety of ways. Predators include scyphomedusae and other hydromedusae as well as some fishes and sea turtles. A few hydromedusae are parasitic (some larval Narcomedusae may parasitize other hydromedusae) and they in turn harbor few parasites or commensals (some larval trematodes and actinians). Abnormalities, such as in the number of radial canals, are fairly common.

*Plate 36*

DEVELOPMENT: The fertilized egg usually develops into a ciliated, mouthless planula larva. Planulae are typically club-shaped with distinct anterior and posterior ends. They are free-swimming for a period ranging from several hours to several days. The next stage of development depends upon the type of life history manifested by the medusa. In species with alternating polyp and medusa stages, the planula settles to the bottom and metamorphoses into a hydroid; in species without, the planula metamorphoses either directly into a free-swimming medusa or into an actinula-like larva which develops into a medusa. Certain species of hydromedusae (*Cytaeis tetrastyla, Bougainvillia niobe*) may form additional medusae by asexual budding.

REFERENCES: The works of MAYER (1910a, b), RUSSELL (1953) and KRAMP (1959, 1961) are particularly useful references.

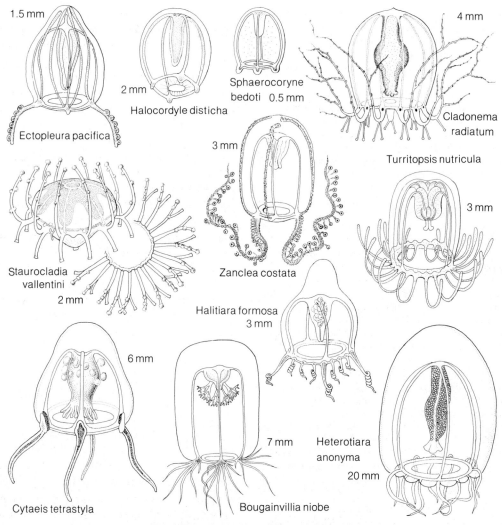

**40 ANTHOMEDUSAE (Hydromedusae 1)**

The most comprehensive reports dealing specifically with the hydromedusae of Bermuda include the papers by FEWKES (1883) and BIGELOW (1938), as well as the zooplankton investigations by MOORE (1949) and MORRIS (1975).

*Plate 40*

O. **ATHECATA** (ANTHOMEDUSAE): Hydrozoa with umbrella margin entire; radial canals present. Gonads on manubrium, infrequently extending along radial canals; ocelli present or absent; statocysts lacking.

S.O. **CAPITATA:** Anthomedusae with simple mouth; gonads typically in 1 or more continuous rings about manubrium.

F. **TUBULARIIDAE:** Capitata with or without exumbrellar nematocyst tracks; manubrium entirely within subumbrellar cavity, oral tentacles lacking. Radial canals 4; marginal tentacles 1-4, simple, solitary; ocelli lacking. (1 sp. from Bda.)

***Ectopleura pacifica*** Thornely (=*E. minerva*): Pear-shaped; apical projection prominent; exumbrella with 8 longitudinal rows of nematocysts; tentacles 2, opposite, each with 6-9 abaxial swellings. Neritic.

F. **HALOCORDYLIDAE:** Capitata with reduced medusae. Radial canals 4; marginal tentacles 4, permanently rudimentary; ocelli present or absent. (1 sp. from Bda.)

***Halocordyle disticha*** Goldfuss: Ellipsoidal, with thin mesoglea; mouth simple, manubrium tubular, bearing gonads; ocelli lacking. A reduced and short-lived medusa, liberated from hydroid in greatest numbers at dusk. Neritic.

F. **SPHAEROCORYNIDAE:** Capitata without oral tentacles; exumbrella with or without longitudinal rows of nematocysts; manubrium typically cruciform in cross section. Radial canals 4; marginal tentacles 2-4; abaxial ocelli present. (1 sp. from Bda.)

***Sphaerocoryne bedoti*** Pictet: Young medusa thimble-shaped, mesoglea thin, exumbrella with scattered nematocysts. Ring canal present; manubrium short, oral tentacles and lips lacking; marginal tentacles rudimentary, tentacle bulbs 4. Neritic.

F. **CLADONEMATIDAE:** Capitata with knobbed oral tentacles. Radial canals variable in number, some simple, others bifurcated; marginal tentacles branched, adhesive organs present; ocelli present. (1 sp. from Bda.)

***Cladonema radiatum*** Dujardin: Dome-shaped, with varying number of radial canals; manubrium with 4-5 capitate oral tentacles. Marginal tentacles 8-10, each with several branches; basal branches with adhesive organs. Neritic, creeping and swimming.

F. **ELEUTHERIIDAE:** Capitata with ring of nematocysts on umbrella margin; oral tentacles lacking. Radial canals variable in number, branched or unbranched; marginal tentacles bifurcating, adhesive organs present; ocelli present (1 sp. from Bda.)

***Staurocladia vallentini*** (Browne): Flattened; marginal tentacles up to 24, lower branch with adhesive organ, upper branch with median nematocyst clusters. Neritic, creeping, fairly common on *Thalassia*.

F. **ZANCLEIDAE:** Capitata with 4 radial canals; when present, marginal tentacles 2 or 4, with stalked nematocyst capsules. Exumbrellar nematocyst patches, oral tentacles, and ocelli present or absent. (2 spp. from Bda.)

***Zanclea costata*** Gegenbaur: Thimble-shaped, with nematocyst patches of varying length between tentacle bulbs and apex; ocelli lacking. Predominantly neritic, although its hydroid is common to abundant on pelagic *Sargassum*.

S.O. **FILIFERA:** Anthomedusae with mouth provided with lips, lobes, oral tentacles or nematocyst clusters. Gonads in longitudinal masses on manubrium.

F. **CLAVIDAE:** Filifera with mouth drawn into 4 lips fringed with a row of nematocysts. Radial canals 4, simple; marginal tentacles solid, numerous; adaxial ocelli present. (1 sp. from Bda.)

*Turritopsis nutricula* McCrady: Thimble-shaped, with relatively thin mesoglea; endodermal cells of radial canals highly vacuolated above manubrium. Marginal tentacles not in clusters. Neritic.

F. **CYTAEIDIDAE:** Filifera with simple, flask-shaped manubrium with unbranched oral tentacles. Radial canals 4; marginal tentacles 4, solid; ocelli absent. (1 sp. from Bda.)

*Cytaeis tetrastyla* Eschscholtz: Bell-shaped; manubrium large, occasionally bearing medusa buds; several oral tentacles inserted about mouth rim. Epipelagic; common.

F. **BOUGAINVILLIIDAE:** Filifera with short manubrium with or without peduncle; oral tentacles simple or branched. Radial canals 4; marginal tentacles either solitary or arising from 4, 8, or 16 marginal bulbs; ocelli present or absent. (2 spp. from Bda.)

*Bougainvillia niobe* Mayer: Thimble-shaped, mesoglea thick, oral tentacles 4, branched 4 times, basal trunk long; manubrium flask-shaped, medusa buds occasionally present, peduncle absent. Gonads 8, adradial; marginal tentacles all alike, in 4 groups; ocelli present. Neritic and epipelagic; common.

F. **PANDEIDAE:** Filifera with manubrium with or without peduncle; mouth with 4 perradial lips, oral tentacles lacking. Radial canals usually 4; marginal tentacles 2, 4, or more, solitary, lacking terminal knob; rudimentary tentacles and ocelli present or absent (3 spp. from Bda.)

*Halitiara formosa* Fewkes: Pyriform, with a solid apical projection; mouth simple, opening cruciform; marginal tentacles of 2 sizes, perradial ones long, intermediate ones short, tightly coiled; ocelli lacking. Neritic.

F. **CALYCOPSIDAE:** Filifera with manubrium lacking peduncle; mouth with 4 lips. Radial canals 4 or 8, simple or branched, centripetal canals occasionally present; marginal tentacles 8 or more, solitary, lacking basal bulbs, terminal knob present; rudimentary tentacles and ocelli present or absent. (1 sp. from Bda.)

*Heterotiara anonyma* Maas: Bell-shaped; mesoglea thick; oral tentacles lacking. Radial canals 4, centripetal canals lacking; marginal tentacles 8-12. Neritic and epipelagic.

**Plate 41**

O. **THECATA** (LEPTOMEDUSAE): Hydrozoa with umbrella margin entire; radial canals present. Gonads on radial canals, occasionally contiguous with manubrium; ocelli infrequent; statocysts, when present, as

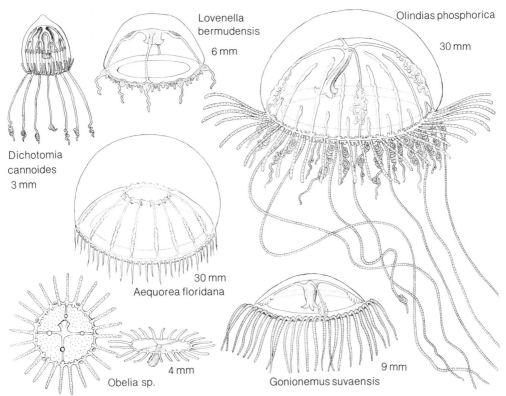

**41 LEPTOMEDUSAE, LIMNOMEDUSAE (Hydromedusae 2)**

marginal vesicles; cordyli sometimes present.

F. **DIPLEUROSOMATIDAE:** Leptomedusae with narrow manubrium base, gastric peduncle present or absent. Radial canals 3, 4 or more, either branched or irregularly arranged; marginal appendages sometimes present, marginal and lateral cirri lacking; ocelli occasionally present, statocysts lacking. (1 sp. from Bda.)

***Dichotomia cannoides*** Brooks: Bell-shaped, with a conical apex. Radial canals 4, each branched dichotomously several times; marginal tentacles up to 50, of different sizes; gonads on radial canals. Neritic.

F. **LOVENELLIDAE:** Leptomedusae with small manubrium, gastric peduncle lacking. Radial canals 4; tentacle bulbs with lateral cirri, excretory pores lacking; marginal vesicles closed. (2 spp. from Bda.)

***Lovenella bermudensis*** (Fewkes): Hemispherical; manubrium short, fairly wide; gonads oval, on radial canals near manubrium. Marginal tentacles 8; up to 6 rudimentary bulbs between successive tentacles; cirri beside tentacles and some rudimentary bulbs; marginal vesicles 16 or more, each with 1 concretion. Neritic.

F. **AEQUOREIDAE:** Leptomedusae with broad manubrium, distinct gastric peduncle lacking. Radial canals numerous, sometimes branched; tentacle bulbs with excretory pores, lateral and marginal cirri lacking; marginal vesicles closed. (2 spp. from Bda.)

***Aequorea floridana*** (L. Agassiz): Bowl-shaped; mesoglea thick centrally; subumbrellar surface lacking rows of gelatinous papillae; manubrium about 1/5 width of umbrella. Radial canals 16, unbranched; marginal tentacles 80-100, tentacle bulbs fusiform, abaxial spurs and excretory papillae present; marginal vesicles numerous. Neritic.

F. **CAMPANULARIIDAE:** Leptomedusae with small manubrium, gastric peduncle lacking. Radial canals 4; excretory pores and cirri lacking; gonads surrounding radial canals; marginal vesicles closed, 8 or more. (2 distinct spp. from Bda.)

***Obelia*** spp.: Flat; mesoglea thin, velum reduced; manubrium short, quadrate, gonads round, midway along radial canals. Marginal tentacles 16 or more, short; marginal vesicles 8, each with one concretion. Neritic. Medusae of the various species of *Obelia* are virtually inseparable morphologically.

O. **LIMNOMEDUSAE:** Hydrozoa with umbrella margin entire; radial canals present. Gonads on manubrium, or radial canals, or both; ocelli present or absent; statocysts, if present, as enclosed sensory clubs.

F. **OLINDIASIDAE:** Limnomedusae with 4 or 6 unbranched radial canals, centripetal canals present or absent. Gonads on radial canals, or on both radial canals and manubrium; marginal vesicles internal, ocelli lacking. (2 spp. from Bda.)

***Olindias phosphorica*** (Delle Chiaje): Hemispherical; mesoglea fairly thick. Radial canals 4, centripetal canals present; marginal tentacles numerous, primaries with and secondaries without terminal adhesive pads, marginal clubs present; statocysts spherical, normally 2 at base of each primary tentacle. Neritic.

***Gonionemus suvaensis*** Agassiz & Mayer: Flatter than a hemisphere; mesoglea fairly thick. Gonads on distal portion of radial canals; radial canals 4, centripetal canals absent; marginal tentacles 40-70, with rings of nematocysts, adhesive pads rudimentary; statocysts 16, spherical. Neritic.

**Plate 42**

O. **ACTINULIDA:** Free-living solitary Hydrozoa with entirely ciliated body and solid, filiform tentacles. Development direct; no asexual reproduction. Interstitial in marine sand.

F. **HALAMMOHYDRIDAE:** Actinulida with conical body with up to 32 tentacles in 2 alternating girdles, endodermal statocysts and an adhesive organ, all at the aboral end. Sexes separate; gonads in the walls of the gastric cavity. (1 sp. from Bda.)

***Halammohydra*** sp.: With 8 long and 8 short tentacles, and adhesive papillae. In coarse clean sand in- and offshore, 5-10 m; uncommon.

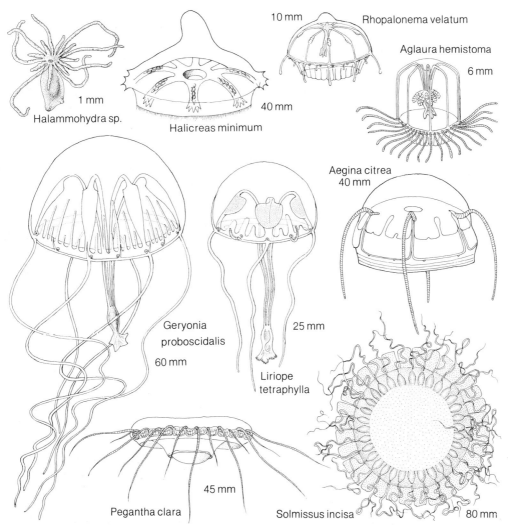

**42 ACTINULIDA, TRACHYMEDUSAE, NARCOMEDUSAE (Hydromedusae 3)**

O. **TRACHYMEDUSAE:** Hydrozoa with umbrella margin entire; radial canals present. Gonads typically on radial canals; sensory clubs free or enclosed.

F. **HALICREATIDAE:** Trachymedusae with broad, circular manubrium; peduncle lacking. Radial canals 8 or more, broad, centripetal canals lacking; marginal tentacles numerous, lacking adhesive discs, proximal portion flexible, distal portion stiff. (3 spp. from Bda.)

*Halicreas minimum* Fewkes: Disc-shaped, with a conical apical projection; periphery with 8 clusters of gelatinous papillae. Radial canals 8; gonads on radial canals; marginal tentacles as many as 640; statocysts 3-4/octant. Bathypelagic.

F. **RHOPALONEMATIDAE:** Trachymedusae with narrow manubrium; peduncle present or absent. Radial canals 8 or more, narrow, centripetal canals lacking; marginal tentacles morphologically uniform throughout, lacking adhesive discs. (6 spp. from Bda.)

***Rhopalonema velatum*** Gegenbaur: Hemispherical, with an apical knob; gonads 8, oval, on radial canals. Marginal tentacles of 2 sizes, 8 long club-shaped ones and up to 24 very small ones; velum broad; statocysts adjacent to tentacle bases. Epipelagic; fairly common.

***Aglaura hemistoma*** Péron & Lesueur: Thimble-shaped with apex flattened; mesoglea thin; gastric peduncle slender, manubrium small; gonads 8, sausage-shaped, on gastric peduncle near manubrium. Marginal tentacles numerous, all alike; statocysts 8, club-shaped. Epipelagic; fairly common.

F. **GERYONIIDAE:** Trachymedusae with narrow manubrium; peduncle present. Radial canals 4-6, centripetal canals present; marginal tentacles of 2 kinds, hollow and solid. (2 spp. from Bda.)

***Geryonia proboscidalis*** (Forskål): Hemispherical; mesoglea fairly thick; gastric peduncle long, conical, manubrium small. Radial canals, gonads, and lips 6; gonads broad. Marginal tentacles of 2 types, 6 long ones with rings of nematocysts and 6 small ones with adaxial nematocyst clusters; statocysts 12. Epipelagic.

***Liriope tetraphylla*** (Chamisso & Eysenhardt): Hemispherical, mesoglea fairly thick; gastric peduncle long, conical, manubrium small. Radial canals, gonads and lips 4; gonads broad, variable in shape. Marginal tentacles of 2 types, 4 long ones with rings of nematocysts and 4 small ones with adaxial nematocyst clusters; statocysts 8. Epipelagic; very common.

O. **NARCOMEDUSAE:** Hydrozoa with lobed umbrella margin. Radial canals absent; gonads on walls of manubrium; sensory clubs free.

F. **AEGINIDAE:** Narcomedusae with interradial, divided gastric pouches. Primary perradial marginal tentacles inserting between pouches, secondary tentacles present or absent on umbrella margin. (2 spp. from Bda.)

***Aegina citrea*** Eschscholtz: Hemispherical; mesoglea thick apically, margin with 4-6 lobes; manubrium large, circular; gastric pouches 8-12, rectangular. Marginal tentacles 4-6, issuing from exumbrella halfway between apex and margin; statocysts numerous. 0-1,000 m or more.

F. **SOLMARISIDAE:** Narcomedusae with circular manubrium, gastric pouches lacking. Marginal tentacles issuing from umbrella opposite periphery of manubrium. (1 sp. from Bda.)

***Pegantha clara*** R. P. Bigelow: Flattened, mesoglea thick, exumbrella smooth; manubrium large; peripheral canal system present, canals narrow. Marginal tentacles 20-40; marginal lobes 20-40, tongue-shaped, each with 3-5 statocysts and long otoporpae. Epipelagic.

F. **CUNINIDAE:** Narcomedusae with perradial, undivided gastric pouches. Marginal tentacles inserting opposite center of pouches, secondary tentacles on umbrella margin lacking. (1 sp. from Bda.)

***Solmissus incisa*** (Fewkes): Flattened, exumbrella smooth, mesoglea soft, fragile. Marginal tentacles, lobes, and gastric pouches 20-40; gastric pouches oval; lobes rectangular, each with 2-5 statocysts; peripheral canal system and otoporpae lacking. Bathypelagic.

D. R. CALDER

## ORDER SIPHONOPHORA

CHARACTERISTICS: *Colonial, holoplanktonic HYDROZOA; highly polymorphic, with several types of asexual polypoid and sexual or asexual medusoid individuals connected by a stem.* Siphonophore colonies vary from several millimeters to 50 m or more in length. Although some are for the most part colorless, others are variously pigmented. *Hippopodius hippopus* responds to touch by turning milky white, and becomes transparent again only after an interval of 15-30 minutes. A number of genera, including *Agalma, Hippopodius, Diphyes,* and *Abylopsis,* have species known to be luminescent. The movements of siphonophores vary; some swim slowly or merely float, but many others are actively propelled by their muscular swimming bells. *Physalia* floats at the surface and is propelled by wind and water currents. Although the various individuals of a siphonophore colony are capable of independent activity, coordinated movement is also evident. Nearly 150 species are recognized, about 45 of which have been reported from Bermuda. Of these, 13 are included here.

OCCURRENCE: Free-swimming or floating, from the epipelagic to the abyssopelagic zone. A number undergo significant vertical migrations diurnally. Fewer species normally exist in the upper 200 m than in deeper waters, although the number of individuals is often high near the surface. Some species vary greatly in abundance from season to season.

Colonies are very fragile and are best collected by SCUBA divers using jars or plastic bags. Specimens may be obtained from trawl and plankton tows but are generally fragmentary. *Physalia* can be removed from the water using tongs or dipnets.

Several siphonophores are capable of stinging humans, and contact with *Physalia* tentacles (which may extend 50 m!) can be extremely painful. Affected skin areas should immediately be cleared of adhering tentacles (use a knife or tweezers to scrape or pluck them off), then treated with meat tenderizer, alcohol or sun lotion to prevent further nematocyst discharge and relieve pain. In case of severe stings a doctor should be consulted (for detailed information on medical aspects see HALSTEAD 1978).

IDENTIFICATION: Often difficult because colonies are complex, polymorphic and usually fragmentary when collected by routine methods. Microscopic examination is usually necessary for positive identification.

Colonies consist of a variety of polymorphic polypoid and medusoid individuals ("persons") that arise by budding, either directly or indirectly, from an asexual, juvenile, oozooid polyp. In Cystonectae and Physonectae look for the float (pneumatophore), a 2-walled sac that is partly lined with chitin. It contains a gas gland and is either closed or provided with a sphincter muscle. The stem is a budding zone of varying length that gives rise to secondary polyps and medusoids. Gastrozooids, palpons, bracts and gonophores are budded off from a section of the stem called the siphosome. Nectophores are usually formed from another area of the stem (nectosome). Polypoid forms include the gastrozooids and palpons. Gastrozooids, specialized for feeding, possess a mouth and a single tentacle. The tentacles are very contractile and bear lateral branches or tentilla, each of which is equipped with a subterminal nematocyst battery. The protozooid is the primary gastrozooid located at the oral end of juvenile siphonophores as well as some mature Cystonectae and Physonectae. Palpons are reduced gastrozooids with a simple tentacle. Gonodendra are complexes consisting of gonopalpons and gonophores; nectophores may also be present. Bracts, regarded as medusoid forms by some authors and as polypoids by others, have thick mesoglea and provide buoyancy and protection. Part of the common gastric cavity may be occluded in some mature bracts and is termed the phyllocyst. Medusoid forms include the gonophores and nectophores. The gonophores vary in their development from medusiform to styloid. Swimming bells or nectophores have radial canals and an umbrella, velum, and ring canal, but lack gonads, tentacles, sense organs and a manubrium. Other parts of the nectophore include the subumbrellar cavity (nectosac), the area of the velar opening (ostium), the ventral cavity housing the retracted stem (hydroecium), and a diverticulum of the common gastric cavity (somatocyst). Nectophores vary greatly in shape, sculpturing, size and arrangement. The vestigial nectophores of *Physalia* are referred to as "jelly-polyps." Finally, various growth phases and groups of polymorphic "persons" deserve mention. Stem groups (cormidia) comprise aggregations of various individuals such as gastrozooids, palpons, bracts, gonophores, gonodendra and nectophores. Cormidia are generally not liberated in Cystonectae and Physonectae, but they may break loose and become free-living when fully developed in the Calycophorae, and are referred to as eudoxids. The polygastric phase in Calycophorae consists of the nectophores, stem and attached cormidia. Free medusoids may be liberated in some Calycophorae, and all 3 suborders have basically different larval forms.

Siphonophores are extremely delicate and must be relaxed, fixed and preserved with care. Anesthetize in a dilute solution of magnesium chloride, and fix with a small amount of neutralized formalin and Zenker's fluid. Preserve in either a 5% solution of formalin in seawater, or 5% neutralized formalin.

BIOLOGY: Individual gonophores are dioecious, but each colony bears gonophores of both sexes. An exception is *Physalia*, entire colonies of which are either male or female. Little is known about siphonophore growth rates and life-span. Pneumatophores of young *Physalia* may increase several centimeters in length per month, and the age of some large specimens has been estimated at about 1 yr. One of the more puzzling aspects of siphonophore behavior is pneumatophore rolling in *Physalia*, which has been interpreted by some as a means to keep the float moist. Crustaceans and fishes constitute major prey organisms. Food is captured with the tentacles and ingested by the gastrozooids. Predators probably include certain planktivorous vertebrates. Parasites known to infest siphonophores include a few Protozoa and Trematoda, as well as an amphipod. An interesting association exists between *Physalia* and the man-of-war fish, *Nomeus gronovii*. *Physalia* provides shelter and probably food for *Nomeus* which in turn may lure prey fishes into the tentacles of the siphonophore. It is unclear how *Nomeus* avoids becoming prey as well.

**Plate 36**

DEVELOPMENT: Where known, fertilized eggs develop into rounded or elongated planula larvae. Subsequent larval differentiation is more or less distinct in each of the 3 suborders. In the cystonect larva of *Physalia*, the primary gastrozooid and an aboral pneumatophore appear, along with a siphosomal budding zone occurring on 1 side of the pneumatophore base only. In physonects, a larval bract, later shed, normally provides buoyancy until the pneumatophore develops. No pneumatophore occurs in calycophores, but a precocious larval nectophore is formed, which may be lost. In both physonect and calycophore larvae, the primary gastrozooid develops at the oral end, and nectosomal and siphosomal budding zones appear, usually on opposite sides of the stem.

Larvae may be collected with plankton nets. Young *Physalia* with developed

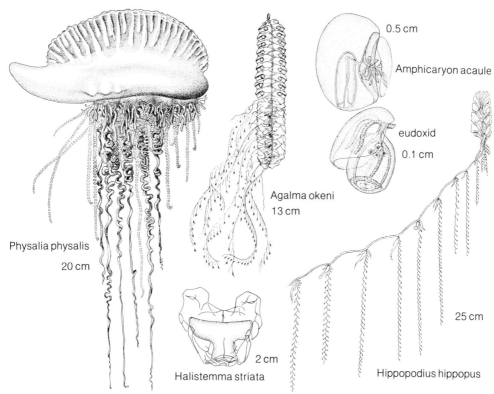

**43 CYSTONECTAE, PHYSONECTAE, CALYCOPHORAE 1 (Siphonophores 1)**

pneumatophores may be taken in neuston nets.

REFERENCES: A detailed overview of systematics and morphology is given by TOTTON & BARGMANN (1965). TOTTON (1960) and MACKIE (1960) provide detailed accounts of the natural history, morphology, and behavior of *Physalia*.

Among the more important contributions to our knowledge of Bermudian species are the studies of BIGELOW (1918), TOTTON (1936), MOORE (1949) and GRICE & HART (1962). An appreciation for the abundance and variety of siphonophores in Bermuda waters may be gained from the bathysphere observations of BEEBE (1934).

## Plate 43

**S.O. CYSTONECTAE:** Siphonophora with a pneumatophore; budding zone on one side of pneumatophore base only; cormidia usually not liberated; bracts lacking.

F. **PHYSALIIDAE:** Cystonectae with large, horizontal pneumatophore. (1 sp. from Bda.)

*Physalia physalis* (L.) (Portuguese man-of-war): Colonies in mirror images (enantiomorphic); pneumatophore to 30 cm long. Secondary buds developing in 2 anterior zones, forming complex cormidia with gastrozooids, palpons, gonodendra, gonophores, nectophores and jelly-polyps; tentacles unbranched, to 50 m or more in length when fully extended. Pelagic, floating at the surface; sometimes associated with *Nomeus* and the juveniles of other fish species. Often beached in large numbers. A severe stinger!

**S.O. PHYSONECTAE:** Siphonophora with a pneumatophore; larvae with siphosomal and nectosomal budding zones normally on opposite sides of pneumatophore; cormidia usually not liberated; bracts present.

**F. AGALMIDAE:** Physonectae with long stems; nectophores arranged in 2 tiers on either side of nectosome. (5 spp. from Bda.)

*Agalma okeni* Eschscholtz: Pneumatophore small, ovoid, reddish brown apically, nectosome dodecagonal, with up to 18 nectophores on either side; bract with distal border very thick, obliquely truncated, faceted; nectosac Y-shaped in top view; tentilla tricornute. 0-200 m. Feeding mostly nocturnal, principal prey believed to be vertically migrating crustaceans and fishes, but epipelagic copepods may also be captured. Common.

*Halistemma striata* Totton & Bargmann: Nectophores with 4 lateral ridges descending at an angle of about 30° from the vertical; ridge at ostial end bifurcating; upper branch extending just below ostial chromatophore, lower branch scarcely visible laterally without staining; stem end with prominent lateral wedge-processes having marked articulation cavities and bosses. Type material collected at depth of about 1,530 m by W. Beebe; known only by its nectophores.

**S.O. CALYCOPHORAE:** Siphonophora without a pneumatophore; cormidia normally liberated as free-swimming eudoxids; bracts usually present.

**F. PRAYIDAE:** Calycophorae with nectophores usually opposite; nectophores rounded, oval, or subcylindrical, mesoglea very thick. (6 spp. from Bda.)

*Amphicaryon acaule* Chun: Nectophores 2, decidedly unequal in size, vestigial nectophore embraced by persistent nectophore; vestigial nectophore with 4 distinct, simple radial canals, nectosac present but not opening to exterior. Eudoxids with a pair of lateral hydroecial canals in bract. 0-800 m. Taken regularly.

**F. HIPPOPODIIDAE:** Calycophorae with a succession of up to 12 or more alternating, closely appressed nectophores; stem retracted up among nectophores; cormidia not liberated as eudoxids. (2 spp. from Bda.)

*Hippopodius hippopus* (Forskål): Nectosome about 2 cm long; definitive nectophores horseshoe-shaped, to 19 mm wide, mesoglea thick, dorsal knobs 4, rounded, forming an arc above ostium. Gastrozooids and gonophores arising together on stem, mature gonophores breaking away; bracts lacking. 0-800 m. Present in varying numbers throughout the year.

**Plate 44**

**F. DIPHYIDAE:** Calycophorae usually with 2 similar, typically pyramidal nectophores, one behind the

# SIPHONOPHORA

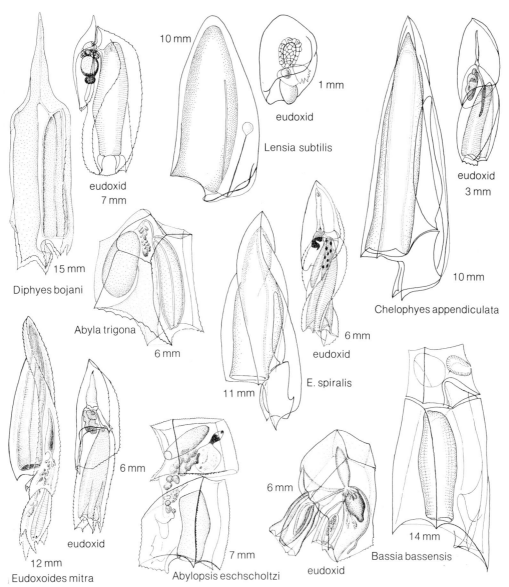

**44 DIPHYIDAE, ABYLIDAE (Siphonophores 2)**

other; somatocyst lacking in posterior nectophore; posterior nectophore occasionally vestigial or suppressed. (18 spp. from Bda.)

***Diphyes bojani*** (Eschscholtz): Anterior nectophore 14 mm long, slender, apex pointed; longitudinal ridges 5, frequently expanded apically; nectosac gradually narrowing apically; hydroecium extending along 1/3 of nectosac length; somatocyst fusiform; mouth plate with small teeth on dorsal side. Posterior nectophore slender, 10 mm long; lateral hydroecial teeth and basal edges serrated. Eudoxid 7 mm long; bract shield-shaped, its axis parallel with that of anterior nectophore. 0-1,000 m; common, maximum numbers in winter.

***Lensia subtilis*** (Chun): Anterior nectophore 11 mm long, with 5 non-crested ridges; somatocyst globular, with a long stalk. Posterior nectophore 11 mm long, truncate proximally, with 5 ridges; lateral radial canals sigmoid. Bract about 1 mm long, rounded, basal cavity shallow, phyllocyst oval; gonophore about 2 mm long, becoming detached and free-swimming with characteristic yellow pigment. 0-800 m; common, maximum numbers in summer.

***Chelophyes appendiculata*** (Eschscholtz): Anterior nectophore with 3 ridges at apex; baso-dorsal and lateral teeth lacking; somatocyst and ventral facet not twisted. Posterior nectophore with basal end of ventral ridges ending in pronounced teeth, 1 tooth 1/3 shorter than the other. Bract with deep cavity enclosing peduncle of gonophore; phyllocyst cylindrical, apex tapering. 0-1,300 m. Known to undergo marked vertical migrations diurnally. Found in abundance at surface during moonlit nights in vicinity of Bermuda by MOORE (1949). Common.

***Eudoxoides mitra*** (Huxley): Genus with polygastric phase usually having 2 nectophores; anterior nectophore pentagonal, with complete dorsal ridge; ostial teeth inconspicuous, mouth plate divided. Posterior nectophore with lateral radial canals forming an apical loop.—Species with anterior nectophore 12 mm long, not twisted; longitudinal ridges 5, serrate, becoming smooth at apex; apex of hydroecium truncated; somatocyst short, pear-shaped. Posterior nectophore with characteristic apico-dorsal notch. Eudoxid 8 mm long; bract with cone-shaped phyllocyst; gonophore pedicel long; gonophore with 2 dorso-lateral teeth, mouth plate with concave edge and 2 lateral teeth. 0-1,000 m. Abundant at surface on moonlit nights; known to undergo marked vertical migrations diurnally (MOORE 1949). One of the commonest diphyids in Bermuda waters.

***E. spiralis*** (Bigelow): Genus as above.—Species with anterior nectophore 11 mm long, twisted; longitudinal ridges 5, twisted; apex of hydroecium pointed; somatocyst cylindrical. Posterior nectophore not produced. Eudoxid 6 mm long; bract with long, cylindrical phyllocyst; gonophore pedicel relatively short; gonophore spirally twisted. 0-1,100 m. Known to undergo marked vertical migrations diurnally. Common in winter.

F. **ABYLIDAE:** Calycophorae with a small anterior nectophore and a large, heteromorphic, posterior nectophore; hydroecial cavity of anterior nectophore closed in on ventral side, forming a tube. (7 spp. from Bd.).

***Abyla trigona*** Quoy & Gaimard: Anterior nectophore with heavily and irregularly serrated ridges; apico-dorsal facet sharply turned upward from insertion of lateral ridges to transverse apical ridge; lateral protrusions sharp, prominent; facets depressed below surrounding ridges. Posterior nectophore with 6-8 teeth on comb; basal margin of right ventral wing with 2 rows of teeth; ostial teeth heavily serrated. Eudoxids known but as yet indistinguishable from those of certain other species of *Abyla*. 0-200 m. Taken regularly.

***Abylopsis eschscholtzi*** (Huxley): Anterior nectophore with a ridge formed by junction of 2 lateral facets; bases of lateral radial canals not forming an upward loop; somatocyst extending to ventral side of hydroecium; apical region of nectosac extending upward between somatocyst and dorsal facet. Posterior nectophore with width nearly equal to length; radial canals 4, normally arranged. Bract with lateral ridges of dorsal facet sloping toward apical facet; basosagittal ridge long. 0-600 m. Common.

***Bassia bassensis*** (Quoy & Gaimard): Anterior nectophore with a ridge formed by junction of 2 lateral facets; somatocyst not extending to ventral side of hydroecium; apical region of nectosac not extending upward between somatocysts and dorsal facet. Posterior nectophore with suppressed median dorsal ridge; basal rim of hydroecium and basal end of left ventral flap serrated. Bract with an apical ridge; phyllocyst lacking ventro-lateral branches. 0-1,000 m. Common, winter and spring.

D. R. CALDER

## Class Scyphozoa (Jellyfish)

CHARACTERISTICS: *CNIDARIA with usually large medusae carrying lappets on their margin; polyps small or reduced, with 4 funnel-shaped septa incompletely dividing the stomach into 4 pouches.* The majority of Scyphomedusae are large (2-40 cm or more) and often of robust cartilaginous consistency, whereas the polyp stage when present is very inconspicuous (a few millimeters). Their delicate colors (orange, pink and purple prevail) and the graceful rhythmic contraction of the umbrella that propels them through the water have to be appreciated by observation in the field.

Of the 5 orders, one (the small, sedentary Stauromedusae) is not represented in Bermuda. The systematics is somewhat complicated by the difficulty of correctly assigning polyps to medusae (cf. HYDROZOA). Of over 200 species, about 10 have been reported from around Bermuda; 7 are included here.

OCCURRENCE: Exclusively marine, although some (e.g., *Aurelia*) venture into brackish water. The majority is found in coastal waters where medusae often drift in enormous numbers near the surface, especially on cloudy days and at twilight. The polyps, rarely noted because of their inconspicuous size, are attached to hard substrates (preferably overhangs or cave ceilings), often on or in other encrusting organisms (sponges, branching corals, ascidians).

Collect polyps individually (or as clumps of associated organisms) with chisel and hammer by diving. Large medusae can be dipped out of the water with a bucket; smaller ones are sampled with plankton nets.

A number of jellyfish can inflict painful stings on contact; in some cases even swimming in their proximity (e.g., over *Cassiopea*) may produce a stinging sensation. Contact with Cubomedusae can cause serious, occasionally fatal anaphylactic reactions. However, most of the really virulent species occur in Australian waters, whereas Atlantic species such as *Carybdea alata* are comparatively mild stingers.

Jellyfish stings should be treated by immediately removing tentacles that adhere to the skin with the use of sand, bathing towels or other available materials, with subsequent application of alcohol, sun lotion or meat tenderizer (to inhibit further nematocyst discharge). In cases where pain spreads beyond the area of the sting, or is accompanied by headache, nausea, fever, etc., a doctor's assistance should be sought without delay.

IDENTIFICATION: For live observation, medusae should be placed in sufficiently large aquaria; polyps have to be examined under a low-power microscope.

The saucer- to helmet-shaped umbrella of medusae is composed of an outer (exumbrella) and inner (subumbrella) layer which enclose the massive, gelatinous to cartilaginous mesoglea (middle layer) and the gastrovascular cavity (stomach). The mouth opens at the base of the manubrium (mouth tube), which hangs down from the subumbrella and is usually drawn out into 4 or 8 often frilled oral arms. The margin of the umbrella is scalloped to form lobes (lappets), each of which carries a sense organ (rhopalium). There are also marginal tentacles, which can be short and very numerous (e.g., *Aurelia*), long and few (e.g., *Pelagia*) or absent. Nematocysts are concentrated on the tentacles and the manubrium. Polyps can be solitary or colonial, and in some species are protected by a chitinous peridermal tube.

For fixation of large medusae use buffered formaldehyde (4%), after anesthesia in magnesium chloride. For histology, fix small specimens or tissue samples in Bouin's, then transfer into alcohol.

BIOLOGY: With few exceptions sexes are separate. Eggs and sperm develop in the wall of the stomach and are eventually expelled through the mouth opening; it is during or after this journey that fertilization occurs. In some (all?) Cubomedusae, however, the male actively attaches spermatophores (sperm packets) to the tentacle of a female. The life cycle of large forms may span more than 1 yr. Most species feed on plankton (including ctenophores, salps, and other medusae) that gets caught on the tentacles or oral arms, or on the sticky subumbrella. *Cassiopea* rests upside down in calm shallow bays trapping small animals in its frilly oral arms; in addition, its symbiotic algae (zooxanthellae) probably provide it with supplementary nutrition from photosynthesis. It is believed that few animals (sea turtles?) prey on scyphomedusae, but a number of taxa (particularly crustaceans and juvenile fish) associate with them for shelter. Scyphopolyps are grazed on by nudibranch gastropods. A few species (*Pelagia*, *Carybdea*) are luminescent.

**Plate 36**

DEVELOPMENT: In the typical life cycle the fertilized egg develops into a free-swimming planula larva which settles eventually to metamorphose into a polyp (called scyphistoma). At certain times of the year the scyphistoma (which can generate additional polyps by asexual budding) produces medusae (ephyrae) either by direct transformation or by transverse fission (strobilation). Ephyrae may develop either 1 at a time (monodisc strobilation, as in *Cassiopea*), or many immature ephyrae are stacked like saucers before they swim away (polydisc strobilation, as in *Aurelia*). The ephyra eventually grows into the adult medusa. In some cases, ephyrae are not liberated but become sexually mature while still connected with the polyp. In some pelagic species (e.g., *Pelagia*), each planula grows directly into an ephyra.

Ephyrae are commonly caught in inshore plankton hauls. To obtain planulae, place a mature *Aurelia* or *Cassiopea* medusa upside down in a shallow dish so that it is barely covered with seawater; after a while planulae will appear (as a brown "slime") at the margin of the umbrella.

# SCYPHOZOA

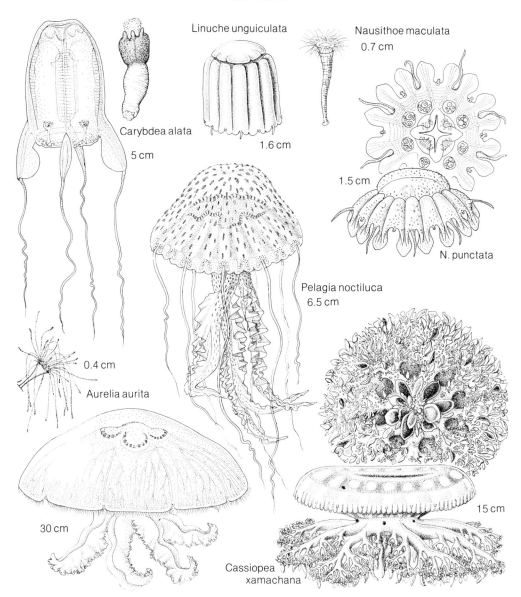

**45 SCYPHOZOA (Jellyfish)**

*Plate 45*

REFERENCES: For a general introduction consult HYMAN (1940); classical monographs (containing most of Bermuda's species) are by MAYER (1910c) and RUSSELL (1953). A more recent synopsis is by KRAMP (1961). Reproduction has been summarized by CAMPBELL (1974), and new observations on the life cycle have been added by THIEL (1962) for *Aurelia*, and WERNER (1974, 1975) for Cubomedusae and Coronatae. HALSTEAD (1978) deals with accidents involving Scyphozoa.

MOORE (1949) and MORRIS (1975) give scant ecological data on Bermuda's species.

**O. CUBOMEDUSAE** (Box jellies, sea wasps): Scyphozoa having cuboid medusae with simple margin and 4 marginal tentacles; velarium (subumbrellar fold containing gastrovascular canals) present. Confined to warm oceans. (1 sp. from around Bda.)

*Carybdea alata* Reynaud: Medusa with 4 simple tentacles with widened proximal parts; bell high-pyramidal. Medusae occasionally found in offshore surface waters; polyp not recorded from Bermuda.

O. **CORONATAE:** Scyphozoa having generally smallish medusae with a horizontally constricted bell; margin with lappets and tentacles. Polyps (originally described as *Stephanoscyphus*) simple or branching, usually with a peridermal tube; often in association with other sedentary organisms (especially sponges). Medusae mostly meso-or bathypelagic (4 spp. reported from around Bda; none of them in both medusa and scyphistoma stages.)

*Nausithoë maculata* Werner: Medusa unknown. Polyp solitary, with about 50 tentacles and white spots at the tentacle bases. In shallow inshore cave entrances on encrusting Ascidiacea; probably not uncommon.

*N. punctata* Kölliker: Medusa with dome-shaped central disc, 8 gonads and 8 marginal tentacles. Exumbrella finely dotted. Colorless to greenish, gonads red or yellow. Polyp colonial, forming a branching growth within sponges, with the polyps projecting out of the host's oscula. The medusae may be encountered offshore, occasionally in fair numbers. The polyp has not been found in Bermuda yet.

*Linuche unguiculata* (Schwartz): Medusa thimble-shaped; brownish, with 8 short tentacles; lappets short, rounded, and at an angle to the margin. Polyp unknown. Occasionally found in inshore waters in early summer, sometimes in enormous numbers. When swimming, the medusa spins clockwise owing to the configuration of the lappets.

O. **SEMAEOSTOMAE:** Scyphozoa having generally large medusae with a bowl-shaped bell; angles of mouth opening developed into curtain-like lips. Stomach with radial canals extending to bell margin. (About 4 spp. from around Bd.).

*Aurelia aurita* (L.) (Moon jelly): Medusa massive, saucer-shaped, with 8 simple marginal lobes and many small marginal tentacles. With 4 long, unbranched oral arms. Polyp with 24 very long tentacles when fully grown; strobilation mostly of the polydisc pattern. Common inshore; polyps on overhangs and the ceilings of intertidal undercuts.

*Pelagia noctiluca* (Forskål): Medusa with 8 marginal tentacles, 16 lappets, and 16 radiating stomach pouches. Exumbrella beset with numerous orange to purple clusters of nematocysts. Development direct (the planula turns into an ephyra, bypassing the polyp stage completely). Offshore; sometimes (esp. during water months) drifting inshore in large numbers.

O. **RHIZOSTOMAE:** Scyphozoa having generally large medusae without marginal tentacles. Oral arms branched, with many openings ("secondary mouths"); original mouth lost. Stomach with many anastomosing radial and ring canals. Mostly in warm, shallow seas. (1 sp. from Bda.)

*Cassiopea xamachana* Bigelow (Upside-down jellyfish, Cabbage-head jellyfish): Medusa with 4 pairs of elaborately branched oral arms. Umbrella flat, saucer-shaped, with a well-defined central depression (acting as a sucker) on the exumbrella. Color variable, mostly greenish gray-blue, the greenish color due to zooxanthellae embedded in the mesoglea. Polyp slender; strobilation of the monodisc type. Medusae are found, upside-down and usually in large congregations, on the muddy bottoms of inshore bays and ponds.

W. STERRER

## Class *Anthozoa* (Corals, anemones)

CHARACTERISTICS: *Exclusively polypoid, solitary or colonial CNIDARIA. Oral end expanded into oral disc which bears the mouth and one or more rings of hollow tentacles. Stomodeum well developed, often with 1 or 2 siphonoglyphs. Gastrovascular cavity compartmentalized by radially arranged mesenteries. Mesoglea a mesenchymal or fibrous connective tissue.* Adult body sizes range from Scleractinia of 2 mm diameter to whip-like Antipatharia 7 m long. Colonies of Scleractinia may also reach several meters in diameter and weigh several tons. Most anthozoans, however, measure between 3 and 50 cm. Most orders possess some kind of inorganic supporting structure, such as a calcareous or horny axis or microscopic calcareous sclerites distributed throughout the tissue. Some gorgonians and pennatulids have both supporting axes and sclerites, whereas the Actiniaria, Corallimorpharia, Zoanthidea and Ceriantharia have no such supporting structure. Anthozoans are among the most colorful animals in the reef environment. Unfortunately the pigments are rarely retained in preserved specimens and, although distinctive in some species, color is not usually a reliable taxonomic character. Anthozoans are predominantly sedentary animals, often firmly attached to the substrate. The only locomotion involved in most life cycles is in the planktonic larval stage. However, many Actiniaria and Ceriantharia can move if exposed to unfavorable conditions. Actiniaria can creep along on their pedal discs at 8-10 cm/hr, pull themselves by their tentacles, move by peristalsis through loose sediment, float in currents, and even swim by coordinated tentacular motion.

Both subclasses are represented in Bermuda. Because the orders are so diverse morphologically, they are often discussed separately. In some classifications the anthozoan orders are grouped into 3 (not the 2 considered here) subclasses, splitting off the Ceriantharia and Antipatharia into a separate subclass, the Ceriantipatharia. Corallimorpharia are sometimes considered a suborder of Scleractinia. Approximately 6,500 species of Anthozoa are known. Of 93 species reported from Bermuda, 76 are included here; the remaining 17 are deep-water or rare shallow-water species.

OCCURRENCE: Throughout the world oceans from the Arctic to the Antarctic, from the intertidal to hadal depths (10,700 m), and in temperatures of $-1°$ to $30°C$. In general, more species occur in the shallow-water tropical and subtropical regions associated with coral reefs, but virtually all benthic marine habitats are exploited. Full oceanic salinity is usually required, but some scleractinians and gorgonians can tolerate brackish water of 1.7% salinity. Some zoanthids and actinians thrive in polluted areas. Substrate type is an important factor governing distribution; some species require mud, others sand or hard substrates, and others are epizoic or epiphytic. Anthozoans are usually patchy in distribution, often abundant in coral reef habitats where conditions of substrate, light and current combine to create highly productive areas. Densities of 27-30 colonies of Scleractinia or Gorgonacea per square meter are not unusual in reef environments, whereas antipatharians rarely exceed 2-3 colonies/m$^2$.

Most intertidal and shallow-water Anthozoa can be collected with the aid of a hammer and chisel while snorkeling or SCUBA diving. A wooden spatula or flat bivalve shell is good for detaching anemones. Burrowing actinians, cerianthids and pennatulids may be obtained by sieving of sand and mud, preferably after the animals have been spotted in the expanded condition. It is best to collect cerianthids at night or at dusk, when most species expand. If kept moist, and the temperature is controlled, anemones may live for several days and even survive shipment by mail. Deep-water (over 70 m) species are usually collected fortuitously by trawling and dredging, or selectively by submersible.

Some anthozoans are commercially or pharmacologically important. Some gorgonians and zoanthids contain compounds (prostaglandins and palytoxin, respectively) that are biologically very active and of pharmaceutical value. Care must be taken when collecting the zoanthid *Palythoa* in late summer because it contains a highly toxic substance (palytoxin) which is very irritating to the skin and open cuts, causing severe pain and long-lasting blisters. Actinians, some scleractinians and cerianthids are popular aquarium animals. Finally, because anthozoans constitute most of the mass of coral and patch reefs, they contribute to storm protection, tourist recreation and sand production, not to mention the habitats provided for innumerable other animals, many of which are eaten by humans. The taking of corals for souvenirs is strongly discouraged, and prohibited by law in certain areas (Coral Reef Preserves).

IDENTIFICATION: Because of the morphological diversity within the class, identification relies on a great number of characters and methods.

The basic structural unit of anthozoans is the polyp, which may exist alone or in a colonial arrangement, all contiguous polyps tracing their ancestry through asexual budding to a single founder polyp. The polyp usually has a short, squat, cylindrical body (column) with a flattened oral end (oral disc). The mouth occupies the center of the oral disc and is surrounded by several rings (cycles) of hollow tentacles. The mouth leads through a short, longitudinally ridged tube (stomodeum) into the gastrovascular cavity (coelenteron). The hollow tentacles are continuous with this cavity. Sometimes the stomodeum is flattened, with one or both of the narrow edges modified into flagellated grooves (siphonoglyphs) that propel water into the polyp. The gastrovascular cavity is radially partitioned by thin lamellae (mesenteries, septa) on which the gonads develop. The inner, free edges of the mesenteries bear distinct, thickened rims, the (mesenterial) filaments, which perform the functions of digestion, absorption and excretion. These filaments generally have the form of a simple cord, often very sinuous in the lower part of the polyp. In Ceriantharia the filaments in the upper part of the polyp are bilobed, in Actiniaria and Zoanthidea trilobed. Mesenteries (except in Ceriantharia) are arranged in cycles of different order. They are termed perfect or complete if they are attached to the column, oral disc, base and stomodeum; imperfect or incomplete if they do not reach the stomodeum. The inner, free edges of the imperfect mesenteries (mesenterial filaments) are sometimes very sinuous and trilobed. Most anthozoan polyps are bilaterally symmetrical, the axis being determined by the siphonoglyphs or the flattened stomodeum. Two mesenteries may occur symmetrically on both sides of the axis (coupled) and/or directly adjacent (in pairs). The space between the mesenteries of a pair is called an endocoel, the space between 2 pairs an exocoel. A directive mesentery pair flanks a siphonoglyph. In some elongate actinians there may be 8 well-developed, fertile macromesenteries alternating with 8 smaller sterile micromesenteries. Nematocysts, the characteristic "stinging capsules" of Coelenterata, are found on the tentacles, stomodeum, column and mesenterial filaments of all Anthozoa. The fine structure of nematocysts as well as the cnidom (types of nematocysts in a species or higher taxon) are important criteria for classification. The body wall of an anthozoan is composed of 3 layers: an outer ectoderm (epidermis), an inner endoderm (gastrodermis) and an intermediate mesoglea, composed of a mesenchymal or fibrous connective tissue. The mesoglea of a colonial anthozoan is collectively termed coenenchyme. Many species contain symbiotic zooxanthellae (dinoflagellates) that enable many Scleractinia to be reef-building (hermatypic) in shallow, well-lit waters; the other, non-reef-building (ahermatypic) Scleractinia, i.e., without zooxanthellae, are able to live in dark, cold deeper waters.

The following terms pertain to features of individual orders rather than the entire class.

All gorgonaceans have a central supporting structure, usually called the axis in Scleraxonia and the central core in Holaxonia. The axis is composed of horny material and usually fused sclerites; the central core has a horny outer layer (cortex) surrounding a chambered core but sclerites are not present. Surrounding the cortex the inner and outer rinds, concentric layers of mesoglea (coenenchyme), make up most of the colony. Within the coenenchyme numerous tiny calcareous skeletal elements (sclerites) lend support to the coenenchyme. Sclerites occur in many different sizes and shapes. One of the simplest kinds is the spindle, a straight, monaxial structure pointed at both ends. Scales are thin, flat sclerites and plates are thicker scales. Monaxial, elongate sclerites that are enlarged at one end (the head) are called clubs. Leaf-clubs have heads ornamented with foliate processes whereas wart-clubs have only low, blunt protuberances on their heads. Capstans are monaxial spicules with 2 whorls of tubercles at either end and terminal tufts. Quadriradiates, also called "butterfly spicules", have 4 terminal processes, usually radiating in the same plane. The portion of the expanded polyp that projects above the surface of the branch is called the anthocodium; it can be retracted into a tubular basal neck zone (calyx), which may support a transverse ring of sclerites (collaret). Directly above the collaret are groups of vertically arranged sclerites, peaking at each tentacle. The peaks are called points and the collaret and points are collectively termed the crown and points. Above the points are the 8 tentacles.

Pennatulaceans are composed of 1 large, axial primary polyp that consists of an anchoring peduncle and a distal rachis, from which the secondary polyps arise. Secondary polyps consist of siphonozooids, controlling water circulation, and autozooids, the typical food-gathering polyps.

Many species of Actiniaria have a muscle band (marginal sphincter) that encircles the upper part of the column. Each mesentery bears a longitudinal muscular band, the retractor. Basilar muscles are present in species with a broadly attached pedal disc. The column can usually be divided into the scapus (column proper) and the scapulus, a histologically differentiated region at the level of the sphincter. Sometimes there is a capitular region above the scapulus, separated by a fold of tissue (collar or parapet), and a groove (fossa) that occurs between the collar and the base of the capitulum. The scapulus and capitulum are invariably smooth, but the scapus may bear a variety of special structures, such as vesicles, hollow, nematocyst-bearing outgrowths of the gastrovascular cavity; acrorhagi, vesicle-like nematocyst batteries occurring on the parapet or in the fossa; pseudotentacles, large, branched, inflated, vesicle-like structures bearing acrorhagus-like nematocyst batteries; warts, adhesive, generally brightly colored, button-shaped structures similar to vesicles; and tenaculi, adhesive, solid mesogleal papillae also used to attach foreign particles to the column. The column may also be pierced by tiny, regularly arranged pores (cinclides) that provide for rapid water expulsion during a sudden contraction and for the emission of acontia. Acontia are long, spirally coiled threads generally below the mesenterial filaments, thought to function in defense.

The calcareous skeleton of scleractinian coral colonies (corallum) is made up of individual units (corallites) produced by individual polyps. The calice is the oral surface of a corallite. Some of the growth forms of colonial Scleractinia include cerioid—closely appressed prismatic corallites, which usually share fused walls; plocoid—slightly more widely spaced, cylindrical corallites with separate walls; phaceloid—parallel or nearly parallel, laterally free corallites; and meandroid—meandering rows of confluent corallites with walls only between rows. In meandroid coralla, rows of corallites (valleys) alternate with slightly elevated walls (collines); the colline may be grooved on top by an ambulacrum. Between each pair of mesenteries is a thin calcareous septum (sometimes called scleroseptum to distinguish it from a mesentery). Each septum is composed of numerous radiating rods (trabeculae) arranged in a plane. The trabeculae of each septum are usually grouped into one or more fan-shaped patterns (fan systems). Usually the trabeculae are parallel and closely adjacent (solid septum), or there is space between the trabeculae (fenestrate septum). Small rods or bars (synapticulae) may connect opposed faces of adjacent septa. If the upper margin of a septum projects above the calicular edge it is termed an exsert septum. Small vertical lobes or pillars on the lower inner edges of the septa are termed pali or paliform lobes. The columella is a calcareous axial structure found in the calices of many species. It is quite variable in shape, including twisted or straight rods, straight or labyrinthiform lamellae, and spongy or solid masses. In colonial Scleractinia the calcium carbonate deposited between corallites is called coenosteum. It may be solid or vesicular; in the latter case, small thin plates called dissepiments form perpendicular to the growth gradient, forming small cavities in the coenosteum. Dissepiments may also occur within the corallites of both colonial and solitary corals.

In the Corallimorpharia, the tentacles on the oral disc are sometimes segregated into peripheral marginal tentacles and more central discal tentacles; the latter are often branched. Some of the unbranched tentacles have knob-like tips (acrospheres) heavily laden with nematocysts.

The mesenteries in Zoanthidea are both paired and coupled, with alternating complete and incomplete mesenteries, except for the 4th through 6th mesenteries on either side of the dorsal directive. Most zoanthids have 2 consecutive imcomplete mesenteries in the 5th and 6th positions (the brachycnemic arrangement), but some have 2 adjacent, complete mesenteries in the 4th and 5th positions (the macrocnemic arrangement).

Ceriantharia also have tentacles of 2 kinds: short labial tentacles surrounding the mouth, and larger peripherally arranged marginal tentacles. Mesenteries are arranged in couples, not pairs, which one after another develop in the multiplication chamber situated opposite the single siphonoglyph. Successive couples are arranged in a regular bilateral pattern of repeating duplets (suborder Penicillaria) or quartets (suborder Spirularia) of mesenteries. Mesenterial filaments may bear both craspedonemes, thread-like, branched, and often bunched processes, or a single aboral acontioid, a short, thick, acontium-like thread.

The horny skeleton (axis) of Antipatharia is often covered by numerous small spines. If a colony consists primarily of a single main stem it is called monopodial but even monopodial colonies may bear pinnules (symmetrically arranged, simple or branched ramifications of smaller branch diameter). If the pinnule branches dichotomously once or twice, secondary and tertiary pinnules, respectively, result. There are no calcareous elements in the mesoglea.

Characters used in the classification of the Alcyonaria include nature of the skeleton if any, arrangement of polyps, branching pattern, terminal branch diameter, sometimes color and, most importantly, the shape and distribution of sclerites. The sclerites can be easily separated from the tissue by dropping a small amount of sodium hypochlorite (full-strength commercial bleach) onto a small piece of tissue on a glass slide. Within minutes the organic tissue dissolves and the sclerites may be rinsed, covered with a cover slip, and examined. For more permanent mounts the sclerites should be repeatedly rinsed with water in a small vial, after which an alcoholic suspension of spicules should be pipetted onto a slide, allowed to dry, and mounted in balsam or a synthetic resin.

To identify the non-skeleton-bearing Zoantharia (Actiniaria, Corallimorpharia, Ceriantharia, Zoanthidea), both the internal and external anatomy must be examined. This is generally done on the basis of preserved specimens. Menthol or magnesium chloride may be used to relax specimens prior to preservation. The general internal anatomy is best studied on the basis of a transverse cut of the column. To determine the character of the (marginal) sphincter a longitudinal cut is required in addition. To study the general anatomy of Ceriantharia it is enough to make a longitudinal cut of the body. It may also be important to examine the types and distribution of nematocysts, though this is not necessary for identifying Bermuda species. The nematocysts are prepared by making a squash preparation of a tiny piece of (preferably living) tissue in a drop of seawater. Examine with a microscope (preferably with interference contrast) under oil immersion ($1,000\times$).

The classification of the Scleractinia relies entirely on characteristics of the corallum. The tissue should be removed by soaking the specimen in sodium hypochlorite, followed by thorough rinsing. It may be necessary to cut through or break a corallum to examine its internal structure.

Identification of the Antipatharia requires examination of the colony form, axis spination, and external characteristics of the polyp.

Haematoxylin and eosin and Gomori Trichrome Stain are usually good histological stains for anthozoan tissue.

Fixation and Preservation: In general, all anthozoans can be fixed in 6-10% formalin (mixed with seawater) and preserved in 70% ethanol. Exceptions are Actiniaria and Zoanthidea, which retain their colors better in formalin. It is usually desirable to fix the specimen in the expanded condition. For gorgonians and zoanthids a 7.5-8.0% solution of magnesium chloride in fresh water is effective for narcotization. The magnesium chloride is repeatedly substituted for half of the water in the container until most of the seawater has been replaced. The animal is then killed with concentrated formalin. Menthol crystals dropped on the water surface are effective in relaxing many Actiniaria and Ceriantharia. Scleractinia are usually preserved dry, as are some gorgonians and antipatharians. Bouin's fixative may be used for histology; however, neither Bouin's nor formalin should be used in cases where calcareous structures are to be preserved.

BIOLOGY: Sexual reproduction is either dioecious or hermaphroditic; protandrous hermaphroditism is usual in monoecious forms. Gonads are often limited to certain mesenteries. Fertilization is either internal or external. Periods of sexual activity for most species are unknown. The gorgonian *Plexaura homomalla* is sexually active in June and July; however, some tropical Scleractinia are fertile year round, but on a lunar periodicity. All orders except Actiniaria and Ceriantharia include species that produce colonies by asexual reproduction. The original founder polyp, a product of sexual reproduction, may produce colonies consisting of millions of asexually budded polyps. When budding occurs the original polyp usually remains intact and one additional polyp is produced, either within the original tentacular ring (intratentacular budding) or outside of it (extratentacular budding). In the case of parricidal budding, however, the original polyp is destroyed, giving rise to numerous daughter polyps. Although the Actiniaria do not produce colonies, they are capable of asexual reproduction by pedal laceration and longitudinal and transverse fission. Transverse fission is also known to occur in Ceriantharia. Growth rates and longevity of anthozoans are also poorly known. One tropical shallow-water gorgonian has an annual growth rate of 0.1-4.0 cm, whereas a deep-water species is known to average 0.9 cm/yr. Branching reef Scleractinia may grow 4.5-27 cm/yr, massive species 0.3-1.9 cm/yr, and deep-water ahermatypes about 0.6 cm/yr. Antipatharians can grow 2-10 cm/yr. Some deep-water gorgonians and antipatharians reach sexual maturity at 12-13 yr and may live for over 70 yr. Actinians have been kept in aquaria for over 70 yr, and it is not unreasonable to surmise that some massive colonial corals are centuries old.

Anthozoans obtain nourishment by a variety of methods. Most are suspension feeders, preying on zooplankton and small crustaceans, fish, jellyfish, etc., by using their tentacles and extruded mesenterial filaments. Both structures have abundant nematocysts that, when discharged, are capable of immobilizing and adhering to small animals. Cilia and mucus aid in the capture and transport of food to the mouth. In this manner many anthozoans are also known to ingest particulate organic matter (detritus), including bacteria, in the same manner. Dissolved organics are quickly assimilated, but their role in nutrition is not fully understood. Some anthozoans may also receive nourishment from a symbiotic relationship. All orders except the Pennatulacea, Ceriantharia and Antipatharia include species that possess symbiotic dinoflagellate algae (zooxanthellae) in their endodermal tissue. These symbiotic species are especially abundant in tropical reef habitats. The zooxanthellae receive protection and waste products from the anthozoan, whereas the advantages to the anthozoan are variable and not fully understood. Zooxanthellae undoubtedly promote calcification in Scleractinia; in general, zooxanthellae probably provide some nourishment for the anthozoan in the form of dissolved organics, e.g., vitamins, glucose and amino acids. Some gorgonians seem to depend entirely on nutrition from the zooxanthellae; they die if placed in darkness, even if provided with food.

Although formidably armed with nematocysts for protection, Anthozoa are not immune to predation. Some of the most common predators are gastropods, polychaetes, echinoids, asteroids, pycnogonids and fishes.

There are numerous associations, both parasitic and symbiotic, of Anthozoa with other animals, e. g., parasitic copepods in most gorgonian gastrovascular cavities; pycnogonid cysts on gorgonians; and *Millepora* encrustation of branching corals. Other less destructive examples include associations of Actiniaria with ciliate protozoans, amphipods, shrimps, crabs and fishes; Zoanthidea with sponges; gorgonians with ophiuroids and crinoids; and gorgonians and scleractinians with polychaetes, which alter the coral branches to form their tubes.

REFERENCES: For general information see HYMAN (1940) and BAYER & OWRE (1968). General references to the following orders include: Alcyonacea, Gorgonacea, and Pennatulacea (BAYER 1956; BAYER & WEINHEIMER 1974; CAIRNS 1976); Actiniaria and Corallimorpharia (CARLGREN 1949; DEN HARTOG 1980); Scleractinia (WELLS 1956); Zoanthidea (HADDON & SHACKLETON 1891, CARLGREN 1923); Ceriantharia (CARLGREN 1912; DEN HARTOG 1977); and Antipatharia (OPRESKO 1972).

For a general reference to Bermuda's Anthozoa see VERRILL (1900a, 1901a, 1907b). Specific references to the Bermudian fauna include Alcyonaria (BAYER 1961; VERRILL 1900a, 1907b); Actiniaria (VERRILL 1899, 1900a, 1907b); Scleractinia (VERRILL 1900a, 1901b, c, 1907b; LABOREL 1966; WELLS 1972); Corallimorpharia (VERRILL 1907b, DEN HARTOG 1980); Zoanthidea (VERRILL 1900a, 1907b); Ceriantharia (VERRILL 1901a). Antipatharia are previously unreported from Bermuda.

S.CL. **ALCYONARIA** (=OCTOCORALLIA): Colonial Anthozoa with 8 pinnate (feathered) tentacles and 8 unpaired mesenteries. (All orders but the Stolonifera, Telestacea and Coenothecalia are represented in Bda.)

*Plate 36*

*Plate 46*

DEVELOPMENT: The fertilized anthozoan zygote undergoes cleavage, resulting in a stereo- or coeloblastula. Gastrulation proceeds by delamination and invagination, or ingression, producing a ciliated planula, the common planktonic larva. The planula may drift in the plankton for weeks, during which time it forms its first mesenteries. A ring of tentacles is formed, usually after the planula has settled. The pelagic larvae of zoanthids are called zoanthella (or Semper's larva) and zoanthina. The former usually develop into species of *Palythoa*, the latter into *Zoanthus*. The planulae of Ceriantharia float on the surface and are therefore often collected in surface plankton tows. These larvae pass through a cerinula stage of 6 mesenteries.

O. **ALCYONACEA** (Soft corals): Alcyonaria with thick and encrusting, lobate, or erect and arborescent attached colonies consisting of very long polyps that usually extend from the base of the colony to the uppermost branches. (Of approximately 1,000 species only 14 are known from the western Atlantic; 1 has been collected from Bda.)

F. **NIDALIIDAE:** Alcyonacea with simple or divided, stiff, cylindrical branches. Coenenchyme rigid, densely packed with tuberculate spindles. Anthocodial armature in form of crown and points.

***Nidalia occidentalis*** Gray (Dandelion coral): Clavate colonies with polyps containing flat scales with scalloped edges. Granulated platelets abundant in pharyngeal walls. Spindles have diameters of about 1/6 their lengths. Colonies white, pale orange or yellowish brown. Lower shelf and upper slope depths. (Color Plate 5.12.)

O. **GORGONACEA** (Sea fans, Sea whips): Alcyonaria with arborescent (rarely lobate or encrusting), attached colonies consisting of polyps with uniformly short gastrovascular cavities. Specialized axial structures usually present. (Of approximately 1,200 species, 23 are known from Bda; 20 are included here.)

S.O. **SCLERAXONIA:** Gorgonacea with axial structure composed of free or fused spicules.

F. **BRIAREIDAE:** Scleraxonia with axis of separate spicules; axis perforated by gastrodermal canals throughout the branches and not separated from cortex by boundary canals. (1 sp. from Bda.)

***Briareum polyanthes*** (Duch. & Mich.): Genus with large (up to 0.8 mm) tuberculate spindles and 3-armed bodies.—Species forms incrustations on rocks, dead coral, and other living gorgonians. Polyps very large, purplish brown. At few nearshore localities.

S.O. **HOLAXONIA:** Gorgonacea with a distinct central axis composed of horny material or a mixture of horny and calcareous substances.

F. **PLEXAURIDAE:** Holaxonia with a soft, cross-chambered central core. Cortex of axis loculated. Sclerites often over 0.20 mm long; clubs usually present. Branches stout. (14 spp. from Bda.)

***Plexaura homomalla*** (Esper) (=*Plexauropsis tricolor* Stiasny, 1935): Genus with stout tree-like colonies having thick, bushy branches. Outer layer of rind with sclerites composed mostly of spindles and clubs; inner layer of spiny spindles that are usually purple.—Species with end branches 2.5-5.0 mm wide. Inner rind contains purple capstans, middle rind composed of spindles, and outer rind with large asymmetric leaf-clubs having serrate leaves with distinct transverse collaret. Colonies to 35 cm tall; dark brown. On patch reefs and outer reefs. Most active spawning VI-VII. Growth rate: 0.13-4.2 cm/yr, average 2.0 cm/yr. Predators: gastropods *Cyphoma gibbosum* (Flamingo tongue) and *Simnia*; also labrid fish.

***P. flexuosa*** Lamouroux (=*P. edwardsi* sensu Stiasny, 1935): Genus as above.—Species with end branches 2.5-4.5 mm wide. Inner rind contains purple capstans and short rods, middle rind composed of short spindles, outer rind large leaf-clubs with serrate folia. Anthocodia without

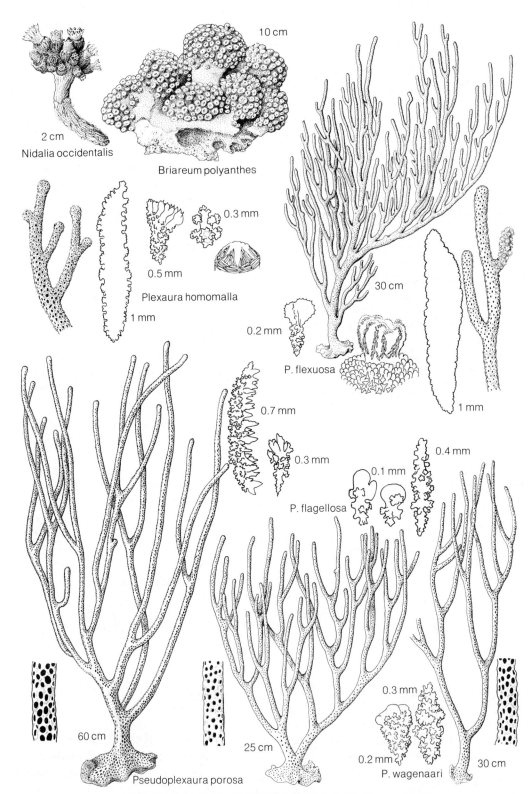

**46 ALCYONACEA, GORGONACEA 1 (Soft corals 1)**

collaret. Colonies to 40 cm tall. Color variable: yellow, brown, purple and reddish purple; commonly purple in Bermuda. Primarily a clear-water (patch reef and outer reef) species. Spawning VI-VII. The dominant octocoral species in shallow water (to 3 m). (Color Plate 5.8-10.)

***Pseudoplexaura porosa*** (Houttuyn) (=*Plexauropsis bicolor* Verrill, 1907; *P. crassa* sensu Verrill, 1907): Genus with outer rind of smooth leaf-clubs or smooth-headed wart-clubs, spiny spindles, and capstans, all colorless. Middle rind with white or purple spindles. Polyps lacking sclerites and fully retractile, resulting in gaping, elliptical pores on branch.—Species with calycular apertures separated by distances smaller than their own diameters (hence the porous appearance of dry specimens!). Apertures 1.0-1.5 mm wide. Leaf-clubs of outer rind large (to 0.4 mm long) and coarsely sculptured; spindles also large (to 1 mm long), often unilaterally spinose. Colonies to 225 cm tall (!), yellow, brownish or reddish purple. Extremely common. Near shore, patch and outer reefs, 1-15 m. (Color Plate 5.3, 4.)

***P. flagellosa*** (Houttuyn) (=*Plexaura esperi* Verrill, 1907): Genus as above.—Species with calycular apertures usually separated by distances greater than their own diameters. Apertures 0.5-1.0 mm wide. Leaf-clubs of outer rind rarely exceed 0.1 mm in length and have rounded folia, usually not globose; spindles slender, rarely over 0.4 mm long, with simple sculpture. Colonies to 100 cm tall; purple. Inner and outer reefs, 5-15 m.

***P. wagenaari*** (Stiasny): Genus as above.—Species with calycular aperture separated by distances greater than their own diameters. Apertures 0.5-1.0 mm wide. Leaf-clubs of outer rind rarely exceed 0.2 mm and have globose heads; spindles stout, usually less than 0.5 mm long, with complicated sculpture. Colonies to 30 cm tall. Rose, gray, light greenish gray or purple. Inner and outer reefs, 5-15 m.

## Plate 47

***Eunicea fusca*** Duch. & Mich.: Genus with tree-like colonies and bushy branches with prominent, tubular calyces. Outer rind with small, colorless clubs and purple, warty spindles; middle rind with large colorless, white or purple spindles.—Species with terminal branches 3 mm or less in diameter. Outer rind contains small (rarely over 0.1 mm) wart-clubs; middle rind has spindles about 1 mm long. Colonies rarely above 50 cm tall; gray.

***E. tourneforti*** Milne Edwards & Haime (=*Euniceopsis atra* Verrill, 1901a; *Euniceopsis tourneforti* sensu Verrill, 1907): Genus as above.—Species with outer rind containing leaf-clubs and wart-clubs to 0.15 mm long; middle rind with stout, long (to 1.5 mm)

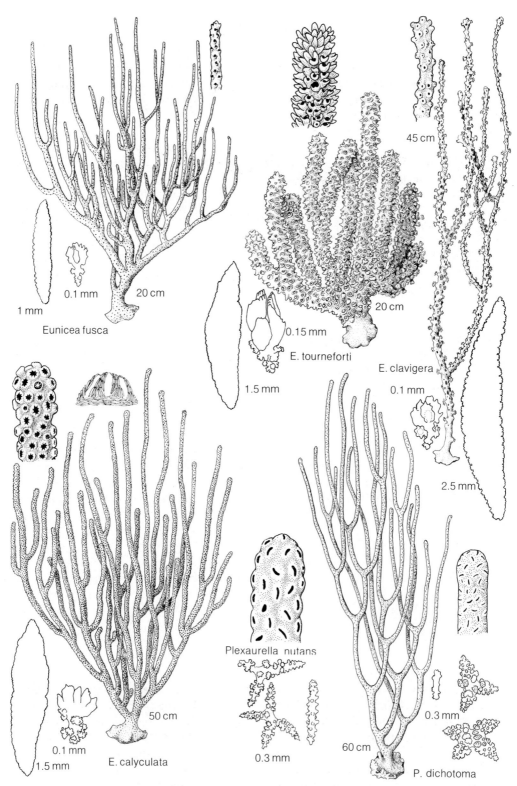

**47 *Eunicea*, *Plexaurella* (Soft corals 2)**

spindles. Colonies broad, candelabrum-shaped. The typical form has terminal branches 10-15 mm in diameter and well-developed lower calycular lips. *E. tourneforti* f. *atra* Verrill, 1901a, has terminal branch diameters of 6-10 mm and poorly developed lower calycular lips. Colonies to 61 cm tall; dark gray, blackish brown, or black. Most common on outer reefs and reef slope. (Color Plate 5.15, 16.)

***E. clavigera*** Bayer: Genus as above.—Species with outer rind containing small leaf-clubs about 0.1 mm long; middle rind with very long (to 3 mm) spindles. Colonies straggly, not candelabrum-shaped. No collaret. Colonies rarely larger than 50 cm; brown.

***E. calyculata*** s. s. (Ellis & Solander) (=*E. grandis* Verrill, 1900a; *Euniceopsis grandis* sensu Verrill, 1907): Genus as above.—Species with outer rind containing small wart-clubs, rarely longer than 0.15 mm; middle rind with large (to 2 mm) white spindles resembling rice grains. Colonies not candelabrum-shaped. Distinct collaret present. Branches long and stout, 8-16 mm in diameter. To 1 m tall; yellowish brown. On patch reefs and nearshore; never abundant.

***Plexaurella dichotoma*** (Esper): Genus with rind containing numerous quadriradiates ("butterfly spicules"). Erect, dichotomously branched colonies of thick, furry appearance. Calycular apertures slit-like.—Species with weakly armed polyps, the rods only 0.05-0.07 mm long. Rind containing numerous stout tri- and quadriradiates both to 0.5 mm long. Colonies to 100 cm tall; yellowish brown. On outer, inner and patch reefs.

***P. nutans*** (Duch. & Mich.): Genus as above.—Species has strongly armed polyps with stout rods about 0.3 mm long. Rind containing tri- and quadriradiates with slender arms, length of sclerites to 0.45 mm. Colony sparsely branched, to 100 cm tall; gray or brown. On inner and outer reefs.

## Plate 48

***Muricea laxa*** Verrill: Genus with arborescent, densely branched colonies. Branches hard and prickly, with numerous, close-set, tubular or shelf-like, lower calycular rims (edges).—Species with calyces having long spindles that are spiny proximally and smooth distally, forming prominent terminal spikes in the calyx. Branching lateral; end branches long, thin (2-3 mm in diameter) and flexible. Colonies 25-30 cm tall; gray, bluish white or yellow. Usually found in deeper water of outer reef, but sometimes as shallow as 2 m. (Color Plate 5.5-7.)

***M. muricata*** (Pallas): Genus as above.—Species with tuberculate outer rind spindles lacking smooth terminal spikes. Branching lateral in 1 plane; end branches short and thick (4.5-6.0

# KINGDOM ANIMALIA

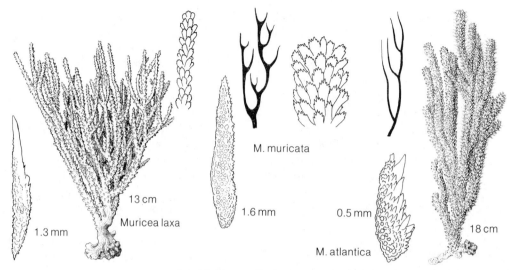

**48 *Muricea* (Soft corals 3)**

mm in diameter). Axis conspicuously flattened at points of branching. Rarely over 30 cm tall; white, beige or yellow. Common on patch and outer reefs.

***M. atlantica*** (Kükenthal) (=*M. muricata* sensu Verrill, 1907): Genus as above.—Similar to above species, except that axis is not flattened at points of branching. Also, outer rind spindles sometimes have strong spines on one side. Rarely over 50 cm tall; white or yellow. Outer reefs. (Color Plate 5.13, 14.)

## Plate 49

F. **GORGONIIDAE:** Holaxonia with a soft, cross-chambered central core. Sclerites rarely over 0.15 mm long; no clubs. Cortex of axis slightly loculated. (3 spp. from Bda.)

***Pseudopterogorgia americana*** (Gmelin) (=*Gorgonia americana* sensu Verrill, 1907): Genus with pinnate branching, twigs round or slightly flattened; branches do not anastomose. Two types of sclerites predominate: straight spindles and C-shaped scaphoids.—Species very slimy to the touch when alive (when dead, the mucus usually causes the branches to stick together). Scaphoid sclerites tuberculate on the concave side and echinulate on the convex side. Colonies to 100 cm tall; pale yellow or light purple. Common on inner and outer reefs.

***P. acerosa*** (Pallas) (Sea plume): Genus as above.—Species not slimy in life. Scaphoids weakly curved, more or less smooth on the convex side. Colonies to 180 cm tall. Color of live colony light purple, purple-red or light yellow becoming white upon drying.

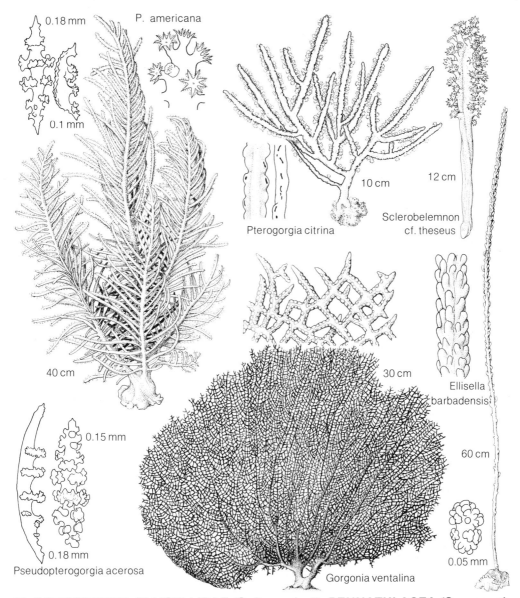

**49 GORGONIIDAE, ELLISELLIDAE (Soft corals 4), PENNATULACEA (Sea pens)**

Mostly nearshore; often host to snails (*Cyphoma*), the shrimp *Tozeuma* and brittle stars.

***Pterogorgia citrina*** (Esper) (=*Gorgonia citrina* sensu Verrill, 1900a, 1907): Branching lateral; end branches slender, stiff, and strongly flattened. Branches do not anastomose. Slit-like calyces arranged along the 2 narrow edges of the branches. Scaphoids, spindles, and anthocodial rods present in coenenchyme. Colonies to 45 cm tall; yellow, sometimes with purple edges. Primarily inshore. (Color Plate 5.1, 2.)

*Gorgonia ventalina* L. (= *G. flabellum* sensu Verrill, 1907): Branches anastomose, forming uniplanar, reticulate, fan-shaped colonies. Branches round or slightly compressed in the plane of the fan. Very small calyces located in 2 rows along edges of branches. Scaphoids, spindles, and anthocodial rods present in coenenchyme. Colonies to 180 cm tall and 150 cm wide; purple, rarely yellow or brown. Restricted to outer and patch reefs. (Color Plate 5.17.)

F. **ELLISELLIDAE:** Holaxonia with a solid, calcified, flexible, horny core. Spicules characteristically dumbbell-shaped or clubs. (3 spp. from Bda.)

*Ellisella barbadensis* (Duch. & Mich.): Whip-like, unbranched colonies with biserial or multiple lateral bands of calyces. Whips to 2 m long and usually less than 8 mm in diameter; vermilion red when alive. Most common on the fore-reef slope (50 m and more). (Color Plate 14.5.)

O. **PENNATULACEA** (Sea pens): Alcyonaria without branches and not firmly attached to substrate. Primary polyp elongate proximally, forming a stalk which anchors the colony in mud. Secondary polyps originate from distal rachis. (Of approximately 300 described species, only 1 is known from Bda.)

S.O. **SESSILIFLORAE:** Pennatulacea with secondary polyps arising directly from rachis, not united near their bases by ridge-like or leaf-like structures.

F. **KOPHOBELEMNIDAE:** Sessiliflorae with bilaterally oriented polyps on rachis, leaving a dorsal streak bare. Colonies clavate, with a well-developed axis.

*Sclerobelemnon* cf. *theseus* Bayer: Slender, elongate, slightly clavate colonies with autozooids arranged in about 9 irregular longitudinal rows. Sclerites are small oval or slightly constricted platelets with serrated ends, and flat scales resembling double-bitted axe heads. Colonies to 15 cm tall; orange or yellow. At depths of 50-60 m. (Color Plate 5.11.)

S.CL. **ZOANTHARIA** (= HEXACORALLIA): Colonial or solitary Anthozoa usually with more than 8 simple (not pinnate) tentacles. Mesenteries both paired and unpaired. (All recent orders of this subclass are represented in Bda.)

O. **ACTINIARIA** (Sea anemones): Solitary Zoantharia with hexamerously arranged mesentery pairs. After first 6 pairs of mesenteries, additional pairs are formed in all exocoels. Each tentacle corresponds to one intermesenterial space. No skeleton. (Of approximately 800-1,000 species in this order, 17 are known from Bda, 15 are included here. Of the 5 superfamilies, the Abasilaria and Mesomyaria are not represented in Bda.)

**Plate 50**

SUP.F. **BOLOCEROIDARIA:** Actiniaria without marginal sphincter muscle. Column divided into scapus (often bearing vesicles) and scapulus. Acontia absent.

F. **BOLOCEROIDIDAE:** Boloceroidaria with tentacles separated from the coelenteron by a thin, centrally perforated diaphragm and a sphincter, which cause the tentacles to break off easily at this partition. Basilar muscles absent. (2 spp. from Bda.)

***Bunodeopsis antilliensis*** Duerden (=*B. globulifera* Verrill, 1900a): Scapus provided with sessile or stalked vesicles of variable shape and development (small and simple to large and highly compound), and variegated in color, usually green, brown or white. Scapulus, oral disc, and tentacles smooth and transparent. Mouth occasionally surrounded by a blue ring. With 20-40 tentacles, variable in length and densely speckled with fine white dots (nematocyst batteries); usually expanded only at night. Diameter of base rarely exceeds 15 mm. Common on dead coral, living sponges, mangrove roots, and *Thalassia* leaves. With zooxanthellae. Asexual reproduction by pedal laceration common. A strong stinger. (Color Plate 6.1.)

F. **ALICIIDAE:** Boloceroidaria with tentacles in open connection with the coelenteron. Basilar muscles weak. (1 sp. from Bda.)

***Lebrunia danae*** Duch. & Mich. (Brown anemone; B Gill-bearing anemone): Scapus smooth, with 4-8 highly expandable, dendritic pseudotentacles that bear conspicuous, semiglobular, white to bluish nematocyst batteries. Scapus and pseudotentacles usually brown (zooxanthellae). Oral disc has 96 tentacles, speckled with fine white dots (nematocyst batteries). Diameter of pedal disc to 5 cm, that of the expanded crown of pseudotentacles often more than 25 cm. Found in reef habitats, the base attached in holes or cracks among coral and rocks. Expands pseudotentacles during daytime (photosynthesis), and tentacles at night (feeding). A severe stinger. In deeper waters (10 m) often associated with the shrimp *Thor amboinensis*. (Color Plate 9.11.)

SUP.F. **ACONTIARIA:** Actiniaria with a mesogleal sphincter. Acontia always present.

F. **AIPTASIIDAE:** Acontiaria with a very weak sphincter. Column divided into scapus and scapulus. Scapus smooth, with cinclides. Tentacles often long. Six, rarely 8 pairs of perfect, fertile mesenteries. (2 spp. from Bda.)

***Bartholomea annulata*** (Lesueur) (=*Aiptasia annulata* sensu Verrill, 1900a, 1907) (Ringed anemone): Tentacles with distinct white to faint blue, incomplete annulations. Common in bays, inlets and on coral reefs, among rocks, stones and corals. Also common in mangrove areas, but usually not on the mangrove roots. No asexual reproduction. With zooxanthellae. (Color Plate 6.5.)

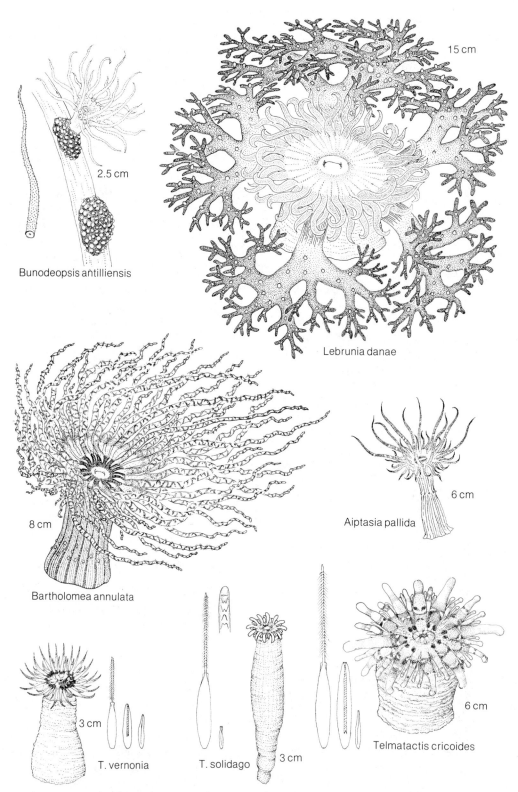

## 50 BOLOCEROIDARIA, ACONTIARIA (Sea anemones 1)

***Aiptasia pallida*** (Verrill) (=*A. tagetes* sensu Verrill, 1900a, 1907) (Pale anemone; B White-speckled anemone): Tentacles long and smooth. Upper part of scapus with 1 or 2 (rarely 3) cycles of slightly elevated cinclides. Scapus pale brown, semitransparent. Scapulus, tentacles and oral disc often speckled with pale, cream-colored or bluish dots. Pedal disc irregular in outline, about 0.5-3.0 cm in diameter. Common, found predominantly in protected bays and inlets among stones and algae, on sponges and often on mangrove roots. Pedal laceration common, often resulting in the formation of large, unisexual clones. With zooxanthellae. (Color Plate 6.4.)

F. **ISOPHELLIIDAE:** Acontiaria with a distinct sphincter. Column divided into long scapus, often with an investment, and a narrow, smooth scapulus. With or without tenaculi and cinclides. Mesenteries divided into macro- and micromesenteries. Without zooxanthellae. (4 spp. from Bda.)

***Telmatactis cricoides*** (Duch.) (=*Phellia americana* Verrill; *Phellia rufa* sensu Verrill, 1901a, 1907; not *Phellia rufa* Verrill, 1900a; *T. clavata* Duch. & Mich): Genus with elongate cylindrical column. Scapus without tenaculi and cinclides but with a distinct investment. More than 1 acontium per mesentery.—Species with column to 10 cm long. Scapus orange-red with a brown or reddish brown investment. Oral disc with up to 80 tentacles; those of the 2 innermost cycles are relatively large and distinctly clavate, whereas those of the outer cycles are smaller and less distinctly clavate or blunt. Pedal disc to 3 cm in diameter. Retractor muscles of mesenteries never kidney-shaped in cross section. Many acontia. Common on reefs and in bays, usually under stones and corals, hanging upside down.

***T. solidago*** (Duch. & Mich.) (=*Phellia simplex* Verrill, 1901a): Genus as above.—Species with elongate to vermiform column to 6 cm long, but generally smaller. Scapus and investment sand colored to dirty yellow ochre. Oral disc with 24-48 short, blunt tentacles that are never capitate. Tentacles white, with conspicuous, dark W-shaped marks. Pedal disc to 8 mm in diameter. Retractor muscles of mesenteries distinctly kidney-shaped in cross section. Few acontia. Found on reefs and in bays, usually under stones dug in sand, the base loosely attached to solid substrate.

***T. vernonia*** (Duch. & Mich.) (=*Phellia rufa* Verrill, 1900a): Genus as above.—Species with a slightly conical column to 5 cm tall. Scapus orange-red with a brown to reddish brown investment. Scapulus often brightly colored (yellow, soft green or lilac). Oral disc with 36-80 acute tentacles, often dark or with 1 or 2 dark sectors. Base to 2 cm in diameter. Retractor muscles of the mesenteries distinctly kidney-shaped in cross section. Many acontia. Common on reefs and in bays, in both sandy and rocky bottoms. Usually found in crevices or under stones. (Color Plate 6.11.)

## Plate 51

**SUP.F. ENDOMYARIA:** Actiniaria with an endodermal, occasionally reduced sphincter. Acontia absent.

**F. ACTINIIDAE:** Endomyaria with a distinct sphincter (rarely reduced) and with an adherent pedal disc with well-developed basilar muscles. Column divided into scapus and a narrow capitulum. More than 6 pairs of mesenteries are perfect. (8 spp. from Bda.)

***Actinia bermudensis*** (McMurrich) (B Red anemone): Column smooth and low, deep red to brownish. Fossa contains 0-24 blue, globular acrorhagi. Oral disc with 96-140 rather short, acute tentacles. Tentacles without pattern and often more brightly colored than the rest of the body. Two pairs of directive mesenteries and 2 distinct siphonoglyphs. Diameter of base to 4 cm. Intertidally in holes and under stones, often in rather exposed localities. Viviparous; asexual reproduction by longitudinal fission has occasionally been reported. (Color Plate 6.2.)

***Pseudactinia melanaster*** (Verrill) (=*Anemonia elegans* Verrill, 1901a, 1907; *Actinia melanaster* sensu Verrill, 1901a, 1907; *Anemonia sargassensis*; *Anemonia antilliensis*) (B Dark-star anemone): Column smooth, low, fawn colored to reddish brown. Variable number (sometimes none) of cream colored acrorhagi on parapet. Oral disc with 30-200 tapering tentacles, usually with acute tips. Oral disc with a distinct, stellate pattern of alternating pale and dark stripes. No directive mesenteries or siphonoglyphs. Diameter of base varies from 0.5 to 4.0 cm. Subtidally and intertidally on exposed shores, in pools, on and under stones and in holes. A small form of this species occurs on floating *Sargassum*. Asexual reproduction by longitudinal fission common. (Color Plate 6.3.)

***Condylactis gigantea*** (Weinland) (=*C. passiflora* sensu Verrill, 1901a; *Ilyanthopsis longifilis* sensu Verrill, 1907) (B Purple-tipped anemone): Column smooth, short-cylindrical to trumpet-shaped; color bluish gray, yellowish or brick-red. Acrorhagi absent. Up to 150 long, tapering, greenish or brownish tentacles with a conspicuous design of pale, densely arranged, ruffle-like striae. Tentacle tips blunt or slightly swollen, often rose or purple (less frequently blue or bright green). Diameter of expanded oral disc and tentacles to 30 cm; of base to 8 cm. Juveniles often with knobby tentacles. Common in shallow water (less than 10 m) usually on coral reefs, rubble flats and *Thalassia* fields. With zooxanthellae. Often associated with shrimp (e.g., *Periclimenes anthophilus, Thor amboinensis*), crabs (e.g., *Stenorhynchus seticornis, Mithrax* spp.), and a va-

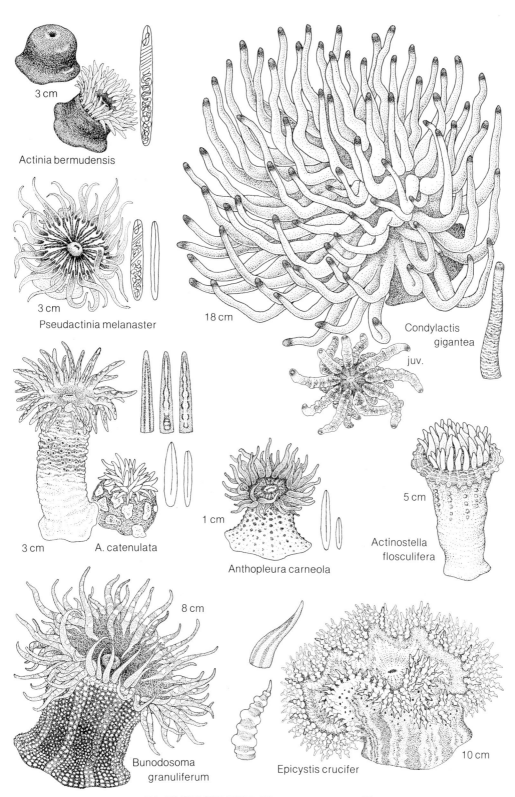

## 51 ENDOMYARIA (Sea anemones 2)

riety of fishes (e.g., juvenile wrasses, *Apogon* spp.). (Color Plate 9.12.)

***Anthopleura carneola*** (Verrill) (=*Bunodactis stelloides* var. *carneola* Verrill, 1907; *A. varioarmata*): Genus with acrorhagi and scapus covered with adhesive warts. Sphincter often circumscript.—Species with dirty-green, pink or wine-colored column, which is entirely covered with bright green, yellowish or reddish warts. Lower part of column often pale. Variable number of cream colored to faint reddish acrorhagi in the fossa. Oral disc with 30-60 short, acute tentacles, with or without a faint pattern. Diameter of base to 2 cm. Common in the upper parts of the intertidal zone under stones, in holes or crevices. Columnar warts often encrusted with shell fragments. Asexual reproduction by longitudinal fission.

***A. catenulata*** (Verrill) (= *Actinoides pallida* sensu Verrill 1900a; *Bunodactis stelloides* var. *catenulata* Verrill, 1907): Genus as above.—Species with elongate milk-white to pale pink colored column. Warts of the same colors restricted to the upper part of the column. Oral disc with 20-43 tentacles, which have a conspicuous color pattern. Oral disc usually with a conspicuous green ring around the mouth. Diameter of base to 1.5 cm. Common under stones and in crevices in tide pools; also found in sand, the base attached to buried stones. Upper part of column often covered by shell fragments adhering to the warts. No asexual reproduction.

***Bunodosoma granuliferum*** (Lesueur): Column short and cylindrical, completely covered with small vesicles and marked by a pattern of alternating pale and dark longitudinal bands. Parapet modified into small, marginal lobes that often bear cream or pinkish, globular acrorhagi on their aboral face. Oral disc with 96 (rarely more) brown, orange, purple to almost black tentacles with opaque, grayish or white cross bars above. Diameter of base to 5 cm. Found intertidally on exposed shores, but usually in sheltered niches (under stones, in holes); not common. (Color Plate 6.12.)

***Actinostella flosculifera*** (Lesueur) (=*Phyllactis flosculifera*; *Asteractis expansa*; *Actinactis flosculifera* sensu Verrill 1900a; *Asteractis flosculifera* sensu Verrill 1899, 1907) (Sand anemone): Column elongate when expanded, with a broad, grayish to yellowish green, disc-shaped collar, provided with numerous, radially arranged, short, often branched protuberances. Scapus often with a flamed pattern of cream and bluish gray or red; the upper scapus is darker and provided with longitudinal rows of distinct warts. Oral disc with 48 tentacles, often green, yellow or red with cream colored crossbars above. Expanded collar to 10 cm across; base to 4 cm. With zooxanthellae. On sandy flats, in bays and inlets, and on reefs; rarely seen. Often buried in the sand, the base attached to a solid substrate, the expanded collar resting on the sand. Viviparous.

F. **PHYMANTHIDAE:** Endomyaria with or without a weak sphincter and with an adherent pedal disc with well-developed basilar muscles. Column not divided into regions. Oral disc wide; divided into a distinct, peripheral, tentaculate region and a central region with radial rows of wart-like protuberances (discal tentacles). (1 sp. from Bda.)

*Epicystis crucifer* (Lesueur) (=*Phymanthus crucifer*) (B Cross-barred anemone): Column short and trumpet-shaped, with a flamed pattern of cream and red. Numerous longitudinal rows of 3-6 conspicuous purple warts in the marginal region, each row ending in a conical marginal wart. Tentacles short, up to 384 in number. Oral disc to 15 cm across; base to 8 cm in diameter. Two varieties occur in Bermuda: a form with elevated crossbars on the tentacles and with predominantly gray (sometimes yellow, green or dark purple) oral disc and tentacles (Color Plate 6.9); and a form with smooth tentacles, a brown oral disc, and tentacles with yellow to faint orange, radial-longitudinal stripes. (Color Plate 6.10.) Intermediates also occur. With zooxanthellae. In shallow water on reefs, rocky shores and stony-sandy flats. Often in sand, the base attached to stones; uncommon.

O. **SCLERACTINIA** (=MADREPORARIA) (Stony corals): Solitary or colonial Zoantharia very similar to the Actiniaria but with a calcareous skeleton. (Of approximately 2,500 living species, 34 are known from Bda; 25 are included here. All 5 suborders are represented in Bda.)

### Plate 52

S.O. **ASTROCOENIINA:** Scleractinia with septa composed of a few simple trabeculae. Corallites small (less than 3 mm in diameter); polyps rarely with more than 12 tentacles. Colonial.

F. **ASTROCOENIIDAE:** Astrocoeniina with beaded septal margins and poorly developed coenosteum. Phaceloid to cerioid colonies formed by extratentacular budding. (1 sp. from Bda.)

*Stephanocoenia michelinii* Milne Edwards & Haime (=*Plesiastrea goodei* Verrill, 1900a, 1901b, 1907) (B Small-eyed star coral): Corallum massive, cerioid to plocoid, often encrusting. Calices 2-3 mm in diameter, containing 24 septa. Twelve paliform lobes arranged before the 1st and 2nd cycles of septa. Columella styliform. Brown. Rare on outer reefs; more common in open, inshore waters; 2-5 m.

F. **POCILLOPORIDAE:** Astrocoeniina having septa with smooth inner margins or reduced to spines. Well-developed, solid coenosteum. Plocoid and usually

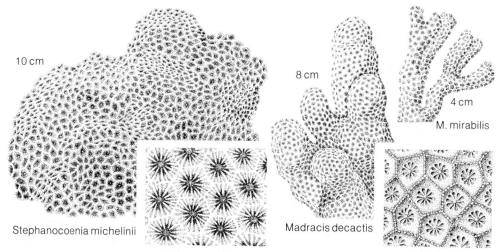

**52 ASTROCOENIINA (Stony corals 1)**

ramose colonies formed by extratentacular budding. (3 spp. from Bda.)

***Madracis decactis*** (Lyman) (B Ten-rayed star coral): Genus with well-developed septa arranged in groups of 6, 8 or 10. Columella styliform and prominent.—Species with encrusting, massive, nodular or clavate corallum. Calices about 2 mm in diameter, with 10 septa each. Pali absent. Green or brown. Common on inner reefs and in inshore waters, especially on vertical reef edges; 1-4 m.

***M. mirabilis*** (Duch. & Mich.): Genus as above.—Species with bushy colonies, to 2 m in diameter. Branches 5-9 mm in diameter, with blunt tips. Calices 1-2 mm in diameter, with 10 septa each. Pali absent. Pale cream to bright yellow. Very common in 0.5-6.0 m, especially on the vertical edges of inshore reefs.

**Plate 53**

S.O. **FUNGIINA:** Scleractinia with septa composed of numerous trabeculae in fenestrate arrangement. Synapticulae present. Corallites usually larger than 2 mm in diameter; polyps usually with more than 12 tentacles. Colonial and solitary.

F. **AGARICIIDAE:** Fungiina with solid septothecal walls. Solitary or colonial, the latter condition resulting from intratentacular budding. (1 sp. from Bda.)

***Agaricia fragilis*** Dana (B Hat coral, Shade coral): Corallum pedicellated, with broad (rarely over 15 cm in diameter), thin, saucer-shaped fronds. Calices only on upper side of frond. Up to 24 septa per calice. No columella. Chocolate- or purple-brown. Very common in shaded areas such as shallow caves; inner and outer reefs, and inshore waters; 1-15 m.

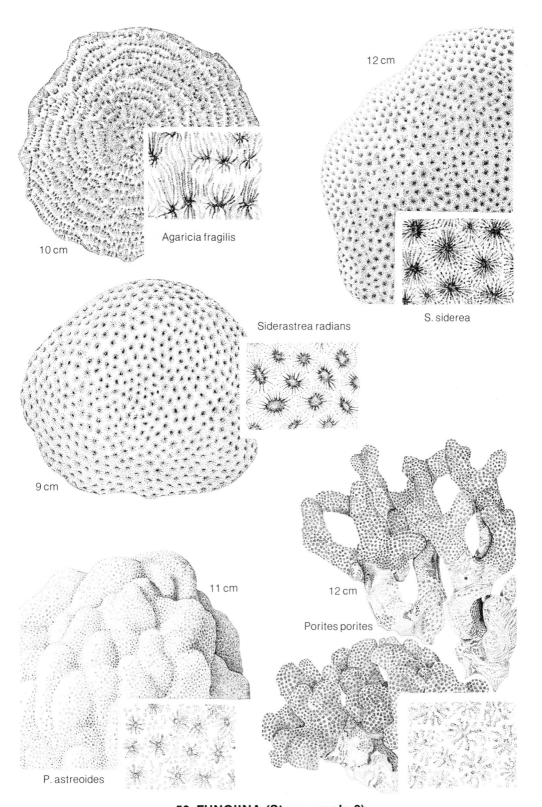

## 53 FUNGIINA (Stony corals 2)

**F.** **SIDERASTREIDAE:** Fungiina with slightly porous, synapticulothecate walls. Usually colonial, formed by both extra- and intratentacular budding. (2 spp. from Bda.)

*Siderastrea radians* (Pallas): Genus with extratentacular budding and cerioid corallum.—Species with spheroidal or hemispherical corallum to 30 cm in diameter. Calices 2.5-3.5 mm in diameter, with less than 48 septa/calice. Inner edges of septa almost vertical. Greenish to brown. Common on inner and outer reefs and in inshore waters, including mud flats where specimens may be partially buried in the mud and periodically exposed at low tide; 0-10 m.

*S. siderea* (Ellis & Solander): Genus as above.—Species with encrusting or hemispherical, cerioid coralla to 100 cm in diameter. Calices 4-5 mm in diameter, with 50-60 septa. Inner edges of septa slope gently (about 45°) toward the papillose columella. Light reddish brown. On inner and outer reefs and in inshore waters; fairly common; 0-10 m.

**F.** **PORITIDAE:** Fungiina with extremely porous walls and septa. Septa composed of 3-8 loosely united trabeculae. Colonial, formed by extratentacular budding. (2 spp. from Bda.)

*Porites porites* (Pallas) (=*P. polymorpha* Verrill, 1901b): Genus with 2 cycles of septa (12) and closely united corallites.—Species with cerioid, ramose corallum consisting of thick clumps or irregular, stout branches. Calices about 2 mm in diameter; septa poorly defined. Light brown to purple. Common in open inshore waters; 0.5-3.0 m.

*P. astreoides* Lamarck: Genus as above.—Species with cerioid, flat to hemispherical corallum, often covered with small bumps. Calices about 1.5 mm in diameter. Septa poorly defined. Yellowish brown. Very common in inner and outer reefs and open inshore waters; 0.5-15 m. (Color Plate 6.14.)

**S.O.** **FAVIINA:** Scleractinia with septa composed of numerous trabeculae in laminar arrangement; septal edges dentate. Synapticulae absent. Corallite diameter usually greater than 2 mm. Colonial and solitary.

### Plate 54

**F.** **FAVIIDAE:** Faviina with exsert septa composed of 1 or 2 fan systems producing a more or less regularly dentate inner margin. Colonies formed by both intra- and extratentacular budding. (All 5 spp. from Bda included.)

*Favia fragum* (Esper) (B Small star coral, Golf ball coral): Small, plocoid, pebble-like coralla or encrustations, usually less than 10 cm in diameter. Corallites mono-, di-, or tricentric. Calicular diameter 5-6 mm. Yellowish brown. Common on inner reefs, in open inshore waters, and in

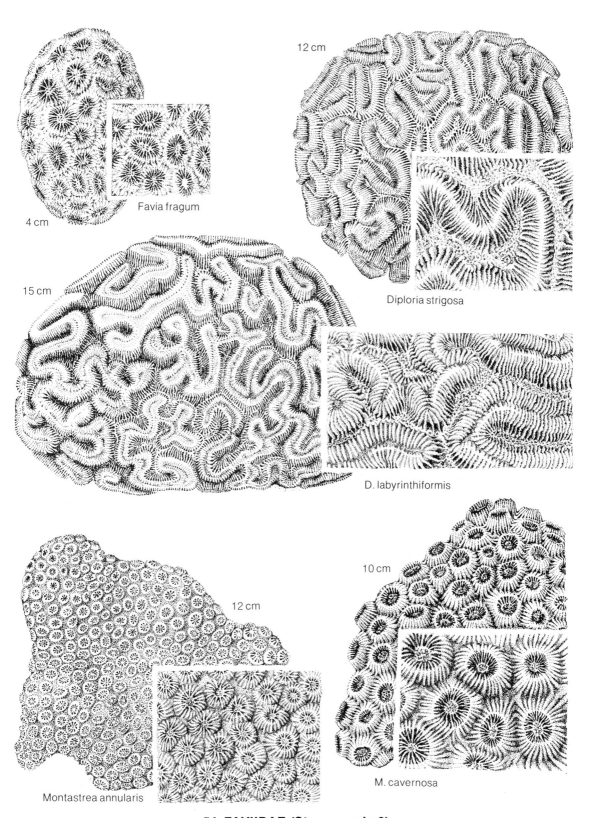

## 54 FAVIIDAE (Stony corals 3)

tide pools; 1-6 m. Planulation VI-VIII.

***Diploria strigosa*** (Dana) (=*Meandra cerebrum* sensu Verrill, 1901b, 1907) (Brain coral): Genus with meandroid, sinuous series of corallites. Paliform lobes absent. Trabecular linkages between corallite centers.—Species with hemispherical, spheroidal, or encrusting coralla to 200 cm in diameter. Valleys about 6 mm wide; collines not grooved. With 15-20 septa/cm. Yellow to greenish brown. Very common in inner and outer reefs and even in muddy bays; 1-8 m.

***D. labyrinthiformis*** (L.) (=*Meandra labyrinthiformis* sensu Verrill, 1901b, 1907) (Brain coral): Genus as above.—Species with hemispherical coralla reaching 200 cm in diameter. Valleys about 5 mm wide; collines distinctly grooved by ambulacra. With 14-17 septa/cm. Yellow or brown. Very abundant on the outer reefs and some inshore water; 1-30 m.

***Montastrea annularis*** (Ellis & Solander) (=*Orbicella hispidula*; *Orbicella annularis* sensu Verrill, 1900a, 1901b, 1907) (Small star coral): Genus with plocoid coralla and costate coenosteum. Columella spongy.—Corallum of species shaped as massive boulders, upright pillars, or encrustations to 200-300 cm in diameter. Calices 2-4 mm in diameter; 24 septa/calice. No paliform lobes; columella well developed. Yellowish to brown. Very common on inner and outer reefs and in open inshore waters; 1-30 m.

***M. cavernosa*** (L.) (=*Orbicella cavernosa* sensu Verrill, 1900a, 1901b, 1907) (Great star coral): Genus as above.—Corallum of species shaped as massive boulders, platy fronds, or encrustations to 200 cm in diameter. Calices 5-11 mm in diameter; 48 septa/calice. Columella well developed. Brown or green. Restricted to the open clear waters of the inner and outer reefs; 3-30 m. (Color Plate 6.13.)

**Plate 55**

F. **RHIZANGIIDAE:** Faviina with septa composed of 1 fan system that produces an irregular septal margin. Colony formation by extratentacular, stoloniferous budding, with individual polyps often losing their original connections. (3 spp. from Bda.)

***Astrangia solitaria*** (Lesueur): Solitary corallites or quasicolonies formed by asexual budding from basal stolons. Stolons often subsequently disrupted, giving the appearance of exclusively solitary corallites. Corallites cylindrical, 5-20 mm tall, 3-4 mm in diameter. With 36 septa; no paliform lobes. Common but cryptic; found on undersurfaces of rocks or other corals; 1-3 m. Ahermatypic. (Color Plate 6.7.)

***Colangia immersa*** Pourtalés: Solitary corallites or quasicolonies formed by asexual budding from a common basal stolon. Corallites cylindrical, to 10 mm tall, 6-7 mm in diameter.

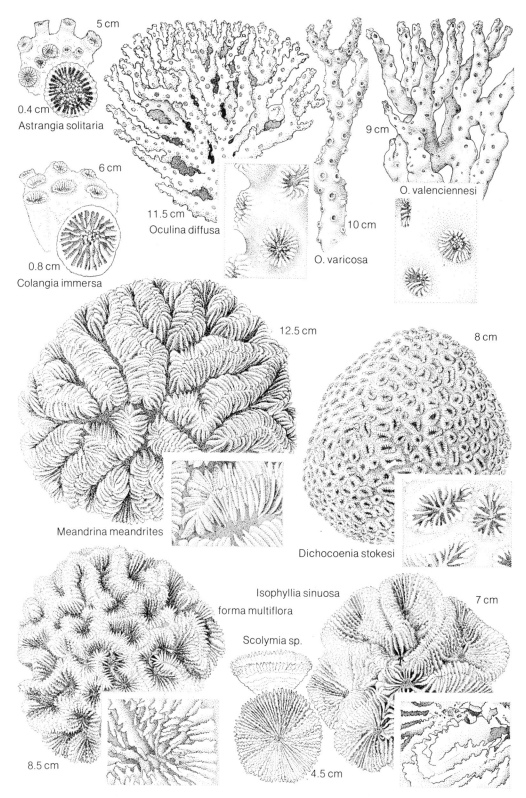

## 55 RHIZANGIIDAE-MUSSIDAE (Stony corals 4)

With 48 septa; those of 1st cycle much more exsert than others. Twelve paliform lobes. Green, with clear tentacles. Rare; found on undersides of platy corals; 3-95 m. Ahermatypic.

F. **OCULINIDAE:** Faviina with exsert septa composed of 1 fan system that produces a smooth or minutely dentate septal margin. Coenosteum dense. Colony formation by extratentacular budding. (4 spp. from Bda.)

***Oculina diffusa*** Lamarck (Ivory coral, Bush coral): Genus with well-developed columella of twisted processes. Pali arranged in an irregular crown before first 1 or 2 cycles of septa. No axial corallites.—Species have ramose, compact colonies rarely exceeding 30 cm in diameter. Branches to 10 mm in diameter. Calice 3-4 mm in diameter; 24 septa/calice. Pale yellow. Very common on inshore reefs and some inshore waters, especially in areas of high sedimentation; 1-3 m.

***O. varicosa*** Lesueur (Large ivory coral, Tree coral): Genus as above.—Species with open, irregularly branched, arborescent coralla to 40 cm tall. Branches long, crooked, to 50 mm in diameter. Calices usually raised on mounds. Calicular diameter 3-4 mm. Rare; found in some inshore waters; 12-25 m.

***O. valenciennesi*** Milne Edwards & Haime (Ivory coral, Tree coral): Genus as above.—Species with open, irregularly branched arborescent colonies to 40 cm tall. Branches long, 12-20 mm in diameter. Calices low, often sunken in the coenosteum, surrounded by a low ridge. Calicular diameter 3-5 mm. Yellow. Common in some inshore waters; 2-20 m.

F. **MEANDRINIDAE:** Faviina with exsert septa composed of 1 fan system that produces a smooth or finely dentate septal margin. Well-developed endothecal dissepiments. Solitary, or colonial by intratentacular budding. (2 spp. from Bda.)

***Meandrina meandrites*** (L.): Massive, rounded or flat boulders reaching to 100 cm in diameter (rarely more than 30 cm in Bda); smaller coralla unattached. Meandroid. Septa 6-8/cm, with smooth inner edges. Columella lamellar. Yellow, brown, or white. Moderately common in open, clear waters such as outer patch reefs; 1-10 m.

***Dichocoenia stokesi*** Milne Edwards & Haime: Heavy, rounded or flat coralla reaching to 50 cm in diameter. Calices mono- or polycentric, arranged in a Y-shape or in long (to 50 mm) meandroid valleys. Septa 10/cm, with smooth inner edges. Columella trabecular. Yellow or brown. Rare; on outer reefs; 7-10 m.

F. **MUSSIDAE:** Faviina with exsert septa composed of more than 2 fan systems that produce large, coarse, septal dentations. Well-developed endothecal dissepiments. Solitary, or colonial by

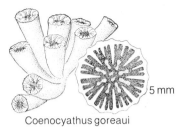
Coenocyathus goreaui

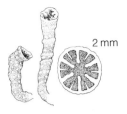
Guynia annulata

Rhizopsammia bermudensis

## 56 CARYOPHYLLIINA, DENDROPHYLLIINA (Stony corals 5)

*Plate 56*

intratentacular budding. (2 spp. from Bda.)

***Isophyllia sinuosa*** (Ellis & Solander) (=*I. multiflora* Verrill, 1901b; *Mussa anectens* Verrill, 1901c, 1907; *Mussa roseola* Verrill, 1907; *Mussa dipsacea* sensu Verrill, 1907; *Mussa fragilis* sensu Verrill, 1907; *I. fragilis* sensu Verrill, 1901b; *I. dipsacea* sensu Verrill, 1901b) (Rose coral): Small meandroid coralla usually less than 20 cm in diameter. Septa 7-9/cm, with inner edges bearing coarse teeth. Columella spongy. Valleys 20-25 mm wide. Form *multiflora* is smaller, with narrower valleys (12-15 mm wide) and has 11-12 septa/cm. Both forms occur in a variety of colors: green, white, lavender, brown and variegated. Very common on outer reefs and in inshore waters; 0.5-5 m. (Color Plate 6.8.)

***Scolymia*** sp.: Solitary, subcylindrical, firmly attached coralla to 7 cm in diameter. Calice round, with 4 or 5 cycles of septa. Septa and costae coarsely dentate. Columella large and trabecular. (The specimens collected from Bda are too small to identify as to species.) Rare, on outer reefs.

S.O. **CARYOPHYLLIINA:** Scleractinia with septa composed of numerous trabeculae in laminar arrangement; septal edges smooth. Synapticulae absent. Corallite diameter usually greater than 2 mm. Mostly solitary.

F. **CARYOPHYLLIIDAE:** Caryophylliina with solid septothecal walls. Solitary or colonial by intra- or extratentacular budding. (Of 7 spp. known from Bda the only shallow-water species is included here.)

***Coenocyathus goreaui*** Wells: Small bushy colonies of intertwined cylindrical corallities. Corallites to 20 mm long and 4-6 mm in diameter. Usually 32 septa arranged octamerally. Eight pali. Pale pink. Rare, known only from cavities in reef rock. Ahermatypic.

F. **GUYNIIDAE:** Caryophylliina with epithecal wall penetrated by regular rows of pores. Exclusively solitary and ahermatypic. (1 sp. from Bda.)

***Guynia annulata*** Duncan: Corallum cylindrical, attached basally or along its side. Extremely small; calicular diameter 1 mm, lenght to 10 mm. Septa 12 or 16. Columella consists of one twisted lath. Inconspicuous, and extremely rare in shallow water. Known depth range 3-653 m.

S.O. **DENDROPHYLLIINA:** Scleractinia with irregularly perforate septa composed of numerous trabeculae; septal edges smooth. Synapticulae present. Corallite diameter usually greater than 2 mm. (Contains only 1 family, Dendrophylliidae, and only 1 sp. is known from Bda.)

***Rhizopsammia bermudensis*** Wells: Small colonies, formed by budding of corallites from an encrusting base. Corallites cylindrical, 6-8 mm in diameter, 5-10 mm tall. Septa 48/calice. Septa and theca porous. Bright orange. Rare; known only from cavities in reef rock. Ahermatypic (Color Plate 6.6.)

## Plate 57

O. **CORALLIMORPHARIA** (=ASCLEROCORALLIA) (Coral anemones, False corals): Solitary or colonial Zoantharia with hexamerously arranged mesentery pairs. More than one tentacle corresponds to each intermesenterial space. No skeleton. (Of approximately 35 spp. worldwide, 3 spp. are known from Bda).

F. **CORALLIMORPHIDAE:** Corallimorpharia with simple, retractile tentacles, bearing conspicuous, globular acrospheres. Spirocysts always present. (1 sp. from Bda.)

***Corynactis parvula*** Duch. & Mich. (Jewel anemone): With 50-100 tentacles arranged in radial rows of 3-7 tentacles each, which increase in size toward the margin. Outer tentacles longest, often exceeding the diameter of the oral disc. Diameter of base and oral disc about 5-8 mm, that of the expanded crown of tentacles to 25 mm. Color variable: body often orange to brown, acrospheres bright orange to red. On reefs, under stones and in holes among dead coral; rare. Often several specimens are clustered as a result of asexual reproduction by pedal laceration. (Color Plate 7.5.)

F. **DISCOSOMATIDAE:** Cup- or disc-shaped, often firm Corallimorpharia with simple or dendritic discal tentacles lacking acrospheres. Tropical shallow-water forms, invariably with zooxanthellae. (2 spp. from Bda.)

***Discosoma sanctithomae*** (Duch. & Mich.) (=*Actinotryx sanctithomae* sensu Verrill, 1900, 1907; *Rhodactis sanctithomae*): Genus as the family.—Species with oral disc to 5 cm in diameter, and with a narrow, naked marginal zone, generally bordered by distinct but tiny marginal tentacles which often have white acrospheres. Discal tentacles usually well developed and distinctly dendritic. No more than 3-7 discal tentacles in the principal radial rows, the total number not exceeding 300.

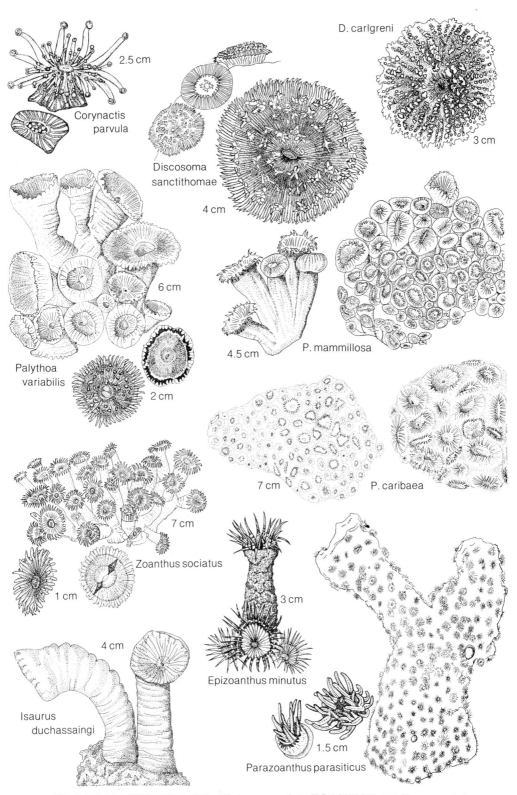

**57 CORALLIMORPHARIA (False corals), ZOANTHIDAE (Sea mats)**

Column greenish or brown; tentacles often iridescent; inside of stomodeum white. Very common on and around coral reefs, often forming extensive, brightly colored mats. (Color Plate 7.12, 13.)

***D. carlgreni*** (Watzl) (=*Rhodactis carlgreni*; *Paradiscosoma carlgreni*): Genus as above.—Species with a smaller and more rigid body than *D. sanctithomae*. No naked marginal zone. Marginal tentacles generally smaller and the margin often drawn out into small, blunt or trifid lobes. Discal tentacles wart-like to short and dendritic, up to 20 in the principal radial rows. Total number of discal tentacles may reach over 1,000. Color variable: variegated, green with brown, orange or red; some discal tentacles may be conspicuously white or yellow; inside of stomodeum usually yellow. On and around coral reefs; uncommon. (Color Plate 7.6, 7.)

O. **ZOANTHIDEA** (Colonial anemones, Sea mats): Solitary and colonial Zoantharia with paired but no hexamerously arranged mesenteries. All mesentery pairs after the first 6 are formed in the ventro-lateral exocoels. No skeleton, but sometimes calcareous encrustation in mesoglea. (Of approximately 300 spp., 9 are known from Bda; 7 are included here.)

F. **ZOANTHIDAE:** Zoanthidea with 1 or 2 mesogleal sphincter muscles and a brachycnemic arrangement of mesenteries. (7 spp. from Bda.)

***Palythoa variabilis*** (Duerden) (=*Protopalythoa grandis* Verrill, 1900): Genus with encrusted mesoglea and single mesogleal sphincter muscle.—Species with polyps that may arise singly or be connected by a lamellar basal coenenchyme in groups of 4 or 5. Large colonies cover to 0.5 $m^2$ of substrate. Length of column and width of oral disc of large polyps are 5 cm and 4 cm, respectively. Column light brown to white, depending on the degree of encrustation and concentration of zooxanthellae. Oral disc dark brown, light brown, slightly greenish or variegated, with large white areas. Tentacles light brown, numbering up to 82, but usually about 68. Common on rocks in shallow water; occasionally found in small colonies at depths to 30 m. Often found with *Zoanthus sociatus*. (Color Plate 7.3, 4.)

***P. mammillosa*** (Ellis & Solander) (=*P. grandiflora* Verrill, 1900a, 1901a, 1907): Genus as above.—Species with variable colony form depending on habitat. On rocks exposed to wave action, the coenenchyme is about 3 cm thick; colonies are 8-12 cm across, and polyps are separated by channels about 1 cm wide. On sand, the coenenchyme is usually not thicker than 1 cm and the polyps are spaced several centimeters apart, usually with only the oral discs visible. Expanded adult polyp 15 mm long with a 13 mm wide oral disc. Column and disc ocher, never variegated. There are usually 44-48 tentacles in fully grown polyps. Moderately common on rocks in shallow water.

***P. caribaea*** Duch. & Mich. (=*P. mammillosa* sensu Verrill, 1901a): Genus as above.—Species usually forming extensive colonies with numerous closely arranged polyps. Proximal ends of polyps joined in a basal coenosteum about 1 cm thick. When fully expanded, the distal portions of the polyps rise about 4 mm from the basal coenenchyme but when contracted the surface of the colony is almost flat. The colony is light ocher. Number of tentacles increases with body size (from 20 to 44). Common on most patch reefs and nearshore, often found growing on dead areas of coral heads. (Color Plate 7.8.)

***Zoanthus sociatus*** (Ellis & Solander) (=*Z. proteus* Verrill, 1900) (Green sea mat): Polyps squat, elongate, or trumpet-shaped. Basal coenenchyme laminar, covering the surface of the substrate to which the colony is attached, or reticulate, composed of a network of stolons that joins the colony on several separate substrates. A typical colony covers about 10 cm$^2$. Mesoglea lacks encrustations. Length of column and width of oral disc of expanded polyp are 6 mm and 3 mm, respectively. Color variable, but most polyps have a bluish green or yellowish green oral disc. The pigment of the disc is often arranged in a series of concentric circles and there may be a dark triangular patch at each corner of the slit-shaped mouth. Tentacles light green, usually 46-50. Column smooth, but often polyps at edge of a colony have numerous protuberances on the mid-column. Column light green, often bluish towards the distal end. Common in most of the shallow-water bays, sometimes intertidally. Often found with *Palythoa variabilis*. (Color Plate 7.1.)

***Isaurus duchassaingi*** (Andres): Polyps occur singly or in small clusters. Column usually bent, with tubercles on the convex side and a smooth concave side. During the daytime the oral disc and tentacles are normally infolded, leaving a small aperture at the center of the closed oral disc. Length and width of the column of a large solitary specimen are 7 cm and 1 cm, respectively. Clustered polyps are smaller. Column light brown, orange, white or chartreuse, sometimes translucent because of very thick non-encrusting mesoglea. Oral disc and tentacles light brown. Large polyps generally have 44-46 tentacles. Filamentous green algae and other debris often adhere to the upper warty surface of the column. On reefs and rocks; 0-20 m; very rare. (Color Plate 7.2.)

F. **EPIZOANTHIDAE:** Zoanthidea with a single mesogleal sphincter muscle, macrocnemic arrangement of mesenteries, and encrusted body wall. (1 sp. from Bda.)

***Epizoanthus minutus*** Duerden: Colonies vary in size, but usually 10-20 polyps joined by thin stolons constitute 1 colony. Length of column and width of oral disc of a large, fully expanded polyp are 10 mm and 4 mm, respectively. Color of polyp dependent on the type of mesogleal encrustation, but usually light brown. The oral disc is darker brown with a few radiating white lines. Tentacles are brown with dark brown bands and white tips. Adult polyps have 32-40 tentacles. Sphincter muscle weak, located in several small cavities near the center of the mesoglea. Found attached to the undersurface of rubble in shallow wave-exposed locations; rare. (Color Plate 7.11.)

F. **PARAZOANTHIDAE:** Zoanthidea with a single endodermal sphincter muscle, macrocnemic arrangement of mesenteries, and encrusted body wall. (1 sp. from Bda.)

*Parazoanthus parasiticus* (Duch. & Mich.): Polyps solitary or colonial, joined in groups of 3 or 4 by thin stolons. Length of column to 4 mm, width of oral disc to 4.5 mm. Column encrusted with fine calcareous and siliceous materials that give it a greenish white color. Oral disc and tentacles light brown. Up to 28 tentacles per polyp. In shallow water on reefs and in enclosed bays; exclusively on sponges, such as *Niphates erecta* and *Callyspongia vaginalis*. (Color Plate 7.9.)

## Plate 58

O. **CERIANTHARIA** (Tube anemones): Exclusively solitary Zoantharia with unpaired mesenteries. After first 6 mesenteries, all subsequent mesenteries are formed in multiplication chamber opposite siphonoglyph. Two tentacles correspond to each intermesenterial space. No skeleton; tube dwellers. (Of approximately 100 species, 1 adult sp. is known from Bda.)

S.O. **PENICILLARIA:** Ceriantharia with mesenteries arranged in duplets. Craspedonemes absent. Older, fertile mesenteries bear acontioids. (Only 1 family, Arachnactidae, in the suborder.)

*Arachnanthus nocturnus* den Hartog ( = *Cerianthus natans*? Verrill, 1901a): Tube encrusted with sand or gravel. Body to 30 cm long, yellowish brown (sometimes with dark brown streaks). Upper part of body usually white. Marginal tentacles brown with 3-6 pale crossbars on the upper surface and fine green streaks between their insertions. Labial tentacles pale brown, without crossbars. Siphonoglyph connected with about 1/3 of the total number of mesenteries. In shallow, sandy bays (often just below tide level) where its tubes are attached to stones. Locally common, but easily overlooked because it is usually retracted into its tube during daytime. Juveniles, often without a tube, may be found under stones. (Color Plate 7.10.)

O. **ANTIPATHARIA** (Black corals): Colonial Zoantharia with 6, 10 or 12 unpaired mesenteries and 6 tentacles. Internal axis keratin-like. Colonies firmly attached to substrate. (Of approx-

# ANTHOZOA

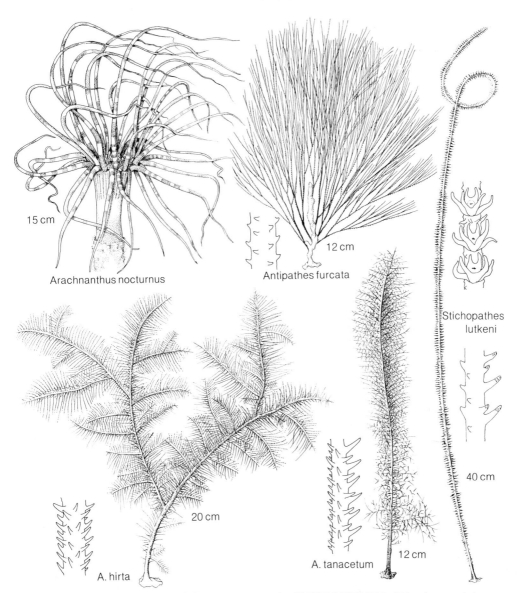

**58 CERIANTHARIA (Tube anemones), ANTIPATHARIA (Black corals)**

imately 175 spp., 5 are known from Bda; 4 are included here.)

F. **ANTIPATHIDAE:** Antipatharia with simple (non-pinnate), non-retractile tentacles.

*Antipathes furcata* Gray: Genus with sparsely to densely branched colonies. Branches simple or pinnulate with simple or bifid spines. Polyps oval.—Species with fan-shaped colonies to 40 cm tall. Branches not pinnulate. At 40-60 m, growing upright on a gently sloping bottom; not uncommon.

*A. hirta* Gray (=*A. picea*): Genus as above.—Species with

sparsely branched, monopodial colonies to 80 cm tall. Pinnules arranged biserially in 4-6 longitudinal rows. Tertiary pinnules rare. Not uncommon at 40-60 m on gently sloping bottom.

***A. tanacetum*** Pourtalès: Genus as above.—Species with a monopodial colony with pinnules arranged in 4-6 rows along the length of the axis. Tertiary pinnules common. Not uncommon below 50 m.

***Stichopathes lutkeni*** Brook: Monopodial colony without pinnules (a naked rod). Polyps arranged on 1 side of the axis throughout its length. Colonies straight, but sometimes coiled near free end. To 4 m long but not more than 1 cm in diameter. Most common between 30 and 60 m in areas with great vertical relief.

S. Cairns, J. C. den Hartog
& C. Arneson

## Phylum Ctenophora (Comb-jellies)

Characteristics: *Extremely transparent, gelatinous, mostly pelagic, bisymmetric METAZOA with 8 rows of comb-like ciliary plates; without nematocysts.* Generally 1-10 cm (rarely to 150 cm). Most species are so transparent that only the iridescent flashes emanating from the beating combs reveal their presence.

The phylum is thought to be an early offshoot from the ancestral medusoid cnidarian. Of about 90 species known, many of the pelagic ones are cosmopolitan; 5 species from the Bermuda region have been chosen to illustrate 5 of the 6 orders.

Occurrence: Marine; some species in brackish water. The great majority is pelagic, drifting in swarms with surface currents; some representatives of the order Platyctenida are adapted for benthic and even parasitic life and found on encrusting algae, mangrove roots, sedentary animals and other substrates.

Collect small forms with a plankton net outfitted with a large-volume bucket; to obtain intact specimens of the more fragile species, individuals should be caught in glass jars by SCUBA.

Identification: Only observation in the open water can fully reveal the delicate beauty of these organisms. Species are identified under the dissecting microscope, preferably live, in a glass dish.

Note the 2 planes of symmetry (one through the tentacles, the other perpendicular). The generally 8 rows of combs (plates made of fused cilia), on the aboral pole, toward the statocyst to which they are connected by delicate rows of cilia. The movement of the combs always starts at, and is directed toward, the aboral pole, propelling the animal through the water mouth first. Most species have 1 or more pairs of delicate tentacles, beset with adhesive bodies (colloblasts), with which they catch prey; in Cydippida these retract into tentacular sheaths. The mouth opens into a stomach from which a system of canals extends into the tentacular sheaths and under the comb rows; the latter canals also house the gonads. In Lobata, the laterally compressed body bears 4 short auricles and 2 often extensive lobes that are used in locomotion; in the ribbon-shaped Cestida, 4 comb rows are reduced. The benthic Platyctenida are extremely flattened, and have reduced comb rows.

Fix Cydippida and Beroida, preferably after anesthesia with magnesium chloride, in Flemming or a mixture of mercuric chloride (100 g) and glacial acetic acid (3 ml) in seawater (300 ml). After 15-30 min, carefully replace fixative with fresh water, and transfer to 70% alcohol by stages, starting with 30%. Most Lobata and Cestida fragment completely when exposed to preservative.

BIOLOGY: Hermaphrodites; each meridional canal produces both male and female gametes. Fertilization takes place in the seawater (except in a few species with brood protection). Many species produce gametes twice: as larvae, and again after reaching full growth. All ctenophores are predators; swimming mouth forward, *Pleurobrachia* trails its enormously long sticky tentacles to catch small Crustacea, Chaetognatha, planktonic larvae, etc. Ctenophores are preyed upon by scyphomedusae and fishes and sometimes parasitized by larval sea anemones and Hyperiidae. Many species are bioluminescent.

**Plate 36**

DEVELOPMENT: As a rule, the gastrula develops into a free-swimming ovoid cydippid larva (never a planula!) which already shows anlagen of most of the organ systems. Only gradually the larva changes into the shape of the (often greatly modified, e.g., *Cestus*) adult. It is not yet possible to identify larvae to species; the duration of development is unknown. Larvae feed without using their tentacles, i.e., directly with the mouth. The ability for regeneration is well developed in the phylum; some Platyctenida (*Vallicula*) are known to reproduce asexually by segregation of peripheral tissue that subsequently develops into a normal animal.

REFERENCES: For an account of northwestern Atlantic species see MAYER (1912).
Bermuda's pelagic species have not been systematically investigated. A few are listed in FEWKES (1883) and MORRIS (1975). RANKIN (1956) and FREEMAN (1967) deal with the benthic *Vallicula*.

**Plate 59**

CL. **TENTACULIFERA:** Ctenophora with 1 pair of tentacles, or many small secondary tentacles.

O. **CYDIPPIDA:** Tentaculifera with egg-shaped body, without lobes; tentacles retractile into pouches.

*Pleurobrachia pileus* Vanhöffen (= *P. rhododactyla*): Body barely compressed; tentacles with numerous, simple, filamentous side branches. To 2 cm; tentacles may extend 15-20× body length.

O. **LOBATA:** Tentaculifera with moderately compressed body and 2 large oral lobes; tentacles not in pouches.

*Mnemiopsis leidyi* A. Agassiz: Pear-shaped, with deep lateral furrows; with smooth outer surface. To 10 cm.

O. **CESTIDA:** Tentaculifera with ribbon-shaped body (through extreme compression of tentacular axis), without lobes. Paired tentacles and 4 comb rows reduced; secondary tentacles present.

*Cestum veneris* Lesueur: Subtentacular canals arching upward toward aboral pole. To 150 cm. Regular in the Gulf Stream; sometimes drifting inshore.

O. **PLATYCTENIDA:** Tentaculifera with greatly flattened body (depression of the oral-aboral axis); comb rows reduced or absent in adult.

*Vallicula multiformis* Rankin: Body folded in 2 along tentacular axis;

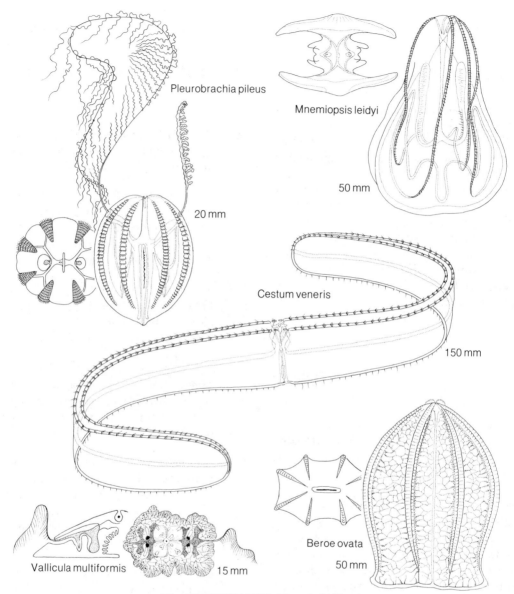

## 59 CTENOPHORA (Comb jellies)

tentacular sheaths anchor-shaped. Viviparous. Greenish; to 1.8 cm; tentacles may reach many times the length of the body (!) when expanded. In shallow, sheltered locations (such as mangrove ponds), crawling on algae, mangrove roots, Bryozoa and Ascidiacea. Sometimes in large numbers. May be misidentified as Polycladida.

CL. **NUDA:** Ctenophora without tentacles and lobes.

O. **BEROIDA:** Nuda with miter-shaped compressed body.

*Beroe ovata* Chamisso & Eysenhardt (=*B. punctata*): Laterally compressed; meridional canals anastomose in adults. Young animals spotted brown-green. To 12 cm. A common warm-water species; feeds on other Ctenophora, engulfing all or bite-size pieces with special modified oral macrocilia.

W. STERRER

# BRANCH BILATERIA

CHARACTERISTICS: *METAZOA with bilateral symmetry (at least in the larval stage), and with a cellular layer (mesoderm) between the outer (ectoderm) and the inner (endoderm) layer. All 3 layers differentiated into organs.*

Traditionally, Bilateria have been subdivided into 2 groups: Protostomia and Deuterostomia. Protostomia (Platyhelminthes, Nemertina, "Aschelminthes", Priapulida, Sipuncula, Echiura, Pogonophora, Annelida, Arthropoda, Tardigrada, Mollusca and Lophophorata) are mainly characterized by a nervous system consisting of a yoke-shaped brain straddling the foregut, and paired ventrolateral nerves; furthermore the larval mouth opening generally turns into the adult mouth, and the anus opens secondarily. Deuterostomia (Chaetognatha, Echinodermata, Hemichordata and Chordata) mostly have an unpaired dorsal nerve chord, and the larval mouth becomes the anus of the adult. Because the exact definition and even the validity of the concept continue to be controversial, protostomes and deuterostomes are here rather considered as major trends within the Bilateria. Similarly, the widely accepted use of the origin and organizational level of body cavities for the grouping of phyla (acoelomates, pseudocoelomates and coelomates) has received renewed criticism in the light of fresh evidence (as most recently expressed in the textbook by RE-MANE et al. 1980). Within the conservative framework of this field guide, therefore, the following terminology is used in a purely descriptive manner:

"without body cavity"—no body cavity present other than the digestive tract; mesoderm fills space between outer and inner body layer;
"with incomplete body cavity"—body cavity present but incompletely lined with mesoderm;
"with complete body cavity (=coelom)"—body cavity completely lined with mesoderm.

The system as used in this book recognizes 28 phyla of Bilateria of which 24 are represented in Bermuda (see Contents). Of two commonly recognized superphyla only one (Lophophorata) is used here; the term Aschelminthes, which embraces a motley assemblage of more or less worm-shaped phyla (Gnathostomulida, Gastrotricha, Nematoda, Nematomorpha, Rotifera, Kinorhyncha, Acanthocephala and Loricifera), is not employed. Phyla not represented in Bermuda are the predominantly freshwater-dwelling Nematomorpha, the recently described interstitial marine Loricifera (KRISTENSEN 1983), the terrestrial Onychophora and the endoparasitic Pentastomida.

## Phylum Platyhelminthes
(Flatworms)

CHARACTERISTICS: *Generally worm-shaped or flattened BILATERIA; body surface covered by cilia or a false, intracellular cuticle. Anus usually lacking; space between internal organs and body wall completely filled with loose tissue (parenchyma).*

All 3 classes are represented in Bermuda: the most free-living Turbellaria (p. 198), and the parasitic Trematoda (p. 203) and Cestoda (p. 205). A recently proposed system (EHLERS 1984) abolishes the class Turbellaria as a monophyletic taxon.

## Class Turbellaria (Free-living flatworms)

CHARACTERISTICS: *Mostly free-living, thread- to leaf-shaped PLATYHELMINTHES with cilia usually covering entire body surface.* Mostly of microscopic dimensions (0.5-2 mm; only some Polycladida to several centimeters), turbellarians are generally colorless or whitish; some of the larger polyclads, however, can be colorful and ornate. Small species glide or swim smoothly by means of their cilia, whereas large Polycladida are able to swim by undulating contractions of the body musculature.

Of 12 orders, the predominantly freshwater-inhabiting Temnocephalida, Tricladida and Lecithoepitheliata, and the small marine group Haplopharyngida are not included here. Of approximately 6,000 known species only about 50 have been described from Bermuda so far; 17 are included here as examples.

OCCURRENCE: In marine, freshwater and moist terrestrial habitats. Generally abundant in sand, mud and plant and animal epigrowth on rocky bottoms; rarely pelagic. Greatest abundance in the lower intertidal and shallow (1-100 m) subtidal but some do extend into the deep sea.

Turbellaria living in plant and animal epigrowth are best collected by diving, with nylon bags put over epigrowth under water. For separation of specimens from substrate samples climate deterioration or the magnesium chloride method can be used. Sand- and mud-living Turbellaria are collected with sediment, in containers with lids. Sand Turbellaria are extracted with the magnesium chloride method or seawater-ice method. Mud forms are best extracted with the Swedmark method.

IDENTIFICATION: Live material is needed for field identification by squeeze preparation. Positive identification often requires phase contrast microscopy and histological sectioning.

Of special importance in identification are the following: 1) Foregut (=pharynx) structure. The mouth opening may lead directly into the digestive parenchyma (=Pharynx O). A simple ciliated tube as foregut is called pharynx simplex. A ring-shaped fold in such a foregut tube is called pharynx plicatus; it can be disc-shaped (as in some Polycladida) or elongated and tube-like (as in some Proseriata) but is always enclosed in the surrounding pharynx cavity. The pharynx bulbosus may be a ring-shaped, downward-pointing (pharynx rosulatus) or barrel-shaped, forward-pointing (pharynx variabilis or doliiformis) muscular organ that is sharply separated from the surrounding parenchyma (filling tissue). 2) Presence, absence and special structure of the statocyst. 3) Male and female parts of the genital system. The gonads may be unpaired, in a single pair or in many pairs (then called follicular). The yolk in the female gonad is either produced in the egg cell itself (then called ovaries) or as part of the germative cells in so-called yolk glands. The male gonoducts usually end in a penis organ, which may be provided with a stylet. The female system often includes structures (bursal organs) functioning in storage of a partner's sperm and during copulation. Male and female gonoducts may join in a common gonopore. Accessory glandular organs are common.

For fixation, Bouin's fixative gives generally good results. Whole mounts can be made after squeeze preparation by using Meixner's technique.

BIOLOGY: Turbellaria are hermaphrodites with internal fertilization by copulation. Eggs are laid singly or in small cocoons (5-1,000 eggs). The life-span is usually 1-6 months; some live 1 yr or more. Most Turbellaria are active predators or scavengers, but detritivores and diatom feeders are common as well. Turbellaria, Nematoda and Copepoda are common prey items. Predators include Nemertina and Nematoda. Sand and mud forms are likely to be eaten by macrobenthic deposit feeders. Parasites in Turbellaria include Sporozoa and Orthonectida. Many Turbellaria have unicellular algae as symbionts (e.g., zooxanthellae).

# TURBELLARIA

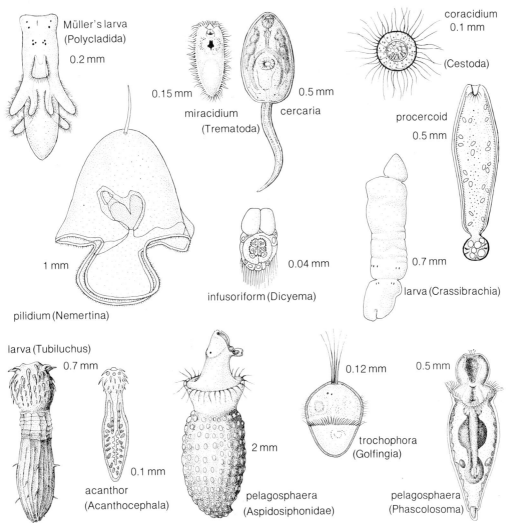

**60 PLATYHELMINTHES-POGONOPHORA (Lower worms): Development**

## Plate 60

DEVELOPMENT: Spiral cleavage; with few exceptions (Müller's larva of Polycladida) development is direct.

REFERENCES: HYMAN (1951a) still stands out as the best English summary of the class. For more recent references see EHLERS (1985).

Bermuda's Turbellaria, particularly the microscopic species that make up the majority of the fauna, are very insufficiently known. See HYMAN (1939) for Acoela and Polycladida, RIEGER (1977) for some Macrostomida, and KARLING (1978) for several other orders.

## Plate 61

**O. CATENULIDA:** Fragile, filiform Turbellaria with dorsal ♂ gonopore; mostly with a pharynx simplex, a ciliated gut, a single statocyst; ♀ gonad as ovary. Most catenulids live in fresh water; the few marine forms are found in reducing sediments. (1 sp. from Bda.)

***Retronectes*** sp.: With pharynx simplex and statocyst. Both body ends narrowing to a tip. In intertidal or shallow subtidal sand rich in organic detritus; rare.

O. **NEMERTODERMATIDA:** Small ovoid to filiform Turbellaria with a statocyst containing 2 symmetrically arranged statoliths; ♀ gonad as ovaries. Exclusively marine, in subtidal sands and muds. (1 sp. from Bda.)

***Flagellophora*** sp.: With frontal mouth opening and broom organ. In shallow subtidal sands; rare.

O. **ACOELA:** Small droplet-, leaf- or thread-shaped Turbellaria without distinct gut lumen, almost always with 1 statocyst with a single statolith; usually with pharynx O and ♀ gonads as ovaries. Except for 2 genera exclusively marine, in all main habitats. (5 spp. from Bda.)

***Amphiscolops bermudensis*** Hyman: With 2 mouth pieces on bursal organ. Posterior end with characteristic mid-dorsal lobe. With characteristic pattern of light-refracting granules and numerous zooxanthellae. Fast swimmers, in algae and on *Thalassia* leaves; not uncommon.

O. **MACROSTOMIDA:** Small Turbellaria with ciliated pharynx simplex and usually sac-like gut; without statocyst and with ♀ gonads as ovaries. Common in sandy bottoms and vegetation. (4 spp. from Bda.)

***Paramyozonaria bermudensis*** Rieger: Gut with muscle ring. With complicated ♂ stylet. Slightly brownish. In shallow subtidal sand; not uncommon.

O. **PROLECITHOPHORA:** Small Turbellaria with pharynx plicatus or pharynx variabilis and sac-like gut; without statocyst; ♀ gonad with yolk glands. Common in vegetation in the shallow subtidal. (4 spp. from Bda.)

***Plagiostomum girardi bermudensis*** Karling: Mouth rostral, common gonopore caudal. Rostral end with dark brown pigment spot between 2 eyes, tail pointed. In muddy sand overgrown by green algae, from where they like to swim up into the water; not uncommon.

O. **PROSERIATA:** Elongated Turbellaria with pharynx plicatus and gut with lateral diverticles; ♀ gonads follicular and with yolk glands. Mostly with characteristic statocyst. Especially on sandy bottoms (2 spp. from Bda.)

***Pseudominona dactylifera*** Karling: With statocyst. ♀ germative cells behind pharynx. Tail 3-pronged, with adhesive organs. Quickly moving and haptic. On exposed beaches and shallow subtidal; common.

***Polystyliphora*** sp.: Without statocyst; tail region with numerous stylet-bearing accessory organs. Slow moving; on shallow subtidal sandy bottoms; rare.

O. **RHABDOCOELA:** Small Turbellaria with pharynx bulbosus and sac-like gut; without statocyst, ♀ gonad

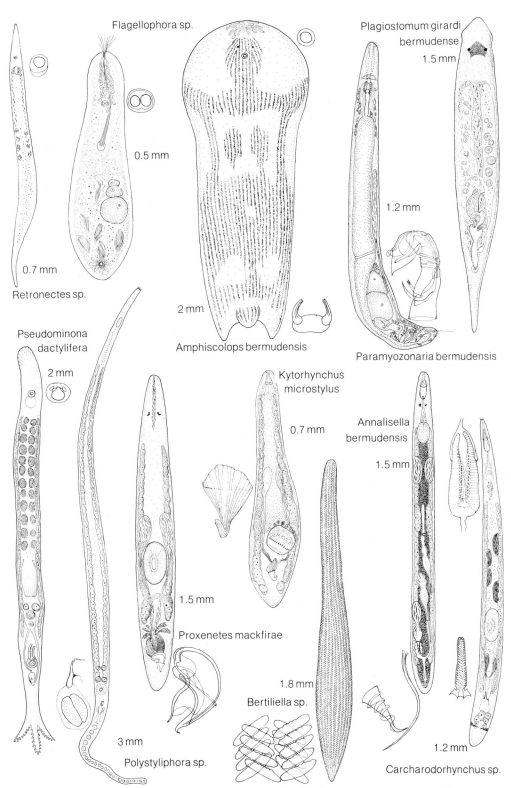

**61 CATENULIDA-RHABDOCOELA (Flatworms 1)**

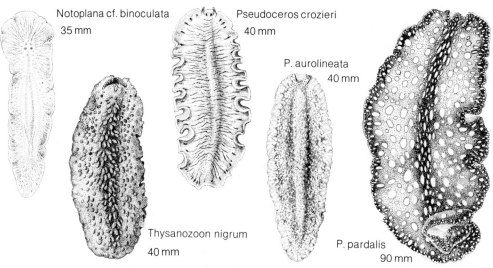

**62 POLYCLADIDA (Flatworms 2)**

with yolk glands. Common in all main habitats. (Of 3 suborders, Dalyellioida have not been recorded from Bda. About 20 spp. from Bda.)

**S.O. TYPHLOPLANOIDA:** Rhabdocoela with mostly ventrally pointing pharynx bulbosus.

*Proxenetes mackfirae* Karling: With complex ♂ stylet and small eyes. In shallow sandy habitats with growth of green algae and adjacent mangrove areas; not uncommon.

*Kytorhynchus microstylus* Rieger: With triangular invaginated pit at rostral tip. Without eyes and with funnel-shaped ♂ stylet. Dark brown. In shallow subtidal sand; common.

**S.O. KALYPTORHYNCHIA:** Small, elongated Rhabdocoela with pharynx bulbosus pointing ventrally or rostrally. With a rostrally located proboscis for prey capture. Common in sand; often haptic and highly contractile.

*Bertiliella* sp.: Body wall with a skeleton made of spicules in longitudinal rows. Whitish-iridescent in reflected light, slow moving. In clean subtidal sand; very rare.

*Annalisella bermudensis* Karling: With conical proboscis and 2 characteristic stylets. Thread-like, slightly yellowish, with 2 eyes. In coarse sand with green algae or *Thalassia*; not uncommon.

*Carcharodorhynchus* sp.: With forceps-like proboscis with small teeth. With characteristic ♂ stylet; yellowish brown. In shallow subtidal sand; common.

**Plate 62**

**O. POLYCLADIDA:** Usually large, flat Turbellaria with pharynx plicatus; gut with many diverticula. Without yolk glands and statocyst. Exclusively marine. (About 15 spp. from Bda.)

**S.O. ACOTYLEA:** Polycladida without a median sucker behind the ♀ pore; tentacles when present of the nuchal type (originating in the neck region).

***Notoplana* cf. *binoculata*** (Verrill): Anteriorly rounded, posteriorly tapered, with thin, undulated edges. Pale flesh-colored to whitish. Cerebral ocelli form 2 distinct round clusters; sometimes a row of small ocelli anteriorly. Common under stones in the intertidal; very active, disappearing readily into crevices when disturbed.

**S.O. COTYLEA:** Polycladida with a sucker behind the ♀ pore; tentacles when present of the marginal type.

***Thysanozoon nigrum*** Girard: Elongated-oval; velvety black above, with numerous rounded papillae. Common in moderately calm areas on ascidians (usually *Eudistoma olivaceum*). Able to swim.

***Pseudoceros crozieri*** Hyman (Tiger flatworm): Genus generally without dorsal papillae.—Species color varying from whitish to orange (resulting from ingested ascidians); dorsal surface marked with black wavy cross lines, some marginally ending in black spots. Body margin white. Not uncommon on ascidians, especially *Ecteinascidia turbinata*. (Color Plate 15.12.)

***P. aurolineata*** Verrill: Genus as above.—Species with purplish-fawn mottled coloration on a whitish background. Under stones and on ascidians; uncommon.

***P. pardalis*** Verrill (Leopard flatworm): Genus as above.—Species broad, with bright yellow spots on a dark purple background. Subtidally under stones; rare.

R. M. RIEGER & W. STERRER

## Class Trematoda (Flukes)

CHARACTERISTICS: *Flattened or cylindrical, parasitic PLATYHELMINTHES with ventral adhesive organs.* Generally 1 mm-3 cm (rarely to 7 m!), colorless; locomotion is a sluggish inchworm crawl; some larval stages swim.

Now divided into 3 biologically distinct orders (sometimes considered classes): Monogenea, Digenea and Aspidogastrea. More than 6,000 species are known; of Digenea alone, over 50 species have been recorded from Bermudian fishes. Four species are given here as examples.

OCCURRENCE: Monogenea are largely ectoparasites on the skin and in the mouth and gill chambers of fishes and Crustacea. Adult marine Digenea and Aspidogastrea are endoparasitic in the intestinal tract or blood vessels of fishes and turtles; their larval stages are found in intermediate hosts (mostly Mollusca), or are briefly free-swimming.

Collect adults by screening skin, gills and viscera of suspected fresh hosts under a dissecting microscope. Larvae can be extracted from gastropods by subjecting the host to stress from stagnant seawater.

Although a number of species are known to cause diseases in humans (e.g., schistosomiasis), no species that may infest humans has been reported from Bermuda.

IDENTIFICATION: External features (such as number, position and shape of suckers) distinguish the orders; in Monogenea sclerotized components (penis, vagina, clamps and hooks) are important. Microanatomy (serial sectioning) and stained whole mounts are indispensable for species identification.

Fix in Bouin's or 4% neutral formaldehyde; preserve in 75% alcohol.

BIOLOGY: Predominantly hermaphrodites, with cross-fertilization by copulation, rarely self-fertilization. Endoparasites are facultative anaerobes.

**Plate 60**

DEVELOPMENT: Monogenea develop via 1 free-swimming larva, and without an intermediate host. In the complete digenean life cycle, a free-swimming miracidium hatches from the egg, enters an intermediate host, and develops into a redia or a sporocyst, which in turn produces free-swimming larvae called cercariae. The latter infest either another intermediate host, in which they become metacercariae (often clearly visible in the skin of the host fish), or the final host, and end up in the host's gut or blood vessels where they develop into adults. The sequence of sexual and asexual reproduction, involving up to 4 hosts, results in enormous numbers of progeny, which ensure infestation of the usually highly specific hosts.

REFERENCES: The vast literature is condensed in DAWES (1946) and YAMAGUTI (1963a, 1971).

Bermuda's species have been reported by LINTON (1905, 1907), BARKER (1922), RAECKE (1945), LEBOUR (1949), HANSON (1950), REES (1970), CONE & BEVERLEY-BURTON (1981), RAND & WILES (1985) and RAND et al. (in press).

**Plate 63**

O. **MONOGENEA:** Trematoda with a large, often armed posterior adhesive organ, and a pair of anterior suckers flanking the mouth. Largely ectoparasites on skin and gills; only 1 host. (Example:)

*Microcotyle incisa* (Linton): Slender; with black pigment spots on dorsal surface; posterior adhesive organ with rows of about 45 clamps each. On gills of gray snapper, porgies and hogfish.

O. **DIGENEA:** Trematoda with typically 2 suckers, 1 around the (terminal) mouth, the other ventrally. Exclusively endoparasites with 2-4 hosts. (Examples:)

*Alcicornis carangis* MacCallum: Mouth ventral, behind mid-body. No ventral sucker; apical disc with 7 tentacles. In the gut of *Caranx ruber*.

*Lepidapedon trachinoti* Hanson: Oral sucker subterminal, ventral sucker small; genital pore near ventral sucker. In the gut of *Trachinotus* sp.

O. **ASPIDOGASTREA:** Trematoda with a large, subdivided adhesive organ covering most of the ventral surface. Mostly endoparasites in fishes; with 1 or 2 hosts. (1 sp. from Bda.)

*Lobatostoma ringens* ( = *Aspidogaster ringens*) (Linton): Ventral disc elliptical, with about 42 disclets around the border, and a median longitudinal ridge. In the gut of *Halichoeres radiatus*.

W. STERRER

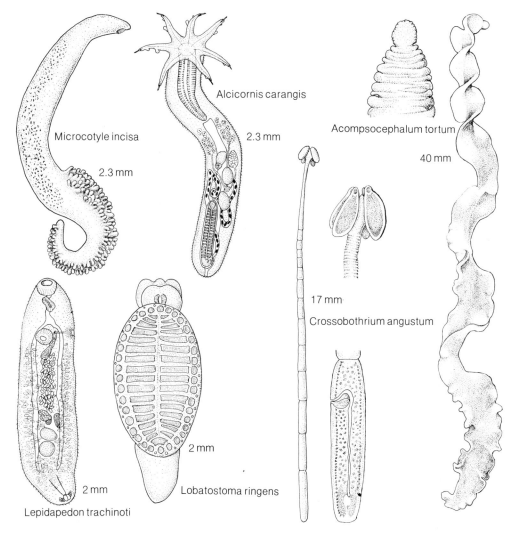

**63 TREMATODA (Flukes), CESTODA (Tapeworms)**

## *Class Cestoda* (Tapeworms)

CHARACTERISTICS: *Elongated, flattened endoparasitic PLATYHELMINTHES with nonciliated, cuticularized epidermis; without a digestive tract.* Colorless or white, 1 mm–15 m (!).

Of 2 subclasses the first (the trematode-like Cestodaria) is not known from Bermuda; the second (Eucestoda) comprises the bulk of the species. Of about 3,400 described species, only 10 have been recorded from Bermuda, mainly from sharks; 2 are given as examples, illustrating 2 of the 10 orders of Eucestoda.

OCCURRENCE: Endoparasites in the gut of vertebrates (some larval stages in Crustacea and Acari, and the body cavity and musculature of Osteichthyes). Marine Teleostei harbor few adult but many larval species, whereas the reverse is true for sharks (because of their higher position in the food chain).

IDENTIFICATION: Under a dissecting microscope, specimens with undamaged (!) anterior end can be identified to order; species identification usually requires whole mounts and serial sectioning.

Note the scolex (head), with a varying number of grooves, suckers, tentacles and/or hooks; and the strobila (body), made up of linearly arranged proglottids (identical body sections carrying the reproductive organs).

Fix in hot (60-70°C) Bouin's or 4% formaldehyde, preferably under a slide to obtain stretched specimens.

BIOLOGY: Hermaphrodites; self-fertilization is common, i. e., one proglottid copulates with another of the same worm. Generally, posterior, fertilized proglottids separate from the organism, die and are discharged with the host's feces, liberating millions of eggs. Adults, sometimes in considerable numbers, adhere to the host's gut epithelium by means of the scolex armature, taking up food through the body wall. Anaerobic metabolism is predominant. Comparatively little is known about marine species.

### Plate 60

DEVELOPMENT: When an egg has been liberated, or ingested by a host, an oncosphaera (a larva with movable hooklets, called coracidium when it is ciliated) hatches, bores its way into the blood system, and develops into a cysticercus, plerocercus, procercoid or plerocercoid, often settling in striated muscle. When the intermediate host is eaten by a predator, the larva ends up in the gut of its primary host where it grows into an adult.

REFERENCES: For monographs see WARDLE & McLEOD (1952) and YAMAGUTI (1959).

Bermuda's species are listed in LINTON (1889, 1907) and REES (1969).

### Plate 63

O. **PSEUDOPHYLLIDEA:** Eucestoda with scolex (when present) provided with 2 flat sucking grooves. Primary hosts from among all vertebrate classes, often with 2 intermediate hosts. (Example:)

*Acompsocephalum tortum* (Linton): Without a scolex (!) and attachment organs, and without external segmentation; 4-7 cm long. In *Synodus intermedius*. Displays a striking corkscrew movement when exposed to seawater.

O. **TETRAPHYLLIDEA:** Eucestoda with scolex provided with 4 flat suckers that may be stalked, armed or divided. Primary hosts are exclusively sharks; little is known about intermediate hosts. (Example:)

*Crossobothrium angustum* (Linton) (=*Orygmatobothrium angustum*): Suckers on short stalks, oval and unarmed. Mature segments 5× as long as wide; to 2 cm. In *Carcharhinus*.

W. STERRER

## Phylum Mesozoa

CHARACTERISTICS: *Minute, elongate, parasitic BILATERIA consisting of an axis of 1 to a few reproductive cells surrounded by a single layer of ciliated cells.* From 0.05 mm to 7 mm and colorless-transparent, they swim slowly by means of cilia.

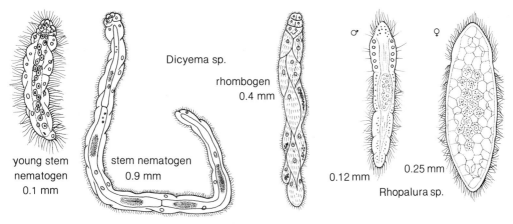

**64 MESOZOA**

Their relationships (to Protozoa, Platyhelminthes or Aschelminthes) remain obscure. Of about 50 species in 2 orders, 2 are given as examples.

OCCURRENCE: Parasites of marine invertebrates. Dicyemida are found, often in masses, in the kidneys of Cephalopoda, whereas Orthonectida occur in Turbellaria, Nemertina, Polychaeta, Ophiuroidea, a.o.

Collect the parasites by taking samples of host tissue and body fluids.

IDENTIFICATION: Use a high-power microscope with phase contrast to study squeeze preparations of live animals.

Distinguish—in Dicyemida—between the 2 cell layers, and the body regions (a "head" made up of few large cells, and a "trunk"). The various developmental stages (see below) are extremely difficult to tell apart.

Preserve in Bouin's fluid for histology, in osmium tetroxide for electron microscopy. Smears of infected host tissue can be impregnated with silver nitrate.

*Plate 60*

BIOLOGY AND DEVELOPMENT: Life cycles are complex and incompletely understood. For Dicyemida it has been shown that young cephalopods contain nematogens that multiply asexually via vermiform larvae. As the host matures, nematogens turn into rhombogens which, by way of differentiations (called infusorigen) within their axial cells, produce eggs and sperm. The fertilized egg develops into a rounded infusoriform larva, which is passed out with the host's urine and presumably requires 1 (unknown) intermediate host to complete the life cycle. In Orthonectida, males and females are released by the host simultaneously, whereupon the male discharges sperm near the genital opening of the female. The characteristic ciliated larva infects a new host, then develops into an irregular syncytium that eventually produces sexual males and females.

REFERENCES: For an introduction see STUNKARD (1954) and McCONNAUGHEY (1963). The life cycle of Dicyemida was most recently described by McCONNAUGHEY (1951); morphology and ultrastructure are dealt with by KOZLOFF (1969, 1971; Orthonectida) and RIDLEY (1969; Dicyemida).

Bermuda's Mesozoa have yet to be studied.

*Plate 64*

**O. DICYEMIDA:** Elongate, completely ciliated Mesozoa with complex life cycle in which nematogens are followed by rhombogens; the latter produce a larva sexually. (Example:)

***Dicyema*** sp.: Nematogen long, worm-shaped; head consisting of 8 cells. In the renal cavity of *Octopus*; usually in very large numbers.

O. **ORTHONECTIDA:** Spindle-shaped, annulated, often incompletely ciliated Mesozoa; ♂ and ♀ developing from irregular syncytia within the host. (Example:)

***Rhopalura*** sp.: ♀ no more than 10× longer than wide. In ophiuroids and other invertebrates.

W. STERRER

## Phylum Nemertina
(=Rhynchocoela) (Ribbon worms)

CHARACTERISTICS: *Worm-shaped BILATERIA with ciliated epidermis, an anus, a circulatory system, and an evaginable proboscis contained in a tubular cavity (rhynchocoel) above the gut.* Body length 1 mm to 30 m (!), usually very contractile. Frequently with bright colors or patterns. They crawl sluggishly, often tying themselves into a knot when disturbed.

Of 4 orders, 2 (Palaeonemertini and Bdellomorpha) have not yet been found in Bermuda. Of approximately 800 species about 20 are known from Bermuda, and 8 are included here.

OCCURRENCE: Mainly marine; bottom dwelling, but several genera are bathypelagic. Relatively few taxa occur in freshwater and terrestrial habitats. Some species are commensals or parasites of other marine invertebrates (e.g., decapod crustaceans and bivalves). The benthic nemerteans have a cryptic mode of life and hide in crevices and among plant tangles and epizoic growth, or are part of the infauna of soft bottoms.

The larger littoral species can be collected by turning over stones and corals and checking their undersides as well as the sediment they were lying on; manual sorting of coral debris may also produce some specimens. However, for more efficient collecting, particularly for quantitative purposes, it is necessary to obtain samples of hard substrates or aquatic plants and extract the nemerteans with the climate deterioration method. For extraction from sand samples the magnesium chloride method is recommended. To separate nemerteans from mud samples, small portions of the sediment should be examined under a dissecting microscope. Subtidal collecting is best done by SCUBA, and on soft bottoms also with dredge or core samplers.

IDENTIFICATION: The external characters of live specimens together with morphological features of the proboscis and its armature suffice for determining the species listed herein. It has to be noted, however, that the systematics of nemerteans in general is based primarily on details of the internal organization, and for exact identification the study of histological sections is frequently indispensable.

Body and head shape, color patterns, and the presence or absence of cephalic slits (rather deep, longitudinal indentations along the lateral sides of the head), cephalic grooves (shallow, transverse indentations demarcating the head from the remainder of the body), and a caudal cirrus (a delicate, filamentous appendage at the posterior end of the body) can easily be observed with a dissecting microscope. The latter is also adequate for discerning the configuration (i.e., either a single tube or branching tubules) of the everted proboscis, and unless the animal has voluntarily evaginated this organ it can be induced to do so by poking it with a needle or by adding a few drops of alcohol to the water in which it is kept. The proboscis emerges from a pore near the tip of the head, and the proboscis armature (if present) is located at the anterior end of the fully everted proboscis. To distinguish between monostiliferous (1 stylet on a pyriform base) and polystiliferous (several stylets on a crescent-shaped base) armature it is necessary to establish the number of the minute (20-100 μm long) stylets by examining the proboscis under a cover slip with high-power optics, preferably phase contrast.

Fixation should always be preceded by narcotization (chloral hydrate or menthol) to avoid excessive contraction or autotomization. For histology fix with Bouin's and transfer specimen after several days to 70% alcohol for preservation.

BIOLOGY: Most nemerteans are dioecious but the majority display no external sexual dimorphism; a few species are hermaphroditic. Except for the rare instances of viviparity, fertilization is external; some nemerteans deposit their eggs in mucous cocoons. Asexual reproduction by fragmentation and regeneration occurs in some taxa. Life expectancy is from 1 to several years. Nemerteans are carnivorous and predators as well as scavengers. To catch live animals, the proboscis is everted and coiled around the prey or, in Hoplonemertini, the proboscis stylets are used to inflict a wound into which a toxic secretion is injected. The diet ranges from Protozoa to small fishes, and the food is swallowed whole. Predators are crabs and fishes; parasites are common (Sporozoa, Ciliata, orthonectid Mesozoa, larval stages of Trematoda, and Nematoda).

### Plate 60

DEVELOPMENT: Direct or indirect. The indirect type is characteristic for Heteronemertini and involves either a helmet-shaped, free-swimming pilidium larva or a Desor larva which remains confined within the egg membrane. In both instances the young worm develops inside the larva, which undergoes a complicated metamorphosis.

REFERENCES: For general orientation see GIBSON (1972) and HYMAN (1951a); for a systematic overview consult COE (1943).

Nemerteans from Bermuda have been treated by VERRILL (1900e), COE (1904, 1936, 1944, 1945) and WHEELER (1942). It would seem that Bermuda's nemertean fauna, though not particularly diverse, is still insufficiently known.

### Plate 65

**O. HETERONEMERTINI:** Nemertina with mouth underneath brain; proboscis not armed with stylets.

*Gorgonorhynchus bermudensis* Wheeler: Body soft, to 12 cm long, with posterior portion flattened and wider than anterior region. With cephalic slits and caudal cirrus. Proboscis branched. Reddish orange, with spatulate head somewhat paler. Common under stones in shallow water. Capable of swimming.

*Cerebratulus leidyi* Verrill: Slender and cylindrical in anterior portion and much flattened behind. Head tapering to acute tip. With cephalic slits and caudal cirrus. Proboscis not branched. To 20 cm or more in length. Deep red without markings. Fairly common in sand under stones in shallow water.

*Lineus albocinctus* Verrill: Genus with cephalic slits but without caudal cirrus; proboscis not branched.— Species slender, with spatulate head. Ventral side whitish, dorsal side greenish brown with about 30 white transverse bands and a white median spot near tip of head. To 7 cm long.

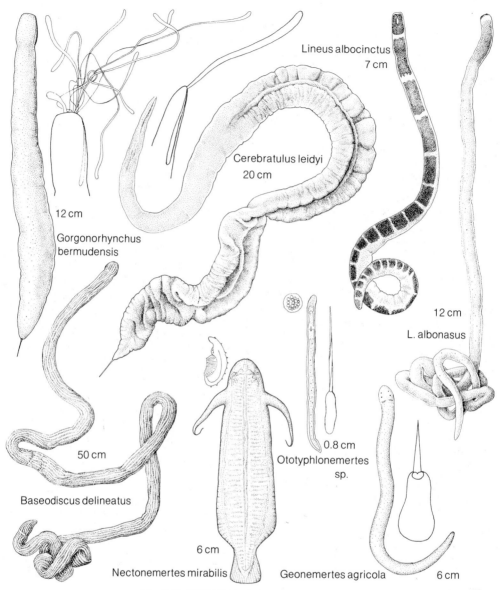

**65 NEMERTINA (Ribbon worms)**

Under stones and in crevices of rocks and dead corals in shallow water.

***L. albonasus*** Verrill: Genus as above.—Species with anterior portion of head yellowish white with 4 eyes faintly shining through; remainder of body brownish red anteriorly, becoming light red toward posterior end. To 12 cm long and frequently twisted into irregular mass. Under stones and in crevices of rocks and dead corals in shallow water.

***Baseodiscus delineatus*** (Delle Chiaje): Very slender, to 1 m long. Head rounded and demarcated from remainder of body by shallow transverse groove. Without cephalic slits and without caudal cirrus; proboscis not branched. Numerous small eyes along lateral margins of head. Brownish yellow with frequently interrupted, laterally anastomosing longitudinal, brown stripes. Common under stones and in crevices of rocks and dead corals in shallow water.

O. **HOPLONEMERTINI:** Nemertina with mouth anterior to brain; proboscis armed with 1 or more stylets.

*Nectonemertes mirabilis* Verrill: Broad, flattened, with thin lateral and caudal fins; to 6 cm long. Sexually mature ♂ with a pair of lateral tentacles in head region. Proboscis with several stylets on crescent-shaped base. Translucent red, orange or pink, except tentacles and fins, which are colorless. Well adapted for swimming. Occasionally found in plankton samples from 200 to 2,000 m depth.

*Ototyphlonemertes* sp.: Very slender, to 0.8 cm long. Without eyes but with 2 statocysts in brain region, each containing a statolith composed of numerous granules. Single proboscis stylet with spiral grooves; stylet base elongate and of irregular shape. Semi-transparent white, with orange tinge in brain region. Common in clean subtidal sand of sheltered beaches.

*Geonemertes agricola* (Willemoes-Suhm): Slender, to 6 cm or more in length. With 4 eyes on rounded head. Single proboscis stylet on pyriform base. Pale gray, whitish, yellowish, orange or brownish. Hermaphroditic and viviparous; with embryos in VI-VII. Terrestrial, together with earthworms in humid localities under stones, decaying wood, and leaf litter; common along shores of mangrove swamps and on adjacent hillsides.

E. KIRSTEUER

## Phylum Gnathostomulida
(Jaw worms)

CHARACTERISTICS: *Free-living wormshaped acoelomate BILATERIA with a muscular pharynx usually provided with paired jaws and an unpaired cuticular basal plate. Without an anus. The epidermis is monocilated (1 long cilium per cell).* Body length 0.2-3 mm; most are colorless, some are bright red. Locomotion is a slow ciliary gliding.

The systematic position is still doubtful, for they share characteristics with Platyhelminthes and Aschelminthes. Approximately 50 species described so far; however, many further species can be expected. About 10 species are known from Bermuda; 4 are included here.

OCCURRENCE: Exclusively marine, with 1 genus penetrating into the brackish groundwater. The typical habitat is sand with a high amount of organic matter (but never mud!), as found intertidally in sheltered bays, or subtidally between reefs or patches of turtle grass. One locality often produces several species, and many specimens.

For qualitative purposes, sand collected by dredge, SCUBA, or, in the intertidal, with a spade (always take sand to at least 10 cm depth, never just the surface!) is kept in a bucket with little overlying seawater. After days to months (!), often accompanied by hydrogen sulfide production (unpleasant odor!), animals migrate to the sample surface and can be extracted using the magnesium chloride method (a mesh size of 65 μm is recommended for the filter).

IDENTIFICATION: Live animals can be examined under a cover slip. Good high-power optics (phase-contrast) are mandatory.

Genera can be distinguished on the basis of body and head shape, basal plate and jaws, and the number and arrangement of the very delicate sensory bristles and cilia on the head. Species differ mostly in the structure, number of teeth, etc. of basal plate and jaws, as well as in details of the reproductive system (sperm, bursa).

Preserve whole mounts with formaldehyde-glycerol; for histology with warm Bouin's or glutaraldehyde.

BIOLOGY: Hermaphrodites with internal fertilization; life cycle possibly annual. The little we know about feeding indicates that the complex pharyngeal apparatus is used to graze on the surface film of sand grains, including bacteria, blue-green algae, and fungal hyphae. Very low oxygen consumption, possibly anaerobic metabolism. Encystment in some species.

DEVELOPMENT: Direct; 1 egg at a time is attached to a sand grain. The juvenile hatches without jaws but with the central piece of the basal plate developed.

REFERENCES: The most comprehensive recent treatment is by STERRER (1972, 1982); for systematic discussion see also AX (1985) and STERRER et al. (1985).
For species from Bermuda see STERRER & FARRIS (1975), STERRER (1976, 1977) and FARRIS (1975, 1977).

## Plate 66

O. **FILOSPERMOIDEA:** Gnathostomulida with filiform sperm; without a bursa, vagina or penis stylet. Without paired sensory organs on the head. Body usually very elongated, head pointed and slender, not delimited from body. (3 spp. from Bda.)

*Haplognathia* cf. *rosacea* Sterrer: Bright red. Jaws solid, tweezer-like, with few teeth. Basal plate delicate, shield-shaped. Not common, in fine sand.

O. **BURSOVAGINOIDEA:** Gnathostomulida with a bursa and often a vagina; sperm not filiform. With paired sensory organs on the head. Body elongate to fairly plump; head more or less rounded, and delimited from body.

S.O. **SCLEROPERALIA:** Bursovaginoidea with a cuticular bursa. Usually with a ♂ stylet consisting of concentrically arranged rods. Testes paired, sperm small. (4 spp. from Bda.)

*Problognathia minima* Sterrer & Farris: Very small (300 μm) and plump, with short rounded head. Basal plate wide, with 35-49 teeth on anterior, concave edge. Jaws lamellar, with 2 rows of delicate teeth. Regular in sheltered locations with fine sand and detritus.

*Gnathostomula tuckeri* Farris: With a well-defined head and tail. Basal plate with pronounced anterior and lateral wings. Jaws lamellarized, with 3 rows of teeth. The most eurytopic species, typically found in fine sand with detritus, but also encountered (in smaller numbers) in clean and sometimes coarse sand.

S.O. **CONOPHORALIA:** Bursovaginoidea with a soft bursa. Penis without a cuticular stylet. Testis usually unpaired, sperm large ("conuli"). (3 spp. from Bda.)

*Austrognathia microconulifera* Farris: Elongated without a tail. Basal plate with 3 anterior lobes, and 13 posterior teeth. Jaws with 2 short rows of teeth. Conuli small (10 μm).

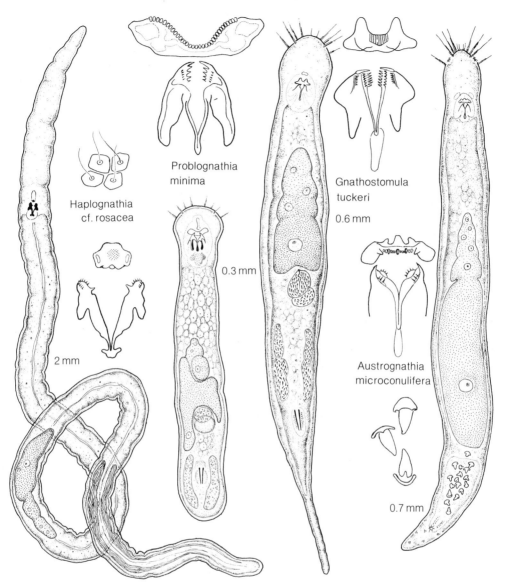

**66 GNATHOSTOMULIDA (Jaw worms)**

Not uncommon in fine to medium sand with detritus.

W. Sterrer

## Phylum Gastrotricha

CHARACTERISTICS: *Minute, short to elongate, free-living BILATERIA with cilia on head and ventral surface only, and mostly with paired adhesive tubules. Cuticula often differentiated into scales and spines.* Among the smallest metazoans (0.1-3 mm), colorless or (rarely) red, they glide over the substrate by means of ciliary action, or creep like inchworms, using adhesive tubules for attachment.

Of some 350 described species, about half are marine. Bermuda's fauna has not been treated systematically as yet; 7 species are included here as representatives of the 2 orders.

OCCURRENCE: Macrodasyida occur exclusively in marine and brackish habitats, whereas Chaetonotida are also found in fresh water. With rare exceptions, marine species live interstitially between sand grains, from the coastal groundwater to several hundred meters depth, reaching their highest diversity in clean, coarse subtidal sand. They can be locally abundant, with densities of 400 specimens/cm$^3$ of sediment.

Collect sand samples with a shovel at tide level, with a hand-held net or bucket by SCUBA, or with a dredge. For quantitative samples use a core sampler. In the laboratory, animals can be first concentrated by means of the climate deterioration method, then extracted with the magnesium chloride or seawater ice technique.

IDENTIFICATION: The study of live animals in squeeze preparation with a high-power phase contrast microscope is essential. Because animals can attach themselves rapidly to any surface, they should be transferred quickly, by means of a finely drawn pipette.

Note body shape and proportions, and number and arrangement of adhesive tubules and cuticular elements. The latter can occur as simple, keeled lamellar, or stalked scales; as spines; or as single- or multi-pronged hooks. Mouth and anus are terminal (rarely subterminal); the intestine is straight and consists of an anterior muscular portion (pharynx) and a posterior gut. The reproductive organs consist of paired or unpaired ovaries and testes (the latter are lacking in many Chaetonotida); in some species there is also an organ for sperm transfer (rarely provided with a cuticular stylet) and a sperm-storing bursa.

Preserve in formaldehyde-glycerol for long-term storage of squeeze preparations; for histology in Bouin's, and for electron microscopy in glutaraldehyde and osmium tetroxide.

BIOLOGY: Hermaphrodites; or parthenogenetically reproducing females only.

In at least some hermaphroditic species (Macrodasyidae) copulation results in reciprocal cross insemination in which sperm is transferred externally (!) from the male pores to a caudal "penis" before being injected into the partner. Only few, large eggs are produced at a time; the life-span is probably a few weeks. Many species are highly gregarious, especially prior to copulation. Most species feed on microalgae (particulary diatoms) and Protozoa, and fall prey to Turbellaria and other meiofauna predators.

DEVELOPMENT: Direct, from eggs that are usually attached to sand grains.

REFERENCES: The classical mongraph is by REMANE (1936); for an introduction see HYMAN (1951b), and the more recent and specialized papers by RUPPERT (1978a, b; reproductive biology) and HUMMON (1975; ecology).

Some species from Bermuda have been described by SCHOEPFER-STERRER (1969, 1974).

## Plate 67

O. **MACRODASYIDA:** Usually elongate Gastrotricha with pharyngeal pores and with many adhesive tubules anteriorly, laterally and posteriorly. Hermaphrodites; exclusively marine.

F. **MACRODASYIDAE:** Elongate Macrodasyida without a defined head region and without dorsal spines or scales; body tapered posteriorly to form a tail. Often abundant in sheltered tidal and subtidal sand.

*Macrodasys* sp.: Tail short; anterior tubules arranged in a transverse half-circlet. In fine to coarse sand, intertidally to subtidally; rather ubiquitous and abundant.

*Urodasys nodostylis* Schoepfer-Sterrer: Tail very long; anterior

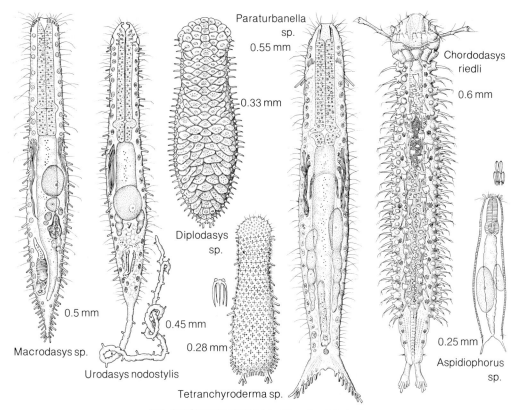

**67 GASTROTRICHA (Gastrotrichs)**

tubules arranged in longitudinal rows. With a cuticular vagina mouthpiece and a proximally constricted ♂ stylet. In fine, clean to detritus-rich subtidal sand; not uncommon.

F. **THAUMASTODERMATIDAE:** Mostly compact Macrodasyida with 2 adhesive pedicles posteriorly. Cuticle complex, forming spines and scales. Very diverse in coarse subtidal sand.

*Diplodasys* sp.: Short, blunt, flattened; with very large subterminal mouth. Cuticle as large flat scales; laterally with 4-angled spines. In coarse to fine subtidal sand; occasional.

*Tetranchyroderma* sp.: Short, with large mouth. Cuticle as 5-pronged hooks. In clean, fine subtidal sand; fairly common.

F. **TURBANELLIDAE:** Ribbon-shaped Macrodasyida with well-defined head often bearing lateral tentacles. Anterior tubules arranged as "hands", posterior tubules on paired lobes. Mostly in subtidal, fine sand.

*Paraturbanella* sp.: Head without tentacles, body not covered with scales or spines. With a lateral pair of 2 long unequal tubules in anterior pharynx region. In fine to coarse sand with detritus, intertidal to subtidal; occasional.

*Chordodasys riedli* Schoepfer-Sterrer: Head with a pair of long tentacles;

body covered with spines and scales. With a chordoid organ behind the anus. In subtidal coarse sand; rare.

**O. CHAETONOTIDA:** Usually short, cylindrical Gastrotricha without pharyngeal pores, and typically with only few tubules located posteriorly. Freshwater and marine. Of many species only ♀ specimens are known.

**F. CHAETONOTIDAE:** Short Chaetonotida with rounded head and somewhat inflated trunk drawn posteriorly into a furca. Cuticle smooth, or with scales and spines. Ventral ciliation not as cirri. Mostly in clean, fine tidal and subtidal sand.

*Aspidiophorus* sp.: Body covered with small stalked scales that join together to form an outer armor. In fine subtidal sand with detritus; fairly common.

W. STERRER

## Phylum Nematoda (Roundworms)

CHARACTERISTICS: *Free-living or parasitic, worm-shaped BILATERIA. Body not divided into regions. Epidermis without cilia, covered by a cuticle. Only longitudinal somatic muscles; protonephridia lacking.* Body length of free-living marine species from 0.1 to 5 mm (parasitic species to 8 m!). With few exceptions colorless. Movement through snake-like writhing, in some cases on stilt bristles.

Of the 2 subclasses Phasmidia and Adenophorea only the latter—containing the marine free-living nematodes—is treated here. About 5,000 marine free-living species have been described so far. Of the several hundred species that can be expected in Bermuda only about 50 have been identified; 7 are included here as examples.

OCCURRENCE: Nematodes occur in every benthic marine habitat, generally in large numbers of species and individuals. Especially sheltered intertidal sand and sediment-rich algal assemblages may yield enormous numbers.

Sand, gathered by dredge, corer or by hand, should be kept in a bucket for some days (to concentrate organisms); the surface layer is then elutriated with warm (50°C) seawater or fresh water, or with the magnesium chloride method. Algae and other tangled substrates should be washed in magnesium chloride, or treated with the climate deterioration method. Mud samples may necessitate the sugar flotation method.

IDENTIFICATION: Live or preserved animals are examined under a cover slip with high-power optics. Interference contrast (Nomarski) is recommended.

The structure of the cuticle (smooth, striated, dotted, etc.), the number and arrangement of cephalic setae, the shape of lateral sense organs (amphids), the structure of the buccal cavity (teeth, stylets, jaws) and the foregut (esophagus), and the number of gonads are used to distinguish higher taxa. Special characters of the male copulatory apparatus (spicula, gubernaculum, supplements) and relative proportions of the various characters define genera and species.

Fix in 4% formaldehyde in seawater; preserve whole mounts, unstained or stained with cotton blue, in pure glycerol after slow evaporation.

BIOLOGY: Dioecious, with internal fertilization; most species are oviparous, 1-10 eggs are deposited at one time. The life cycle is completed on the average within about 20 days, but may be extremely short (2-3 days) or long (over 1 year). Most species feed on detritus and/or bacteria, fungal hyphae or microalgae, some are carnivorous. Predators are all carnivorous meiofauna species, especially Turbellaria, and small shrimp. Non-selective deposit feeders are also likely to ingest them. Some species occur in sediments with little or no oxygen, some may be obligate anaerobes. Several associations with microorganisms are known.

DEVELOPMENT: Direct; juveniles resemble adults; 4 molts during growth.

REFERENCES: A recent comprehensive account can be found in GRASSÉ (1965), and a complete bibliography is given by GERLACH & RIEMANN (1973-1974).

Although some parasitic species from Bermuda fishes were described as early as 1907 by LINTON, free-living nematodes were first studied by COULL (1970; ecology), and subsequently by WIESER et al. (1974; ecophysiology), GERLACH (1977; ecology) and WIESER & SCHIEMER (1977; ecophysiology). New species descriptions are by JENSEN & GERLACH (1976, 1977), OTT (1977) and HENDELBERG (1977).

## Plate 68

O. **MONHYSTERIDA:** Adenophorea with smooth or annulated cuticle; amphids circular.

*Paramonohystera wieseri* Ott: Buccal cavity simple, cup-shaped; esophagus without terminal bulb; only anterior ovary developed (Fam. Monhysteridae). Amphids weakly cuticularized, vesicular; annulation fine; labial and cephalic setae very long; living animals brown to black. In the deeper anoxic layers of intertidal fine sand, probably an obligate anaerobe.

O. **DESMODORIDA:** Adenophorea with generally conspicuously annulated cuticula, head mostly with strong cephalic helmet; amphids spiral.

*Eubostrichus dianae* Hopper & Cefalu: Cuticle without longitudinal ridges; buccal cavity small, unarmed; esophagus with small round terminal bulb; amphids a large simple spiral close to the anterior end (Subfam. Stilbonematinae). With pairs of hollow modified setae in cervical and postanal region; body covered by filamentous (to 100 μm long) microorganisms, giving the animal a furry appearance. Extremely abundant in all sheltered subtidal and intertidal sands, in and close to anoxic sediment layers. (The association with microorganisms is typical for all members of the subfamily.)

*Cyttaronema reticulatum* (Chitwood): Cuticle with conspicuous annulation and longitudinal ridges; well-developed cephalic helmet; amphids a simple spiral, often drawn out longitudinally; buccal armature reduced (Fam. Ceramonematidae). Ten cephalic setae in 2 separate circles, cuticle vacuolated. On all subtidal sandy bottoms.

O. **CHROMADORIDA:** Adenophorea with striated, punctated or elaborately ornamented cuticle; amphid spiral or slit-like.

*Nannolaimoides decoratus* Ott: With punctate cuticle; amphids spiral with many turns; anterior portion of buccal cavity with 12 cuticular rods (rugae), posterior portion mostly with teeth (Fam. Cyatholaimidae). With a row of simple supplements pre-anally in ♂; spicula large, gubernaculum simple. Abundant in intertidal sheltered fine sand. With a remarkably high temperature tolerance.

O. **ENOPLIDA:** Adenophorea with smooth cuticle and generally pouch-like amphids.

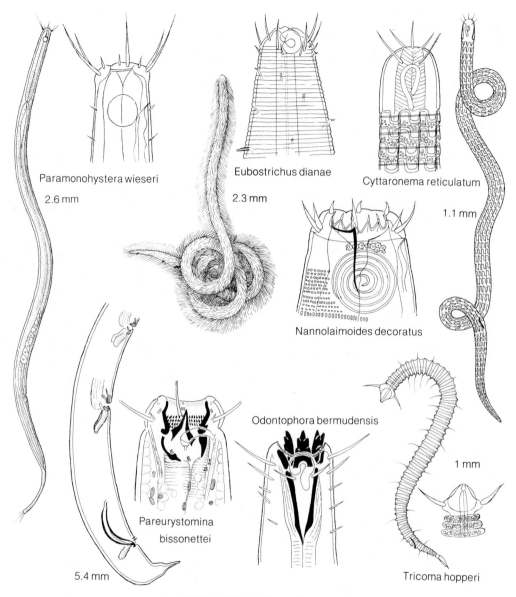

**68 NEMATODA (Roundworms)**

***Pareurystomina bissonettei*** Hopper: Buccal cavity not embedded in esophageal tissue, strongly cuticularized, divided into 2 chambers by a transverse ring (Fam. Enchelidiidae). Large subventral tooth and 4 or 5 rows of denticles. Subcuticular vesiculation. Tail abruptly tapered and spicate in posterior extremity. In sheltered inter- and subtidal sand, near *Thalassia* beds.

**O. ARAEOLAIMIDA:** Adenophorea with smooth or finely striated cuticle, amphid horseshoe- or loop-shaped.

***Odontophora bermudensis*** Jensen & Gerlach: With 6 lips; esophagus without terminal bulbus; ♂ without tubular copulatory supplements; with well-armed buccal cavity (Fam. Axonolaimidae). Six large eversible odontia (jaws) with toothed edges and large complicated apophyses. Posterior portion of buccal cavity conical, deep. In sheltered intertidal sand, below the oxidized top layer.

O. **DESMOSCOLECIDA:** Adenophorea with conspicuous cuticular rings; amphids vesicular. Locomotion caterpillar-like on stilt bristles.

*Tricoma hopperi* Timm: Cuticular annules without spines (Fam. Desmoscolecidae). Concretion rings (77-81) separated by interzones bearing no distinct annules. Red pigment spot within rings 12-14. On subtidal soft bottoms.

J. OTT

# Phylum Rotifera
(Wheel animalcules)

CHARACTERISTICS: *Minute, elongate to sac-shaped BILATERIA with an anterior ciliated organ; body covered by a cuticula, pharynx provided with complex jaws.* Among the smallest metazoans (0.04-3 mm), rotifers are transparent or opaque, mostly colorless, and move either by creeping leech-like over the substrate, or by swimming.

Of about 1,500 species in 3 orders (Seisonidea, Bdelloidea and Monogononta), only 3 species are known from Bermuda so far and are given here as examples.

OCCURRENCE: The majority of species live in fresh water; only some 50 are strictly marine, although many freshwater species venture into marine and brackish areas. Most marine species are free-living on algae or between sand grains, and some are pelagic; however, a number are epizoic (on crustaceans, particularly the gills, and on holothurians) or parasitic (in oligochaetes and hydroids).

Collect likely substrate, and extract specimens by means of the climate deterioration method (from algae) or the magnesium chloride method.

IDENTIFICATION: Live specimens should be observed in squeeze preparation, possibly after anesthesia with magnesium chloride.

Note body regions and proportions, cuticle thickness and sculpture, the presence and shape of foot and toes, and the structure of the ciliary crown (corona). Other identification criteria are the number of ovaries and eyes, the shape and number of spines, and particularly the structure of the jaw apparatus (mastax) which is essentially composed of 1 unpaired and 8 paired elements.

Preserve in formaldehyde-seawater, or for histological purposes in Bouin's. For studying the mastax, whole mounts should be embedded in polyvinyl-lactophenol and examined under a high-power phase contrast microscope.

BIOLOGY: Sexes are separate; males are smaller than females and often somewhat degenerate, or are lacking altogether (in Bdelloidea). Copulation is by hypodermic impregnation (the male injects sperm under the cuticle of the female). Only a few eggs are produced, and are brooded in some species. The life-span is only a few weeks. Rotifers either feed on suspended particles or are predators on protozoans and other meiofauna which they trap, or grasp with their mastax.

DEVELOPMENT: Direct; juveniles resemble adults (males are sexually mature) at hatching.

REFERENCES: Marine Rotifera are poorly known in general, and no systematic study has been made of

220  SIPUNCULA

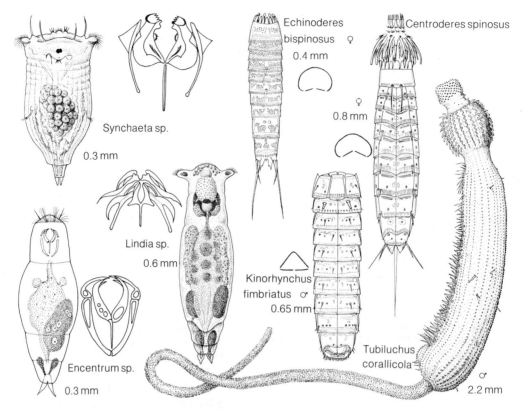

**69 ROTIFERA (Wheel animalcules), KINORHYNCHA (Kinorhynchs), PRIAPULIDA (Priapulids)**

Bermuda's species. For an overview see HYMAN (1951) and the classical treatment by REMANE (1929). A key to marine genera is given by THANE-FENCHEL (1968), and illustrations of plankton species can be found in YAMAI (1971).

### Plate 69

**O. MONOGONONTA:** Swimming or sessile Rotifera with only 1 ovary. (Examples from the suborder Ploima:)

***Synchaeta*** sp.: Delicate, urn-shaped, with 4 bristles anteriorly; foot with 2 toes. Mastax large, heart-shaped. In inshore plankton, abundant in late summer. A lively swimmer.

***Encentrum*** sp.: Body soft, tapered at both ends. Anteriorly with a short dorsal rostrum; corona pointing obliquely downward. Foot short, with 2 toes. Without eyes. In fine sublittoral sand inshore; locally not uncommon.

***Lindia*** sp.: Body short, cylindrical; foot rudimentary, with 2 soft toes. With 1 median eye. Mastax with flattened anterior elements, some with fine teeth. In fine sublittoral sand inshore; rare.

W. STERRER

## Phylum Kinorhyncha

CHARACTERISTICS: *Minute, free-living BILATERIA composed of 13 segments, the 1st of which forms an eversible head, and the 2nd (neck) consists of plates that cover the head when it is withdrawn into the trunk. The chitinous cuticle is differentiated into plates, spines and setae; cilia are lacking.* Less than 1 mm in length and colorless, they pull themselves forward by repeated eversion of the head.

Of the 84 species described from adults, 4 have been recorded from Bermuda; 3 are illustrated here.

OCCURRENCE: In marine and brackish water sediments, occasionally associated with algae or various sedentary invertebrates such as bryozoans and sponges. Some are intertidal but most are subtidal, occurring at depths of several thousand meters. They are often both common and abundant representatives of meiobenthic communities.

Collect the upper few centimeters of fine or mixed fine and coarse sediment, add sufficient seawater so as to form a slurry when mixed thoroughly, then introduce a stream of fine bubbles from an aquarium air stone and allow the sediment to settle for a few minutes. Specimens can then be gathered by touching the surface film with a piece of bond paper, which is immediately removed and its wetted surface with adherent organisms washed into a dish or fine mesh net by a wash bottle.

IDENTIFICATION: A good microscope, preferably with phase or interference contrast, is essential for studying live specimens in squeeze preparation.

Note body shape and cross section, the armature of the head (recurved spines) and the 9 anteriorly directed oral styles that surround the terminal mouth opening. The 11 trunk segments are variously divided into tergal (dorsal) and sternal (ventral) plates, which may bear spines, setae, adhesive tubes and sensory and/or glandular spots.

Kinorhyncha may be fixed in either 70% ethyl alcohol or 10% formalin. Judicious, well monitored rinsing of specimens for a few minutes in fresh water may promote extrusion of the head; an isosmotic magnesium chloride solution is also helpful in relaxation. Fixed specimens can be stored in anhydrous glycerin, or double-cover slip whole mounts may be made using Hoyer's medium colored with Lugol's iodine.

BIOLOGY: All are dioecious but sexes are not always distinctive externally. In Homalorhagida, males deposit a spermatophore on the caudal end of the female. Kinorhynchs are thought to feed primarily on unicellular algae, bacteria and possibly detritus and, in turn, are a source of nutrients for shrimp and other bottom-feeding organisms.

DEVELOPMENT: Kinorhyncha hatch singly as hyaline juveniles having essentially all 13 adult segments. They undergo a series of juvenile stages that may not closely resemble the definitive adult until the last several molts have occurred.

REFERENCES: For general reference see HIGGINS (1982a).

The first published report of kinorhynchs from Bermuda is by COULL (1970), whose collection has only recently been identified and described by HIGGINS (1982b).

## Plate 69

O. **CYCLORHAGIDA:** Second (neck) segment consists of a ring of 14-16 plates. Usually with articulated mid-dorsal, lateral and midterminal spines (at least in juvenile stages). Trunk generally round to rounded triangular in cross section.

S.O. **CYCLORHAGAE:** Cyclorhagida with 1st (and sometimes 2nd) trunk segment consisting of complete ring of cuticle.

F. **ECHINODERIDAE:** Cyclorhagae with midterminal spines only in first 3 of 6 juvenile stages; only lateral terminal and accessory lateral terminal spines (♀) in the adult. (2 spp. from Bda.)

***Echinoderes bispinosus*** Higgins: Adult trunk length 245-285 μm. Middorsal spines fragile (easily broken, therefore may be indicated only by scars), on segments 6 and 8. Lateral spines on segments 4, 7-11, often pressed tightly to trunk segment. Lateral terminal spines long, 41-56% of trunk length. With distinct perforation sites (indications of cuticular hairs). Distinctive muscle scars lateral and ventrolateral on segment 4. In mixed fine and coarse sediments, inshore, to 14 m; abundant.

F. **CENTRODERIDAE:** Cyclorhagae with midterminal spine in addition to 1 or 2 pairs of lateral terminal spines. Second trunk segment divided ventrally into 2 sternal plates. Middorsal and lateral spines usually present on most segments.

***Centroderes spinosus*** (Reinhard): Adult trunk length 390-500 μm. Midterminal spine longer than adjacent 2 pairs of lateral terminal spines. Prominent lateral spine originating near ventral midline at margin on segment 3, smaller spines (probably adhesive tubes) on lateral margin of sternal plates 4, 7, 9 and 10 (♀). Robust lateral spines on segments 10 and 11. Middorsal spines on segments 3-11, 13 in ♀ and 3-13 in ♂ and juvenile. The 12th middorsal and subdorsal spines are crenulate and very flexible in ♂. Juveniles lack subdorsal spines but have the middorsal spine on segment 12. In mixed fine and coarse subtidal sediments; common.

O. **HOMALORHAGIDA:** Second (neck) segment consisting of 6-8 plates. First trunk segment consisting of vaulted tergal plate and 1-3 variously divided sternal plates. Remaining trunk segments with 2 sternal plates. Trunk triangular in cross section. No articulated lateral or middorsal spines except for lateral and, less commonly, a midterminal spine in some genera.

F. **PYCNOPHYIDAE:** Homalorhagida with lst trunk segment with completely separated single sternal and 2 adjacent episternal plates. (1 sp. from Bda.)

***Kinorhynchus fimbriatus*** Higgins: Adult trunk length 592-697 μm. Trunk without lateral terminal spines. Terminal tergal border with distinct fringe. In mixed fine and coarse sediment subtidally; abundant.

R. P. HIGGINS

## Phylum Priapulida

CHARACTERISTICS: *Worm-shaped, free-living BILATERIA with a cuticular epidermis; retractable proboscis (introvert) provided with hooks. With a spacious coelom and a straight intestine.* Length 0.2-6 cm (rarely 20 cm). Transparent, greenish or reddish.

The phylum comprises only 10 species; the only species reported from Bermuda belongs to the larger of the 2 orders (Priapulimorpha).

OCCURRENCE: Marine to brackish water; on muddy to sandy bottoms, from tide level to the deep sea. With few exceptions in cold water (e.g., circumpolar); sometimes abundant.

Collect with a light dredge; sediment is then treated with the Swedmark method.

IDENTIFICATION: Between slide and cover slip, anesthetized in magnesium chloride, under a phase contrast microscope.

Note the division in introvert (retractable proboscis), trunk and tail; note further the superficial segmentation, and the cuticular structures covering the body.

Fix in formalin-glycerol 1:1 (for whole mounts), in warm Bouin's for histology.

BIOLOGY: Poorly known. Sexes are separate; eggs and sperm are liberated into the seawater where fertilization occurs. Animals burrow through the sediment by contractions of the body. They are predacious, feeding on soft-bodied, slow-moving prey, particularly Polychaeta. Numerous bottom-feeding fishes are their enemies.

### Plate 60

DEVELOPMENT: The hatching larva is unciliated and already bears hooks on the introvert. It can be found in the same sediment as the adults.

REFERENCES: A recent summary is by VAN DER LAND (1970).
For Bermuda's only species see COULL (1970), and KIRSTEUER & RÜTZLER (1973).

### Plate 69

*Tubiluchus corallicola* van der Land: Hooks on the introvert arranged in 20 longitudinal rows. Adults with comb-shaped pharyngeal teeth, and with a long tail. ♂ with numerous ventral setae on abdomen. Larvae display radial symmetry. Intestine often greenish. Very abundant in shallow mud and muddy sand.

W. STERRER

## Phylum Acanthocephala
(Spiny-headed worms)

CHARACTERISTICS: *Worm-shaped, endoparasitic BILATERIA with a retractable proboscis provided with hooks, and with a cuticularized epidermis. Without a digestive tract. Length mostly 1-2 cm (rarely 0.15-50 cm).*

Of uncertain systematic relationship, in some features (proboscis) similar to Kinorhyncha and Priapulida. Of the 3 orders, only the Palaeacanthocephala are represented in Bermuda. A total of 500 species are known; of 3 reported from Bermuda, 1 is given as an example.

OCCURRENCE: Adults are parasites in vertebrates, whereas Insecta and Crustacea act as intermediate hosts.

Collect adults by screening, under a dissecting microscope, the opened gut of fresh fish. Immature stages can be collected from the outside (!) of viscera and mesenteries of intermediate hosts.

Acanthocephala are not known to infest humans.

IDENTIFICATION: In seawater, under a dissecting microscope, gross morphological features can be discerned; species identification often requires dissection or microanatomy.

Note body dimensions, and the shape, number and arrangement of hooks on the proboscis.

Preserve in 4% neutral formaldehyde; for histology in Bouin's.

BIOLOGY: Dioecious; males are usually smaller than females. Fertilization is by copulation. Large numbers of eggs are continuously produced. Life-span is 1 to several years. Considerable numbers of adults can be found in a single host, attached to the gut epithelium by means of the everted proboscis. Food is taken up through the body wall.

## Plate 60

DEVELOPMENT: When the fertilized egg is eaten by an intermediate host, the acanthor (a hook-bearing larva) hatches and drills its way into the host's body cavity where it encysts. The final host, having fed on the intermediate host, digests the cyst, and the young worm takes up residence in the gut lumen.

REFERENCES: The phylum has received monographic treatment by YAMAGUTI (1963b).
For Bermudian species see LINTON (1907) and REES (1970).

## Plate 70

*Echinorhynchus medius* (Linton): Body elongated; anterior end and proboscis slightly deflected. With 22 vertical rows of 20 hooks each. To 4-5 cm long. Adults in *Mycteroperca venenosa*; immature stages in other groupers, snappers and porgies.

W. STERRER

## Phylum Sipuncula (Peanut worms)

CHARACTERISTICS: *Worm-shaped BILATERIA with a narrow anterior part (introvert) retractable into a thicker posterior trunk. Body wall cuticularized, often bearing papillae and hooks, but never bristles. Coelomic cavity spacious, unsegmented.* From a few millimeters to more than 30 cm in length, sipunculans are largely robust worms, white, pinkish or brown in color. They burrow in sand and rock by slow peristaltic contractions, and are rather helpless when removed from their substrate.

Of 320 currently recognized species, only 10 have so far been reported from Bermuda; 7 are included here.

OCCURRENCE: As members of the marine benthos, some species burrow in sand or mud, whereas others occur in gravel, in crevices or under rocks. A few species inhabit discarded mollusk shells and, in the vicinity of coral reefs, many species bore into dead coral or other calcareous rock. Inhabitants of shells and rock-dwelling species may occur in densities as great as 200-4,000/[m[wu2]. Recorded depths for sipunculans range from the high intertidal to the abyssal plains. Many species are cosmopolitan.

Collect sand-burrowing sipunculans in shallow waters with shovel and sieve and in deeper waters by dredge. Small species live near the surface of the sand and larger species burrow as deep as 1-1.5 m. Rock-dwelling species must be extracted from their burrows by fracturing rocks (use hammer and chisel) and completely exposing the burrows. Specimens cannot be pulled from their burrows without danger of rupture.

IDENTIFICATION: Genera of local sipunculans may be recognized by an examination of external features with a low-power microscope; determination of species most often requires dissection to reveal the internal anatomy.

Distinguish between the smooth or variously sculptured trunk region and the introvert, which is frequently ornamented with spines or hooks. The mouth is located terminally at the introvert, either inside a tentacular crown or ventral to it. The anus opens dorsally, usually near the trunk-introvert border, and there are nephridiopores in ventrolateral position on the anterior part of the trunk. The intestine is looped and usually coiled with descending and ascending portions. The inner longitudinal muscle layer of the body wall is either continuous or arranged in bundles. One to 4 introvert retractor muscles are attached to the body wall at specific levels. Important generic characters are the relative smoothness of the body wall, i.e., the development and dispersal of surface papillae, the presence of the longitudinal and transverse furrows, and the elaboration of anterior and posterior cuticular thickenings on the trunk. The shape of tentacles and their arrangement relative to the mouth are also used as generic characters. Species-specific characters include numbers of longitudinal muscle bundles, number and position of attachment of introvert retractor muscles, and presence and attachments of small muscles fastening the gut to the body wall. External features useful in species identification are overall body size, relative length of introvert and trunk, presence and structure of hooks and spines on introvert, and form and distribution of papillae over the body.

Identification is facilitated if, before preservation, specimens are relaxed so that introvert and tentacles are extended. Place rock-dwelling species in 10% ethanol in seawater and sand-burrowers in 7.5% magnesium chloride. Gentle pressure applied to the trunk during relaxation will sometimes cause a retracted introvert to be extended. For systematic collections specimens should be preserved in 5% formalin (or 5% formalin made up in 70% ethanol) for 24 hr and later transferred to 70% ethanol for storage.

BIOLOGY: Most sipunculans are dioecious. Gametes begin their development in the ovary or testis usually located at the base of the ventral retractor muscles. At an early stage oocytes or spermatocytes are released into the coelom where they undergo most of their development as freely floating cells. Before spawning they are accumulated by the nephridia and are shed via the nephridiopores during simultaneous spawning of males and females into the seawater where fertilization occurs. The life-span of individual sipunculans has been estimated to be as long as 25 years. There have been few observations on feeding in sipunculans, but several patterns have been proposed. Many sand burrowers are believed to engulf sand as they burrow, digesting from it the organic matter. Species with well-developed tentacles may be ciliary-mucus feeders, extending tentacles from burrows for prolonged periods while collecting suspended particles from the surrounding water. Species dwelling in burrows in rock extend the introvert over the surface of the rock, presumably scraping off detritus and ingesting it. Species inhabiting coral rock excavate their own burrows over a long period of time by a combination of chemical secretions and abrasive body movements. Burrows are frequently shared with bivalves and polychaetes, and parasites include copepods and sporozoans. Many fish species are predators of sipunculans, as are some gastropods. Sipunculans may occur in great numbers in the dead basal portions of coral colonies, which they weaken by their boring activity; thus they are believed to contribute significantly to the natural erosion of the reef.

### Plate 60

DEVELOPMENT: Some species develop directly, with the embryo emerging from egg envelopes as a small crawling worm. Most species, however, have indirect development, with a trochophore and an independently feeding larva (pelagosphaera) that may live for long periods in the plankton before undergoing metamorphosis into a juvenile worm. Pelagosphaera larvae, many of them still unidentified, have been found in surface plankton in all the major east-west currents of the North and South Atlantic Oceans and, with an estimated duration of several months, are thought to survive transport across the ocean.

REFERENCES: A compilation of the systematic literature on all described species is found in the monograph by STEPHEN & EDMONDS (1972). Other useful general references are HYMAN (1959), RICE & TODOROVIC (1975, 1976) and GIBBS (1977).

No systematic study of the group in Bermuda has been published since VERRILL'S (1900a, 1901a) accounts.

## Plate 70

F. **SIPUNCULIDAE:** Usually large, cylindrical Sipuncula. Tentacular fold or numerous tentacles surround mouth. Longitudinal integumentary coelomic canals or coelomic sacs extend between muscle bands in body wall. Longitudinal muscle layer of body wall separated into bands. Retractor muscles usually 4; nephridia 2. (3 spp. from Bda.)

*Sipunculus norvegicus* Danielssen: Integument appears in rectangles formed by transverse and longitudinal furrows. Tentacular fold surrounds mouth and is elaborated to form tentacles. Flat triangular papillae arranged as scales on relatively short introvert. Relatively long trunk frequently with bulbous expansion. Well separated transverse and longitudinal muscle bands. Longitudinal integumentary coelomic canals between muscle bands. With 24 longitudinal muscle bands. Post-esophageal accessory intestinal loop. Intestine attached to body wall by numerous mesenteries. Fairly large species (to 90 mm). Sand burrowers in the intertidal and shallow subtidal of sheltered bays; not uncommon.

*Siphonosoma cumanense* (Keferstein): Numerous filiform tentacles surround mouth. Longitudinal muscles divided into bands; circular muscles anastomosing without clearly defined bands. Coelomic sacs extend into body wall. Dorsal contractile vessel with villi. Intestinal spindle muscle arises anteriorly from 3 roots and attaches to body wall posteriorly. Two nephridia with characteristic semilunar nephrostomal lips. Transverse mesenteric dissepiments extending into coelom. Longitudinal muscle bands 18-21. To 190 mm. Burrows in sand in the intertidal and shallow subtidal of sheltered bays; not uncommon.

F. **GOLFINGIIDAE:** Small to medium-sized Sipuncula. Tentacles, commonly digitiform or filiform, usually surround the mouth. Skin usually smooth but small papillae may be present. Body wall musculature continuous, i.e., not separated into longitudinal or circular bundles. With 1-4 introvert retractor muscles, and 1 or 2 nephridia. (2 spp. from Bda.)

*Golfingia elongata* (Keferstein): Elongate, to 50 mm in length. Trunk 4-5× length of introvert. Simple tentacles surround mouth. Hooks on anterior introvert arranged in rows. With 2 pairs retractor muscles, joining far anteriorly in introvert. Ventral retractors arise mid to anterior 1/3 of trunk; dorsal retractors arise just posterior to anus. Intestine with numerous coils not attached posteriorly. Two fixing muscles attach intestinal spiral and rectum to body wall. With 2 nephridia, not attached to body wall for most of length. In sand around turtle grass and in rock.

F. **ASPIDOSIPHONIDAE:** Small to medium-sized Sipuncula with a thickened cuticular shield or calcareous knob at anterior extremity of trunk and usually a posterior cuticular shield. Partial circle of short tentacles. In most genera, introvert extends at angle from trunk, ventral to anterior shield. Longitudinal musculature of

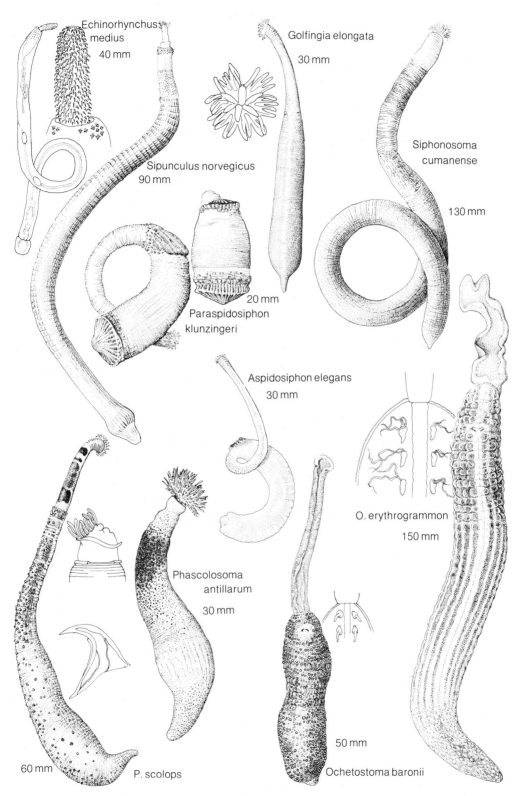

**70 ACANTHOCEPHALA (Spiny-headed worms), SIPUNCULA (Peanut worms), ECHIURA (Spoon worms)**

body wall continuous or arranged in bands. Introvert retractor muscles usually 1 or 2. Two nephridia. (3 spp. from Bda.)

***Aspidosiphon elegans*** (Chamisso & Eysenhardt): Body slender; trunk to 35 mm. Hooks on anterior introvert with 2 points, occurring in regular rows. Prominent red-yellow spines on posterior introvert. Anterior shield oval, composed of numerous irregularly arranged platelets. Posterior shield conical or discoid. Longitudinal musculature of body wall continuous. Two retractor muscles soon join to form 1; they arise at posterior 1/3 of trunk. In coral rock intertidally to subtidally; often numerous.

***Paraspidosiphon klunzingeri*** (Selenka & De Man): Trunk to 30 mm. Hooks on anterior introvert with single points, arranged in rows; spines on posterior introvert. Anterior shield oval with radiating furrows; posterior shield cone-shaped with basal ridge, furrows radiating from apex. Longitudinal musculature arranged in up to 40 bundles. One introvert retractor muscle with broad base arising in posterior trunk. Rectal caecum with many lobes. In subtidal rock; uncommon.

F. **PHASCOLOSOMATIDAE:** Sipuncula of medium size in which tentacles are arranged in a crescent dorsal to the mouth. Papillae usually concentrated at anterior and posterior extremities of trunk; comprised of small platelets. Longitudinal musculature of body wall usually in bands. (2 spp. from Bda.)

***Phascolosoma antillarum*** Grübe & Oersted: Genus with longitudinal muscles divided into separate bundles which often anastomose.—Species with thick trunk; length to 30 mm, about 4× width. Body covered with flat, dark brown papillae, concentrated and larger at anterior and posterior trunk. Tentacles long and filiform. Introvert short, about 1/2 length of trunk; no hooks. Longitudinal musculature divided into approximately 30 bands in mid-trunk region, considerable anastomosis of bands. Contractile vessel with villi. In intertidal rock; common.

***P. scolops*** (Selenka & De Man) (=*Physcosoma varians* of VERRILL 1900a): Genus as above.—Species with fusiform trunk to 40 mm long, 6-9× as long as wide; frequently with scattered brown pigment spots. Papillae dome-shaped. Short digitiform tentacles. Introvert narrow, as long as or longer than trunk; dorsally with dark brown pigment bands. Rows of hooks on anterior introvert; hooks with central clear streak without expansion and basal triangle on convex side. Longitudinal muscle bands 20-21; some anastomosis. Contractile vessel simple, no villi. In intertidal rock; very common.

M. E. RICE

## Phylum Echiura (Spoon worms)

CHARACTERISTICS: *Subcylindrical, sac-shaped BILATERIA with spacious, unsegmented coelom; usually with a pair of ventral setae and a highly extensible proboscis that cannot be retracted into the body.* Body length 0.5-15 cm (rarely more); proboscis of some (non-Bermudian) species to 150 cm when extended. Color reddish-brown, green or transparent. Echiurans burrow by means of peristaltic contractions and are sluggish animals when removed from their substrate.

Of about 130 species in 4 families, only 2 (representing 1 family, Echiuridae) have been recorded from Bermuda to date; both are included here.

OCCURRENCE: Exclusively marine, from the intertidal to the abyss. Generally found in U-shaped burrows in sand or mud, under rocks or in rock crevices of protected habitats. Intertidal and shallow subtidal burrowing specimens may be collected with a shovel and sieve, specimens from deeper water may be dredged or airlifted. Species living in rock or coral must be carefully extracted with a hammer, chisel and forceps. Species forming burrows in soft substrates are difficult to obtain intact. It is advantageous to begin collecting by identifying both burrow entrances. This can be accomplished by allowing the worm to pump a small quantity of tracer dye (e.g., fluorescein) from one burrow aperture to the other. Careful excavation and gentle screening of the sediment between the 2 openings may then proceed.

IDENTIFICATION: Internal and external characters are necessary for the identification of most echiurans. Use a low-power microscope for live observation, and for the study of the internal anatomy of dissected specimens.

Note the position of mouth (ventral to the base of the proboscis), setae (behind the mouth) and anus (terminal). The proboscis has the shape of a ventrally open gutter or spoon (hence the common name!); in some (non-Bermudian) species it is terminally forked. A dorsal dissection of relaxed and preserved specimens exposes the important features of the internal anatomy. These include absence or presence and number of longitudinal muscle bands in the body wall, number of nephridia, morphology of the nephrostome and anal vesicles, and the presence or absence of a rectal caecum.

Before fixation echiurans should be relaxed with magnesium chloride. Fixation should be in 5% formalin for 24 hr followed by transfer to 70% alcohol for storage. The proboscis is an important taxonomic structure that is often broken off or lost during collection. Though possibly detached, it should be preserved along with the trunk.

BIOLOGY: Sexes are separate and, in Echiuridae, indistinguishable externally (in another family, a dwarf male is carried on or in the female!). Most of gametogenesis takes place while the gametes are floating freely in the coelomic fluid. Older gametes are acccumulated in the nephridia until spawning. Fertilization is external (except in 1 family where eggs are fertilized internally just before passing out of the nephridiopores at spawning). With the exception of 1 genus that feeds by trapping small particles in a mucous net, echiurans are deposit feeders. With the aid of the proboscis, sediments surrounding the burrow aperture are transported to the mouth and ingested. After the organic or bacterial component of the sediment is digested, cleaned sediment is returned to the environment as small fecal pellets. Echiurans can turn around in their burrow and hence feed from either opening. Coprophagy is common. Some polychaetes, decapods and mollusks are known commensals; Protozoa and Platyhelminthes are parasites.

DEVELOPMENT: A trochophore larva is produced 1-2 days after fertilization. An echiuran trochophore may be distinguished from a polychaete trochophore by the absence of eyes and the presence of green pigment spots or bands. Further, late echiuran trochophores lack the segmentation present in late polychaete trochophores.

REFERENCES: For general orientation see the monograph by STEPHEN & EDMONDS (1972), and the collection of recent studies in RICE & TODOROVIC (1976).

VERRILL (1901a) reported 1 species from Bermuda.

### Plate 70

***Ochetostoma baronii*** (Greeff) (= *Thalassema baronii* of VERRILL 1901a): Genus having proboscis without bifurcation; longitudinal muscles in bands; oblique muscles in fascicles; nephrostome with long, spirally coiled lips.—Species with trunk to 80 mm length; trunk dark green, proboscis lighter green; 17-19 longitudinal muscle bands; 2 pairs of nephridia; anal vesicles are long brown tubes with short branching outgrowths. Found intertidally and subtidally in cracks and crevices of rocks, or in sand under stones.

***O. erythrogrammon*** Leuckart & Ruppell: Genus as above.—Species large, body to 160 mm long. Contracted proboscis about 1/3 body length, easily shed. Color reddish brown. Longitudinal muscles gathered into 14 bands, most prominent ventrally. Oblique muscles in transverse fascicles. Muscle bundle arrangement causes contracted specimens to have regular rows of raised "blisters" over the body. With 3 pairs of nephridia; nephrostomes with long, spirally coiled lips. Anal vesicles are long unbranched tubes with scattered ciliated funnels. Forms deep burrows in intertidal muds; large mounds indicate burrow openings. Not uncommon but difficult to collect.

J. F. PILGER

## Phylum Pogonophora
(Beard worms)

CHARACTERISTICS: *Tube-living, worm-shaped BILATERIA without mouth, alimentary canal or anus. Hair-thin (!) body composed of short forepart, very long trunk and short opisthosoma. First coelomic segment cephalic, with 1 or more tentacles; 2nd segment includes posterior forepart and all of trunk; several very short segments make up the opisthosoma. With short toothed bristles on trunk and opisthosoma. Body length from 10 mm to 1 m (!); diameter 0.05-6.0 mm. Tube chitinous; white, yellow, brown or black; often ringed, sometimes segmented.*

Previously classified with Deuterostomia but probably related to Annelida, Echiura and Sipuncula. Approximately 130 species described; 2 are known from near Bermuda. Both belong to the order Athecanephria; some representatives of the other order (Thecanephria) occur in the northwest Atlantic.

OCCURRENCE: Marine, in mud or muddy sand (rarely in decaying wood, etc.). In depths greater than 300 m in the tropics, shallower (to 25 m in Norway) toward the poles. Pogonophora are typical of continental slopes and some deep trenches.

Collection by dredge, grab or small trawl, followed by hand sorting or sieving with water. Keep cold, examine or fix as soon as possible.

IDENTIFICATION: Live animals can be extracted from their tubes by a combination of dissection, squeezing and pulling, but for identification it is important to have both animal and tube as complete as possible (the opisthosoma is almost always lost during collection).

Note tube size; color of various parts of the body; arrangement of papillae; number of tentacles; arrangement, size and shape of bristles (setae); shape and size of spermatophores.

Fix the animal inside its tube, with 4% neutral formaldehyde, and transfer to 70% alcohol for storage. Small fixed specimens can be examined under a cover slip in 50% glycerin. Larger ones must be at least partly dissected from their tubes.

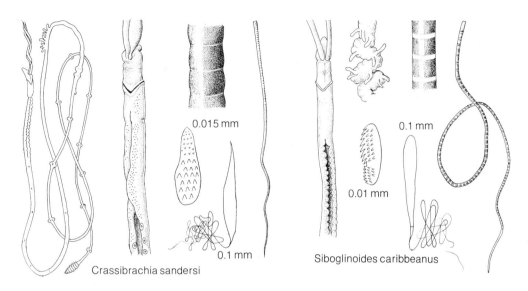

## 71 POGONOPHORA (Beard worms)

BIOLOGY: Dioecious, with slight external dimorphism: genital apertures of male are at anterior end of trunk, of female in mid-trunk region, anterior to girdles. The anterior part of the trunk becomes full of spermatophores or oocytes, which may be seen easily through the body wall. Spermatophores are provided with long filaments; when released by the male into water they must drift with currents and be picked up by nearby females. The site of fertilization has not been established. Tubes are partly buried in the seafloor, with the anterior end projecting into the water. The animal is capable of considerable contraction and extension and can move up and down inside the tube. Feeding is partly by epidermal absorption of dissolved organic compounds, and partly by symbiotic relationship with internal chemoautotrophic bacteria. The function of tentacles is chiefly respiratory.

### Plate 60

DEVELOPMENT: Known only for a few species: eggs are incubated inside the anterior part of the tube of the female; a crawling/swimming larva emerges which settles almost immediately and starts to burrow and form a tube.

REFERENCES: IVANOV (1963) has written a monograph on the phylum. A recent review article is by SOUTHWARD (1971a).

Descriptions of species from the Bermuda area are found in SOUTHWARD (1968, 1971b, 1972).

### Plate 71

O. **ATHECANEPHRIA:** Pogonophora with sac-shaped 1st coelom; its 2 coelomoducts are short and wide apart. Spermatophores spindle- or cigar-shaped.

F. **OLIGOBRACHIIDAE:** Athecanephria with 2-12 tentacles; pear-shaped (tube-forming) glands of the anterior part of the trunk are in irregular lateral bands, not separate papillae.

*Crassibrachia sandersi* Southward: With 2 fat tentacles, no pinnules. Brittle black tube about 0.2 mm in diameter. Spermatophore spindle-shaped. (Note that the illustration given is of an idealized small pogonophore rather than of this species.) Depth range 2,000-5,000 m.

F. **SIBOGLINIDAE:** Athecanephria with 1 or 2 tentacles; pear-shaped glands of anterior part of trunk usually in 2 rows of separate papillae.

*Siboglinoides caribbeanus* Southward: With 2 tentacles with pinnules; 4 apparent rings of bristles or setae on midregion of trunk. Tube 0.10 mm in diameter, with yellow-brown rings. Spermatophore cigar-shaped. Found at 1,000 m.

E. C. SOUTHWARD

# Phylum Annelida
(Segmented worms)

CHARACTERISTICS: *Worm-shaped BILATERIA with a true coelom; body consisting of a single preoral lobe (prostomium) and a linear series of basically similar segments.*

Note body segmentation, which can be fairly uniform, or regionally specialized (particularly the head region), or superimposed by secondary annulation (as in Hirudinea), or reduced. The body is usually covered with a non-chitinous cuticle. External characters such as parapodia (fleshy foot-like lateral appendages) and setae (chitinous bristles), where present, also follow the segment pattern as do features of internal anatomy such as the coelom, musculature and the nervous, circulatory and excretory systems.

The phylum is divided into 4 classes: Polychaeta (p. 232), Oligochaeta (p. 258), Hirudinea (p. 263) and Myzostomaria (=Myzostomida). The latter class, parasites of echinoderms, has not yet been recorded from Bermuda, nor have Branchiobdellida, a group of leech-like ectoparasites of crayfish, sometimes considered a separate (5th) class of Annelida. Oligochaeta and Hirudinea share the absence of parapodia and paired antennae, and the presence of a clitellum, a permanent or temporarily (during phases of sexual activity) thickened body region; they are, therefore, frequently grouped together as a subphylum, Clitellata. Another group, Archiannelida, was formerly thought to represent a separate class but is now considered to be an assemblage of aberrant Polychaeta of uncertain affinities.

## Class Polychaeta (Bristle worms)

CHARACTERISTICS: *Usually marine ANNELIDA with numerous basically similar segments. Segments provided with developed or reduced parapodia. Parapodia with fascicles of many setae of diverse, often specific, ornamentation and structure. Sexes usually separate and gonoducts, where present, are simple. Head (prostomium) usually with sensory or feeding structures.* Body length varies greatly from 0.5 mm to 3 m. Many are brightly colored. Some move freely over or through substrates; others are confined to tubes or permanent burrows.

In the past the class Polychaeta was divided into the subclasses Errantia and Sedentaria. However, these groupings are no longer considered to have any taxonomic status and are used now only as a convenient tool to divide the large class into 2 general groupings. Several attempts have been made to arrange the many polychaete families into superfamilies and orders. Because no one scheme has received widespread acceptance, we choose to use the more traditional scheme of treating the polychaetes by family. The Archiannelida, previously considered an order of Polychaeta but now regarded as an assemblage of 5 families that are not particularly related to each other except by their interstitial mode of life, have been treated here as an appendix. There are approximately 8,000-10,000 species in about 5,000 genera in nearly 80 families. The Bermuda polychaete fauna is comprised of at least 35 families, 99 genera and 151 species; of these, 34 families, 67 genera and 74 species are treated herein.

OCCURRENCE: Ubiquitous, mostly marine, benthic; in the intertidal and subtidal regions (often to depths greater than 1,000 m), on floating objects (e.g., wood, *Sargassum*) and in the plankton. They may burrow in sand or mud or into coral rock, or live in tubes attached to hard substrates (e.g., rocks, pilings, *Thalassia*) or buried in sand, mud or under rocks. They are commonly a part of fouling communities (hydroids, sponges, bryozoans, algae, etc.) attached to hard substrates.

Collecting techniques vary with substrate and habitat. Soft substrates (sand and mud): small quantities (about 100 cm$^3$ are extracted with seawater or 7% magnesium chloride, and large quantities by sieving. Hard substrates: break apart coral rock or wood or turn over rocks (be sure to replace rocks as you found them!). Other habitats: break apart *Thalassia* root systems; treat fouling communities with climate deterioration; for planktonic species use plankton nets.

IDENTIFICATION: A good dissecting microscope is essential, and a compound microscope is frequently required for specific identification. The external characters of live or fixed specimens are usually sufficient for species identification.

Several external characters are important (although all are not always required for an accurate identification): (1) Body regionation: note if the body is divided into 2 or more regions (usually marked by a change in the structure of the parapodia, types of setae, presence or absence of branchiae, etc.); if so, the number of segments or setigers (segments bearing setae) in each region, particularly the anterior region, is often useful; the anterior region is sometimes modified to form a lid (operculum) that serves to close the open end of the tube and thus protect the occupant. (2) Head (prostomium): the prostomium is the most anterior region of the body; note its shape and the types, number and location of any associated sensory structures; these may include eyes (presumed photoreceptors that consist of pigment cups with or without lenses, or of aggregations of pigmented granules; eyes may be present on other areas of the body also), a nuchal organ (a sensory structure that is usually ciliated and eversible), palps (structures that are considered to function in food handling and, possibly, tasting; palps are sometimes associated with the peristomium, see below), and one or more tentacles (processes of variable shape that have a sensory and/or food-gathering function). (3) First segment (peristomium): the peristomium is the segment immediately posterior to the prostomium and may consist of a single segment or represent a series of fused segments; it bears the mouth opening and sometimes one or more tentacular cirri (processes with sensory function). (4) Parapodia (1 pair per segment): these are fleshy projections from the lateral body surface that usually consist of a notopodium (dorsal portion) and neuropodium (ventral portion), each with its associated setae, aciculum (a short internal rod that supports a parapodial lobe), cirrus (a sensory projection from the dorsal or ventral margin that is often long and tapered) and lobes; note if the parapodia are biramous (notopodium and neuropodium present) or uniramous (usually the notopodium is reduced or absent); also note the development and shape of the dorsal and ventral cirri (occasionally 1 or both may be absent). (5) Gills (branchiae): these are respiratory structures sometimes associated with some or all of the parapodia; note their shape and distribution throughout the body. (6) Setae: bristle-like structures that project from the parapodia; note if they are simple (unjointed), compound (jointed) or specialized to form a hook (a long simple seta with the free tip recurved to form 1 or more teeth and with the terminal portion sometimes invested by a transparent hood) or uncinus (a plate-like seta provided with teeth of variable size and shape and occurring with other uncini, side by side, in rows, like stacked coins, with the teeth exposed at the body surface); also note if any setae are subacicular (situated in the region of a neuropodium that is ventral to the aciculum and the neurosetae). (7) Pygidium; this is the most posterior region of the body that usually bears the anal opening and may bear (anal) cirri. (8) Color pattern: this is helpful but often fades rapidly after fixation.

The most useful internal structure is the proboscis

(often everted in fixed specimens). This is the most anterior region of the gut that is protrusible through the mouth opening and used for burrowing and/or food gathering. It may be essential for the accurate identification of genera and species in some families (e.g., Nereididae). Note the presence and development of any associated jaws and also note if soft papillae or paragnaths (hardened conical structures) are present on the surface of the proboscis.

Fixation with 10% formalin (=4% formaldehyde) neutralized with borax ($Na_2B_4O_7$) for at least 24 hr provides good results (never use ethyl alcohol as a fixative!). The specimens can be left in the formalin solution for several weeks if necessary. For best results relax the specimens in 7% magnesium chloride for at least 1/2 hr, or in 1-2 drops of oil of cloves per 500 ml of seawater for about 10 min, before fixing. Following fixation, the specimens should be rinsed once or twice in fresh water and then transferred to 70% ethyl alcohol. The alcohol should be changed at least once to ensure full strength.

BIOLOGY: The majority of species are dioecious and reproduce sexually; however, hermaphroditic species are known, as well as species capable of asexual reproduction. Sexual dimorphism is rarely exhibited. Fertilization may be external, with eggs and sperm released into the seawater, or internal, with species using a variety of copulatory mechanisms.

In the families Syllidae, Nereididae and Eunicidae sexual reproduction is sometimes accompanied by the phenomenon of epitoky. This process results in a sexual form (epitoke) that differs in several characteristics from the non-sexual form. Changes may include loss or rearrangement of internal organs and musculature, development or enlargement of sensory structures on the prostomium and modification of setal types and parapodial structure along all or part of the body. Segments that contain eggs or sperm are usually the most modified. Epitokes usually swim to the water surface to swarm. This spawning behavior of sexual individuals helps to ensure the likelihood of fertilization when the eggs and sperm are shed into the seawater. In many species (e.g., *Odontosyllis enopla*) swarming is closely linked to a specific phase of the moon (lunar periodicity).

Life-span varies greatly, from several months to a few years. Polychaetes display a wide range of feeding types and include carnivores, herbivores, filter feeders and deposit feeders. Predators include crabs and other Crustacea, fishes, and other polychaetes. Parasitic species are usually associated with Crustacea or other polychaetes. Parasites of polychaetes include Sporozoa, Trematoda, Cestoda, Nematoda, Copepoda, and other Polychaeta.

**Plate 72**

DEVELOPMENT: Typically, species with external fertilization have indirect development that includes 1 or more free-swimming larval stages. Species with internal fertilization usually display some degree of direct development whereby 1 or all of the free-swimming larval stages are lost. The earliest recognizable larval stage is the trochophore. This top-shaped larva uses 3 bands of cilia for swimming and has not yet developed segmentation. The next stage, the metatrochophore, shows external signs of segmentation, usually in the form of segmentally arranged ciliary bands or precursors of setal sacs. The 3rd stage, the nectochaeta, is a free-swimming larva with several segments and usually well-developed setae. The development and appearance of the nectochaeta larva vary considerably depending upon family. A gradual metamorphosis into the adult form usually occurs for those nectochaetae that remain free-swimming for a long period of time (up to several weeks),

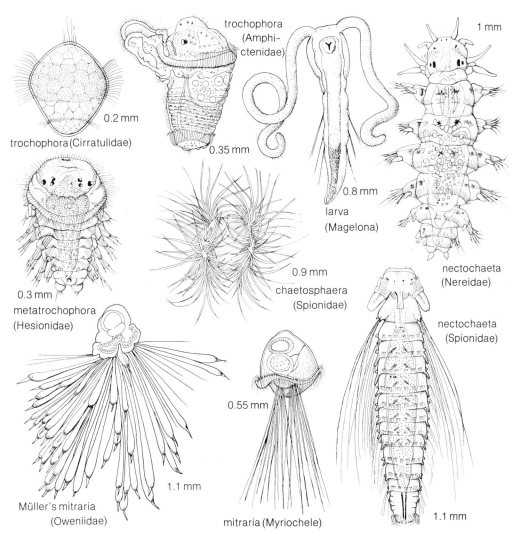

**72 POLYCHAETA (Bristle worms): Development**

whereas those that are free-swimming only for a short period of time (few hours to a few days) usually settle out of the seawater and undergo a more rapid metamorphosis. Among specialized forms of the nectochaeta stage are the mitraria larvae of the family Oweniidae and the chaetosphaera larvae of several species in the families Polynoidae and Spionidae. All larval stages can be captured from the plankton with a medium to fine-mesh net.

REFERENCES: For a general introduction see DALES (1963). Information on feeding and reproduction is given by FAUCHALD & JUMARS (1979) and SCHROEDER & HERMANS (1975), respectively. For identification of families and genera see FAUCHALD (1977). Other useful works include FAUVEL (1923, 1927) and DAY (1967). Archiannelida have been summarized by JOUIN (1971).

At present, there is no comprehensive treatment available for Bermuda polychaetes. Literature on the group is scattered over many years and by several authors, most notably WEBSTER (1884), VERRILL (1900e, 1901a, 1902a), TREADWELL (1917, 1936,

1941), WESTHEIDE (1973) and STERRER & ILIFFE (1982).

## Plate 73

F. **POLYNOIDAE** (Scale worms): Polychaeta with body dorsoventrally flattened, with relatively long parapodia. Prostomium usually with 3 antennae. Flattened scales (elytra) on dorsal surface, in generically distinct distributions. Setae all simple, in notopodia and neuropodia. Eversible proboscis with 4 jaws and fringe of papillae at distal margin. Most probably carnivores or scavengers. (4 spp. from Bda.)

*Halosydna leucohyba* (Schmarda): With 18 pairs of elytra; to 25 mm. With 3 prostomial antennae equally developed, lateral ones on extension of anterior prostomial margin. Elytra with characteristic medial ovals of gray pigment. Among rocks and cobbles; infrequent.

F. **CHRYSOPETALIDAE:** Polychaeta with body relatively narrow, fragile, dorsoventrally flattened; to 25 mm. Prostomium with 3 antennae, with or without dorsoposterior extension (caruncle). Notosetae (paleae) extending medially to dorsal midline, simple, usually flattened, overlap those on next parapodium like shingles. Neurosetae compound. Most likely carnivores. (3 spp. from Bda.)

*Bhawania goodei* Webster: Body with over 200 segments; about 25 mm. Caruncle absent. Paleae symmetrical, broad to narrow, with rounded tips, giving shiny golden-yellow appearance. Among rocks and cobbles; rare.

*Paleonotus elegans* (Bush in Verrill): Body with about 40 segments; about 11 mm. Caruncle present. Paleae symmetrical to asymmetrical, with pointed tips, imparting shiny light yellow aspect. Among rocks and cobbles; rare.

F. **AMPHINOMIDAE:** Polychaeta with body nearly square in cross section, elongate or short and nearly oval. Prostomium with 1-5 antennae. Caruncle present. Setae usually white, in rounded bundles; branchiae usually branching, in tight clumps. Called "fireworms" owing to the pain accompanying the breaking off of the setae in the flesh of the unwary handler. Carnivores and scavengers. (4 spp. from Bda.)

*Hermodice carunculata* (Pallas): To 300 mm. Caruncle nearly as wide as long, extending to about setiger 4, with transverse folds. General aspect brownish red with a green fringe. In subtidal depths, associated with hard substrata; not uncommon. Feeds on corals. (Color Plate 8.14.)

*Eurythoe complanata* (Pallas): To 250 mm. Caruncle rather short, consisting of single longitudinal ridge and pair of narrower folds on each side. General aspect flesh-red. Under rocks in shallows; common. (Color Plate 8.8.)

F. **ALCIOPIDAE:** Planktonic Polychaeta with body short to long, with segmental organs at bases of parapodia. Prostomium with 4 or 5 antennae and 2 extremely large, spherical, orange or red eyes. With 3-5 pairs of tentacular cirri. Parapodia uniramous, with foliaceous dorsal and ventral cirri. Setae simple, compound or both. Pygidium with anal cirri. Proboscis eversible, without teeth or jaws but

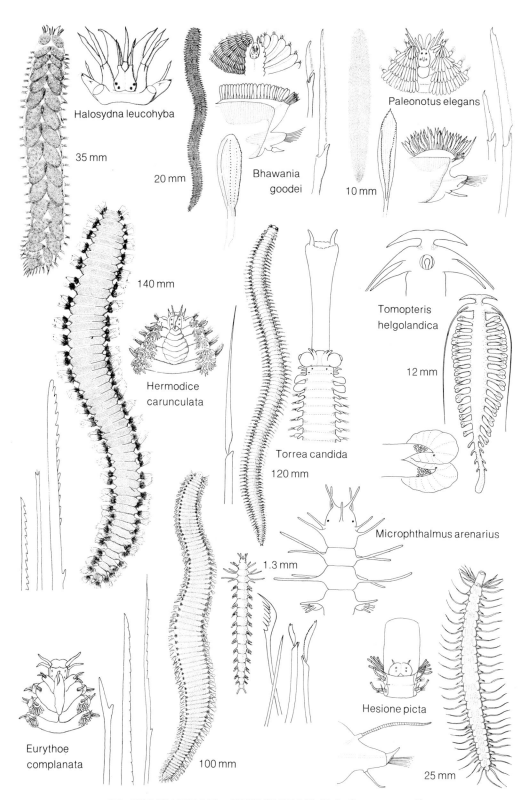

**73 POLYNOIDAE—HESIONIDAE (Bristle worms 1)**

often with papillae or lateral horns. Carnivorous. (2 spp. from Bda.)

***Torrea candida*** (Delle Chiaje): Body elongate, to 250 mm. Setae all compound, with long pointed blades. Segmental organs darkly pigmented. Dorsal surface often with segmental pigment stripes. Proboscis with lateral horns. Planktonic in the open ocean; uncommon.

F. **TOMOPTERIDAE:** Planktonic Polychaeta with body short, transparent, with few segments. Prostomium fused with first 2 segments; with 2 antennae, 1 pair of eyes and 1 pair of nuchal organs. With 1 or 2 pairs of tentacular cirri; 2nd pair often as long as body. Parapodia well developed anteriorly, rudimentary posteriorly, without setae but with 1 pair of membranous pinnules often with associated glands. Proboscis short, without teeth or jaws. Carnivorous. (1 sp. from Bda.)

***Tomopteris helgolandica*** Greeff: To 12 mm. With 1 pair of tentacular cirri about 2/3 as long as body. Tail with 3-4 rudimentary parapodia. Planktonic in the open ocean; uncommon.

F. **HESIONIDAE:** Polychaeta with short fragile body. Prostomium with 2 or 3 antennae. Peristomium with 2-8 pairs of tentacular cirri. Notopodia sometimes reduced or lacking. Neurosetae compound; notosetae, if present, simple. Dorsal cirri slender, long. Most likely carnivorous. (5 spp. from Bda.)

***Hesione picta*** Müller: Body round in cross section, to 50 mm. Prostomium with 2 antennae. Peristomium with 8 pairs of tentacular cirri. Parapodia with neuropodia only. Neurosetae compound, with short to long blades. Body light colored with transverse brown bands. Among rocks; uncommon.

***Microphthalmus arenarius*** Westheide: To 1.3 mm long and 0.2 mm wide. Prostomium with 5 antennae and 2 eyes. With 2 pairs of tentacular cirri on each of first 3 segments. Parapodia well developed. Setae simple, pointed and comb-like and compound with short hooked blades. Pygidium with anal plate and 2 anal cirri. Intertidally in exposed sandy beaches; common.

## Plate 74

F. **SYLLIDAE:** Polychaeta with body usually small and slender. Prostomium with 3 antennae and 1 pair of palps; palps separate or fused for part or all of length. With 1 or 2 pairs of tentacular cirri on peristomium. Barrel-shaped proventriculus separating esophagus and gut at about setigers 4-12. Parapodia usually uniramous, setae simple or compound. At sexual maturity parapodia sometimes biramous. Dorsal cirri smooth, articulated, or bead-like. Carnivores, diatom and surface deposit feeders. (32 spp. from Bda.)

***Odontosyllis enopla*** Verrill (Bermuda fireworm): To 35 mm. Palps fused at base only. With flap overlying posterior area of prostomium. Pharynx with many small teeth. With compound setae with short blades. Dorsal cirri smooth. Hardly ever collected on the shallow, protected gravelly bottoms on which it presumably lives, this species displays lunar periodicity in its spawning behavior. Females appear at the surface and attract males, both emitting bright greenish light. Swarming occurs during most months of the

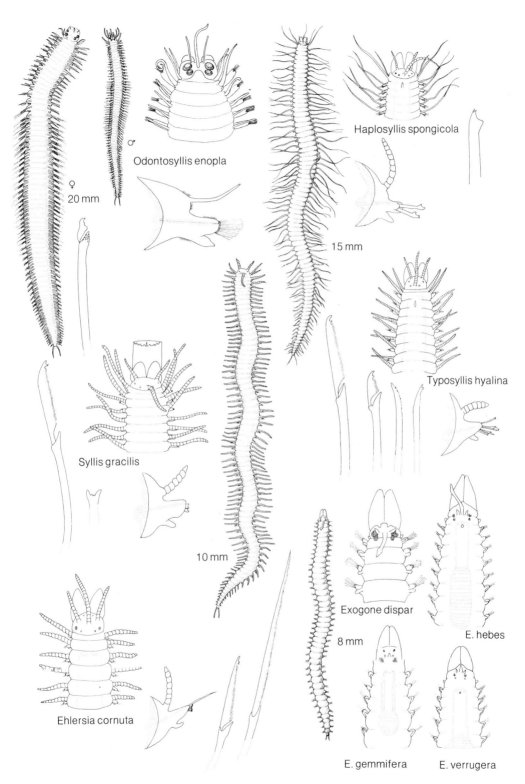

## 74 SYLLIDAE (Bristle worms 2)

year but is particularly spectacular in summer, with its peak on the 3rd night after the full moon, 56 min after sunset (MARKERT et al. 1961). (Color Plate 8.2.)

*Haplosyllis spongicola* (Grube): To 20 mm. Palps free throughout. Setae simple, heavy tridentate hooks. Dorsal cirri articulated. Usually associated with sponges, but also found among rocks; fairly common.

*Syllis gracilis* Grube: To 25 mm. Palps free throughout. Anterior and posterior setae compound, with long blades. Setae of middle parapodia simple, stout, bidentate, Y-shaped; dorsal cirri bead-like. Associated with rocks; uncommon.

*Typosyllis hyalina* (Grube): To 16 mm. Palps free throughout. Parapodia with compound setae throughout, with short to long blades. Posterior parapodia also with simple setae. Dorsal cirri bead-like. Associated with rocks and sand; common.

*Ehlersia cornuta* (Rathke): To 10 mm. Palps free throughout. Parapodia with compound setae only, with long to very long blades. Dorsal cirri bead-like. Associated with sand, rocks and *Thalassia*, common.

*Exogone dispar* (Webster): Genus with palps fused throughout. Prostomium with 3 antennae. Dorsal cirri small, papilliform.—Species to 8 mm. Median antenna long, laterals short. Setiger 2 with dorsal cirri. Associated with sand, rocks, and *Thalassia*; common.

*E. hebes* (Webster & Benedict): Genus as above.—Species to 8 mm. Median antenna long, laterals short. Setiger 2 without dorsal cirri. Associated with rocks; uncommon.

*E. verrugera* (Claparède): Genus as above.—Species to 8 mm long. All 3 antennae short, subequal. Setiger 2 with dorsal cirri. Associated with sand, rocks and *Thalassia*; uncommon.

*E. gemmifera* Pagenstecher: Genus as above.—Species to 8 mm. All 3 antennae short, subequal. Setiger 2 without dorsal cirri. In sand; uncommon.

### Plate 75

F. **NEREIDIDAE:** Polychaeta with elongate body. Prostomium with 1 pair of antennae, 1 pair of palps, and usually 2 pairs of eyes. Peristomium with 2 or 4 pairs of tentacular cirri. Eversible pharynx with 1 pair of jaws and variable number of paragnaths on 1-8 areas of pharyngeal surface; paragnaths sometimes lacking and pharyngeal surface smooth or with papillae. Parapodia usually biramous, setae simple or compound; blades of compound setae short or long, hooked or pointed. Mostly omnivorous. (7 spp. from Bda.)

*Nereis riisei* Grube: To 120 mm. With conical paragnaths on pharyngeal surface. Setae all compound, with long, pointed and short, hooked blades. Associated with *Thalassia* and rocks; uncommon.

*Perinereis anderssoni* Kinberg: To 120 mm. With conical paragnaths and transverse bars on pharyngeal surface; latter on dorsal proximal surface of everted pharynx. Setae all compound; notosetae all with long, pointed blades; neurosetae with long, pointed

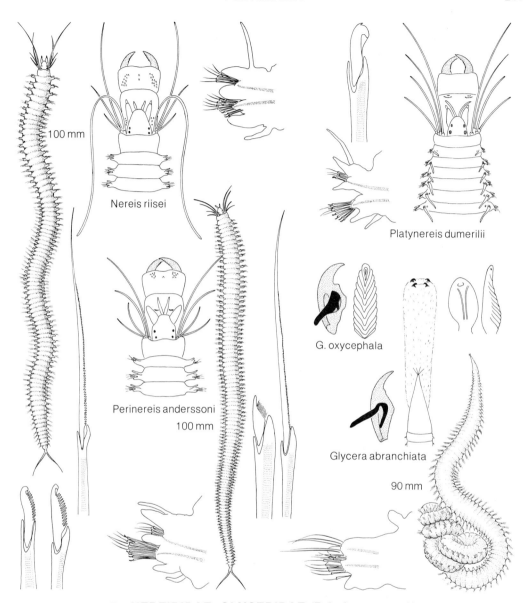

**75 NEREIDIDAE, GLYCERIDAE (Bristle worms 3)**

and short, hooked blades. Associated with rocks and *Thalassia*; fairly common.

***Platynereis dumerilii*** (Audouin & Milne Edwards): To 60 mm. With very long peristomial tentacular cirri. Without conical paragnaths but with rows of very small comb-like bars; setae all compound, with long, pointed and short, hooked blades; some notosetae of middle region with recurved tips. Associated with rocks; uncommon.

**F. GLYCERIDAE:** Polychaeta with elongate body and tapering prostomium with 4 minute tentacles at tip. With 4 heavy black jaws at end of everted pharynx; jaws provided with ailerons (supports); with characteristic sensory and/or secretory "organs" on pharyngeal surface. Parapodia either biramous or uniramous throughout. Mostly carnivores. (2 spp. from Bda.)

***Glycera abranchiata*** Treadwell: Genus with all parapodia biramous except on first 2 segments.—Species to 100 mm. Presetal parapodial lobe incised and bilobate. With 2 kinds of pharyngeal organs. Jaws with 2 ailerons each. General aspect red. In sand and *Thalassia*; common.

***G. oxycephala*** Ehlers: Genus as above.—Species to 50 mm. Presetal parapodial lobe entire and unilobate. With 1 kind of pharyngeal organ. Jaws with 1 aileron each. General aspect white. In sand; fairly common.

### Plate 76

F. **ONUPHIDAE:** Polychaeta with narrow, relatively long body. Prostomium with 2 frontal and 5 occipital antennae. With or without dorsolateral peristomial tentacular cirri. Parapodia with neurosetae only, with dorsal cirri, with or without branchiae. Setae simple and compound; former long and slender, small and finely dentate, and subacicular hooks; latter with long or short blades or with obscured junction with setal base. Tubicolous; scavengers. (1 sp. from Bda.)

***Mooreonuphis jonesi*** Fauchald: Small, to 20 mm. With short, frontal prostomial antennae. With dorsolateral peristomial tentacular cirri. Branchiae simple and unbranched or lacking. Tubes of sand-size particles. In sand; also associated with *Thalassia*; very common.

F. **EUNICIDAE:** Polychaeta with long to extremely long body. Prostomium with 1-5 occipital antennae, lacking frontal antennae. Parapodia with neurosetae only, with dorsal cirri, with or without branchiae. Setae simple and compound; former long and slender, small finely dentate, and heavy subacicular hooks; latter with long and short blades. Mostly carnivores, some scavengers or detritus feeders. (17 spp. from Bda.)

***Marphysa sanguinea*** (Montagu): To 200 mm. With 5 occipital antennae. Peristomium lacking dorsolateral tentacular cirri. Branchiae beginning on about setiger 20; subacicular hooks from about setiger 30. In muddy sand and *Thalassia*; common.

***Eunice vittata*** (Delle Chiaje): Genus with 5 occipital antennae, and with dorsolateral peristomial tentacular cirri.—Species to 60 mm. Branchiae beginning on about setiger 30; subacicular hooks from about setiger 28. Associated with rocks; common. (Color Plate 8.1.)

***E. cariboea*** Grube: Genus as above.—Species to 70 mm. Without branchiae. Subacicular hooks beginning on about setiger 24. Up to 700 segments long. Associated with rocks; common.

***Lysidice ninetta*** Audouin & Milne Edwards: To 100 mm. With 3 occipital antennae. Peristomium lacking dorsolateral tentacular cirri. Without branchiae. Subacicular hooks beginning on about setiger 20. Associated with rocks and *Thalassia*; uncommon.

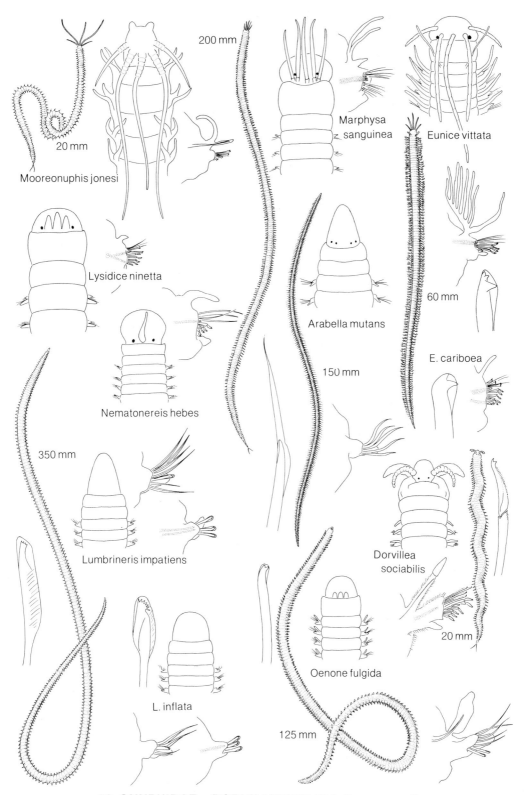

**76 ONUPHIDAE—DORVILLEIDAE (Bristle worms 4)**

*Nematonereis hebes* Verrill: Small and thread-like, to about 25 mm. With 1 occipital antenna. Peristomium lacking dorsolateral tentacular cirri. Without branchiae. Associated with sand, rocks and *Thalassia*; common.

F. **LUMBRINERIDAE:** Polychaeta with superficial aspect of iridescent earthworm. Prostomium lacking eyes. With short uniramous parapodia. Setae simple, pointed and hooded, and compound with hooded blades. Carnivorous, herbivorous and surface deposit feeders. (2 spp. from Bda.)

*Lumbrineris impatiens* (Claparède): Genus without antennae or branchiae.—Species to 350 mm. Prostomium somewhat pointed, length:width = 1.5:1. In sand; common.

*L. inflata* (Moore): Genus as above.—Species to 60 mm. Prostomium rounded, length:width = 1:1. Associated with rocks and *Thalassia*; common.

F. **ARABELLIDAE:** Polychaeta with superficial aspect of iridescent earthworm, frequently darkly pigmented. Prostomium lacking antennae; eyes sometimes present. With prominent uniramous parapodia. Setae simple and pointed and/or blunt spines; no hooded setae. Most likely carnivores. (1 sp. from Bda.)

*Arabella mutans* (Chamberlin): To 150 mm. Prostomium somewhat pointed, length:width = about 1.5:1, with eyes near posterior margin. In sand, rocks and *Thalassia*; very common.

F. **LYSARETIDAE:** Polychaeta with elongate body. Prostomium with 1-3 occipital antennae. Notopodia with large foliaceous dorsal cirri. Setae simple or compound, with bidentate blades. Possibly carnivores. (1 sp. from Bda.)

*Oenone fulgida* (Savigny): At least 125 mm long. Prostomium with occipital antennae. Setae all simple, pointed and stout, bidentate. Subtidal, associated with rocks; uncommon.

F. **DORVILLEIDAE:** Polychaeta with short body. Prostomium with 1 pair of palps and 1 pair of antennae. With well-developed neuropodia; notopodia 2-parted, cirriform. Setae simple, pointed and/or forked, as well as compound, with pointed or hooked blades. Carnivores and possibly some herbivores. (4 spp. from Bda.)

*Dorvillea sociabilis* (Webster): To 20 mm. With beaded antennae and smooth palps. Setae simple, with obscure bidentate tips, and compound, with long bidentate blades; lacking forked simple setae. Associated with rocks; rare.

**Plate 77**

F. **ORBINIIDAE:** Polychaeta with body elongate, slender, with moderate number to many segments, divided into short thoracic region with lateral parapodia and long abdominal region with dorsal parapodia. Prostomium small, pointed or truncate, without appendages. With 1 or 2 buccal segments without setae. Parapodia well developed, biramous. Branchiae usually on some thoracic and most abdominal segments. Proboscis eversible, without teeth or jaws. Burrowing deposit feeders. (4 spp. from Bda.)

*Naineris laevigata* (Grube): Genus with a pair of subdermal eyes; only buccal segment without setae.—Spe-

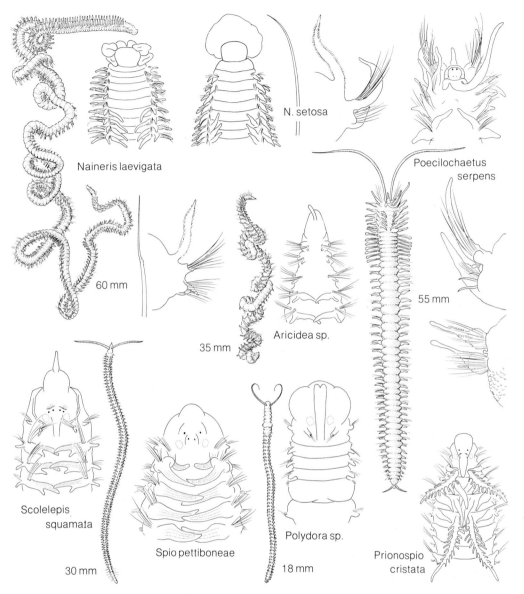

**77 ORBINIIDAE—SPIONIDAE (Bristle worms 5)**

cies to 60 mm long and 3 mm wide. Prostomium somewhat rounded. Thoracic neuropodia with slender pointed setae and stout acicular setae producing light brownish color. Proboscis with numerous lobes. In intertidal and shallow subtidal sand; common.

***N. setosa*** (Verrill): Genus as above.— Species to 95 mm long and 3-4 mm wide, not sharply divided into thoracic and abdominal regions. Prostomium truncate. Thoracic neuropodia with numerous slender pointed setae producing golden silky appearance, without stout acicular setae. Proboscis saclike. In intertidal and shallow subtidal sand; very common.

F. **PARAONIDAE:** Polychaeta with small, thread-like body, usually with numerous segments. Prostomium conical, with nuchal slits and sometimes with single median antenna and eyespots. Parapodia biramous, poorly developed posteriorly. Pair of cirriform branchiae on variable number of anterior segments. Setae all simple, slender and pointed, with some variously developed specialized forms on posterior segments in most species. Proboscis eversible, without teeth or jaws. Burrowing deposit feeders. (2 spp. from Bda.)

*Aricidea* sp.: Body slender, to 35 mm long and 1 mm wide. Prostomium with single median antenna and with eyespots. With 30 pairs of branchiae, beginning on setiger 4. Specialized neurosetae of posterior segments consisting of bidentate hooks with terminal hair. In intertidal and shallow subtidal sand; infrequent. Easily overlooked because of small size.

F. **POECILOCHAETIDAE:** Polychaeta with elongate, posteriorly tapering body. Prostomium subrectangular to rounded, with 4 eyes, 1 antenna and usually a nuchal organ with 1-3 elongate lobes. Peristomium reduced, with 1 pair of long grooved palps (easily lost). First segment usually with long setae extending anteriorly. Parapodia well developed, biramous, with lateral sense organs between rami. Setae all simple, slender and pointed, spinous, plumose or acicular. Branchiae sometimes present on middle and posterior parapodia. Suspension feeders or surface deposit feeders. (1 sp. from Bda.)

*Poecilochaetus serpens* Allen: To 55 mm. Nuchal organ well developed, with elongate median and 2 shorter lateral lobes. Parapodia of setiger 1 with long notosetae and neurosetae. Parapodia of setigers 2-3 with stout acicular neurosetae. Parapodia of setigers 7-13 with flask-shaped dorsal and ventral cirri. In intertidal and shallow subtidal sand; uncommon.

F. **SPIONIDAE:** Polychaeta with slender, short to long body, with moderate number to many segments, not obviously divided into regions. Prostomium variably shaped, rounded, tapered, bifurcate, sometimes with eyes, occipital tentacle or frontal horns. Peristomium often surrounding prostomium, with 1 pair of grooved palps inserted at sides of prostomium (easily lost). Parapodia moderately developed, biramous. Notosetae simple, usually slender and pointed; hooded hooks in some species. Neurosetae simple, hooded hooks on middle and posterior segments. Branchiae associated with variable number of segments. Pygidium variably shaped, sometimes with cirri. Mostly surface deposit feeders. (13 spp. from Bda.)

*Scolelepis squamata* (Müller): To 30 mm long and 1 mm wide. Prostomium pointed anteriorly, with 2 pairs of eyes, without appendages. Branchiae beginning on setiger 2, continuing to near posterior end. Bidentate neuropodial hooded hooks beginning on setigers 27-40. Pygidium cushion-shaped. Restricted to clean sand; common.

*Spio pettiboneae* Foster: To 11 mm long and 1 mm wide. Prostomium rounded anteriorly, with 2 pairs of eyes, without appendages. Branchiae beginning on setiger 1, continuing to near posterior end. Tridentate neuropodial hooded hooks beginning on setigers 12-16. Pygidium with 4 blunt anal cirri. In intertidal and shallow subtidal clean sand; not uncommon.

*Polydora* sp.: To 18 mm long and 1 mm wide. Prostomium bifid anteriorly, without eyes or appendages. Setiger 5 modified, wider and longer than others, with stout acicular setae and slender pointed setae. Branchiae beginning on setiger 8, continuing to near posterior end. Bidentate neuropodial hooded hooks beginning on setiger 7. Pygidium flange-shaped, with 3 lobes. In clean sand; common. Other species of this genus are common in fouling communities (e.g., on buoys and boats).

*Prionospio cristata* Foster: To 10 mm long and 1 mm wide. Prostomium bluntly rounded anteriorly, with 2 pairs of eyes, without appendages. Branchiae 4 pairs, beginning on setiger 2; 1st and 4th pairs pinnate, 2nd and 3rd pairs strap-like. Membranous crest across dorsal surface of setigers 7-9. Multidentate notopodial hooded hooks beginning on setigers 21-37; multidentate neuropodial hooded hooks beginning on setigers 11-12. Pygidium with 3 anal cirri. In intertidal and shallow subtidal sand; uncommon.

## Plate 78

F. **MAGELONIDAE:** Polychaeta with slender, elongate body divided into anterior thoracic region of 9 setigers and longer abdominal region. Prostomium spatulate, without eyes or appendages, sometimes with frontal horns. Peristomium reduced, with pair of long papillated palps inserted ventrally at sides of mouth. Parapodia biramous, moderately developed. Branchiae absent. Setae all simple, slender and pointed on thoracic segments, and hooded hooks on abdominal segments; setiger 9 sometimes with specialized setae. Proboscis eversible, without teeth or jaws. Surface deposit feeders. (1 sp. from Bda.)

*Magelona* sp.: To 73 mm long and 0.5 mm wide. Prostomium without frontal horns. Setiger 9 without specialized setae. Bidentate hooks on abdominal segments. In clean sand, barely subtidal; uncommon.

F. **CHAETOPTERIDAE:** Tubicolous Polychaeta with elongate body divided into 3 regions. Anterior region of 9-15 setigers, dorsally flattened, with uniramous parapodia; setiger 4 with special stout acicular setae. Middle and posterior regions with biramous parapodia. Prostomium inconspicuous, often surrounded by collar-like buccal segment, usually with 1 pair of long grooved palps. Sometimes with 1 pair of tentacular cirri posterior to palps. Branchiae absent. Primarily filter feeders using 1 or more mucous bags secreted by the parapodia of the middle region. Surface deposit feeding also may be important for those species that possess long paired palps. Tubes are very characteristic and often species-specific. (3 spp. from Bda.)

*Mesochaetopterus minutus* Potts: Short, slender, to 15 mm. Prostomium with 1 pair of eyes. Buccal segment with 1 pair of long palps. Without tentacular cirri. Anterior region with 10-13 setigers; middle region with 2 setigers with simple flattened notopodia. Tube fragile, constructed of fine sand. Gregarious species, often occurring in dense masses in sandy areas; common.

*Spiochaetopterus costarum oculatus* Webster: Slender, to 60 mm. Prostomium with 1 pair of eyes. Buccal

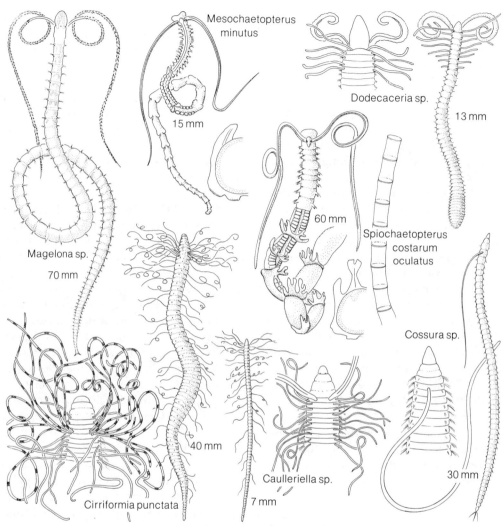

**78 MAGELONIDAE—COSSURIDAE (Bristle worms 6)**

segment with 1 pair of long grooved palps. Without tentacular cirri. Anterior region with 9 setigers; middle region with 20-24 setigers with trilobed notopodia. Tube horny, ringed. In intertidal and shallow subtidal sand; uncommon.

F. **CIRRATULIDAE:** Polychaeta with body essentially cylindrical, sometimes flattened posteriorly, with few to numerous segments. Prostomium small, without appendages, fused to peristomium, followed by 2 poorly defined segments lacking setae. Pair of grooved palps, or 2 groups of grooved tentacular filaments, at posterior margin of peristomium or apparently originating from 1 or more anterior setigers. Parapodia biramous, poorly developed. Variable number of segments with branchial filaments. Setae simple, slender and pointed and/or acicular. Proboscis eversible, without teeth or jaws. Surface deposit feeders. (14 spp. from Bda.)

*Cirriformia punctata* (Grube): To 40 mm long and 2-4 mm wide. Prostomium bluntly conical. Grooved tentacular filaments usually on setigers 3 and 4. Branchial filaments on large number of segments throughout length of body. All parapodia with slender pointed setae in both lobes and with slender curved hooks beginning in notopodia of setigers 8-13 and in neuropodia of setigers 5-7. Color in life brown speckled with black; tentacular filaments yellow with black bars, branchial filaments brownish or blackish. In shallow water with rocks; occasional. (Cf. Color Plate 8.7.)

*Dodecaceria* sp.: To 13 mm long and 0.7 mm wide. Prostomium blunt, forming hood over mouth. With 1 pair of grooved palps. Branchial filaments on a few anterior segments. Setae on anterior segments slender and pointed, middle and posterior segments also with 2 or 3 spoon-shaped or distally curved hooks in each lobe. Color in life yellowish green or blackish. Boring in corals, and washed from rocks and sponges in shallow waters; not uncommon.

*Caulleriella* sp.: To 7 mm, with very crowded anterior segments. Prostomium with blunt tip and 2 dark red eyes near posterolateral margin. Pair of grooved palps on dorsal surface of setiger 1. With branchial filaments on several segments throughout length of body. Slender pointed setae in all notopodia and anterior neuropodia; 7-9 bidentate hooks in all neuropodia and in notopodia from about setiger 25. In shallow water with sand or *Thalassia*; uncommon.

F. **COSSURIDAE:** Polychaeta with small, thread-like body. Prostomium without appendages and usually with eyes. With 1-2 segments without parapodia following prostomium. Single median tentacle arising from dorsal surface on one of anterior segments. With 1 or 2 kinds of simple setae. Pygidium with anal cirri. Proboscis eversible, without teeth or jaws. Burrowing deposit feeders. (1 sp. from Bda.)

*Cossura* sp.: With 1 kind of simple seta, and 1 peristomial segment. Median tentacle arising on anterior margin of setiger 3. Simple setae with paired transparent wings along length. In subtidal sand or sandy mud; infrequent. Easily overlooked because of small size.

## Plate 79

F. **OPHELIIDAE:** Polychaeta with fusiform, short body with few segments and usually with ventral and lateral grooves. Prostomium without appendages but with 1 pair of eversible nuchal organs and 1 pair of subdermal eyes. Eyespots often present along sides of body. Setae simple, slender and pointed. Pygidium usually elongate, tubular, with anal cirri. Proboscis eversible, without teeth or jaws. Burrowing deposit feeders. (3 spp. from Bda.)

*Armandia maculata* (Webster): Body with ventral groove, to 25 mm long and 2.5 mm wide, with 28-31 segments. Lateral eyespots between bases of most parapodia. Prostomium sharply pointed. Parapodia biramous. Ciliated strap-like branchiae on all parapodia except 1st and sometimes last 1-3. Pygidium tubular, with marginal papillae and long midventral cirrus. In sand; fairly common. Burrows rapidly.

*Polyophthalmus pictus* (Dujardin): Body with ventral groove, to 25 mm, with about 28 segments. Lateral eyespots on middle segments. Prostomium rounded. Parapodia poorly developed, with single bundle of long pointed setae, except on last 1-3 segments. Branchiae absent. Pygidium with short tube and marginal papillae. Brown bars on dorsum. In mud, muddy sand or *Penicillus*; fairly common.

F. **CAPITELLIDAE:** Polychaeta with elongate, reddish body, earthworm-like in appearance, divided into short thoracic region and longer abdominal region. Prostomium small, without appendages but with 1 pair of nuchal slits and sometimes eyespots. Parapodia reduced and poorly developed, setae emerging from pockets in body wall in thoracic region. Setae simple, slender and pointed, and hooded hooks. Branchiae sometimes present. Proboscis eversible, without teeth or jaws, papillated. Burrowing deposit feeders. (6 spp. from Bda.)

*Notomastus latericeus* Sars: To 300 mm. Thoracic region of peristomium and 11 segments with pointed setae in both parts of parapodia; long abdominal region with hooded hooks in both parts of parapodia. Prostomium with eyespots. Rudimentary branchiae as triangular projections above abdominal neuropodia. Lateral organs between abdominal notopodia and neuropodia. In intertidal and shallow subtidal sand; very common.

*Capitella capitata* (Fabricius): Small, to 40 mm long and 1 mm wide. Thoracic region of 9 segments with pointed setae in both parts of parapodia of segments 1-5 or 1-6; segment 7 variable, with pointed setae and/or hooded hooks; ♀ with hooded hooks in parapodia of segments 8-9; ♂ with stout genital hooks on dorsal surface of segments 8-9 and with hooded hooks ventrally; abdominal region with numerous segments and hooded hooks in both parts of parapodia. Color in life bright red. Intertidally in muddy areas; fairly common but easily overlooked because of small size.

*Dasybranchus lunulatus* Ehlers: To 100 mm long and 4 mm wide. Thoracic region of peristomium and 13 segments with pointed setae in both parts of parapodia; long abdominal region with hooded hooks in both parts of parapodia. Prostomium with eyespots. Retractile branchial filaments above middle and posterior neuropodia. Lateral organs between notopodia and neuropodia. Color in life bright red. In intertidal sandy or muddy areas; fairly common.

F. **ARENICOLIDAE:** Polychaeta with elongate body divided into 2 or 3 regions. Prostomium small, without appendages. Parapodia biramous, with neuropodial ridges and stout notopodial lobes. Branching branchiae on some segments. Setae simple, slender and pointed, and hooked. A posterior body region made up of segments without setae is sometimes present. Proboscis eversible, without teeth or jaws. Deposit feeders, often enriching sediment by pumping water into L-shaped burrow. (2 spp. from Bda.)

*Arenicola cristata* Stimpson (Lugworm): Body cylindrical, to 250 mm long and 15 mm wide, divided into 2 regions; anterior thoracic region of 17 segments with branching branchiae on segments 7-17 and posterior region of segments lacking setae. Notosetae spinous; neurosetae simple hooks, often

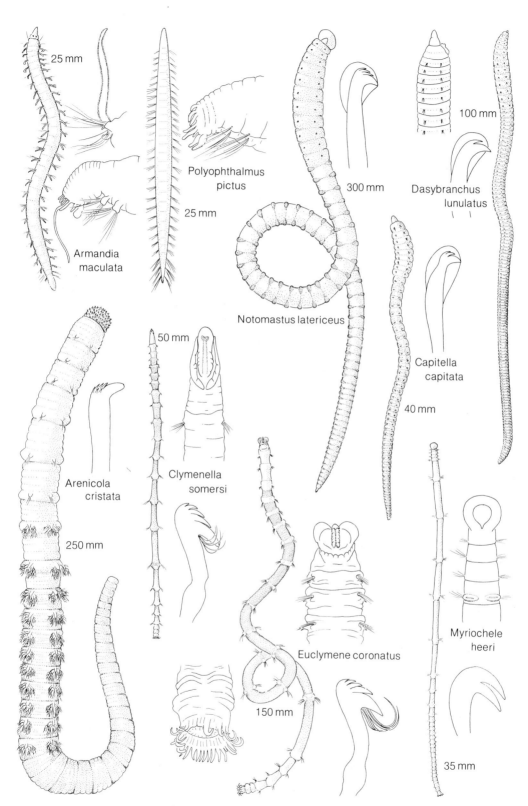

**79 OPHELIIDAE—OWENIIDAE (Bristle worms 7)**

absent on segments 1-2. Color in life olive greenish yellow to brownish; branchiae red. In muddy sand; fairly common. ♀ attaches long, brownish mucous mass containing eggs to mouth of burrows. (Color Plate 8.9.)

F. **MALDANIDAE:** Tubicolous Polychaeta with cylindrical body with few greatly elongate segments. A posterior body region of segments without setae is usually present. Prostomium poorly defined, without appendages, fused to peristomium, sometimes with a cephalic plate and often surrounded by raised flange. One pair of nuchal grooves present. Parapodia moderately developed, neuropodia as raised ridges. Notosetae simple, slender and pointed; neurosetae simple hooks. Pygidium variously shaped, often with anal cirri. Proboscis eversible, without teeth or jaws. Tube sandy. Deposit feeders. (2 spp. from Bda.)

*Clymenella somersi* Verrill: To 50 mm long and 1 mm wide, with 18 setigerous segments and 2 preanal segments lacking setae; middle and posterior segments elongate. Cephalic plate well developed, surrounded by raised flange. Nuchal grooves straight. Well-developed collar on segment 4. Pygidium funnel-shaped, with ring of about 20 anal cirri. Color in life light reddish-orange. In intertidal and shallow subtidal sandy areas; common.

*Euclymene coronatus* Verrill: To 150 mm long and 5 mm wide, with 19 setigerous segments and 2 preanal segments lacking setae; middle and posterior segments elongate. Cephalic plate well developed, surrounded by raised flange. Nuchal grooves straight. Neurosetae of segments 1-3 stout, acicular; neurosetae of remaining segments simple hooks with 5 teeth and well-developed "beard" below teeth. Pygidium funnel-shaped, with ring of about 30 anal cirri. Color in life banded bright red. In intertidal and shallow subtidal sandy areas; common. (Color Plate 8.13.)

F. **OWENIIDAE:** Tubicolous Polychaeta with cylindrical body of about 30 segments. Middle segments elongate, posterior segments short. Prostomium with or without appendages and eyespots. Notosetae slender and pointed; neurosetae in large fields as minute bidentate hooks. Pygidium with or without appendages. Filter feeders or deposit feeders. (1 sp. from Bda.)

*Myriochele heeri* Malmgren: To 37 mm long and 1.5 mm wide. Prostomium without appendages. Neurosetae with 2 teeth arranged one above the other. In intertidal or shallow subtidal sand; uncommon. Easily overlooked because of small size.

**Plate 80**

F. **SABELLARIIDAE:** Tubicolous Polychaeta with short to long body divided into 4 regions: 1st region of 2 segments with uniramous parapodia; 2nd region of 3-4 segments with biramous parapodia; 3rd region of numerous segments with biramous parapodia and branchiae; 4th region of few rudimentary segments lacking setae, reflected forward over 3rd region. Anterior end crowned with operculum of stout golden setae (paleae). Prostomium indistinct. Numerous oral filaments ventrally on anterior end. One pair of grooved palps at base of opercular cleft. Setae all simple, slen-

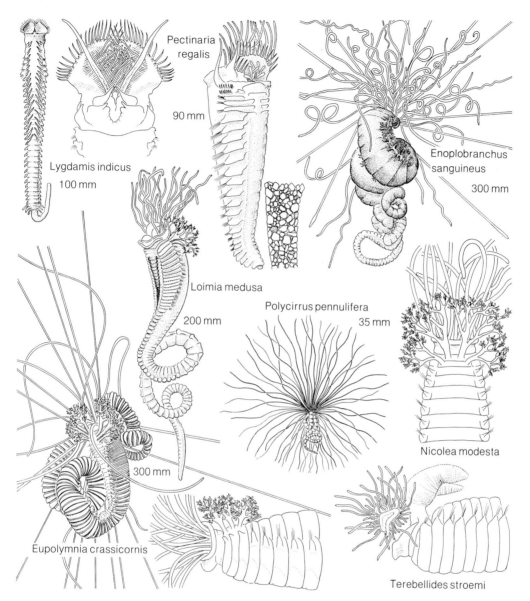

**80 SABELLARIIDAE—TRICHOBRANCHIDAE (Bristle worms 8)**

der and pointed, and short, hooked (uncini). Filter feeders. (1 sp. from Bda.)

***Lygdamis indicus*** Kinberg: Elongate, to 110 mm long and 10 mm wide. Anterior end with slanting operculum crowned with numerous paleae. Stout, dark recurved hooks on dorsal surface of region 2. Attached to hard subtidal bottoms associated with sandy areas; rare. Tubes of sand-size particles of limestone and shells. (Color Plate 8.5.)

F. **PECTINARIIDAE:** Tubicolous Polychaeta with short to long body, tapering posteriorly, divided into 4 regions: anterior region consisting of broad opercular plate dorsally, with fan of stout spines anteriorly, indistinct prostomium with cephalic veil with marginal papillae, numerous buccal tentacles around mouth; segments 3 and 4 each with 1 pair of lamellated branchiae; 1 pair of tentacular cirri on buccal segment and segment 2; 2nd region of 3 segments with uniramous parapodia; 3rd region of 12-15 segments with biramous parapodia; 4th region of segments forming flattened plate often reflected against tube and with stout setae at its base. Setae all simple, slender and pointed and short, hooked (uncini). Subsurface deposit feeders. (1 sp. from Bda.)

*Pectinaria regalis* Verrill (Icecream cone worm): Elongate, to 90 mm long and 13 mm wide. Opercular plate with 11-14 pairs of golden paleae. In intertidal and shallow subtidal sandy areas; uncommon. Slightly curved tubes of sand-size pieces of limestone, open at both ends. (Color Plate 8.3, 4.)

F. **TEREBELLIDAE:** Usually tubicolous Polychaeta with body divided into 2 regions. Anterior end with numerous long, grooved contractile tentacles. Thoracic region with biramous parapodia; branchiae usually 1-3 pairs on segments 2-4, variably shaped, dendritic, single filaments or club-shaped. Abdominal region of numerous segments with notopodia reduced or absent. Setae all simple, slender and pointed, and short, hooked (uncini). Pygidium without anal cirri. Mostly surface deposit feeders. (11 spp. from Bda.)

*Enoplobranchus sanguineus* (Verrill): Body elongate, tapered, to 350 mm, bright red in life. Without branchiae on segments 2-4. Abdominal region with palmately branched notopodia with small tufts of setae at tips of branches. Tube absent, gallery dweller, i.e., in cavity in substrate, without opening. Possibly subsurface deposit feeder. In intertidal and shallow subtidal sand and muddy sand; common. (Color Plate 8.6.)

*Loimia medusa* (Savigny): Elongate, to 200 mm, flesh-colored or white in life, with an elongated, blood-red triangle ventrally. With 3 pairs of dendritically branched branchiae on segments 2-4. Tube of sand and shell fragments. In intertidal and shallow subtidal sand; also on rocks. Very common.

*Eupolymnia crassicornis* (Schmarda): Elongate, to 300 mm, cream-white with brown transverse stripes. With 3 pairs of dendritically branched branchiae on segments 2-4. Tube of coarse gravel and limestone. In intertidal and shallow subtidal sand and under rocks; common. (Color Plate 8.12.)

*Nicolea modesta* Verrill: Short, to 15 mm. With 2 pairs of dendritically branched branchiae on segments 2 and 3. Tube constructed of sand. In intertidal and shallow subtidal sand; common.

*Polycirrus pennulifera* Verrill: Relatively short, to 35 mm, bright red in life. Branchiae absent. Tube of sand. In intertidal and shallow subtidal sand; common.

F. **TRICHOBRANCHIDAE:** Tubicolous Polychaeta, body with 2 regions. Anterior end with numerous long, grooved, contractile tentacles. Branchiae either a single pair of filaments on segments 2-4 or single branchia with stout trunk and 4 lamellated lobes. Surface deposit feeders. (2 spp. from Bda.)

***Terebellides stroemi*** Sars: Relatively long, to 70 mm. Single branchial trunk arising from segments 2-4 and with 4 lamellated lobes. Tube of mud. In subtidal and intertidal muddy areas; rare.

## Plate 81

F. **SABELLIDAE:** Tubicolous Polychaeta with cylindrical body divided into short thoracic region with slender pointed notosetae and short hooked neurosetae (uncini), and longer abdominal region with setal types reversed in position. Prostomium indistinct; anterior end with crown of bipinnate radioles (ciliated feeding tentacles), sometimes with eyespots and small projections on outer surfaces. Operculum absent. Lateral and pygidial eyespots sometimes present. Tube membranous, usually reinforced with mud. Filter feeders. (10 spp. from Bda.)

***Sabella melanostigma*** Schmarda: To 20 mm long and 2 mm wide. Thoracic region with about 11 setigerous segments, abdominal region with numerous segments. Well-developed collar around bases of radioles, notched dorsally and ventrally. Radioles with paired eyespots along length, without external projections. Patches of purple pigment on radioles, collar and bases of thoracic neuropodia. Tube membranous, reinforced with mud. Intertidal or in shallow subtidal areas on rocks, mangrove roots or other hard objects; common. (Color Plate 8.10.)

***Branchiomma nigromaculata*** (Baird): To 50 mm long and 4 mm wide. Thoracic region with 8 setigerous segments, abdominal region with numerous segments. Well-developed collar around bases of radioles, interrupted dorsally. Radioles with numerous paired eyespots along entire length and with paired projections on outer surfaces. Body with numerous specks of black pigment. Tube membranous, reinforced with mud. On submerged hard substrates; common.

***Hypsicomus elegans*** (Webster): To 50 mm long and 3 mm wide. Thoracic region with 8 setigerous segments; notosetae of 1st thoracic segment short, arranged in an antero-lateral diagonal row; succeeding thoracic segments with longer notosetae arranged in tufts; abdominal region with numerous segments. Collar at base of tentacular crown with shallow notch dorsally. Radioles united by a web along basal 1/4, deeply pigmented purplish brown. Purplish brown stripe on midventral surface. Tube leathery. In hard substrates such as coral.

***Megalomma lobiferum*** (Ehlers): To 60 mm long and 6 mm wide. Thoracic region with 8 setigerous segments, abdominal region with numerous segments. Well-developed collar around bases of radioles. Many radioles with single large purple eye near tips, without external projections. Tube covered with sand and shell fragments. In shallow subtidal sandy areas; infrequent.

F. **SERPULIDAE:** Tubicolous Polychaeta in calcareous tubes usually on hard substrates. Body cylindrical, divided into short thoracic region with slender pointed notosetae and short hooked neurosetae (uncini), and longer abdominal region with setal types reversed in position. Prostomium indis-

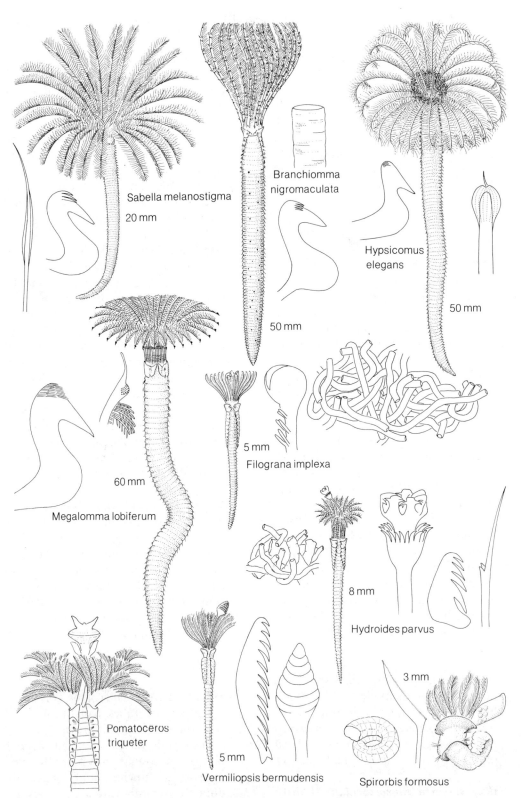

**81 SABELLIDAE, SERPULIDAE (Bristle worms 9)**

tinct; anterior end with crown of bipinnate radioles (ciliated feeding tentacles) one of which is usually modified as a stalked operculum. Second segment usually expanded to form foliaceous collar. Filter feeders. (7 spp. from Bda.)

*Filograna implexa* Berkeley: Small gregarious worms often forming large intertwining masses of tubes. To 5 mm long and 0.3 mm wide. Thoracic region with up to 12 segments, abdominal region with about 30 segments. Opercula poorly developed, sometimes 2 radioles slightly modified distally forming cup-like opercula. Collar setae simple, slender and pointed and abruptly bent with few teeth at base of winged expansion. Color in life pink. On hard substrates in shallow water; fairly common.

*Hydroides parvus* (Treadwell): To 8 mm long and 0.5 mm wide. Thoracic region with 7 segments, abdominal region with numerous segments. Opercular stalk slender, smooth. Opercular funnel with about 25 spines; opercular crown with 6 large curved spines, each with 1 pair of small lateral spines at base of curved part. Notosetae of setiger 1 simple, slender and pointed, and bayonet-like, each with 2 coarse teeth at base of blade. Subsequent thoracic notosetae slender and pointed; neurosetae of thorax and notosetae of abdomen are uncini with 5 teeth. Abdominal neurosetae comb-like. Tube curved, circular in cross section. On hard substrates in shallow water; common.

*Pomatoceros triqueter* (L.): To 70 mm. Thoracic region with 7 segments, abdominal region with numerous segments. Opercular stalk slender, with paired lateral wings at base of crown. Opercular crown plate-like, with or without few unbranched horns. Radioles usually banded blue or red. Offshore, on coral in shallow water; fairly common. (Color Plate 8.11.)

*Vermiliopsis bermudensis* (Bush): To 5 mm long and 0.3 mm wide. Thoracic region with 7 segments, abdominal region with about 25 segments. Opercular stalk slender, without wings. Operculum cone-shaped, with several concentric rings. Notosetae of setiger 1 slender and pointed. Uncini with numerous teeth. Tube with prominent spine over anterior opening. On hard substrates in shallow water; fairly common.

*Spirorbis formosus* Bush: Very small, coiled, to about 3 mm. Thoracic region with 3 segments, abdominal region with about 9-11 segments. Operculum well developed, flattened, often containing incubating embryos in adults. Tube coiled anti-clockwise when viewed from above. On *Sargassum* floating in open water and occasionally on rocks; common.

### Plate 82

**APPENDIX: ARCHIANNELIDA:** Variously reduced free-living Polychaeta; parapodia small or absent, setae simple (rarely compound) or lacking. Pharynx usually a muscular bulb. With few exceptions they live in the interstices of marine sand and move by means of cilia. (Although they are still poorly known in Bermuda, of 5 families only the Polygordiidae are yet unrecorded.)

**F. SACCOCIRRIDAE:** Elongated, many-segmented Archiannelida with retrac-

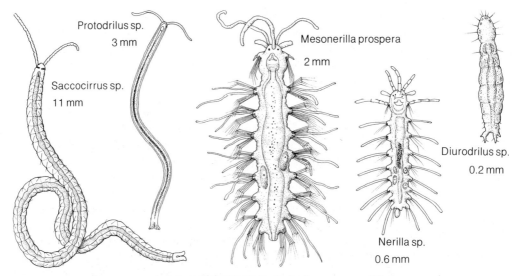

**82 ARCHIANNELIDA (Bristle worms 10)**

tile parapodia and setae; with 1 pair of movable cephalic tentacles.

***Saccocirrus*** sp.: With about 60 segments; pygidium with 2 rounded lobes. Without eyes. In clean subtidal sand, often common.

F. **PROTODRILIDAE:** Elongated, many-segmented Archiannelida without parapodia; setae reduced or absent; with 1 pair of movable cephalic tentacles.

***Protodrilus*** sp.: Body orange; without eyes. Pygidium with 2 adhesive lobes. In clean subtidal sand; common.

F. **NERILLIDAE:** Small Archiannelida with 7-9 segments, most with simple or compound setae. With up to 3 tentacles and 1 pair of palps.

***Mesonerilla prospera*** Sterrer & Iliffe: With 9 segments and compound setae; each parapodium with 2 long cirri. Head with 2 palps, 3 tentacles, and eyes. On walls of marine caves; locally not uncommon.

***Nerilla*** sp.: With 9 segments and simple setae; each parapodium with 1 cirrus. Head with 2 palps, 3 tentacles, and eyes. In clean subtidal sand; common.

F. **DINOPHILIDAE:** Very small Archiannelida with few segments and without parapodia, setae and cephalic tentacles. Ciliation prominent, often in segmental rings.

***Diurodrilus*** sp.: With strong ventral ciliation, but no ciliary rings. In clean sublittoral sand; rare.

M. L. JONES, S. L. GARDINER,
M. E. PETERSEN (CIRRATULIDAE)
& W. STERRER (ARCHIANNELIDA)

## Class Oligochaeta

CHARACTERISTICS: *Free-living, regularly segmented, cylindrical ANNELIDA with a clitellum; without parapodia, but each segment normally with 2 lateral and 2 ventral bundles of few setae. Head region lacking appendages. Mostly small (2-25 mm, exceptionally to 90 mm)* and transparent-whitish to blood colored, marine oligochaetes present the familar habitus of the earthworm. They move snake-like or by means of peristalsis; some can attach themselves to the substrate.

Of 3 orders, only Haplotaxida contain marine representatives. About 200 marine oligochaetes are known; of about 40 species reported from Bermuda, 10 are included here.

OCCURRENCE: Originated from limnic and terrestrial ancestors, most marine oligochaetes live in muddy or sandy sediment, in decaying seaweeds or between pebbles in intertidal shores or shallow sublittoral reaches where they can attain population densities of several million per square meter. Some tubificids, however, have recently been reported from deep-sea samples. Most marine oligochaetes are free-burrowing macrobenthic or interstitially wriggling meiobenthic forms. As predominantly littoral animals they tolerate fluctuations in temperature and salinity quite well; they usually prefer, however, the uppermost, well oxygenated sediment layers.

Macrobenthic forms are sampled with dredges, large corers or by hand, and gently sieved out with 500 μm screens. Meiobenthic worms from sands are usually elutriated or thoroughly decanted after anesthetization in magnesium chloride. Seaweed and coarse debris can be treated with the climate deterioration method. In fixed samples (neutralized formaldehyde) quantitative sorting will be facilitated by staining with Rose Bengal. Collecting from mud samples may require flotation methods.

IDENTIFICATION: Reliable identification requires the use of a very good dissecting and compound microscope (preferably with Nomarski interference contrast). Because most critical characteristics are internal and minute, species identification is extremely difficult and requires, apart from expertise, fully mature specimens with well-filled spermathecae, developed clitellar glands, and complete number of segments.

Note external characters of systematic relevance: color, length, diameter (in relaxed worms), number of segments (segments are usually given roman numerals). Record the number of setae per segment and per bundle and their shape: Enchytraeidae with simply pointed needles, Tubificidae mostly with furcate crochets (bifid—with 2 hooks; trifid—3 hooks). The position of the clitellum (a region of thickened segments), and of pores associated with spermathecae (pouches for sperm storage after copulation) and genital organs are important, as are most other structures associated with reproduction. The sperm, produced in 1 or 2 paired testes, are often stored in 1 or 2 seminal vesicles (sperm sacs), large pouches derived from the septa in the respective segment. After maturation, the sperm masses are transferred by relatively long paired sperm ducts, whose anterior part is usually a massive sperm funnel tapering into a slender vas deferens which, at its opening, is often surrounded by a muscular penial bulb. In tubificids, the posterior part of the duct is dilated to form an atrium into which 1 or 2 prostate glands open. In many tubificid species, a distinct eversible penis is developed, often supported by a cuticularized penial sheath. After mutual copulation of the hermaphroditic worms, the partner's sperm is stored in a paired (seldom unpaired) spermatheca, an ectodermal pouch usually opening ventrolaterally.

Observation of live worms in a drop of seawater should always precede fixation. Animals are narcotized in magnesium chloride and fixed immediately, for 48 hr, in a solution of 10% neutral formalin in seawater, then stored in a more dilute formalin solution (5%) or in 85% ethyl alcohol. Preserved species with a diameter as small as 1 mm can be dissected, preferably after 3 hr in 50% ethanol, by opening the anterior half of the body by a dorsal longitudinal cut. In smaller species not transparent enough for direct examination under the microscope (many tubificids), the solid parts of the genital organs can be made visible by clearing with oil of cloves. Preserved enchytraeid specimens that lack internal hard structures should be stained in a carmine stain (borax carmine, Ranvier's picrocarmine) and stored in alpha-terpineol or Bergamot oil if not whole-mounted in Canada balsam. In difficult cases, sectioning is necessary.

BIOLOGY: Hermaphrodites with mutual internal fertilization; 1 to about 8 eggs are deposited in cocoons excreted by the clitellum and attached to sand grains or plants. Most species feed on detritus, diatoms and microorganisms on sediment particles. Main predators of oligochaetes are turbellarians, nemerteans, some polychaetes, small fish and shorebirds. Worms often contain gregarines in their coelomic cavity and seminal vesicles, and holotrichous ciliates in their gut as parasites or commensals.

DEVELOPMENT: Direct; in smaller enchytraeids hatching occurs after about 14 days, time to maturity is 3-5 weeks. There may be 3-5 generations per year. In many tubificids generation time is much longer (1-2 yr); worms often die after cocoon deposition.

REFERENCES: General overviews are by STEPHENSON (1930) and BRINKHURST & JAMIESON (1971). A report on selected biological aspects in aquatic oligochaetes is by BRINKHURST & COOK (1980), the biology and ecology of marine Oligochaeta is reviewed in GIERE & PFANNKUCHE (1982). For species identifications consult BRINKHURST & JAMIESON (1971), BRINKHURST (1982), JAMIESON (1971) and NIELSEN & CHRISTENSEN (1959).
   Bermuda's species have been reported or described by ERSÉUS (1979a,b), GIERE (1979) and LASSERRE & ERSÉUS (1976).

## Plate 83

O.   **HAPLOTAXIDA:** ♂ funnels at least 1 segment anterior to ♂ pore(s).

S.O.  **TUBIFICINA:** Haplotaxida with ♂ pore(s) in the segment following the testes.

F.   **TUBIFICIDAE:** Tubificina with spermathecae in segment X, close to genital segments; atria mostly present, ♂ pores in XI; dorsal and ventral setae frequently dissimilar. (About 30 spp. from Bda.)

*Phallodrilus leukodermatus* Giere: Genus mainly marine; without hair setae, generally with penial setae. ♂ efferent ducts paired in XI; sperm ducts ciliated, joining atrium apically or entally, but never in its ectal half. Penial structures mostly absent. Two pairs of often stalked prostates, anterior pair entering atrium near entrance of sperm ducts, posterior pair attached to ectal atrial end. Spermathecae paired, in X.—Species with about 65 segments; cuticle secondarily annulated owing to regular skin constrictions. Mouth lacking, gut reduced; 2 or 3 bifid setae per bundle; setae with distal ligament, 2 hooked penial setae. Living specimens shiny white due to subcuticular bacteria. Worms thin and thread-like, rather stiff, slow; in sand around shores and reefs.

*Aktedrilus monospermathecus* Knöllner: With true penes and unpaired middorsal spermatheca. Prostomium elongate, contains conspicuous globular coelomocytes. With (2)3 to 4(5) bifid setae per bundle anteriorly, (1)2 to 3(4) setae posteriorly. Gray to slightly reddish in life. Ubiquitous mainly in intertidal, unsorted sands, usually in the upper 10 cm of sediment.

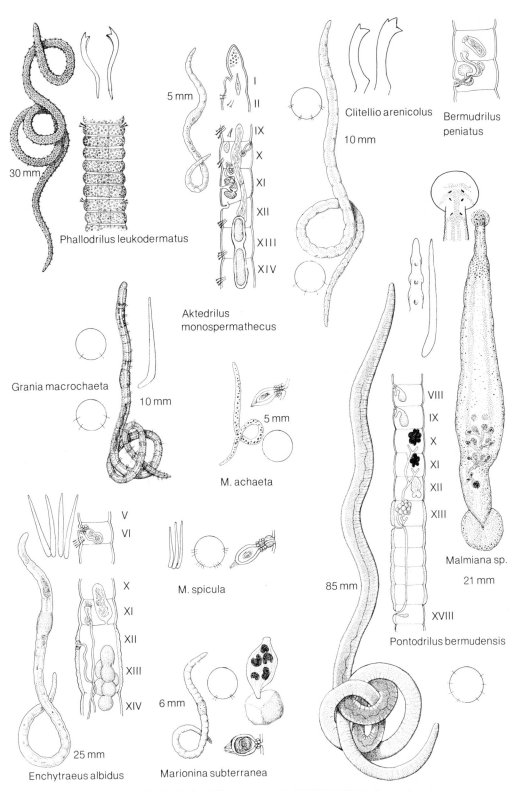

## 83 OLIGOCHAETA (Oligochetes), HIRUDINEA (Leeches)

*Clitellio arenicolus* (Pierantoni): ♂ sperm ducts tubular; prostate glands diffuse or absent, penes and spermathecae present; with spermatophores; no coelomocytes. With 40-60 segments; anterior setae bifid, trifid or blunt, 2 per bundle. Atria with diffuse covering of prostate cells. Subtidally in coarse sand.

*Bermudrilus peniatus* Erséus: Hair setae absent, with penial setae in XI; sperm ducts dilated, nonciliated, containing sperm and entering ental ends of atria; 1 large prostate attached to ental part of each atrium. Length 3.0-3.9 mm; 40-41 segments; diameter about 0.12 mm with 2-4 somatic setae (all bifid) anteriorly, posteriorly 2 setae per bundle. Penial bundle contains straight setae with ectal ends clubbed bearing a thin, hooked tooth. Penial and spermathecal pores paired; cuticularized penis in copulatory sac; oval spermatheca thin-walled. In medium to coarse coral sands.

F.  **ENCHYTRAEIDAE:** Tubificina with spermathecae 5 segments anterior to genital segments; no atria; spermathecae (of prime importance for identification) in V, ♂ pores in XII; no penes, but penial bulbs; dorsal and ventral setae usually the same. (About 10 spp. from Bda.)

*Enchytraeus albidus* Henle: Relatively large, of whitish appearance. With straight setae; brain rounded posteriorly, blood often faintly yellowish; unbranched "spongy" tubes of peptonephridia (excretory organs) in II and IV with 3 pairs of primary septal glands united dorsally, no ventral lobes (secondary glands). Sperm funnel 5-8× longer than wide, with collar of same width; sperm duct very long, extending into XX. With 6-12 mature eggs at a time; thin ectal duct of spermatheca covered with dense layer of glands; ampulla of spermatheca of varying shape, often with dilatations. Very common, ubiquitous. Typical for the upper littoral shore, under decaying seaweed, and for semiterrestrial soils rich in detritus and wrack. Also in garden soil and limnic habitats.

*Grania macrochaeta* (Pierantoni): Whitish-transparent, slim; 8-15 mm; 45-70 segments; length of setae about 70-80 μm. Epidermis striated, with thick cuticula and many glands. Setae large, straight, of typical shape, only 1 per bundle, totally absent in anteriormost segments; ventrally present from IV or V, dorsally from about XVIII onward. Primary septal glands never united dorsally, secondary glands of varying number; no true peptonephridia (excretory organs). Dorsal blood vessel arises behind clitellum. Seminal vesicle large (often reaching XVII); sperm funnel several times longer than wide; sperm duct long and coiled. Spermathecae with distinct muscular duct and well set-off egg-shaped ampulla, duct lacking all glands, its ectal opening latero-dorsally; sperm arranged in balls or scattered. Lives interstitially in coarse subtidal sand. Owing to its thick, glandular epidermis and stiff movements it superficially resembles nematodes.

*Marionina subterranea* (Knöllner): Genus without peptonephridia; mostly straight setae with ental hook; blood colorless.—Species small, whitish, 22-29 segments; no dorsal bundles, 2 setae per bundle;

septal glands united dorsally, with small lobes in IV and V, and large ones in VI. Sperm funnel 1.5× longer than wide, with a collar half its width; sperm duct short, coiled in XII. Large seminal vesicle, often reaching VIII. Ectal duct of spermatheca short, orifice surrounded by glands; ampulla ovoid or somewhat square. Abundant and ubiquitous, intertidally and around high-water line in medium sand, under wrack. Body surface rather sticky owing to glands; often adheres with its anal end to particles by excretion of viscous coelomic fluid. Deposits cocoons on sand grains.

*M. spicula* (Leuckart): Genus as above.—Species whitish, 27-30 segments, relatively stout. With 2-4(5) setae per bundle; septal glands IV and V united dorsally (at V with ventral lobes), at VI only separate ventral parts developed. Sperm funnel 2× longer than wide, often with lobed outline, collar half its width; sperm duct short, 2-3× length of funnel. No seminal vesicle. Spermatheca with distinct duct covered with small glands, 1 large gland at ectal orifice, ampulla oval. Intertidally or along high-water line in near-surface detritus-rich sand. Adheres with posterior end to substrate.

*M. achaeta* Lasserre: Genus as above.—Species intensely white in reflected light (spotted dark in transmitted light) owing to large disc-shaped coelomocytes, especially in posterior segments. Setae completely reduced. Septal glands all united dorsally, those at V and VI with large ventral lobes. Sperm funnel 4× longer than wide, its collar of same width; sperm duct long and coiled in XII; 2 seminal vesicles present. Spermatheca with distinct duct covered with glands, ampulla oval. In subsurface sand, preferably on the upper part of the beach.

S.O. **LUMBRICINA:** Haplotaxida with ♂ pores usually several segments behind last testes; 1 pair of ovaries in XIII. With 8 setae per segment.

F. **MEGASCOLECIDAE:** Lumbricina with spermathecae in VIII + IX; ♂ pores in XVIII, in conjunction with prostates. (1 sp. from Bda.)

*Pontodrilus bermudensis* Beddard: Large (70-100 mm) but thin (1-3 mm), pigmentless-grayish to slightly reddish (blood). With 100-120 segments; 8 regular rows of setae, ornamented (=slightly serrated) at their tips; gizzard rudimentary. One pair of testes each in X and XI; seminal vesicles extend from anterior wall of XI and XII. ♂ pores open, combined with those of paired tubular prostates, in 2 ventral grooves between ventro-lateral lips. ♀ pores in XIV. With 2 pairs of large, oval spermathecae with thin and long ducts, with a tubular diverticulum at their orifice; spermathecal pores intersegmentally in VII/VIII and VIII/IX, in line with latero-ventral setal row. Clitellum from XIII to XVII, annular. This unique "marine earthworm" is common around high-water line under pebbles or seaweed in sand rich in mud and detritus. Gut usually filled with sand grains.

O. Giere, C. Erséus & P. Lasserre

### Class Hirudinea (Leeches)

Characteristics: *Dorso-ventrally flattened ANNELIDA with secondary external annula-*

tion and clitellum. Body of 34 segments, without parapodia and usually without setae; 1st and last segments modified as suckers. From 1 cm to 30 cm in length, leeches are transparent-colorless to opaque, the coloration ranging from black to olive-green or red, often with striped or spotted patterns. "Leech-like" crawling is typical; some are able to swim with undulating body movements.

Of 4 orders, the Acanthobdellida, Gnathobdellida and Pharyngobdellida are not known from Bermuda. The class contains about 500 species; about 20 marine species are known from the northwestern Atlantic, and only 1 from Bermuda.

OCCURRENCE: Cosmopolitan; mostly aquatic in marine, brackish and freshwater habitats, some terrestrial. Although a few are predacious, the majority are blood-sucking ectoparasites, mostly of vertebrates (esp. fishes and turtles); they may attach to decapod crustaceans presumably as a means of dispersal. Leeches ectoparasitic on fishes are found mainly in mouth and gill chamber, and on the ventral body surface, and fins of the host.
Specimens are collected by carefully screening likely hosts.

IDENTIFICATION: Because the pigmentation, eyes and ocelli are important diagnostic features but susceptible to dissolution by preservatives, specimens should first be studied alive and under a dissecting microscope.
Note the anteriorly tapered body with the discoid or cupped, terminal suckers. The oral sucker is usually the smaller one, and bears 2-6 highly developed eye spots; these rarely occur on the caudal sucker.
Prior to fixation, small amounts of ethyl alcohol should be added to the water to relax the specimens so they can be straightened. Preserve in 70-80% ethanol and store in the dark. To observe the internal anatomy, a longitudinal incision can be made in the dorsal body wall; small specimens can be stained with hematoxylin or borax carmine, dehydrated in glacial acetic acid and mounted in Canada balsam.

BIOLOGY: Hermaphroditic; copulation involves a mutual exchange of spermatophores (which in some species are transferred by means of hypodermic impregnation), whereupon sperm are carried to the ovaries where fertilization occurs. The clitellum secretes a cocoon that may be attached to the host or deposited in the water, beneath stones and on vegetation. Most leeches have an annual cycle.

DEVELOPMENT: Direct, without free larval stages.

REFERENCES: For a brief systematic overview see STUART (1982). Marine leeches of the Atlantic are treated by APPY & DADSWELL (1981), KHAN & MEYER (1976), MEYER & KHAN (1979) and SAWYER et al. (1975). In the absence of a systematic study on Bermuda species, the latter paper especially may be of help in preliminary identification.

## Plate 83

O. **RHYNCHOBDELLIDA:** Hirudinea with a proboscis but without jaws; circulatory system distinct from the coelom. Strictly aquatic.

F. **PISCICOLIDAE** (Fish leeches): Subcylindrical Rhynchobdellida with usually more than 3 annuli per segment.

*Malmiana* sp.: Tegument smooth; annulation obscure; oral sucker large, wider than neck; caudal sucker discoid, equal to or only slightly larger than maximum body width. Found on the half beak *Hemirhamphus bermudensis*.

T. G. RAND

## Phylum Tardigrada (Water bears)

CHARACTERISTICS: *Microscopic, free-living BILATERIA with 4 pairs of segmented legs with claws, rod-shaped adhesive discs or round suction discs often inserted on the foot via toes. Without circulatory or respiratory structures.* With a size of 0.05-1.7 mm (most marine species 0.1-0.3 mm) they are among the smallest Metazoa. The herbivorous tardigrades move with a slow, bear-like gait.

Because of similarity with Aschelminthes (especially Rotifera, Nematoda and Kinorhyncha) the phylogenetic position of Tardigrada is still somewhat doubtful; there are, however, strong arguments (especially the segmentation of the nervous system, and the structure of sense organs) for a close relationship with the annelid-arthropod complex. Of the 3 orders, only 1 is reported from Bermuda; Mesotardigrada are only known from a hot sulfur spring and Eutardigrada have few a marine representatives. Of about 500 species, 56 are marine; 25 are known from Bermuda (mostly recent, undescribed collections) of which 5 are included here.

OCCURRENCE: Freshwater (including moist plants such as mosses) and marine. Marine tardigrades are found from the intertidal to the deep sea, with the highest abundance in supra- to sublittoral sand, algae, detritus, or as ectoparasites; the highest species diversity is found in subtidal coralline sand or shell gravel.

Collecting of live beach specimens is done by shocking large samples of sediment with fresh water and thereafter transferring the animals back to seawater. Live subtidal tardigrades can be collected by the seawater-ice method, or by shaking the substrate in magnesium chloride and subsequent filtration through a 65 μm screen.

IDENTIFICATION: Most genera can be sorted or recognized under a dissecting microscope (50-63×). Mounting under a cover glass and examination with phase contrast or interference contrast microscope is necessary for further investigation.

The key characters for the identification of marine heterotardigrades are mostly based on differences in shape of toes and claws, and in cephalic appendages that consist of club- or dome-shaped chemoreceptors (clavae) and hair- to spike-like mechanoreceptors (cirri). Further distinguishing characters are the buccal apparatus and the genital system (especially the seminal receptacles).

For permanent slides marine forms are preserved in 2-3% buffered formalin after freshwater shocking; staining with osmium tetroxide vapor is recommended. Fixed animals should be transferred to a 1:10 glycerin/absolute alcohol solution, which is then evaporated to glycerin after a few days. The cover glass should be sealed by epoxy or Murrayite. For histology, fixation in trialdehyde or glutaraldehyde is recommended.

BIOLOGY: One marine species is hermaphroditic; all others are dioecious. Sexes are not always easily distinguished externally but the males can be smaller and have longer clavae. Internal fertilization by copulation is present in nearly all marine species, preceded by a complex courtship behavior; sperm are stored in 1 or 2 seminal receptacles. Heterotardigrades lay only few eggs (1-8) at a time. The life-span varies according to temperature; arctic species can live more than 1 yr, temperate-tropic species only a few months. As a reaction to desiccation or low temperatures the life cycle of tidal species can be lengthened by cryptobiosis, a state of extremely low metabolism. Furthermore, true marine species can pass into a dormant state, encystation, which may last several years. Cyclomorphosis (change in body shape over several generations) is present in at least 1 genus. Most heterotardigrades are herbivorous, feeding by piercing plant cells with a stylet-armed buccal apparatus and sucking the contents by a strong muscular pharyngeal bulb with armature. Sedentary ectoparasitism is found in 2 cases (on echinoderms and barnacles). True carnivorous tardigrades are perhaps more common than we know today; one species is known to feed on bryozoans. Predators may be Turbellaria, Nematoda or Crustacea, but no observations have been made. Sporozoa and yeasts are very common and often fatal parasites of marine tardigrades.

DEVELOPMENT: Direct, or indirect via a modified larva. Juveniles often have a smaller number of claws or toes. The adult stage is reached after 3-8 molts, but molting continues after the animals are mature. In heterotardigrades defecation only happens in connection with molts.

REFERENCES: The most recent monographs are by RAMAZZOTTI (1972), GREVEN (1980) and MORGAN (1982). For marine species see POLLOCK (1976) and RENAUD-MORNANT (1982).
Bermuda's species are treated in RENAUD-MORNANT (1971).

## Plate 84

O. **HETEROTARDIGRADA:** Up to 11 cephalic appendages differentiated in clavae and cirri, legs segmented and foot with toes or only claws. The pharyngeal bulb has calcium carbonate-encrusted placoids never separated in bar-like structures. Gonopore always preanal.

S.O. **ARTHROTARDIGRADA:** Heterotardigrada usually with a median cephalic cirrus, legs with 3-5 telescopic segments. Foot with toes or claws directly inserted on the foot. (21 spp. from Bda.)

F. **STYGARCTIDAE:** Arthrotardigrada with dorsal segmental plates. No digitation on the foot. With 2-4 claws directly inserted on the 1st foot segment (tarsus). Complete set of cephalic appendages (secondary clavae always present). Caudal plate with 2 cirri E. Eyes and stylet supports absent. A pair of seminal receptacles is present. No secondary sexual dimorphism.

*Parastygarctus sterreri* Renaud-Mornant: Adult body length 135-175 μm. Head divided into 2 parts. Dorsal cuticle thickened with 3 unpaired segmental plates (extended into a single lateral expansion on each side) between the cephalic and caudal plates. Cephalic plate with 2 lateral processes. Caudal plate without spikes. Primary and secondary clavae both elongated. Each leg with 4 claws. The 2 central claws with filamentous accessory spines. Found in coarse coralline sediments, 4-8 m; common.

F. **HALECHINISCIDAE:** Diverse Arthrotardigrada without dorsal segmental plates. Each leg has 2-4 digitations with a claw. Stylet supports can be absent. A pair of ventral seminal receptacles.

*Florarctus antillensis* van der Land: Body length of ♀ 200-300 μm, of ♂ 130-150 μm. Sexual dimorphism especially in the primary clavae. Secondary clavae are fused as are H-shaped plates surrounding the mouth cone. Body dorso-ventrally flattened. The 5 wing-like extensions (alae) with latero-posterior supporting hooks. Caudal ala 4-lobed. Dorsal cuticle with indistinct segmentation. The 2-clawed larvae have small alae. Foot with the 2 lateral toes having hook-shaped supporting structures (peduncles) and lateral claws with outer pseudoseg-

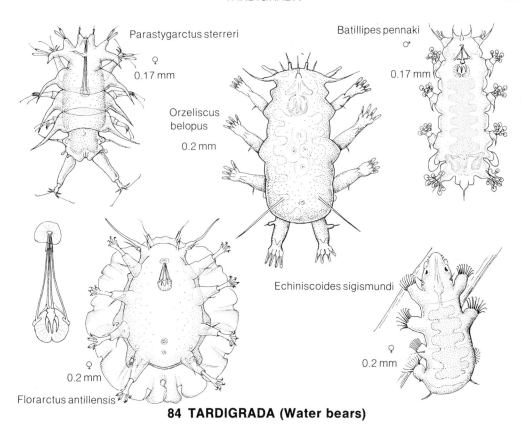

**84 TARDIGRADA (Water bears)**

ments (beak of bird-like claws); central claws with accessory spines. In subtidal coralline sand (1-8 m); very common.

F. **ORZELISCIDAE:** Arthrotardigrada without dorsal segmental plates. Each leg has 4 digitations in adults and 2 in larvae. The toes without claws but with elongated rod-like adhesive structures. Stylet supports present. Can be hermaphrodites; a pair of seminal receptacles present.

***Orzeliscus belopus*** Bois-Reymond Marcus: Adult body length 200-250 μm. All cephalic appendages present. Primary clavae slightly constricted, secondary clavae large flat papilla on the ventral side near the 3-lobed mouth cone. The cephalic cirri in 3 or 4 segments. Three 1st leg appendages spine-like, 4th leg appendages papillae with small spines. Seminal receptacles located in latero-ventral projections. Stylet supports small and curved. Large placoids in the pharyngeal bulb. Eyes absent. In intertidal to subtidal coralline sand (1-8 m); very common.

F. **BATILLIPEDIDAE:** Arthrotardigrada without dorsal segmental plates. Adult has 6 toes, larva has 4 toes of different length on each leg. The toes without claws, but with a disc-shaped suction structure. Stylet supports present. Seminal receptacles absent or only one present. Caudal spike present or absent.

***Batillipes pennaki*** Marcus: Adult body length 175-225 μm. The lateral borders of the body with 2 folds in each segment. A single constriction present on primary clavae, secondary clavae absent or indistinct. Cephalic cirri long. All leg spines present, but reduced in length. Caudal spike arising from a single-lobed base. The 2nd toe of legs 1-3 and the 2 middle toes of legs 4 are reduced; the suction discs insert directly on tarsus. The 3 placoids in the pharyngeal bulb with large knobs. Eyes absent. In intertidal sand; very common.

S.O. **ECHINISCOIDEA:** Heterotardigrada usually without a median cephalic cirrus, legs stumpy without telescopic segments. With claws directly inserted on the foot (no digitations). (3 spp. from Bda.)

F. **ECHINISCOIDIDAE:** Echiniscoidea without dorsal segmental plates. Black eye pigment. Adult with 3-13 claws on each leg. Stylet supports absent. Cephalic cirri small to reduced. Primary clavae round papilla similar to 4th leg appendages. Cirri A and E similar. Secondary clavae dome-shaped.

***Echiniscoides sigismundi*** (Schultze): Adult body length 150-400 μm; animals from Bermuda are small (150-200 μm), perhaps belonging to a new subspecies. With 7-10 claws on each leg. Legs 4 normal, with 1 claw less than legs 1-3. Flat secondary clavae placed on large projections lateral on the head. The 1st and 2nd leg appendages absent. Seminal receptacles absent. Anus placed posteriorly, with 2 triangular lateral plates. Cuticle smooth or with a fine punctuation. Found in filamentous algae near the high-tide level; abundant.

R. M. Kristensen & W. Sterrer

## Phylum Arthropoda

CHARACTERISTICS: *Segmented BILATERIA with jointed appendages and a chitinous exoskeleton; segments often modified and fused to form body regions.*

The skeleton may be thin and soft (as in Copepoda), or thick and hard (e.g., in beetles). It can be further strengthened by calcareous deposition, as particularly in larger Crustacea. The skeleton may be composed of individual plates (and jointed tubes in the case of appendages), or the plates of several segments may be fused to form a rigid carapace (e.g., in crabs). Externally, body segmentation can be further blurred by the presence of a bivalved shell (e.g., Ostracoda, Cirripedia), or a parasitic mode of life.

The 10 living classes (see PARKER 1982) can conveniently be grouped into 4 subphyla: Chelicerata (p. 268), Crustacea (p. 277), Uniramia (p. 381) and Pentastomida, the last (parasites of vertebrates) not represented in Bermuda. Because they share the possession of mandibles, Crustacea and Uniramia have sometimes been united as Mandibulata; however, crustacean mandibles are primitively biramous and directed backwards rather than uniramous and directed downwards as in Uniramia. Arthropod relationships are discussed by MANTON (1977), and development and phylogeny have been treated by ANDERSON (1973). Information on air-breathing marine arthropods is presented by CHENG (1976).

### Subphylum Chelicerata

CHARACTERISTICS: *Terrestrial and aquatic ARTHROPODA without antennae, and with chelicerae and pedipalpi as oral appendages. Body divided into prosoma and opisthosoma; eyes, when present, usually not compound (except in Merostomata).*

The chelicerae, often in the form of a pincer, are the main food-handling appendages, followed by variously shaped and used pedipalpi, and usually 4 pairs of walking legs. Respiratory organs, when present, are book lungs, tracheal tubes, or gills.

Of 3 classes, Merostomata (horseshoe crabs) do not occur in Bermuda. The 2 others, Arachnida (p. 269) and Pycnogonida (p. 275) are represented by at least some orders.

## Class Arachnida (Spiders, mites, scorpions, etc.)

CHARACTERISTICS: *Mostly terrestrial CHELICERATA with tracheal system or/and book lungs (in small forms the respiratory system may be reduced). Opisthosomal appendages when present highly modified.*

This large class contains 11 orders of which only 3 have some marine affinities: Pseudoscorpiones (p. 269), Acari (p. 270) and Araneae (true spiders), the last without marine representatives in Bermuda.

## ORDER PSEUDOSCORPIONES
(=Chelonethida) (False scorpions)

CHARACTERISTICS: *Small terrestrial ARACHNIDA with prosoma carrying a pair of small chelicerae, a pair of long, chelate pedipalpi, and 4 pairs of legs. Opisthosoma clearly divided into 11 or 12 segments and broadly connected to prosoma.* Body size less than 7 mm, coloration brownish. Pseudoscorpions can run backward as well as forward. Several species are phoretic; i.e., they attach themselves to larger Arthropoda such as flies, probably as a means of dispersal.

Of about 1,300 species known, mainly from the tropics, only 1 has been described from Bermuda.

OCCURRENCE: Pseudoscorpions live in the upper layer of the soil, in crevices under bark or in rocks, in nests of birds and mammals, in caves, etc.; a small number of species prefer the seashore.

Collecting is done by hand, or by treating substrate samples in a Berlese apparatus.

IDENTIFICATION: Specimens fixed in 70-75% ethanol should be studied under a microscope. Specimens can be made transparent by warming in lactic acid.

Important taxonomic criteria are the structure of carapace and tergites; presence and number of eyes; and particularly the structure of chelicerae and pedipalpi. Each chelicera bears several setae some of which are grouped together into a flagellum; the movable finger is provided with silk glands, and with a serrated membrane (serrula) on its inner side. The pedipalpi usually contain venom glands that open at the tip of 1 or both fingers.

BIOLOGY: Sexes are separate, without a distinct sexual dimorphism. The sperm transfer is indirect; a stalked spermatophore is deposited on the substrate, with or without courtship. The life-span can exceed more than 3 yr. Their food consists chiefly of various arthropods, in particular Collembola. The large chelate pedipalps are used for capturing the prey, which is narcotized by poison of the pedipalps. Sometimes pseudoscorpions occur in human habitations, especially in libraries (but they are not dangerous to humans), where they hunt tiny insects such as book lice. Silk is used to weave chambers for molting, breeding and hibernation.

DEVELOPMENT: The female commonly carries 3-40 eggs in a sac on the underside of her abdomen. The hatched embryos remain in this sac feeding, with a special pump-organ, on a secretion from the female. After molting the young leave the sac and remain for a few days in the silken breeding chamber together with the female. In general it takes several months from hatching to maturity.

REFERENCES: For general information see CHAMBERLIN (1931), VACHON (1949) and BEIER (1963).
 The first record of pseudoscorpions in the intertidal region of Bermuda was recently published by MAHNERT & SCHUSTER (1981).

## Plate 85

**F. OLPIIDAE:** Opisthosoma more or less slender; tergites mostly undivided; 4 eyes. Chelicerae relatively small, mobile "finger" without "teeth".

*Pachyolpium atlanticum* Mahnert & Schuster: Tergites undivided, number of tergite bristles exceedingly high (about 17-20 at the tergites in the middle and posterior part of opisthosoma); eyes large, 2 on each side. Size about 2 mm. Dorsal shield of prosoma (carapace) and pedipalps brown, tergites olive-brown; juvenile stages generally lighter. Common in the intertidal crevices on the rocky shore. Females with egg sacs were observed in summer.

R. SCHUSTER

## ORDER ACARI (Mites and ticks)

CHARACTERISTICS: *ARACHNIDA with variable or reduced segmentation; body often subdivided by sutures or furrows. Adults mostly with 4 pairs of walking legs.* Body size is usually small, 0.2-7 mm (engorged female ticks may reach 30 mm); coloration ranges from brownish to bright colors. Though some aquatic mites are able to swim, most marine forms crawl over the substrate; terrestrial forms are capable of fast running.

Because the higher classification of Acari is presently in flux a simplified system is presented here. Of this morphologically and ecologically very diverse group, 30,000 species have so far been described. Only a few are associated with the marine environment; of a probably much larger number occurring in Bermuda, 12 species are listed here.

OCCURRENCE: The majority is air breathing and lives in various terrestrial biotopes, especially in the soil where mites are the most common arthropods; soil mites are also a characteristic component of the intertidal fauna. All ticks and many species of terrestrial and freshwater mites, at least as larvae, are parasites; some of them are dangerous to humans because they may transmit diseases. The marine mites (Halacaridae) live on algae, in colonies of hydroids, bryozoans, barnacles and sponges, and in sediments.
 Collecting of intertidal soil mites is best done by Berlese apparatus but it is also possible to collect larger species by hand with a moistened brush. The best method to get halacarids is to place the substrate into a series of sieves, the upper with 1-2 mm apertures, the lowest with 0.06 mm, and wash with a strong jet of water. A 0.06 mm sieve will retain nearly all mites; most of the halacarids will pass a 1 mm sieve, but the residue of this sieve ought to be examined as well. Sediment samples can also be mixed with water 3-4 times, stirred vigorously, and the supernatant water poured through a 0.06 mm sieve. The seawater ice-technique will give results of lower efficiency.

IDENTIFICATION: A microscope is necessary for studying morphological details. Soil mites, especially strongly sclerotized forms, become sufficiently cleared after warming in concentrated lactic acid. The mite is transferred into a small quantity of this liquid on a depression slide and covered with a glass cover slip. If the slip is then gently pushed with a fine needle the mite begins to roll, and can be observed in various positions. If necessary the mite should be dissected in lactic acid. The gnathosoma (region of oral orifice and mouthparts) is torn off with a sharp needle. The mites are then kept in pepsin (at 40°C) or in lactic acid (at 50-60°C) for several hours. By slight pressure with a rounded needle the gut content is forced out. Halacarids are washed and mounted in Hyrax or glycerin jelly. Halacarids should not be treated with alkaline solutions. Soil mites are best preserved in alcohol 70-75%; halacarids can be preserved in alcohol or in glycerin-glacial acetic but not in formalin.

Note the apparent lack of a clear body division or segmentation. The abdomen is broadly fused with the thorax to form the idiosoma (body), which may be covered with a sclerotized carapace. The gnathosoma (head region) bears the mouthparts, which can be variously specialized for grasping, crushing or piercing. In some groups, the chelicerae of the male are modified for sperm transfer (spermadactyls). The walking legs generally consist of 6 segments (coxa, trochanter, femur, patella, tibia and tarsus, the last bearing claws). The tracheal system, where present, opens to the outside via stigmata that are located dorso-laterally, ventro-laterally, anteriorly on the mouthparts, or ventrally on the leg bases. Stigmata may be provided with grooves (peritreme). The shape of sclerotized plates, structure of genital openings, and presence and distribution of eyes, bristles and trichobothria (sensory bristles) provide further identifying characteristics.

BIOLOGY: Most species are dioecious; some with a distinct sexual dimorphism (e.g., Gamasina, Uropodina), others without (e.g., the majority of Oribatei). Parthenogenesis is rare except in a few groups. Sperm transfer is highly diverse and ranges from copulation to several methods using spermatophores. Stalked spermatophores, often deposited in connection with a complex courtship behavior, are produced, among others, by Halacaridae. In Gamasina and Uropodina, males use their chelicera (which bears an appendage, the spermadactyl) for the implantation of the sac-like spermatophore. The number of eggs in the halacarids is low, generally 10-20 per female per year; in soil mites the numbers vary much more. Most of the marine mites have a 1-yr life-span, with only 1 generation per year, whereas soil mites often have more than 1 brood/yr. The remarkable range of acarian food habits is evident in the diversity of their mouthparts. The chelicerae are generally adapted for biting, chewing, piercing or sucking. In some cases the pedipalps are raptorial organs. Many of the free-living mites are predators. The food of the non-predacious soil mites consists of decaying organic material, fungal hyphae and spores, excrements of other soil animals, etc.; in the littoral fringe, green and blue-green algae may be an important food source. The marine Rhombognathinae are algivorous. All other halacarids are supposed to be carnivorous; few species become parasites.

Phoretic behavior is known from juvenile stages of various groups. Some soil mites, including also a few littoral species, are able to produce silk chambers that may be found occasionally in littoral crevices. Periodic flooding with seawater is the limiting ecological factor for intertidal, air-breathing soil mites. Although they have a high tolerance, they become immobile when covered with seawater; in this physical state oxygen consumption seems to be very low.

DEVELOPMENT: The life cycle runs from the egg through a larval stage and 1, 2 or 3 nymphal stages to the adult; larvae have 3 pairs, nymphae 4 pairs of legs. In general, juvenile stages show a weak sclerotization and a reduced bristle count. Sometimes their habit is quite different from those of the adults.

REFERENCES: For general orientation see KRANTZ (1978) and BAKER & WHARTON (1959); for a systematic and ecological overview about Halacaridae of the Atlantic consult NEWELL (1947).
  Information about ecology and biology of littoral soil mites, including examples from Bermuda's fauna, is given by SCHUSTER (1979). A treatise about Bermuda's Acari does not exist, but records of Halacaridae were recently published by BARTSCH (1978, 1983), MONNIOT (1972b) and BARTSCH & ILIFFE (in press); new littoral soil mites were described by FAIN & SCHUSTER (1983).

## Plate 85

**S.O. MESOSTIGMATA:** Stigmata lateral, with grooves (peritreme); eyeless; a pair of horn-like processes (corniculi) at the terminus of the hypostome.

**COHORT GAMASINA:** Body (but not chelicerae) covered with dorsal shield(s). Chelicera of ♂ with a spermadactyl; ♀ with a large ventral plate covering the genital orifice. Terrestrial.

F. **RHODACARIDAE:** Gamasina with body covered by 1 or 2 shields; leg II of ♂ spined; peritreme conspicuous. (1 sp. from Bda.)

*Hydrogamasus* sp.: One large dorsal shield that is fused with the ventral shield in the anal region; tibia of leg I with 3 ventral bristles; leg II of ♂ with 1 relatively blunt spine; long bent spermadactyl. Color light brown (juvenile stages white or yellowish); size (length of the dorsal shield) 0.56-0.64 mm. Rocky intertidal; common.

**COHORT UROPODINA:** Body mostly flattened; chelicera can be retracted into the body. Roundish genital orifice of ♂ in the center of sternal region, orifice of ♀ covered by a large plate of various shapes. Terrestrial.

F. **UROPODIDAE:** Uropodina with dorsal shield usually entire; fovae pedales present, body well sclerotized. (2 spp. from Bda.)

*Deraiophorus* sp.: Prominent "shoulders", with peritreme; reticulate ornamentation; on the lateral border prickly bristles, inserted on cone-like protuberances. Size 0.73-0.75 mm, color yellowish brown. Slow-moving. On the rocky coast in crevices of the intertidal region.

**S.O. PROSTIGMATA:** Stigmata near the base of chelicerae or lacking; the rest of morphological features highly diverse.

F. **HALACARIDAE:** Prostigmata with tracheal system reduced; stigmata absent. Usually with 4 sclerotized dorsal plates: an anterior predorsal plate, a pair of ocular plates and a posterior postdorsal plate. Surface of plates often ornamented with a coarse panelling and raised costae. Within demarcated areolae, the cuticular layers of the plates are pierced by fine pores (canaliculi). ♂ with a dense corona of bristles around genital opening; ♀ with only a few genital bristles. Aquatic. (17 spp. from Bda.)

*Rhombognathus* sp.: Body dark green. Palpi very small. Ventral plates fused to a ventral shield. The 2 claws with palmate process with minute teeth. Third segments of

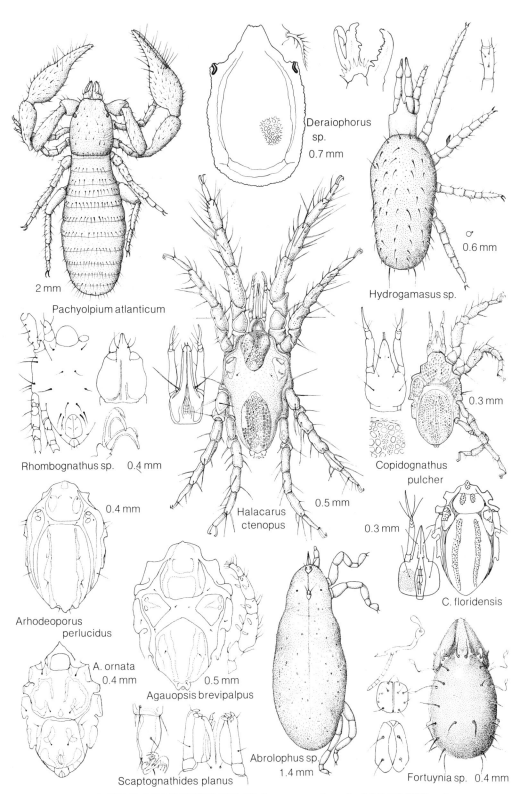

## 85 PSEUDOSCORPIONES (False scorpions), ACARI (Mites)

legs III and IV with ventral setae. Size 0.4 mm. Algivorous. Abundant on algae at low water edge.

*Halacarus ctenopus* Gosse: Idiosoma elongated. Predorsal plate often with process. Palpi 4-segmented; 3rd segment medially with spine. Patella on legs I and II nearly as long or longer than femur or tibia. Leg I with long spines. Ocular and postdorsal plates present. Predorsal and ocular plates with a pair of large pores. Size 0.5 mm. Carnivorous. Common in the sublittoral fringe and below.

*Copidognathus pulcher* (Lohmann): Genus with slender palpi; 3rd segment without setae. Tibia on leg I with 3 ventral setae.—Species with porose panels on costae of postdorsal plate; plate distinctly panelled outside porose areas. Gnathosoma with 3 pairs of long ventral setae. Size 0.3 mm. Common on algae at low water edge and below.

*C. floridensis* Newell: Genus as above.—Species with 2 elevated semidomes on predorsal plate. Costae on dorsal plates with rosette pores (an ostium surrounded by 4-10 canaliculi). Fourth segment of palpi as long as 2nd. Size 0.3 mm. Common on algae at the low water edge and below.

*Arhodeoporus perlucidus* Bartsch: Palpi slender, 4-segmented; 3rd segment without seta. Patella on leg I with 4 ventral setae. Ocular plates posteriorly drawn out into tail-like projections. Size 0.3 mm. Common in sublittoral sediments.

*Agauopsis brevipalpus* (Trouessart): Genus with short, tapering palpi slightly longer than rostrum. Leg I with heavy spines.—Species with H-shaped elevated area on predorsal plate. Leg I femur with 4, patella with 2, tibia with 3, and tarsus with 1 blunt spine. Size 0.5 mm. Common in the eulittoral and sublittoral.

*A. ornata* (Lohmann): Genus as above.—Species with porose areas on ventral plates. Size 0.4 mm. Common on algae.

*Scaptognathides planus* Monniot: Claws on leg I with teeth arranged umbrella-like. Palpi distally with 4 spines. Rostrum slender. Dorsal plates panelled. Size 0.2 mm. In very coarse, clean sand between reefs.

F. **ERYTHRAEIDAE:** Sac-like Prostigmata; prosoma dorsally with a sclerotized median crest (crista metopica) bearing trichobothria; legs arranged in 2 groups, I + II and III + IV. Terrestrial. (2 spp. from Bda.)

*Abrolophus* sp.: Body elongated, covered with large numbers of short bristles; lateral, small eyes. Crista with 2 sensillary areas, each with a pair of long trichobothria. Size about 1.4 mm; color (dark) gray, velvety, legs and gnathosoma reddish (in alcohol the red color disappears in time). Running fast on the surface of rocks in the eu- and supralittoral zone.

S.O. **ORIBATEI:** Weakly or strongly sclerotized; a large dorsal shield (notogaster) covers the idiosoma; stigmata near the base of legs II and III. Terrestrial, only a few species aquatic (fresh water).

F. **FORTUYNIIDAE:** Oribatei with a special canal system laterally between the trichobothrium and the base of legs III and IV; genital plate with 5 bristles. (1 sp. from Bda.)

*Fortuynia* sp.: End of sensillus thickened; anal plates with 2 bristles; tarsus with 1 sickle-shaped large claw. Sexual dimorphism conspicuous: ♂ with a large lateral cone and 2 pairs of long bristles with broadened ends, ♀ without. Color blackish, surface glossy, size about 0.4 mm; locomotion slow. Juvenile stages grayish brown, with a flat shield-like dorsum. In the intertidal zone, especially in crevices; abundant.

R. Schuster & I. Bartsch

## Class Pycnogonida (=Pantopoda)
(Sea spiders)

CHARACTERISTICS: *Spider-like CHELICERATA with short body divided into proboscis, appendage-bearing cephalothorax and reduced, 1-segmented abdomen. Each segment of cephalothorax extended laterally to articulate with ambulatory legs, usually in 5 (more rarely 4 or 6) pairs.* With a body length of 1-10 mm (rarely to 60 mm) and a leg span of 10-50 mm, most shallow-water species, drab-colored (rarely with green or red pigment) and sluggishly stalking, are easily overlooked.

No orders are recognized; about 500 recent species belong to 8 accepted families. Of 4 families and 8 species (including 2 endemics) recorded from Bermuda, 5 species are considered herein.

OCCURRENCE: Exclusively marine and mostly benthic (some are planktonic or parasitic, but not in Bermuda), to 6,000 m; most diverse in antarctic waters. They are usually found clinging to prey species, especially hydroids, but also gorgonians, anemones, bryozoans and sponges. Often very abundant on floating *Sargassum* and in fouling communities.

Collect specimens by gathering likely substrate, if necessary by dredging, and scan under a dissecting microscope, isolating specimens with the help of needles.

IDENTIFICATION: Place specimen under a dissecting microscope to assess key characters (species identification may require higher magnification).

Note the generally elongated, narrow, segmented body, the 4 eyes situated on the dorsal surface of the neck, and the appendages. The latter consist of a pair of often clawed (subchelate) cheliphores =chelicerae), a pair of palps, a pair of egg-carrying legs (ovigers), and 4-6 pairs of walking legs. Characters important for identification are the number of legs and the proportions of their segments, the presence or absence of palps and cheliphores and their segment numbers, the structure of the ovigerous legs, and the claws at the ends of the ambulatory legs.

Using dissecting needles, tease individuals apart and free from substrate, then slowly add ethanol to seawater without disturbance until animal ceases to move. Fix in 10% buffered formalin and after a day or two transfer to 70% ethanol for storage. Handle the animals very carefully because ambulatory legs, especially in long-legged species, tend to fall off readily.

BIOLOGY: Nearly all pycnogonids are dioecious, and sexes can be readily distinguished because ovigerous legs are either reduced or absent in females, and eggs are clearly visible within the ambulatory legs (!) of ripe females. During mating, ventral surfaces are opposed, and as the female extrudes eggs the male uses its ovigerous legs to remove them and form them into balls; evidently fertilization occurs at this time, and then the male carries the developing embryos. Most pycnogonids are carnivorous, though food preferences are

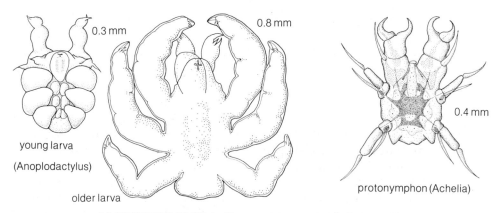

**86 PYCNOGONIDA (Sea spiders): Development**

generally unknown. Certain pycnogonids are known to eat hydroids, anemones, bryozoans and red algae. The palps and cheliphores serve to manipulate food to the mouth. Some pycnogonids probably fall prey to fishes and other bottom feeding animals.

### Plate 86

DEVELOPMENT: Cleavage follows a typical arthropodan pattern, its completeness varying with the amount of yolk present, which differs greatly among genera. Upon hatching, some species are completely formed miniature adults, but most emerge as a protonymphon, a larva bearing 3 pairs of appendages and superficially resembling a crustacean nauplius. Never independent, the protonymphon clings to its parent's cheliphores until it assumes adult form after a series of molts.

REFERENCES: The best recent summary of the class is the little book by KING (1973). Classical treatments are those by DOHRN (1881) and HEDGPETH (1943); tropical Atlantic fauna is covered by STOCK (1974).

Most species known or expected from Bermuda are dealt with by HEDGPETH (1948). The only report devoted exclusively to the pycnogonid fauna of Bermuda, now out of date, is that of COLE (1904); additional species have been recorded by VERRILL (1900b) and GILTAY (1934).

### Plate 87

F. **ENDEIDAE:** Pycnogonida without cheliphores and palps; ovigers (of 7 segments) only in ♂. Legs long, ending in well-developed propodi with auxiliary claws. (1 sp. from Bda.)

***Endeis spinosa*** (Montagu): Slender, often greenish species; legs about twice as long as body. With 6-30 mm, this is the largest species in Bermuda, commonly found on turtle grass and floating *Sargassum*.

F. **AMMOTHEIDAE:** Pycnogonida with generally subchelate cheliphores, 6-10-segmented palps, and 9-10-segmented ovigers in both sexes. (2 spp. from Bda.)

***Achelia gracilis*** Verrill: Propodus with basal spines and auxiliary spines; cheliphores 2-segmented. Trunk segments not completely separated, body circular, abdomen reduced. Palp 7-segmented. Common in fouling communities (e.g., on buoy chains).

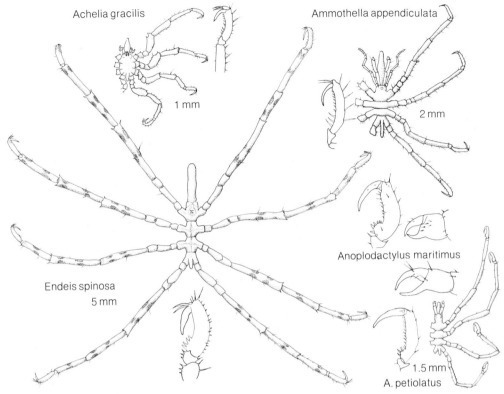

**87 PYCNOGONIDA (Sea spiders)**

*Ammothella appendiculata* (Dohrn): Propodus well developed, with basal spines and auxiliary claws. Trunk segmentation distinct; proboscis large, elliptical or pyriform; no clubbed spines on basal leg segments. Common in fouling material (e.g., on buoy chains).

F. **PHOXICHILIDIIDAE:** Pycnogonida with 2-segmented cheliphores; ovigers (of 5-9 segments) only in ♂. Palps absent. (4 spp. from Bda.)

*Anoplodactylus maritimus* Hodgson (=*A. parvus* Giltay): Genus with cephalic segment extended forward as neck; proboscis cylindrical; abdomen not markedly elongate.—Species without prominent projections on end of femur; 2nd oviger segment about $3 \times$ as long as wide. Specific locality unknown.

*A. petiolatus* (Krøyer): Genus as above.—Species with long setae on legs, and prominent tubercles on lateral processes; 2nd oviger segment about $5 \times$ as long as wide. Common on *Sargassum*.

J. Markham

### Subphylum Crustacea

CHARACTERISTICS: *Mostly aquatic ARTHROPODA with usually branched (biramous) limbs, 2 pairs of antennae, and with mandibles. Trunk often protected by a carapace or paired valves; head often fused with some thoracic somites to form a cephalothorax.*

The biramous limb typically consists of a basal protopodite of 2 pieces: coxopodite (coxa) and basopodite (basis); the latter carries an outer (expododite) and inner branch (endopodite). This basic pattern can be variously modified, especially in adults, to form mouthparts; grasping, walking and swimming legs; or copulatory structures. Instead of true gills (epipodites = processes of the coxa), the abdominal legs or—in very small forms—the whole body surface may have respiratory function. There are 2 different kinds of eyes: compound lateral eyes, often on stalks, and simple median eyes (nauplius eyes).

Of 9 classes, 5 are represented in Bermuda: Branchiopoda (p. 278), Ostracoda (p. 280), Copepoda (p. 288), Cirripedia (p. 299) and Malacostraca (p. 305). The sediment-dwelling, minute Cephalocarida and Mystacocarida and the parasitic Branchiura have not been recorded yet, nor have the recently discovered, cave-dwelling Remipedia. In older systems, the subphylum (or class) was divided into Malacostraca ("higher" Crustacea) and Entomostraca ("lower" Crustacea), the latter a heterogeneous assemblage of what are now treated as separate classes.

## Class Branchiopoda

CHARACTERISTICS: *Small, free-living CRUSTACEA in which the carapace may form a dorsal shield or a bivalve shell, or be entirely absent. Trunk limbs vary greatly in number, but are generally of uniform, leaf-like structure.* They are small (0.5-2 mm), transparent to opaque (with a black eye and yellow gut) animals that move in a jerky way by means of their 2nd antennae.

Of the 4 orders, 2 live exclusively in fresh water (Notostraca: tadpole shrimps, and Conchostraca: clam shrimps) and 1 in highly saline water (Anostraca: fairy shrimps); only the Cladocera, common in freshwater lakes and streams, have a few marine species. Of about 7 marine species known, 5 have been recorded from Bermuda; 4 are presented here.

OCCURRENCE: The few marine species are planktonic and of worldwide distribution. Bermuda's species occur in surface or near-surface oceanic water (*Evadne*) or in inshore lagoons and bays (*Podon, Penilia*). They form a significant if sporadic portion of marine zooplankton, being particularly abundant when the water column becomes stratified (May-October).

Collecting is done with plankton nets of 0.1-0.5 mm mesh opening.

IDENTIFICATION: Examine preserved whole animals in small dishes or on slides under dissecting or compound microscope. For differentiation beyond whole body features, appendages have to be removed with fine needles and examined individually.

Note the bivalved carapace that generally encloses the trunk but not the head and the large 2nd antennae, and often terminates posteriorly in a spine. In some species, carapace and trunk are fused to form a brood chamber, and the legs are exposed. The number of setae on the (small) endopodite are species-specific.

Fixation is generally by 4% formaldehyde in seawater.

BIOLOGY: Sexes are separate; males are smaller than females. Reproduction follows 2 modes. Under favorable environmental conditions (warm water, sufficient food), females produce thin-shelled ("summer") eggs that, unfertilized, develop parthenogenetically into females for several generations, resulting in a rapid population increase. Eventually, males appear, and fertilized ("resting") eggs are produced via copulation. Fertilized eggs, only a few per clutch and large, are sealed into a saddle-shaped capsule (ephippium) in which they can weather adverse conditions. The life-span of Cladocera is proba-

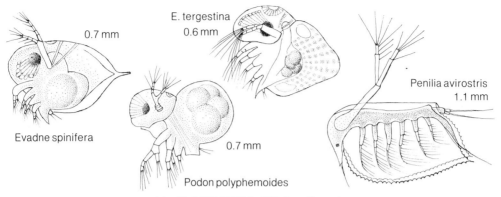

**88 CLADOCERA (Water fleas)**

bly only a few months. Feeding by filtering small plankton and particles, they are in turn an important source of food for other plankton organisms.

DEVELOPMENT: Direct, from yolk-rich eggs; hatching juveniles are similar to adults.

REFERENCES: For general taxonomy see APSTEIN (1901). More recent taxonomy and ecology can be found in RAMMNER (1939) and GIESKES (1971).
MORRIS (1975) provides data on the distribution of some Bermudian species.

## Plate 88

O. **CLADOCERA:** Laterally compressed Branchiopoda with 4-6 pairs of trunk limbs. Carapace bivalved, without hinge or adductor muscle, fused with 2 or more thoracic somites, and not covering the head; sometimes reduced and serving only as a brood chamber.

F. **POLYPHEMIDAE:** Cladocera with carapace not enclosing trunk and limbs; with 4 pairs of trunk limbs, all more or less compressed. Carapace and body fused into prominent brood chamber. (4 spp. from Bda.)

*Evadne spinifera* Müller: Genus without a pronounced saddle depression between head and brood chamber.—Species with needle-like projection of posterior margin of brood pouch. With 2 setae on 3rd endopodite. Occupies surface waters of Sargasso Sea, occasionally inshore.

*E. tergestina* Claus: Genus as above.—Species with bluntly rounded posterior brood chamber; slight depression common between head and brood chamber. With 3 setae on 3rd endopodite. Primarily restricted to open-ocean waters.

*Podon polyphemoides* Leuckart: Genus with distinct depression between head and brood chamber; the latter round in profile. - Species with 3 setae on 1st exopodite. In estuarine and inshore waters; in ocean waters only as strays.

F. **SIDIDAE:** Cladocera with carapace completely enclosing trunk and limbs; with 6 pairs of trunk limbs, all similar and foliaceous. (1 sp. from Bda.)

*Penilia avirostris* Dana: Eye small. Carapace margins dentate; posteroventral corners acute. Rostrum pointed and projected ventrally. Common in inshore waters.

B. F. Morris

### Class Ostracoda (Mussel shrimps)

CHARACTERISTICS: *Minute CRUSTACEA with a bivalved carapace and unsegmented body. With 5-7 pairs of appendages capable of being withdrawn completely within the carapace when the valves are closed by an adductor muscle that traverses the body and is attached to the inside of each valve, generally near the center.* The carapaces of most Ostracoda are 0.5-2.0 mm long (rarely to 30 mm) and of dull colors (hues of brown, occasionally yellow or red). Locomotion is a jerky swimming, or scuttling and plowing over and through the sediment.

Compared to other crustacean groups the Ostracoda of Bermuda are relatively unknown, which probably accounts for the apparent absence of 1 of the 5 orders (Cladocopida) and of 2 of the 5 families of Myodocopida. Of well over 2,000 described living species, at least 120 are known to occur in Bermuda waters; 20 are included here.

OCCURRENCE: Marine and freshwater; only a few species in moist terrestrial habitats. Marine forms live in or on all types of substrates—mud, sand, vegetation and hard bottom—and at abyssal as well as intertidal depths. A few species are parasitic or commensal on other organisms (mostly Crustacea).

Collect living specimens by dragging a net or trawl with a fine mesh along the bottom for epibenthic forms, and through the water for planktonic forms. They may also be obtained from sediment collected with a dredge or grab sampler, and small numbers may be found in core samples. In Platycopida, Podocopida and Cladocopida, whose shells are generally strongly calcified, empty carapaces and separated valves representing the remains of dead animals, or those that have molted, are generally more abundant than living animals. However, living animals are more abundant than empty shells in Myodocopida and Halocyprida, whose shells are not strongly calcified, or disintegrate rapidly. The living animals may be concentrated by swirling the sample in a tray and then decanting the water through a fine mesh screen or net, where most of the living Ostracoda will be retained. If interest is only in the strongly calcified forms, these may be concentrated from dried sediment by various flotation techniques.

IDENTIFICATION: The orders and many of the families and genera may be recognized by the external appearance of the carapace, but internal examination of the carapace and/or appendages is necessary for identification of most species. For removing the soft parts from the carapace it is generally expedient to put the specimen in a drop of glycerin on a depression slide, and to separate and remove the soft parts with fine needles while viewing the specimen under a dissecting microscope. A stereoscopic dissecting microscope is generally sufficient for examining details of the carapace such as its hinge and adductor muscle scars, which play an important part in the classification of the Podocopida, but a compound microscope with magnification $100\times$ to $500\times$ is needed for study of appendages in transmitted light.

The ostracode carapace consists of 2 valves, right and left, connected along the dorsal margin by a flexible hinge. The outer lamella of the carapace is calcified in most orders; the peripheral part of the inner lamella is calcified only in Podocopida. The resulting pocket between the outer and calcified inner lamella is called the vestibule, and radial pore canals may connect it with the edge of the valve. The uncalcified part of the inner lamella is continuous with the chitinous exoskeleton of the soft body and appendages.

In well-calcified ostracodes and especially in Podocopida, the hinge is modified for more rigid articulation, and several hinge types are recognized. Adont is the simplest type, in which the ridgelike edge of the smaller valve fits into a groove on the larger valve. Merodont describes a 3-part wavy (crenulate) hinge with a series of denticles on the smaller valve that fit into tiny depressions (loculae) on the larger valve; generally the anterior and posterior regions of this hinge have larger and more prominent denticles or loculae than the median region. In Cytherideidae the median region shows anterior/posterior differentiation. Amphidont hinges have 4 parts, in the smaller valve consisting of anterior tooth, anteromedian socket, posteromedian bar and posterior tooth; the hinge of the larger valve has complementary elements. Loxoconchidae have a unique gongylodont hinge of alternating very tiny denticles and loculae, with a superimposed size gradient from very large connected loculae at the anterior end (in the smaller valve) to large connected denticles at the posterior end. Macrocyprididae have a unique 5-part hinge consisting in the smaller valve of anterior and posterior crenulate teeth, anteromedian and posteromedian crenulate sockets, and median bar.

The attachment areas of the adductor musclefibers are small scars whose number and arrangement are of great taxonomic importance. Aggregate scar patterns consist of a great many scars closely assembled in a rather large circular area medially. A biserial scar pattern consists of 2 vertical rows of 6-11 scars each, usually located dorsally. A rosette scar pattern has fewer scars in a smaller circular area, but arrangement of individual scars may be apparently random. Discrete scar patterns have 4-8 individual scars arranged in a variety of taxonomically conservative configurations. One or 2 frontal scars and 2 mandibular scars may be located anterior and anteroventral to the adductor scars, respectively.

Preserve animals with soft parts (appendages) in 70% alcohol or in buffered 10% formalin. Empty valves and calcified carapaces are stored dry in paper micropaleontological slides.

BIOLOGY: Most marine species appear to be dioecious, but in many species adult males do not live long after copulation and, therefore, are less abundant than adult females in most samples. Eggs are brooded within the postero-dorsal part of the carapace by members of the Myodocopida, Platycopida and a few Halocyprida, but in the Podocopida, Cladocopida and Halocyprida, many species deposit their eggs on the substrate or attach them to plants. Little is known about the life-spans of marine Ostracoda, but apparently some live as long as several years. Unlike Conchostraca, a new set of valves appears at each molting. Ostracoda manifest a variety of food preferences: some are scavengers, living on dead plant or animal matters; others prefer a diet of living plants or animals. Although many Ostracoda, because of their small size, are ingested accidentally by other animals browsing on plants or burrowing in bottom muds, they also form an important part of the diet of fishes, as well as of other invertebrates. Ostracodes may be hosts to parasites such as Ciliata, Nematoda and Copepoda.

## Plate 89

DEVELOPMENT: The eggs of Podocopida hatch as nauplii but in the Myodocopida and in some Halocyprida, development continues within the egg and the hatching animal has the 5 anterior appendages. All forms pass through 4 or more larval stages before reaching maturity, but molting discontinues at maturity for most species.

REFERENCES: More information about the terminology, taxonomy, ecology and study methods for Ostracoda may be found in MOORE (1961) and MORKHOVEN (1962, 1963). The best modern compilation of soft-part anatomy and biology is by HARTMANN (1966, 1967, 1968, 1975).

Bermudian species have been reported in papers by BRADY (1880), VAN DEN BOLD (1963), MADDOCKS (1969, 1973, 1976), DEEVEY (1968), and KORNICKER (1981).

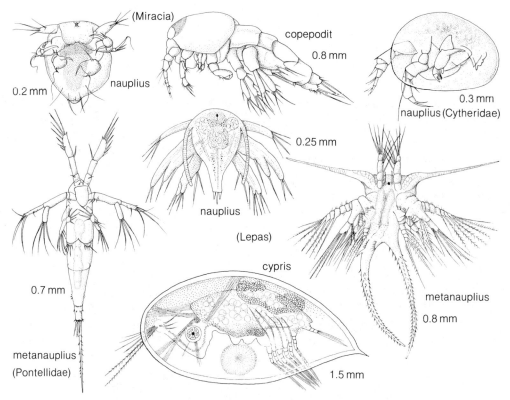

**89 "ENTOMOSTRACA" (Lower crustaceans): Development**

*Plate 90*

O. **MYODOCOPIDA:** Benthic and pelagic marine Ostracoda with basically oval carapaces, some plain, others highly ornamented, with varying degrees of calcification. Inner lamella uncalcified. With 7 paired appendages and well-developed caudal lamellae; 2nd antenna with large basal article and well-developed 9-jointed exopodite with long bristles adapted for swimming or burrowing, or both; none of the limbs adapted for walking. Medial and lateral eyes and bellonci organ (sensory organ attached to median eye) usually present. (Of 5 families, only 3 are known from Bda so far.)

F. **RUTIDERMATIDAE:** Myodocopida generally with small anterior rostrum. Mandible of ♀ and juvenile ♂ with stout claw on endopodial joints 1 and 2 forming pincers. All furcal claws separated from lamella by suture. Carnivorous. (1 sp. from Bda.)

***Rutiderma sterreri*** Kornicker: Carapace strongly calcified with horizontal ribs and radiating riblets; anterior with small rostrum, posterior with projecting caudal process. On algae, sea grasses and sediments; common.

F. **SARSIELLIDAE:** Myodocopida generally without anterior rostrum on ♀ and juvenile ♂, but with small rostrum on adult ♂,

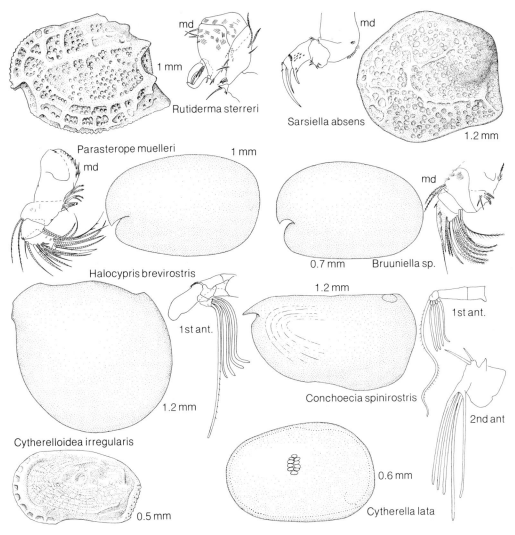

**90 MYODOCOPIDA, HALOCYPRIDA, PLATYCOPIDA (Mussel shrimps 1)**

which are generally sparse. Mandible of ♀ and juvenile ♂ of most species with stout claw on endopodial joints 1-3. Distal furcal claw never separated from lamella by suture. Carnivorous. (1 sp. from Bda.)

***Sarsiella absens*** Kornicker: Carapace of ♀ oval in lateral view with short rounded caudal process and with posterior part of each valve forming large bulge. In sand in shallow water; uncommon.

F. **CYLINDROLEBERIDIDAE:** Myodocopida with oval to elongate carapace with rounded posterior and anterior ends in lateral view, and generally with deep incisura below middle of anterior margin. Species collected in Bermuda all belong in subfamily Cylindroleberidinae, which have smooth carapace, and mandibles usually with 3 well defined endopodial joints with a single claw present on 3rd joint, or rarely fused endopodial joints with total of 3 claws. Flaps along postero-dorsal part of body thought to represent gill structures. All furca claws separated from lamellae by sutures. Filter feeders. (2 spp. from Bda.)

***Parasterope muelleri*** (Skogsberg): Carapace generally tumid (swollen) in lateral view with narrower anterior end. Mandible with 3 well-defined endopodial joints, and with single claw on 3rd joint. On shallow water sediments and algae; uncommon.

***Bruuniella*** sp.: Carapace in lateral view with widest part near middle. Mandible with fused endopodial joints bearing 3 claws. On shallow rubble bottoms; uncommon.

O. **HALOCYPRIDA:** Planktonic marine Ostracoda with thin, weakly calcified elongate carapaces having straight dorsal hinge. Inner lamella uncalcified. With 7 paired appendages, and well-developed caudal lamellae with distal claw located on anterior margin of lamella dorsal to 2nd claw. Eyes absent. (Of 2 families, 1 is known from Bda.)

F. **HALOCYPRIDIDAE:** Halocyprida having short 1st antennae with distal part bending downward; furcal lamella with long distal claw on anterior margin and 6-7 claws on ventral margin decreasing in size gradually. (42 spp. from the Sargasso Sea off Bda.)

***Halocypris brevirostris*** (Dana): The 2 terminal, often fused, segments of the 1st antenna with 5 setae. Basal segment of endopodite of 2nd antenna without a tubercle on its anterior margin. Carapace without distinct rostrum. Offshore, throughout the year in the upper 500 m; occasional.

***Conchoecia spinirostris*** Claus: The 2 terminal, often fused segments of the 1st antenna with 5 setae. Basal segment of endopodite of 2nd antenna with a tubercle on its anterior margin. Carapace in lateral view with small but distinct rostrum and small concave incisura. Offshore; the most abundant species in surface waters.

O. **PLATYCOPIDA:** Benthic marine Ostracoda with well-calcified, laterally compressed, asymmetric oblong carapace with straight ventral margin. Right valve larger than and overlapping left, hinge not differentiated from rest of margin. Inner lamella uncalcified. Muscle-scar pattern biserial or aggregate. Most frequent above 200 m, especially in shallow lagoons. (Only living family:)

F. **CYTHERELLIDAE:** Platycopida with a biserial muscle-scar pattern located dorsally, consisting of 2 vertical rows of 6-11 small scars. The soft body shows apparent vestigial segmentation of the posterior region and a biramous antenna. Six elaborate appendages are adapted for burrowing and filter feeding; the second thoracic leg is missing. (5 spp. from Bda.)

***Cytherella lata*** Brady: Smooth oblong carapace without external sulcus or ornament. Fairly common in inshore and lagoon sediments.

***Cytherelloidea irregularis*** (Brady): Robust carapace ornamented with fine tracery of tiny pits, deep muscle-scar pit, deep fossae and thick peripheral ridge around the anterior margin, and thick denticulate posterior marginal ridge. Common in reef sediments near the edge of the platform.

### Plate 91

O. **PODOCOPIDA:** Benthic marine and freshwater Ostracoda with well-calcified carapace, usually with convex dorsal and concave ventral margin, differentiated hinge, calcified inner lamella, discrete muscle-scar pattern, and sexually dimorphic shape. Five cephalic and 2 thoracic appendages are variously adapted for feeding, cleaning, clasping, walking, swimming, burrowing, sensory or secretory functions. Of 5 superfamilies, the fresh- and brackish-water Darwinulacea have not been reported from Bermuda. At least 10,000 species are living, of which perhaps half have been described to date. The podocopid ostracode fauna of Bermuda, which has never been monographed, comprises more than 70 species, many of which are new.

SUP.F. **SIGILLIACEA:** Podocopida with smooth ovoid carapace, left valve larger than and overlapping right valve tightly all around, aggregate muscle-scar pattern of 20-40 scars and merodont hinge. Posterior region of the soft body with vestigial segmentation. (Only living family: Sigillidae.)

***Saipanetta brooksi*** Maddocks: Small, asymmetric, egg-shaped, brown carapace, wider than high. Interstitial in coarse sand offshore; very rare.

SUP.F. **CYPRIDACEA:** Podocopida with smooth ovate to elongate-subtriangular carapace, adont or 5-part (macrocyprid) hinge and 5 or 6 adductor muscle scars.

F. **MACROCYPRIDIDAE:** Cypridacea with elongate carapace showing pronounced overlap of right valve over left, 5-part hinge, and 2-part rosette adductor muscle-scar pattern. Exclusively marine. (2 unnamed spp. from Bda.)

***Macrocyprina*** sp.: Small, length nearly 3× height, brown or transparent with 3 opaque white spots. Common in lagoon and near-reef sediments.

F. **PONTOCYPRIDIDAE:** Cypridacea with smooth, white or yellow, ovate to subtriangular carapace, with adont hinge and 5 adductor muscle scars. Exclusively marine. (At least 10 spp. in Bda.)

***Propontocypris*** sp.: Fragile yellow subtriangular carapace with very numerous normal pore canals. Infrequent on plants and coralline detritus in the lagoons.

F. **CYPRIDIDAE:** Cypridacea with smooth ovate to elongate-subtriangular carapace, with left valve overlapping right inconspicuously, adont hinge and 6 adductor muscle scars. Diverse and abundant in freshwater, brackish and shallow marine environments worldwide. (At least 4 spp. in Bda.)

*Triangulocypris laeva* (Puri): Elongate-oval carapace with deep anterior and posterior vestibules and short straight radial pore canals, brown color in fresh specimens. Abundant in most shallow lagoon and reef environments.

*Thalassocypria* sp.: Fragile, transparent, laterally compressed, elongate-ovate carapace with narrow calcified inner lamella, shallow vestibules, inconspicuous radial pore canals, and topmost adductor muscle scar divided. Abundant in brackish inland ponds.

SUP.F. **BAIRDIACEA:** Podocopida with robust, smooth or punctate, asymmetric (left over right) carapace with characteristic lemon-seed shape, 4 or 8 adductor muscle scars. Especially diverse, abundant and characteristic of tropical reef and carbonate lagoon habitats. (Only living family: Bairdiidae, with at least 18 spp. in Bda.)

*Glyptobairdia coronata* (Brady): Large, thick-shelled, opaque white, very asymmetric, densely pitted carapace, ornamented with raised ridges (a bar within a ring). Tiny denticles on the antero-dorsal and postero-dorsal edges of the right valve fit into corresponding sockets beneath the hinge of the left valve. Common on coral reef-masses and associated detritus.

*Paranesidea* sp.: Rounded subtriangular, faintly pitted carapace with small marginal spines; the carapace of fresh specimens is mottled white, brown and transparent and may have conspicuous brown external setae. Common in near-reef and lagoon habitats.

SUP.F. **CYTHERACEA:** Podocopida with smooth to elaborately ornamented carapaces, with 4 adductor muscle scars arranged in a vertical row, and often with complex hinges. The predominant ostracodes in most benthic marine environments. (At least 40 spp. in Bda, many of them new.)

F. **CYTHERIDEIDAE:** Cytheracea with smooth or pitted, ovate-oblong carapace and 3- or 4-part modified merodont hinge. Many species are notably euryhaline and abundant in brackish or hypersaline waters; in some cases their geographic distribution follows the flyways of migratory water fowl.

*Cyprideis* sp.: Large white ovoid carapace, smooth except for narrow V- or Y-shaped grooves around normal pore canals. Abundant in brackish inland ponds.

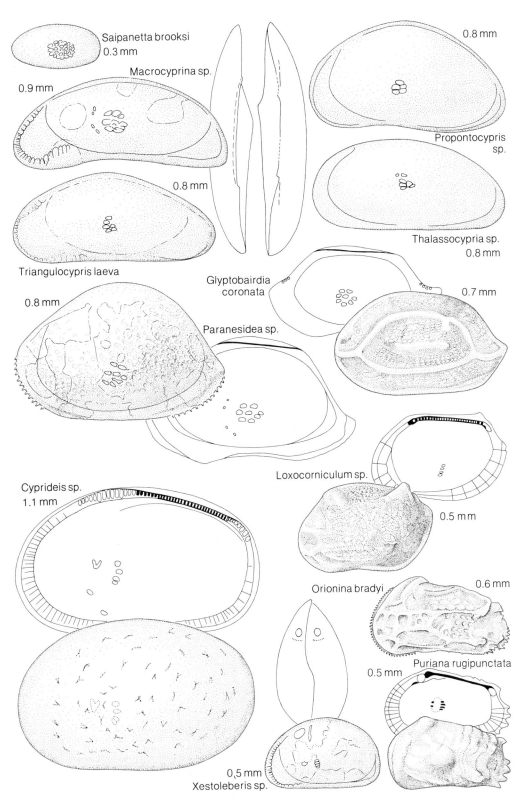

**91 PODOCOPIDA (Mussel shrimps 2)**

F. **LOXOCONCHIDAE:** Cytheracea with subrhomboidal, densely pitted carapace with gongylodont hinge. Very common in shallow coastal marine waters, especially on plants.

*Loxocorniculum* sp.: Prominent postero-dorsal and postero-ventral tubercles, less prominent anterior eye tubercle and 2 anterior ridges, remainder of surface densely but delicately pitted. Extremely abundant in all lagoon and near-reef habitats.

F. **HEMICYTHERIDAE:** Cytheracea with elongate quadrangular carapace outline, amphidont hinge and divided median adductor muscle scars. Very diverse and abundant in shallow marine environments.

*Orionina bradyi* van den Bold: Subtrapezoidal lateral outline with obliquely rounded anterior margin, dorsal margin sloping posteriorly, tiny anterior and posterior marginal spines; external ornament is a distinctive pattern of ridges and fossae, with a median ridge that connects with the dorsal ridge to form a posterodorsal loop, and a ventrolateral ridge that splits and then recombines anteriorly. Common in reef and lagoon habitats.

*Puriana rugipunctata* (Ulrich & Bassler): Small, very thick-shelled, opaque white carapace, with a distinctive ornament of broad rounded ridges and narrow grooves. Common in reef and lagoon environments.

F. **XESTOLEBERIDIDAE:** Cytheracea with small, egg-shaped carapace with flattened venter and swollen posterior, with prominent clear eye spot and white eye scar just behind it. Very common in shallow marine environments, especially on plants.

*Xestoleberis* sp.: Carapace smooth, mottled white and transparent, with deep vestibules and short simple radial pore canals. Extremely abundant in reef and lagoon environments.

R. F. MADDOCKS & L. S. KORNICKER

## Class Copepoda

CHARACTERISTICS: *Mostly subcylindrical (except parasitic forms) CRUSTACEA with 6 pairs of trunk limbs; without a carapace, compound eyes or abdominal appendages.* Mostly small (0.5-2 mm) and transparent (but sometimes brightly colored—often red—or iridescent), they swim or crawl jerkily by means of their legs.

Of 7 orders, only 2 (the parasitic Monstrilloida and Lernaeopodoida) have not been reported from Bermuda as yet. Of 7,500 known species, about 540 (!) have been described from the Bermuda area, mostly from the plankton and shallow sediment bottoms; 27 are included here as examples.

OCCURRENCE: Many Copepoda occur in fresh water, but the majority are marine. They are the most abundant component of the zooplankton, reaching maximum density beneath the well-lit (euphotic) zone and often undergoing daily vertical migrations. Benthic species are found on fine sediments and algae, or interstitially in sand. Parasitic species, sometimes highly modified, are primarily associated with marine fishes and mammals, less frequently with invertebrates.

Collecting of free-swimming species is best done with plankton nets; benthic species may be sampled with dredge or grab, with subsequent extraction by the magnesium chloride method. Parasitic species are collected by carefully examining potential hosts, in particular the mouth and gills.

IDENTIFICATION: Species determinations are generally made after dissection of specimens (by means of insect needles) and transfer to slides for high-magnification examination. Species differences are often minute, and superficial examination without careful study of original literature only allows gross identification.

Body divided into anterior (prosome) and posterior region (urosome) by a major articulation whose position does not (except in Calanoida) coincide with the border between cephalothorax and abdomen. The (median) eye can be provided with cuticular lenses. Note genital pores on 1st abdominal (=genital) segment, and anus on the last (=anal) segment. Appendages comprise 1st antenna ($A_1$), often long and geniculate (kneed); 2nd antenna ($A_2$); mouthparts: mandibles (Md), maxillae ($Mx_1$, $Mx_2$), and maxillipeds (Mxp); 5 pairs of swimming legs ($P_1$-$P_5$), the last pair often modified as copulatory stucture in males; and the furca (tail fork) consisting of 2 rami (branches). All measurements refer to length not including the rostrum, antennae and caudal rami.

Fix in 4% formaldehyde-seawater. After detailed study of skeletal features, dissected specimens can be transferred into glycerol-water (1:1) or polyvinyl alcohol for permanent storage.

BIOLOGY: Sexes are separate and usually dimorphic; reproduction is strictly sexual. Typically, the male transfers sperm packaged in a spermatophore to the female. Eggs are laid free and singly, or held in clusters contained in 1 or 2 egg sacs and cemented to the female. Life expectancy is probably up to a year. By far the majority of Copepoda are filter feeders, with mouth parts adapted to strain the water current, produced by antennae and limbs, for food particles (e.g., phytoplankton). Other feeding specializations range from raptorial predation to piercing and sucking, browsing and parasitic habits. Planktonic forms are the major food source for many other animals, including commercially important fish species (because of their grazing habit copepods have been called the "cattle of the sea"). Some planktonic species (e.g., *Oncaea*) are bioluminescent. Copepods may serve as intermediate hosts for Trematoda, Cestoda and Nematoda, and may be overgrown by Suctoria or diatoms.

### Plate 89

DEVELOPMENT: The hatching larva (nauplius) typically passes through 6 naupliar and 5 copepodite stages, gradually approaching the shape of the adult. Larval stages are still undescribed for many species, including parasites where the number of stages is often reduced.

REFERENCES: There is no recent treatment of the class. For the planktonic species a good account is by OWRE & FOYO (1967); an older though useful reference is WILSON (1932). For general biology see MARSHALL & ORR (1955). Harpacticoida have received monographic treatment by LANG (1948).

Bermuda's benthic species have been dealt with by WILLEY (1930, 1935), COULL (1969, 1970) and VOLKMANN (1979a, b); planktonic groups by ESTERLY (1911), MOORE (1949), GRICE & HART (1962), HERMAN & BEERS (1969), MORRIS (1975) and DEEVEY & BROOKS (1977). For parasitic groups, still poorly known in Bermuda, see CRESSEY (1967), YEATMAN (1957) and ILLG (1958). Larval stages of plankton forms are described in BJÖRNBERG (1972).

### Plate 92

O. **CALANOIDA:** Copepoda with last thoracic segment firmly connected with preceding segment but movably articulated with 1st abdominal segment. $P_5$ always on prosome; similar to preceding pair or reduced or absent in ♀; always present in ♂ and modified as copulatory organ. One or neither $A_1$ (♂) geniculate. Eggs deposited free or carried in a single cluster. Free-living, mostly pelagic. (About 400 spp. from Bda and the Sargasso Sea.)

F. **PARACALANIDAE:** Calanoida with $P_5$ (♀) minute, uniramous, 2-segm. $P_1$ endopodite 1-2-segm., all others 3-segm. Head fused with thorax; 4th and 5th segments fused. Abdomen of ♀ 4-segm., of ♂ 5-segm. Left $P_5$ (♂) long, 5-segm.; right 2-segm. Right $A_1$ (♂) not geniculate. (5 spp. of *Paracalanus* from Bda.)

*Paracalanus parvus* (Claus): $P_5$ (♀) with long and slender medial terminal setae. Rostrum with 2 thin filaments. Common in inshore waters and on reef platform.

F. **PSEUDOCALANIDAE:** Calanoida with $P_5$ (♀) uniramous, 2-3-segm.; variable in ♂. Endopodite of $P_1$ 1-segm., of $P_2$ 1-2-segm., of $P_3$ 2-3 segm.; $P_4$ 3-segm. Abdomen 4-segm. in ♀, 5-segm. in ♂. Head fused with 1st thoracic segment; 4th and 5th thoracic segments fused. Right $A_1$ (♂) not geniculate. (8 spp. of *Clausocalanus* from Bda.)

*Clausocalanus furcatus* (Brady): $P_5$ (♀) 3-segm.; right $P_5$ (♂) unsegm. Genital segment (♀) shorter than each of the following 2 segments. Furca twice as long as wide. A near-surface oceanic species.

F. **CENTROPAGIDAE:** Calanoida with $P_5$ (♀) biramous, 3-segm. Abdomen in ♀ often asymmetric, 3-segm.; 4-5-segm. in ♂. Swimming legs usually all with 3-segm. rami. Head separate from thorax; 4th and 5th segments not fused. Right $A_1$ (♂) a geniculate grasping organ. (5 spp. of *Centropages* from Bda.)

*Centropages violaceus* (Claus): Posterior corners of thorax rounded. Genital segment in ♀ without lateral spines, but with ventral swelling. $P_5$ (♂) with curved finger-like processes (left very long; right shorter, with globose base). Pale lavender. An epipelagic oceanic species.

F. **PONTELLIDAE:** Calanoida with head separate from thorax; 4th and 5th segments mostly fused. Head bluntly triangular, often with lateral hooks. Posterior corners of thorax acutely produced. Eyes usually present, large, with up to 2 pairs of cuticular lenses. Abdomen of ♀ 1-3-segm., of ♂ 4-5-segm. $A_1$ (♂) strongly geniculate and globose. Endopodite of $P_1$ 2-3-segm., of $P_{2-4}$ 2-segm. Right $P_5$ (♂) chelate, often large. (4 spp. of *Pontella* and 2 spp. of *Calanopia* from Bda.)

*Pontella atlantica* (Milne Edwards): Left of posterior thorax corners larger than right. Abdomen of ♀ 3-segm. and strongly asymmetric; genital segment with dorsal boss and right ventral hook. Eye with 1 pair of cuticular lenses. Right $P_5$ (♂) much larger than left and strongly chelate, with large cusps on basal segment and finger. Cephalothorax blue. In oceanic surface waters.

*Calanopia americana* Dahl: Head without lateral hooks; rostrum stout, bifurcated. Cuticular lenses absent. Abdomen of ♀ 2-segm.; posterior thoracic corners symmetrical. $P_5$ (♂) of equal size. Transparent to opaque. In lagoon and inshore waters; near

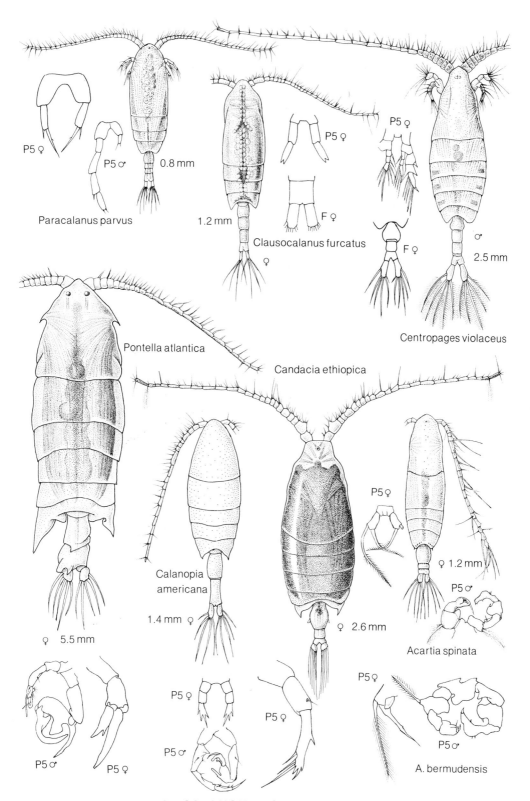

**92 CALANOIDA (Copepods 1)**

bottom during the day, migrating toward the surface at night.

F. **CANDACIIDAE:** Calanoida with $P_5$ (♀) uniramous, 3-segm. $P_{1-4}$ endopodites 2-segm. Abdomen of ♀ 3-segm. Blunt head separate from thorax, 4th and 5th segments fused. Posterior corners of thorax pointed and usually asymmetrical. (8 spp. of *Candacia* from Bda.)

***Candacia ethiopica*** (Dana): Genital segment in ♀ elongated on left side. Right posterior margin of thorax in ♂ with a bifid spine-like protrusion, and right side of genital segment with a rounded tubercle and spine-like point. Thorax and appendages distinctively chocolate-brown. Epipelagic-oceanic.

F. **ACARTIIDAE:** Calanoida with head separate from thorax; 4th and 5th segments fused. Abdomen of ♀ 3-segm., of ♂ 5-segm. $P_{1-4}$ all with 3-segm. exopodite and 2-segm. endopodite. $P_5$ uniramous. (4 spp. of *Acartia* from Bda.)

***Acartia spinata*** Esterly: Genus with $P_5$ (♂) 4-5-segm. on right, 5-segm. on left; $P_5$ (♀) 2-3-segm., with terminal segment usually claw- or awl-shaped.—Species with posterior thoracic corners of ♀ symmetric, rounded. Numerous small spines along posterior margins of last thoracic and of abdominal segments. $P_5$ (♀) 3-segm., with fused basal segments. Rostral filaments present. $A_1$ (♀) with a spine on segments 2 and 3. Found in shallow water near edge of platform.

***A. bermudensis*** Esterly: Genus as above.—Species with posterior thoracic corners of ♀ symmetric, rounded. Numerous delicate spinules along posterior margins of last thoracic and of abdominal segments. $P_5$ (♀) with blunt flap on inner margin. Rostral filaments absent. $A_1$ (♀) without spines. Primarily in shallow inshore waters.

### Plate 93

O. **CYCLOPOIDA:** Copepoda with last thoracic segment movably articulated with the preceding, and firmly united with the 1st abdominal segment which it resembles in size and form. Cephalothorax much wider than abdomen and more depressed. $P_5$ in both sexes small and unmodified, or vestigial. $A_1$ with no more than 15 segments and moderate in length; both or neither geniculate in the ♂. Eggs generally carried in paired sacs attached to lateral or subdorsal (but never ventral) surface. Includes free-swimming, commensal and parasitic species. (About 60 spp. from Bda.)

F. **OITHONIDAE:** Slender, elongate Cyclopoida with moderately widened cephalothorax and narrow sublinear abdomen; the latter 4-segm. in ♀, 5-segm. in ♂. Head separate from 1st thoracic segment. $A_1$ 10-15-segm. $P_5$ small, conical, with a long apical and a long anterior basal seta. (8 spp. of *Oithona* from Bda.)

***Oithona nana*** Giesbrecht: Genus with rami of $P_{1-4}$ all 3-segm.—Species with narrow, truncate forehead; rostrum absent. $A_1$ short, reaching to 3rd or 4th thoracic segment. $P_4$ with 1,1,2 spines on outer edge of 3 exopodite segments. Common on reef platform, most abundant inshore.

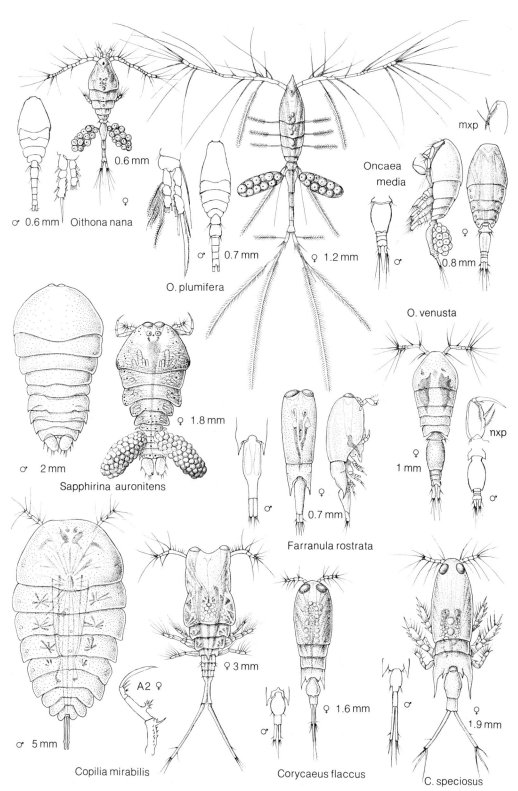

## 93 CYCLOPOIDA (Copepods 2)

***O. plumifera*** Baird: Genus as above.—Species with acutely pointed forehead; rostrum present. $A_1$ long, extending to 4th abdominal segment, bearing plumose setae. $P_5$ with plumose seta as long as urosome. $P_4$ with 0,0,1 spines on outer edge of exopodite segments. Plumose setae often bright orange. Oceanic, in surface waters.

F. **ONCAEIDAE:** Cyclopoida with elongate-elliptical cephalothorax, strongly vaulted dorsally. Head separate from 1st thoracic segment. Abdomen with 4 segments in ♀, 5 in ♂. $A_1$ 6-segm. Mxp prehensile, moderately chelate, with strong spine-like terminal segment. (6 spp. of *Oncaea* from Bda.)

***Oncaea media*** Giesbrecht: Genus with genital segment enlarged, cylindrical, esp. in ♂.—Species with genital segment as long as rest of abdomen. Caudal rami longer than anal segment, less than $3 \times$ as long as wide. Basal segment of Mxp with 2 slender bristles. In nearshore and inshore waters.

***O. venusta*** Philippi: Genus as above.—Species genital segment longer than rest of abdomen. Caudal rami longer than anal segment, about $4 \times$ as long as wide. Basal segment of Mxp with 2 strong spinous setae. Body strongly chitinous; thorax and abdomen purple. In near-surface oceanic waters.

F. **SAPPHIRINIDAE:** Flattened Cyclopoida, sexes dissimilar. Head separate from 1st thoracic segment. Cuticular lenses present. $A_1$ short, to 6 segments; $A_2$ prehensile, with terminal claw. Mxp claw-like. All rami of legs 3-segm. (15 spp. of *Sapphirina* and 5 spp. of *Copilia* from Bda.)

***Sapphirina auronitens*** Claus: Genus with leaf-like body. Prosome 5-segm.; urosome 5-segm. in ♀, 6-segm. in ♂. Genital segment in ♀ subdivided dorsally. Caudal rami flattened, leaf-like.—Species with cephalothorax slightly wider than abdomen. $P_4$ endopodite 2/3 of the length of exopodite. $A_1$ 5-segm., 2nd segment in ♂ as long as remainder of antenna. Heavily pigmented and iridescent. In near-surface oceanic waters.

***Copilia mirabilis*** Dana: Genus with pronounced sexual dimorphism: ♀ rather slender, cylindrical; ♂ flat, leaf-shaped. Head fused with 1st thoracic segment in ♀, separate in ♂. Urosome 3-segm. in ♀, 5-segm. in ♂. Caudal rami slender, elongate. $A_1$ 6-segm. $P_4$ endopodite 1-segm., all others 3-segm.—Species with posteriorly widened cephalothorax in ♂. Cuticular lenses only in ♀, well separated. $A_2$ (♀) with dentate 1st segment, and thorn-like spine on 2nd; $A_2$ (♂) with short spine on 2nd segment, this segment equal in length to 1st. In oceanic near-surface waters.

F. **CORYCAEIDAE:** Slender Cyclopoida with only slightly widened cephalothorax. Posterior prosome corners acutely produced backward. Ventral surface of thorax with cone- or beak-like projection. Two large cuticular lenses anterodorsally. Urosome 2-3-segm., caudal rami often narrow and elongate. $A_1$ 6-segm.; $A_2$ 3-segm., prehensile. $P_4$ endopodite 1-segm., all other rami of swimming legs 3-segm. (13 spp. of *Corycaeus* and 5 spp. of *Farranula* from Bda.)

***Corycaeus flaccus*** Giesbrecht: Genus with 3rd and 4th thoracic segments not fused.—Species with pear-shaped, 1-segm. abdomen in ♀. In both sexes, both basal setae of $A_2$ equal in length. $A_2$ (♂) with dentate inner distal border. Caudal rami shorter than abdomen in ♂; furca in ♀ about equal to length of abdomen. In ocean near-surface waters and in lagoon.

***C. speciosus*** Dana: Genus as above.—Species with 2-segm. abdomen in ♀. Caudal rami long, slender, strongly divergent. $A_2$ (♀) with one basal seta about twice the length of the other. Inner distal border of $A_2$ (♂) ending in single large tooth. Setae of legs and caudal rami orange. Primarily in near-surface oceanic waters.

***Farranula rostrata*** (Claus): Genus with strong beak-like ventral projection on thorax. $A_2$ 4-segm., prehensile. $P_4$ endopodite lacking; all other rami of swimming legs 3-segm. Abdomen 4-segm.—Species with smoothly tapering abdomen in ♀, about 4 times as long as caudal rami. Caudal rami only twice as long as wide. Thorax blue. In near-surface oceanic waters.

**Plate 94**

**O. HARPACTICOIDA:** Generally elongate-cylindrical Copepoda with 5th thoracic segment firmly attached to 6th; movable articulation between 4th and 5th. $A_1$ short, usually less than 9-segm., both prehensile in ♂. $P_1$ usually prehensile, next 3 pairs natatory. $P_5$ never natatory, usually lamellar and 2-segm. Primarily free-living benthic species. (With about 70 spp. the group is still insufficiently known in Bda.)

**F. ECTINOSOMIDAE:** Fusiform or cylindrical Harpacticoida with head rectangular or attenuated in front, fused with 1st thoracic segment and produced anteriorly into lamellate rostrum. Abdomen 4-segm. in ♀, 5-segm. in ♂. Eyes absent. $A_1$ usually short, with few segments. $P_1$ endopodite with 2 or 3 segments, all other rami 3-segm. $P_5$ lamelliform, ending in strong spinous setae. Egg sac single. (Only 1 sp. of *Ectinosoma* from Bda.)

***Ectinosoma dentatum*** Steuer: $A_1$ very short, much widened proximally. Apical seta of caudal rami usually less than half of body length. $P_1$ endopodite 3-segm. Basal process of $P_5$ reaching middle of exopodite; setae as in figure. Common in silty sand of inshore waters.

**F. TISBIDAE:** Somewhat flattened Harpacticoida with distinct anterior and posterior body regions. Cephalothorax 5-segm.; abdomen 4-segm. (♀) or 5-segm. (♂). Eyes present or absent. $A_1$ slender, with 8-9 segments, geniculate in ♂. Both rami of legs 3-segm. $P_5$ with poorly developed basal process, exopodite elongate. Egg sac single. (11 spp. of *Tisbe* from Bda.)

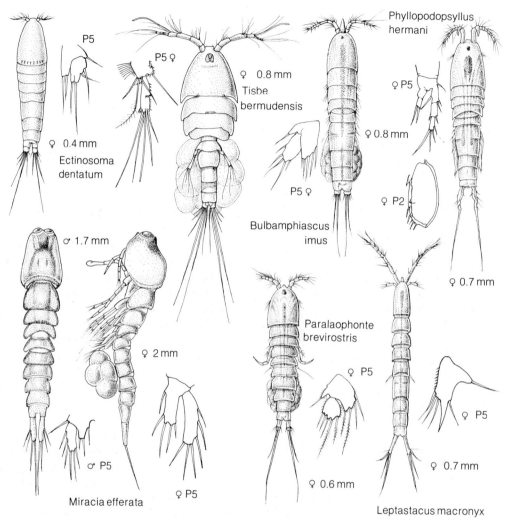

**94 HARPACTICOIDA (Copepods 3)**

***Tisbe bermudensis*** Willey: Eyes present; head segment moderately wide, tapering anteriorly; rostrum short, wide. $A_1$ 9-segm., 4th segment half the length of the 3rd. $P_5$ (♀) with moderately long exopodite, with 5 long setae clustered at the summit of the distal lamella. Common in algae and debris (such as the masking weeds on sea urchins).

F. **DIOSACCIDAE:** Variously shaped but never flattened Harpacticoida. Anterior body region usually not well defined. Rostrum well defined at base. $A_1$ short to moderately long, with 8-9 segments (rarely 5), geniculate (♂). $P_1$ endopodite 2-3-segm., $P_2$ endopodite (♂) transformed into copulatory organ. Both rami of $P_{2-4}$ 3-segm. $P_5$ 2-segm., flattened, larger in ♂. With 2 egg clusters. (1 sp. of *Bulbamphiascus* from Bda.)

***Bulbamphiascus imus*** (Brady): $P_4$ endopodite 3-segm. First segment of $P_2$

exopodite with an inner seta; distal segment of $P_3$ endopodite with 5 setae. Caudal rami as wide as long. Burrowing in sandy-silt inshore sediments.

F. **TETRAGONICIPITIDAE:** Slender, cylindrical Harpacticoida; body regions not defined. Head fused with 1st thoracic segment. Rostrum articulated with head. Caudal rami distinctively shaped. $A_2$ exopodite 2-segm. Mxp prehensile. $P_4$ endopodite with less than 3 segments. (4 spp. of *Phyllopodopsyllus* from Bda.)

*Phyllopodopsyllus hermani* Coull: Rostrum very short, rounded. $A_1$ 9-segm., with claw-shaped process on 2nd segment. $P_4$ with long 3-segm. exopodite and short 2-segm. endopodite; the latter with 1 plumose terminal seta. $P_5$ (♀) exopodite foliaceous, forming a brood pouch. Only 1 furcal seta (♀), not bulbous at the base. Rather ubiquitous, although most abundant in well-sorted, medium to coarse (0.5-1 mm) subtidal sand.

F. **CYLINDROPSYLLIDAE:** Very elongated, uniformly cylindrical Harpacticoida; body regions not defined. Head fused with 1st thoracic segment. $A_2$ exopod at most 2-segm. Mxp prehensile; other mouthparts much reduced. $P_4$ exopodite 3-segm., elongate; endopodite short, slender, 1-2-segm. $P_5$ a single plate. With 2 egg sacs. (1 sp. of *Leptastacus* from Bda.)

*Leptastacus macronyx* (T. Scott): Mxp large, terminal claw with accessory spinules. $P_{1-4}$ endopodite 2-segm. Last segment of $P_4$ with 2 inner setae. $P_5$ one fused segment, with long, triangular, acute projection. A common interstitial species in well-sorted fine beach sand.

F. **LAOPHONTIDAE:** Cylindrical Harpacticoida with posterior margins of segments more or less protruding and armed with spinules. Mxp well developed, prehensile. $P_1$ endopodite longer than exopodite, both 2-segm.; endopodite ending in strong claw. $P_{2-4}$ of swimming type; exopodites 3-segm., shorter endopodites 2-segm. $P_5$ variable. Egg sac usually single. (2 spp. of *Paralaophonte* from Bda.)

*Paralaophonte brevirostris* (Claus): Rostrum bilobular at tip. $A_1$ 6-segm. Caudal rami not twice as long as broad. $P_1$ exopodite 3-segm. $P_5$ 2-segm., basal segment carrying 4 setae and extending past middle of exopodite; exopodite with all setae arising from truncated end portion. Interstitially in sand of inshore basins.

F. **MACROSETELLIDAE:** Slender, cylindrical Harpacticoida; head fused with 1st thoracic segment; abdomen 4-segm. (♀) or 5-segm. (♂). Caudal rami slender, cylindrical, longer than last 2 abdominal segments combined. $A_1$ 8-9-segm., geniculate in ♂. $P_1$ endopodite 2-segm.; $P_2$ endopodite (♂) 2-segm., other rami of swimming legs 3-segm. $P_5$ 2-segm. Egg sac single. (1 sp. of *Miracia* from Bda.)

*Miracia efferata* Dana: Body tapered gradually toward posterior. Two large cuticular lenses on forehead. $A_1$ 8-segm.; $A_2$ exopodite

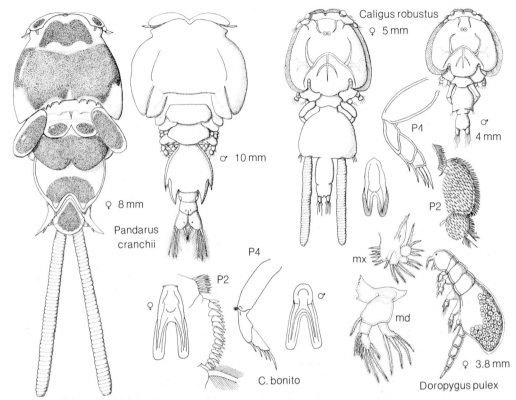

**95 SIPHONOSTOMATOIDA, NOTODELPHYOIDA (Parasitic copepods)**

small, 1-segm. $P_5$ 2-segm., elongate. Color a dark metallic greenish blue, yellowish along margins. In near-surface oceanic waters.

**Plate 95**

**O. SIPHONOSTOMATOIDA:** Dorsoventrally flattened Copepoda with segment bearing $P_5$ free. Mouth in form of a tube containing stylet-shaped mandibles. Eggs produced serially as egg strings. Parasitic as adults.

**F. PANDARIDAE:** Siphonostomatoida usually with dorsal plates. Head fused with 1st thoracic segment. Thoracic segments 2-4 free. Genital segment conspicuous. Abdomen of 1 or 2 segments with or without dorsal plates. Oral area with or without dorsal plates, and with or without adhesion pads. $A_1$ 2-segm. Md with 10-12 apical teeth. Mxp with terminal claw. $P_{1-4}$ biramous. $P_5$ reduced. (1 sp. from Bda.)

*Pandarus cranchii* Leach: Dorsal thoracic plates present on segments 2-4. Plates on segment 2 separate, extending laterally beyond tip of plate of segment 3. Tip of caudal ramus extends to or beyond abdominal dorsal plate. ♀ with heavy brown-black pigment, ♂ without. Ectoparasite on carcharhinid sharks.

F. **CALIGIDAE:** Siphonostomatoida with frontal lunules present or absent. Prosome broad, flattened dorsoventrally. First 3 thoracic segments fused with head, 4th free. Sternal furca present or absent. Postantennal process (maxillary hook) present or absent. $P_4$ usually uniramous, rarely biramous or absent. Immature forms attach to host by means of frontal filament. (5 spp. from Bda.)

***Caligus robustus*** Bassett-Smith: Genus with frontal lunules; postantennal and postoral processes present; sternal furca present. Endopod of 1st leg absent or reduced; 2nd and 3rd legs biramous, each ramus with 3 segments; 4th leg uniramous. Body form rounded without elongations or winglike processes.—Species with $P_2$ endopodite with large patches of heavy spinules on last 2 segments. $P_4$ exopodite 3-segm.; each segment nearly as wide as long, last segment with 3 setae, each only slightly longer than segment. Base of furca narrow. Parasitic on fish, especially Carangidae.

***C. bonito*** Wilson: Genus as above.—Species with 1st segment of $P_2$ endopodite with small patch of fine spinules at outer distal corner, 2nd segment with stout spinules along outer edge. $P_4$ exopodite 2-segm.; terminalmost seta longer than other 4. Tines of ♀ sternal furca parallel. Primarily a parasite of scombrid fishes; in Bermuda, however, collected from the hogfish.

O. **NOTODELPHYOIDA:** Copepoda with brood sac within body cavity of ♀. $A_2$ with prehensile terminal claw. Associated with invertebrates, mainly ascidians. (1 sp. from Bda.)

***Doropygus pulex*** Thorell: Body elongate with conspicuous brood sack. Rostrum well developed. $A_2$ 8-segm. Md exopod with 4 setae. $Mx_1$ endopod with 2 setae, exopod with 4 setae. Second segment of $Mx_1$ with setae rather than hook. Caudal ramus about 4.5 times as long as wide. In ascidians; in Bermuda recorded from *Styela plicata*.

B. F. MORRIS & R. CRESSEY

### Class Cirripedia (Barnacles)

CHARACTERISTICS: *Imperfectly segmented, sedentary (!) CRUSTACEA with body enclosed in a partly or wholly bivalved carapace, and with thoracic appendages modified as filtering apparatus; or amorphous parasites.* From 0.5 to 5 cm in size (rarely to 80 cm), most species are an often abundant yet usually drab-colored and rather shapeless component of the macrofauna whose crustacean features are not immediately apparent. Movement of adults is confined to the rhythmic beating of filtering appendages and the opening and closing of valves.

Of 4 orders, the Ascothoracica (parasitic in Decapoda) are not known from Bermuda. Of over 600 species known, only 19 have been definitely recorded from Bermuda; 15 are included here.

OCCURRENCE: Though all species are marine, some (especially Thoracica) spend much time out of the water owing to their preference for the intertidal. The majority live attached to hard surfaces such as rocks, pilings, or floating objects; some burrow into calcareous rocks, or are embedded in corals or sponges, and a few encrust crustaceans, turtles or whales.

Collecting of Balanomorpha is best done at low tide with hammer and chisel (detach complete specimens on a piece of substrate!), the deep-water species by dredging.

Barnacles are of considerable economic importance because of their abundance and rapid growth on ships and resulting inefficiency and maintenance costs.

IDENTIFICATION: Live specimens placed in seawater under a dissecting microscope should be allowed to open the shell and display the cirri and the soft flaps of colored tissue on either side of the opening. Dead or preserved specimens usually need dissection: cirri and oral appendages are pulled off and mounted on microscope slides with glycerin jelly or a synthetic water-miscible mounting medium. Shell and opercular plates are cleaned in 10% potassium hydroxide or Clorox for a few hours, then rinsed, brushed free of remaining tissue, and mounted dry in cell slides. It is sometimes necessary to grind sections of the shell to observe tubular structure.

Note the difference between the elongated goose barnacles, divided into stalk (peduncle) and body (capitulum); and the conical acorn barnacles with a broad, membranous or calcareous attachment disc (basis). The capitulum of most Lepadomorpha carries 5 plates: an unpaired carina, and the paired terga and scuta (the latter nearer the stalk). Balanomorpha have a rigid carapace whose walls (parietes) are made up of the unpaired opposed rostrum and carina, flanked by a varying number of paired lateral plates. The apical opening is guarded by 2 pairs of movable valves, terga and scuta, which together form the operculum. Some plates display a raised growth center (umbo) from which deposition of minerals proceeds in concentric layers. Appendages consist of mouthparts (of which the shape of the paired mandibles and maxillae are important taxonomic characters), thoracic limbs (cirri; usually 6 pairs, the first 3 often smaller and modified as maxillipeds), and a sizable penis.

Fix in 5% formaldehyde-seawater, buffered to pH 7 or 8, or 70% ethanol, the latter being best for long term storage and handling. Histology requires prior dissection or else Bouin's fluid or dilute formaldehyde containing EDTA (ethylenediamine tetracetic acid, sodium salt) for decalcification. Narcosis with menthol crystals or magnesium chloride will relax the opercular plates but cirri are never fully protruded because extension depends on muscular contraction to produce hydraulic pressure.

BIOLOGY: Except for a few species with dwarf males, most species are cross-fertilizing hermaphrodites, 1 specimen acting as male inseminating another that has just cast its outer exoskeleton and is about to lay eggs (note: barnacles are probably the only fully sessile animals with copulation!). Life expectancy varies from 1 to 20 or more years according to climate, tide level occupied or predation. Balanomorpha and Lepadomorpha feed on plankton and detritus of all sizes, captured by the large cirri that are swept through the water like a net, or else filtered off inside the mantle cavity by a current of water passing through a fine filter formed by the hairs on the small cirri. Predators are gastropods, crabs, and shallow-water fishes such as wrasses and parrot fishes which graze on the rocks; wreck fishes feed on the Lepadomorpha growing on floating objects. The commonest parasites are isopods found inside the shell, close to the substrate, and taking the place of the ovary. Diptera are often associated with *Chthamalus* in Bermuda.

### Plate 89

DEVELOPMENT: Egg masses develop inside the shell cavity into the 1st nauplius stage (with 3 pairs of swimming appendages) which is then shed into the water and passes through to stage VI while feeding on phytoplankton. After a few days or weeks there follows a short-lived non-feeding stage with a bivalve shell (cypris) which resembles an ostracod (but lacks 2nd antennae, and has only 6 pairs of similar legs). The cypris seeks out a suitable place to settle, attaches itself by means of its 1st antennae, and metamorphoses into the adult, the preoral body portion becoming the stalk or part of the base. *Tesseropora* and a few others incubate larvae to the cypris stage before release.

Larval stages of common species can be taken with small plankton nets close to the shore or among seaweed.

REFERENCES: A systematic overview is given by NEWMAN et al. (1969) and NEWMAN & ROSS (1976); for a broad orientation it is still necessary to consult DARWIN (1851-1854) and PILSBRY (1907, 1916). A monograph of Acrothoracica is by TOMLINSON (1969), and of *Lithotrya* by CANNON (1947).

Bermudian species are included among others in VERRILL (1901a), PILSBRY (1907-1916), CROZIER (1916), HENRY (1958), ZULLO et al. (1972), HENRY & McLAUGHLIN (1975), SOUTHWARD (1975) and NEWMAN & ROSS (1977).

## Plate 96

O. **THORACICA:** Permanently attached Cirripedia with a mantle usually strengthened by calcareous plates, and with 6 pairs of biramous thoracic appendages. (Of the 3 suborders, the Verrucomorpha have not been recorded from Bda.)

S.O. **LEPADOMORPHA** (Goose barnacles): Thoracica with body differentiated into capitulum and stalk. (8 spp. from Bda.)

F. **SCALPELLIDAE:** Lepadomorpha with umbones of scuta and carina above the middle of the plates or apical; a basal whorl of smaller plates below the principal five; stalk or peduncle scaly.

*Lithotrya dorsalis* (Ellis): Shell with 8 valves, umbones apical, with finely crenated growth lines; scuta, terga and carina large; rostrum and 1 pair of latera smaller; peduncle covered with small scales, those of the upper row crenated; peduncle terminating in a calcareous cup or row of discs. Deeply embedded in coral limestone with only the tips of the valves visible at the mouth of the burrow; from low intertidal to several meters, possibly deeper.

F. **LEPADIDAE:** Lepadomorpha with 5 or less plates on capitulum; 1st maxilla with step-like cutting edge.

*Lepas anatifera* L. (Common goose barnacle): Genus with 5 plates; carina extending up between the terga.—Species large, with only faintly striated valves, and with an umbonal tooth in the right scutum. Commonly found on drifting objects (wood, bottles, buoys).

*L. pectinata* Spengler: Genus as above.—Species small, with strongly striated valves, often profusely spinose; carina contracted just above the fork; occludent margin and umboapical ridge very close. The most common goose barnacle on *Sargassum*.

*Conchoderma virgatum* (Spengler): With 5 reduced plates on a conspicuously striped capitulum; without ear-like projections behind the terga. Rare; on pelagic fishes, turtles and whales, and drifting objects.

F. **POECILASMATIDAE:** Lepadomorpha with 5 variously reduced plates on capitulum; cutting edge of 1st maxilla not step-like.

***Octolasmis forresti*** (Stebbing): Small, with 5 incompletely calcified plates, scutum composed of 3 slender branches. In the gill cavity of *Panulirus*.

S.O. **BALANOMORPHA** (Acorn barnacles): Thoracica without stalk, with box-like shell; wall of 4-8 plates, bilaterally symmetrical, sometimes indistinguishably fused together; aperture usually closed by operculum of 4 plates hinged in pairs opening down the midline. (10 spp. from Bda.)

F. **CATOPHRAGMIDAE:** Balanomorpha with wall of 8 or 6 plates, having 1 or more whorls of supplementary plates; mandible tridentoid.

***Catophragmus imbricatus*** Sowerby: Shell with 8 major plates and several rows of small imbricated plates at base. A surf-loving form common on the outer reefs, found together with *Tesseropora*.

F. **CHTHAMALIDAE:** Balanomorpha with wall of 8, 6 or 4 plates, without basal whorl of supplementary plates; mandible tridentoid or quadridentoid. Basis commonly membranous; 3rd cirri unspecialized, labrum concave or nearly straight, without central notch.

***Chthamalus angustitergum thompsoni*** (Henry): Wall of 6 plates; rostrum complex and overlapped on either side by the 2 lateral plates. Tergum short; pectinate setae on cirrus II without basal guards. Small cirri hairy. One of the commonest intertidal species, possibly a recent immigrant.

F. **CORONULIDAE:** Balanomorpha with wall of 6 or 8 plates, parietes tubular; basis membranous; opercular plates absent or reduced. Attached to other animals.

***Xenobalanus globicipitis*** (Steenstrup): Shell reduced to a small star-shaped body forming the attachment to the skin of the host; rest of body elongated into pseudopeduncle and capitulum; without opercular plates. A cosmopolitan form found on whales and porpoises.

F. **TETRACLITIDAE:** Balanomorpha with a wall of 4 (sometimes 6) plates; parietes usually porous; basis commonly membranous. Inferior margin of mandible pectinate or serrate; cirrus II and III with specialized setae.

***Tesseropora atlantica*** Newman & Ross: Shell wall white, composed of 4 plates each containing a single row of square pores. Radii moderately developed, orifice toothed; basis membranous. A "relict" species known only from Bermuda and the Azores. Restricted to wave-swept rocks and the outer edges of the reefs at low tide level.

F. **CHELONIBIIDAE:** Balanomorpha with wall of 8 or 6 plates; opercular plates smaller than orifice and only weakly articulated; sheath extending to base, forming compartment for rest of body; shell parietes porose, with 1 row of tubes but often much filled up with incomplete subsidiary laminae; basis membranous. (Only genus:)

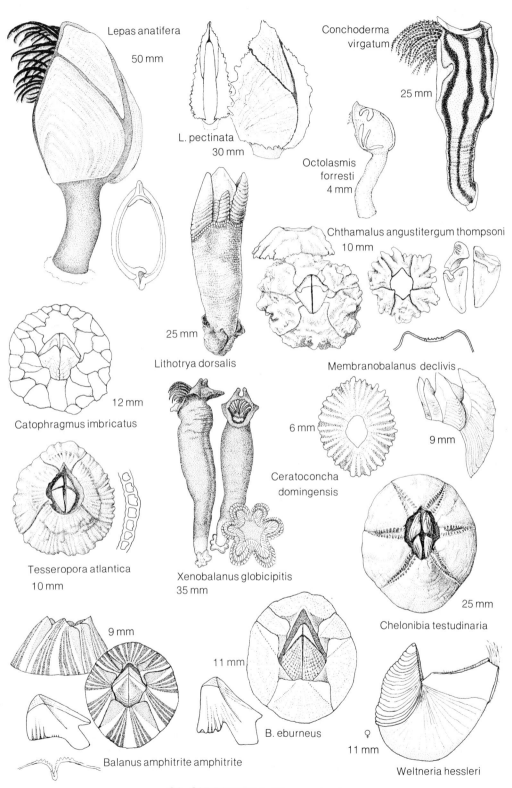

## 96 CIRRIPEDIA (Barnacles)

*Chelonibia testudinaria* (L.): Shell only superficially attached, not embedded in the skin of the turtle. Radii well-marked, with crenulated edges, forming a distinct star-shaped arrangement. Parietes with tubes partly filled by many involuted incomplete laminae, robust in form yet light in weight. On the carapace of sea turtles.

F. **ARCHAEOBALANIDAE:** Balanomorpha with wall of 6 to 4 plates; parietes usually not porous; basis often calcareous.

*Membranobalanus declivis* (Darwin): Walls of 6 plates, thin and weakly united; basis membranous; rostrum boat-shaped, twice as long as the other plates; spur of tergum very short and broad; cirri rather specialized. Always found embedded in keratose sponges (esp. *Spheciospongia othella*), cirri protruding through a pore maintained in the sponge tissue.

F. **PYRGOMATIDAE:** Balanomorpha with a wall of 4 plates or wholly concrescent; basis usually calcareous. Embedded in living corals.

*Ceratoconcha domingensis* (des Moulins): Wall of 4 plates; opercular valves fully developed; tergum with single large crest for depressor muscle. Embedded in the skeleton of *Porites astreoides*, to which its growth is synchronized. Shell covered by living coral tissue except for tip from which the cirri protrude; possibly commensal rather than semi-parasitic. (Color Plate 6.14.)

F. **BALANIDAE:** Balanomorpha with wall of 6 or 4 plates; parietes with tubes in single row, sometimes with supplementary tubes at base; basis calcareous, usually also with tubes.

*Balanus amphitrite amphitrite* Darwin: Genus with wall of 6 plates; rostral plate broad, overlapping the lateral plates on either side; cirrus III small and specialized, like cirri I and II; labrum with deep V-shaped notch in center.—Species with basically white shell, each plate bearing several vertical purple stripes, irregularly arranged; flaps of tissue on either side of opening white or yellowish white, crossed by 3 black or dark purple bands. Labrum with 10 or more small teeth on each side, running down into notch. Uncommon, represented by a few large specimens at low tide or sublittoral inshore.

*B. eburneus* Gould: Genus as above.—Species with white or dirty white shell, without colored stripes; operculum appearing triangular compared with preceding species; outer surface of scutum crossed by longitudinal striations as well as parallel growth ridges, giving cross-hatched effect; flaps surrounding opening in operculum speckled black or brown on a white or yellow ground. Not uncommon in sublittoral inshore.

O. **ACROTHORACICA:** Small Cirripedia with soft carapace, without calcareous plates; cirri usually reduced. Male dwarf and structurally reduced. Burrowing into calcareous substrates such as limestone, mollusk shells, coral and barnacles. (No shallow-water species recorded for Bda as yet; 1 example from deep water:)

***Weltneria hessleri*** Newman: With 5 pairs of biramous terminal cirri; caudal appendages present. With a calcified rostral plate. In burrows in foraminiferal chalk at 1,000 m.

A. J. SOUTHWARD

## Class Malacostraca
(Higher crustaceans)

CHARACTERISTICS: *Generally robust CRUSTACEA with 8 thoracic and 6 (rarely 7 or 8) abdominal somites, all bearing paired appendages. First antennae usually biramous; eyes often stalked.*

In most Malacostraca, the first 1-3 pairs of thoracic appendages are modified as mouthparts (maxillipeds); in the following pairs, the endopodites are generally produced to form cylindrical walking legs (pereopods). The anterior abdominal appendages (pleopods), often biramous and foliaceous, are used for swimming, carrying eggs (♀), transferring sperm (♂), or for gas exchange; only rarely are they reduced (♀ of Cumacea and Mysidacea) or lacking (Caprellidae). The 6th pair of abdominal appendages (uropods) is often flattened to form a tail fin together with the tail spine (telson).

Of 15 orders, 10 are represented in Bermuda: Leptostraca (p. 305), Stomatopoda (p. 306), Euphausiacea (p. 310), Decapoda (p. 312), Mysidacea (p. 358), Mictacea (p. 361), Cumacea (p. 362), Tanaidacea (p. 364), Isopoda (p. 366) and Amphipoda (p. 372). The small interstitial, groundwater or cave-dwelling Anaspidacea, Stygocaridacea, Bathynellacea, Thermosbaenacea and Spelaeogriphacea have not been recorded yet. In older systems, Mysidacea and Euphausiacea were grouped together as Schizopoda.

Most systems now distinguish 6 superorders: Phyllocarida (Leptostraca), Hoplocarida (Stomatopoda), Syncarida (Anaspidacea, Stygocaridacea, Bathynellacea), Eucarida (Euphausiacea, Decapoda), Pancarida (Thermosbaenacea) and Peracarida (Mysidacea, Cumacea, Spelaeogriphacea, Tanaidacea, Isopoda, Amphipoda). The new order Mictacea, only just decribed from marine caves in Bermuda, has also been assigned to the superorder Peracarida. In all Peracarida, the ♀ is characterized by extensions of anterior pereopods (oostegites) which form a brood pouch (marsupium).

## ORDER LEPTOSTRACA

CHARACTERISTICS: *Small MALACOSTRACA with a bivalved (but unhinged) carapace enclosing thorax and part of abdomen and produced anteriorly into a movable rostral plate. With 8 thoracic segments carrying 8 pairs of foliaceous legs, all similar; and with 8 abdominal segments. Eyes stalked.* Species range in size from 5 to 40 mm, and are colorless-translucent or lightly pigmented. All species apparently swim well, but some burrow into the loose benthic substrate.

Of about 11 species in 5 genera, 1 is known from Bermuda.

OCCURRENCE: Exclusively marine, in subtidal habitats on many coasts of the world including the Arctic and Antarctic. Some live in mud, some in algae and others burrow. Still others are bathypelagic.

Collect specimens by washing algae or mangrove roots. They are also taken from the bottom of rocks found on sand and mud or obtained by benthic dredging.

It is possible that some species indicate polluted conditions because they appear to be especially abundant in foul water and in places where water is muddy from channel dredging.

IDENTIFICATION: Binocular microscope for whole animals and compound microscope for details of dissected appendages placed in glycerin under a cover slip.

Note the hinged rostral plate, the adductor muscle that closes the carapace valves, and the 8 short thoracic segments; of the 8 (including the telson) abdominal segments the first 4 bear swimming legs, the next 2 bear reduced appendages, and the last one is provided with a furca.

Preserve in 70% alcohol. About 3% glycerin can be added to prevent drying for long-term storage.

BIOLOGY: Sexes are separate; males (which occur more rarely than females) are usually more slender and have longer antennae than females. Fertilization is internal. The number of progeny can be at least 50. Growth is by molting; the life-span is unknown. Most are recorded to be filter feeders, stirring up the bottom sediment and filtering the water current produced by the thoracic legs. Very few details are known about their life history.

DEVELOPMENT: Direct; eggs develop under the carapace until juveniles are released.

REFERENCES: For a systematic account including fossil forms see ROLFE (1969).
VERRILL (1923) reviewed literature on the species from Bermuda.

### Plate 98

*Paranebalia longipes* (Willemoes-Suhm): Thoracic legs extend well beyond edge of carapace. Eyestalks elongate and denticulate. ♂ with antenna modified. Subtidal, under stones, on red mangrove roots, especially in muddy places with rich, sometimes foul decayed matter. Common.

G. A. SCHULTZ

## ORDER STOMATOPODA
(Mantis shrimps)

CHARACTERISTICS: *Elongate MALACOSTRACA with eyes and triramous antennules borne on movable stalks anterior to the carapace. With 5 pairs of subchelate (clawed) mouthparts (2nd pair enlarged as raptorial claws), and 3 pairs of biramous pereopods.* These small (2 cm) to very large (more than 30 cm) crustaceans are often colorful, their robust bodies decorated with ridges and spines. Locomotion is by walking, sometimes swimming, but they are capable of rapid backward swimming by means of telson and uropods.

Of the more than 300 species described, some 70 occur in the western Atlantic, and 8 of these are known from Bermuda; 7 are included here.

OCCURRENCE: Exclusively marine, predominantly tropical, most species in shallow water. All stomatopods live in burrows, some constructing permanent burrows on level bottoms, others preferring reefs and rough bottoms where existing cavities are used.

Stomatopods may be taken by using a small trawl, by dredging, by digging, by using dip nets or push nets, night lights or by hand. By using the latter method, they may be found by turning rocks or breaking coral clumps. Sometimes they can be attracted to the mouth of their burrow with bait (oyster or shrimp). Some night-active species can be collected with hand nets or push nets on grass flats.

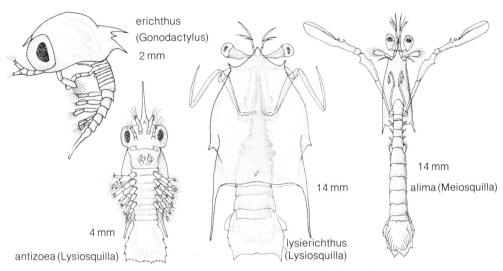

## 97 STOMATOPODA (Mantis shrimps): Development

IDENTIFICATION: Stomatopoda are differentiated exclusively through external characters.

Note the short carapace, which does not cover the 3 posterior segments of the thorax. The 5 pairs of appendages of the covered segments (maxillipeds) bear claws, whereas the 3 pairs of walking legs (pereopods) do not. The abdomen is strongly depressed, smooth or longitudinally carinate. The telson is large, flattened, variously decorated and spined. Strong uropods with ventral spines are present. Familial characters are based on posterior maxillipeds and telson; generic and specific characters include eye shape and size, and details of shape, carination and armature of various parts of carapace and appendages.

Great care should be exercised in handling live stomatopods. They can inflict painful wounds with their raptorial claws or their spined uropods (hence the Bermudian names "split-thumb", "split-toe"!).

Prior to fixation a drop of oil of cloves in a finger bowl is an excellent narcotizing agent. Fix in 10% neutralized formalin or in 70% ethyl alcohol; larger specimens may require injection. Store in alcohol rather than formalin to avoid permanent damage to the integument.

BIOLOGY: Sexes are separate. Copulation is the rule, often preceded by elaborate courtship. Eggs are laid in a rounded mass that is carried by the female in her maxillipeds. Apparently the female does not feed or leave her burrow while tending the eggs. Few observations are available on reproductive biology. Gonodactylidae lay few eggs, 500-800 at a time, whereas egg masses of Squillidae may contain up to 50,000 eggs. Stomatopoda, especially Gonodactylidae, are aggressive animals, the behavior of which can readily be studied in the laboratory. Some species actively forage in daytime; others rarely leave the burrow. All Stomatopoda appear to be predators; they consume mollusks, fishes, worms and other crustaceans. They are in turn preyed upon by fishes.

## Plate 97

DEVELOPMENT: Stomatopod larvae hatch in 2 different forms. The larvae of lysiosquilloids hatch as an antizoea, which lacks pleopods and has uniramous antennules and 5 pairs of biramous thoracic appendages. The pleopods develop in progression from front to rear. Other stomatopods hatch as a pseudozoea, with 4 or 5 pleopods, biramous antennules, and 2 pairs of uniramous thoracic appendages. Antizoeas develop into an erichthus stage, a stage with functional chelae, named after the genus; thus the erichthus of a *Lysiosquilla* is a lysioerichthus. Most pseudozoeae also develop into an erichthus, also named from their corresponding genera, e.g., gonerichthus (*Gonodactylus*). Squillid pseudozoeae develop into an alima, which differs from erichthus larvae in having 4 rather than 2 denticles between the submedian and intermediate marginal teeth on the telson. The complete larval development from hatching to post-larva is known for but 1 West Indian *Gonodactylus* species. The larva passes through 3 stages in the burrow before entering the plankton. There are 4 pelagic stages preceding the molt to postlarva, at which time the larva leaves the plankton to begin its benthic existence. In *G. oerstedii* the duration of larval life, from hatching to metamorphosis, is about 35 days. Late larval stages can be regularly found in plankton and will, unless damaged, feed readily on bits of shrimp or clam. Two of the species, *Alima hyalina* and *Meiosquilla lebouri*, were first recorded from Bermuda on the basis of planktonic stages reared to postlarvae (GURNEY 1946).

REFERENCES: Adult stomatopods from the western Atlantic have been treated in detail by MANNING (1969). Information on larvae has been summarized by GURNEY (1946), who included observations on some species from Bermuda. MANNING & PROVENZANO (1963) and PROVENZANO & MANNING (1978) reported on the larval development of *Gonodactylus*.
DINGLE (1964, 1969a, b) and DINGLE & CALDWELL (1969) studied various aspects of the behavior of *Gonodactylus bredini* from Bermuda, and DENNELL (1950b) reported *Lysiosquilla*.

## Plate 98

**F. GONODACTYLIDAE:** Body smooth dorsally, lacking longitudinal carinae; tubular, compact. Propodus of 4th maxilliped longer than broad, not beaded or ribbed ventrally. Dactylus (distal part) of raptorial claw articulating subterminally. Telson with sharp median carina or broad median ridge, no more than 2 intermediate marginal denticles present.

***Gonodactylus bredini*** Manning (=*G. oerstedi* p.p.) (B Split-thumb, Split-toe): Genus with dactylus of claw inflated, unarmed.—Species with intermediate marginal teeth of telson not widely separate from submedians. Dorsal surface of carinae of telson lacking erect tubercles. This species occurs in 6 distinct color phases, from a mottled gray to green, brown and black; all without conspicuous patches of black pigment. To 7.5 cm. Intertidal and subtidal in rocks, coral or on reefs. Common.

***G. spinulosus*** Schmitt: Genus as above.—Species with intermediate marginal teeth of telson widely separate from submedians. Dorsal surface of some carinae on telson with tubercles. Color variable, mottled blue or green, with conspicuous patches of dark pigment on 6th thoracic and 1st abdominal somites. To 5.5 cm. Sublittoral, in *Porites*, rock or on reefs. Uncommon.

***Pseudosquilla ciliata*** (Fabricius): Genus with 2 spines on basal prolongation of uropod; dactylus of claw slender, with 3 teeth.—Species without apical spine on rostral plate. Cornea cylindrical. Telson with median and 3 pairs of dorsal carinae. Color variable, usually mottled green, juveniles longitudinally striped. Carapace lacking paired eyespots, 6th thoracic and 1st abdominal somites and telson each with dark spot laterally. To

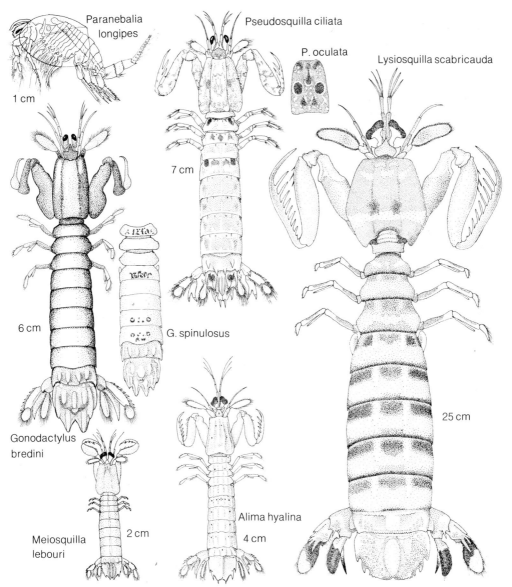

**98 LEPTOSTRACA, STOMATOPODA (Mantis shrimps)**

9 cm. An active species on sand, grass beds, and coral reefs. Usually shallow sublitoral; uncommon.

***P. oculata*** (Brullé): Genus as above. —Species with apical spine on rostral plate. Cornea broadened. Telson with median and 4 pairs of dorsal carinae. Background color mottled with white spots, carapace with pair of dark circles surrounded by lighter ring. To 12.5 cm. Usually occurs deeper than *P. ciliata*, on coral reefs.

F. **LYSIOSQUILLIDAE:** Body smooth dorsally, lacking longitudinal carinae, depressed, loosely articulated. Propodus of 4th maxilliped broader than long, conspicuously beaded or ribbed ventrally. Telson without distinct median carina.

***Lysiosquilla scabricauda*** (Lamarck) (Queen mantis shrimp): Dactylus of raptorial claw with 8-11 teeth, fewer in adult females. Terminal abdominal somite and telson coarsely tuberculate. Body conspicuously banded. To 30 cm. Usually in burrows in soft bottom, rare. Edible.

**F. SQUILLIDAE:** Body depressed, compact, longitudinally carinate. Propodus of 4th maxilliped longer than broad, not beaded or ribbed ventrally. Telson with distinct median longitudinal carina, 4 or more intermediate marginal denticles present.

***Alima hyalina*** Leach: Lateral process of 5th thoracic somite (1st exposed somite) bilobed, posterior lobe rounded. Dactylus of raptorial claw with 6 teeth. Mandibular palp absent. Four epipods present. Submedian teeth of telson with fixed apices. With 2 rounded lobes between spines of basal prolongation of uropod. Color whitish, with black spots on surface. To 4.5 cm. In sand or grass beds; uncommon.

***Meiosquilla lebouri*** (Gurney): Lateral process of 5th thoracic somite single, obscure. Dactylus of claw with 4 teeth. Mandibular palp absent, 2 epipods present. Submedian teeth of telson with movable apices. With 1 lobe between spines of basal prolongation of uropod. Color light, marked with lines and patches of black pigment. To about 2 cm. Habitat unknown.

R. B. Manning

## ORDER EUPHAUSIACEA (Krill)

CHARACTERISTICS: *Shrimp-like MALACOSTRACA with carapace fused to and covering the entire cephalothorax but not the gills, and with 8 pairs of mostly uniform thoracic appendages none of which are modified as maxillipeds; eyes stalked.* Length 5-30 mm (rarely up to 95 mm); transparent or with red pigment. Locomotion is by movement of the abdominal pleopods which results in a steady swimming.

Of the 2 families only 1 is represented here; the other (Bentheuphausiidae) comprises only 1 species. Approximately 90 (often cosmopolitan) species are known; 4 of about 10 species commonly found off Bermuda are included here.

OCCURRENCE: Pelagic, but some species are found near shore. Most species prefer great depths (hundreds to thousands of meters); larvae are usually found closer to the surface.

Collecting may be done with a coarse plankton net, preferably at night.

IDENTIFICATION: Adults are easily distinguished on the basis of external characters (use a dissecting microscope).

Genera differ in the number and differentiation of thoracic appendages (the 7th and 8th are often reduced; and the 2nd, the 3rd or both frequently produced); species (♂) are best distinguished by the inner rami of the 1st and 2nd pair of pleopods which function as a copulatory organ. Note the exposed gills. Lens-shaped luminescent organs (photophores) are usually found on the eyestalks, the base of the 2nd and 7th thoracic appendage, and under each of the first 4 segments of the abdomen. Luminescence can be observed in a darkened aquarium.

Preserve in 80% alcohol or 40% formaldehyde.

BIOLOGY: Sexes are separate; sperm are transferred to the ♀ as spermatophores. Eggs are liberated into the seawater, or

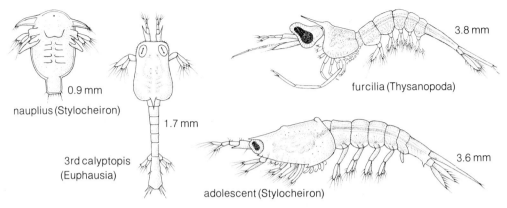

**99 EUPHAUSIACEA (Krill): Development**

temporarily retained under the thorax. Life-span is probably 1-4 yr. Most species are filter feeders, the first 6 thoracic appendages forming a "food basket"; only a few (*Stylocheiron*) are predators, capturing Chaetognatha with a specialized 3rd thoracic appendage. Many euphausids live in enormous swarms that undergo vertical migrations, approaching the surface at nighttime. They are of economic importance as food to humans, fishes and seabirds, and they constitute the staple diet of baleen whales.

## Plate 99

DEVELOPMENT: There are 4-5 larval stages: the nauplius and metanauplius develop into a calyptopis (with body divided, but eyes still under carapace), then a furcilia which still uses its 1st antennae, and an adolescent (formerly cyrtopia, or late furcilia) which already uses its pleopods for swimming.

REFERENCES: Classical treatments can be found in expedition reports, e.g., SARS (1885); recent biological monographs are by MAUCHLINE & FISHER (1969) and MAUCHLINE (1980); for taxonomy see also BODEN et al. (1955) and MAUCHLINE (1971a, b).

Some of Bermuda's species and their larvae are treated in LEBOUR (1949); for distribution see MOORE (1949), MORRIS (1975) and GRICE & HART (1962).

## Plate 100

F. **EUPHAUSIIDAE:** Euphausiacea with 1st and 2nd pairs of pleopods (♂) modified as copulatory organs, and with photophores.

***Thysanopoda aequalis*** Hansen: With 8th pair of thoracic legs reduced, 7th pair smaller than preceding; abdominal segments without dorsal spines. Rather frequent in 100-500 m.

***Euphausia brevis*** Hansen: With 7th and 8th pairs of thoracic legs quite reduced, 6th pair similar to 5th; lateral margin of carapace with 2 pairs of denticles. The commonest euphausid around Bermuda; surface to 300 m.

***Thysanoëssa gregaria*** G. O. Sars: With 2nd pair of thoracic legs greatly produced, the last 2 segments armed with bristles; without a mediodorsal keel on abdomen. With brilliant red

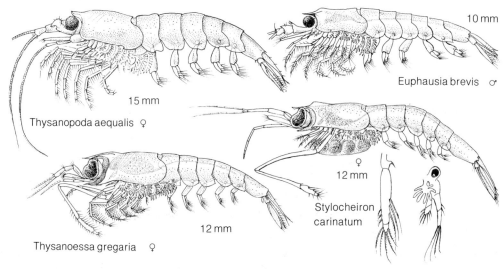

### 100 EUPHAUSIACEA (Krill)

pigment dorsally and laterally. Common, surface to 200 m.

***Stylocheiron carinatum*** G. O. Sars: With 3rd pair of thoracic legs greatly produced, its terminal segments forming an incomplete claw (subchela). Carapace with distinct dorsal keel. Upper part of eyes narrow and with enlarged cones. Colorless. Common above 200 m.

W. STERRER

ORDER DECAPODA (Shrimps, lobsters and crabs)

CHARACTERISTICS: *Generally robust MALACOSTRACA in which the 3 anterior pairs of thoracic appendages function as mouthparts and are morphologically distinct from the 5 posterior pairs. Exopod of the 2nd maxilla (scaphognathite) expanded anteriorly and posteriorly.* Though the size range within the order is impressive (from a few millimeters in length to a leg span of several meters), most Bermudian species are in the 2-20 cm bracket, with only the spiny lobster reaching much larger dimensions (50 cm). Few animal orders display such a diversity of appearance, from the tiny transparent *Lucifer* to the graceful long-tailed shrimps, the sluggish locust lobsters and the cumbersome crabs. Color and patterns often simulate the substrate (mottled red, brown and gray), or have signal function (as in cleaning shrimps), in which case they are bright (e.g., white and red). Locomotion is in 3 different ways: forward swimming by action of the abdominal appendages, as in shrimps, especially pelagic species; rapid evasive movements backward, produced by flexure of the powerful abdominal, muscles as in shrimps and lobsters; and multi-directional walking (often sideways in crabs) on thoracic appendages. In some members of most infra-orders (especially the portunid crabs), 1 or more pairs of pereopods are flattened and used as swimming, or sculling, appendages.

Traditionally (and still for practical purposes), long-tailed Decapoda (Macrura) have been divided into Natantia (shrimps) and Reptantia (lobsters); the more recent system followed here does not use these categories. Thalassinoidea, in some systems, are not included in Anomura. Of about 9,000 living species known, about 275 have been recorded from Bermuda; 115 are included below.

OCCURRENCE: Most common in the sea, where they are found in both benthic and pelagic situations from the surface to 5,500 m. Highest diversity on shallow hard and sandy bottoms. Some families are adapted to fresh water (e.g., crawfish) or a terrestrial existence (Coenobitidae, Gecarcinidae). Although the majority of local species does not occur in large congregations (as do some commercially exploited prawns elsewhere), a few shrimps (e.g., *Palaemon northropi*) often form small schools, and spiny lobsters can be found crowding into reef crevices by the dozens.

Collecting of all pelagic forms and larvae can be done by means of a plankton net. Species inhabiting hard bottoms can be caught by carefully overturning rocks (using SCUBA and hand net if necessary), or by cracking collected rocks open by means of a hammer and chisel. A dredge or push net should be used on sediments and grass beds; species occupying burrows or mud flats can be extracted with a suction device called a "yabby pump". Benthic shrimps can be trapped in a wide mouthed gallon jar baited with decaying fish and covered with a wire-mesh funnel; deepwater species (esp. hermit crabs) are lured into baited wire traps.

Other than the spiny lobsters (*Panulirus argus* and *P. guttatus*) which are caught in traps during lobster season (!), no species is of any commercial importance in Bermuda. Scyllaridae and several crabs (e.g., *Grapsus grapsus*) are occasionally eaten.

IDENTIFICATION: Because of the morphological diversity of decapods, diagnostic characters vary from family to family and often from genus to genus. External features of preserved specimens, however, are sufficient for identification, and with experience, probably most of the species will eventually be recognizable in life. Color or color patterns can be very useful in identifying some forms (e.g., Palinuridae, Diogenidae, Paguridae, *Calappa*), but should be used with caution on others, because variability within the species is not well known (this applies especially to Alpheidae), or the color in individuals may change considerably according to the background, the time of the day or other factors that now defy explanation (see Scyllaridae, Porcellanidae, Grapsidae and Majidae).

## Plate 101

The body is composed of 2 parts: a uniform cephalothorax and a segmented abdomen. The former is covered dorsally by a carapace whose shape, regions and ornaments (such as spines, grooves, sutures) and extensions (e.g., rostrum, or head spine) are important diagnostic features. In particular, the position of longitudinal lines (linea thalassinica: close and parallel to dorsal midline, and linea anomurica: below the lateral margin) is characteristic for certain taxa. The eyes, consisting of stalk and cornea, sit in variously shaped orbits, sometimes covered by the carapace. Ventrally, the epistome (region anterior to the mouth) and the sternum (region between walking legs) are of importance. Gills are enclosed in lateral branchial chambers, some attached to the body wall, some to the coxae of the legs (podobranch). Gill numbers and structure - dendrobranchiate (tree-shaped), phyllobranchiate (lamellar) and trichobranchiate (filamentous) -are used to distinguish natantian infraorders. The cephalothorax carries 3 groups of external appendages: 2 pairs of antennae, consisting of peduncle (basal part) and flagella (whips), the 2nd pair often provided with an antennal scale; mouthparts comprising an upper lip, mandibles, 2 pairs of maxillae and a lower lip, assisted by 3 pairs of maxillipeds; and 5 pairs of walking legs (pereopods). Pereopods usually consist of 7 parts: coxa, basis, ischium, merus, carpus (sometimes subdivided), propodus and dactylus. The latter 2 can form an incomplete claw (subchela) where the dactylus bends back against the propodus, or a complete claw (chela) where the dactylus moves against a fixed extension of the propodus. The abdomen, often flexed temporarily (shrimps, lobsters) or permanently (crabs) under the cephalothorax, typically consists of 7 somites that

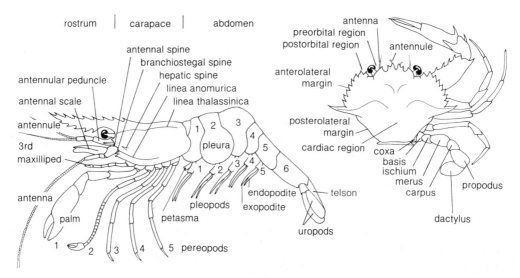

**101 DECAPODA (Decapods): Schematic**

can be broadened laterally into flaps (pleura), or fused (as in hermit crabs). Somites carry paired swimmerets (pleopods); the last pair (uropods) often forms a tail fan together with the tail spine (telson). The specific shape of the anterior pleopods is an important character for species and sex determination in several groups. In male Penaeidea, the mesial (innermost) sides of the proximal segments (protopodites) of the anterior pair of pleopods is broadened into a complicated organ (petasma) used for sperm transfer. Female Penaeidea possess an external pocket (thelycum) on the ventral side of the thorax which functions as a receptacle for sperm. In most Caridea there are two lappets—appendix interna and appendix masculina—arising from the mesial margin of the endopod of the second pleopod; only the appendix interna is present in females. In many of the true lobsters and crabs, as well as some of the anomurans, the two anterior pairs of pleopods are modified as copulatory organs in the males, whereas the female pleopods are adapted to support the eggs. In most female spiny lobsters, the posterior pair of pereopods terminate in short pincers, which are used in tending the eggs, whereas these legs are not chelate in males. The sexes of most adult true crabs can be determined easily by the shape of the abdomen—subtriangular in males, oval in females. In the absence of secondary sexual characters in reptant decapods, the sex can usually be ascertained by the position of the genital openings—on the proximal joint of the 5th (last) pair of pereopods in males and in a corresponding position on the 3rd pair in females. Sizes given in plates are the lengths of average mature specimens, measured from the tip of the rostrum to the end of the telson (in all long-tailed forms); and the maximum width of the cephalothorax (in Brachyura).

Before preservation note colors of live specimens (color photography!) for they fade rapidly, especially in shrimps. If possible, specimens should be narcotized with oil of cloves or magnesium chloride, or thoroughly chilled or allowed to die in stagnant water before being immersed in preservative, in order to avoid the loss of appendages by autotomy. Specimens may be preserved initially in either 75% ethyl alcohol or 10% buffered formalin. Large specimens, such as spiny lobsters, should be injected with formalin to ensure thorough fixation. For permanent storage, at least 70% ethyl alcohol is preferred.

BIOLOGY: Probably all decapods are dioecious, although some caridean shrimps and a few anomurans are protandrous hermaphrodites in varying degrees, changing from males to females during growth. In many of the shrimps, the males are smaller than the females, but in some of the carideans, as well as in many of the lobsters and crabs, they are larger, often considerably so. Fertilization is preceded by copulation, often possible only when the female is in the soft-shell stage after molting; or the male may deposit adhesive sperm on the ventral surface of the female (as in *Panulirus*) or in a sperm storage organ (thelycum) on the ventral surface (as in Peneaidea), in which case fertilization is presumably external; or the sperm may be deposited in a seminal receptacle inside of the female gonopore, leading to internal fertilization (as in the true crabs). The number of eggs produced by a female at any one time varies from a single egg in some of the caridean shrimps to millions in some of the crabs and lobsters. Eggs are carried in clusters on the pleopods of the female until hatching. Growth is discontinuous and regulated by molt frequency, itself a function of environmental conditions, e.g., food availability. Though data are still scarce, it is estimated that it takes *Panulirus argus* 4-5 yr to grow from a postlarva to breeding size (ca. 22 cm length, ca. 650 g weight); it may reach more than 8 kg in weight and probably several decades in age. Most smaller species probably have a much shorter life-span. Complex behavior is displayed by some, such as pre-copulatory courtship, invitation to fishes for cleaning (e.g., *Periclimenes*), or the enigmatic migration into deep water by *Panulirus*, in which many animals may move in an unbroken line. Some species are associated with sea anemones, either seeking refuge between the tentacles (shrimps of the genera *Periclimenes*, *Thor*), or—as in the case of various pagurid hermit crabs—by encouraging anemones to live on the shells they inhabit. Most decapods are probably fortuitous and omnivorous feeders, capturing live prey when available but apparently scavenging most of the time. Some of the more sedentary species (Callianassidae, Porcellanidae, Hippidae) are plankton or detritus feeders. In turn, all but the largest forms are prey for numerous fishes, cephalopods (*Octopus*), other decapods and seabirds.

The chief parasites are other Crustacea: all groups are subject to branchial, less frequently abdominal, epicaridean Isopoda; the reptant forms are attacked by even more harmful rhizocephalan Cirripedia (which have not been recorded from Bermuda). Some crabs are regularly infested with Sporozoa, and stages of Trematoda-Digenea. Commensal Cirripedia (*Octolasmis*) are commonly found attached to the gill of *Panulirus*.

## Plate 102

DEVELOPMENT: Postembryonic development runs the full gamut from a complete series of larval stages—beginning with a nauplius—in Penaeidea, through variously abbreviated series to direct development, without metamorphosis. Most decapods bypass the initial nauplius stage and hatch as a zoea, with 6 pairs of appendages and a segmented abdomen. The shape of the zoea is usually distinctive for larger taxa: the shrimp zoea often carries thorns on the abdomen; Anomura are characterized by paired thorns and Brachyura by unpaired horns on the carapace. Specialized types of zoeae are called a mysis in Penaeidea, acanthosoma in Sergestidae, eretmocaris in Lysmatidae and phyllosoma in Palinuroidea. The succeeding basic stage is a postlarva, which is known as a mastigopus in Penaeidea, and as a megalopa in the true crabs. Larval life in the plankton usually takes 1-3 months (but possibly as long as 18 months in some Palinuroidea!). The fact that a comparatively large number of decapod species have never been recorded as ovigers (egg-bearing) in Bermuda, together with the high occurrence, in plankton around Bermuda, of decapod larvae that cannot be assigned to local species, may indicate that many species do not breed here, but have their populations replenished by larvae

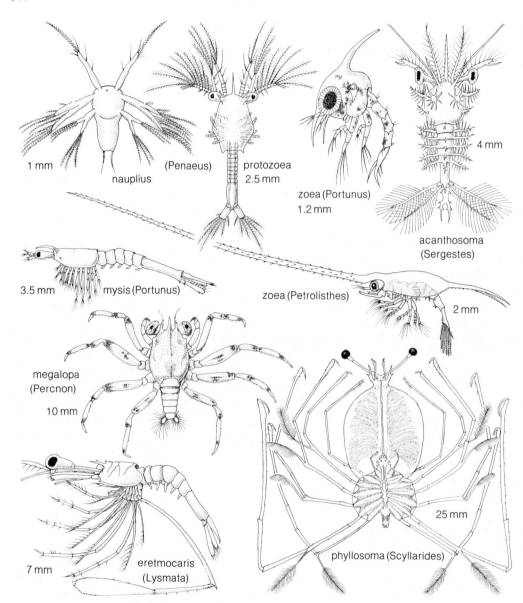

**102 DECAPODA (Decapods): Development**

drifting in via the Gulf Stream. Larvae can be collected with a fine-mesh plankton net, or attracted to a light and then dipnetted. Larvae can also be hatched off from egg-bearing females. Methods for rearing larvae have been described by RICE & WILLIAMSON (1970) and INGLE & CLARK (1977).

References: For a broad systematic overview see GLAESSNER (1969) and ABELE & FELGENHAUER (1982). Most of the shrimps covered in the present volume are included in CHACE (1972); some Penaeidea have been reviewed by PERÉZ FARFANTE (1969, 1971). Most lobsters can be found in the field guide by OPRESKO et al. (1973); HOLTHUIS & ZANEVELD (1958) provide information on the biology of some species. Keys to the western Atlantic Callianassidae are given in BIF-

FAR (1970); there has been no general study of the Axiidae of the region. The literature on the western Atlantic hermit crabs is still rather scattered, because the known species are currently undergoing revision and expansion; for recent references see PROVENZANO (1959, 1960) and McLAUGHLIN (1975). Keys to the western Atlantic galatheids are included in CHACE (1942) but they are in need of updating, and those to the porcelain crabs may be found in HAIG (1956). The four monographs by RATHBUN (1918, 1925, 1930 and 1937) are still basic to any study of the brachyuran crabs of the western Atlantic; supplemental references are the publications by WILLIAMS (1965) on the decapods of the Carolinas, by CHACE & HOBBS (1969) on the freshwater and terrestrial species, by HOLTHUIS (1958) on the genus *Calappa* and by WILLIAMS (1974) on the genus *Callinectes*.

The basic works specifically on Bermudian decapods are those by VERRILL (1908a, b; 1922), but their nomenclature is now outdated. MARKHAM & McDERMOTT (1980) give an up-to-date list of Bermuda's Decapoda. *Sargassum*-inhabiting decapods are listed in MORRIS & MOGELBERG (1973), and some cave shrimps have been described by HART & MANNING (1981). SUTCLIFFE (1957) provides information on the biology of *Panulirus*; HAZLETT & WINN (1962a, b) have reported on sound production in various Decapoda, KENSLEY (1980) described an *Axiopsis*, and MARKHAM (1977) made observations on *Calcinus*. Larval stages from Bermuda are particularly well known thanks to GURNEY (1936, 1942) and LEBOUR (1941, 1944, 1950a, b); palinuroid larvae are described by ROBERTSON (1969).

## Plate 103

I.O. **PENAEIDEA:** Decapoda with lateral flap (pleuron) of 2nd abdominal somite not overlapping that of 1st. Third pair of legs chelate (except in *Lucifer*), but not much more robust than preceding pairs. ♂ with petasma on 1st pleopods. Gills dendrobranchiate (tree-shaped).

F. **PENAEIDAE** (Prawns): Penaeidea with carapace unarmed in posterior 2/3 of dorsal midline. Scale-like projections arising from base of eyestalk and from mesial margin of basal segment of antennular peduncle. All legs well developed. Third and 4th pairs of swimmerets (pleopods) with 2 branches (endopod and exopod). At least 15 conspicuous gills on each side. (7 spp. from Bda.)

***Penaeus duorarum*** Burkenroad (Large prawn): Rostrum with 1 or more teeth on ventral margin. Carapace without lateral suture. Telson without fixed lateral spines. Petasma symmetrical, with distal portion of ventral costa armed along free border with minute spines. Reaches 270 mm in length. Prefers rather firm bottoms of mud and silt with coral sand and mixture of mollusk shells, usually in depths of 10-35 m (occ. to 330 m). Adults active only at night, remaining buried during the daytime except on cloudy days or when water is turbid, but immature specimens can be found at any time. Spawning probably occurs throughout the year.

***Metapenaeopsis smithi*** (Schmitt): Rostrum with dorsal teeth only. Carapace without lateral suture. Telson with posterior pair of lateral spines fixed, not movable. Petasma asymmetric; with distoventral projection cleft by deep sinus into 2 long, subequal lobes. Color pattern a mottled brown-yellow. Found on a variety of bottom types, usually in shallow water, occasionally to more than 350 m. Most active at night. WHEELER (1937) noted that these shrimps come to the surface at Bermuda about 1 hr after sunset at the time of the new

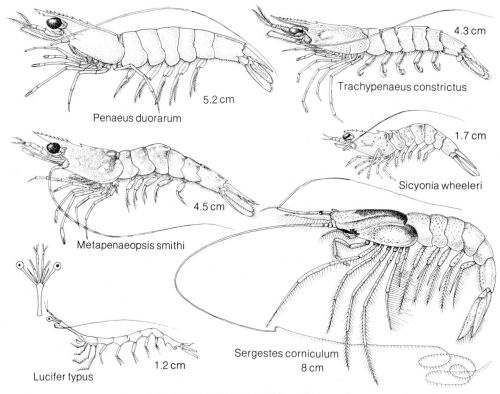

**103 PENAEIDEA (Whip shrimps)**

moon, reaching maximum numbers on the 2nd and 26th nights of the lunar month.

***Trachypenaeus constrictus*** (Stimpson): Rostrum with dorsal teeth only. Carapace with longitudinal lateral suture. Telson with lateral spine movable. Petasma symmetrical. Female with thelycum pubescent, lips of transverse groove strongly convex, not truncate. Male with sternal elevation between coxae of 5th legs goblet-shaped, constricted posteriorly, rather than simply triangular. Prefers sandy or mud and shell bottoms in high- salinity water, to a depth of 55 m. Spawning probably occurs throughout the year.

F. **SICYONIIDAE:** Penaeidea with at least 1 strong tooth in posterior half of dorsal midline of carapace. No scale-like projections arising from eyestalk or antennular peduncle. All legs well developed. Third and 4th swimmerets (pleopods) with single branch (endopod lacking). Only 11 conspicuous gills on each side. (1 sp. from Bda.)

***Sicyonia wheeleri*** Gurney (Wheeler's rock shrimp): Carapace with only 2 teeth in dorsal midline, largest situated in posterior half. Buttressed tooth (antennal spine) below orbit. Dorsal carina on 2nd abdominal somite entire, not transversely incised. Fifth abdominal somite without tooth or sharp angle at posterior and of dorsal carina. Anterior leg without spine on 2nd or 3rd segments. Probably prefers sandy bottoms in 1-42 m. Active at night; probably buried during the daytime, well camouflaged owing to its mottled coloration.

F. **SERGESTIDAE:** Penaeidea with carapace unarmed in posterior 2/3 of dorsal midline. No scale-like projections arising from eyestalk or antennular peduncle. Two posterior pairs of legs usually reduced or absent. Third and 4th pairs of swimmerets (pleopods) with 2 branches (endopod and exopod). No more than 7 conspicuous gills on each side, sometimes absent. Pelagic. (19 spp. from around Bda.)

***Lucifer typus*** Borradaile: Genus with extremely compressed body, its depth several times its thickness. Fourth and 5th pair of pereopods absent; the others not chelate. No gills. - Eyes of species more than 1/2 as long as "neck". In offshore areas within a few meters of the surface; not uncommon.

***Sergestes corniculum*** Krøyer: Genus not with extremely compressed body. Fourth and 5th pair of pereopods present albeit reduced. No gills.—Species with a thorn near base of exterior uropod; eyestalk about as long as antennular peduncle. Larvae are common in offshore waters, within a few meters of the surface.

I.O. **CARIDEA:** Decapoda with lateral flap (pleuron) of 2nd abdominal somite overlapping those of 1st and 3rd somites. Third pair of legs simple, not chelate. Males without petasma on 1st pleopods. Gills phyllobranchiate (leaf-shaped).

*Plate 104*

F. **PASIPHAEIDAE:** Caridea with rostrum reaching little if at all beyond eyes. Eyes exposed and freely movable, not unusually long and slender. Third maxilliped not unusually broadened to form operculum covering anterior mouthparts. Two anterior pairs of legs subequal and chelate, with slender, pectinate fingers; 5th segment (carpus) of 2nd leg not subdivided. (5 spp. from Bda.)

***Leptochela bermudensis*** Gurney: Rostrum an unarmed, normal anterior extension of carapace. Carapace completely unarmed and without excavation at anterolateral angle (branchiostegal sinus), dorsal surface with 3 longitudinal ridges in breeding females only; orbital margin entire, not serrate, without tooth at suborbital angle. Fifth abdominal somite without dorsal elevations or posterior tooth. Sixth somite with transverse, carinate ridge near anterior end of dorsal surface and long, posteriorly directed spine near posterior end of ventrolateral margin. Telson with 1 pair of dorsolateral spines in addition to anterior mesial pair, posterior margin with minute paired mesial spines between mesial pair of 5 pairs of large posterior spines. Has been collected at the surface under a light at night in water as shallow as 6 m, but it has also been taken between more than 1,000 m and the surface.

F. **DISCIADIDAE:** Caridea with rostrum reaching little at all beyond eyes. Eyes short and stout, exposed and freely movable. Third maxilliped not unusually broadened to form operculum over anterior mouthparts. First pair of legs robust, much stouter than other legs, movable finger compressed, semicircular, deeply recessed in slit in penultimate segment when flexed. Fifth segment (carpus) of 2nd leg not subdivided. (2 spp. from Bda.)

***Discias atlanticus*** Gurney: Rostrum narrow with subparallel margins in dorsal aspect. Sixth abdominal somite nearly twice as long as 5th, and slightly longer than telson. First taken in plankton at night in shallow inshore waters just prior to a new moon, but since found in depths to 50 m.

F. **RHYNCHOCINETIDAE:** Caridea with rostrum overreaching antennal scale and strongly dentate. Eyes stout, exposed and freely movable. Third maxilliped not unusually broadened to form operculum over anterior mouthparts. First pair of legs more robust than 2nd pair. Fingers of chelae of both pairs bearing long marginal spines forming kind of basket when flexed. Fifth segment (carpus) of 2nd leg not subdivided. (1 sp. from Bda.)

***Rhynchocinetes rigens*** Gordon: Rostrum incompletely articulated with carapace, movable in adults. Fourth and 5th abdominal somites with acute lateral tooth dorsal to posterior margin of lateral flap (pleuron). Coloration mottled red and white. Nocturnal and probably herbivorous; in daytime in deep crevices in riprap and in caves from low tide line to a depth of 15 m.

F. **PALAEMONIDAE:** Caridea with variable rostrum. Eyes exposed, usually freely movable. Third maxilliped not unusually broadened to form operculum over anterior mouthparts. Second pair of legs with 5th segment (carpus) not subdivided, more robust than anterior pair. (13 spp. from Bda.)

***Palaemon northropi*** (Rankin): Rostrum dentate. Carapace with 2 spines (antennal and branchiostegal spines) on or near anterior margin, with longitudinal suture (branchiostegal groove) between them. Antennal scale well developed. Third maxilliped with lash (exopod) arising from 2nd segment (basis). Second pair of legs elongate, subequal. Walking legs with distal segment (dactyl) simple, undivided. Very transparent, with few thin brown transverse stripes. Very common, occurring in schools in shallow water, tide pools and around pilings. (Color Plate 9.8.)

***Leander tenuicornis*** (Say): Rostrum dentate. Carapace with 2

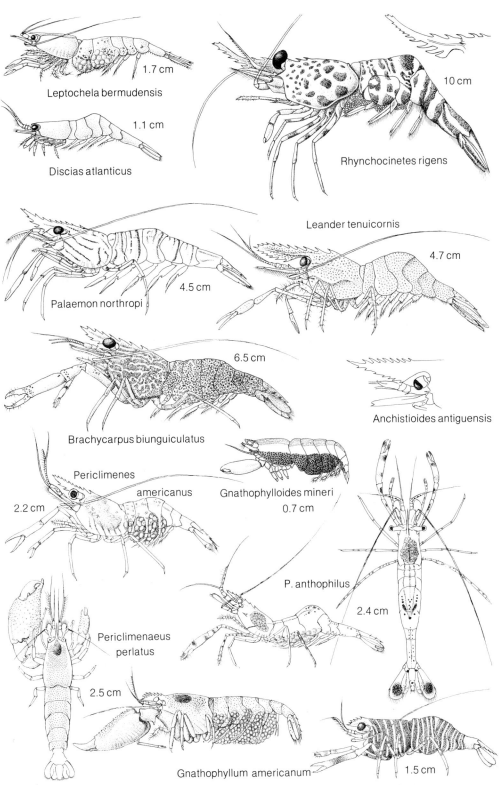

**104 PASIPHAEIDAE—GNATHOPHYLLIDAE (Shrimps)**

spines (antennal and branchiostegal spines) on or near anterior margin, without longitudinal sinus (branchiostegal groove) between them. Antennal scale well developed. Third maxilliped with lash (exopod) arising from 2nd segment (basis). Second pair of legs elongate, subequal. Walking legs with dactyl simple, undivided. Color variable; usually yellowish brown, with brown dots. On grass flats in shallow water and, especially, associated with *Sargassum* in the open sea.

***Brachycarpus biunguiculatus*** (Lucus): Rostrum dentate. Carapace with 1 (antennal) spine on anterior margin, longitudinal suture (branchiostegal groove) below it, and 2nd (hepatic) spine near posterior end of latter. Antennal scale well developed. Third maxilliped with lash (exopod) arising from 2nd segment (basis). Second pair of legs elongate, subequal. Walking legs with dactyl terminating in 2 teeth. Coloration a mottled green and brown. Occurs in most littoral and sublittoral habitats to a depth of at last 140 m. (Color Plate 9.10.)

***Anchistioides antiguensis*** (Schmitt): Rostrum dentate. Carapace armed with single (antennal) spine near lower margin of orbit, strong turbercle postero-dorsal to orbit, without longitudinal suture (branchiostegal groove). Antennal scale well developed. Third maxilliped without lash (exopod). Second pair of legs elongate, subequal. Walking legs with dactyl terminating in 2 unequal teeth. Maximum total length about 25 mm. Tends to swarm at the sea surface an hour after sunset during the time of the new moon, specifically on the 2nd and again on the 26th night of the lunar month (WHEELER 1937). Habitat during daylight hours not yet known.

***Periclimenes americanus*** (Kingsley): Genus with single (antennal) spine on anterior margin and 2nd (hepatic) spine posterior and slightly ventral to 1st. Antennal scale well developed. Third maxilliped with lash (exopod) arising from 2nd segment (basis). Second pair of legs elongate, subequal.—Species without longitudinal suture (branchiostegal groove) on carapace. Walking legs with dactyl simple, undivided. Spine on ventral surface between anterior pair of legs. Colorless, with whitish "backbone" pattern medio-dorsally. Very common in most shallow water habitats; also recorded from nearly 75m.

***P. anthophilus*** Holthuis & Eibl-Eibesfeld: Genus as above.—Species with longitudinal suture (branchiostegal groove) present, at least anteriorly. Walking legs with dactyl terminating in 2 uncqual teeth. No spine on ventral surface between anterior pair of legs. Very transparent, with distinctive white, pale purple and reddish brown markings. Found in intimate association with the sea anemone *Condylactis gigantea*. Known to clean fishes, which it invites to the "cleaning station" by rhythmically waving its white antennae (SARGENT & WAGENBACH 1975). (Color Plate 9.12.)

***Periclimenaeus perlatus*** (Boone): Rostrum dentate dorsally. Carapace armed with single, strong (antennal) spine near lower angle of orbit, longitudinal suture (branchiostegal groove) short and indistinct. Antennal scale well developed. Third maxilliped with lash (exopod) arising from 2nd segment (basis). Second pair of legs massive, unequal. Walking legs with dactyl terminating in 2 teeth. Lives in sponges (e.g., *Aplysina fistularis*).

F. **GNATHOPHYLLIDAE:** Caridea with short, dentate rostrum. Eyes exposed and freely movable. Third maxillipeds greatly expanded to form operculum over anterior mouthparts. Second pair of legs with 5th segment (carpus) not subdivided, more robust than 1st pair. (2 spp. from Bda.)

***Gnathophylloides mineri*** Schmitt: Antero-lateral angle of carapace not reaching beyond level of antennal spine. Second leg with carpus broader than long. Dorsal surface porcelaneous white with fine longitudinal stripes, sides dark brown. Common (but almost invisible owing to its color pattern) between the spines of the urchin *Tripneustes ventricosus*.

***Gnathophyllum americanum*** Guérin-Ménéville (Bumblebee shrimp): Posterior tooth of dorsal series on rostrum placed anterior to level of posterior margin of orbit. Antero-lateral angle of carapace distinctly overreaching (antennal) spine at ventral orbital angle. Second leg with 5th segment (carpus) distinctly longer than broad. Walking legs with dactyl terminating in 2 teeth. Color pattern composed of dark brown and cream transverse stripes. Sometimes associated with sea urchins (*Lytechinus*) and sponges. Littoral zone to 50 m; rare.

**Plate 105**

F. **ALPHEIDAE** (Snapping shrimps): Caridea with rostrum much reduced and not dentate. Eyes usually partially or entirely covered by carapace, with little freedom of lateral movement. Third maxillipeds rarely expanded to form operculum over anterior mouthparts. Second pair of legs with 5th segment (carpus) subdivided, considerably weaker than 1st pair. (More than 24 spp. from Bda.)

***Alpheopsis labis*** Chace: Genus with triangular movable plate or scale at posterolateral angle of 6th abdominal somite lateral to basal segment of tail fan (uropod). Eyes largely covered by carapace but not concealed from anterior view. Epipods on at least 2 anterior pairs of legs.—Species with anterior margin of carapace tapering to single median (rostral) tooth. Carapace smooth, without carinae. First legs unequal, chelae not triangular in

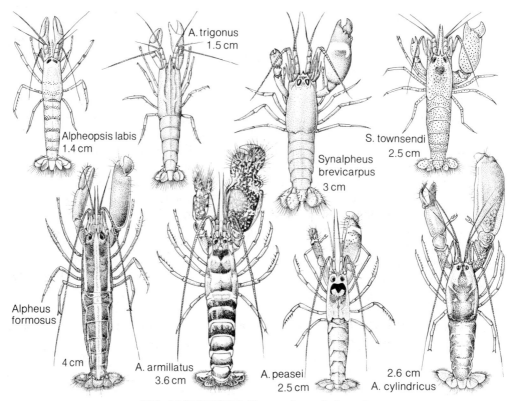

**105 ALPHEIDAE (Snapping shrimps)**

cross section. Color white-opaque, sometimes with transverse red stripes. Littoral and sublittoral, in and under coral rubble; not uncommon. (Color Plate 9.4.)

***A. trigonus*** (Rathbun): Genus as above.—Species with anterior margin of carapace tridentate. Carapace tricarinate dorsally. First legs subequal, chelae triangular in cross section. Color white opaque, sometimes with transverse red stripes. Littoral and sublittoral, in and under coral rubble; not uncommon. (Color Plate 9.2.)

***Synalpheus brevicarpus*** (Herrick): Genus without a movable plate or scale at postero-lateral angle of 6th abdominal somite lateral to basal segment of tail fan (uropod). Eyes covered by carapace and concealed from all but antero-ventral view by deflexed front. Movable finger of major 1st chela with molar-like tooth fitting into socket in fixed finger. Movable finger of minor 1st chela without prominent fringe of curved hairs on extensor surface. Legs without epipods.—Species with anterior margin of carapace armed with 3 subequal, acutely triangular teeth. Overall color yellowish; chelae orange. May occupy any habitat that offers protective cover, such as sponges, eroded dead coral, worm rock, abandoned gastropod shells, and cover provided by stones and coral on grass flats, to a depth of 68 m.

**S. *townsendi*** Coutière: Genus as above.—Species with anterior margin of carapace armed with 3 prominent teeth, lateral pair much broader than median and tapering to slender, sharp tips. Whitish, with many small reddish-brown spots. May be found in numerous habitats offering concealment, such as grass flats with clumps of *Porites*, but especially in eroded dead coral, to a depth of 100 m.

***Alpheus formosus*** Gibbes: Genus without a movable plate or scale at postero-lateral angle of 6th abdominal somite lateral to basal segment of tail fan (uropod). Eyes covered by carapace and concealed from all but antero-ventral view by deflexed front. Movable finger of major 1st chela with molar-like tooth fitting into socket in fixed finger. Epipods on at least 2 anterior pairs of legs.—Species with anterior margin of carapace armed with prominent flattened rostrum and spine on each orbital hood. Paired dorsal longitudinal (adrostral) grooves partially delimited posteriorly. Major chela entire, without marginal notch or sinus. Third and 4th legs with dactyl simple, not biunguiculate; 4th segment (merus) without distal tooth on flexor margin; and 3rd segment (ischium) bearing movable spine on lateral surface. Carapace dark brown, with light brown stripe along dorsal midline; legs blue. May be found in any habitat offering concealment, from intertidal to 40 m.

**A. *armillatus*** H. Milne Edwards: Genus as above.—Species with anterior margin of carapace with single (rostral) point supported by median ridge. Paired dorsal longitudinal (adrostral) grooves usually abruptly delimited posteriorly. Ventral margin of major chela with marginal notch or sinus. Third and 4th legs with dactyl simple, not biunguiculate; 4th segment (merus) armed with distinct distal tooth on flexor margin; and 3rd segment (ischium) bearing movable spine on lateral surface. With brown-olive transverse stripes; chelae usually with white spots. Occurs in various habitats, such as under stones and oyster bars, in the interstices of coral rock and on grass flats, in littoral and sublittoral depths. (Color Plate 9.5.)

**A. *peasei*** (Armstrong): Genus as above.—Species with anterior margin of carapace with rostral point supported by median ridge and spine on each orbital hood. Paired dorsal longitudinal (adrostral) grooves not abruptly delimited posteriorly. Ventral margin of major chela with marginal notch or sinus. Third and 4th legs with dactyl biunguiculate; 4th segment (merus) armed with distinct distal tooth on flexor margin; and 3rd segment (ischium) without movable spine on lateral surface. Coloration pale green, with characteristic white mark behind eyes. Commonly found on, under, or in the interstices of rocks and dead coral, occasionally in sponges, to a depth of 7 m or more.

***A. cylindricus*** Kingsley: Genus as above.—Species with anterior margin armed with 3 broad flat teeth. Paired dorsal longitudinal (adrostral) grooves lacking. Major chela smoothly rounded, without marginal notch or sinus. Third and 4th legs with dactyl biunguiculate; 4th segment (merus) without distal tooth on flexor margin; and 3rd segment (ischium) without movable spine on lateral surface. Color pattern a vivid red on white; chelae reddish-brown. Frequents rough bottoms and often lives in canals of loggerhead sponges; littoral to a depth of 82 m. (Color Plate 9.7.)

## Plate 106

F. **HIPPOLYTIDAE:** Caridea with variable rostrum. Eyes exposed and freely movable. Third maxilliped not unusually broadened to form operculum over anterior mouthparts. Second pair of legs with 5th segment (carpus) subdivided into 2 or more articles, more slender than anterior pair. (12 spp. from Bda.)

***Hippolyte zostericola*** (Smith): Genus with 5th segment (carpus) of 2nd leg composed of 3 articles. Antennular peduncle without subtriangular movable plate overhanging base of flagellum. Third maxilliped with lash (exopod) arising from 2nd segment (basis). Rostrum not greatly expanded ventrally; supraorbital spine present.—Species with rostrum usually armed with 2 dorsal teeth. Carapace armed with single (antennal) spine on anterior margin, and a hepatic spine posteroventral to orbit. Fifth abdominal somite without spines on posterior margin. Walking legs with dactyl terminating in 3 long spines. Coloration often an intense green, eyes white. Usually found on grass flats (e.g., *Halodule*). (Color Plate 9.6.)

***H. coerulescens*** (Fabricius): Genus as above.—Species with rostrum usually with single inconspicuous tooth on dorsal and ventral margins. Carapace armed with branchiostegal spine, overreaching anterior margin. Fifth abdominal somite with a pair of strong spines on posterior margin. Walking legs with dactyl terminating in 2 long spines. Body banded with brownish yellow in such a way that it seems to be broken up into 2 parts, each of which looks like a vesicle of *Sargassum*—the substrate on which it is fairly common.

***Latreutes fucorum*** (Fabricius): Rostrum with deep ventral blade projecting posteriorly between bases of antennules, anterior margin subtruncate and spinulose. Carapace smooth except for single tooth in dorsal midline posterior to base of rostrum, (antennal) spine near lower margin of orbit, and few (branchiostegal) spines on antero-lateral margin; no (supraorbital) tooth above orbit. Antennular peduncle without subtriangular movable plate overhanging base of flagellum.

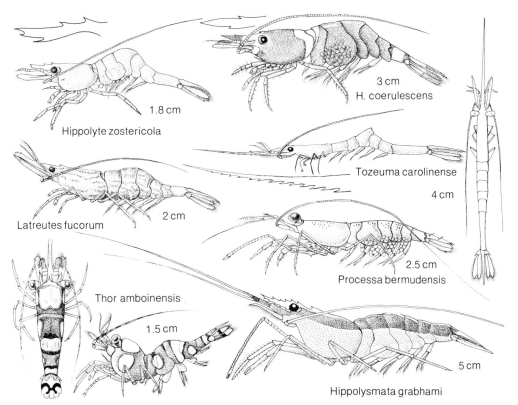

**106 HIPPOLYTIDAE, PROCESSIDAE (Shrimps)**

Antennal scale with blade tapering, not expanded distally. Third maxilliped with lash (exopod) arising from 2nd segment (basis). Fifth segment (carpus) of 2nd legs composed of 3 articles. Walking legs with dactyl terminating in 2 teeth. Brownish yellow with white markings. Common in floating *Sargassum*, but also found in shallow turtle grass areas.

***Tozeuma carolinense*** Kingsley (Arrow shrimp): Rostrum very long and unarmed dorsally, moderately expanded and dentate ventrally but ventral blade not projecting posteriorly between bases of antennules. Carapace smooth except for 2 (antennal and pterygostomian) spines on anterior margin and (supraorbital) spine above orbit. Antennular peduncle without subtriangular, movable plate overhanging base of flagellum. Third maxilliped without lash (exopod) arising from 2nd segment (basis). Fifth segment (carpus) of 2nd legs composed of 3 articles. Transparent green with white markings. Usually found on grass flats (e.g., *Syringodium*), where it often swims in a vertical position, as well as among alcyonarians, to a depth of 75 m. (Color Plate 9.9.)

***Thor amboinensis*** (De Man): Rostrum not overreaching eyes, not expanded ventrally. Carapace smooth and unarmed except for single (antennal) spine on anterior margin, without (supraorbital) spine above orbit. Antennular peduncle bearing subtriangular, movable plate overhanging base of flagellum. Third maxilliped with lash (exopod) arising from 2nd segment (basis). Fifth segment (carpus) of 2nd legs composed of 6 or 7 articles. Chocolate-brown, with large white patches, possibly mimicking the color pattern of the anemone *Lebrunea danae* with which it is usually associated (less commonly with *Condylactis gigantea*); shallow subtidal offshore. (Color Plate 9.11.)

*Hippolysmata grabhami* Gordon (Red-backed cleaner shrimp): Rostrum elongate, dentate, not expanded ventrally. Carapace smooth, armed with 2 (antennal and pterygostomian) spines on anterior margin, without (supraorbital) spine above orbit. Antennular peduncle without subtriangular, movable plate overhanging base of flagellum. Third maxilliped with lash (exopod) arising from 2nd segment (basis). Fifth segment (carpus) of 2nd legs composed of 17 to 23 articles. Walking legs with dactyl armed with series of prominent spines on flexor margin. With pairs of bright red longitudinal stripes and a median white stripe dorsally. Usually found in crevices and excavations in coral reefs to a depth of at least 21 m. They have been observed cleaning a variety of fishes, including moray eels. Very rare. (Color Plate 9.3.)

F. **PROCESSIDAE:** Caridea with short rostrum, unarmed except for subdistal dorsal tooth forming asymmetrically bifid, setose tip. Eyes exposed and freely movable. Third maxilliped not expanded to form operculum over anterior mouthparts. Second pair of legs with 5th segment (carpus) multiarticulate, more slender than 1st pair. (2 spp. from Bda.)

*Processa bermudensis* (Rankin): Carapace completely smooth and unarmed, without (antennal) spine on anterior margin below orbit. Lateral flap (pleuron) of 5th abdominal somite without distinct postero-lateral tooth. First pair of legs asymmetric, one chelate, the other simple, neither with lash (exopod) arising from 2nd segment (basis). Second legs asymmetric, right with 19-29 articles in 5th segment (carpus), left with 13-15. Active at night on grass flats in shallow water.

### Plate 107

I.O. **STENOPODIDEA:** Decapoda with lateral flap (pleuron) of 2nd abdominal somite not overlapping that of 1st. Third pair of legs chelate, much more robust than preceding pairs. ♂ without petasma on 1st pleopods. Gills trichobranchiate (filamentous). (Only family: Stenopodidae; with 2 spp. of *Stenopus* from Bda.)

*Stenopus hispidus* (Olivier) (Banded coral shrimp): Rostrum unarmed ventrally. Carapace and abdomen covered with forwardly curved erect spines. Antennal scale with 2 or 3 rows of spinules on dorsal surface.

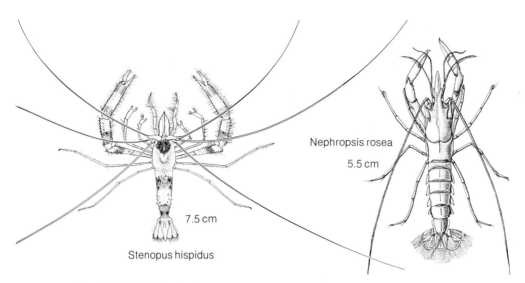

**107 STENOPODIDEA (Coral shrimps), ASTACIDEA (Lobsters)**

Mesial branch of tail fan (uropod) with 2 dorsal ridges. Red and white banding on carapace and 3rd pair of legs. Commonly occurring in pairs in shaded coral crevices and caves, usually on the ceiling near the entrance, also in any large holes or man-made objects such as automobile tires and buckets, frequently in 2-4 m (but recorded to a depth of 120 m). This is the largest of the known clean shrimps. It does not climb onto the fishes that it cleans but it advertises its presence by waving its white antennae and picks at fishes that are thus lured into its cave. (Color Plate 9.1.)

**I.O. ASTACIDEA:** Decapoda with subcylindrical carapace; rostrum and abdomen well developed. Frontal portion of carapace not fused with epistome. Antennae with 5-segmented stalk and scale. Third maxilliped pediform. First 3 pereopods chelate. Abdominal pleura well developed. Genital openings coxal.

**SUP.F. NEPHROPOIDEA** (True lobsters): Astacidea with carapace with dorso-median longitudinal suture or with simple or spinous ridge extending from caudal margin of carapace at least to base of rostrum. Sternal plate between 5th pereopods fused with that between 4th pair. Podobranchiae of first 3 pereopods with discrete branchial and epipoditic parts. First pleopod of ♂ without individual sperm conduit. (Of the several families which make up this superfamily—including the true lobsters and crawfish—only one, Nephropidae, is represented, with 1 rare deep-sea species.)

*Nephropsis rosea* Bate: Rostrum with 1 pair of lateral spines. A post-supraorbital spine present behind each supraorbital spine. Carpus of 2nd cheliped shorter than palm; carpus of 3rd cheliped half as long as chela. Color pale red or orange-red, ventral surface darker, appearing two-tone. Originally described from off Bermuda in 1,262 m. Usually in depths between 550 and 750 m; rare.

**I.O.** **PALINURA:** Decapoda with cylindrical or dorso-ventrally compressed carapace; rostrum not prominent but commonly spinose, fused laterally with epistome. Antennal stalk with 5 segments. Maxilliped pediform. Abdomen well developed.

**SUP.F.** **PALINUROIDEA:** Palinura with cylindrical or dorso-ventrally compressed carapace, mostly without projecting rostrum. Base of antenna fused with epistome and lateral carapace margin. No antennal scale. First 4 pereopods without chelae. Abdominal pleura well developed. Telson and uropods only partly calcified.

### Plate 108

**F.** **PALINURIDAE** (Spiny lobsters): Palinuroidea with cylindrical carapace without lateral keels. Antennal flagella long and strong. Rostrum usually absent. Strong supraorbital spines usually present. (4 spp. from Bda.)

*Justitia longimanus* (H. Milne Edwards) (Long-armed spiny lobster): Supraorbital spines serrate dorsally. First legs elongate and heavier than remainder (especially in males), with dactylus closing against a projection of the propodus. Abdomen with transverse grooves, not squamiform sculpture. First pereopods conspicuously banded red and white, body brick-red, variously striped and spotted with yellow. On reef slopes in depths beyond 20 m; rare. (Color Plate 10.5.)

*Panulirus argus* (Latreille) (Spiny lobster, B Bermuda lobster): Genus with dorsally rounded carapace covered with numerous spines and nodules. Rostrum absent.—Species with transverse dorsal groove on each abdominal somite. Color very variable, tail with 4 major yellow spots, 2 on each side of 2nd and 6th abdominal somites. Legs longitudinally striped. In rocks and on reefs, low tide to about 90 m; usually in shallow water. An important food item, the Bermuda lobster is protected from any fishing during part of the year.

*P. guttatus* (Latreille) (Spotted spiny lobster, Star lobster, Guinea chick lobster): Genus as above.—Species with transverse dorsal groove on each abdominal somite. Color green, blue or brown, covered with cream spots, latter extending onto legs; propodus of legs longitudinally striped. In shallow water on outer reefs and jetties, and in caves; locally common. (Color Plate 10.11.)

*P. laevicauda* (Latreille) (Smooth-tailed spiny lobster): Genus as above.—Species without transverse grooves on abdominal somites. Color variable, background green, yellow or shades of purple, each abdominal somite with posterior line of small white spots, tail fan spotted, legs longitudinally striped. In deeper water, between 25 and 50 m; rare.

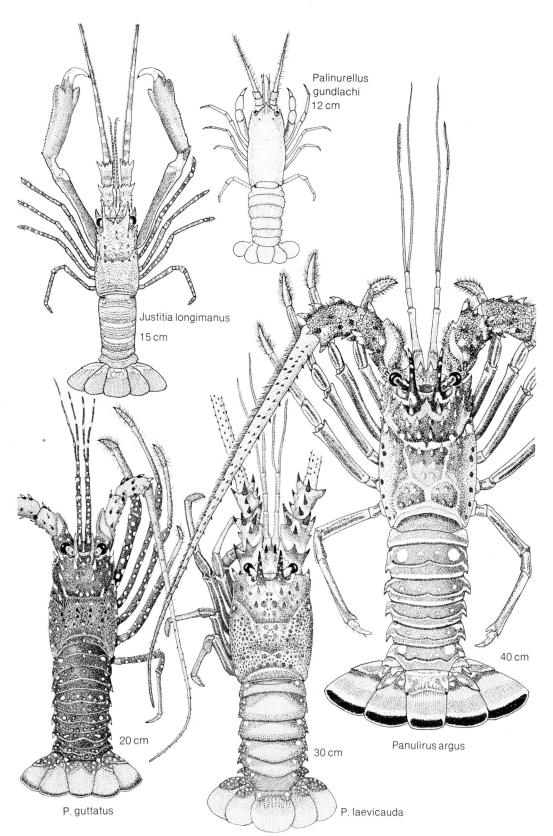

**108 PALINURIDAE (Spiny lobsters), SYNAXIDAE**

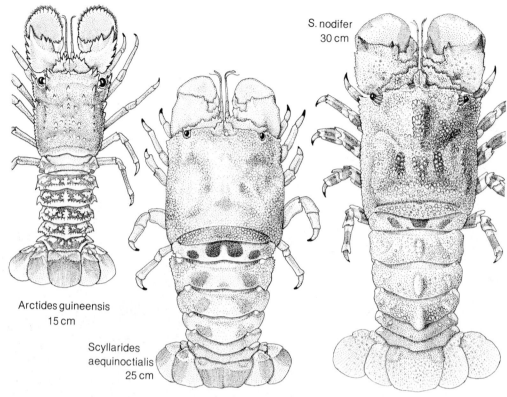

**109 SCYLLARIDAE (Locust lobsters)**

F. **SYNAXIDAE:** Palinuroidea with subcylindrical, slightly flattened carapace, with distinct median rostrum, lacking dorsal spines or ridges. Antennae slender. Eyes in orbits, lacking strong supraorbital spines. (1 sp. from Bda.)

*Palinurellus gundlachi* von Martens: Body hairy, lacking spines or tubercles. First legs largest. Color orange. Usually found on or near reefs, in depths to about 35 m; very rare.

**Plate 109**

F. **SCYLLARIDAE:** Palinuroidea with dorso-ventrally more or less flattened carapace with sharp lateral margins. Antennae short, flagella replaced by plates with dentate or lobulate margins. Orbits in anterior margin; no supraorbital spines. (3 spp. from Bda.)

*Arctides guineensis* (Spengler): Anterior margin of antennae and lateral margin of carapace cut into teeth. Abdomen conspicuously sculptured, with scalloped transverse grooves. Color brownish to bright red, irregularly mottled with yellow. On outer reefs; rare. (Color Plate 10.6.)

*Scyllarides aequinoctialis* (Lund): Genus with forward edge of antenna and lateral margins of carapace smooth, not cut into teeth.—Species with 2nd, 3rd and 4th abdominal somites round, without distinct median carina. First abdominal somite with 4 round, red spots, the middle 2 more or less joined together. Overall color terra-cotta, orange-brown or reddish brown, irregularly marked with red. From a few meters to deeper water, on reefs and rock ledges. Not uncommon.

***S. nodifer*** (Stimpson): Genus as above.—Species with 2nd through 4th abdominal somites with a distinct median ridge. First abdominal somite with 3 round, red spots, 1 on midline. Color very variable, cream to dull brown variously ornamented with red. Off outer reefs, in deeper water.

I.O. **ANOMURA:** Decapoda with carapace not fused with epistome. Last thoracic sternite free. Third maxilliped narrow. Third pereopods lacking chelae, 5th modified in shape and position. Abdomen not strongly calcified, mostly reduced in length.

## Plate 110

SUP.F. **THALASSINOIDEA** (Mole shrimps, Mud shrimps): Anomura with well-developed abdomen but more or less reduced pleura. First pereopods usually chelate.

F. **AXIIDAE:** Thalassinoidea with rostrum and cervical groove on carapace; without linea thalassinica. Antennular flagella well developed. First pereopods with large chelae, 2nd with small chelae. (1 sp. from Bda.)

***Axiopsis serratifrons*** A. Milne Edwards: Rostrum triangular, toothed laterally. Eyes well developed, free, pigmented. Antennular spine long. Chelae unequal, asymmetric, fingers shorter than palm. Telson with spinulose dorsal carinae. ♂ chestnut-brown to black, ♀ pale brown or olive. Common in burrows without mounds on clean coarse sand in reef areas. (Color Plate 10.1.)

F. **CALLIANASSIDAE:** Lobsterlike Thalassinoidea with a long, cylindrical, weakly calcified body. Cervical groove and linea thalassinica usually present. Rostrum flattened, usually a blunt projection on midline. Eyes small, pigment reduced. Antennal peduncle with 5 joints. Antennal scale usually vestigial or absent. First pereopods usually unequal, greatly enlarged, carpus and palm of stronger leg usually enlarged. Second pereopods smaller, equal, chelate. Third pereopods simple. Abdomen greatly elongated. Tail fan well developed. Uropods lacking transverse broad appendages on posterior 3 abdominal segments. (2 spp. from Bda.)

***Callianassa branneri*** (Rathbun): Genus with short claws; first 2 pairs of pleopods unlike subsequent 3 pairs.—Species with 3 rounded projections on front. Ischium and merus of major cheliped serrate ventrally.

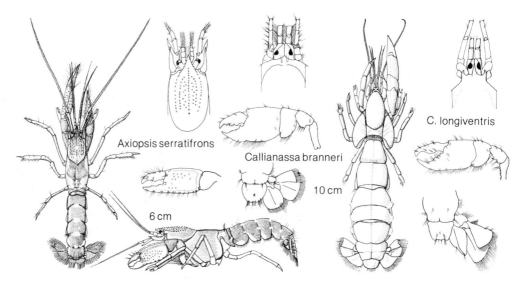

**110 THALASSINOIDEA (Mole shrimps)**

Uropodal endopod squarish. Body off-white, sometimes with pinkish tinge. In burrows, intertidal and subtidal on sand beaches and grass flats. (Color Plate 10.10.)

*C. longiventris* A. Milne Edwards: Genus as above.—Species with 3 spines on front. Ischium and merus of major cheliped spinous ventrally. Uropodal endopod triangular. Brightly colored, with orange and dark brown on chelae, articular membranes red. Pereopods yellow-orange. In burrows on sand in grass beds near reefs, in depths of 2-5 m.

## Plate 111

**SUP.F. COENOBITOIDEA** (Hermit crabs): Anomura with generally cylindrical or dorso-ventrally compressed body; abdomen usually well developed, symmetric or asymmetric; occasionally with abdomen reduced and with tergites moderately well calcified. Eyes usually well developed; antennal scale usually moderately well developed. First pereopods chelate, subequal, or left usually considerably larger than right; 2nd pereopods simple. Third maxillipeds approximate basally. Telson reduced, generally with transverse suture; uropods usually asymmetric, with well-developed rasps.

F. **COENOBITIDAE** (Land hermit crabs): Coenobitoidea with unequal chelipeds; fingers moving in vertical plane. Carapace moderately well calcified; rostrum reduced or absent. Antennular peduncles very long, flagella terminating abruptly and bluntly. Flagella of 2nd and 3rd maxillipeds reduced. (1 sp. from Bda.)

*Coenobita clypeatus* (Fabricius): Antennular peduncles 5× length of eyestalks, latter compressed laterally. Left cheliped massive. Left 2nd pereopod with propodus smooth and laterally compressed. Color of left chela varying from blue to bright purple, with dactyl yellow-orange or orange. On the high sand hills and in the woods some distance from the shores; uncommon in Bermuda, probably owing to the lack of suitable shells.

F. **DIOGENIDAE:** Coenobitoidea with equal or subequal chelipeds, or left considerably larger than right. Third maxillipeds with or without crista dentata but, when present, lacking accessory tooth. With few exceptions, no paired pleopods in either sex. (12 spp. from Bda.)

***Allodardanus bredini*** Haig & Provenzano: Left cheliped massive, dactyl and fixed finger corneous-tipped; palm covered with very small granules tipped with minute corneous spinules. Lateral faces of 2nd and 3rd pereopods covered with minute corneous-tipped spinules; ventral surfaces of each dactyl and propodus with row of corneous spines. Shield and cephalic appendages, 4th and 5th pereopods and coxae of first 3 pereopods bright yellow or yellow-orange; meri, carpi and propodi of chelipeds and ambulatory legs pink; dactyls yellow. Very common in deep water (100-200 m). (Color Plate 10.7.)

***Dardanus venosus*** (H. Milne Edwards): Left cheliped of genus considerably larger than right.—Species with lateral surface of palm bearing numerous scales, each with 1 to 6 tubercles; carpus with continuous fringe of setae along distal margin. Ventral margin of dactyl of left 3rd pereopod with rounded ridge and series of tubercles and adjacent tufts of setae proximally; lateral surface with vertical tuberculate ridges interrupted dorsally by shallow groove. Left chela red or occasionally deep crimson with red, pink or occasionally purple tubercles. Ambulatory legs with broad bands of orange-red on propodi, carpi and meri. In shallow reefs; uncommon.

***D. insignis*** (de Saussure): Genus as above.—Species with spinose ridges on lateral surface of palm. Lateral surface of dactyl and propodus of left 3rd pereopod without median longitudinal carina. Generally whitish, with transverse orange-red bands on chelae and legs; eyestalks with alternate red and white banding. In deep water (100-200 m); common. (Color Plate 10.9.)

***Calcinus tibicen*** (Herbst): Genus with left cheliped much larger than right; dactyl and fixed finger with calcareous tips.—Species with palm of left cheliped unarmed and devoid of setae. Propodus of left 3rd pereopod with longitudinal furrow on lateral face. Chelipeds red-brown to maroon, often also tinged with purple, fingertips white. Dactyls of ambulatory legs white or yellow. Lower intertidal (tide pools) and subtidal; not uncommon, but never particularly abundant.

***C. verrilli*** (Rathbun): Genus as above.—Species with row of spines on dorsal margin of palm of left cheliped. Propodus of left 3rd pereopod without longitudinal furrow on lateral face. General body color purple with red spots. A common shallow subtidal species that is endemic to Bermuda. May also be sendentary, occupying vermetid shells (MARKHAM 1977). (Color Plate 10.2.)

***Clibanarius anomalus*** A. Milne Edwards & Bouvier: Genus with equal chelipeds, and with dactyls of ambulatory legs shorter than propodi.—Species with ocular acicles widely separated basally. Eyestalks dorsally and antennular peduncles orange; ambulatory legs generally uniformly orange or yellow-orange. Very abundant in deep water (100-300 m).

***C. tricolor*** (Gibbes): Genus as above.—Species with ocular acicles approximating basally. Eyestalks blue; antennular peduncles orange. Dactyls orange proximally, white or yellow generally, with dark tips; proximal portions of propodi also orange, segments otherwise pale to dark blue. Extremely common in shallow water. Females are ovigerous V-X.

***C. antillensis*** Stimpson: Genus as above.—Species with ocular acicles approximating basally. Eyestalks greenish blue; antennular peduncles similarly colored, but with orange basally. Ambulatory legs with broad white stripe on the lateral faces of each of the 3 distal segments. Uncommon, usually occurring subtidally.

SUP.F. **PAGUROIDEA** (Hermit crabs): Anomura with generally dorsoventrally compressed body; abdomen well developed, asymmetric, and frequently coiled, or reduced, folded under cephalothorax. Eyes usually well developed; antennal scale usually moderately well developed. First pereopods chelate, subequal, right usually considerably larger than left; 2nd pereopods simple. Third maxillipeds widely separated basally. Telson reduced or absent; if present, with or without transverse suture; uropods, if present, usually asymmetric and with well-developed rasps.

F. **PAGURIDAE:** Paguroidea with usually unequal chelipeds, right considerably larger than left. Third maxillipeds with crista dentata, with or without 1 or more accessory teeth. Sometimes with paired appendages on the 1st or first 2 abdominal segments in the males or on the 1st abdominal segment in females. Sometimes males with sexual tubes. (7 spp. from Bda.)

***Pagurus brevidactylus*** (Stimpson): Rostrum broadly triangular, rounded or obsolete. Ocular peduncles moderately long, slender; ocular acicles with terminal margins produced into several spinose projections. Left chela with longitudinal row of strong spines on dorso-lateral face in proximity of dorso-lateral margin. Eyestalks mottled pink with

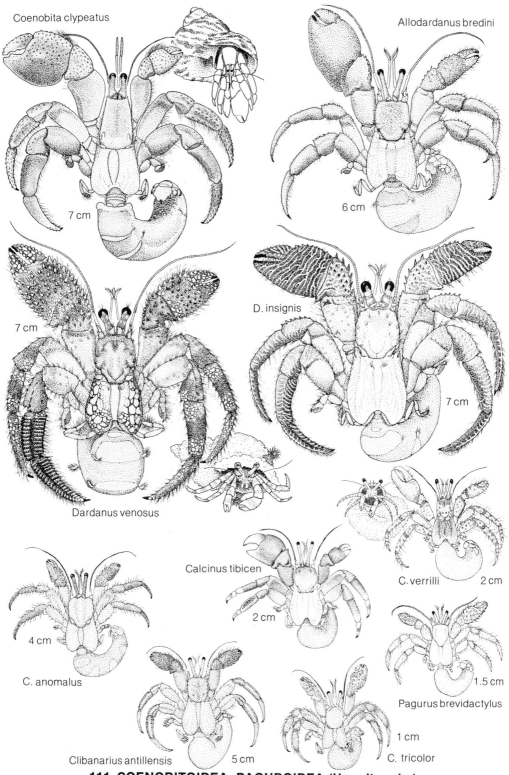

**111 COENOBITOIDEA, PAGUROIDEA (Hermit crabs)**

brown and green black stripes. Chelipeds white with dorsal longitudinal stripes of green or brown. Ambulatory legs white with dorsal, lateral, mesial and ventral longitudinal stripes of green or brown-black and darker broad transverse band medially on each segment; lateral faces also with median longitudinal stripe of red or red-brown. Not uncommon in low intertidal, particularly in exposed rocky areas where the water circulation is good.

## Plate 112

SUP.F. **GALATHEOIDEA:** Anomura with well-developed carapace, generally with distinct lateral edges and marked with linea anomurica; rostrum well developed. Eyestalks short and stout. First pereopods chelate. Abdomen more or less reduced, with well-developed pleura; telson subdivided by sutures.

F. **GALATHEIDAE:** Galatheoidea with shrimp-like body, carapace elongate; rostrum triangular or styliform. Abdomen curved ventrally but not bent under thorax. (3 spp. from Bda.)

*Munida simplex* Benedict: Carapace ovate, with transverse elevated, ciliated ridges, and with several marginal spines. Rostrum long, acute. Antero-lateral spine elongate, followed by 6 smaller spines. Pale yellow with pink markings. In deep water (100-200 m); rare.

F. **PORCELLANIDAE** (Porcelain crabs): Galatheoidea with crab-like body, carapace broad, regions poorly defined; frontal margin broad, often subdivided into 3 lobes; median or rostral lobe usually broad. Antennal peduncle usually directed posteriorly. Chelipeds broad and usually compressed. Abdomen bent under thorax. (1 sp. from Bda.)

*Petrolisthes armatus* (Gibbes): Carapace granulate and plicate; usually with 1 epibranchial spine. Anterior margin of carpus of chelipeds with 3 or 4 wide-set teeth. Ambulatory legs each with row of spines on anterior margin of merus. Color variable, usually mottled gray to brown; light blue animals common in the local population. Abundant near the low water mark under rocks. (Color Plate 10.3, 4.)

SUP.F. **HIPPOIDEA** (Mole or sand crabs): Anomura with ovate or square carapace, usually relatively smooth, with well-marked lateral edges. Rostrum small or wanting. Eyes lack distinct orbits. First pereopods simple or subchelate. Abdomen bent under thorax; uropods present.

F. **ALBUNEIDAE:** Hippoidea with subquadrangular carapace, without postero-lateral extensions. First pereopods subchelate and flattened; 2nd to 4th with distal segments curved and also flattened. Telson oval and foliaceous. (1 sp. from Bda.)

*Albunea paretii* Guérin: Carapace with 11 or 12 spines on each side of median rostral tooth.

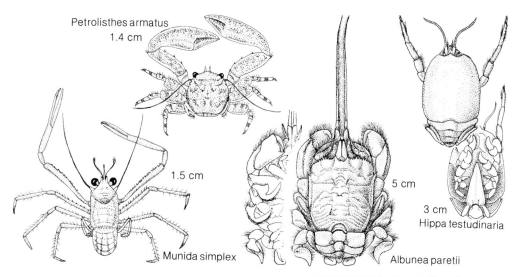

**112 GALATHEOIDEA (Porcelain crabs), HIPPOIDEA (Sand fleas)**

Antennular and antennal peduncles long; eyestalks also quite long. Burrowing in sandy beaches; rare.

F. **HIPPIDAE:** Hippoidea with ovoid carapace, with posterolateral extensions covering posterior pereopods. first pereopods simple. Telson lanceolate. (1 sp. from Bda.)

*Hippa testudinaria* (Herbst) (=*H. cubensis* (de Saussure)): Carapace broad, depressed; anterior margin sinuated, shallowly trilobate, median projection very short. Antennae and eyestalks short. White, with brown marble markings. Not uncommon; however, special effort is needed to catch specimens for they bury themselves in the sand at and below the low tide level. (Color Plate 10.8.)

I.O. **BRACHYURA** (True crabs): Short, compact Decapoda with epistome fused to carapace, and last thoracic segment fused with others. Symmetrical, very reduced abdomen permanently flexed under carapace, and not used in swimming; uropods rare and never biramous. Outer pair of mouthparts (3rd maxillipeds) broad; 1st pair of legs (pereopods) chelate and more conspicuous than the rest. Brachyura may be divided into 5 sections: Gymnopleura, Dromiacea, Oxystomata, Brachyrhyncha and Oxyrhyncha, all of which are represented in Bermuda.

*Plate 113*

SECT. **GYMNOPLEURA:** Brachyura with anterior part of sternum broad and tapering posteriorly. Female genital openings on coxae of 3rd pair of legs rather than on sternum. Last pair of legs located dorsally.

F. **RANINIDAE** (Frog crabs): Gymnopleura with elongated carapace not covering all of the flexed abdomen. Third maxillipeds long and narrow. Palm usually very flat and finger more or less deflected so that dactylus is usually applied against anterior border of palm. (1 sp. from Bda.)

***Symethis variolosa*** (Fabricius): Anterior half of carapace eroded. Chelae elongated with palms swollen (atypical for family!); fingers long and slender. Dactyls of walking legs sickle-shaped. Color white with splotches of light brown and pink. From shallow (9 m) to deep water (100 m). Probably confined to sandy bottoms offshore; rare.

SECT. **DROMIACEA:** Brachyura with carapace elongate to subglobose or subquadrate with narrow front. Last 1 or 2 pairs of legs dorsal or subdorsal and smaller than others. ♀ genital openings on coxae.

F. **DROMIIDAE** (Sponge crabs): Dromiacea with subglobular carapace, rarely flattened. Last pairs of legs short, subdorsal and with small hook-like nails. Sixth segment of abdomen generally with rudimentary uropods. (2 spp. from Bda.)

***Dromia erythropus*** (G. Edwards) (Large sponge-carrying crab): Carapace wider than long, with postero-lateral margins converging; covered with stiff dense hairs; 4 conspicuous conical antero-lateral spines on carapace, with low blunt tooth between 2nd and 3rd. Last 2 pairs of legs have articulating spines forming a chelate condition with dactyls. Hairs dark brown, surface of carapace whitish. Probably frequent in deeper water (50 m). Covers itself with living sponges held by the last 2 pairs of legs. Goose barnacles (*Lepas* sp.) have been found around mouthparts.

***Dromidia antillensis*** Stimpson (Small sponge-carrying crab): Carapace only slightly broader than long, with lateral margins of posterior 2/3 nearly parallel; covered with stiff dense hairs; 4 antero-lateral spines rather pointed; the posterior one located at lateral sulcus. Last pair of legs much longer than penultimate pair. Color highly variable, ranging from white or gray to shades of green, yellow or red; fingers with dark bases and white tips. No females reported. A rarely collected deep-water species. Carries living sponges, colonial ascidians or zoanthid polyps over its carapace.

SECT. **OXYSTOMATA:** Brachyura with mouth region triangular in outline, prolonged forward to form a gutter; excurrent branchial channels at middle of mouth region.

F. **LEUCOSIIDAE:** Oxystomata with circular, oval or polygonal carapace. External mouthparts completely enclosing mouth cavity. Incurrent branchial openings on either side of endostome (i.e., to sides of mouth area). Eyes and

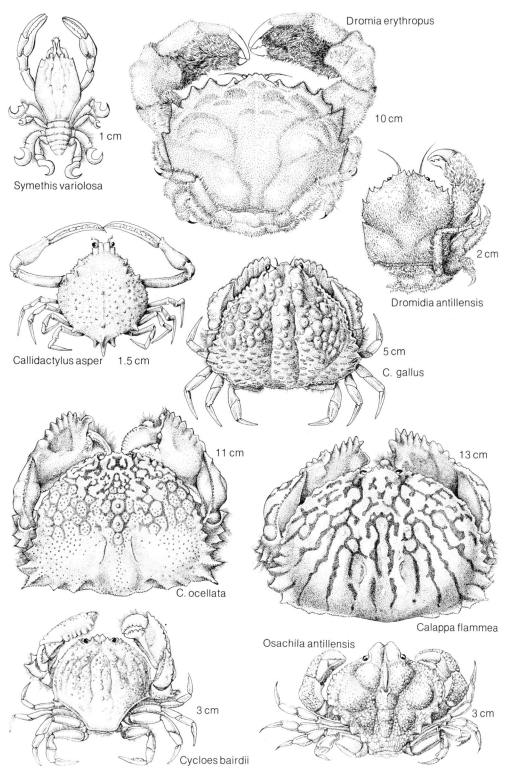

**113 GYMNOPLEURA (Frog crabs), DROMIACEA (Sponge crabs), OXYSTOMATA (Box crabs)**

orbits very small. Chelipeds Sequal. (2 spp. from Bda.)

***Callidactylus asper*** Stimpson: Carapace rounded, nearly as broad as long; posterior half with 7 spines, and anterior half with 3 on either side. Color white. On patch reefs; rare. (Color Plate 11.9.)

F. **CALAPPIDAE** (Box crabs): Oxystomata with carapace usually distinctly wider than long. Outer mouthparts not completely closing the buccal cavity. Incurrent branchial openings in front of bases of chelipeds. (6 spp. from Bda.)

***Calappa flammea*** (Herbst) (Flaming box crab): Genus with postero-lateral region of carapace expanded and dentate.—In the species, these wing-like projections not distinctly set off from antero-lateral margin; teeth not very acute. Surface of carapace lightly granular. Carapace gray to pale yellowish, with purplish brown interlacing bands on anterior half becoming oblique longitudinal stripes posteriorly. Buried in sand of shallow, sheltered bays, but probably more common in deeper water; inconspicuous as only anterior edge projects above sand. Largest species of the genus and the most common. May be infected with intestinal Sporozoa (BALL 1951). Ovigerous crabs never recorded. (Color Plate 11.13.)

***C. ocellata*** Holthuis (Ocellated box crab): Genus as above.— Species with postero-lateral wing-like projections of carapace set off from the antero-lateral margin; teeth with slender pointed tips. Granulation of carapace fairly coarse. Anterior half of carapace red with large number of white, often ocellated spots and a few white streaks, forming a compact reticulated red pattern. Buried in sand; apparently uncommon.

***C. gallus*** (Herbst) (Yellow box crab): Genus as above.—Species with postero-lateral wing-like projections of carapace not distinctly set off from antero-lateral margins; postero-lateral teeth blunt; a deep hollow on either side of mid- region in anterior 1/3. Posterior 1/3 of carapace covered with short transverse granulated lines; anterior 2/3 with circular, tuberculated swellings. Upper parts of body and legs orange to orange- brown, becoming brighter on front of chelae; under parts dull yellow. In shallow sandy bays and in deeper water.

***Cycloes bairdii*** Stimpson: Carapace somewhat heart-shaped, slightly broader than long, with a small postero-lateral spine behind middle of carapace. Pale yellow or whitish with lemon-yellow spots in irregular rows, and many small bright red spots laterally. Chelipeds and walking legs bright yellow, spotted and banded with scarlet. In sandy shallow bays.

***Osachila antillensis*** Rathbun: Carapace angular, surface uneven with distinct broad protuberances, front protruding to form a bilobed rostrum, and postero-lateral margins thickened with rounded lobes. Yellowish white, with light rusty splotches on carapace and same color bands on legs; tips of fingers chocolate brown. A deep-water species, commonly collected in baited traps, at 200-360 m. Ovigerous females not observed, but crabs recorded with well-developed ovaries and sperm in receptacles. (Color Plate 11.3.)

SECT. **BRACHYRHYNCHA**: Brachyura with anteriorly broad carapace with rostrum absent of very reduced; shape oval, round or squarish, usually broader than long. Orbits mostly complete.

## Plate 114

F. **PORTUNIDAE** (Swimming crabs): Brachyrhyncha with usually transversely oval carapace, with 5 to 9 anterolateral teeth (includes outer orbital and lateral spine); widest at the last antero-lateral tooth (lateral spine). Last pair of legs usually adapted for swimming, the dactylus broad and flattened into an oval paddle. (15 spp. from Bda.)

*Portunus anceps* (de Saussure) (White crab): Genus with 9 antero-lateral teeth on carapace, the first 8 of which are roughly of uniform size. A distinct medio-distal spine on the wrist (carpus) of cheliped.—Species with antero-lateral margins forming an arc of a circle with center near the posterior margin; carapace pubescent with several indistinct, arching, granulate transverse ridges; 6 frontal teeth, including inner orbitals, latter blunt and shorter than next pair; submedian teeth short and smaller than inner orbitals; antero-lateral teeth small, acute, curved forward, with lateral spine sharp, slender and about as long as space occupied by 4 preceding teeth. Carapace white with light brown mottling; undersides of chelipeds with light brown stripe running from base to near tip of immovable finger; appendages other than chelipeds have little pigment. On sandy bottom of shallow bays. Ovigerous crabs in IV and IX-X, which probably bracket a summer spawning season.

*P. sayi* (Gibbes) (Sargassum swimming crab): Genus as above.—Species with antero-lateral margins forming an arc of a circle with center near the posterior margin of carapace; the latter very convex, smooth and glossy; 6 frontal teeth, including inner orbitals; 2 submedian teeth smaller but on a line with next pair; most antero-lateral teeth relatively blunt and not curved forward. Chocolate brown or purplish with olive-green or brown shading and irregular white spots; orange margins on spines of chelipeds. The most common portunid among the *Sargassum* fauna (MORRIS & MOGELBERG 1973). Most specimens in this algal community are juveniles. Ovigerous crabs found VII-IX. (Color Plate 11.1).

*P. ordwayi* (Stimpson) (Silvery-clawed crab): Genus as above.—This species and the three that follow (*P. depressifrons, spinimanus* and *sebae*) have an-

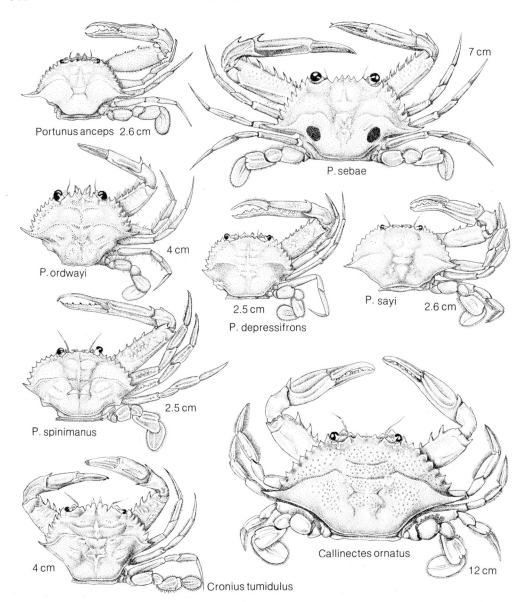

**114 PORTUNIDAE (Swimming crabs)**

tero-lateral margins of carapace forming an arc of a circle with center near the middle of cardiac region (i.e., not near posterior margin); elevations granulate and depressions pubescent, with conspicuous, curving, transverse ridges; 6 frontal teeth with middle pair deeply separated and advanced beyond others; lateral spine about as long as space occupied by 2 preceding teeth. Smooth silvery or iridescent area on the outer surface of chelipeds. Carapace and legs finely variegated and mottled with red, yellowish brown and gray, producing a general reddish brown effect; pale orange beneath; deeper orange on chelipeds and legs, and chelae deep red-brown above; hairs on appendages deep red. Probably frequent on sandy bottoms of shallow bays.

**P. depressifrons** Stimpson: Genus as above.—Species with antero-lateral margins of carapace forming an arc of a circle with center near the middle of cardiac region; carapace uneven, pubescent, and with indistinct transverse ridges; 6 frontal teeth, with submedian teeth somewhat smaller and slightly more advanced than next pair, and inner orbitals a little behind line of other 4 teeth; lateral spine scarcely longer than 1 in front. Carapace irregularly mottled with sand colors (light and dark gray); chelipeds and swimming legs same but paler; first pair of walking legs bright purple or deep blue, next 2 pairs lighter. One of the most common members of the genus, living on sand in shallow water. Ovigers in IV.

**P. spinimanus** Latreille: Genus as above.—Species with antero-lateral margins of carapace forming an arc of a circle with center near the middle of cardiac region; carapace finely granulate and pubescent, with prominent, curved, coarsely granulate transverse ridges; 8 frontal teeth; lateral spine about $2\times$ longer than other teeth and usually curved forward. Yellowish or reddish brown pubescence; ridges of carapace, spines of chelipeds, fingers and tips of legs reddish brown; antero-lateral teeth reddish at base and white at tips. Usually reported as juveniles within *Sargassum* community, although not common.

**P. sebae** (H. Milne Edwards) ("Four-eyed" swimming crab): Genus as above.—Species with antero-lateral margins of carapace forming an arc of a circle with center near the middle of cardiac region; with a large, round, red spot on each postero-lateral slope; 2 submedian frontal teeth more prominent than the outer 2 pairs; antero-lateral teeth deeply cut and curving forward; lateral spine tends to curve forward. An erect spine on the base (basis) and a postero-distal spine on arm (merus) of swimming legs. Carapace and legs lightly but conspicuously pubescent; legs, chelipeds and spines fringed with larger red hairs; fingers red. Probably uncommon; on sand in shallow water.

***Cronius tumidulus*** (Stimpson): Antero-lateral teeth unequal, alternating large and small. Body and legs brick-red to purplish. In shallow water on sandy bottom; probably uncommon.

***Callinectes ornatus*** Ordway: Genus without an internal spine on the carpus of chelipeds; abdomen of males broad at base, narrow distally, and roughly an inverted T in shape.—Species with submedian pair of frontal teeth small or somewhat rudimentary; next lateral pair more prominent. Antero-lateral teeth (exclusive of outer orbital and lateral spine) progressively more pointed laterally; first 5 teeth with posterior margins longer than anterior margins; last 2 teeth with margins approximately equal in length. Lateral spine trends forward, about as long as space occupied by 3 preceding teeth. Adult ♂ with dull olive to dark brown carapace, usually with a large, poorly defined roundish spot of orange or orange-red on each side posteriorly. Lateral spines and antero-lateral teeth maroon, light blue or whitish, with white tips. Chelipeds proximally similar to carapace, spotted with blue or purple, joints red, fingers mostly purple, tipped with red. Other legs bright blue above, red at joints. ♀ similar, except upper surface of chela more violet; fingers with white or fuchsia colored teeth. Of 5 species of this genus found in Bermuda, this is the only common one. On sandy bottom in shallow water; easily caught with dip net at low tide in small enclosed bays.

### Plate 115

F.  **XANTHIDAE** (Mud crabs): Brachyrhyncha with transversely oval or hexagonal carapace, almost always broader than long; front broad; anterior margin of buccal frame not covered by external mouthparts; antennules folding obliquely or transversely. Habitats diverse. (32 spp. from Bda.)

*Micropanope spinipes* (A. Milne Edwards): Antero-lateral borders shorter than postero-lateral borders; 3rd and 4th anterolateral teeth larger than 1st, 2nd and 5th (lateral spine). Chelipeds unequal, with distinct spines on inner edge of wrists; major chela with large basal tooth on movable finger. Walking legs very slender. Preserved ♂ pale buff above and on legs; fingers of chelae brown, with color on immovable finger not running onto palm. Probably relatively common, in shallow bays living in crevices of old corals. Ovigerous ♀ in VI.

*Actaea setigera* (H. Milne Edwards) (Hairy crab): Carapace wide, ovoid, regions well defined; covered with short, stiff, yellow hair and granules; antero-lateral border divided into 4 lobes that do not project beyond the general outline of the carapace. Chelipeds hairy and granulose; fingers black, color extending onto palm of full-grown ♂, but not ♀. Body and legs are reddish brown to purplish red. Under stones in shallow bays and on patch reefs to 50 m; common. (Color Plate 11.2.)

*Platypodia spectabilis* (Herbst) (Calico crab): Surface of carapace lobulated and granulated, with unique color pattern. Hands of chelae compressed and surmounted by a sharp crest. Carapace dusky orange to deep orange-red; legs and undersides orange. Irregular paired patches of pale orange to cream, each bordered by black or dark purple; black spots within the patches or small black-bordered light circles within patches. Fingers of chelae chocolate brown. Found frequently in bays and on patch reefs, among rocks, corals

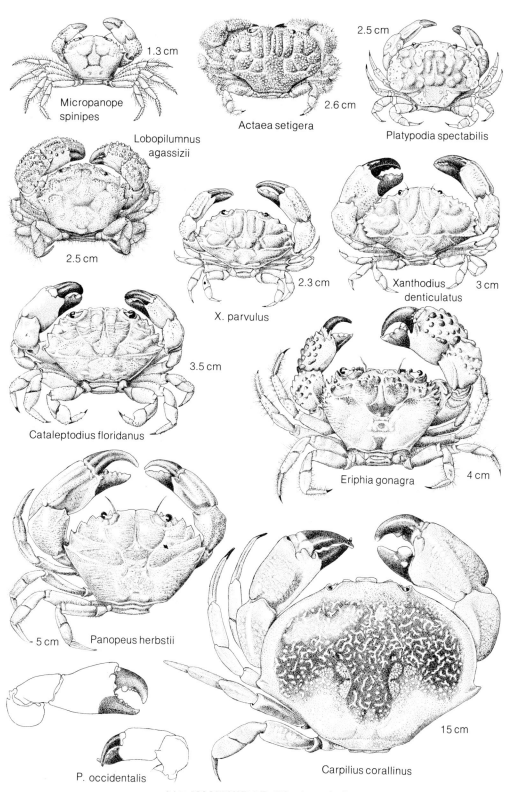

## 115 XANTHIDAE (Mud crabs)

and sponges, a colorful background into which it blends perfectly. (Color Plate 11.11.)

***Lobopilumnus agassizii*** (Stimpson): Carapace swollen and deeply lobulate anteriorly; front with 2 rounded lobes; antero-lateral margin cut into 3 large teeth behind the orbital tooth; they are roughened by numerous spinules or tubercles. Chelae with sharp tubercles pointing forward, arranged largely in longitudinal lines. Carapace, cleaned of calcareous silt, is tannish with brown patches; ventral side whitish with similar patches; brown bands on legs. Brown fingers, color not running onto palm. Common; found under stones and dead corals on reefs to at least 30 m, and close to rocky shores. Ovigerous crabs in IV-V and X.

***Xanthodius denticulatus*** (White): Genus with antero-lateral margin lobate or dentate; orbits subcircular.—Species with 9 (or 10) antero-lateral teeth behind outer orbital; antero-lateral margin continued well behind widest part of carapace; surface deeply sculptured. Fingers not spooned. Carapace reddish, purplish red or salmon; variable pattern of darker patches. Undersides whitish with some pale brown spots on abdomen. Black color of fingers extending slightly onto palm. Probably fairly common, under stones in shallow bays and among dead corals on reefs. LEBOUR (1944) found ovigers in VII, from which she obtained the first zoea.

***X. parvulus*** (Fabricius) (= *Leptodius americanus*): Genus as above.—Anterior 2/3 of carapace with deep grooves outer orbital tooth low and inconspicuous; remaining antero-lateral teeth broad, shallow, with subacute tips; last tooth directed outward or a little backward. Wrist and upper part of hand granulate, with longitudinal ridges on hand; fingers spooned. Color tan, mottled with brown; fingers black with color extending onto palm. Not common; in lower intertidal under stones. Ovigerous in VI- VII.

***Cataleptodius floridanus*** (Gibbes) (= *Leptodius floridanus*): Anterior 2/3 of carapace with regions well defined by deep grooves, and antero-lateral borders cut into 4 strong teeth in this genus.—In the species, a transverse, raised, granulate line runs inward across branchial region just behind tip of last anterolateral tooth. Fingers spooned at tips. Color tan, mottled with brown. Fingers brown to black with white tips and teeth; color extending onto palm. Usually in shallow water under stones or in crevices of dead coral on reef; often in lower intertidal under stones, in association with *Panopeus herbstii*; Common. Ovigerous crabs IV-VIII; mature ovaries in X.

***Eriphia gonagra*** (Fabricius) (Warty crab): Antero-lateral borders of carapace much shorter than postero-lateral, and meeting the latter at an open angle; orbits widely separated from antennae; carapace roughened anteriorly and antero-laterally with tubercles. Hands and wrists with large rounded tubercles; major dactylus with large rounded tooth at base. Brightly colored but variable; carapace and legs with dark markings on a light yellow background; bright yellow around external maxillipeds; tubercles of chelipeds and fingers chocolate brown. Common species, living under loose stones and dead masses of coral on reefs and in tide pools on rocky shores. Ovigerous crabs observed in summer and autumn.

***Panopeus herbstii*** H. Milne Edwards (Common mud crab): Genus with moderately wide and moderately convex carapace; 5 distinct antero-lateral teeth, the first 2 of which may be coalescent; large basal tooth on dactylus of major chela.—Species with color of immovable finger extending onto palm. One dark red rounded spot on the inside of each maxilliped. Generally brownish in color dorsally and mottled with various shades of brown; lighter below. Fingers of chelae black. Largest and most common member of the genus in Bermuda; usually in burrows under stones in lower half of intertidal zone, also subtidally under rocks and dead corals. An intestinal sporozoan is parasitic in this species (BALL 1951).

***P. occidentalis*** de Saussure: Genus as above.—Species with groove parallel to distal margin on wrist of chelipeds. Dark color of immovable finger not continued onto palm. One light red elongated spot on the inside of each external maxilliped. Color variable, generally brownish mottled dorsally, lighter below. Usually more subtidal than *P. herbstii*; under stones and dead coral in bays at 1-3 m; occasionally intertidal. A host for an intestinal sporozoan (BALL 1951). Ovigerous in IV and VIII.

***Carpilius corallinus*** (Herbst) (Coral crab, Queen crab): No antero-lateral teeth except for small blunted lateral spines. Major chela with large molar-like tooth on immovable finger. Carapace brick-red, with fine white or yellowish meandering lines; underparts yellowish; legs veined with brown; fingers of chelae brown. A relatively deep-water species, caught in fish traps. Not common.

## Plate 116

F. **GRAPSIDAE** (Shore crabs): Brachyrhyncha with usually more or less quadrangular carapace; lateral borders either straight or convex; orbits at or near antero-lateral angles; front usually quite broad and depressed in all Bermudian species except *Percnon* and *Plagusia*. Fast-moving crabs, living among rocks or other shelter in the intertidal to shallow subtidal, and also in floating *Sargassum*. Best known major group of brachyurans in Bermuda; all breed here. (13 spp. from Bda.)

***Cyclograpsus integer*** H. Milne Edwards: Carapace more than 4/5 as long as wide and somewhat rectangular; regions poorly marked and almost smooth; no antero-lateral teeth. Tan to light brown with anterior half of carapace darker; legs banded with tan. Common but not abundant; in the high intertidal (splash) zone, under rocks or beach wrack. Not as agile or pugnacious as most grapsids. Ovigers not recorded but no doubt breeds; females with mature gonads and full seminal receptacles observed in II, X and XII.

***Planes minutus*** (L.) (Gulf weed crab, Sargassum crab): Carapace about as long as wide; subquadrate in young, trapezoidal in medium-sized crabs, and laterally convex in old individuals; surface smooth except for faint lines on posterolateral parts. Flattened legs; fringes of hairs on first 3 pairs are adaptations for swimming. Color pattern of carapace extremely variable; generally brown or brownish yellow with irregular lighter markings; often with a small white spot on each side or a large white region on the anterior half. Very common; found primarily in the floating *Sargassum* community; may attach to sea turtles or floating debris; also lives temporarily in moist beach wrack. Ovigers found mainly III-X but occasionally in other months. (Color Plate 11.10.)

***Pachygrapsus transversus*** (Gibbes) (Mottled shore crab): Genus with fine transverse striations anteriorly on the carapace and oblique ones posterolaterally; broader than long.—Species with front slightly more than 1/2 as wide as carapace; median depression on front; sides convergent. Upper surface of movable finger smooth. Ground color of carapace and legs dull olive green or yellowish brown with irregular mottlings of darker brown or dark olive. One of the most common intertidal crabs; under rocks and in crevices, mainly in upper half of intertidal zone; may occur in Sbeach wrack. Where it occurs with *P. gracilis*, it is 10 times more abundant. About 25% of adult crabs may harbor the branchial isopod, *Leidya bimini*. Is host for an intestinal sporozoan (BALL 1951). Crabs with dull gray coloration are massively infected with metacercariae of digenetic trematodes. Ovigers found throughout the year, but peak in V-X.

***P. gracilis*** (de Saussure): Genus as above.—Species with carapace with strongly convergent sides; striations more distinct than in *P. transversus*; front nearly straight and about 2/3 as wide as carapace. Upper surface of movable finger tuberculated. Generally dark in color, with very little mottling. Under stones in the upper intertidal, where it is much less common than *P. transversus*. Ovigers found IV-V and XI.

***Geograpsus lividus*** (H. Milne Edwards): Carapace squarish and widening posteriorly; front not greatly deflexed, about 2/5 width of carapace. Last 3 segments of walking legs with long slender hairs. Color pattern of carapace

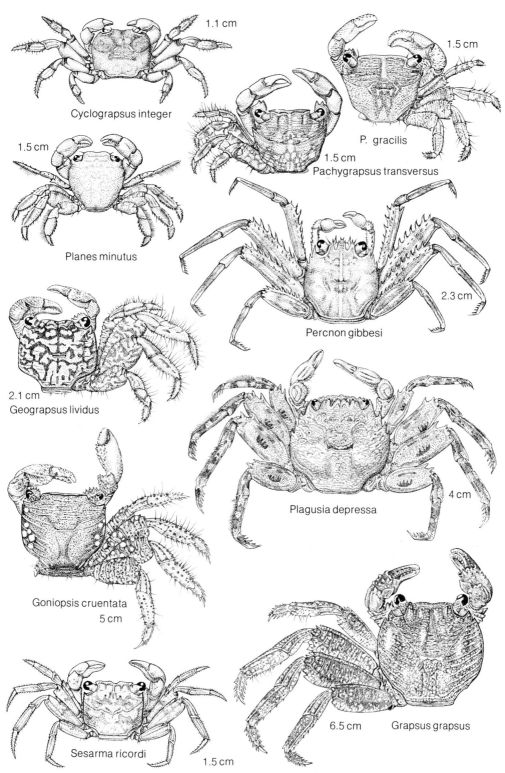

**116 GRAPSIDAE (Shore crabs)**

characteristic; ground color bluish green or yellowish, with variable dark brown irregularly anastomosing patches interspersed with some solitary spots of irregular shapes; similar variable marking on walking legs; chelipeds with less blue-green, and mainly light orange with brighter orange joints. Frequent in some areas but very secretive; lives mainly in the upper intertidal under loose rocks and debris. Oviger in IX.

***Goniopsis cruentata*** (Latreille) (Mangrove crab): Carapace squarish, widening very slightly posteriorly; front sharply deflexed, about 1/2 width of carapace; distinct antero-lateral tooth posterior to outer orbital. Scattered spiniform tubercles on chelae. The most colorful grapsid; carapace dark brown with small greenish, red or purple markings, and circular white spots laterally; legs reddish above, variegated with purplish black and white spots; chelipeds similarly colorful becoming white at tips. Common intertidally; in burrows among tangled roots of mangroves or other vegetation and under rocks or debris. Ovigers found VIII. (Color Plate 11.12.)

***Sesarma ricordi*** H. Milne Edwards: Carapace squarish, sides nearly parallel; surface flattened; front strongly deflexed but can be seen from dorsal view; upper edge of front deeply lobed. Carapace with tan ground color mottled with dark brown to form a variable pattern; legs irregularly banded with dark brown. A common but inconspicuous, fast-moving crab found mainly at or above the high-water mark among rocks and beach wrack, and in the grasses above this level on fairly dry soil. Ovigers in VI.

***Percnon gibbesi*** (H. Milne Edwards) (Flat crab, Spray crab): Carapace very flat, disc-like, longer than wide, and covered with minute stiff hairs; 4 antero-lateral teeth (including outer orbital); front deeply cut and spiny. Chelipeds distinctly larger in male. Carapace mottled with dark brown pattern on light green background; thin median white or pale blue stripe; walking legs broadly banded with dark brown; chelipeds dusty orange; undersides of body pale blue, legs pale pink. Common under rocks at the mean low-water level; may be more abundant in crevices below this level but more difficult to capture. Breeds IV-XI; megalopae found under rocks at low tide level. (Color Plate 11.5.)

***Plagusia depressa*** (Fabricius): Carapace with convex sides giving a nearly circular shape, highly tuberculated and hairy with deep grooves marking regions; 4 antero-lateral teeth; front deeply notched. Wrists of chelipeds without broad flattened spine. Carapace and legs reddish purple with blood red and dark purple mottling and banding; longitudinal dark bands on chelipeds; undersides yellowish white. Common but inconspicuous, found in the same exposed rocky regions as *Grapsus*, but nearer the low-water mark,

**COLOR PLATE 1: CHLOROPHYTA (Green algae).** 1 *Acetabularia crenulata*; 2 *Neomeris annulata*; 3 *Valonia macrophysa*; 4 *Halimeda monile*; 5 *Halimeda incrassata*; 6 *Anadyomene stellata*; 7 *Caulerpa verticillata*; 8 *Caulerpa sertularioides*; 9 *Avrainvillea nigricans*; 10 *Penicillus dumetosus*; 11 *Ulva lactuca*; 12 *Cymopolia barbata*; 13 *Codium decorticatum*. (Sponsored by *The Bank of Bermuda Ltd.*)

**COLOR PLATE 2: PHAEOPHYTA (Brown algae), RHODOPHYTA (Red algae).** 1 *Stypopodium zonale*; 2 *Dictyota divaricata*; 3 *Amphiroa fragilissima*; 4 *Sargassum bermudense*; 5 *Padina vickersiae*; 6 *Sargassum fluitans* (with *Halopteris* sp.); 7 *Dasya baillouviana*; 8 *Liagora farinosa*; 9 *Galaxaura obtusata*; 10 *Laurencia obtusa*; 11 *Ceramium byssoideum*; 12 *Bostrychia montagnei*; 13 *Halymenia floresia*. (Sponsored by L. P. Gutteridge Ltd.)

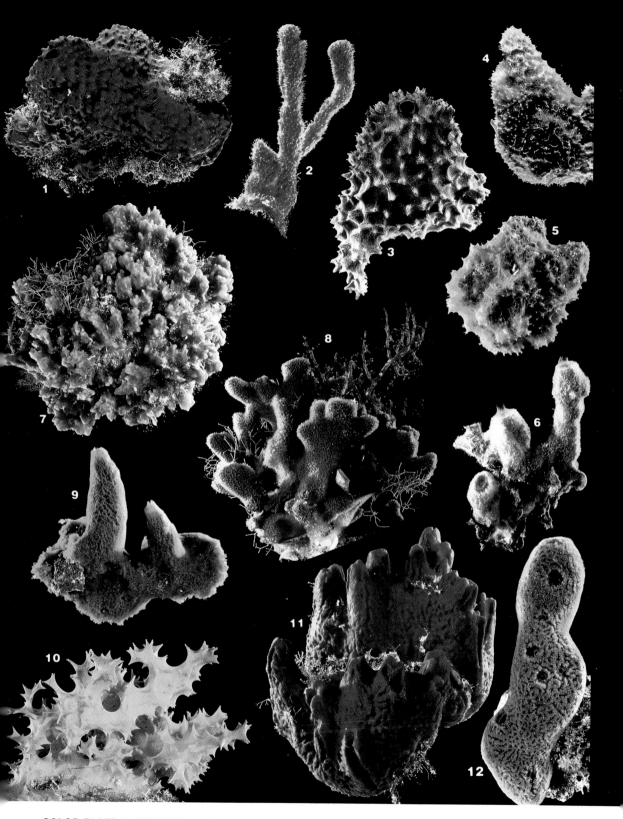

**COLOR PLATE 3: PORIFERA 1 (Sponges).** 1 *Pseudoceratina crassa*; 2 *Haliclona monticulosa*; 3 *Ircinia felix*; 4 & 5 *Dysidea etherea*; 6 *Dysidea janiae*; 7 *Lissodendoryx isodictyalis*; 8 *Haliclona molitba* (with the red algae *Laurencia papillosa* and *Amphiroa fragilissima*); 9 *Amphimedon viridis*; 10 *Aplysilla longispina*; 11 *Tedania ignis*; 12 *Niphates erecta*. (Sponsored by The Trustees of the Bermuda Biological Station.)

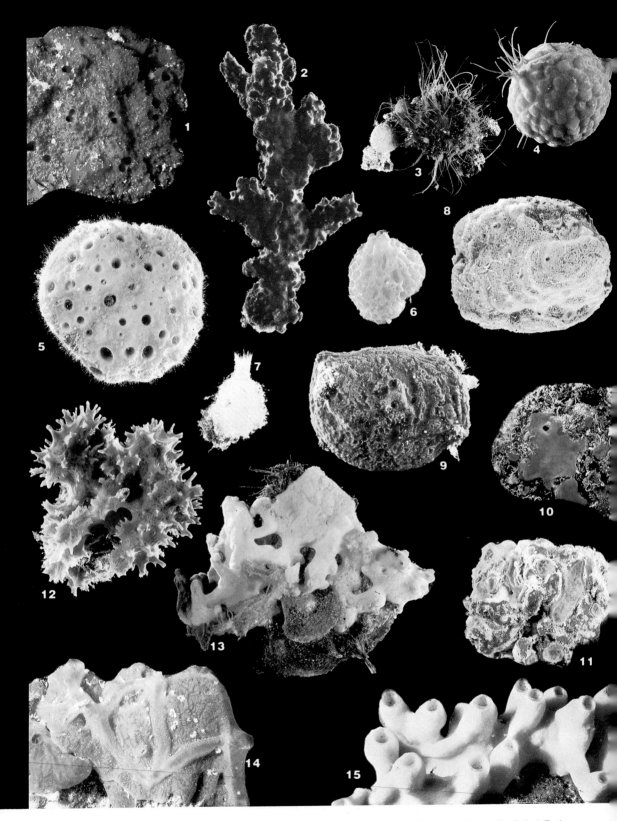

**COLOR PLATE 4: PORIFERA 2 (Sponges).** 1 *Pseudaxinella explicata*; 2 *Homaxinella rudis*; 3 & 4 *Tethya actinia*; 5 *Cinachyra alloclada*; 6 *Clathrina coriacea*; 7 *Sycon ciliatum*; 8 & 9 *Cliona lampa*; 10 *Cliona dioryssa*; 11 *Cliona caribbaea*; 12 *Ulosa ruetzleri*; 13 *Terpios aurantiaca* (on *Isognomon alatus*); 14 *Spirastrella mollis*; 15 *Leucetta microraphis*. (Sponsored by *Shell Company of Bermuda Ltd.*)

**COLOR PLATE 5: ANTHOZOA 1 (Octocorals).** 1 & 2 *Pterogorgia citrina*; 3 & 4 *Pseudoplexaura porosa*; 5-7 *Plexaurella nutans*; 8-10 *Plexaura flexuosa*; 11 *Sclerobelemnon* sp. cf. *S. theseus*; 12 *Nidalia occidentalis*; 13 & 14 *Muricea atlantica*; 15 & 16 *Eunicea tourneforti*; 17 *Gorgonia ventalina* (with *Cyphoma gibbosum*). (Sponsored by *The Bank of Butterfield Ltd.*)

**COLOR PLATE 6: ANTHOZOA 2 (Sea anemones and stony corals).** 1 *Bunodeopsis antilliensis* (on *Halodule bermudensis*); 2 *Actinia bermudensis*; 3 *Pseudactinia melanaster*; 4 *Aiptasia pallida*; 5 *Bartholomea annulata*; 6 *Rhizopsammia bermudensis*; 7 *Astrangia solitaria* (with *Homotrema rubrum*); 8 *Isophyllia sinuosa*; 9 & 10 *Epicystis crucifer*; 11 *Telmatactis vernonia*; 12 *Bunodosoma granuliferum*; 13 *Montastrea cavernosa*; 14 *Porites astreoides* (with *Ceratoconcha domingensis*). (Sponsored by Mr. William de V. Frith.)

**COLOR PLATE 7: ANTHOZOA 3 (False corals, sea mats, tube anemones).** 1 *Zoanthus sociatus*; 2 *Isaurus duchassaingi*; 3 & 4 *Palythoa variabilis*; 5 *Corynactis parvula*; 6 & 7 *Discosoma carlgreni*; 8 *Palythoa caribaea*; 9 *Parazoanthus parasiticus* (on *Niphates erecta*); 10 *Arachnanthus nocturnus*; 11 *Epizoanthus minutus* (with *Darwinella rosacea*); 12 & 13 *Discosoma sanctithomae*. (Sponsored by *Mr. and Mrs. Gerard M. Ives.*)

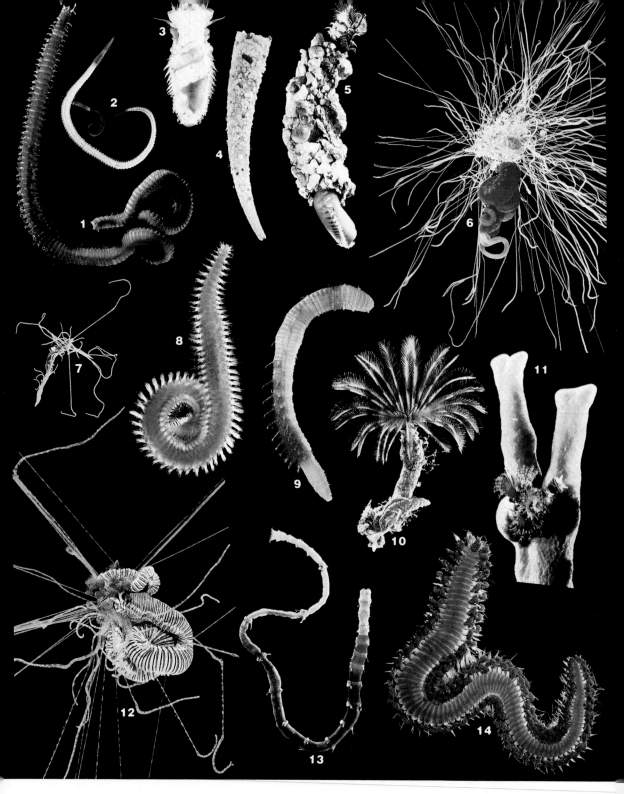

**COLOR PLATE 8: POLYCHAETA (Bristle worms).** 1 *Eunice vittata*; 2 *Odontosyllis enopla* (♂ & ♀); 3 & 4 *Pectinaria regalis*; 5 *Lygdamis indicus*; 6 *Enoplobranchus sanguineus*; 7 *Cirriformia punctata*; 8 *Eurythoe complanata*; 9 *Arenicola cristata*; 10 *Sabella melanostigma*; 11 *Pomatoceros* sp. (on *Millepora alcicornis*); 12 *Eupolymnia crassicornis*; 13 *Euclymene coronatus*; 14 *Hermodice carunculata*. (Sponsored by *The Trustees of the Bermuda Biological Station*.)

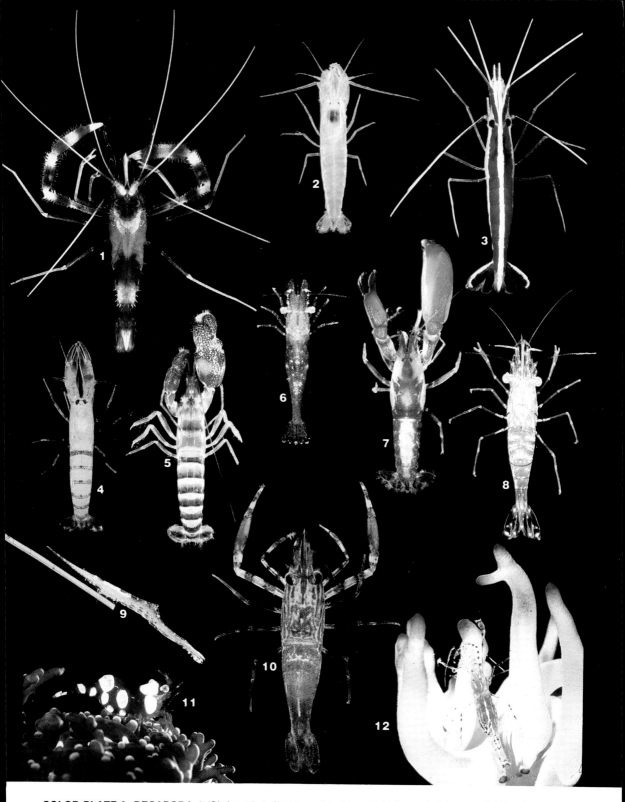

**COLOR PLATE 9: DECAPODA 1 (Shrimps).** 1 *Stenopus hispidus*; 2 *Alpheopsis trigonus*; 3 *Hippolysmata grabhami*; 4 *Alpheopsis labis*; 5 *Alpheus armillatus*; 6 *Hippolyte zostericola*; 7 *Alpheus cylindricus*; 8 *Palaemon northropi*; 9 *Tozeuma carolinense* (on *Syringodium filiforme*); 10 *Brachycarpus biunguiculatus*; 11 *Thor amboinensis* (on *Lebrunia danae*); 12 *Periclimenes anthophilus* (on *Condylactis gigantea*). (Sponsored by *Sir James and Lady Pearman in memory of Eugene Charles Pearman*.)

**COLOR PLATE 10: DECAPODA 2 (Spiny lobsters, mole shrimps, hermit crabs a.o.).** 1 *Axiopsis serratifrons*; 2 *Calcinus verrilli*; 3 & 4 *Petrolisthes armatus*; 5 *Justitia longimanus*; 6 *Arctides guineensis*; 7 *Allodardanus bredini* (with *Parathelges* sp.); 8 *Hippa cubensis*; 9 *Dardanus insignis*; 10 *Callianassa branneri*; 11 *Panulirus guttatus*. (Sponsored by *Dr. W. Redwood Wright*.)

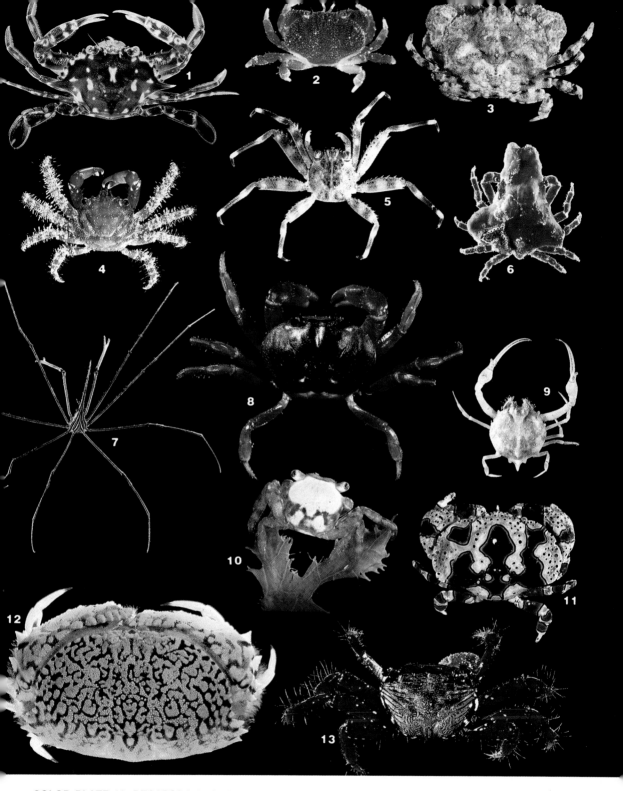

**COLOR PLATE 11: DECAPODA 3 (Crabs).** 1 *Portunus sayi*; 2 *Actaea setigera*; 3 *Osachila antillensis*; 4 *Mithrax forceps*; 5 *Percnon gibbesi*; 6 *Macrocoeloma trispinosum nodipes* (with *Tedania ignis* and *Botrylloides nigrum*); 7 *Stenorhynchus seticornis*; 8 *Gecarcinus lateralis*; 9 *Callidactylus asper*; 10 *Planes minutus* (on *Sargassum* sp.); 11 *Platypodia spectabilis*; 12 *Calappa flammea*; 13 *Goniopsis cruentata*. (Sponsored by Mr. and Mrs. William T. Kemble.)

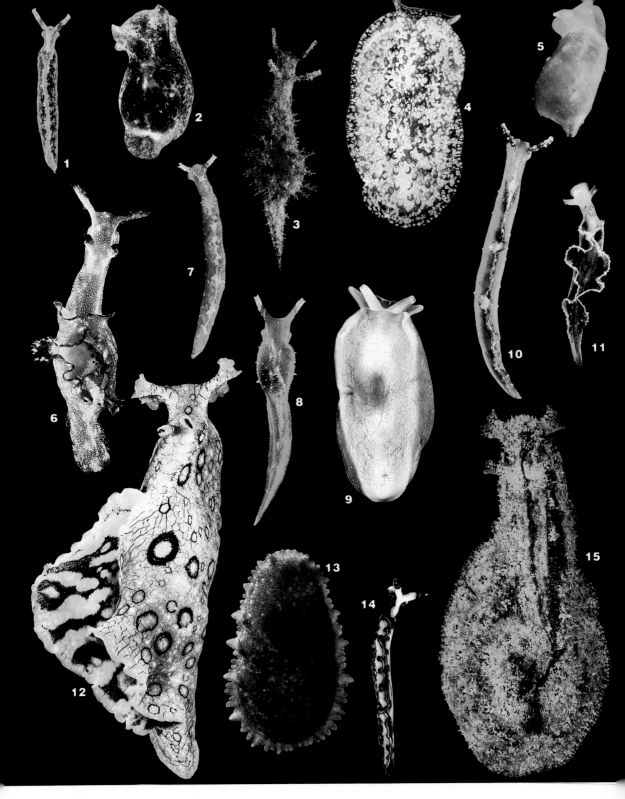

**COLOR PLATE 12: GASTROPODA 1 (Sea hares and other sea slugs).** 1 *Elysia papillosa*; 2 *Haminoea antillarum*; 3 *Stylocheilus longicauda*; 4 *Pleurobranchus areolatus*; 5 *Volvatella bermudae*; 6 *Aplysia parvula*; 7 *Elysia tuca*; 8 *Oxynoe antillarum*; 9 *Berthella agassizi*; 10 *Elysia subornata*; 11 *Elysia flava*; 12 *Aplysia dactylomela*; 13 *Onchidella floridana*; 14 *Elysia picta*; 15 *Dolabrifera dolabrifera*. (Sponsored by The Bank of Bermuda Ltd.)

**COLOR PLATE 13: GASTROPODA 2 (Sea slugs).** 1 *Cyerce antillensis*; 2 *Chromodoris clenchi*; 3 *Aegires sublaevis*; 4 *Cratena pilata*; 5 *Scyllaea pelagica* (on *Sargassum* sp.); 6 *Bosellia mimetica*; 7 *Glaucus atlanticus*; 8 *Tritoniopsis frydis*; 9 *Cyerce cristallina*; 10 *Hypselodoris zebra*; 11 *Costasiella ocellifera*; 12 *Gymnodoris* sp.; 13 *Chromodoris bistellata*; 14 *Okenia zoobotryon*; 15 *Favorinus auritulus*; 16 *Spurilla neapolitana*. (Sponsored by *Dr. Rudolph F. Nunnemacher*.)

**COLOR PLATE 14: ECHINODERMATA (Echinoderms).** 1 *Diadema antillarum*; 2 *Echinoneus cyclostomus*; 3 *Ophioderma appressum*; 4 *Ophiomyxa flaccida*; 5 *Asteroporpa annulata* (on *Ellisella* sp.); 6 *Ophionereis reticulata*; 7 *Lissothuria antilliensis*; 8 *Synaptula hydriformis* (on *Caulerpa sertularioides*); 9 *Ocnus surinamensis*; 10 *Isostichopus badionotus*; 11 *Goniaster tessellatus*; 12 *Coscinasterias tenuispina*; 13 *Oreaster reticulatus*; 14 *Asterinopsis pilosa*. (Sponsored by Appleby, Spurling and Kempe Ltd.)

**COLOR PLATE 15: ASCIDIACEA 1 (Sea squirts).** 1 *Aplidium exile*; 2 *Trididemnum savignyi*; 3 & 4 *Diplosoma listerianum*; 5 *Didemnum* sp.; 6 *Ecteinascidia conklini*; 7 *Distaplia corolla*; 8 *Eudistoma olivaceum*; 9 *Eudistoma obscuratum*; 10 *Eudistoma clarum*; 11 *Ascidia curvata*; 12 *Ecteinascidia turbinata* (with *Pseudoceros crozieri*); 13 *Clavelina picta*; 14 *Polyclinum constellatum*; 15 *Lissoclinum fragile*; 16 *Perophora formosana*; 17 *Clavelina oblonga*. (Sponsored by *Electronic Communications Ltd.—The K&J Group*.)

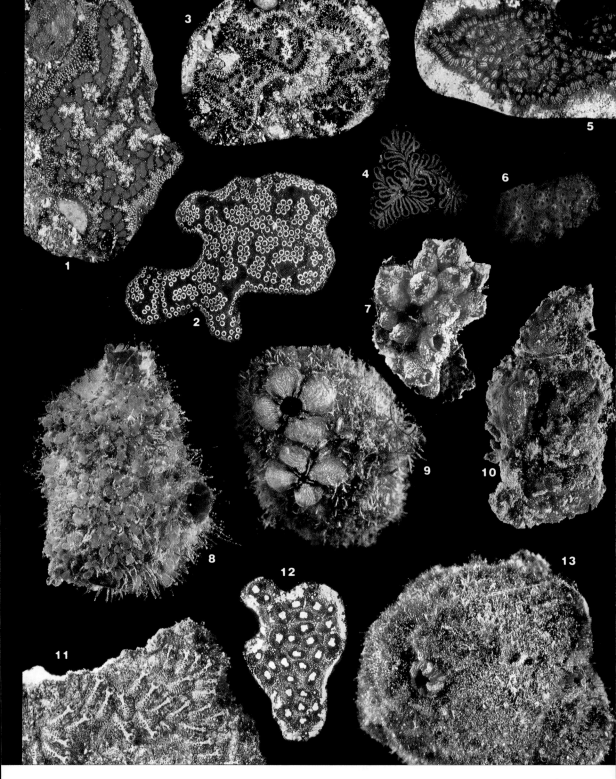

**COLOR PLATE 16: ASCIDIACEA 2 (Sea squirts).** 1 & 2 *Botryllus planus*; 3-6 *Botrylloides nigrum*; 7 *Stolonica sabulosa*; 8 *Microcosmus exasperatus* covered by *Perophora formosana*; 9 *Styela plicata*; 10 *Polyandrocarpa tincta*; 11 & 12 *Symplegma viride*; 13 *Pyura torpida*. (Sponsored by *The Bank of Butterfield Ltd.*)

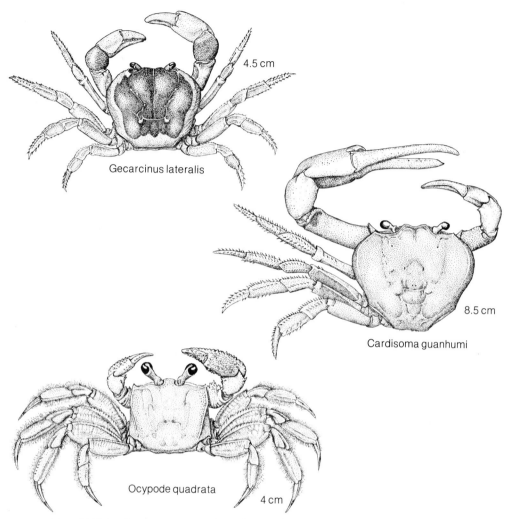

**117 GECARCINIDAE (Land crabs), OCYPODIDAE (Ghost crabs)**

and in algae-covered tide pools. Oviger recorded in XI.

***Grapsus grapsus*** (L.) (Cliff crab, Red shore crab, Sally Lightfoot): Carapace with convex sides giving a nearly circular shape. Wrist of cheliped armed with broad flattened sharp-tipped spine. Color variable, but carapace and legs mainly dark red to reddish brown with pale blue markings. Lives throughout the intertidal on steep, barren, rocky, exposed cliffs; also on gently sloping rocky areas if protected by vegetation. Ovigers recorded in IV-V and VIII (probably brackets main breeding season).

### Plate 117

F. **GECARCINIDAE** (Land crabs): Brachyrhyncha with transversely oval carapace, antero-lateral borders strongly arched, surface very convex, front deflexed, eyestalks not very large; chelipeds of ♂ large, powerful and unequal. (2 spp. from Bda.)

***Gecarcinus lateralis*** (Fréminville) (Common land crab): Fronto-orbital border more than 1/2 width of carapace. Carapace deep reddish brown or plum; often a wide lighter band around lateral and posterior regions; a pair of white spots close behind orbits and another pair near the midposterior region; legs light grayish brown; chelipeds darker and reddish. Occurs in deep burrows, well above high tide level; very abundant along the sandy, vegetation-covered dunes, but also atop rocky cliffs. Leading a largely nocturnal life, they usually emerge from their burrows in the evening, especially after rainfall. During VII-VIII, crabs migrate to the ocean to release zoeae. Young crabs, 7-9 mm long, recorded in IV. (Color Plate 11.8.)

***Cardisoma guanhumi*** Latreille (Great land crab): Fronto-orbital distance about 2/3 of maximum carapace width in adult ♂, nearly 3/4 in ♀. Chelipeds massive and unequal in both sexes. Carapace of large crabs pale gray, becoming bluish gray on margins and on the legs; intermediate sized crabs may be bright blue; chelipeds pale gray to white. In deep burrows in low-lying muddy regions above high-water mark; wander out at night. Rare.

F. **OCYPODIDAE:** Brachyrhyncha with front of carapace usually narrow and somewhat deflexed, orbits occupying the whole anterior border outside of the front. Eyestalks long and slender. One cheliped of ♂ often greatly enlarged. Amphibious, burrowing, commonly gregarious crabs, particularly abundant in the tropics. (1 sp. from Bda.)

***Ocypode quadrata*** (Fabricius) (Ghost crab): Carapace quadrilateral, broader than long. Eyestalks large and club shaped. Chelipeds well developed and unequal in both sexes. Color of carapace gray, grayish white, pale yellow, or yellowish white imitating color of beach sand; some yellow markings below. Common on sandy beaches where it lives in burrows near to or a distance above high-water mark. Mainly nocturnal. Records of ovigers in late VII.

**Plate 118**

SECT. **OXYRHYNCHA:** Brachyura with anteriorly narrowed body, and usually with a distinct rostrum; postero-lateral regions well developed. Epistome usually large. Antennules usually folded longitudinally.

F. **MAJIDAE** (Spider crabs): Oxyrhyncha with a triangular, tear-drop or pyriform carapace; orbits generally more or less incomplete; hooked hairs usually present. Chelipeds usually not massive or powerful; walking legs often long and thin, giving rise to common name. Generally sluggish crabs found in a wide variety of habitats. (20 spp. from Bda.)

***Epialtus bituberculatus*** H. Milne Edwards: Carapace with oblong rostrum slightly bilobed at tip; angular to rounded antero-lateral and postero-lateral lobes; slightly wider across postero-lateral region, although this may be less obvious in females. Carapace with delicate pastels of pink, purple, green and brown; same for legs except that the last pair is whitish. Conspicuous white patch in center of carapace and light lateral and posterior borders. A frequent but inconspicuous crab found under rocks and dead coral where it may blend in with colorful sponges and tunicates. Found in shallow bays and in deeper water on sandy bottom. Oviger in IV.

***Acanthonyx petiverii*** H. Milne Edwards: Carapace with broad, somewhat triangular bilobed rostrum; rounded antero-lateral lobes and 2 pairs of smaller postero-lateral lobes; may be tubercles on midline of ♂. Dark fawn-color. Abundant among algae in rocky pools and in deeper water. May have algae on carapace. Ovigerous females found IV.

***Podochela riisei*** Stimpson: Carapace elongate, pyriform, uneven, and with tufts of hairs; rostrum broad, rounded and deeply excavated below. Body light brown with darker edges, or brick-red; legs lighter, grading to whitish; chelipeds whitish to nearly transparent. Found among hydroids on pilings and in rocky areas down to 55 m. May have ascidians, bryozoans or algae attached to body and legs.

***Stenorhynchus seticornis*** (Herbst) (Arrow crab): Carapace triangular, smooth, and much longer than broad; rostrum very slender, tapered, flattened and with lateral spines. All legs very long and slender; chelipeds larger in males. Ground color gray with stripes of brown or black, diverging obliquely from median line to the posterior margin (giving effect of nested series of inverted V's); legs reddish brown with bright red joints; fingers bluish purple. In shallow water among corals and in deeper water to 180 m, sometimes associated with *Condylactis gigantea*; uncommon. (Color Plate 11.7.)

***Microphrys bicornutus*** (Latreille): Carapace pyriform, uneven, very tuberculate, small but conspicuous spine at each postero-lateral angle; horns of rostrum generally divergent. Usually covered with algae, bryozoans, sponges or sediment. Carapace dull yellowish brown; chelipeds grayish white with small round purplish spots. Very common subtidally on rocky shores, reefs, and under stones and dead coral in shallow sandy bays; down to 30 m. Breeds mainly from IV to XI, but ovigers also found in other months.

***Macrocoeloma subparallelum*** (Stimpson): Genus with triangular carapace, not armed with series of strong spines on lateral margin; rostrum with 2 strong horns.—Species with carapace covered with short pubescence;

row of 5 protuberances between postero-lateral spines; rostral horns separated by U-shaped sinus and nearly parallel. Brownish yellow when cleaned of adherent algae. Found in shallow water; uncommon. Oviger collected in IX.

***M. trispinosum nodipes*** (Desbonne) (Decorator crab): Genus as above.—Species with swollen carapace with rounded prominences, and covered with velvety pubescence. Distinct broad postero-lateral spines and 1 posterior spine; rostral horns diverge. Body usually covered with growth of sponge and algae, but when cleaned is a reddish brown. Common, sluggish species; in shallow bays to 20 m. Ovigers recorded from IV through midsummer. This subspecies (*nodipes*) appears to be the only variety in Bermuda. (Color Plate 11.6.)

***Mithrax forceps*** (A. Milne Edwards) (Common spider crab): Genus with carapace ovate or oblong-ovate, broader than long or slightly longer than broad; antero-lateral margins usually with 4, sometimes 3, teeth behind orbit; postero-lateral margin sometimes having spines; front with rostral horns either pointed or truncate. Long strong chelipeds, especially in ♂.— Carapace of the species with conspicuous, smooth, oblique grooves; 4 acute antero-lateral teeth (not including post-orbital tooth); rostral horns short, wide, truncated, and with broad V-shaped median notch. Large teeth on fingers of ♂ chela. Carapace uniform yellowish brown to greenish brown or terra-cotta in largest crabs; a pale yellow medial dorsal stripe in young; legs may be banded. Most abundant shallow-water member of the genus. Easily collected in shallow sandy bays under coral rocks and in crevices of dead coral; from low tide mark to 30 m. Shells of serpulid polychaetes may attach to carapace and legs. Breeds throughout much of the year but especially from II to XI; becomes ovigerous at about 10 mm width. (Color Plate 11.4.)

***M. pleuracanthus*** Stimpson: Genus as above.—Carapace of species conspicuously tuberculate; rostral horns shorter and wider than in *M. hispidus*; notch between horns shallower than in latter. Body yellowish white, with blotches of bright red; legs yellowish white, blotched or barred with red; chelae light red with pale tips. In shallow water; rare.

***M. hispidus*** (Herbst) (Large red spider crab): Genus as above.— Carapace of species somewhat tuberculate, rostral horns short, narrow and separated by conspicuous U-shaped notch; 3rd and 4th antero-lateral teeth pointed and curved forward. Postero-lateral border with a smaller tooth higher on carapace. Chelipeds unequal in male. Deep brownish red or terra-cotta in color, brighter on chelipeds and darker on walking legs; latter may have bright red bands at joints. A sluggish crab, fairly common in deeper water but also found in shallow bays. Often

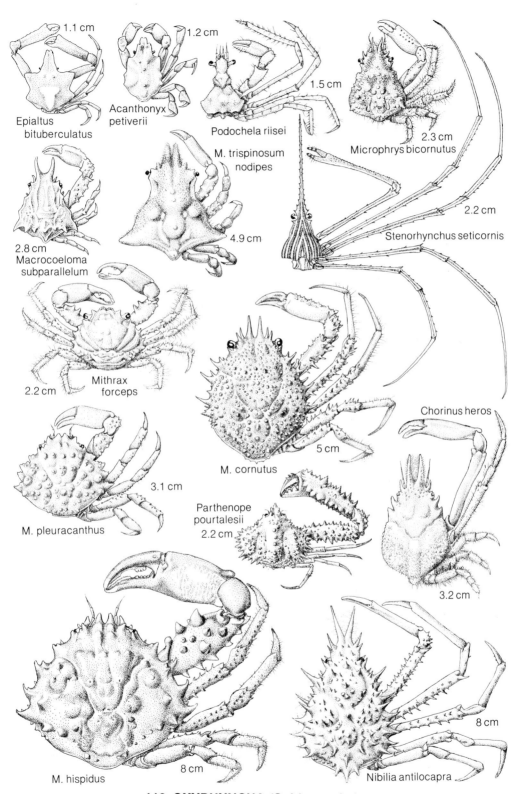

**118 OXYRHYNCHA (Spider crabs)**

taken in fish traps set outside reefs.

***M. cornutus*** de Saussure (Red spider crab): Genus as above.— Carapace of species very spiny, covered with short woolly pubescence, as are the legs; anterolateral teeth not all on same level, and main teeth have secondary teeth; rostral horns long, tapering and divergent. Color bright red, yellowish or rosy. Deep-water species, from 55 to 420 m; collected in baited traps. Ovigers VIII-X.

***Chorinus heros*** (Herbst): Carapace oval with 2 long hairy rostral horns. Chelipeds of ♂ long and stout, naked and smooth. Color in life unknown. Rare; probably only in deeper water.

***Nibilia antilocapra*** (Stimpson): Carapace pyriform and spiny; rostral horns divergent; conspicuous spines projecting from around posterior margin. Arms of chelipeds spiny. Whitish in preserved condition. In deep water (365 m); rare.

F. **PARTHENOPIDAE:** Oxyrhyncha with chelipeds usually much longer and heavier than other legs. Eyes usually retractile within well-defined orbits. Antennules folded somewhat obliquely. (4 spp. from Bda.)

***Parthenope pourtalesii*** (Stimpson): Carapace broadly triangular, convex, spiny and tuberculate; deep grooves separate branchial areas from other regions; broader than long. Chelipeds strongly spinous along edges. Color purplish red with bands of buff on all legs; tips of fingers brown. In deep water (180 m); rare.

F. A. CHACE, JR., J. J. MCDERMOTT, P. A. MCLAUGHLIN & R. B. MANNING

## ORDER MYSIDACEA
(Opossum shrimps)

CHARACTERISTICS: *Shrimp-like MALACOSTRACA usually with stalked eyes; carapace fused only to the 3 anterior somites but covering most of the thoracic region; thoracic limbs with well-developed exopods and endopods. Uropods large, flattened, leaf-like, forming a tail fan together with the telson.* Mysids are slender, usually 5-25 mm long (2-180 mm); the exoskeleton is usually flexible, only rarely stiff and mineralized. Over a light background color there is a pattern resulting from branched, dark chromatophores, but deep-sea species are often red. Most mysids are steady swimmers using the exopods of the thoracic limbs, or the pleopods. Escape movements are performed by sudden flexing of the abdomen and the tail fan against the thorax.

Of approximately 800 species described, 10 are known from Bermuda waters; 8 are included here.

OCCURRENCE: Mostly marine; some live in brackish, some in fresh water. The majority occur in nearshore coastal waters, but species have been found in all regions of the ocean, to a depth of 7,210 m (!).

For collecting, a hand net can be used along the shore. Sampling on level bottoms can be performed from a boat with a light detritus sledge or a D-net (tow net with D-shaped mouth mounted on a light sled with broad runners), and in open water by plankton nets or mid-water trawls with fine-meshed net. Many mysid species rise towards the surface at night and can be collected by the combined use of a light source and a hand net or plankton net. Species living in association with sponges corals etc., are best caught by SCUBA and hand net. Mysids are easily sorted from plankton samples. A useful method to extract them from sediment samples is by repetitive washing with seawater, shaking, decanting and filtering through a fine-meshed sieve.

IDENTIFICATION: A good dissecting microscope is mandatory, a compound microscope useful.

Presence or absence of gills, biramous pleopods in the female and statocysts, and the shape of the pleural plates are important characters. Species differ mostly in the shape of the margins of the carapace, the antennal scale, the eye, the uropods and the telson.

For identification and study of morphology fix in 2-4% neutralized formalin-seawater. For storage transfer to 2% formalin solution or 70-80% alcohol. For general histology fix in Carnoy or Bouin.

BIOLOGY: All mysids are dioecious. External sexual dimorphism exists: adult females have a brood pouch; adult males have a setiferous lobe on the distal part of the antennular peduncle and a paired copulatory organ at the base of the 8th thoracic limb. All females (except Lophogastrida) have reduced pleopods; most males have at least some biramous pleopods. Mysids copulate, and fertilization takes place in the brood pouch. The number of progeny is very variable (1-350). Brood size within a species at any time of the year is related to the body length of the female. Brood volume or weight is approximately 10% of the body volume or weight of the female. Life-span ranges from some weeks in warm waters to more than 7 yr in cold, deep waters. Mysids rather unselectively get their food by filter feeding and/or picking up large particles. Many mysids form shoals, swarms or schools and a number of species perform diurnal vertical migrations. They serve as an important food item for some of the larger invertebrates, many fishes, shorebirds and some marine mammals. Mysids are parasitized by different types of Protozoa, Cestoda, Copepoda and Isopoda.

DEVELOPMENT: Direct, in the brood pouch. Incubation time ranges from 4 days in warm waters to more than 10 months in cold, deep waters. The young differ very little from the adult.

REFERENCES: For general orientation and systematic overview see TATTERSALL & TATTERSALL (1951) and MAUCHLINE (1980).

For species from Bermuda see SARS (1885), VERRILL (1923), TATTERSALL (1951), TATTERSALL (1955) and BRATTEGARD (1969).

## Plate 119

**S.O. LOPHOGASTRIDA:** Mysidacea with gills on some or all of the thoracic limbs; both sexes with biramous pleopods. Endopod of uropod with statocyst.

**F. LOPHOGASTRIDAE:** Lophogastrida with distinct pleural plates on abdominal somites. (2 spp. from Bda.)

*Gnathophausia* cf. *ingens* (Dohrn): Anterior part of carapace produced into an elongate, spear-shaped, often denticulate rostrum. Large (to 150 mm !); color red. Bathypelagic.

**S.O. MYSIDA:** Mysidacea without gills; pleopods of ♀ generally reduced, not natatory.

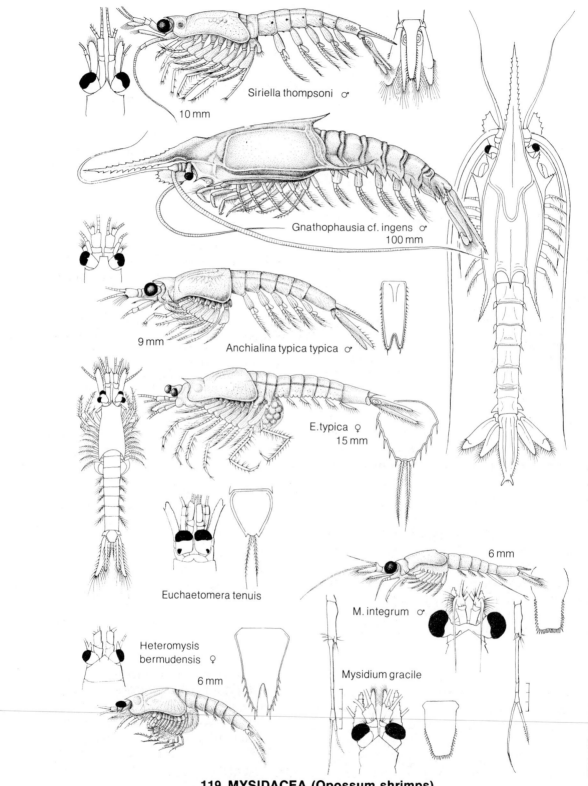

**119 MYSIDACEA (Opossum shrimps)**

**F. MYSIDAE:** Mysida with penultimate segment of thoracic endopods divided into subsegments; usually 2 or 3 pairs of brood lamellae (oostegites); pleopods of ♀ generally reduced to small, unsegmented plates. Statocyst present. (8 spp. from Bda.)

*Siriella thompsoni* Milne Edwards: Exopod of uropod shorter than endopod and with well- marked distal suture, proximal segment armed with spines. Antennal scale 5-6 times as long as broad. Oceanic, epipelagic. Collected offshore at night with light.

*Anchialina typica typica* (Krøyer): Exopod of uropod undivided, posterior margin of carapace transverse. Neritic and oceanic. Collected at night with light.

*Euchaetomera tenuis* G. O. Sars: Genus with functional ocelli of eye divided into 2 distinct areas.—Species with lateral margins of telson unarmed. Oceanic, mesopelagic.

*E. typica* G. O. Sars: Genus as above.—Species with lateral margins of telson armed with 6 or 7 equally spaced spines. Oceanic, mesopelagic.

*Mysidium gracile* Dana: Genus with antennal scale elongate, setose all around; 4th pleopods (♂) with endopod rudimentary, exopod 3-4-segmented, penultimate segment with very long seta, ultimate segment with long seta; telson short.—Species with apex of telson slightly emarginate, the 3 distal segments of the 4th pleopod (♂) of equal length. In shallow water, where it swarms among gorgonians and corals, often in shade.

*M. integrum* W. M. Tattersall: Genus as above.—Species with apex of telson transverse or slightly convex, the 3 distal segments of the 4th pleopod (♂) of unequal length. In shallow water; swarms among mangrove and around gorgonians and corals, often in shade.

*Heteromysis bermudensis* G. O. Sars: Third thoracic endopod larger than any other, pleopods (♂) all rudimentary, telson cleft. Color pinkish white. In shallow water, between sedentary organisms; probably associated with sea anemones.

T. BRATTEGARD

## ORDER MICTACEA

CHARACTERISTICS: *Small MALACOSTRACA with small lateral carapace folds covering only bases of maxilla 1, maxilla 2 and maxilliped. Antenna 1 with 2 flagella; antenna 2 with small scale. Maxilliped without epipod; uropod biramous, rami 2-5-segmented. Female with ventral thoracic marsupium. Only 3-3.5 mm long and whitish-opaque when alive, they swim slowly and steadily except for a rapid darting escape response.* Of only 2 species known, 1 has been described from Bermuda.

OCCURRENCE: Marine, in caves or in the deep sea; the cavernicolous species lives in the water column of caves with only little water movement.

Collect with a hand net or, to avoid damage, individually by means of suction bottles or vials.

IDENTIFICATION: A dissecting microscope is necessary to identify species.

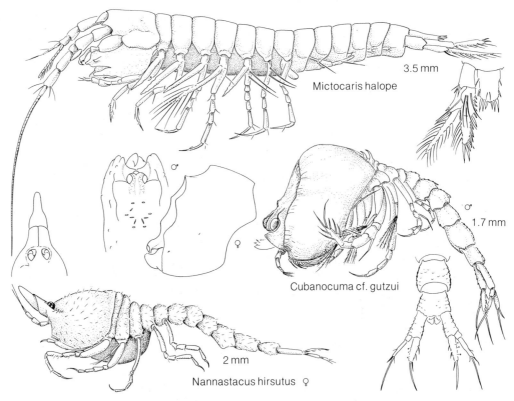

**120 MICTACEA (Mictaceans), CUMACEA (Cumaceans)**

Note the slender, cylindrical body, the pereopods 1-5 with natatory exopods (in ♀ with oostegites) without branchiae; ♂ pereopod 7 with penes on coxa. The pleopods 1-5 are reduced, uniramous, 1-segmented (except for ♂ pleopod 2 which is enlarged, 2-segmented).

Preserve in 70% alcohol.

BIOLOGY: Sexes are separate; nothing is known about life-span, feeding, predators, etc.

DEVELOPMENT: The embryo, flexed dorsally, is brooded in the marsupium and released at the manca stage (i.e., resembling the adult but lacking pereopod 7).

REFERENCES: First discovered in Bermuda, the order has been established by BOWMAN et al. (1985), and the species described by BOWMAN & ILIFFE (1985).

**Plate 120**

*Mictocaris halope* Bowman & Iliffe: Rostrum narrowly triangular; eyestalks pyriform, without visual eleSments. Pereon slightly longer than pleon, pereonites subequal in length. In the water column of marine caves, not uncommon. Known only from Bermuda.

W. STERRER

## ORDER CUMACEA

CHARACTERISTICS: *Small MALACOSTRACA with short, bulging carapace fused to first 3 or 4 (rarely 6) thoracomeres, and set off distinctly from the long, slender abdomen. Pleopods usually absent in females, or reduced in males. Uropods rod-like. Eyes unstalked, often fused dorsally.* One to 2 mm (rarely to 35 mm) long and mostly whitish, yellow or reddish brown, they either glide smoothly and rapidly over the substrate, or burrow into it.

Of some 900 known species in 8 families, only 4 species are known from Bermuda of which 2 are considered here.

OCCURRENCE: With few exceptions, cumaceans are marine and benthic, living buried in muddy and sandy bottoms during daytime, but swarming into the plankton at night to molt and mate. Though not abundant in Bermuda, they are known to inhabit suitable substrates (which they select by grain size) in amazing densities (17,000/0.1 $m^2$!) The majority of species occur in shallow water (intertidal to 200 m), but many have a wide depth range that extends into the deep sea.

Collect sediments with a dredge, grab or corer, or—in caves—stir up the fine sediment and then sieve animals out with a hand-held fine-meshed net. In the laboratory animals are extracted by elutriation, or by washing with an isotonic magnesium chloride solution.

IDENTIFICATION: A dissecting microscope set at high power may be adequate, but it is usually necessary to mount specimens on microscope slides (preferably depression slides to prevent crushing) and use a compound microscope. A vital stain (eosin, borax carmine, etc.) may be used to make details more visible. Specimens are often covered by sediment particles and need to be cleaned for identification.

The carapace and the 3-5 uncovered thoracomeres may be smooth, or variously ornamented with grooves, ridges or spines. Rostrally, the carapace is produced into 2 lobes that join (but never fuse) to form a pseudorostrum. The abdomen (pleon) consists of 6 somites, the 5th usually being the longest. The 1st antenna (antennule) is usually short, the 2nd rudimentary in females, but very long in mature males. The mandible lacks a palp. There are 2 maxillae, and the first 3 thoracopods are modified as maxillipeds. The remaining 5 pairs of thoracopods (pereopods 1-5) are simple, cylindrical; all except the last may have an exopod. Up to 5 pairs of pleopods may be present in males; none in females. The number of articles of the inner ramus (endopod) of the uropods is of taxonomic value, as is the presence or absence of a telson.

Preserve in 70% ethanol in 10% buffered formalin, preferably after removal from the sediment.

BIOLOGY: Little is known. Sexes are separate, and differ mainly in the presence of the brood pouch in the female, and the long 2nd antennae in the male. At copulation (probably at night, during swarming), the female molts and releases 10-100 eggs into the marsupium. The life-span is probably 1 yr (more rarely 4 yr). When buried in the substrate, cumaceans usually assume the shape of a U, with only the pseudorostrum and uropods showing. They feed by grazing off sand grains, by selectively sorting mud for organic particles, or prey on Foraminifera and interstitial Crustacea. They are probably important as food for small bottom-dwelling fishes.

DEVELOPMENT: Eggs develop in the marsupium and, after 1-3 months, hatch as juveniles that resemble adults except for the lack of the last pair of pereopods.

REFERENCES: A good general introduction is by KAESTNER (1970); other useful references are by WATLING (1979), JONES (1969, 1976), CALMAN (1912) and SARS (1900).

Except for 1 species recorded by VERRILL (1923) and recent collections from caves (mostly new species), Bermuda's Cumacea remain unstudied.

## Plate 120

F. **NANNASTACIDAE:** Without a free, independent telson; endopod of uropod 1-segmented. ♂ without pleopods.

*Nannastacus hirsutus* Hansen: With 2 separated groups of eyes; molar process of mandibles thick, truncate. Carapace short and wide; pseudorostrum long, curved upwards. Carapace and abdomen without spines, but covered with setae. Found only once; in algae.

*Cubanocuma* cf. *gutzui* Bacescu & Muradian: With 1 group of eyes on a lobe that is particularly prominent in the ♂. Molar process of mandibles styliform, sharp. Carapace high, anteriorly truncate; pseudorostrum inconspicuous. ♂ with 4 exopodites. On silty bottoms in inland seawater caves; not uncommon.

J. Markham & W. Sterrer

## ORDER TANAIDACEA

CHARACTERISTICS: *Small, elongate-cylindrical MALACOSTRACA with eyes on immovable stalks; carapace short, fused to the 2 anterior somites. First pair of pereopods (= gnathopods) chelate, 2nd adapted for burrowing, other 5 for walking.* From 2 to 6 mm (rarely to 25 mm) and drab yellowish, greenish or brownish, they swim occasionally; most often they scurry over soft substrates into which they can burrow rapidly.

Of some 350 species described, 16 are known from Bermuda of which 5 are included here.

OCCURRENCE: Almost exclusively marine and benthic, from the littoral to 6,000 m. Many live buried in mud or hide among plants or sedentary animals, some spin mucous tubes in the sediment. Although encountered regularly, they never occur in large numbers.

Collect likely substrates; shake algae or samples of sedentary organisms (tunicates, sponges); or treat them with the climate method. For extraction from sand use the magnesium chloride method; muddy sediments should be treated by elutriation, or left to settle until animals can be picked individually from the sediment surface.

IDENTIFICATION: Microdissection and study of appendages under a compound microscope are mandatory for species identification.

Note the short carapace, the structure of the 2 pairs of antennae, and the presence or absence of eyes. The first thoracic limb is a maxilliped, the second an often massive cheliped, and the following 6 pereopods are uniramous walking legs. In the female, pereopods 1-4 (or pereopod 4 only) develop oostegites which form the marsupium (brood pouch). There are no more than 5 free pleomeres; a 6th is fused to the telson. Pleopods may be reduced in females; uropods are usually rod-shaped. The paired male gonopores open on 1 or 2 cones (see diagnoses of orders!) on the last (8th) thoracic sternite.

Preserve in 5-10% buffered formalin, then transfer into 70% ethanol for storage.

BIOLOGY: Little is known. Sexes are separate, though there is sex reversal in some species. After copulation, fertilized eggs develop in the female's marsupium. The life-span is probably short for males, which lack mouthparts and do not feed as adults. Most species are filter feeders that also ingest larger particles and small invertebrates (e.g., nematodes).

DEVELOPMENT: After 2-3 weeks a manca stage hatches; it closely resembles the adult except for lacking the full set of appendages which it acquires through successive molts.

REFERENCES: A brief general introduction is by KAESTNER (1970). SIEG & WINN (1979) have presented a key to the currently recognized families, which can be used in conjunction with classical treatments by SARS (1896) and RICHARDSON (1905), and the identification guide by GOSNER (1971).

Bermuda's fauna is only spottily known through an initial survey by RICHARDSON (1902), and subsequent reports by GREVE (1974) and GARDINER (1975).

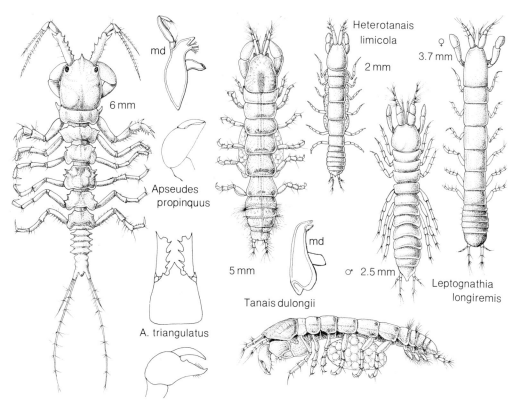

**121 TANAIDACEA (Tanaids)**

## Plate 121

**S.O. MONOKONOPHORA:** Tanaidacea with biflagellate antennule and palp-bearing mandible. Pleon usually narrower than pereon. In the ♂ a single genital cone; in the ♀ marsupium of 4 pairs of oostegites. (Of 8 families, 1 is represented in Bda, with 5 spp.)

**F. APSEUDIDAE:** Monokonophora with convex body; pereopod 1 flat and more expanded than others; other pereopods lacking setae around distal dactyl spine; 1st maxilla with palp.

***Apseudes propinquus*** Richardson: Genus with 6 pleomeres, 5 pairs of unarticulated biramous pleopods, and exopodites on both pairs of gnathopods; antennules usually similar in both sexes.—Species with 3 spines on medial margin of basal segment of antenna; rostrum long, acute. All pereomeres with ventral spines; pleomeres 1-5 laterally pointed. Both rami of uropod long. On sand in shallow bays, but also to 120 m; not uncommon.

***A. triangulatus*** Richardson: Genus as above.—Species with 4 spines on medial margin of basal segment of antenna; rostrum spearpoint-shaped. On algae in shallow bays, probably not uncommon. Endemic.

**S.O. DIKONOPHORA:** Tanaidacea with uniflagellate antennule and palpless mandible. Pleon usually as broad as pereon. In the ♂ 2 genital cones; in the ♀ marsupium of 1 or 4 pairs of oostegites. (Of 8 families, 5 are represented in Bda, with 11 spp.)

F. **TANAIDAE:** Dikonophora with 2nd antenna having no more than 7 joints; mouthparts of ♂ not reduced; uropods uniramous. Marsupium of 1 pair of oostegites.

*Tanais dulongii* Audouin (=*T. cavolinii, T. tomentosus*): With 4 pleomeres and 3 pairs of pleopods. Eyes well developed. Pleomeres fringed with stiff bristles. Whitish, variegated dorsally with brown. In inshore waters on dead coral.

F. **PARATANAIDAE:** Dikonophora with much-reduced antennules and antennae; maxillules and maxillae reduced in ♀, absent in ♂; marsupium of 4 pairs of oostegites; exopodites of pleopods thickly setose marginally.

*Heterotanais limicola* (Harger) (=*Leptochelia limicola*): Eyes distinct; with 6 pleomeres and 5 pleopods. Uropods biramous, with outer ramus small and inner long. Gnathopod of ♀ normal, of ♂ nearly subchelate. Moderately common in algae amoung fouling material, as on buoy chains.

F. **LEPTOGNATHIDAE:** Dikonophora with antennae and antennules of relatively few articles; mouthparts quite well developed, with basis of maxilliped partly fused medially; gnathopod with coxa articulating with mediolateral margin of basis. Marsupium of 4 pairs of oostegites.

*Leptognathia longiremis* (Lilljeborg): Eyes lacking. Mandibles small, feeble. Inner ramus of uropod more than 3 times as long as outer. In fine sand and shell in inshore waters.

J. MARKHAM

ORDER ISOPODA (Pill bugs, wharf lice, fish lice)

CHARACTERISTICS: *Usually dorso-ventrally flattened MALACOSTRACA without a carapace; with 7 thoracic segments and 7 (or 5: Gnathiidea) pairs of cylindrical legs. Abdomen with 5 pairs of foliaceous legs functioning as gills. Eyes unstalked.* Of small size (1-35 mm; but some deep-sea species to 45 cm!) and brown-gray mottled coloration, most species move over the substrate, often rapidly (*Ligia*), by means of thoracic legs; most can swim, some only short distances; a few are adapted to planktonic life.

Of 9 suborders, the semiparasitic Gnathiidea and the mostly groundwater-dwelling Microcerberidea and Phreatoicidea are not represented in Bermuda. Of some 9,000 species (among Crustacea, second only to Decapoda!), about 100 are known from the Caribbean, which suggests that the 50 species known from Bermuda (of which 25 are included here) are only a partial record.

OCCURRENCE: Present in most terrestrial, freshwater and marine habitats, from the beach to the deep sea, reaching particularly high densities on fine-sediment bottoms, in plant associations and in beach wrack.

Collecting is done by washing algae, sponges, mangrove roots, etc., or by sifting sand and mud. Littoral species are picked individually from crevices with forceps. Parasitic species are often conspicuous for they deform the host (e.g., shrimps); or they can be collected from the skin, fins, gills and throats of fish.

Because of the damage certain species do to man-made wooden structures, and as parasites of food fish and crustaceans, they are of considerable economic interest.

IDENTIFICATION: Use a low-power microscope for whole animals, and a compound microscope for details of appendages after dissection in a glycerin-filled depression slide by means of insect pins glued to large tooth picks.

Body (thorax) and abdomen (pleon) shape distinguish suborders. All species have 7 conspicuous body segments at least in some stage of their life cycle. A varying number of the primitively 6 abdominal segments can be fused, and combined with the telson (tail plate) to form a pleotelson. The head, always fused with the 1st thoracic segment, carries the (usually shorter) 1st antennae and the 2nd antennae, composed of basal portion (peduncle) and whip (flagellum). Mouth parts consist of mandible, 1st and 2nd maxilla, and 1 maxilliped. The (5)-7 thoracic appendages (pereopods) are uniramous; their basal segment (coxa) is often enlarged laterally to form wing-shaped extensions that, in most groups, fuse more or less indistinguishably to the tergites (coxal plates). Medially the bases of the anterior pereopods carry flat extensions (oostegites) that together form a brood pouch (marsupium). The abdominal appendages (5 pairs of foliaceous pleopods and 1 pair of flattened or elongate-conical uropods) are biramous. The 2nd pair and sometimes the 1st are sexually dimorphic. In Valvifera, uropods form covers over all pleopods. In parasitic groups, the main body axis of the female can be bent and the legs can be reduced to hooks.

Preserve in 70% alcohol, adding 3% glycerin to prevent drying during long-term storage.

BIOLOGY: Sexes are separate except in some highly specialized parasites. In Epicaridea, usually the first specimen infesting a host becomes a female; later arrivals become males. In most Epicaridea the tiny male lives on the body of the female or in her brood pouch. Fertilization is internal, and mating is often accomplished when the female molts. Eggs (a dozen in small free-living species, several thousand in parasites) develop in the brood pouch. Growth is by molt and apparently continuous throughout life, except for species parasitic on crustaceans. Life-spans of 3-4 yr have been recorded for terrestrial species, but those of marine species are unknown. Many species, when disturbed, roll into a ball (hence "pill bug"). Food preferences for most species are unknown, but omnivorous scavengers and herbivores dominate. Anthuridea live in burrows of their own making from which they extend their bodies to catch food. Isopods are preyed upon by many invertebrates, fishes and birds.

DEVELOPMENT: Direct. The hatching stage is a postlarva (manca stage) with 6 pairs of legs; the 7th pair forms at the first molt just after release from the marsupium in most species. In the crustacean parasites the juvenile stage is planktonic for a short time.

REFERENCES: For general introduction see KAESTNER (1970) and the monograph by SCHULTZ (1969).

RICHARDSON (1902) did the first comprehensive study on Bermuda's species (marine and terrestrial); SCHULTZ (1972) reviewed the terrestrial species from Bermuda, including those that live near the sea, and BOWMAN & MORRIS (1979) clarified some synonymies. New species have been described by MARKHAM (1979; crustacean parasites), SCHULTZ (1979; from deep water), and SKET (1979) and BOWMAN & ILIFFE (1983) from marine caves.

## Plate 122

S.O. ANTHURIDEA: Isopoda more than 7× as long as broad. Head about as wide as remainder of body. Abdomen with uropods and pleotelson (or true telson in some) large and flattened, with many fringing setae. Second antennae longer in ♂ than in ♀. Benthic. Most live in holes in mud or sand, others on vegetation. (3 spp. from Bda.)

***Paranthura infundibulata*** Richardson: The 5 abdominal segments together shorter than posterior thoracic segment. Under rocks in 1-2 m.

**S.O. FLABELLIFERA:** Isopoda with head much narrower than thorax, often partially enveloped in 1st thoracic segment. With 7 free thoracic segments. Abdomen with 1-5 free segments plus a broad, sometimes flat, proportionately large, conspicuous pleotelson. The most commonly encountered suborder. (11 spp. from Bda.)

***Colopisthus parvus*** Richardson: Abdomen with 1 (sometimes 2 can be distinguished in dorsal view) free segments plus a pointed pleotelson. Frontal margin of head straight with slight medial protrusion between antennae. In algae.

***Eurydice littoralis*** (Moore): Posterior margin of pleotelson with 4 moderately long teeth and long plumose setae. Abdomen narrower than thorax. Antennae attached to frontal margin of head. In sand and plankton (!); also washed from vegetation.

***Paracerceis caudata*** (Say): Pleotelson of ♂ split (2 "tails"); of ♀ with large medial bump (without split). Uropods long and spine-like. Rolls into imperfect ball. Washed from vegetation, on rocks and in and among encrusting organisms; common.

***Dynamenella perforata*** (Moore): Pleotelson of ♂ with small slotted hole in posterior margin. Pleotelson of ♀ smooth, posterior margin broadly rounded. Uropods broad and paddle-like. Rolls into near-perfect ball. On piles among encrusting organisms, and in mantle groove of *Chiton tuberculatus*.

***Limnoria tuberculata*** Sowinski: Pleotelson flat with 3 tubercles arranged in triangle towards anterior edge. Body margins more or less parallel. In wooden structures and driftwood in tubes of very small diameter (when compared to holes of the clam *Teredo*). Of considerable economic importance because of their destructive habit.

***Exocorallana quadricornis*** (Hansen): ♂ with 4 "horns" on head between eyes, ♀ without (or with only slight indications). Thorax and abdomen with segments well defined. Pleotelson with medial dorsal groove; posterior margin pointed and edge fringed with long setae. In sponges.

***Cymothoa oestrum*** (L.): Head enclosed in 1st thoracic segment. Eyes and antennae small. Abdominal segments partially enveloped laterally by last thoracic segment. Pleotelson large and flat with no fringing setae. Legs end in hooks. Parasite in gills and mouth of jacks (Carangidae).

***Alcirona krebsii*** Hansen: Eyes large. Three abdominal segments show in dorsal view. Pleotelson flat with rounded posterior margin fringed with setae. Anterior legs with 1 large and 4 small claws. Parasite on the skin of fishes including

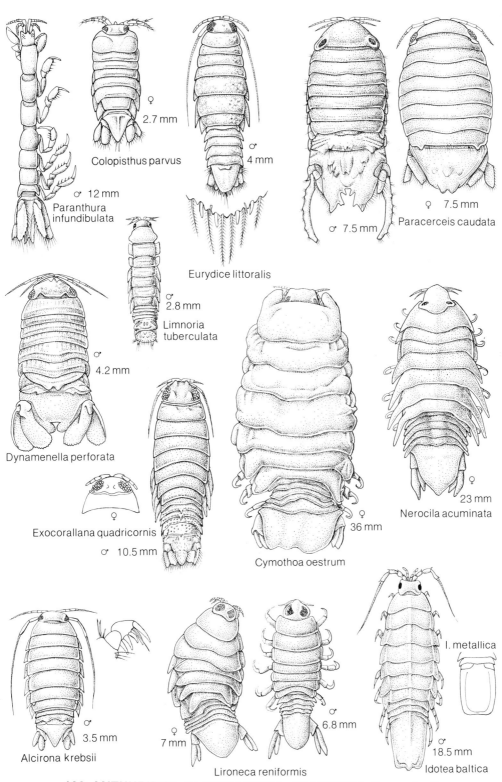

**122 ANTHURIDEA, FLABELLIFERA, VALVIFERA (Isopods 1)**

groupers; sometimes found not on a host.

***Nerocila acuminata*** Schioedte & Meinert: Edges of thoracic segments 4-7 and edges of coxal segments 4-7 long and pointed. Broad flat pleotelson. Uropods long, with both rami ending in a point. Ectoparasite; no record of the host fish from Bermuda.

***Lironeca reniformis*** Menzies & Frankenberg: Main body axis of ♀ bowed; ♂ straight. Eyes large. Coxal segments show in dorsal view. Legs with hooked ends. Caught by seine in shallow water; host fish unknown.

**S.O. VALVIFERA:** Isopoda with uropods forming valves or covers for pleopods. Mandibular palp absent. First antennae with all articles of flagellum generally fused. Most species live in the deep sea; several are planktonic. (2 spp. from Bda.)

***Idotea baltica*** (Pallas): Genus with more or less convex body margins, with 2 completely free abdominal segments, and a long pleotelson.—Species with acuminate (tapering to a point) posterior margin of pleotelson. Washed from vegetation, and in plankton.

***I. metallica*** Bosc: Genus as above.—Species with straight or truncate posterior margin of pleotelson. In floating Sargassum; uncommon.

## Plate 123

**S.O. ASELLOTA:** Generally dorsoventrally flattened Isopoda. Thoracic segments usually distinct. Pleotelson generally shield-shaped. Sexual dimorphism in some subgroups, especially in 1st pair of legs which, in the ♂, are subchelate and large. ♂ with 3 parts (operculate pleopods) to cover other pleopods in abdominal cavity; ♀ with 1 operculate pleopod. Mostly in algae and encrusting organisms, including the cavities of sponges. Many are epibenthic; some are infaunal. (5 spp. from Bda.)

***Stenetrium stebbingi*** Richardson: Antennae, legs and uropods long. First legs of ♂ large and subchelate. Antero-lateral margins of head acutely pointed. Washed from algae.

***Carpias bermudensis*** Richardson: Antennae, legs and uropods long. First legs of ♂ subchelate and large. (Intermediate types between mature and immature males have been described as separate species in Bermuda.) Washed from algae near shore, and from floating *Sargassum*.

***Joeropsis rathbunae*** Richardson: Antennae, legs and uropods short. Uropods incorporated into margin of pleotelson. Eyes located dorso-laterally. Washed from algae.

**S.O. EPICARIDEA** (=BOPYRIDEA): Parasitic Isopoda with distinct sexual dimorphism: ♀ tiny to large, body symmetric, asymmetric or without definite form; antennae and eyes, if present, short and inconspicuous; legs short and modified for attachment to host (sometimes absent); body segments often difficult to recognize. ♂ tiny, usually with thoracic segments distinct; abdominal segments variously indicated or not indicated. Parasites

# ISOPODA

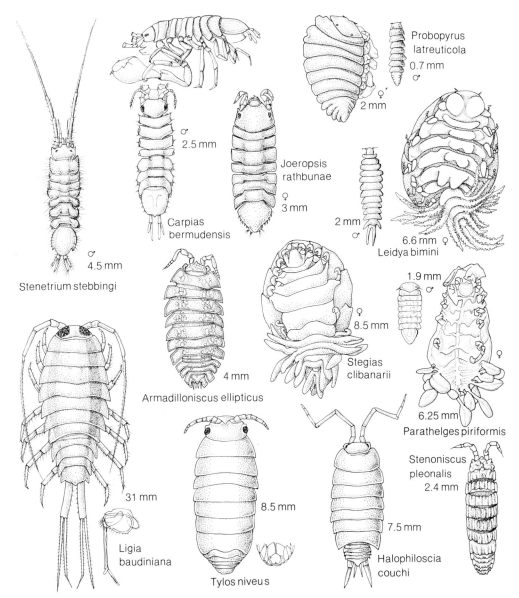

**123 ASELLOTA, EPICARIDEA, ONISCOIDEA (Isopods 2)**

on and in gill chamber of many crustaceans including other isopods, sometimes within body of crabs. (13 spp. from Bda.)

***Probopyrus latreuticola*** (Gissler): Body flat, main axis bowed. Abdominal segments only partially indicated in dorsal view. ♂ with abdominal segments indicated laterally only. In the gill chambers of the shrimp *Latreutes fucorum*.

***Leidya bimini*** Pearse: Very irregular in shape. Blind. Thorax with large bosses dorsally; legs short. Abdomen with lace-like extensions on segments. ♂ with all abdominal segments complete, and with 2 long extensions on posterior segment. In the gill chamber of the crab *Pachygrapsus transversus*.

***Stegias clibanarii*** Richardson: Irregular in shape. Anterior thoracic segments bend in semicircle around base of head. Abdominal segments with long extensions that have smooth (not lace-like) margins. ♂ unknown. In the gill chamber of the hermit crab *Clibanarius tricolor*.

***Parathelges piriformis*** Markham: Very irregular in shape with anterior part of body narrower than posterior part. Extensions on abdominal segments with stem; i.e., narrow at base and broadened apically. In the gill chamber of the hermit crab *Pagurus brevidactylus*.

**S.O. ONISCOIDEA:** Terrestrial Isopoda with 7 pairs of walking legs, 1 pair of conspicuous antennae and an abdomen with 5 segments plus a pleotelson. Most species live in moist habitats on land, others in the littoral zone, even in seawater. Species other than those included here are occasionally found in the intertidal zone, especially at night at low tide, when they feed. (16 spp. from Bda.)

***Tylos niveus*** Budde-Lund: Body highly arched, legs short; rolls into an almost perfect ball. Ventral extensions of abdominal segments approximate each other along most of inner margin. In seaweed, in crevices and under detritus of the intertidal zone.

***Ligia baudiniana*** Milne Edwards (Wharf louse, Slater): Uropods very long; antennae very long; eyes large. Pleopod 2 (♂) with elongate endopod with secondary small extension on apex (usually folded laterally). On seawalls, piles, mangrove roots and under rocks, detritus and seaweeds almost anywhere in the intertidal zone, especially on rocky shores. Runs rapidly and swims away quickly when disturbed.

***Armadilloniscus ellipticus*** (Harger): Body outline elliptical. Dorsoventrally flattened, with short antennae. Clings to the bottom of rocks and boards in the sandy intertidal.

***Halophiloscia couchi*** (Kinahan): Antennae long, with 3 flagellar articles; uropods long; abdomen abruptly narrower than thorax. In the maritime drift high on the beach under leaf litter; sometimes in caves.

***Stenoniscus pleonalis*** Aubert & Dollfus: Body highly arched and highly calcified; antennae short; uropods indistinct. Very passive, feigning death when sifted from more or less dry beach sand which it resembles in consistency and color.

G. A. SCHULTZ

## ORDER AMPHIPODA

CHARACTERISTICS: *MALACOSTRACA without a carapace; typically with a laterally compressed, slightly arched body, a well-defined head, a thorax of 7 segments, each bearing a pair of uniramous walking legs, and an abdomen of 6 segments. Eyes sessile; gills at the inner base of pereopods.* Body length from 0.1 to 28 cm, with Bermudian forms usually less than 1 cm. Body colors are variable and cryptic. Those living on algae are green, dark red or brown, often with white or black spots. Infaunal and pelagic forms are usually white or cream colored to translucent. Some species are able to change color to match the substrate. Flexion of the abdomen and rapid beating of the pleopods provides propulsion for swimming. Walking is usually accomplished by pulling with the forward perepods and by pushing with the rear pereopods. Stout antennae such as found on many tube dwellers may be used to help pull the body forward. Many species ("beach hoppers") are capable of jumping more than a meter by the sudden flexion of the abdomen and by pushing with the uropods.

The nearly 6,000 described species are apportioned to 4 suborders as follows: Gammaridea (85%), Hyperiidea (9%), Caprellidea (6%) and Ingolfiellidea (1%). At least 50 species occur in Bermuda; 22 are included here.

OCCURRENCE: In many freshwater and almost all marine habitats from the supratidal to the hadal, amphipods are common free-living or tubiculous animals on or within most kinds of algae, other animals (including jellyfish, hydroids, coral, sponges, sea urchins, tunicates, fishes and whales), and most mud or sand bottoms. Benthic forms are usually gammarids or caprellids, nearshore plankton forms a mixture of gammarids and hyperiids, and open-ocean forms almost exclusively hyperiids.

Collecting methods differ according to habitat. Algae, hydroids and other material containing amphipods may be shaken in seawater containing formalin, forcing the animals to move into the water, from which they may then be screened. Caprellids generally must be removed from their substrate using forceps. Pelagic amphipods may be collected by plankton net tows; lights held over the water at night often attract unusual benthic and pelagic forms, which may then be netted.

IDENTIFICATION: Accurate identification to genus and species, usually based on male characteristics, necessitates the use of both dissecting and compound microscopes. To avoid loss of mouthparts during dissection, the entire buccal mass of small specimens should be removed from the body and dissected in mounting medium on a glass slide. Body parts should be taken from one side only and mounted in a temporary medium (glycerin or 7% ethylene glycol) or in a permanent medium. Media such as CMC-10 or Hoyer's that do not require dehydration of the specimen before mounting are the most convenient to use. The addition of a small amount of stain to the mounting medium makes it easier to locate small parts on the slide.

Note the generally compressed and arched body with a well-defined head, a thorax of 7 freely articulated segments (pereonites) and an abdomen of 6 segments (pleonites). The abdomen is divided into an anterior pleosome with 3 pairs of pleopods (swimmerets) and posterior urosome with 3 pairs of posteriorly directed uropods. The telson, which can be entire or split into 2 lobes, is a flap over the anus attached to pleonite 6. Besides a pair of eyes the head bears 2 pairs of antennae, each composed of basal part (peduncle) and flagellum; antenna 1 can be provided with an accessory flagellum. Mouthparts consist of epistome, upper and lower lip, mandibles, 1st and 2nd maxillae and maxillipeds; the paired appendages are often provided with a palp. The thorax bears 7 pairs of uniramous walking legs (pereopods), the first 2 pairs of which are often modified for grasping (gnathopods). The numbering system used here for pereopods designates the first 2 legs gnathopods 1 and 2, with the remaining pereopods numbered 3 to 7. Each pereopod has 7 articles of which the 1st (coxa) is often laterally expanded downward to cover part of the remaining leg, and medially to form spoon-shaped processes (oostegites) that enclose the brood pouch (marsupium). Pereopods may be modified for grasping, and termed subchelate (article 7, the dactyl, closing against art. 6), or merochelate or carpochelate, respectively (extensions of art. 4, the merus; or art. 5, the carpus, forming an opposable thumb for the dactyl). Uropods consist of a peduncle (base) and 2 rami (branches).

Preserve specimens initially in 5% buffered formalin in seawater and transfer later to 70% or 95% ethanol for sorting and identification. Ethanol imparts a brittleness to the appendages that greatly facilitates dissection but formalin tends to preserve the animals' natural color. Some species must be relaxed with magnesium chloride or chloral hydrate before killing to avoid loss of antennae or other delicate appendages. Amphipods decay rapidly after death and should be preserved as soon as possible after collection.

BIOLOGY: Sexes are separate although consecutive hermaphrodites are known. Sexual dimorphism is common but not universal. Males are usually larger than females and often possess large, strongly subchelate gnathopods. These are used to carry the female prior to copulation. Eggs are deposited and fertilized in a marsupium where they are held up to several weeks before hatching. Juveniles may remain in the marsupium until after their first molt. The number of eggs in a clutch is generally less than 20, and each female usually reproduces more than once. Growth is rapid at first and decreases with age. Molts occur every 15-30 days and sexual maturity is usually attained by the 5th-7th molt. Although some species are known to live 4-5 yr, most live less than 1 yr. Generations are usually overlapping. Free-living forms are usually herbivores, feeding on macroalgae, but may also be carnivores, omnivores or detritivores. Tube-dwelling, pelagic and free-living infaunal forms may be deposit or filter feeders or may feed on interstitial organisms scraped from the surfaces of sand grains. Predators include invertebrates such as nemertean and nereid worms, birds, and adult and juvenile fishes of many species.

DEVELOPMENT: Except for certain parasitic hyperiids, newly hatched amphipods closely resemble the adults and have no free-living larval stages.

REFERENCES: For a general orientation, systematic overview and introduction to the literature dealing with gammarids see BARNARD (1958, 1969), BOUSFIELD (1973) and SMITH & CARLTON (1975). Caprellid systematics are reviewed by McCAIN (1968), Hyperiidea by HURLEY (1955) and Ingolfiellidea by STOCK (1976).

The review of Bermuda amphipods by KUNKEL (1910) is valuable but out-of-date; some distribution data of plankton species can be found in MOORE (1949), MORRIS (1975) and GRICE & HART (1962).

## Plate 124

S.O. **GAMMARIDEA:** Laterally compressed Amphipoda with 3 pairs of elongate biramous pleopods and 7 pairs of walking legs. Eyes rarely coalesced. Maxilliped with palp. Widely distributed in marine, freshwater and semiterrestrial habitats.

F. **AMPITHOIDAE:** Gammaridea with antenna 1 longer than antenna 2; accessory flagellum present or absent. Lower lip with notched outer lobes; mandible with strong molar, usually bearing palp. Uropods 3 biramous, rami shorter than peduncle, outer armed with 1 or 2 hooks; telson entire, short.

*Cymadusa filosa* Savigny (=*Grubia coei* Kunkel): Antenna 1 with a distinct 1-articulate accessory flagellum; antenna 2, gnathopods 1 and 2 with dense plumulose setae; uropod 1 with large interramal spine; uropod 3 inner ramus longer than outer. Common in algae, dead coral. Intertidal to shallow subtidal.

*Ampithoe rubricata* (Montagu): Antennae 1 and 2 subequal, about 1/2 body length; accessory flagellum absent; peduncle of an-

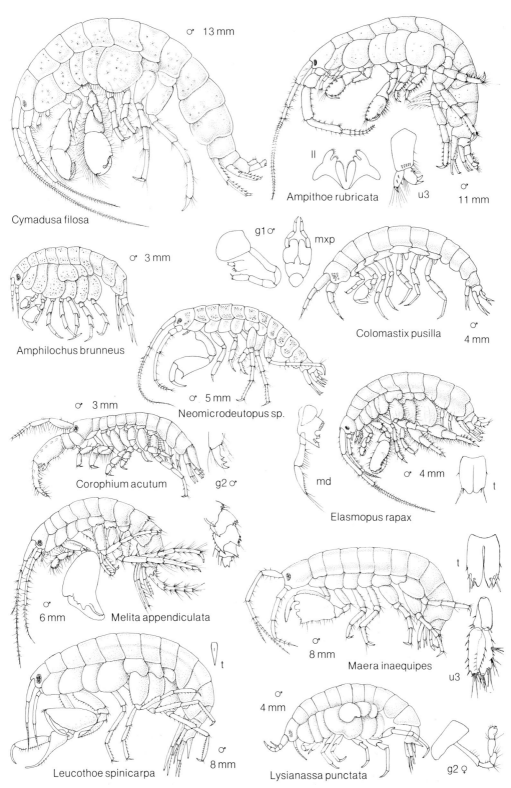

## 124 GAMMARIDEA 1 (Amphipods 1)

tenna 2 stout, twice diameter of antenna 1 peduncle. Uropod 3 peduncle with distal row of 5 stout spines, rami subequal. Moderately common in dead coral, algae. Intertidal to shallow subtidal.

F. **AMPHILOCHIDAE:** Gammaridea with short, subequal antennae; accessory flagellum absent. Mandible usually with 3-articulate palp and strong molar; coxa 1 very small, hidden by following coxae which are usually enlarged; uropod 3 biramous; telson entire.

*Amphilochus brunneus* Della Valle: Gnathopod 2 subchelate with 5th article extending to distal edge of article 6. Uropod 3 long, extending beyond uropods 1 and 2, rami lanceolate, outer shorter than inner; telson slender, triangular. Moderately common in algae, dead coral and mangrove detritus. Intertidal to shallow subtidal.

F. **AORIDAE:** Gammaridea with antenna 1 slightly longer than antenna 2; accessory flagellum long and multiarticulate to absent. Pereopod 7 longer than pereopods 5 or 6; mandible with palp and molar. Uropod 3 biramous or uniramous, rarely projecting beyond uropods 1 and 2; at least 1 ramus of uropod 3 as long or longer than peduncle; telson entire, short.

*Neomicrodeutopus* sp.: Antenna 1 slender, with minute 1-articulate accessory flagellum; antenna 2 stout. Gnathopod 1 carpochelate with long distal tooth on article 5; coxae small, rounded; uropods 3 uniramous, lanceolate. Rare, in algae and dead coral. Shallow subtidal.

F. **COLOMASTIGIDAE:** Gammaridea with short, stout antennae; accessory flagellum absent. Body not strongly laterally compressed, subcylindrical; mandibular palp absent; maxilliped palp slender, inner plate reduced, outer plate large. Uropod 3 biramous; telson entire.

*Colomastix pusilla* Grube: Antenna 1 slightly longer and stouter than antenna 2, both with reduced 3-articulate flagella. Gnathopod 1 dactyl highly reduced on ♂, on ♀ dactyl replaced by a bundle of setae. Uropods all biramous, rami styliform, subequal. Common in sponges and shell/worm tube/sponge associations. Intertidal to shallow subtidal.

F. **COROPHIIDAE:** Gammaridea with antennae variable in length and stoutness; accessory flagellum often absent. Mandible with palp and molar; body subcylindrical, generally depressed; pereopod 7 often longer than pereopods 5 and 6; coxae usually not overlapping. Some or all urosomites often coalesced; uropod 3 short, usually uniramous; telson entire.

*Corophium acutum* Chevreux: Antenna 2 longer than antenna 1, stout; accessory flagellum absent; ♂ with large distal tooth on article 4 of antenna 2. Mandibular palp 2-articulate; anterior margin of articles 2 and 4 on peropods 3 and 4 lack long setae, article 5 very short; ♂ gnathopod 2 dactyl with distinct teeth. Moderately common in coralline algae. Intertidal to shallow subtidal.

F. **MELITIDAE:** Gammaridea with antenna variable in length, rarely stout; accessory flagellum variable in articulation but always present. Mandible always with strong grinding molar and 3-articulate palp; gnathopods usually strongly subchelate, rami of uropods 3 never shorter than peduncle; telson usually deeply cleft. Gnathopod 2 distinctly the larger in the ♂.

*Elasmopus rapax* Costa: Accessory flagellum short, 1-articulate; maxilla 2 inner plate margin bare; inner margin of 3rd article of mandibular palp heavily setose. Uropod 3 with short, stout, spinose rami; telson cleft and with only subapical lateral spines. Common in mangrove detritus, dead coral, coralline algae and sand around *Thalassia*. Intertidal to 30 m.

*Melita appendiculata* (Say) (=*M. fresnelii* Audouin): Accessory flagellum 5-articulate; ♂ gnathopod 2 with very large 6th article, dactyl large, stout. Urosome carnate; uropod 3 biramous with outer ramus very long and spinose, inner ramus scale-like. Very common in dead coral, *Thalassia* detritus, sponges and tunicates. Intertidal to shallow subtidal.

*Maera inaequipes* (Costa): Accessory flagellum 7-articulate; mandibular palp article 3 slender, not heavily setose; distal margin of article 6 of gnathopod 2 with distinct rounded notch. Very common in coral, coralline algae and sand around *Thalassia*. Intertidal to shallow subtidal.

F. **LEUCOTHOIDAE:** Gammaridea with short antennae; accessory flagellum reduced or absent. Mandible without molar; gnathopod 1 carpochelate; gnathopod 2 article 5 elongate. Uropod 3 biramous, rami styliform, shorter than peduncle; telson entire.

*Leucothoe spinicarpa* (Abildgaard): Accessory flagellum 1-articulate; telson long and slender, triangular, tapering to an acute apex, 3 times as long as broad. Moderately common in dead coral, sponges, tunicates (*Ascidia* and *Clavelina*) and algae. Intertidal to shallow subtidal.

F. **LYSIANASSIDAE:** Gammaridea with short antennae; accessory flagellum multiarticulate. Mandible with palp; gnathopod 2, articles 6 and 7 "mitten-shaped"; coxae 1-4 larger than 5-7. Uropod 3 biramous; telson cleft, emarginate or entire.

*Lysianassa punctata* (Costa): Accessory flagellum 4-articulate; uropod 3 rami subequal, styliform and shorter than peduncle; telson entire with shallow apical depression and a short spine on each side. Common on algae; often found associated with urchins buried in clean sand. Shallow subtidal.

## Plate 125

**F. PHLIANTIDAE:** Dorsally depressed Gammaridea with short antennae without accessory flagellum. Mandible without palp; coxae splayed; gnathopods feeble. Uropod 3 usually uniramous, often without rami; telson usually entire.

*Pariphinotus tuckeri* Kunkel: Antennae with bundle of setae at end of each flagellum; maxilla 1 lacks palp; maxillipedal palp 4-articulate. Head with large spatulate rostrum; body ridged dorsally; uropod 3 absent. Rare, in dead coral debris and in detritus around *Padina*. Intertidal to shallow subtidal.

**F. PODOCERIDAE:** Gammaridea with usually long, slender and heavily setose antennae; accessory flagellum often absent. Body often subcylindrical or depressed; urosome strongly depressed, often toothed; pleonite 1 more than twice as long as pleonite 2; coxae small. Uropods 2 and 3 often reduced or absent; telson entire.

*Podocerus* sp.: Accessory flagellum minute, 1-articulate; uropods 3 minute, rami absent; telson small, tapers to a narrow knob with 2 short apical spines. Uncommon, in filamentous algae. Often loses antennae when collected. Shallow subtidal to 10 m.

**F. SYNOPIIDAE:** Gammaridea with usually slender antennae of variable length; accessory flagellum multiarticulate. Eyes coalesced when present; head massive, usually with a downturned rostrum; mandible with molar but poorly developed palp. Gnathopods feeble, usually heavily setose on the posterior margins of articles 5 and 6. Uropods biramous, lanceolate; telson cleft or entire, usually elongate.

*Synopia ultramarina* Dana: Flagella without long hairs; eyes present, coalesced. Telson cleft, much longer than broad, with apicolateral margins serrate, each apex with at least 2 setae. Moderately common in nearshore surface waters at night.

**SUP.F. TALITROIDEA:** Gammaridea with antenna 1 shorter than antenna 2; accessory flagellum absent. Mandible lacks palp but with strong molar. Uropod 3 short, inner ramus reduced or absent; telson cleft or with fused lobes, short.

**F. HYALIDAE:** Talitroidea with antenna 1 longer than peduncle of antenna 2. Maxilliped palp article 4 claw-shaped; lower margins of coxae bare or weakly spinose; telson never entire, usually deeply cleft.

*Parhyale hawaiensis* (Dana) (identified by Kunkel (1910) as *Hyale prevostii*, *H. pontica* and *H. trifolidens*): Palp on maxilla 1 is 1-articulate; article 6 on pereopods 5, 6 and 7 with groups of spines on front and hind margins. Uropod 1 with interramal spine; uropod 3 with minute inner ramus. Very common in

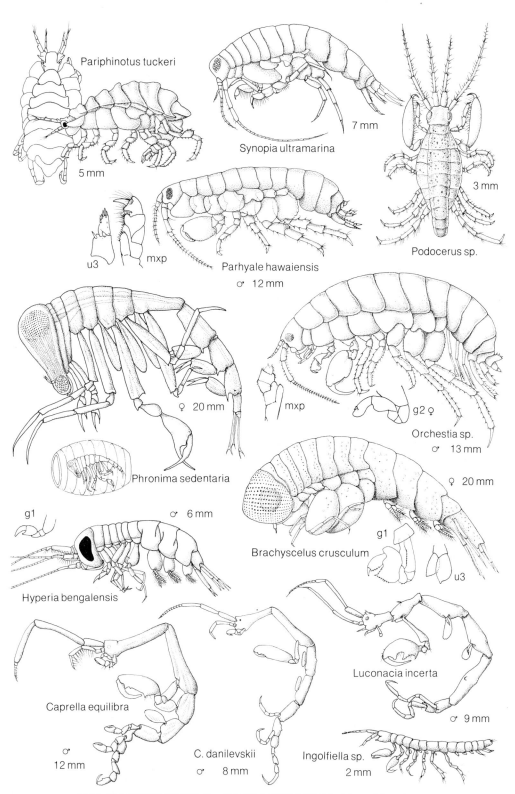

**125 GAMMARIDEA 2—INGOLFIELLIDEA (Amphipods 2)**

dead coral and algae. Intertidal to shallow subtidal.

F. **TALITRIDAE** (Beach hoppers): Talitroidea with antenna 1 shorter than peduncle of antenna 2; antenna 2 with enlarged peduncle. Maxilliped palp not claw-shaped, 4th article reduce or missing; articles 6 and 7 on gnathopod 2 form "mitten" on young and ♀. Uropod 3 uniramous; telson entire, often with apical notch.

*Orchestia* sp.: Pereopod 7 longer than pereopods 5 and 6, articles 4 and 5 not greatly enlarged when compared to articles 6. Uropod 1 extends beyond uropod 3, outer ramus lacks marginal spines; uropod 3 lacks marginal spines; telson with dorsal and apical spines. Extremely common under *Sargassum* beach wrack or buried in sand. Supratidal.

S.O. **HYPERIIDEA:** Usually semitransparent Amphipoda with large head mostly covered by eyes. Coxae small or absent; maxilliped palp absent. Common in open ocean, often associated with jellyfish or pelagic tunicates.

F. **PHRONIMIDAE:** Hyperiidea with antenna 1 very short with 2-articulate flagellum in ♀, multiarticulate in ♂. Mandible without palp; pereopod 5 subchelate. Uropods biramous, lanceolate.

*Phronima sedentaria* (Forskål): No antenna 2 on ♀; short, 2-articulate with terminal setae on ♂. Gnathopods 1 and 2 simple; ♂ 1st pleonite as long as 7th pereonite. Uncommon, offshore to 500 m. Lives in the test of salps, where it also lays its eggs.

F. **HYPERIIDAE:** Hyperiidea with antennae generally long, not reduced. Mandibles with palp. Uropods biramous; telson entire.

*Hyperia bengalensis* (Giles) (=*H. atlantica, Lestrigonus bengalensis*): Gnathopods 1 and 2 weakly carpochelate; article 5 of pereopods 3 and 4 not expanded; pereopods 5 and 6 equal, not elongate. Uropods 3 rami slender, elongate and sharp-pointed; telson longer than wide, narrowing to a subacute apex. Common in offshore waters.

F. **BRACHYSCELIDAE:** Hyperiidea with terminal article of flagellum on antenna 2 short. Gnathopods 1 and 2 carpochelate. Peduncles of uropods 1 and 2 subequal; uropod 3 outer ramus broad.

*Brachyscelus crusculum* Bate: Terminal article of antenna 2 nearly half the length of the preceding article; article 5 on gnathopods 1 and 2 broad and serrate. Uropod 3 outer ramus leaf-like, only slightly broader than inner ramus. Moderately common offshore.

S.O. **CAPRELLIDEA** (Skeleton shrimps): Elongate, skeleton-like Amphipoda with greatly reduced abdomen. Antennae length variable; accessory flagellum absent. Pereonite 1 often fused with head; pereonites 2-4 bear gills; pereonites 3 and 4 bear a conspicuous marsupium in adult ♀; pereopods 3 to 5 are reduced in many genera. Exclusively marine.

F. **CAPRELLIDAE:** Caprellidea with antenna 2 usually shorter than antenna 1. Mandible with molar but without palp; pereonites 3 and 4 with gills but without appendages.

*Caprella equilibra* Say: Genus with flagellum of antenna 2 biarticulate; abdomen of ♂ with pair of appendages and pair of lobes; ♀ with pair of lobes.—Species with strong sharp process extending from venter of pereonite 2 between 2nd gnathopods. Articles 6 and 7 on gnathopod 2 form large subchelate hand. Gills ovate, long axis perpendicular to body. Common on hydroids, bryozoans, colonial ascidians, *Thalassia* and algae. Preys on wide variety of invertebrates. Intertidal to 3,000 m.

*C. danilevskii* Czerniavski: Genus as above.—Species with elliptical gills with long axis parallel to body; without ventral process on pereonite 7. Common on *Thalassia* and *Sargassum*, occasionally in plankton.

F. **AEGINELLIDAE:** Caprellidea with mandible with molar and palp. Gills on pereonites 3 and 4 only; pereonites 3 and 4 with rudimentary, often minute, appendages.

*Luconacia incerta* Mayer (=*Protellopsis stebbingii* Pearse): Antennae lacking dense setae; flagellum of antenna 2 2-articulate; pereopods 3 and 4 2-articulate; pereopod 5 6-articulate and unserted near middle of pereonite 5. Gnathopod 2, article 6 with distinct U-shaped depression on distal margin. Common on mangrove roots, *Sargassum*, *Thalassia*, sponges, hydroids and ascidians.

S.O. **INGOLFIELLIDEA:** Mostly minute, elongate-cylindrical Amphipoda without eyes; pereonites and pleonites disinct, not fused. Gnathopods large, pereopods undifferentiated, pleopods reduced. (1 sp. from Bda.)

F. **INGOLFIELLIDAE:** Ingolfiellidea with 1st pleonite not fused with the head; 2nd gnathopod not chelate. Pleopods rudimentary, uniramous.

*Ingolfiella* sp.: Uropod 2 hardly longer than uropod 1; peduncle of uropod 2 with less than 15 transverse rows of setae. A single specimen collected from an inland marine cave.

S. E. JOHNSON

## Subphylum Uniramia

CHARACTERISTICS: *Mostly terrestrial ARTHROPODA with unbranched (uniramous) limbs, 1 pair of antennae and with mandibles. Head highly sclerotized to form a capsule.*

In addition to the mandibles (which can be adapted for various modes of food uptake) there may be 1 or 2 pairs of maxillae. Respiratory organs, when present, are in the form of tracheal tubes. Eyes occur as single ocelli and compound (faceted) eyes.

The subphylum contains the 5 classes Chilopoda, Symphyla, Diplopoda, Pauropoda and Insecta. Some systems group the first 4, many-legged classes together as Myriapoda, to contrast them against the 6-legged Hexapoda (=Insecta). There is also ongoing discussion regarding the conventional concept—which we have followed here—of uniting all of the various groups of primitively wingless "insects" as the subclass Apterygota of Insecta. In Bermuda, only Chilopoda (p. 382) and Insecta (p. 383) have species with marine affinities.

### Class Chilopoda (Centipedes)

CHARACTERISTICS: *Elongated UNIRAMIA with many (15-180) more or less uniform trunk segments each with 1 pair of legs. First pair modified as maxillipeds with poison claws. Head flattened, with mandibles and 2 pairs of maxillae.* Although terrestrial tropical species may reach more than 25 cm in length, littoral species are much smaller. Yellowish to dark brown; many can run (and bite!) with impressive speed.

Of nearly 3,000 known species in 4 orders, most of them occurring terrestrially in the tropics, about 10 have been recorded in Bermuda; only 1 is common in the littoral fringe and therefore included here.

OCCURRENCE: In dark, moist locations such as rock crevices, under bark and in soil.
Collect littoral species by breaking open intertidal rocks, or by treating pieces of such rock in the Berlese apparatus.
Though the species listed here is harmless to humans, another, larger centipede occurring in Bermuda's terrestrial environment can inflict a reportedly very painful bite.

IDENTIFICATION: A pocket lens or low-power dissecting scope is sufficient.
Note the strong, claw-shaped maxillipeds; the spiracles of the tracheal system that open on the lateral (more rarely on the dorsal) side of the segments; and the genital opening anterior to the anus.

Fix and preserve in 75-80% ethyl alcohol.

BIOLOGY: Dioecious, with little or no external sexual dimorphism. Sperm transfer is indirect; the male deposits sperm drops or spermatophores, generally on silk threads, from where the female picks them up; this is often preceded by courtship behavior. In some orders the female protects the eggs and developing young by curling herself around them. Their life-span can exceed 6 yr. Centipedes are carnivores, feeding mainly on other arthropods as well as worms and mollusks which they kill with their poison claws. Predators (of terrestrial species) are toads, and possibly birds.

DEVELOPMENT: The hatching juvenile may already have the definitive number of segments and legs (as in Geophilomorpha); in other groups the final number is reached after several molts.

REFERENCES: For general orientation see EASON (1964) and HOFFMAN (1982).
Bermuda's centipedes have been treated by CHAMBERLIN (1920).

### Plate 126

O. **GEOPHILOMORPHA** (Soil centipedes): Slender, with 31 to approximately 180 pairs of short legs. Yellow to light brown; without eyes. If disturbed, they coil up or writhe like a snake.

*Hydroschendyla submarina* (Grube): To 32 mm; body yellow, head and posterior end reddish brown. With 47-49 pairs of legs. In the upper intertidal and lower supralittoral region, in porous stone and crevices; widespread but difficult to find and extract.

R. SCHUSTER

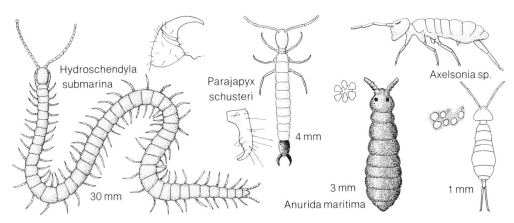

**126 CHILOPODA (Centipedes), APTERYGOTA (Wingless insects)**

## Class Insecta (Insects)

CHARACTERISTICS: *Generally compact UNIRAMIA consisting of relatively few segments organized into 3 regions: head, thorax and abdomen. Head with 1 pair of mandibles and 2 pairs of maxillae (2nd pair fused into a labium). Thorax of 3 segments, each with 1 pair of walking legs; abdomen with rudimentary appendages or none.* Length ranges from 0.2 to 250 mm but the majority of the species are in the 10-50 mm range. Coloration is highly variable, often bright; locomotion is by walking, flying or swimming.

With more than 750,000 described species (in 2 subclasses and about 30 orders) this is by far the largest of all animal classes. Approximately 2,000 species have been reported from Bermuda; of these, 18 are included here as having marine affinities.

OCCURRENCE: The great majority are terrestrial; many are aquatic during at least some stage of development, but fewer than 300 are thought to be obligatorily marine, spending at least part of their life on the surface of or submerged by saltwater. Many more are "marine associates" living in intertidal areas but retreating before the incoming tides or occupying supratidal habitats. Species of the genus *Halobates* are among the very few insects to live permanently on the open ocean.

A light, strong net is required for collecting flying insects, and a sieve of some form is useful in capturing specimens in water. Less active terrestrial forms can be collected with a pair of forceps, an artist's brush moistened with alcohol or an aspirator. Seaweed supports a varied community of arthropods, including insects, and the latter may be recovered by placing samples of seaweed in a standard Berlese funnel with a heat source above (a gooseneck lamp is satisfactory). Specimens retreating from the heat may be collected at the base of the funnel in a small beaker or watch glass containing 75% alcohol. Many winged forms are attracted to lights at night where they may be collected. Larval and pupal forms living in sand and mud may be recovered by various flotation procedures.

Insects as a whole have a mixed reputation as pests of humans, livestock and crops on one hand, and as pollinators, producers of honey, silk and dyes, and as agents of biological control on the other. Some marine insects listed here (*Aedes* spp., *Culicoides* spp., *Tabanus* spp.) can be a nuisance to beach users because of their bloodsucking habits.

IDENTIFICATION: Because of the vast number of insect species, their high diversity and relatively small size, determination of specimens to order is challenging, and placement into families is generally as much as can be hoped for by the nonspecialist. This is particularly true of immature forms. Notwithstanding the above, field determination of most species covered here should be possible if habitat information is considered along with external characters. A good hand lens (10×) is indispensable for field determination of specimens, which are best examined dead, having been killed in alcohol or a killing bottle charged with ethyl acetate or sodium, potassium or calcium cyanide.

The presence of wings, even vestigial ones (wing pads), on thorax segments 2 and 3, automatically places specimens in the subclass Pterygota. Unfortunately, the absence of wings does not mean the specimen falls into the Apterygota because immature forms of many Pterygota, and a considerable number of adult forms in this subclass, are secondarily wingless. Insects with only 1 pair of wings belong to the order Diptera (true flies) and can easily be placed in this taxon. Coleoptera (beetles) have 2 pairs of wings, the outer pair of which (elytra) is horny or leathery, completely concealing the more typical membranous pair and often masking the presence of wings on casual observation. The position in which wings are held at rest is a useful characteristic as are the wing veins, and both can be used in placing specimens into families and lower taxa.

Mouthparts consist of a labrum (upper lip), 1 pair of mandibles, 1 pair of maxillae, and a labium (lower lip; the fused 2nd maxillae). Originally designed for chewing, insect mouthparts have adapted to a wide range of food and hence represent valuable characters for identification. The beetles included here all have mouthparts adapted for biting and chewing. The water strider (*Halobates*) and the water boatman (*Trichocorixa*) have strong stout "beaks" and are generally believed to be predacious fluid feeders; the flies included have mouthparts adapted for sucking or sponging liquids. *Tabanus*, *Aedes*, and *Culicoides* have mouthparts developed for piercing and sucking which they effectively use to withdraw blood from hosts.

Other features used in identification are the shape and number of articles (segments) of antennae and walking legs, particularly the feet (tarsi), and the presence and shape of abdominal appendages such as pincer-like cerci, or jumping organs (as in Collembola).

Most adult and immature insects may be preserved satisfactorily in 75% isopropyl alcohol. Insects may also be preserved dry, preferably pinned through the thorax or, in the case of small specimens, mounted on a card "point" which in turn is supported by a pin. All adult flies, as well as moths and butterflies, should be preserved dry. Slide preparations are appropriate for small, delicate specimens.

BIOLOGY: Insects, with rare exceptions, are dioecious, but do not always exhibit sexual dimorphism. Sperm transfer is by copulation except in some Apterygota where the male deposits spermatophores on the ground for uptake by the female. Most species are oviparous but a considerable number are viviparous. Parthenogenesis is fairly common, with males being completely unknown in a few species. There are even insects that are capable of reproduction in the larval or immature stage (paedogenesis). Life expectancy for adult insects ranges from a single day for such groups as the mayflies and midges, to several years in the case of the queens of certain social insects (e.g., termites and ants). Insects generally spend far more of their life in an immature form than as an adult. As a group, insects feed on almost everything. The species covered here include, as adults, carnivorous scavengers, predators and bloodsuckers. Predators of insects include fish, crabs and birds and other insects, which serve also as important parasites.

### Plate 127

DEVELOPMENT: The Apterygota have a simple (ametabolous) development in which the form of the body remains basi-

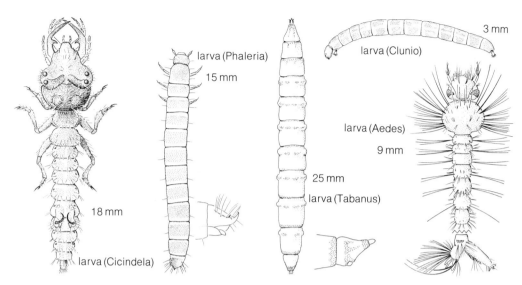

**127 INSECTA (Insects): Development**

cally unaltered from the time of hatching of the egg, apart from the increase in size following each molt.

Most Pterygota undergo a more complex (holometabolous) metamorphosis which, in certain orders including the Diptera and Coleoptera, involves 4 distinct stages: egg, larva, pupa and adult. Dipteran larvae are usually legless, with aquatic forms often provided with elaborate siphon tubes to facilitate breathing as in some mosquitoes. Larvae of the beetles covered here have 3 pairs of clearly distinguishable legs. Insects in the orders Hemiptera and Dermaptera undergo a gradual (hemimetabolous) metamorphosis in which there is generally some difference in outward appearance between immature, or nymphal stages, and the adult insect (particularly with the development of wings) but in which there are no larval or pupal stages.

REFERENCES: For a general introduction see BORRER & DELONG (1971); an extensive synopsis of living taxa is provided by BROWN (1982). Information on marine insects is presented by CHENG (1976); marine Collembola are summarized by STRENZKE (1955).

The only comprehensive work on insects of Bermuda, by OGILVIE (1928), is now outdated but has been supplemented by lists (WATERSTON 1940), card files and collections maintained by the Department of Agriculture and Fisheries as well as articles in the Department's monthly bulletin. Only a few marine taxa have been dealt with so far: midges (WILLIAMS 1956a, b; 1957, 1961; WIRTH & WILLIAMS 1957) and Diplura (NOSEK 1981).

*Plate 126*

**S.CL.** **APTERYGOTA** (Wingless insects): Wingless Insecta; abdomen consisting of 11 distinct segments or less, without normal legs but frequently with variously shaped appendages. Metamorphosis is ametabolous. (Of 4 orders 2 have marine representatives in Bda.)

**O.** **DIPLURA** (Two-pronged bristletails): Slender Apterygota with relatively long antennae consisting of many articles. Eyes lacking. Mouthparts drawn into the head capsule, not immediately visible externally. Cerci either fragile, segmented like an antenna, or unsegmented, pincer-shaped. Body soft, white or yellow in color; size mostly about 1 cm. (1 sp. from Bda.)

F. **JAPYGIDAE:** Diplura with abdomen terminating in pincer-shaped cerci. Body white or yellow, pincers brown or blackish.

*Parajapyx schusteri* Nosek: Length, including antennae and cerci, only 4.5 mm. Body white (sometimes with a black median stripe which is the gut content!), cerci brownish. In crevices of intertidal rocks; infrequent.

O. **COLLEMBOLA** (Springtails): Cylindrical or globular Apterygota with abdomen consisting of only 6 segments. The 4th abdominal segment bears ventrally a forked jumping organ (furca), the 3rd a pair of small appendages (tenaculum) for holding the furca in place, the 1st an unpaired cylindrical ventral tube. Antennae variable in length, mostly of 4 distinct articles. The majority of Collembola have eyes each consisting of maximally 8 ocelli surrounded by a pigmented area. Between the base of the antennae and the eyes there may be a sense organ (post- antennal organ). Mouthparts drawn into the head capsule, not immediately visible externally. Coloration of the body variable, size mostly 1-2 mm. (About 9 spp. from Bda.)

*Anurida maritima* (Guérin): Body cylindrical, with eyes of 5 ocelli on each side of the head. Furca and tenaculum lacking; the post-antennal organ, about as large as 1 eye, consists of 5-10 pearlike tubercles. Color uniformly to blackish blue. Size without antennae maximally 3 mm. Common, sometimes in great numbers, in the intertidal region of the rocky coast, in crevices but also outside on the rock surface.

*Axelsonia* sp.: Body cylindrical, with eyes of 6 large and 2 small ocelli in a dark area on each side of the head. Furca well developed, also tenaculum; post-antennal organ lacking. Color grayish, dorsally dark gray. Size without antennae and furca 1 mm. Between the tide lines on rocks as well as on the air roots of mangrove trees (e.g., in the *Bostrychia* epigrowth). Adults and juveniles run and jump very quickly and are therefore difficult to collect.

### Plate 128

S.CL. **PTERYGOTA** (Winged insects): Insecta mostly with 1 (or 2) pairs of wings on the 2nd (and 3rd) thorax segments; abdomen of adults without appendages. Metamorphosis hemimetabolous or holometabolous.

O. **DERMAPTERA** (Earwigs): Small to medium-sized, slender, flattened Pterygota, with or without wings. If wings are present, the front leathery pair completely hide the membranous, intricately folded hind pair. Integument smooth, tough and generally dark; mouthparts adapted for chewing. Body terminating in a pronounced pair of forceps-like cerci that are used in defense and in the capture of prey; metamorphosis is hemimetabolous. (4 spp. from Bda.)

**S.O. FORFICULINA:** Dermaptera with well-developed compound eyes; forceps strongly developed, body smooth and hard.

**F. LABIDURIDAE:** Forficulina with antennae of 16-30 segments; wings present or absent; 2nd tarsal segment cylindrical and not prolonged distantly beneath 3rd segment.

*Anisolabis maritima* (Gene) (Seaside earwig): Wingless, dark brown with pale legs; 18-24 mm long; antennae with 24 segments. ♂ with asymmetric cerci, the right one more curved than the left; ♀ with tips of cerci crossed. Found regularly in intertidal and supratidal situations, hiding under seaweed, stones and beach wrack.

**O. HEMIPTERA** (Bugs): Small to large Pterygota; a great majority with wings that at rest are held flat over the body; fore wings usually thickened at base, membranous at tip; hind wings completely membranous and shorter than forewings. Mouthparts in the form of a beak adapted for sucking, generally arising from the front part of the head; antennae with 5 or fewer segments, variable in length; tarsi with 3 or fewer segments. Hemimetabolous; immature forms similar to adults but with wings small or absent. (A very large order with over 20,000 described species including many that are aquatic.)

**S.O. CRYPTOCERATA** (Short-horned bugs): Hemiptera with antennae shorter than head, usually concealed in grooves beneath the eyes; ocelli (simple eyes) generally absent. Aquatic or shore-inhabiting forms.

**F. CORIXIDAE** (Water boatmen): Small to medium-sized, generally gray Cryptocerata. Dorsal surface of body flattened; front legs short with single-segmented, scoop-shaped tarsi; hind legs long with stiff hairs and functioning as oars. (About 100 species representing 12 genera have been reported worldwide from coastal and inland saline environments.)

*Trichocorixa reticulata* (Guérin-Ménéville): Small (5-6 mm), highly active aquatic bugs with 4-segmented antennae and a reticulate pattern on the pronotum and fore wings. Very common in brackish ponds and tide-wash pools; apparently able to tolerate wide ranges in salinity. Thought to be largely predacious on planktonic crustaceans and insect larvae.

**S.O. GYMNOCERATA** (Long-horned bugs): Hemiptera with antennae longer than head, and clearly visible from above.

**F. GERRIDAE** (Water striders): Generally dull-colored, predacious Gymnocerata with or without wings, which skate on the water surface. Antennae 4-segmented, eyes large and globular, ocelli absent. The body is covered with a velvet-like hair pile and is generally oval, but in the case of marine species the abdomen is very short and stubby; middle and hind legs are widely separated from the front pair. As far as is known all marine gerrids are wingless. (The family contains about 400 species; with *Halobates* containing 42 described species all of which can be considered marine, with 5 living on the surface of the open ocean.)

***Halobates micans*** (Eschscholtz) (Ocean skater): Small (3.5-6 mm), blue-gray, wingless; abdomen very short; fringe of hairs present on tibia and 1st tarsal segment of middle legs. Often collected offshore, in quiet waters inshore, or on beaches where they are frequently stranded. Appears to feed upon various planktonic crustaceans, insect larvae and fish larvae trapped in the surface film. Eggs are laid, often in great numbers, on floating objects on the ocean, including tar-balls, cork, debris and even on the bodies of seabirds as they rest on the ocean surface.

**O.** **COLEOPTERA** (Beetles): Pterygota with horny or leathery front wings meeting in straight line down the back and covering the membranous hind wings which are the only ones functional in flight. The front wings are often short, occasionally fused or even absent. Antennae generally with 11 or fewer segments. Mouthparts adapted for chewing; tarsi usually 3-5-segmented. Metamorphosis holometabolous. (This is the largest of all insect orders, with close to 300,000 described species. At least 50 species are considered to be obligatorily marine and a great many more are marine associates occurring in the upper tidal or supratidal region.)

**S.O.** **ADEPHAGA:** Coleoptera with 1st abdominal segment divided by hind coxa. Tarsi 5-segmented. Antennae 11-segmented and filiform. Almost always predacious.

**F.** **CICINDELIDAE** (Tiger beetles): Medium-sized (mostly 10-20 mm), elongate, slender, frequently brightly colored beetles of distinctive shape. Head as wide as or wider than pronotum; antennae inserted above base of mandibles; legs long and slender. Very active in flight or when running. Generally found in sandy habitat where they are effective hunters. Larvae live in vertical tunnels, waiting at opening to capture unsuspecting prey. (Only 1 sp. from Bda.)

***Cicindela trifasciata*** LeConte: Iridescent, metallic green with distinctive copper pattern on elytra; approximately 10 mm. Occasionally collected from intertidal and supratidal beach habitats.

**S.O.** **POLYPHAGA:** Coleoptera with 1st abdominal segment not divided by hind coxae; posterior margin extending completely across abdomen. Prothorax generally lacking notopleural sutures.

**F.** **STAPHYLINIDAE** (Rove beetles): Minute to medium-sized, elongate, parallel-sided Polyphaga with very short front wings that leave 3-6 abdominal segments exposed. Antennae 10- to 11-segmented, thread-like to clubbed in form. (A very large family with over 30,000 species, several hundred of which are found in marine habitats. Larvae are similar in form to adults. Both stages are predacious for the most part.)

***Cafius bistriatus*** (Erickson): Elongate, to 8-10 mm; dark brown, with 6 abdominal segments ex-

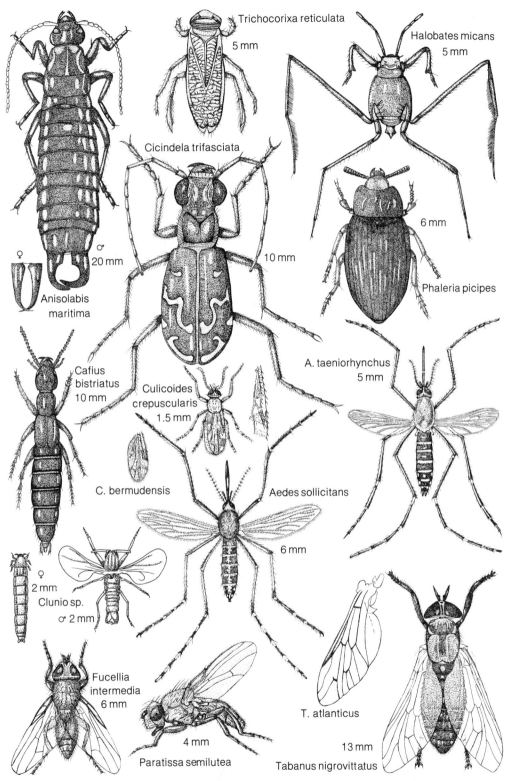

**128 PTERYGOTA (Winged insects)**

posed beyond wings. Anterior tibia with spines on outer edge. Very active both running and in flight. Commonly associated with beached seaweed.

F. **TENEBRIONIDAE** (Darling beetles): Generally brown or black Polyphaga variable in form; almost all with 11-segmented, thread-like or slightly clubbed antennae; tarsi 5-segmented on first 2 pairs of legs, 4-segmented on hind pair; 5 abdominal segments visible ventrally. Larvae long and cylindrical, with 2 distinctive terminal hooks on the abdomen. Nocturnal in habit and generally scavengers.

*Phaleria picipes* Say: Oval, brown, 5-6 mm; with slightly clubbed antennae; tibia of front legs flat, spiny, adapted for digging. Adults and larvae associated with seaweed where they live as scavengers. Common.

O. **DIPTERA** (Flies): Pterygota with 1 pair of membranous wings on 2nd thoracic segment (rarely absent); hind wings modified into small, knob-like structures (halteres) that are critical in governing balance. Compound eyes large, often meeting on top of head; antennae variable, simple or with bristle-like arista. Mouthparts adapted for sucking or lapping; tarsi generally 5-segmented. Holometabolous, with larvae usually legless and maggot-like. (More than 80,000 species described worldwide, with many occupying marine habitats.)

S.O. **NEMATOCERA:** Diptera with simple antennae, apparently with 6 or more segments, plumose in the ♂ of some species. Wing venation variable. Most species are slender and soft-bodied with long legs. Larvae with well-developed head and strong mandibles. Many species of economic importance occur in this suborder including bloodsucking pests (mosquitoes, sand flies, etc.) of humans and livestock.

F. **CULICIDAE** (Mosquitoes): Small (5-10 mm) delicate, soft-bodied Nematocera; wings long and narrow with scales along veins and wing margins; proboscis long; antennae with 14 or 15 segments, plumose in ♂. Adult ♀ bloodsucking; all larvae are aquatic, some having elongate air tubes. (About 2,500 species have been described, with approximately 3% breeding in brackish water; 4 spp. from Bda.)

*Aedes sollicitans* (Walker) (Eastern salt-marsh mosquito): Adults of this genus generally ornamented with white scales producing distinct rings on proboscis and tarsi. Spiracular bristles absent, postspiracular bristles present; the posterior border of the scutellum with 3 distinct lobes; abdomen of ♀ appears pointed. Larvae with short, stout air tube with single pair of vertical hair tufts.—Species with prominent, wide white ring on proboscis and numerous white rings on tarsi; thorax golden brown; abdomen with median longitudinal white stripe. Strong fliers and fierce biters, even in daylight hours. Eggs are laid singly in marshy, coastal areas. Larvae and pupae usually live in potholes or depressions not reached daily by high tide, but can tolerate high salt levels.

*A. taeniorhynchus* (Widermann) (Black salt-marsh mosquito): Genus as above.—A small black and white species that lacks white stripe on abdomen and has narrower white rings on tarsi and proboscis. Biology and behavior similar to that of the preceding species. Breeds most frequently in potholes and temporary ponds in mangroves. Adult ♀ less likely to bite in direct sunlight than *A. sollicitans*.

F. **CERATOPOGONIDAE** (Sand flies, No-see-ums): Minute (usually less than 3 mm) Nematocera with long, usually 15-segmented antennae; wings with only few veins but with radius prominent, forming 1 or 2 radial cells and with costa ending 1/2 or 3/4 way to wing tip. Mostly bloodsucking (only ♀); larvae semiaquatic. (13 spp. from Bda.)

*Culicoides crepuscularis* Malloch: Genus with more or less hairy, generally dark wings with a conspicuous pattern of pale spots; 2 radial cells usually present and costa ending about midway along the wing margin. Larvae elongate, cylindrical and about 1 mm.— Species brownish with distinct pattern on thorax and sharply defined white spots on wings. Larvae live in brackish or saline mud. Adults are annoying biters in localized seashore and marsh areas.

*C. bermudensis* Williams: Genus as above.—Species with thorax dull brown and without pattern; white spots on wings poorly defined. This species is parthenogenic in Bermuda and appears also to be autogenous (i.e., the ♀ is able to produce at least an initial egg batch without a blood-meal).

F. **CHIRONOMIDAE** (Midges): Small, soft-bodied Nematocera with weak mouthparts (adults usually do not feed); antennae slender, with 5-14 segments, densely plumose in the ♂ (except in some marine species of *Clunio*!). Emergence of short-lived adults, often in great numbers and at twilight, may be tied to lunar cycles in marine species.

*Clunio* sp.: Antennae of ♂ 11-segmented, not plumose. This genus shows a remarkable sexual dimorphism: while the ♂ possesses wings and actually swarms, the ♀ is worm-shaped, lacks wings and halteres, and her legs and antennae are often vestigial. The ♂, after having found a ♀ still in her pupal skin, assists her in molting and subsequently mates with her. The ♀ usually dies after laying eggs. In the rocky intertidal, males can be seen, often in large numbers, flying-running over rocks at low tide; females are hidden between algae.

S.O. **BRACHYCERA:** Diptera with 5 or fewer (usually 3) antennal segments, the 3rd sometimes annulated, and with a terminal style but rarely with an arista. Vein $R_s$ usually 3-branched. Generally medium-sized to large, robust flies.

F. **TABANIDAE** (Horseflies and deerflies): Brachycera with 3rd antennal segment elongate and annulated. Branches of veins $R_4$ and

$R_5$ divergent, enclosing wing tip; eyes prominent, meeting dorsally in ♂ but separated in ♀. Strong flyers. ♀ bloodsucking and pestiferous. Larvae aquatic in swamps and marshes. (2 spp. from Bda.)

***Tabanus nigrovittatus*** Macquart (Greenhead, B Doctorfly): Genus with a tooth-like process on 3rd antennal segment; wings without pattern.—Species large (12-14 mm), yellowish brown, with clear wings and metallic green eyes. A common, vicious biter at beaches and rocky shores. Eggs are laid on stems or leaves of marsh vegetation. Larvae develop in mud and require high levels of moisture; they are 20-24 mm long and light brown, with a short siphon on the anal segment.

***T. atlanticus*** Johnson: Genus as above.—A pallid, pubescent, robust species with dusky wings and blue eyes. Infrequent.

**S.O.   CYCLORRHAPHA:** Diptera with 3 antennal segments and arista. $R_s$ 2-branched, head generally with frontal suture.

**F.   EPHYDRIDAE** (Shore flies): Small (3-5 mm), black to brown Cyclorrhapha. Subcostal vein incomplete, costa broken near end of $R_1$ and near humeral cross vein, anal cell absent.

***Paratissa semilutea*** (Loew): Reddish brown, 3-4 mm; face somewhat bulging. Very common along rocky shoreline and on beached seaweed.

**F.   ANTHOMYIIDAE:** Cyclorrhapha with well-developed calypters (lobes at posterior side of wing base; 2nd antennal segment with a suture; $R_5$ cell parallel-sided; vein 2A reaches wing margin, at least as a fold.

***Fucellia intermedia*** Lundbreck (Seaweed fly): Reddish brown, bristly, slender, 6-7 mm. Very common on seaweed and refuse on beaches; often an annoyance to bathers. Larvae develop in seaweed, apparently obtaining some if not all their food from this source.

I. W. HUGHES & R. SCHUSTER

## Phylum Mollusca (Mollusks)

CHARACTERISTICS: *Usually stout BILATERIA mostly with a ventral, muscular foot and a dorsal mantle; the latter is generally provided with a calcareous integument in the form of spicules, plates or a solid shell. Gills (ctenidia) located in a cavity between body and mantle; pharynx commonly with jaws and a lingual ribbon (radula). Heart surrounded by a pericardium (coelomic cavity); skin often with mucus-producing cells.*

Of 8 classes, 6 are represented: Solenogastres (p. 393), Polyplacophora (p. 393), Gastropoda (p. 397), Scaphopoda (p. 459), Bivalvia (p. 460) and Cephalopoda (p. 492). Only the small, vermiform Caudofoveata and the deep-sea dwelling Monoplacophora have not yet been reported from Bermuda. Caudofoveata and Solenogastres are sometimes grouped together as Aplacophora, and these 2 and the Polyplacophora as Aculifera. The most recent systematic summary of the phylum is by BOSS (1982); other useful sources, with particular emphasis on shell morphology, are by ABBOTT (1974) and ABBOTT & DANCE (1982).

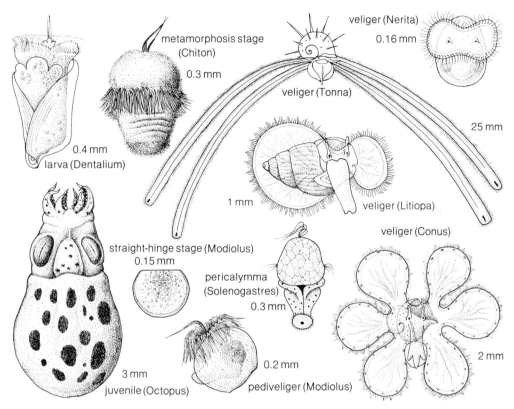

**129 MOLLUSCA (Mollusks): Development**

## Class Solenogastres (Worm mollusks)

CHARACTERISTICS: *Shell-less MOLLUSCA with laterally narrowed vermiform body; mantle covered by a cuticle and embedded calcareous scales and spines. Foot in the form of a ventral groove; mantle cavity subterminal, without ctenidia.* Mostly small (few millimeters to few centimeters) and whitish or pink, solenogasters move slowly by gliding on a mucous trail.

Of about 180 species in 4 orders, only 3 species have so far been found in Bermuda; 1 is included here.

OCCURRENCE: Exclusively marine. The majority occur on sublittoral to hadal, sandy-muddy or clay bottoms; some are found on secondary hard bottoms or colonies of Cnidaria; few are members of the interstitial sand fauna.

Collect sediment with a dredge and extract specimens by magnesium chloride method (if the sediment is sand), or by leaving the sediment undisturbed (and cool, if from deep water!) and then pipetting animals off individually.

IDENTIFICATION: Use a dissecting microscope for observing live animals. Species identification requires dislodging and examining calcareous spicules under the microscope. Further investigation needs microanatomy of preserved specimens.

Note shape and distribution of scales and spines, and characters of the radula (if present). Type of radula and foregut glandular organs are family

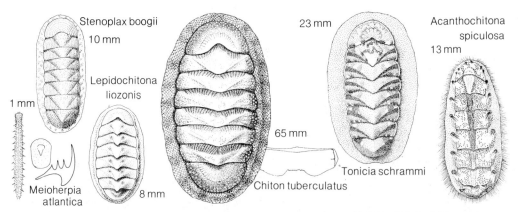

**130 SOLENOGASTRES (Worm mollusks), POLYPLACOPHORA (Chitons)**

characters, whereas other internal organs are used to distinguish genera and species.

BIOLOGY: Without exception hermaphrodites, often protandrous. Fertilization is internal, and several to many eggs are produced. The life-span is unknown. Solenogastres are carnivorous on Cnidaria, rarely on sponges; the stinging nematocysts are ingested and pass the gut without exploding. Food is taken up by the radula and/or by the sucking action of the pharynx.

**Plate 129**

DEVELOPMENT: The free-living larva is called pericalymma; brood protection occurs in a few species.

REFERENCES: For a systematic overview see BOSS (1982) or the older summary by HYMAN (1967). A recent monograph of cold-water species is by SALVINI PLAWEN (1978); some western Atlantic species have been described by TREECE (1979).

A species close to or identical with one found in Bermuda is pictured in RIEGER & STERRER (1975, p. 257), and SALVINI PLAWEN (1985) lists further records from Bermuda.

**Plate 130**

O. **PHOLIDOSKEPIA:** Solenogastres with moderate cuticle; calcareous bodies as scales.

F. **MEIOMENIIDAE:** Pholidoskepia with 3 or more types of scales one of which protrudes laterally; radula bipartite.

*Meioherpia atlantica* Salvini Plawen, Rieger & Sterrer: With a dorso-terminal sensory organ surrounded by 8 triangular scales. Radula with 1 prominent lateral and 3 smaller admedian teeth on each side. In coarse, clean subtidal sand; not uncommon.

L. v. SALVINI PLAWEN

**Class Polyplacophora** (Chitons)

CHARACTERISTICS: *Untorted, bilaterally symmetrical, dorso-ventrally flattened MOLLUSCA with a ventral creeping foot and a dorsal shell composed of 8 separate calcareous valves that are movably articulated.* Species in Bermuda range from a few millimeters to 75 mm; the shell surface may be green, brownish or red. Locomotion is so slow as to be imperceptible; only the pill bug-like curling motion of dislodged specimens is somewhat faster.

Of about 600 species, only 8 have been reported from Bermuda, 2 of which are doubtful or erroneous; 5 species are included here.

OCCURRENCE: Strictly marine and benthic, Polyplacophora occur at all latitudes. Although found predominantly on rocky coasts in the intertidal to subtidal zones, some species have been reported from depths as great as 4,200 m. Chitons are limited to hard substrates, where some species occur exposed whereas others are negatively phototactic, hiding under rocks during the day.

Specimens are best collected by inserting a knife blade or spatula under the broad foot and immediately prying the specimen from the rock. Once the element of surprise is lost, the chiton will clamp down tenaciously and probably be damaged by further efforts to loosen it. A loose specimen will curl up and should therefore be placed in a container of seawater until it can be prepared for study.

IDENTIFICATION: Classification is based largely on characters of the shell valves and girdle, many of which can be observed in living specimens. In order to determine characters of radula and internal shell these should be cleaned of tissue by boiling in a 5-10% solution of sodium or potassium hydroxide followed by a thorough rinsing in distilled water. If the specimen is small, the entire chiton may be boiled until only the shell valves, radula and girdle are left. For larger specimens, preliminary dissection is required. The shell valves should be arranged in the same order as they appeared in the intact animal.

On the live animal note the broad creeping foot, the location of mouth (anterior) and anus (posterior), and the absence of eyes or tentacles. The shell is surrounded by a tough, muscular girdle that may be nude, or ornamented by scales, spicules, spines or hairs. Aesthetes (photoreceptive organs unique to chitons) are located in channels in the shell valves. Shell sculpture, color and valve conformation should be noted. Distinguish between the head valve, the tail valve and the 6 intermediate valves.

Each valve is composed of 4 layers: the periostracum, a thin outer layer of scleroprotein; the tegmentum, an outer, soft, calcareous layer; the articulamentum, an inner, hard, semi-porcelaneous shell layer; and the hypostracum, the lowest ventral calcareous layer, differing from the articulamentum in crystal structure. With few exceptions, the articulamentum extends from beneath the tegmentum as apophyses, and lateral insertion plates that are partly covered by the girdle. The radula may be mounted on a slide and examined under a microscope. It contains 17 teeth per transverse row. The configuration of the 2nd (uncinate) tooth on either side of the central tooth has been used as a taxonomic character by some authors.

Prior to fixation for anatomical studies, a live specimen (in a flat position) should be tied to a tongue depressor, microscope slide or other flat surface. It is helpful to place the tied specimen in seawater, so that it resumes a natural position. The specimen is killed and fixed by immersion in 10% buffered formalin, followed by attenuation to 70% ethanol.

BIOLOGY: With rare exceptions, Polyplacophora are dioecious. Fertilization is external, gametes being shed into surrounding water. Eggs are produced in jelly-like strings or masses that soon break up and are dispersed. Several species brood their young in the pallial grooves. Most chitons are microphagous herbivores, grazing on algae-covered rocks, but several species feed on amphipods and worms. A number of species have been reported to exhibit homing behavior, feeding at night but returning to the same rock by day. Major predators are humans, rats, birds, and other mollusks. Various organisms (mites, crabs, isopods, e.g., *Dynamenella*) may live in the pallial grooves. Parasites include sporozoans and ciliates.

## Plate 129

DEVELOPMENT: Eggs develop by spiral cleavage and hatch after reaching the trochophore stage. Only 7 valves are formed in the late trochophore larva, the anterior valve being added later. Upon metamorphosis, which occurs 15 min to several days after hatching, the larva settles to the bottom, becomes dorso-ventrally compressed and loses its eyes, prototroch and apical tuft.

REFERENCES: For general orientation see ABBOTT (1974) and BOSS (1982).
Of Bermuda's species only *Chiton tuberculatus* has been studied in any detail, originally with emphasis on biological observations (PARKER 1914; CROZIER 1918a, b, 1919, 1920, 1921b, 1922; CROZIER & AREY 1918, 1920; AREY & CROZIER 1919) and more recently on reproduction and biochemistry (SOUTHWICK 1939; KIND & MEIGS 1955; COWDEN 1961, 1963, 1974; HÖGLUND & RAHEMTULLA 1977).

## Plate 130

O. **NEOLORICATA**: Polyplacophora in which the valves contain an articulamentum layer that forms the apophyses and insertion plates.

S.O. **ISCHNOCHITONINA**: Neoloricata in which the girdle does not encroach on the tegmentum. The head valve has only the anterior shell area, the intermediate valves are divided into the median and 2 lateral areas, and the tail valve has a median and a posterior area. The insertion plates are divided into teeth by slits.

F. **ISCHNOCHITONIDAE**: Ischnochitonina with sharply edged, outwardly directed insertion teeth. Tegmentum on intermediate valves divided by diagonal ribs. Shell sculpture usually reticulate-costate, girdle covered by small scales or spicules. Uncinate radular teeth with 1-3 cusps. (3 spp. from Bda.)

*Stenoplax boogii* Haddon (=*S. rugulata* of authors; *Ischnochiton bermudensis* Dall & Bartsch) (Wrinkled chiton): Genus with elongate shell; smooth, granulated or ridged tegmentum; narrow girdle with minute scales. Tail valve larger than the head valve. End valves with 9-15 slits, intermediate valves with 1-3 slits in insertion plates.—Species with brown, pink or red shell 12-20 mm long. All valves sculptured with concentric ridges. Girdle tan to brown. Uncommon; under rocks and rubble, 5 m.

*Lepidochitona liozonis* (Dall & Simpson) (Puerto Rican red chiton): Genus with ovate shell; smooth or finely granular tegmentum; girdle with minute scales. Head valve larger than the tail valve. End valve with 9-12 slits; intermediate valves with 1 slit.—Species with red shell 8-12 mm long. Surface of valves minutely granular. Girdle brownish and unornamented. Uncommon; under rocks and rubble, 5 m.

F. **CHITONIDAE**: Ischnochitonina with well- developed pectination on outside of insertion plates. Variable tegmentum. Girdle may be nude, or have scales, bristles or spicules. Uncinate radular teeth with 3 unequal cusps. (2 spp. from Bda.)

***Chiton tuberculatus*** L. (=*C. squamosus* L., *C. marmoratus* of authors) (Common West Indian chiton): Genus with ovate shell; variable tegmentum; girdle covered by smooth, imbricating scales. Outer edges of insertion plates with tiny, sharp teeth; intermediate valves with 1 or more slits in insertion plates. Interior of valves blue-green.—Species with a greenish shell that may reach 75 mm or more in length. Lateral areas with radiating cords, central areas smooth. Girdle broad, scaly, with bands of green and black. Common; intertidally on rocks.

***Tonicia schrammi*** (Shuttleworth) (Schramm's chiton): Genus with ovate shell. Second valve larger than the rest. Girdle nude or sparsely hairy. Upper surface of valves with aesthetes.—Species with brownish red, mottled shell, 20-30 mm long. Surface of valves glossy. Head valve and central areas of intermediate valves with tiny black aesthetes. Girdle brownish, naked. Moderately common; intertidally on rocks.

S.O. **ACANTHOCHITONINA:** Neoloricata in which the valves are partially to completely buried in the girdle. The tegmentum surface is greatly reduced in area relative to the articulamentum, which is highly developed. Tail valve has a small, triangular, jugal area, 2 adjacent triangular pleural areas and a large posterior area.

F. **ACANTHOCHITONIDAE:** Acanthochitonina with few slits in the insertion plates. Tegmental area narrow. Girdle may be nude or with tufts of long bristles, but never with scales. Gills do not extend the full length of the foot. Uncinate radular teeth always with 3 cusps. (1 sp. from Bda.)

***Acanthochitona spiculosa*** (Reeve) (=*A. astriger* Reeve) (Glass-haired chiton): Genus with elongate shell; tegmentum with nodules arranged in radial rows; girdle naked or hairy with 18 tufts of spicules, 4 on head valve and 1 on either side of each of the other valves.— Species with blackish brown to olive-green shell, 25-40 mm long. Girdle covers most of the valves and has spicules on the lower edge. Rare; under rocks in shallow water.

R. H. JENSEN & M. G. HARASEWYCH

**Class Gastropoda** (Snails, limpets and slugs)

CHARACTERISTICS: *Torted, often asymmetrical* MOLLUSCA *typically with a well-developed head bearing eyes and tentacles, a flat creeping foot, and a shell-secreting mantle. Shell single, apically closed; generally present and usually spirally coiled. Buccal cavity usually with a radula (a lingual ribbon set with teeth).* Gastropods range in size from less than 1 mm to over half a meter. Although most move by slow crawling, some are able to swim, either intermittently (Aplysiidae), as permanent members of the plankton (Gymnosomata, Thecosomata), or as drifters at the surface (Janthinidae, Glaucidae); few are permanently sessile (Vermetidae). Coloration of both the shell and the soft body ranges from colorless and transparent to almost every color and pattern imaginable. The soft body is often slimy to the touch; the shell surface can be rough-sculptured, hairy or glossy.

All 3 subclasses are represented in Bermuda. Of about 39,000 marine gastropod species, approximately 375 have been reported (and another 50 collected but not positively identified) from Bermuda; 177 are included here.

OCCURRENCE: The most successful of the molluscan classes, Gastropoda are abundant in all oceans, from the spray zone to the deep sea, and have invaded—several times independently in the course of evolution—brackish, freshwater and terrestrial habitats. Although the majority of marine species are benthic, others have become adapted to pelagic or parasitic existences. Of the Prosobranchia, Archaeogastropoda are generally restricted to rocky marine substrates, whereas Caenogastropoda occur in a wide range of biotopes ranging from pelagic (e.g., *Janthina*) to sessile (e.g., *Vermicularia*). Opisthobranchia are almost exclusively marine and typically found in rocky areas among algae; however, some minute species (Microhedylidae) are members of the interstitial sand fauna, and 2 entire orders (Gymnosomata, Thecosomata) are pelagic. Most Pulmonata are terrestrial, generally occurring in sheltered areas of high humidity (e.g., under rocks, in decaying vegetation), or they live in fresh water; the few marine species inhabit mangroves (Melampidae) or rocky areas near the tidal zone (Siphonariacea).

Collecting of live gastropods for souvenirs is strongly discouraged, for it has already caused severe depletion of several species. A number of species (*Strombus gigas*, *S. costatus*, *Conus mindanus*, *Oliva circinata* and all Cassidae) are now protected by law in Bermuda waters. Scientific collecting may be done by hand in intertidal areas (under rocks in tide pools, among seaweed) and by means of snorkeling or SCUBA. Small gastropods, especially Opisthobranchia, may be collected by placing substrate samples (algae, sediment, coral rubble) in a small quantity of seawater and waiting for the oxygen to be depleted; the specimens will then congregate at the surface of the water. Pelagic species may be dipnetted (*Janthina*, *Scyllaea*) or caught with a plankton net (*Limacina*, *Clione*). Beach drift and wrack can be collected along the high tide line and carefully examined for specimens. Sand samples taken from pockets on the reef, 1-50 m, often yield many unusually small mollusks (Caecidae, Microhedylidae, *Rhodope*). Dredging and trapping may be done in the deeper waters; dead specimens are frequently inhabited by hermit crabs, which can be lured into baited traps.

Some large species of *Strombus* are an important source of food, particularly in the Caribbean; they have been overfished to near-extinction in Bermuda. Similarly, the disappearance of *Cittarium pica* has been attributed to overharvesting. Some Muricidae produce a purple dye that was highly valued by ancient Mediterranean cultures.

IDENTIFICATION: The external characters of live specimens are usually sufficient for identifying genus and species. Higher taxonomic categories, however, are established on the basis of features of the internal anatomy (radula, nervous system, sexual organs, etc.). The soft parts of shelled forms may be extracted by placing the specimen in water and bringing it to a boil, then cooling, or preferably by alternately freezing and thawing to loosen the columellar muscle. Dissection may be carried out in a water-filled dish with a paraffin bottom to which the soft body can be attached by means of pins. The mantle cavity can be split open medially; subsequent cutting through the base of the mantle cavity reveals the anterior digestive system components. The radula may be extracted by dissecting the proboscis; it may then be boiled in a 5-10% solution of sodium or potassium hydroxide to remove extraneous organic matter. The use of Clorox is not recommended for it tends to obscure finer details by slowly dissolving the radula. For permanence the radula can be mounted in a low-refractive index acid-free medium (glycerin jelly or Turtox CMC9).

The soft body is divided into head, foot and visceral mass, the latter usually covered by the mantle and a shell. The head usually bears eyes (which can be on stalks, e.g., *Strombus* and many Pulmonata) and tentacles: 1 pair (cephalic tentacles) in Prosobranchia, and an additional pair, which may be enlarged for chemoreception (rhinophores), in Opisthobranchia. The mouth is simple or modified as a proboscis. The mantle may be drawn out anteriorly into an inhalant canal (siphon). The foot, in some groups (Anaspidea, Bullomorpha), may be extended into wings (parapodia) which are sometimes used for swimming. The shell-less dorsal body surface (notum) may bear digestive diverticula (cerata) in Nudibranchia, and secondary gills in Doridacea.

As a result of torsion, the nervous system is twisted into a figure 8 and the anus is displaced anteriorly. Among the higher Gastropoda, the Opisthobranchia may exhibit a secondarily derived bilateral symmetry (untwisted nervous system and posterior anus) owing to detorsion, whereas the Pulmonata appear to have a bilaterally symmetrical nervous system owing to anterior concentration of ganglia, but retain the anteriorly displaced anus. The primitive Prosobranchia have paired ctenidia (gills), osphradia (chemoreceptive organs), nephridia (tubular renal organs), and auricles (heart chambers), but in the higher prosobranchs and in all other gastropods, these organs are unpaired. The ctenidium (gill) may be bipectinate (having a row of leaflets on either side of a central axis suspended from the roof of the mantle cavity), monopectinate (a single row of leaflets attached directly to the roof of the mantle cavity), or plicate (longitudinal folds attached to the mantle roof). The osphradium may range in complexity from a longitudinal ridge to a structure resembling a small bipectinate gill. In taxa with internal fertilization, sex can be determined by examining the reproductive organs in the mantle cavity. Males usually possess a prominent penis behind the right cephalic tentacle, whereas in females the capsule gland and albumen gland occupy the right side of the mantle cavity. At the rear of the mantle cavity is the nephridium and, adjacent to it, the pericardium (cavity surrounding the heart), which contains 1 ventricle and 1 or 2 auricles depending on the number of ctenidia. Opisthobranchs are hermaphroditic, and have highly complex reproductive organs; the genitalia are located on the anterior right side.

The radula is particularly useful in diagnosing prosobranchs, which have been divided into a number of groups based in part on radula type. The radula may have a variable number of rows, each with a certain number of teeth. The central tooth in each row is referred to as the rachidian, the major teeth on either side of it are the laterals, and the finer teeth flanking the laterals are called marginals. The radula formula gives the number of teeth of each type per row of the radula. The following radula types can be disinguished in the Prosobranchia: (1) Rhipidoglossate ($\infty:5:1:5:\infty$), i.e., 1 rachidian flanked by 5 laterals and numerous marginals in a fan-shaped arrangement; (2) docoglossate ($3:1:(2+0+2):1:3$), i.e., 4 rachidians (this number may vary), 1 lateral and 3 marginals; (3) taenioglossate ($2:1:1:1:2$), with its derivations; (4) ptenoglossate ($\infty:0:\infty$), i.e., with a long row of numerous, graduated, long, hooked teeth with the outermost the longest; (5) rachiglossate ($1:1:1$ or $0:1:0$); (6) toxoglossate ($1:0:0:0:1$ or $1:0:1$), where the marginals are greatly modified into harpoon-like structures; and (7) gymnoglossate (without a radula). There is a great diversity in the radulae of Opisthobranchia and Pulmonata, and there is currently no radula-based system of higher classification in these groups. However, radular and jaw characteristics are often necessary to identify families, genera and species of opisthobranchs. Other hard structures (gizzard plates, spicules, penial stylet) may also be of use in identification, and may be removed by alkali treatment for study.

## Plate 131

The shell is usually made up of several layers of which the outer (periostracum) is horny and often pigmented, whereas others are of calcite or aragonite, the latter often giving a nacreous (mother-of-pearl) effect. A typical spiral shell consists of nuclear whorls at the apex, which continue into the penultimate whorls (larger whorls of the spire), and finally into the ultimate whorl (body whorl) with the aperture. The shell axis (columella) may be hollow, and open distally into a navel (umbilicus). Bordered by an outer lip and an inner lip (the latter sometimes enlarged into a parietal shield or callus), which may bear folds (plicae), the aperture may be drawn out anteriorly into a siphonal canal, and laterally into an anal canal. The whorls join along the suture; they may bear axial and spiral sculptures such as costae (ribs), cords, nodules, beads and spines. Axial ribs made during major growth intervals are called varices. In most Prosobranchia the soft body also carries an operculum (lid) which seals the aperture when the animal retracts into its shell. The operculum may be of thin-corneous or heavy-calcareous consistency, and bear a

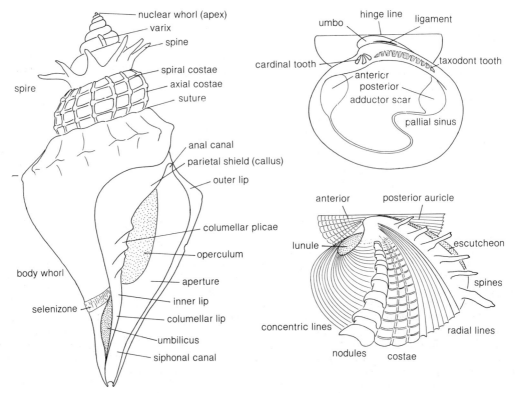

**131 MOLLUSCA (Mollusks): Schematic**

paucispiral (few turns) or multispiral (many turns) sculpture.

For detailed anatomy or histology, the specimen should first be anesthetized by the magnesium chloride method, then fixed in buffered formalin and attenuated to 70% ethanol. To expose all tissue, the shell (if present) should be carefully broken in a vise or clamp after the first 1/2 hr of fixation. Small specimens may be fixed in Bouin's fluid, which will dissolve the shell.

### Plate 132

BIOLOGY: Prosobranchia are mostly dioecious, whereas Opisthobranchia and Pulmonata are hermaphrodites. In some primitive Prosobranchia (e.g., Pleurotomariacea, Fissurellacea) sexual products are released into surrounding seawater where fertilization takes place. All other Gastropoda have internal fertilization by means of copulation and deposit their eggs (several to several thousand, depending on the species) in gelatinous strings or masses. The higher Caenogastropoda enclose their eggs and albuminous fluid in proteinaceous capsules, which are usually attached to the substrate.

A great diversity of feeding mechanisms and diets has evolved within the class. Most Archaeogastropoda, some Opisthobranchia and many Pulmonata are grazing herbivores; several other groups (Turritellidae, Vermetidae, *Crepidula*, Thecosomata) are suspension feeders, trapping small phytoplankton by means of a ciliary current and/or a mucous web. The Nassariidae are among the best-known and most common scavengers. Carnivores have developed diverse feeding mecha-

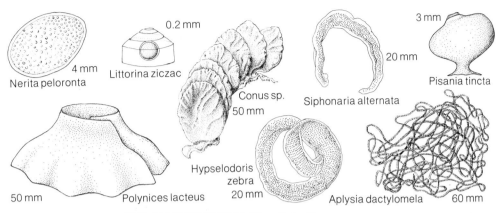

**132 GASTROPODA (Snails): Egg cases**

nisms, from grazing on octocorals (*Cyphoma*) to sucking out colonial ascidians (*Trivia*), drilling through bivalve shells (Naticidae and Muricidae) or harpooning worms, mollusks and fish (*Conus*). Parasitism, both external (e.g., *Melanella* in holothurians) and internal (e.g., worm snails in holothurians), has developed in several lineages. Opisthobranchs are often highly selective, and may feed on only a few closely related prey species, as in the Nudibranchia (Hydrozoa, Bryozoa) and Sacoglossa (algae). A number of species (*Onchidella, Siphonaria*) that live in the rocky intertidal regularly return to their "home scar" after feeding.

Major predators are humans, fish, crabs and other mollusks; intertidal species are eaten by birds and rodents. Shells are often overgrown by epiflora and -fauna (encrusting algae, hydroids, bryozoans); when empty, they provide homes for hermit crabs, and hiding places for other organisms (e.g., the conch fish *Astrapogon*, and *Octopus*). Gastropods are parasitized by ciliates, sporozoans and trematodes.

*Plate 129*

DEVELOPMENT: After fertilization the egg divides by spiral cleavage to form a trochophore larva which develops into a veliger larval stage, easily recognized by the presence of a ciliated, lobed swimming organ (velum). In many species, number and size of lobes increase as the larva grows. The veliger is a suspension feeder. Growth of the larval shell (protoconch), which is often different from but persists in the adult shell, is initiated in the trochophore and continues into the veliger stage where torsion occurs. In most species that are "shell-less" as adults an embryonal shell appears in the veliger, to be cast off at metamorphosis. Metamorphosis can be induced by contact with the substrate (most prosobranchs) or with the adult food (most opisthobranchs). In species with external fertilization all larval stages are free-swimming. In those with internal fertilization, embryos develop within the egg membranes or capsules and may hatch in the veliger stage or even after metamorphosis as free-crawling young.

Well-fed, sexually mature adults will often produce egg masses in the laboratory. These can be easily cultured by regularly supplying fresh, clean seawater until hatching occurs, though antibiotics may be necessary to control bacteria, fungi and ciliates. Veligers are also frequently caught in plankton tows.

REFERENCES: The vast field has most recently been summarized by BOSS (1982) in terms of systematics; HYMAN (1967) continues to be a useful overall reference, and ARNOLD (1965) provides a glossary of terms used in malacology. The major opisthobranch orders have been summarized by THOMPSON (1976) and THOMPSON & BROWN (1984); the pteropods (Gymnosomata, Thecosomata etc.) by SPOEL (1967, 1976), and EALES (1960) revised the genus *Aplysia*. The taxonomic literature on Caribbean Opisthobranchia is extensive: the most useful papers are by MARCUS (1955, 1957, 1977) and MARCUS & MARCUS (1963, 1967, 1970), EDMUNDS (1964) and THOMPSON (1977a); a complete listing of relevant papers is given by MARCUS (1977). Descriptions of reproductive biology of many opisthobranch species occurring in Bermuda are given by CLARK & GOETZFRIED (1978), and diets of Caribbean Ascoglossa are summarized by CLARK & BUSACCA (1978) and JENSEN (1980).

Most species known to occur in Bermuda are listed in the extensive monograph by ABBOTT (1974). Bermuda's marine gastropods have been the subject of only a few recent studies: HAAS (1941, 1943, 1949, 1950, 1952), CLARKE (1959), TURNER (1961), ABBOTT & JENSEN (1967, 1968), WALLER (1973), THOMPSON (1977b, 1981), HARASEWYCH & JENSEN (1979), CLARK (1982, 1984) and SNYDER (1984); COOK (1971, 1976) and COOK & COOK (1981) investigated homing behavior in *Siphonaria*. The older literature has never been summarized and only a few references are given here as examples: HEILPRIN (1889), VERRILL (1901a, 1902a), VERRILL & BUSH (1900), BARTSCH (1911), PEILE (1926) and the many biological observations by CROZIER & AREY (1919a, b), AREY (1937a, b) and AREY & BARRICK (1942). Planktonic gastropods are included in MOORE (1949) and DEEVEY (1971). Eggs and larvae of Bermudian prosobranchs have been described by LEBOUR (1945), larval shells by ROBERTSON (1971).

S.CL. **PROSOBRANCHIA:** Gastropoda with gills contained in the mantle cavity. Visceral mass torted; usually contained in a coiled or cuplike shell. Head with a single pair of tentacles. Sexes usually separate. (Some 20,000 known species in 2 orders, both represented in Bermuda. The system used here unites Mesogastropoda and Neogastropoda as Caenogastropoda.)

O. **ARCHAEOGASTROPODA:** Prosobranchia with 1 bipectinate ctenidium and 1 osphradium in all but the more primitive families where it is paired. Without proboscis or siphon, sometimes without an operculum; with numerous radula teeth in each row. A nacreous inner layer occurs in many shells.

*Plate 133*

SUP.F. **PLEUROTOMARIACEA:** Archaeogastropoda with shells that possess a slit, notch or row of openings. Shell trochiform (a conical spiral), discoidal or ear-shaped. Inner shell layer aragonitic, nacreous, iridescent. Operculum where present small, corneous, multispiral with a central nucleus.

F. **PLEUROTOMARIIDAE:** Pleurotomariacea with medium to large (240 mm) trochiform shells. Exhalant slit with its selenizone (a spiral band of crescentric growth lines on the shell surface generated by a narrow notch or slit) occurring between suture and periphery. Radula rhipidoglossate. Operculum corneous, multispiral. (2 spp. from Bda.)

***Perotrochus quoyanus*** (Fisher & Bernardi) (Quoy's slit shell): Genus without umbilicus. Slit wide, short, situated just above rounded periphery. External part of columella not thickened or curved.—Species to 50 mm in diameter, cream-white with beige brown axial streaks. Sculpturing of raised, fine beading on axial riblets. In deep water; to 80 m. Rare.

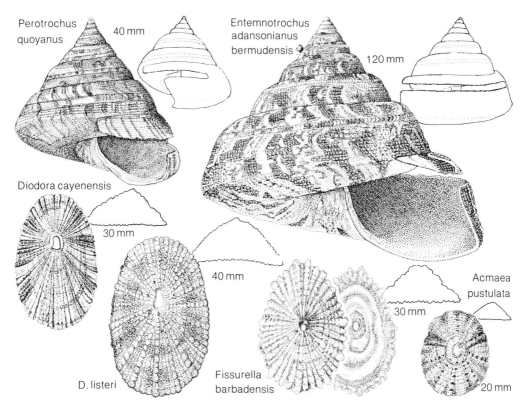

**133 PLEUROTOMARIACEA (Slit shells), FISSURELLACEA (Keyhole limpets), PATELLACEA (Limpets)**

*Entemnotrochus adansonianus bermudensis* (Okutani & Goto) (Adanson's slit shell): Genus with an open, deep umbilicus. Slit long, narrow, 1/3 of the way down between the suture and periphery.—Species to 180 mm in diameter, cream base with terracotta brown indeterminate maculations. In deep water; to 80 m. Rare.

**SUP.F. FISSURELLACEA:** Archaeogastropoda with conical, porcelaneous shell with hole, slit or notch for passage of exhalant current. Horseshoe-shaped muscle scar opening anteriorly. Radula rhipidoglossate.

**F. FISSURELLIDAE:** Fissurellacea with limpet-shaped, bilaterally symmetrical, non-spiral shell (except for nuclear whorls). Perforation for exhalent current at or forward of apex. (9 spp. from Bda.)

*Diodora cayenensis* (Lamarck) (=*Fissurella alternata* Say) (Cayenne keyhole limpet): Genus with orifice forward of the apex. Interior callus squared and excavated behind.—Species to 40 mm. Color variable, white, cream or gray, sometimes with 8 or 9 darkened rays or scattered spots. May have base band of same color. Many radial ribs with each 3rd or 4th enlarged. Ribs finely scaled by concentric threads. Interstices with tiny rectangular pits. Interior callus sometimes edged in black. On underside of rocks or rubble, intertidal to moderate depths. Uncommon.

***D. listeri*** (Orbigny) (Lister's keyhole limpet): Genus as above.— Species to 40 mm, color cream, white or gray, sometimes with weak radial bands. Coarsely sculptured with 38-40 strong round radial ribs, every other enlarged. Concentric threads forming large beads on the ribs. Interstices small, deep squares. Orifice callus outlined or stained blue-black. External markings showing through on the interior. On underside of rocks or rubble; intertidal to moderate depths. Uncommon.

***Fissurella barbadensis*** (Gmelin) (=*F. antillarum* Orbigny, *F. bermudensis* Pilsbry) (Barbados keyhole limpet): Genus with orifice callus same width all around, not truncated or excavated as in *Diodora*. - Species to 35 mm. Shell conical, variable in shape. Exterior color cream, gray-white to pale olive, sometimes with brownish radial markings. Apex orifice round to oval. Numerous radial ribs, of which 10 or 11 are generally stronger. Interior with concentric, white to olive bands. Orifice callus often circled with pinkish-brown line. Common, intertidally on rocks.

**SUP.F. PATELLACEA:** Archaeogastropoda with variable, porcelaneous, depressed conical or cup-shaped, bilaterally symmetrical, limpet-like shells, without perforation, slit or notch. Muscle scar semicircular or horseshoe-shaped, opening anteriorly. Inner layer aragonitic, sometimes iridescent. Radula docoglossate.

**F. ACMAEIDAE:** Patellacea with shells that are low conical, with a non-spiral, conical embryonic whorl. Animal with a free cervical branchial plume (ctenidium) at the left side of the neck; no branchial cordon (a single row of lappets). Eyes present. Aperture with a more or less distinct internal marginal border. Interior never iridescent. (1 sp. from Bda.)

***Acmaea pustulata*** (Helbling) (=*A. punctulata* Gmelin) (Pimple or spotted limpet): Genus with round-oval, porcelaneous, depressed conical shell.—Species to 15 mm. Exterior smooth to finely ribbed, crossed by fine concentric lines. Apex tiny, pointed, slightly forward of center. Color cream to rose, quite often marked with darker rose flecks. May occasionally have a rose-brown ray on each side. Moderately common; on smooth rocks or rubble (i.e., bottles, glass, old porcelain, etc.) in shallow, sheltered waters, to 5 m.

**Plate 134**

**SUP.F. TROCHACEA:** Archaeogastropoda with low to high conical, spiral shells with few to many whorls. Inner layer generally nacreous. Operculum corneous or calcareous. Epipodium (muscular lobe of the foot) highly de-

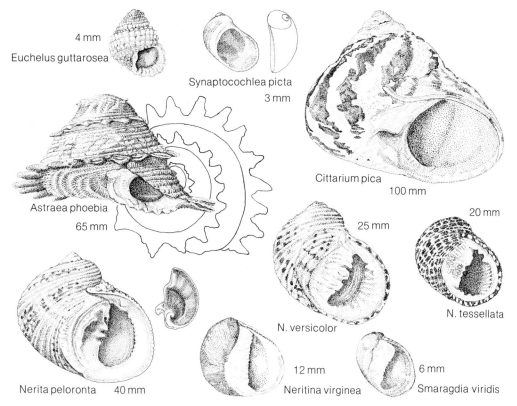

**134 TROCHACEA (Top and star shells), NERITACEA (Nerites)**

veloped. Radula rhipidoglossate. Herbivorous.

F. **TROCHIDAE:** Trochacea generally with broad, conical, top-like shells. Operculum corneous, thin, multispiral. (8 spp. from Bda.)

***Euchelus guttarosea*** Dall (Red-spotted Euchelus): Genus with solid, thick, coarsely sculptured small shells.—Species to 4 mm, white, coarsely noduled; periostracum thin, gray. (The Bermuda form does not have the rose-red markings that are typical of this species.) Nuclear whorl axially ribbed. No umbilicus. Aperture nacreous, 6-8 lirae on outer lip. Uncommon; under rocky rubble, 1-5 m.

***Cittarium pica*** L. (=*Turbo pica*; *Livona pica*) (West Indian top shell; B Wilk, Magpie shell): Shell heavy, conical, 50-150 mm. Color grayish white with black maculations. Aperture round, umbilicus wide and deep. Although the shell is found as fossil only, it is common throughout the island and at times looks freshly dead. Used as a residence by the land hermit crab *Coenobita clypeatus*. Two attempts to reintroduce this species to Bermuda (VERRILL 1902a, and again in 1930, unpublished) failed.

F. **STOMATELLIDAE:** Trochacea with low-spired, few-whorled shells. Not umbilicate. Aperture large, interior non-nacreous. Operculum lacking in most groups. (1 sp. from Bda.)

*Synaptocochlea picta* (Orbigny) (Painted false Stomatella): Genus small, auriform with large aperture. Surface with fine striations. Operculum corneous. —Species 3 mm. Color white to variegated, with reddish brown and/or black markings, to all black. Columella white. Interior of aperture shiny with exterior markings showing through. Moderately common; intertidally to 4 m in rubble and sand.

F. **TURBINIDAE:** Trochacea with globose, few-whorled, strong, solid shells. Exterior surfaces smooth to strongly sculptured. Operculum calcareous. (1 sp. from Bda.)

*Astraea phoebia* Roeding (=*A. longispinum; A. longispina*) (Long-spined star shell): Genus solid, depressed conical, coarsely sculptured, generally umbilicate; spines on periphery; operculum calcareous.—Species 65 mm; with 6 or 7 whorls. Flattened triangular spines on periphery. Aperture round, interior pearly and iridescent. Uncommon; in grassy shallows of quiet waters.

SUP.F. **NERITACEA:** Archaeogastropoda with shells that are conical, mostly globose, with few whorls and low spire. Without nacreous layer; operculum calcareous. Characterized by secondary gills or pulmonary sac. Right nephridium absent. Radula rhipidoglossate.

F. **NERITIDAE:** Neritacea with shells that are globose, turbiniform to patelliform; mostly thick walled. No umbilicus. Inner lip generally thickened by callus, often dentate on inner margin. Operculum with peg-like apophysis for better attachment to foot. (7 spp. from Bda.)

*Nerita peloronta* L. (Bleeding tooth shell): Genus with solid shells. Parietal area flat, with teeth, lirae and/or papillae. Operculum with apophysis.— Species 40 mm, basic shell color yellowish white, color extremely variable; markings, yellow, orange, red, purple or black. Parietal area stained with blood-red marking around 1 or 2 stained teeth. Operculum brown-orange, half of exterior surface smooth, raised and shiny, other half depressed and finely papillose; apophysis reduced. Common on rocky shoreline where there is surf action.

*N. versicolor* Gmelin (Variegated nerite): Genus as above.— Species 25 mm, gray-white with maculations of red and black. Coarsely sculptured with spiral grooves. Parietal area yellowish white, with fine lirae and 4 teeth. Operculum finely papillose, brown-gray; apophysis faint to lacking. Common, associated with above.

*N. tessellata* Gmelin (Tessellated nerite): Genus as above.— Species 20 mm, spotted black and white or all black. Coarsely sculptured with spiral grooves. Parietal area milk-white with weak lirae and papillae. Operculum finely papillose, charcoal colored; apophysis reduced. Very common; congregate in large numbers. Associated with above.

*Neritina virginea* (L.) (Virginia nerite): Genus smooth, glossy, smaller and thinner than *Nerita*. Inner lip of parietal area with weak teeth.—Species 12 mm, glossy, smooth, extremely variable in color (red, brown, gray, black, olive) and pattern (lines, dots, zigzag stripes and/or bands). Parietal area smooth, yellowish white to white with a number of small teeth. Operculum charcoal black, smooth; apophysis prominent. In brackish water; uncommon.

*Smaragdia viridis* (L.) (Emerald nerite; B Green lucky): Genus small, thin, surface smooth and glossy, grass-green with white flecks; may have red or black markings.—Species 6 mm. Operculum greenish white, smooth; apophysis smooth and prominent. Uncommon, in turtle grass areas, 1-5 m. Found in beach drift above such areas.

O. **CAENOGASTROPODA** (= Mesogastropoda + Neogastropoda): Prosobranchia with 1 monopectinate ctenidium, 1 osphradium. With or without an inhalant siphon. Proboscis present. Radula with only up to 7 teeth per row. Shell porcelaneous, nacreous layer absent.

*Plate 135*

SUP.F. **LITTORINACEA:** Caenogastropoda with usually small, conical shells under 2 cm. Operculum corneous. Radula taenioglossate (usually 7 teeth per row). Dioecious.

F. **LITTORINIDAE:** Intertidal marine Littorinacea with small conical shells. Operculum corneous, either paucispiral or multispiral. ♂ with external prong-like penis. Eggs shed freely or laid in floating capsules or jelly masses, or retained in oviduct. Radula with over 1,000 transverse rows of teeth. Proboscis short, mobile, not invaginable. (9 spp. from Bda.)

*Littorina ziczac* (Gmelin) (Zigzag periwinkle): Genus with small, ovate conical shells, without umbilicus. Columella smooth. Operculum thin, paucispiral, color brownish.—Species 20 mm; gray-white with irregular, angular zigzag stripes of brown. With 20-26 microscopic spiral lines on upper whorls. Columella brown. Aperture 1/3 the length of shell. Common on rocky exposed shoreline; intertidal.

*L. mespillum* (Mühlfeld) (Dwarf periwinkle): Genus as above.—Species 6 mm. Solid, globular, with brown periostracum under which the shell is smooth, brown, sometimes with round blackish spots, to cream white with blackish spots. Aperture and columella glossy brown. Umbilicus a narrow slit. A recent (1950-1960) introduction to Bermuda. Un-

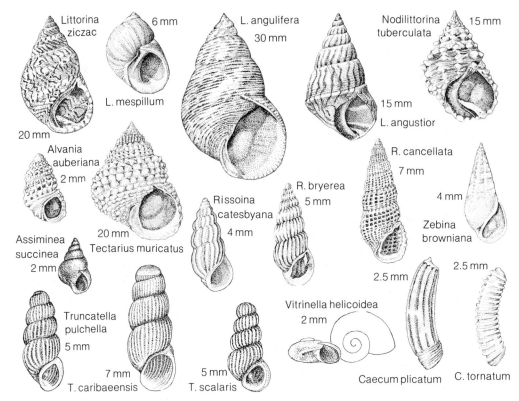

**135 LITTORINACEA (Periwinkles), RISSOACEA (Risso shells)**

common; in tide pools of exposed shores.

***L. angulifera*** (Lamarck) (Angular periwinkle): Genus as above.—Species 30 mm. Thin-shelled, strong with numerous fine spiral lines. Color brownish with lighter maculations. Columella purplish brown. Operculum pale-brown. Common on mangroves or on man-made structures in sheltered waters.

***L. angustior*** (Mörch) (Narrow periwinkle): Genus as above.—Species 6-10 mm. Color similar to *L. ziczac*. Shell carinate at base of whorl; with 6-9 incised spiral lines on upper whorls. Aperture less than 1/2 the length of the shell. Common on rocky exposed shorelines; intertidal.

***Nodilittorina tuberculata*** (Menke) (Prickly winkle): Genus small, solid, low conical with spiral rows of sharpened nodules axially aligned. Columella flattened and dished. Operculum paucispiral.—Species 15 mm, gray-brown. Common on rocky shores, in the splash zone above high tide line.

***Tectarius muricatus*** (L.) (Beaded periwinkle): Genus with small, sturdy shells. Surface sculptured with rounded nodules. No umbilicus. Operculum paucispiral.—Species to 20 mm. Exterior gray, interior dark brown. Surface with 10-11 rows of rounded, evenly spaced nodules. Columella dished. Common; high on rocky shoreline in the upper splash zone.

**SUP.F. RISSOACEA:** Caenogastropoda with generally minute shells, seldom over 10 mm. Usually with siphon, operculum and penis. Radula generally taenioglossate.

**F. RISSOIDAE:** Rissoacea with shells less than 5 mm, smooth or axially sculptured, ovate conic; outer lip simple or slightly thickened. (8 spp. from Bda.)

*Alvania auberiana* (Orbigny) (=*Rissoa minuscula* Verrill & Bush) (West Indian Alvania): Genus globose, solid; sculpturing reticulated; outer lip thickened; umbilicus chink-like, aperture rounded.—Species 2 mm, yellowish white. Base of shell with 3 or 4 spiral cords. Common in sand and rubble, 1-6 m.

**F. RISSOINIDAE:** Rissoacea with shells less than 10 mm, aperture ovate, outer lip thickened and flaring anteriorly. (6 spp. from Bda.)

*Rissoina catesbyana* Orbigny (=*R. bermudensis* Peile; *R. chesneli* (Michaud); *R. scalarella* C. B. Adams) (Catesby's Risso): Genus small, elongate-conical. Operculum yellow, corneous, paucispiral with club-shaped peg on inner surface.—Species to 5 mm, with 11-14 strong, slanting, axial riblets. No spiral striae between ribs. Color glossy white. Prominent tooth inside thickened outer lip. Uncommon; on *Thalassia* in sheltered water.

*R. bryerea* (Montagu) (Caribbean risso): Genus as above.—Species 5 mm with 16-20 strong, slanting axial riblets. Spiral striae on base of shell. Color white or yellowish white. Without tooth inside lip. Common; in shallow water on sand and rubble.

*R. cancellata* Philippi (=*R. pulchra* C.B. Adams; *R. sagraiana* Orbigny) (Cancellate risso): Genus as above.—Species 7 mm. Strongly cancellate. Color white to light tan. Operculum light brown. Common; in shallow water on sand and rubble.

*Zebina browniana* (Orbigny) (=*Rissoina laevissima* C. B. Adams) (Smooth risso): Genus small, lacking axial ribs or spiral threads.—Species 4 mm. Shell glossy smooth, white, rarely brown banded. Thickened outer lip with 2 small nodes within. Common in sand, 1-4 m. (Do not confuse with *Melanella*!)

**F. ASSIMINEIDAE:** Rissoacea with ovately conical or subglobose shells covered with horny epidermis. Operculum corneous, paucispiral. Amphibious. Eat detritus and decaying vegetable matter. Dioecious. (1 sp. from Bda.)

*Assiminea succinea* (Pfeiffer) (=*A. auberiana* Dall non Orbigny; *A. modesta* Lea) (Atlantic Assiminea): Genus with small, smooth, ovate conical shells. Aperture ovate; columellar lip thickened; outer lip thin and sharp.—Species 2 mm, light translucent brown. Umbilicus tiny and chink-like. Common; intertidally on mudflats.

F. **TRUNCATELLIDAE:** Rissoacea with small, cylindrical shells. Adult shells mechanically truncated. Loss of early whorls necessitates an end closure by a domed septum. Aperture oval; peristome blunt, single or double. No umbilicus. Operculum thin, corneous, paucispiral. Dioecious. (4 spp. from Bda.)

***Truncatella pulchella* f. *bilabiata*** (Pfeiffer) (Beautiful Truncatella): Genus with small, ovate-cylindrical shells sculptured by many fine axial ridges. Translucent.—Species to 5 mm, red-brown to pale straw in color. Sculpture ranging from well-formed to much- reduced axial ribs, the former being dominant. Lips whitish, doubled, the inner one thicker. Common under rotting vegetation and in damp, shady recesses above high tide line.

***T. caribaeensis*** Reeve ($=T.$ *subcylindrica* Gray; *T. succinea* C. B. Adams) (Caribbean Truncatella): Genus as above.—Species to 7 mm, with a thin outer lip. (Some workers think this is just a form of *T. pulchella*.) Habitat as above.

***T. scalaris*** (Michaud) ($=T.$ *clathrus* Lowe; *T. piratica* Clench & Turner) (Staircase Truncatella): Genus as above.—Species to 5 mm. Sculptured with strong, solid axial ribs. Faint spiral lines between the ribs. Color dull gray to pale straw. Outer lip double, inner portion greatly thickened. Habitat as above.

F. **VITRINELLIDAE:** Rissoacea with very small to minute, porcelaneous, non-nacreous shells less than 5 mm, varying from low shells with a small raised spire with large umbilicus to high spire and small or closed umbilicus. Operculum corneous, circular, multispiral. Radula taenioglossate. (3 spp. from Bda.)

***Vitrinella helicoidea*** C. B. Adams (Spiral-shaped Vitrinella): Genus with small, discoidal, vitreous shell of about 4 whorls. Spire slightly raised. Deep umbilicus. Aperture oblique, nearly circular.—Species 2 mm, translucent-white, glossy, smooth. Peristome thin, sharp. Uncommon, living under rocks or found dead in beach drift.

F. **CAECIDAE:** Rissoacea with minute, tusk-shaped shells less than 6 mm. Early spiral whorls decollated, broken and closed with a calcareous septum terminating in a sharp, rigid point (mucro). Operculum thin, corneous, multispiral. (8 spp. from Bda.)

***Caecum plicatum*** Carpenter ($=C.$ *obesum* Verrill & Bush; *C. termes* Heilprin) (Plicated or Corrugated Caecum): Genus with curved, tubular shell. Embryonic spiral whorls lost at maturity.—Species to 3 mm with about 15 thick, longitudinal ridges. Axial rings near aperture rapidly thickening to form a strong varix. Color white with tan mottled markings. Mucro low and

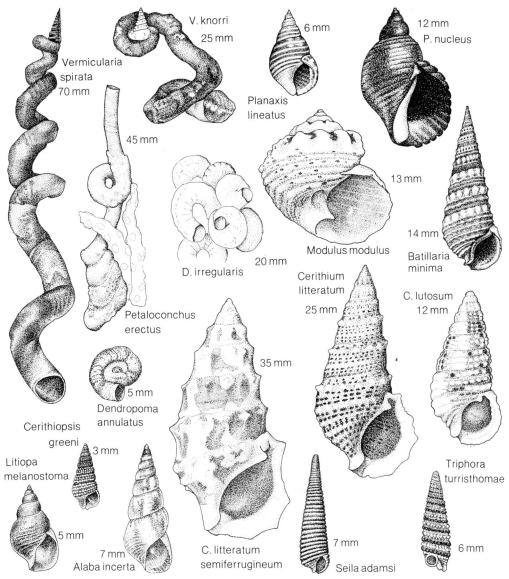

**136 CERITHIACEA (Worm and horn shells)**

pointed. Common in sand and algae, 1-3 m.

**C. tornatum** Verrill & Bush (Bermuda Caecum): Genus as above.—Species 2-2.5 mm with about 20 widely spaced, thick rings. Color white. Mucro large and projecting. Common in sand and algae, 1-3 m.

**Plate 136**

SUP.F. **CERITHIACEA:** Caenogastropoda generally with narrow, pointed shells of many whorls, sometimes provided with a siphonal notch.

F. **TURRITELLIDAE:** Cerithiacea with high, narrow, pointed shells of many whorls, without siphonal notch. Penis lacking; pallial oviduct open. Eggs laid in gelatinous masses. Many species are sedentary and have mucociliary feeding. (2 spp. from Bda.)

***Vermicularia spirata*** (Philippi) (=*Vermetus lumbricalis* L.) (West Indian worm shell): Genus with a turritelloid form of juvenile shell, uncoiling at random as it grows. Operculum corneous, multispiral, dished, without bristles.—Species to 150 mm. Shell brown or orange-brown, nuclear whorls dark brown with 1 or 2 spiral cords. May form colonies of aggregated masses or grow singly on *Oculina* coral, 1-6 m. (For an unusual inshore growth form see GOULD 1968a, b.)

***V. knorri*** (Deshayes) (Florida worm shell): Genus as above.—Species to 50 mm. Similar to *V. spirata*, except that the juvenile whorls are white. The preferred habitat is eroded or irregular limestone rock in quiet shallow waters, 1-3 m.

F. **VERMETIDAE:** Cerithiacea with the spiral juvenile shell cemented to hard substrate (shell or rock). Subsequent shell form resembles the irregular tube of a polychacte worm, sometimes in inextricable masses. Operculum corneous, multispiral, concave. Penis lacking. Pallial oviduct is open. Mucociliary feeding. (5 spp. from Bda.)

***Petaloconchus erectus*** (Dall) (Erect worm shell): Genus with worm-like tubular shell attached to hard substrate. Spiral ridges on inner surface of shell.—Species to 50 mm, white to ivory. Shell attached and randomly coiled for several whorls, then straightening out and projecting upward to 45 mm. Common in rocky, rubble areas in quieter waters to 5 m and offshore to 100 m.

***Dendropoma annulatus*** Daudin (=*D. corrodens* (Orbigny)) (Corroding worm shell): Genus with small, worm-like tubular shell that corrodes a trench in the substrate to which it is attached. Operculum with fringed whorls; underside reddish with a peg in the middle.—Species to 5 mm, dark reddish brown. Common on hard surfaces (rocks, shells, etc.) in intertidal areas to 5 m.

***D. irregularis*** (Orbigny) (Irregular worm shell): Genus as above.—Species with sturdy, rugose, irregularly coiled shell of varying length; diameter to 8 mm; color a light gray-brown to brown. Interior glossy, tan to brown; surface with fine to medium longitudinal ribbing; operculum corneous, multispiral, red-brown, with fringed edge. May grow singly or in compact, confused masses attached to coral, rocks and other shells. An important member of the reef-building community. Prefers to face the open ocean where there is a constant exchange of water. Empty shells are frequently inhabited by the hermit crab *Calcinus verrilli*.

F. **PLANAXIDAE:** Cerithiacea with shells that are ovate-conic with elevated spire. No varices.

Columella flattened and truncated anteriorly. Operculum corneous, paucispiral. Notable for an internal brood pouch. (Only genus: *Planaxis,* with 2 spp. from Bda.)

***Planaxis lineatus*** (da Costa) (Dwarf Atlantic Planaxis): Genus with small, sturdy shells. Small notch at anterior end of aperture. With spiral sculpturing and color patterns. Periostracum fibrous.—Species to 6 mm, solid, glossy-smooth. Cream-white with or without spiral bands of brown, rarely all brown. Sculptured on the juvenile whorls with 4 or 5 thin spiral cords which become obsolete in maturity. Aperture oval; 9 or 10 denticles well within outer lip. Periostracum thin, translucent. Common; generally in colonies under rocks and rubble intertidally.

***P. nucleus*** (Bruguière) (Black Atlantic Planaxis): Genus as above.—Species to 13 mm, solid, very dark brown. May be completely smooth or sculptured with up to 15 strong spiral cords that develop behind the outer lip from suture to base and are faint elsewhere. Aperture oval. Outer lip thickened; 12 or 13 denticles well within. Rare; along exposed rocky shores.

F. **MODULIDAE:** Cerithiacea with short-spired, trochoid-shaped shells. Columella with tooth-like projection at lower end. (1 sp. from Bda.)

***Modulus modulus*** (L.) (Atlantic Modulus): Genus with small, porcelaneous, turban-like shells. Inner lip grooved, ending at base with a sharp denticle.—Species to 13 mm. Top of whorls ribbed or tuberculate, base with 5-8 spiral cords. Narrowly umbilicate. Gray-white with maculations of brown. Operculum corneous, multispiral. Common; on algae in shallow, protected areas.

F. **POTAMIDIDAE:** Cerithiacea with shells that are solid, elongate, conical, many-whorled, with axial, pustulose riblets. Outer lip without denticles. Penis lacking. (1 sp. from Bda.)

***Batillaria minima*** (Gmelin) (=*B. minima rawsoni* Mörch) (False horn shell): Genus with cerithium-like shells. Siphonal canal short. Outer lip smooth within.—Species to 15 mm. Shell elongate, high-conic; sculptured with beaded ribs and uneven spiral threads. Off-white to gray with banded phases, to all black. Aperture narrow. Operculum corneous, round, multispiral. Intertidal, on mud flats in quiet waters, often in enormous numbers.

F. **CERITHIIDAE:** Cerithiacea with elongate, conical, usually strongly sculptured shells that range from small to quite large. Oval aperture, usually with short, bent, siphonal canal. Outer lip flared in the adult. Operculum corneous, paucispiral. Lacks penis. Pallial oviduct open. (23 spp. from Bda.)

***Cerithium litteratum*** (Born) (Lettered horn shell): Genus with small to large, high-conic shells.

Surface smooth to coarsely or finely sculptured with beaded or nodose spiral cords. May have irregularly placed varices. Operculum oval, paucispiral.—Species to 30 mm. Whitish with spiral rows of brownish black tiny rectangles, at times so heavily marked as to appear all black. Sculptured at times with a row of spiny tubercles on the shoulder just below the suture, and occasionally with a second row of lesser tubercles lower on the periphery. Siphonal canal short. On mud and grassy areas; common.

*C. litteratum* f. *semiferrugineum* Lamarck (Semi-rusted horn shell): Genus as above.—A form of the above species. Generally smooth with few to no tubercles. Color all white to white with large, irregular splotches of rust-brown to all rust-brown. In deep waters offshore; quite often brought up in fish pots.

*C. lutosum* Menke (=*C. variabile* C. B. Adams; *C. bermudae* Sowerby; *C. ferrugineum* Say) (Dwarf horn shell): Genus as above.—Species to 12 mm. Shell stocky, not elongate. Color dark blackish brown. Sculptured with 3 or 4 rows of even-sized beads. With 8 or 9 teeth well within outer lip. May have 1 or 2 varices on ultimate whorl. Common under rocks. (Do not confuse with *Batillaria minima*!)

*Litiopa melanostoma* Rang (Brown Sargassum snail): Genus with small, thin, conoid shells. Columella truncated at base end.—Species to 5 mm with 6-7 whorls. Color straw to dark brown, outer lip often bordered internally with darker brown. Aperture elongate, oval. Operculum corneous, thin, paucispiral. Nuclear whorls tiny, smooth; two post-nuclear whorls with axial ribbing; remaining whorls smooth with incised, spiral lines. Pelagic; found on seaweed or washed ashore in beach drift. Common.

*Cerithiopsis greeni* (C. B. Adams) (Green's miniature horn shell): Genus with small, beaded shells; lower half cylindrical, tapering above. Aperture oval, small. Siphonal canal short, slightly recurved, truncated. Operculum corneous, thin, paucispiral.—Species 4 mm, with 9-12 whorls. Apical whorls pale straw, smooth, translucent; the remainder dark brown, with 2-3 spiral rows of large glossy beads. Suture broad, impressed. Uncommon; in shallow waters.

*Alaba incerta* (Orbigny) (=*A. tervaricosa* C. B. Adams; *A. melanura* C. B. Adams) (Varicose Alaba): Genus with small, smooth, translucent shells. Surface with weakly incised spiral lines. Operculum corneous, thin, paucispiral.—Species 7 mm, with 10-12 whorls; may have up to 3 low, rounded varices per whorl from midsection onward. Apical whorls dark. Sutures impressed. Aperture oval, lip thin. Color straw to dark tan. Surface with 12-16 weakly incised spiral lines. Common; in sand, weeds and rubble, 1-3 m, and dead in beach drift.

*Seila adamsi* (H. C. Lea) (Adams' miniature horn shell): Genus with small, elongate-conical shells. Strong spiral cords. Siphonal canal short, truncated. Operculum corneous, thin, paucispiral.—Species 7 mm, with about 12 whorls. Sutures indistinct. Sculptured on each whorl with 3 or 4 elevated spiral cords with fine axial threads between. Color brown. Aperture small, lip thin. Uncommon; to 70 m.

F. **TRIPHORIDAE:** Cerithiacea with small, elongate conical, sinistral shells. Sculptured with spiral cords of fine beads that are axially lined up on each whorl. Aperture small, sub-circular. Siphonal canal tubular, short. Operculum corneous, thin, paucispiral. (4 spp. from Bda.)

***Triphora turristhomae*** (Holten) (=*T. mirabile* C. B. Adams; *T. bermudensis* Bartsch) (Thomas Tower Triphora): Genus with small, slender, sinistral shells. Siphonal canal short, tubular or open, recurved dorsally.— Species to 7 mm, with 10-12 whorls. Nuclear whorls moderately rounded; 1st postnuclear whorl smooth, 2nd and 3rd whorl each with the beginning of a beaded spiral cord to about midsection where a 3rd, lesser beaded cord begins. The last 2 nuclear whorls also have about 25 fine axial threads. Color white, lower row of beading dark straw; entire base slightly darker. Uncommon; in sandy rubble and under stones to 80 m.

**Plate 137**

SUP.F. **EPITONIACEA:** Caenogastropoda with a long row of numerous, similar, graduated teeth on the radula that increase in size laterally; central tooth lacking (ptenoglossate). Protandrous hermaphrodites. Pallial gland secretes viscous purple fluid when irritated. Larval shell reticulated.

F. **JANTHINIDAE:** Epitoniacea with thin, fragile shells. Operculum, eyes and penis lacking. Floats upside down. Modified foot produces an air bubble raft that also carries the egg capsules. Feeds on pelagic Hydrozoa, chiefly *Velella*. Pelagic; circumtropical (2 spp. from Bda.)

***Janthina janthina*** (L.) (=*J. communis* Lamarck) (Common purple sea snail): Genus with fragile, turbinate, purple shell.— Species to 35 mm, shell whorls 2-toned violet or purple on lower half, pale lavender on upper. Aperture angular, terminating with slight projection at base of columella. With slight labial sinus. Uncommon. Sometimes found along beaches after a summer storm.

***J. pallida*** (Thompson) (Pallid Janthina): Genus as above.— Species 20 mm. Shell globose, pale lavender. Aperture rounded, terminating at base of columella in a smooth curve. Without labial sinus. Uncommon; associated with above.

F. **EPITONIIDAE:** Epitoniacea with shells that are usually glossy, all white or white with brown markings; turriculated (a high conical spire); umbilicus absent, or partially covered by an expansion of the inner lip. Many whorls, attached or loosely coiled

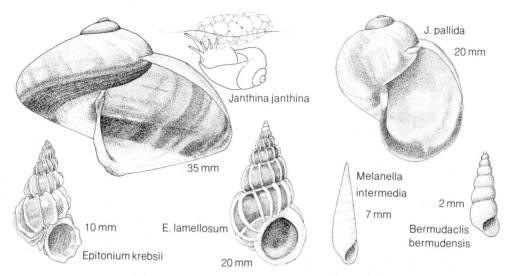

**137 EPITONIACEA (Wentle traps, a.o.), EULIMACEA**

or open and attached by the contact of the blade-like costae. Sculpture varies from species to species. Operculum corneous, paucispiral. Actiniaria are the preferred food. (14 spp. from Bda.)

***Epitonium krebsii*** (Mörch) (= *Scala electa* Verrill & Bush) (Krebs' wentletrap): Genus with white, porcelaneous, high-conic shells. Sculpture of strong to weak, axially bladed costae. Whorls often open, not touching. Generally umbilicate.—Species with glossy, thin but strong, turbinate shell to about 10 mm in length. With 7-9 whorls each with 10-12 costae. Costae connected axially. Umbilicus narrow to wide, and deep; may be partially covered by reflection of inner lip. Uncommon; intertidally to 300 m.

***E. lamellosum*** (Lamarck) (= *Scalaria clathrus* L.) (Lamellose wentletrap): Genus as above.—

F.

Species with glossy, turreted shell about 20 mm. With 11-12 attached whorls. Numerous blade-like costae, each lined up with those of the preceding whorl, giving the appearance of a continuous line to the apex. Color white with brown undertones that show through as maculations or, more generally, in 2 bands that sometimes join together giving an all brown color to the shell. Costae are always white. There is a raised, well-developed spiral thread at the base of shell. Uncommon; 1-70 m.

**ACLIDIDAE:** Epitonicea with small to minute, thin, translucent, elongate-ovate to turreted shells. Surfaces may be smooth, axially ribbed, spirally lirate or reticulated. With or without an umbilical chink. Aperture oval. Operculum corneous, paucispiral. Possess a pair of jaws and radula with very small, hooked teeth. Sexes apparently separate. (1 sp. from Bda.)

***Bermudaclis bermudensis*** (Dall & Bartsch) (Bermuda Aclis): Genus with minute shells, pupoid in outline. A single, smooth, well-rounded nuclear whorl. Post-nuclear whorls rounded, suture depressed. Surface generally with fine reticulated sculpturing. No umbilicus.—Species elongate-conic, white, 2 mm in length. Upper 1/3 of whorl smooth, lower 2/3 with 6 fine raised spiral threads. Entire surface marked with fine axial growth lines. Outer lip thin, surface markings showing through. Uncommon; in shallow water or along drift line on edge of mangrove swamp.

**SUP.F. EULIMACEA:** Caenogastropoda with shell that may be elongate-turriculate or subglobose, or absent in the adult. Anterior siphonal canal absent. There is a trend toward parasitism of echinoderms with subsequent loss of radula and jaws.

**F. MELANELLIDAE:** Eulimacea with small, high-spired, turreted shells; white, occasionally with brown markings, smooth, glossy, with nearly flattened whorls. Sutures shallow, almost indistinct. Nuclear whorls dextral, never tilted or immersed and concealed. Spire sometimes bent to one side. Columella without plications. Operculum thin, corneous, paucispiral. No jaws or radula. Sexes separate; ♂ with a penis. (7 spp. from Bda.)

***Melanella intermedia*** (Cantraine) (Cucumber Melanella): Genus with elongate shells. Not umbilicate. Generally parasitic on holothurians, starfish and sea urchins.—Species to 7 mm; shell narrow, sharply pointed; with 12 or 13 flattened whorls. Glossy white, sometimes tinged with light brown. Aperture narrow, teardrop-shaped. Outer lip thin and sharp. Common; ectoparasitic on *Isostichopus badionotus*; to 15 m.

### Plate 138

**SUP.F. HIPPONICACEA:** Caenogastropoda with low conical to trochoid shells with large ultimate whorl. Operculum thin, corneous, paucispiral. Radula taenioglossate.

**F. FOSSARIDAE:** Hipponicacea with small, solid, spirally ribbed shells. Color white, generally with brown nuclear whorls. Umbilicus wide. Columella nearly straight. (1 sp. from Bda.)

***Fossarus orbignyi*** Fischer (Orbigny's Fossarus): Genus with shells to 10 mm. Turbiniform, umbilicate, spirally corded. Spire short.—Species to 2 mm in length, shell with 4 whorls; nuclear whorls reticulated, translucent, brown, often missing. Sculptured with 5-6 raised, smooth, spiral cords. Interstices axially striated. Umbilicus chink-like, fairly deep. Umcommon; in sand, low tide to 4 m.

**SUP.F. CALYPTRAEACEA:** Caenogastropoda with low spiral or cap-like shells. Operculum generally lacking; when present it is

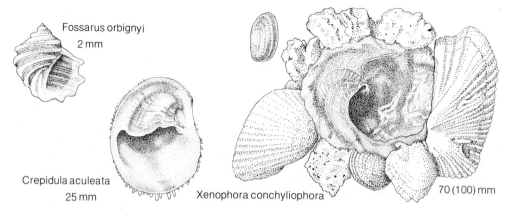

**138 HIPPONICACEA, CALYPTRAEACEA (Slipper and carrier shells)**

corneous, with offset nucleus. Radula taenioglossate.

F. **CALYPTRAEIDAE:** Calyptraeacea with oval, low conical, cap-like shells with a central or posterior beak-like apex. Columella replaced by an internal shelf or other projection for attachment of the soft parts. Practically sessile, with mucociliary feeding. (3 spp. from Bda.)

***Crepidula aculeata*** (Gmelin) (Spiny slipper shell): Genus with limpet-like shell. Apex at posterior end, coiled. Interior shelf covers the posterior of the soft body. No operculum. Exterior smooth or roughened by growth lines or spines.—Species to 25 mm. Surface rough, spinose. Off-white, mottled with reddish brown. Found attached to smooth objects (stones, discarded bottles, other shells, etc.) Shallow water to 30 m; uncommon.

F. **XENOPHORIDAE:** Calyptraeacea with the ability to cement shells, stones and other objects to their shells. Basic shell low conic, trochiform. Umbilicus partially covered or absent. Operculum rounded trapezoid, with lamellar growth lines. (1 sp. from Bda.)

***Xenophora conchyliophora*** (Born) (=*Phorus agglutinans* Montfort not Lamarck; *X. trochiformis* (Born) (Atlantic carrier shell): Genus with strong shells; umbilicus lacking.—Species to 70 mm, with 6 or 7 whorls. Off-white with light brown growth line markings that may be hidden by attached objects but are visible on the base. Aperture large and oblique. Lives a sedentary life spending much time withdrawn within the shell; feeds on microscopic algae. On shallow sand and rubble bottoms; uncommon.

### Plate 139

SUP.F. **STROMBACEA:** Caenogastropoda with small to large and massive shells. Outer lip often expanded, aperture generally with a siphonal canal. Columella callused, without plicae. No umbilicus. Operculum corneous, with apical nucleus, claw-shaped,

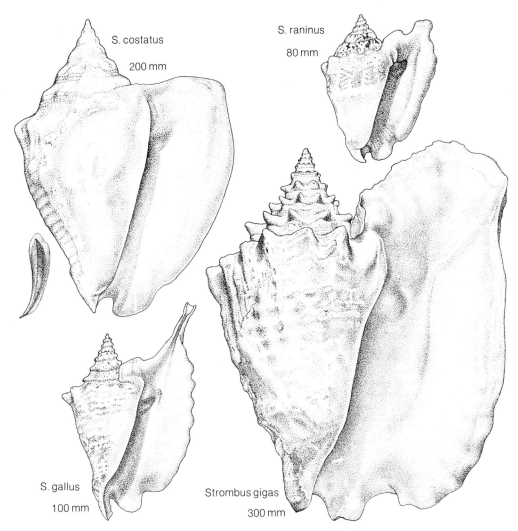

**139 STROMBACEA (Conchs)**

often with serrated edge. Radula taenioglossate. Dioecious.

F. **STROMBIDAE:** Strombacea with solid, low spired to high turreted shells, 8-300 mm. Ultimate whorl takes up the greater portion of the shell. Outer lip expanded and flaring, often with digitations, in the adult. Juvenile shell without flared lip; looks like a cone. The outer lip has an indentation (stromboid notch) at the base to accommodate the right eyestalk. Eyes pedunculated, large, with several orange, yellow, red or black rings; tentacles slender, siphon short, proboscis long, contractile; foot large. Herbivores or detritus feeders. The taking of any live *Strombus* is prohibited by Bermuda law. Predators are hermit crabs, spiny lobsters, fishes, rays, loggerhead turtles and humans. (5 spp. from Bda.)

***Strombus gigas*** L. (Pink conch; B Queen conch, Great conch): Genus with strong shells; surface smooth, may have a row of nodes at shoulder of whorl. Aperture long. Outer lip may become thickened with maturity.—Species to 300 mm. Shell heavy, large flaring lip thickened in the adult. Aperture moderately narrow, the interior generally rose-pink, sometimes yellow to light orange. Nodes on shoulder may be absent, low rounded or extended into blunt spines. Periostracum straw color, corneous, flaking off when dry. Edible and overcollected. May lay as many as 750,000 eggs in a compact mass made up of a long, continous, sand-coated strand. Uncommon; in grass beds, 5-20 m. (For information on biology see RANDALL 1964.)

**S. costatus** Gmelin (=*S. accipitrinus* Lamarck) (Milk conch; B Harbour conch): Genus as above.—Species to about 10 whorls, 200 mm in length. Shell solid, thickened flaring lip in the adult; moderately high, nodeless spire. Exterior color straw, light brown to pinkish orange or mauve, sometimes with brown axial markings. Periostracum straw color, corneous, thin. Aperture milk-white; the columella callus and interior of lip darken to aluminum-gray with age. Edible and overcollected. Prefers the algae beds of quiet inshore waters; 5-20 m.

**S. raninus** Gmelin (Hawk-wing conch): Genus as above.—Species with 9-10 whorls; to 80 mm in length. Juvenile with fine axial ribs crossed with many fine spiral threads, making a reticulated surface. Three low, thickened varices per whorl to the 5th or 6th whorl where varices cease. Shoulder nodes begin on the penultimate whorl, increasing in size to the ultimate whorl where the last 2 nodes on the dorsum become large, thick and blunt. Paralleling the shoulder and widely separated are 2 rows of small, rounded nodes. Lip thickened in the adult; the posterior end with slight extension up to but not beyond the apex. Color a mottled gray-brown to chocolate brown. Aperture narrow, cream colored; interior salmon-pink. Siphonal canal extended and curved. Rare; in grassy shallows to 10 m.

**S. gallus** L. (Rooster-tail conch): Genus as above.—Species with 10-12 whorls, 100 mm in length. Spire moderately elevated; fine axial ribs and spiral threads give a reticulated surface to about the 5th whorl. Spiral threads diminish and ribs become blunt nodes that increase in size with the whorl growth until at the dorsum of the ultimate whorl they are extended and bluntly spinose. Lip flared, blade-like, never thickened; the posterior end with a long, channeled extension much past the apex. Aperture long, narrow, salmon-tan to straw; interior white. Exterior tan, yellow, pink or mauve, maculated with light to dark brown splotches. Siphonal canal quite extended and slightly recurved. Rare; in grassy shallows in reef area, 3-20 m.

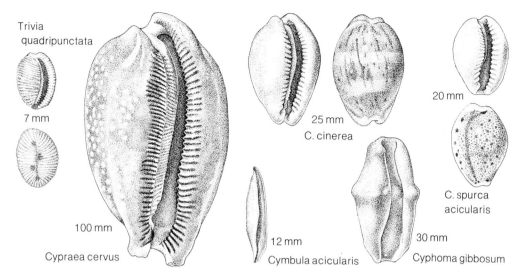

**140 TRIVIACEA, CYPRAEACEA (Cowries, a.o.)**

## Plate 140

**SUP.F. TRIVIACEA:** Caenogastropoda with small shells; spire low conical or immersed (apex concealed by subsequent whorls). Surface smooth and glossy, or with transverse, often bifurcated ribbing. Aperture narrow, with dentition on both sides. No periostracum; no operculum. Feed on and lay eggs in colonial ascidians.

**F. TRIVIIDAE:** Triviacea with ovate-globose shells resembling small cowries, sculptured by many transverse ribs. Aperture narrow, parallel to axis, with dentition on the lips formed by the continuation of the ribbing. With or without sulcus and/or colored blotches on the dorsum. (5 spp. from Bda.)

***Trivia quadripunctata*** (Gray) (Four-spotted Trivia; B Pink lucky): Genus with shell surface decorated with fine to coarse, often bifurcated, transverse ribbing that extends on both sides of the shell from the long, narrow aperture to the mantle line on the dorsum.—Species to 7 mm. Shell with 20-25 fine transverse ribs that are never beaded. Color when alive is bright pink with 2-4 small brown blotches along the mantle line. Mantle pink, orange, light brown or charcoal; papillae branching, white. Common; under stones and loose coral to 10 m.

**SUP.F. CYPRAEACEA:** Caenogastropoda with variably shaped shells: ovoid, globose, pyriform; short or elongate. Spire generally immersed. Surface smooth, glossy, generally patterned. Mantle lobes may be colored, often with tubercles or papillae, plain or frondose. Aperture long, narrow; parallel with the axis, channeled at both ends. Lips inflected, with or without denticles. No operculum although present in the larval stage. No periostracum. Radula taenioglossate.

F. **CYPRAEIDAE:** Cypraeacea with shells that are variable in shape, color and pattern. Smooth surface layer and color pattern deposited by mantle lobes that can extend from both sides to cover the shell, and are completely retractable into the aperture. Aperture with dentition on both sides. Juveniles ("Bulla stage") differ greatly from the adults, resembling elongated bubble shells or wide-mouthed, thin-lipped olive shells; spire not covered. Proboscis completely invaginable. Omnivorous, nocturnal. (3 spp. from Bda.)

*Cypraea cervus* L. (=*C. exanthema* L.; *Trona cervus peilei* Schilder) (Atlantic deer cowrie): Genus with shell that has the spire glaze-covered and concealed in the adult. Aperture the full length of the shell, long and narrow, with denticles on both sides.—Species in Bermuda seldom exceeds 100 mm in length. Shell elongate, oval, inflated; anterior portion of aperture wide. Chestnut brown, speckled with whitish, non-ocellated spots of varying size over the greater portion of the dorsum. Denticles marked with brown. Interior purplish. On sandy bottoms between reefs; rare.

*C. cinerea* Gmelin (Atlantic gray cowrie): Genus as above.—Species oval, inflated; to 25 mm. Color when alive is a deep brown with tiny black flecks, fading to an ashen brown when dead. Base a dark ivory-white. Aperture narrow; space between denticles purplish brown. Uncommon; under rocks and rubble on the reefs, to 60 m.

*C. spurca acicularis* Gmelin (=*C. spurca* L.) (Atlantic yellow cowrie): Genus as above.—Species to 20 mm. Oval, slightly flattened; dorsum covered with many fine, irregular flecks of orange-brown. Base, aperture and denticles ivory-white. Top margin of base with a row of pits that gives the edge a scalloped effect. Uncommon; under rocks and rubble on the reefs, to 60 m.

F. **OVULIDAE:** Cypraeacea with gibbose to transversely subcarinate, ovate to lanceolate shells. May be glossy smooth or with tranverse, incised striations. Spire immersed. Aperture usually narrow, extending the entire length of the shell. Inner lip smooth; outer lip inflected, with or without teeth. Living on octocorals. (7 spp. from Bda.)

*Cymbula acicularis* (Lamarck) (=*Simnia acicularis* (Lamarck) (West Indian Cymbula): Genus with elongate, subcylindrical shells. Usually thin, without adapical umbilication.—Species to 13 mm. Color lavender. Dorsum smooth, glossy, without striations. Aperture fairly wide. Outer lip inflected, slightly thickened, without teeth. Columella smooth, bordered by an off-white, raised axial ridge. Uncommon; attached to the ribs of *Gorgonia ventalina*; 1-20 m.

*Cyphoma gibbosum* (L.) (Flamingo tongue): Genus with glossy,

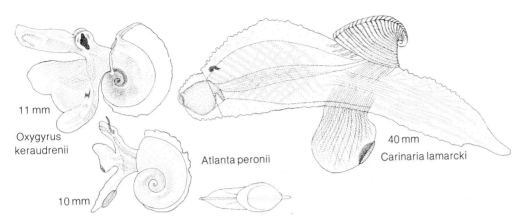

### 141 HETEROPODA (Heteropods)

smooth, oblong, sub-ovate shells. Crossing the dorsum is a swollen ridge that flattens out on the base. Aperture long, narrow, without teeth. Outer lip thick, inflected.—Species to 30 mm. Shell apricot-buff in color with a cream colored patch axially along the dorsum. Mantle flesh colored with closely arranged, yellow-orange, irregularly shaped spots that are ringed in black. Siphon rimmed in black. Foot ivory-white, with detached irregular black lines radiating toward the black-edged margin. Extremely overcollected; must be protected. Common; living on gorgonians, 1-20 m. (Color Plate 5.17.)

### Plate 141

**SUP.F. HETEROPODA** (=CARINARIACEA): Caenogastropoda with a laterally compressed foot. Shell may be whole, reduced or absent. Body light and transparent, adapted for swimming upside down. Without operculum (except in Atlantidae). Large, highly developed eyes that are tubular and projecting. Free-swimming, active predators of medusae, small fish, pteropods and other heteropods.

**F. ATLANTIDAE:** Thin, transparent Heteropoda with small, compressed, planorboid shells bearing a blade-like keel around the periphery that assists in swimming by keeping the shell in a vertical position. Animal can withdraw completely into the shell. Operculate. (7 spp. from Bda.)

***Atlanta peronii*** Lesueur (Peron's Atlanta): To 11 mm. Shell calcareous, glossy-smooth with microscopic growth lines. Spire flat. Apex sharp, not immersed. Color transparent-white, or may be tinged with brown at base of keel on ultimate whorl. Operculum thin, corneous, subtrigonal, but variable in shape. Collected in plankton tows, or occasionally washed up on beaches after storms.

***Oxygyrus keraudrenii*** (Lesueur) (Keraudren's Atlanta): To 10 mm. Calcareous juvenile shell, sculptured by longitudinal, wavy lines, is resorbed in maturity. Corneous adult shell is flexible, colorless or light straw. The keel starts shortly after the beginning of the ultimate whorl and reaches full broad extension at the edge of the aperture. Shell and keel sculptured by numerous fine growth lines. Shallowly umbilicate on both sides with all whorls visible. Apex immersed. Operculum rounded trapezoid, lamellar. Uncommon; associated with above.

F. **CARINARIIDAE:** Heteropoda with a small, nightcap-like, thin and fragile, transparent shell that fits over the visceral mass. (2 spp. from Bda.)

*Carinaria lamarcki* Peron & Lesueur (=*C. mediterranea* Blainville) (Lamarck's Carinaria): Genus pelagic. Shell small, thin, fragile. Apex hooked. Swims rapidly by movement of a modified propodium and steers with a modified metapodium.—Species to 250 mm. Body highly altered, slightly compressed laterally, gelatinous, transparent. Proboscis large, purple. Surface of shell rippled. Rare; might be collected in a plankton net. Shell very seldom found in beach drift.

### Plate 142

SUP.F. **NATICACEA:** Caenogastropoda with low spired, spiral globose to auriform shells, with or without periostracum. Surface smooth or with weak spiral sculpturing. Ultimate whorl large. Umbilicus may be partially or completely closed by a pronounced callus, or open and deep. Operculum corneous or calcareous, paucispiral. Foot large. Prefer a soft substrate where they can burrow below the surface in search of prey. Feed on both gastropods and bivalves by boring a hole in the shell and then inserting the proboscis to rasp out the soft parts with their radula. Dioecious. Eggs laid in the form of "sand collars." (Only family: Naticidae, with 9 spp. from Bda.)

*Polinices lacteus* (Guilding) (Milky moon snail): Genus with solid, glossy, ovate shell; last whorl large, slightly flattened. Umbilicus partially or completely filled with a button-like callus. Periostracum thin, smooth, yellowish. Operculum thin, translucent, corneous, paucispiral, wine-red to amber.—Species to 20 mm, glossy milk-white. In sandy areas, generally burrowing beneath the surface. Common; to 20 m.

*Sinum perspectivum* (Say) (Common baby's ear): Genus with flattened, auriform shells, spire minute. Aperture very large, columella strongly curved. Operculum minute, corneous. Periostracum thin, light brown.—Species to 35 mm. Shell dull white, very flat, sculptured with numerous finely incised spiral lines. Animal may entirely cover the shell. Leaves a wide track as it burrows through the sand in search of food. Uncommon; to 10 m.

*Natica livida* Pfeiffer (=*N. marochiensis* (Gmelin); *Polinices du-*

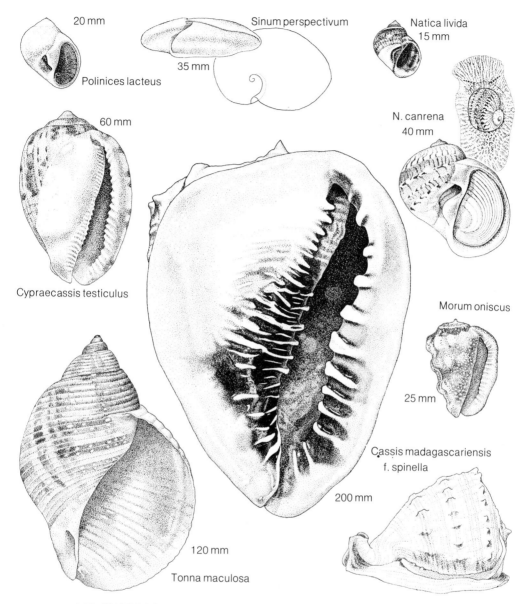

**142 NATICACEA (Moon snails), TONNACEA (Helmets, tuns)**

*plicatus* of authors not Say) (Livid Natica): Genus with globose, low-spired, glossy, smooth, solid shells. Umbilicus partially filled, to completely sealed over, by columellar callus. Aperture large. Operculum calcareous.—Species to 15 mm, gray-brown, with vague darker spiral bands. Operculum white, smooth. Interior of aperture and columella light brown. Umbilicus almost filled in with chocolate-brown callus. In sandy areas, uncommon; to 20 m.

***N. canrena*** (L.) (Colorful Atlantic Natica): Genus as above.—Species to 40 mm. Color pattern quite variable, usually with brown axial zigzag lines on alternating light-brown, brown or cream-white spiral bands. Base of shell, columella, columellar callus and edge of aperture white. Interior of aperture light brown. Umbilicus deep, almost filled by the columellar callus. Operculum calcareous; interior smoothish with fine growth lines; exterior with about 10 incised lines. In sandy areas, common; to 20 m.

**SUP.F. TONNACEA:** Caenogastropoda with small to very large shells with low spire, large apertures and short siphonal canals. Foot powerful. Operculum long, narrow or absent. Prey on echinoderms and other small invertebrates.

**F. CASSIDAE:** Tonnacea with small to very large, strong, porcelaneous shells. Spire low to moderately high. Varices present in most species. Aperture ovate to long and narrow. Parietal shield broad, well developed, the surface smooth, rigid or granulate. Short, curved siphonal canal. Operculum corneous, long and narrow to fan-shaped. Feed on echinoderms. (8 spp. from Bda.)

*Morum oniscus* (L.) (Atlantic wood louse): Genus with small, subcylindrical, knobby shells. Apex papilliform. Parietal shield pustulose. Periostracum thin, grayish.—Species with shell of 3 whorls, 27 mm in length. Base color white with spots or maculations of brown or gray-brown. Nipple-like nucleus white or pink. Aperture narrow. Outer lip thickened, with weak denticles. Uncommon, under rocks or coral slabs just below low tide mark.

*Cassis madagascariensis* f. spinella Clench (Clench's emperor helmet): Genus with large, globose shells. Exterior moderately glossy. Aperture long, narrow. Parietal shield glazed, with many plicae. Outer lip thickened, with heavy denticles.—Species to 300 mm. Exterior cream colored, with 3 widely spaced spiral rows of small blunt nodes starting at the shoulder. Parietal shield deep salmon colored, plicae whitish. Areas between the plicae on the columella and between the heavy denticles on the outer lip stained dark brown. Operculum long, narrow. Uncommon; sandy bottom to 15 m. Protected by law.

*Cypraecassis testiculus* (L.) (Reticulated cowrie helmet): Genus with ovate, short-spired shells. Aperture long, narrow. Columella with numerous fine plicae along its entire length. Operculum small or absent.—Species to 80 mm; color orange-brown; surface strongly reticulated, may have low nodes on shoulder. Parietal shield of the adult cream with a few orange-brown splotches. Outer lip thickened, with numerous denticles. Uncommon; burrows in sand; to 7 m.

**F. TONNIDAE:** Tonnacea with medium to large, thin, strong, ovate shells. Ultimate whorl large. Sculptured with strong spiral ridges; no varices. No

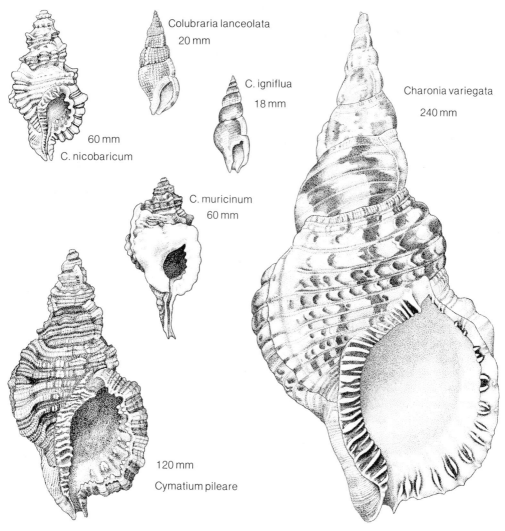

**143 CYMATIACEA (Tritons)**

operculum in the adult. (1 sp. from Bda.)

***Tonna maculosa*** (Dillwyn) (Atlantic partridge tun): Genus with globose shells. Exterior with strong, flattened spiral cords that are crossed by fine axial growth lines.—Species to 135 mm. Color all brown or brown with indefinite mottlings of white. Inside edge of thin lip stained dark brown. Suture impressed. Thin periostracum often worn off. Uncommon; shallow, grassy flats in protected areas.

### Plate 143

**SUP.F. CYMATIACEA:** Caenogastropoda with variably shaped shells. Aperture oval, with a short, open posterior canal or none. Siphonal canal well developed, long or short. Operculum corneous, oval. Shell surface with well-developed sculpturing. Varices usually prominent. Predatory, feeding on echinoderms, worms, gastropods and bivalves. Veliger stage long-lived.

F. **CYMATIIDAE:** Cymatiacea with small to large, strong, often highly sculptured, fusiform shells. No posterior canal. Varices strong, rounded, periostracum hairy. Animal strikingly colored. (13 spp. from Bda.)

***Cymatium pileare*** (L.) (Atlantic hairy triton): Genus with medium-sized shells. Body whorl large. Anterior canal short to long.—Species to 120 mm. Color light orange-brown to deep brown, with light and dark bands that are emphasized on the varices. Surface coarsely sculptured with strong spiral cords that are nodose where crossed by axial ridges. Parietal area orange to dark brown, wih many cream colored plicae. Interior of thickened outer lip orange to brown, with 7 or 8 cream colored, paired denticles that extend into the aperture. Siphonal canal well developed, short, recurved dorsally. Periostracum hairy, brown, thick, axially bladed. Uncommon; to 10 m.

***C. nicobaricum*** (Roeding) (=*C. chlorostomum* Lamarck) (Gold-mouthed triton): Genus as above.—Species to 60 mm. Shell solid, sculptured by deeply incised lines that form 8-10 sets of 2 thin and 1 thick spiral cords that are crossed by axial ridges forming coarse nodes. Color cream to light gray-brown, flecked and lined with rusty brown. Two varices per whorl. Protoconch of 6 whorls, glossy-smooth, straw colored, often missing. Aperture bright orange. Parietal wall with 11-19 cream colored plicae that extend into the siphonal canal. Outer lip with 7-10 single, cream colored denticles that run well into the aperture. Uncommon; to 10 m.

***C. muricinum*** (Roeding)(=*C. tuberosum* Lamarck) (Knobbed triton): Genus as above.— Species to 50 mm, solid, gray to brown, speckled and/or lined with dark brown, may have light spiral band centrally placed on ultimate whorl. Two varices per whorl. Sculptured with finely incised lines crossed by axial ridges forming 3-5 coarse nodes between varices. Interior of aperture dark orange-brown. Parietal callus heavy, white or cream, glossy smooth with several subdued plicae along the columellar wall. Outer lip thickened, white or cream, glossy, with 6 or 7 coarse denticles. Protoconch of 5 whorls, glassy, smooth, straw colored, often missing. Uncommon; grassy shallows and under rocks, to 10 m.

***Charonia variegata*** (Lamarck) (=*C. nobilis* Conrad; *C. tritonis* of authors not L.) (Trumpet triton): Genus with large, high-conic shells; whorls few with widely spaced varices. Sculpturing spiral, with incised lines and wide cords that in some species may be nodulose at the shoulder. Columella colorful, with many plicae. Outer lip crenate and/or denticulated within.—Species to 230 mm. Shell large, heavy; spire high, apex usually missing. Ultimate whorl large; shoulder slightly angular. Color variegated, chocolate-brown, tan and cream, pinkish purple near the apex. Aperture large, oval. Operculum heavy, lamellose; nucleus central. Columella dark chocolate-brown with many thin, raised white plicae. Outer lip with 10-13 paired, short, white denticles, the space between chocolate-brown. Siphonal canal short. Spire whorls usually badly eroded in all but the juveniles. Uncommon; on reefs to 30 m.

F. **COLUBRARIIDAE:** Cymatiacea with small to medium, elongate and moderately slender shells. Varices irregularly spaced or lacking. Sculptured with fine threads or cords that may form beads where crossed by axial growth lines. Aperture elongate-oval. Parietal callus sometimes raised into a columellar shield. Operculum with apical nucleus. Siphonal canal short. (3 spp. from Bda.)

*Colubraria lanceolata* (Menke) (Arrow dwarf triton): Genus with small, slender shells of about 7 whorls. Sculpturing may be beaded, cancellate, or have sinuous axial ribbing. Parietal callus smooth, not wrinkled as in Cymatiidae.—Species to 25 mm. Sculptured with 20-23 fine ribs between varices and many fine spiral threads that give the surface a cancellate appearance. Varices recurved, about 2/3 of a whorl apart. Color bone-white to straw; may be sparsely spotted or blotched with brown. Aperture smooth, glossy, with a short posterior canal. Edge of outer lip crenulated. Uncommon; under rocks intertidally to 10 m.

*C. igniflua* (Reeve) (= *C. swiftii* (Tryon)) (Swift's dwarf triton): Genus as above.—Species with slender shell, to 20 mm. No varices. Sculptured with many fine spiral threads and 15-17 subdued, rounded axial ridges that almost disappear on the ultimate whorl. Color all brown to bone-white with flammules of brown. Nuclear whorls smooth, whitish, often missing in the adult. Aperture white, glossy, with a short posterior canal. Outer lip with 8 or 9 low denticles within. A ridge on the base of the columella and one at the base of the outer lip form a short, channeled, siphonal canal. Common; under rocks intertidally.

## Plate 144

SUP.F. **MURICACEA:** Caenogastropoda with short to long, fusiform to roundish-ovate shells with or without spiral and axial sculpturing. Varices may be nodose, lamellose, foliated, spinose or lacking. Aperture oval. Siphonal canal short to elongated. Operculum corneous, with marginal to terminal nucleus. Mantle with a hypobranchial gland (produces a secretion from which purple dye was made). Predators of other mollusks, particularly bivalves; feed by boring into the shell by chemical and mechanical means.

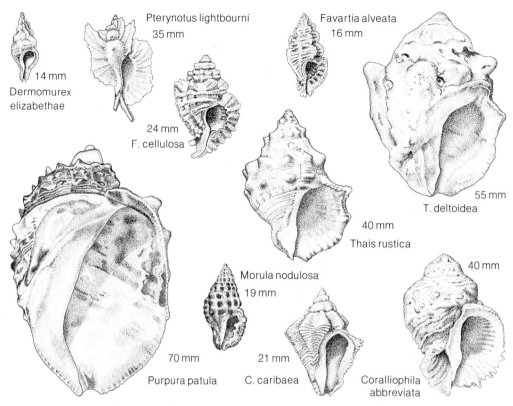

**144 MURICACEA (Rock shells, a.o.)**

F. **MURICIDAE:** Muricacea with generally thick, solid shells varying in shape and size; spire elevated or shortened. Aperture oval to rounded, may have denticles on outer lip. Siphonal canal short to very long, often partially closed. Former varices resorbed in the process of growing. May have a periostracum. (10 spp. from Bda.)

***Dermomurex elizabethae*** (McGinty) (Elizabeth's Aspella): Genus with small, solid, generally high-spired, fusiform shells. Color usually milk-white to yellow-white. A chalky, axially striated or minutely reticulated coating (intritacalx) covers live shells; this is worn off or eroded on dead specimens. May have a thin straw-yellow to buff periostracum.—Species to 18 mm. Shell elongate; each whorl has 3 varices axially aligned with those of preceding whorl and 2 or 3 intervarical nodes or ridges. Aperture glossy, milk-white. Siphonal canal short, almost closed, slightly recurved dorsally. Operculum straw colored. Uncommon; found under old coral slabs to 7 m.

***Pterynotus lightbourni*** Harasewych & Jensen (Lightbourn's Murex): Genus with small to medium-sized, fusiform shells with 3 prominent, smooth to lamellate, wing-like varices per whorl.—Species to 40 mm, milk-white, with 3 faint golden-tan patches on outside of outer lip. Varices broad, thin, translucent. Aperture oval, interior glossy, smooth. Siphonal canal long, open, slightly curved. Operculum, periostracum and soft parts unknown. Trapped in 275-600 m.

***Favartia alveata*** (Kiener) (=*Ocinebra intermedia* C. B. Adams) (Frilly dwarf Murex): Genus with small, broadly fusiform shells. Spire moderately high to high. Color generally gray-white. Aperture oval, interior with subdued lirae on outer wall. Surface with 3-7 low, rounded varices that are crossed by many fine, scaled, spiral cords.—Species to 20 mm. Body whorl with 5-7 raised, rounded, frilly varices. Aperture small, sub-circular. Outer lip finely corrugated. The many spiral cords have axially lamellate scales that give the surface a reticulated look. Unworn specimens have an axially striated coating. Moderately common under rocks in cracks and crevices to 5 m.

***F. cellulosa*** (Conrad) (=*Murex nuceus* Mörch) (Pitted Murex): Genus as above.—Species broadly fusiform, 28 mm gray-white. Surface with 5-7 rounded, frilly varices that occasionally develop short, blunt spines. Scaled spiral cords crossing over the varices give the surface a honeycomb look. Exterior often coated with calcareous algae concealing the sculpturing. Aperture small, round, brown within; outer lip frilly and lace-like. Siphonal canal recurved dorsally. Surface with an axially striated coating. Uncommon; in rocky rubble to 7 m.

F. **THAIDIDAE:** Muricacea with solid, rounded-ovate to fusiform shells without varices but sometimes with knobby or slightly spinose sculpturing. Siphonal canal generally short. Operculum with thickened, polished ridge along the inside of the outer edge. Active predators of barnacles and mussels. (3 spp. from Bda.)

***Thais rustica*** (Lamarck) (Rustic rock shell): Genus with large body whorl; moderately shortened spire; wide, oval aperture; outer lip smooth or with denticles. Columella flattened, often dished.—Species to 35 mm. Color dirty gray, banded with dark brown or purplish brown. Sculptured with 2 spiral rows of blunt, whitened nodes, one row at shoulder, the other just above mid-whorl. Many fine spiral threads. Aperture with white interior, often with exterior colors showing through, particularly on inside edge of lip. Denticles inside lip elongated. Common intertidally along rocky shorelines.

***T. deltoidea*** (Lamarck) (Deltoid rock shell): Genus as above.—Species to 50 mm, heavy, solid, roughly triangular in shape. Sculptured with a row of large blunt spines at shoulder, occasionally a smaller row just above midwhorl. Color grayish white with mottlings of dark brown to purplish brown. Color and sculpturing often obliterated by a heavy calcareous coating, by erosion or by numerous shells of *Spiroglyphus annulatus*. Aperture large, with white, smooth interior; edge of parietal callus often rust-brown; columella tinted with shades of lavender, its base with a ridge that forms the edge of the siphonal canal. Operculum semicircular, dark red-brown. Uncommon; intertidally in exposed areas on rocks to which it clings tenaciously.

***Purpura patula*** L. (Wide-mouthed Purpura): Genus with shells that have a very large aperture; columella smooth, flattened or dished.—Species to 85 mm. Shell ovate-globose; spire short; color dull gray-brown; sculptured with 6 or 7 spiral rows of sharp nodes and numerous incised spiral lines, the nodes becoming much reduced and the surface eroded with age. Aperture very large. Columella flattened, slightly dished, upper portion dark brown. Operculum large, semi-circular, dark red-brown. Rare. Found clinging tenaciously to intertidal rocks exposed to surf action.

***Morula nodulosa*** (C. B. Adams) (Blackberry drupe): Genus with small, fusiform shells. No varices. Surface nodulose. Siphonal canal short, open. Outer lip thin or thick, with denticles on inner surface.—Species to 20 mm. Surface with spiral rows of regularly spaced, small, rounded beads that are lined up forming slightly oblique, nodose, axial ridges, the interspaces of which have 3-5 rows of finely scaled threads. Surface often coated with calcareous growth, obliterating all sculpture. Aperture moderately narrow; outer lip with 4-5 strong, whitish denticles. Common intertidally on and under rocks.

F. **CORALLIOPHILIDAE:** Muricacea with small, sturdy, high- to low-spired shells. Spiral cords fine to coarsely sculptured with flattened spines. May be umbilicate. External color white, often coated with calcareous growths. Interior often soft lavender. Operculum thin, wine-red, dark brown to pale straw; nucleus at marginal edge. Lacks a radula. Feeds suctorially on coral tissues. (6 spp. from Bda.)

***Coralliophila abbreviata*** (Lamarck) (Abbreviated coral shell): Genus with shells that have predominantly spiral sculpture and weak or no axial ribs. Siphonal canal well defined, open, short.—Species to 50 mm; strong, low-spired, round-shouldered, bone-white. Sculptured with many finely scaled spiral cords that become larger and more scaly toward the base. Columella and interior of aperture and siphonal canal may be light orange-brown. Operculum yellow-brown. On corals. Uncommon; to 10 m.

***C. caribaea*** Abbott (Caribbean coral shell): Genus as above.—Species to 25 mm. Shell broadly biconical; may have numerous, subdued, rounded axial ridges. Color bone-white. Aperture elongate-triangular, often tinted lavender. Operculum wine-red. Umbilicus slit-like. At the base of sea fans, to 20 m; uncommon.

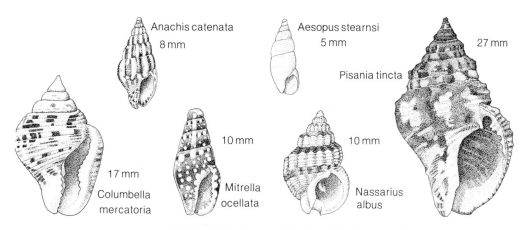

**145 BUCCINACEA (Dove shells, whelks, a.o.)**

## Plate 145

SUP.F. **BUCCINACEA:** Caenogastropoda with sturdy, small to large, oval to fusiform shells. Sculpture well developed but not spinose. Aperture elongate to round, generally with an anterior canal or notch. Inner lip smooth or denticulate. Predators and scavengers.

F. **COLUMBELLIDAE:** Buccinacea with small, fusiform, slender or stocky shells. Surface generally smooth, sometimes with low axial ridges and/or fine spiral cords. Can be quite colorful and patterned. May have a periostracum. Aperture oval to elongate and narrow. Columella and outer lip smooth or with denticles. Siphonal canal short. Not umbilicate. Operculum corneous, small, sickle-shaped or oblong. (11 spp. from Bda.)

***Columbella mercatoria*** L. (=*C. somersiana* Dall & Bartsch) (Common dove shell): Genus with round-shouldered, spirally corded shells with moderate spire. Aperture long and narrow with thickened, denticulated outer lip.—Species to 15 mm, solid, oval, squat. Highly colored, all white, pink, yellow, orange or brown or maculated with white and brown over the base color. Surface with many fine, spiral cords. Periostracum heavy, velvety, grayish. Outer lip white, thick, with about 12 denticles. Intertidally on and under rocks and grazing on algae.

***Anachis catenata*** (Sowerby) (Chain dove shell): Genus with shells that are small, fusiform, with low, rounded, axial ridges; aperture narrow; outer lip thickened, denticulate within.—Species to 6 mm, white, straw to brown, spotted with brown. Sculptured with about 18 low axial ridges. Outer lip slightly thickened, 6 or 7 weak denticles within. Intertidally under rocks.

***Mitrella ocellata*** (Gmelin) (=*Columbella cribraria* Lamarck) (White-spotted dove shell): Genus with small, fusiform shells. Sculpture smooth. Thickened outer lip.—Species to 10 mm. Color light brown to dark black-brown with numerous varying sized white dots, sometimes with white bands. Several incised spiral lines near the base. Aperture less than 1/2 the length of the shell. Outer lip thickened, denticulate within. Columella smooth. Apex often missing. Intertidally under rocks, sometimes in great quantities.

*Aesopus stearnsi* (Tryon) (Stearns' Aesopus): Genus with small, solid, subcylindrical shells sculptured with numerous spiral lines. Aperture small. Columella short. -Species to 5 mm; 5 or 6 whorls; apex blunt. Sculptured with many fine, spiral, axially bladed grooves that appear reticulated under microscopic examination. No axial ridges. Color off-white, may be finely blotched or banded with brown. Uncommon; in shallow water (10 m) dredgings.

F. **BUCCINIDAE:** Buccinacea with small to medium-large shells. Surface smooth to axially or spirally ornamented; no spines. Siphonal canal short to long. Columella smooth to rippled, without plicae. Predators and scavengers. (3 spp. from Bda.)

*Pisania tincta* (Conrad) (Tinted Cantharus): Genus with small, solid, short-spired shells. Sculptured with low, rounded axial ridges and numerous spiral cords. Aperture oval, with a prominent posterior canal.—Species to 25 mm, sturdy, short-spired. Sculptured with 13-15 low axial ridges and numerous spiral cords and threads. Color purple-brown with milk-white markings. Aperture with milk-white interior. Columella with 7-9 low plicae; the uppermost is strong and forms part of the posterior canal. Outer lip thickened, with 12 or 13 denticles. Siphonal canal short. Common; intertidally to 5 m in grassy areas around rocks.

F. **NASSARIIDAE:** Buccinacea with small, solid, ovate to spherical shells sculptured with axial ridges and/or spiral lines. Aperture roundish-ovate. Columella may have a large callus that is formed only in the adult. Outer lip denticulate within. Siphonal canal short and slightly recurved. Operculum round-triangular or claw-shaped. Scavengers. (2 spp. from Bda.)

*Nassarius albus* (Say) (=*Nassa ambiguus* Pulteney) (White basket shell): Genus with small, sturdy, pyriform to spherical shells. Parietal shield generally well developed. Operculum corneous, ovate, small. Very long proboscis.—Species to 12 mm, solid, pyriform. Color generally white, may be light brown, straw or yellow, rarely pink or lavender, sometimes mottled with brown. Sculpture variable, with 10-13 strong, rounded axial ridges crossed by a few to many fine to strong spiral cords. Base of the shell with a deep groove just above the short siphonal canal. Aperture round, interior white, glossy. Columellar callus small, smooth. Outer lip thickened, denticulate within. Common; in grassy shallows and on

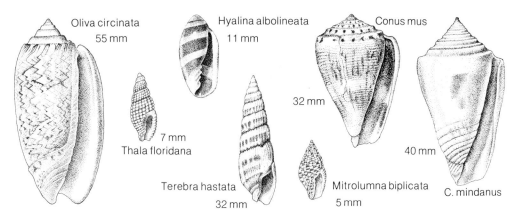

**146 VOLUTACEA (Olives), MITRACEA (Miters), CONACEA (Cones)**

clear sand to 10 m, buried in the sediment.

### Plate 146

**SUP.F. VOLUTACEA:** Caenogastropoda with cylindrical to fusiform shells, surface smooth or sculptured with axial ribs and/or spiral cords. Aperture wide to narrow. Columella usually with plications, the anterior-most being the strongest. Siphonal canal short to long. Operculum often absent. Radula rachiglossate. Predatory, feeding on mollusks and other small marine animals, generally by burrowing in sand and smothering prey with the foot.

**F. OLIVIDAE:** Volutacea with cylindrical, glossy, smooth, colorful, variously patterned shells. Spire short to moderately elevated. Aperture long, narrow; siphonal notch short, oblique. Columella smooth to finely lirate. With or without operculum. Siphon long; foot voluminous. Mantle lobes extend over the sides of the shell. Generally nocturnal, feeding on carrion, small crabs and other invertebrates. (About 8 spp. from Bda.)

***Oliva circinata*** Marratt (= *O. reticularis* of authors not Lamarck) (Netted olive; B Mic Mac): Genus with columella having many low lirae.—Species to 60 mm, elongate, sides slightly swollen at midpoint. Spire low to moderately high. Sutures channeled. Color all cream-white to completely marked with axial rows of purplish brown zigzag markings or small tent-like lines or, rarely, banded with rust-brown or all brown. Aperture long, narrow; pale violet within when fresh. Outer lip thickened, blunt-edged, smooth. Burrowing beneath the surface on sand flats to 30 m; common.

**F. MARGINELLIDAE:** Volutacea with generally small, elongate to conic shells; surface smooth, may be highly polished, rarely with weak axial ornamentation; white to very colorful and patterned. Aperture narrow. Columella with several weak lirae to a few strong folds. Lip thickened, with a rounded margin, smooth to denticulate within. No operculum. With or without a radula. Mantle envelopes the shell. Active carnivores. (About 7 spp. from Bda.)

***Hyalina effulgens*** (Reeves) (=*H. avena*, *H. albolineata* of authors not Kiener; *Volvaria avena* var. *southwicki* Davis) (White-lined Marginella): Genus with small, sturdy, cylindrical, low- spired, glossy shells. Outer lip smooth or weakly denticulate within.— Species to 8 mm; shell moderately elongate, cylindrical. Spire short, nucleus white. Amber colored with 2 white spiral bands. Outer lip smooth and white. Aperture long, narrow. Columella with 4 white folds at the base. Not uncommon; under rubble, intertidal.

SUP.F. **MITRACEA:** Caenogastropoda with a rachiglossate radula and a poison gland. Shell ovate to fusiform. Columella with folds. No operculum. With or without periostracum.

F. **VEXILLIDAE** (=COSTELLARIIDAE): Mitracea with paired accessory salivary glands and a pycnonephridian kidney (primary and secondary lamellae interdigitate). Egg capsules inverted-hemispherical. (1 sp. from Bda.)

***Thala floridana*** (Dall) (=*T. foveata* of authors not Sowerby) (Beaded Florida miter): Genus with small fusiform shells. Sculptured with axial ridges and spiral threads; surface cancellate or beaded. Aperture narrow, lirate within. Columella with 3-6 plicae. Outer lip slightly thickened, denticulate. Siphonal canal short or long. Periostracum thin.— Species to 8 mm, biconical or fusiform. Surface cancellate. Color dark brown, brown spotted, mottled with white or all white. Aperture moderately narrow. Columella with 3 plicae. Outer lip with many small denticles within. Intertidally to 90 m, under rubble or among algae on a hard substrate. Uncommon.

SUP.F. **CONACEA:** Carnivorous Caenogastropoda with variably shaped shells. Proboscis long. Radula toxoglossate (altered to a slender, harpoon-like, venom-injecting tube.)

F. **CONIDAE:** Conacea with small to large, generally conic shells. Spire flat to very high. Aperture elongate, narrow to moderately open. Posterior notch usually deep. Outer lip thin, sharp-edged, fragile, not denticulate within. Columella without plicae. Operculum small, corneous, oval or elongate. Periostracum heavy, opaque to thin, transparent. Carnivores of fish, worms and mollusks. Venomous; no cones toxic to humans known in Bermuda. (6 spp. from Bda.)

***Conus mus*** Hwass (Mouse cone): Genus with cone-shaped shells; narrow, elongate aperture; outer lip smooth, thin.—Species to 40 mm. Shell moderately heavy, dull. Spire low. Body whorl with many fine spiral threads. Shoulder with low white nodes separated by brown blotches. Color drab bluish or brownish gray with light to heavy, irregular brown markings and at midwhorl a lighter, spiral band that shows through into the aperture. Periostracum heavy, opaque. Intertidal; sandy and rubbly bottom under rocks. Uncommon.

***C. mindanus*** Hwass (=*C. agassizi* Dall; *C. bermudensis* Clench) (Bermuda cone): Genus as above.—Species to 50 mm. Shell fairly heavy, glossy. Spire low to moderately high. Whorls above shoulder concave. Nuclear whorls always dark pink. Color all white or white, pale yellow, pink or pinkish lavender mottled with large, irregular pink or pinkish brown splotches. Rarely with 25-30 spiral rows of short, brown and white dashes. A single, wide, lighter, spiral band occurs just below midwhorl. Anterior 1/3 of body whorl with 8-12 widely spaced spiral grooves. In the juvenile and sub-adult stage, the shell may have 5-9 spiral rows of small pustules, may be spirally grooved or a combination of both; this is rarely found in the adult. Operculum thin, oval; brown or tan. Periostracum thin, opaque, rust-brown, often worn off. Found partially buried in a sandy or shell-mud bottom in algal beds. Feeds on annelid worms. Mating season in late summer. Uncommon. Protected by law.

F. **TURRIDAE:** Very small to moderately large, biconical to spindle-shaped Conacea. Outer lip with anal notch or slit. Surface with or without axial ribs and/or spiral threads, cords, carina, beading or incised lines. Siphonal canal short to long. Columella smooth or with plicae. Operculum corneous. Radula toxoglossate or rachiglossate; all possess a venom gland. (15 spp. from Bda.)

***Mitrolumna biplicata*** (Dall) (=*Mitra haycocki* Dall & Bartsch) (Two-plaited miter-like turrid): Genus with small, biconical shells. Surface beaded by the crossing of axial ridges and spiral cords. Aperture narrow. Outer lip slightly thickened, denticulate. Anal notch shallow. Siphonal canal short, wide.—Species to 8 mm. Nucleus white, glossy. Outer lip thin, lirate within. Columella with 2 plicae, the upper one the strongest. Color straw to light brown, often with brown flammules. Uncommon; in beach drift and in dredgings to 15 m.

F. **TEREBRIDAE:** Conacea with elongate, narrow, tapered shells of numerous whorls. Aperture small; siphonal notch short. Outer lip thin, sharp. Shell exterior smooth, with or without spiral incised lines; may be axially ribbed, cancellate, heavily or sparsely beaded. Some forms have a poison injection mechanism similar to the Conidae, others lack a radula. Prey on small invertebrates; bite not toxic to humans. (1 sp. from Bda.)

***Terebra hastata*** (Gmelin) (Shiny Atlantic auger): Genus with smooth, shiny shells with low, rounded axial ridges. Radula with poison gland.—Species to 30 mm. Shell stocky, uppermost whorls taper more rapidly than the lower ones. Exterior glossy, sculptured with numerous rounded axial ribs. Color light straw to light brown with a white band below the suture. Aperture elongate-oval; outer lip thin, sharp; smooth and white within. Columella white, with 2 oblique, low spiral folds. In clean sand offshore, to 20 m.

S.CL. **OPISTHOBRANCHIA:** Partially or fully detorted Gastropoda with a single gill, auricle and kidney; head usually with 2 pairs of tentacles. Shell mostly reduced or lacking; operculum mostly lacking. Hermaphrodites; mostly marine. (The planktonic orders Thecosomata and Gymnosomata are often collectively referred to as pteropods, or sea butterflies.)

O. **BULLOMORPHA** (=CEPHALASPIDEA): Opisthobranphia with cephalic shield; usually with open sperm groove; sensory tentacles sometimes present, but usually without true rhinophores. Shell may be external (spire visible or concealed), internal or absent; operculum present in primitive forms, otherwise absent.

### Plate 147

SUP.F. **ACTEONACEA:** Bullomorpha with external shell, usually large enough to permit complete withdrawal of body; spire external, several whorls visible; aperture does not extend to spire.

F. **ACTEOCINIDAE:** Acteonacea with external shell into which body can completely retract; aperture narrow, but expanded anteriorly; shell cylindroid, with visible nuclear whorl; without operculum. (6 spp. from Bda.)

*Utriculastra canaliculata* Say (=*Acteocina candei*): Genus with glossy shell, elongate aperture, moderately elevated spire, columella with single spiral ridge.—Species glossy, white, with small nucleus; length 6 mm. Height of nuclear whorl and aperture length are variable. Sand or mud bottoms; to 50 m.

F. **APLUSTRIDAE** (=HYDATINIDAE): Acteonacea with globose or oval shell, enlarged aperture, minute nucleus; shell conspicuously banded; head bearing anterior tentaculate processes and lobed posteriorly. (3 spp. from Bda.)

*Hydatina vesicaria* (Lightfoot) (=*H. physis* L.): Genus with globose shell; epipodial lobes absent.—Species 40 mm, with fragile shell; closely spaced spiral brown lines; interior white. Among algae, to 3 m; rare.

*Micromelo undatus* (Bruguière): Genus with flat-spired oval shell, body not fully retractable; cephalic shield with both anterior and posterior lobes.—Species with fragile white shell bearing widely spaced red axial and spiral lines; foot and mantle with bright multicolored iridescence; length 16 mm. Under rocks or among algae, to 30 m.

SUP.F. **PHILINACEA:** Bullomorpha with shell reduced so that body cannot be completely withdrawn,

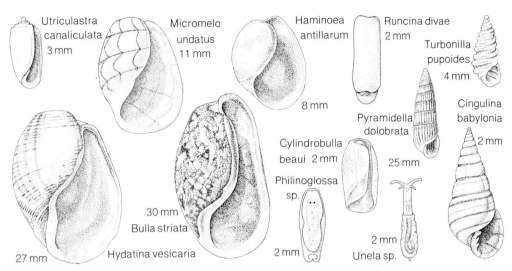

**147 ACTEONACEA—PYRAMIDELLACEA**

internal, or absent; spire involute if present. Cephalic lobes reduced or absent.

F. **RUNCINIDAE:** Philinacea without cephalic tentacles and parapodia; cephalic shield elongated, extending entire length of body. Gills simple, located on posterior right side; gizzard with 4 plates, radular formula 1:1:1. (2 spp. from Bda.)

*Runcina divae* (Marcus): Genus with bilobed rachidian tooth bearing fine denticles; opaline gland usually absent.—Species with variable color, brown, green or reddish, with small white specks; length 3 mm. Larval shell usually retained at posterior tip of body, to left of gill. Relatively common among algae in sheltered shallow areas, to 2 m.

F. **PHILINOGLOSSIDAE:** Philinacea with vestigial shell or shell absent; cephalic shield extending length of body; parapodial lobes vestigial. Gill and gizzard absent;

radula 3:0:3; habitat interstitial. (1 sp. from Bda.)

*Philinoglossa* sp.: Without a shell; body fusiform-squared. With a stomach and a pair of eyes. In shallow, subtidal, clean, coarse sand; not uncommon.

SUP.F. **BULLACEA:** Bullomorpha with external shell, inflated body whorl, aperture extending to region of shell apex; spire sunken below apex; parapodia often present; gizzard plates usually present.

F. **BULLIDAE:** Bullacea with strong, oval shell; columella with callus but without folds. (2 spp from Bda.)

*Bulla striata* Bruguière: Genus with umbilicate apex, mottled coloration of shell, moderate size.—Species with adapical callus; color whitish with brown markings in juvenile, the brown markings increasing with age; shell length 25 mm. Common, especially among sea grasses in shallow water; animals burrowing by day, emerging at night.

F. **ATYIDAE:** Bullacea with fragile, translucent or transparent shell with wide aperture; callus extends length of aperture; parapodia enclose shell when animal is extended. (3 spp. from Bda.)

*Haminoea antillarum* (Orbigny): Genus with very large body whorl; aperture does not extend much beyond apex; shell transparent, yellowish; radula present; gizzard with 3 large and 6 small plates.—Species with smooth shell, apertural lip arising to right of umbilicus; cephalic shield bilobed, body brown with white and orange spots; shell length 15 mm. In shallow water, usually among algae.

F. **CYLINDROBULLIDAE:** Bullacea with thin, flexible shell, with spiral apical slit; aperture ad-apical, with narrow slit to apex; cephalic shield bilobed; radula uniseriate. (1 sp. from Bda.)

*Cylindrobulla beaui* Fischer: Genus with bilobed cephalic shield; aperture ad-apical with narrow slit to apex.—Species to 14 mm, with white to pale straw shell, radular tooth broad with several denticles. To 200 m.

O. **ACOCHLIDIACEA:** Opisthobranchia without shell, gill or cephalic shield; with rhinophores and oral tentacles; with vestigial foot (metapodium) separated from a large, sac-like visceral mass; often with spicules embedded in mantle. Habitat interstitial.

F. **MICROHEDYLIDAE:** Acochlidiacea with mantle spicules; oral tentacles flattened, slightly larger than rhinophores (that may be absent); metapodium short. (1 sp. from Bda.)

*Unela* sp.: Genus with cylindrical rhinophores and flattened oral tentacles; radular formula 1:1:1; penis absent.—Species without spicules or eyes. In shallow, subtidal, clean, coarse sand.

O. **PYRAMIDELLIDACEA:** Opisthobranchia with high-spired, external shell of several whorls, hyperstrophically coiled (seemingly dextral shell on a sinistral animal); operculum present. Proboscis long, with stylet; gill and radula absent. Ectoparasitic on other invertebrates, or rarely detritivorous. (1 family: Pyramidellidae, with 25 spp. from Bda.)

*Pyramidella dolobrata* (L.): Genus with strong, glossy shell, columella with 3 strong spiral folds.—Species to 30 mm, with smooth, white shell bearing 2 or 3 fine, spiral, brown lines. To 55 m.

*Cingulina babylonia* (Adams): Genus with small shell bearing spiral cords or threads and axial riblets of microscopic size; varices absent.—Species with smooth nuclear whorl; body whorl with 3 or 4 spiral cords, remaining whorls with 2 cords; space between cords concave. Length 2 mm. Dredged from shallow water.

*Turbonilla pupoides* Orbigny: Genus with axial ribs, without spiral sculpture; columella with

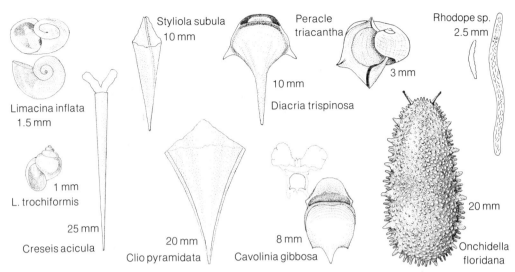

**148 THECOSOMATA (Sea butterflies), RHODOPACEA, ONCHIDIACEA**

single very weak fold.—Species 3 mm, with about 8 flat whorls, flat base; "fat" shape relative to other *Turbonilla*. Depth 15-23 m.

### Plate 148

O. **THECOSOMATA:** Planktonic Opisthobranchia with shell; operculum present or absent, epipodium of foot expanded to form natatory parapodia; with gizzard, jaws and a triseriate radula.

S.O. **EUTHECOSOMATA:** Thecosomata with calcareous shell; lacking proboscis and rostrum; tentacles unpaired, asymmetric.

F. **LIMACINIDAE:** Euthecosomata with small, fragile, trochoid, sinistrally coiled, involute, or planospiral shell; operculum present; mantle cavity dorsal, ctenidium absent. (6 spp. from Bda.)

*Limacina inflata* (Orbigny) (Planorbid pteropod): Genus with thin, smooth, transparent shell, hyperstrophically coiled (seemingly dextral shell on a sinistral animal or, in this case, a seemingly sinistral shell on a dextral animal).—Species to 1.5 mm; shell hyaline, plani-spiral with wide-flaring aperture; outer lip with tooth-like projection; umbilicus deep; brown spots near tooth and umbilicus. Operculum thin, corneous, transparent, slightly conoid and sinistrally coiled. Common; worldwide.

*L. trochiformis* (Orbigny) (Trochiform pteropod): Genus as above.—Species to 1 mm; shell smooth, ultimate whorl with some spiral lines; white to light purple, thicker parts purplish brown; spire elevated; aperture small; columellar aperture concave. Operculum as above; umbilicus narrow, deep.

F. **CAVOLINIIDAE** (=CUVIERIDAE): Euthecosomata with

shell never coiled but bilaterally symmetrical; operculum absent; shell white to brown. (18 spp. from Bda.)

***Creseis acicula*** (Rang) (= *C. recta* Gray) (Needle pteropod): Genus with long conical shell, circular in cross section.—Species to 33 mm; shell light chestnut-brown, slender, straight, elongate conical, tapering to a sharp point. Common; temperate and warm seas.

***Styliola subula*** (Quoy & Gaimard) (Keeled Clio): Genus with acutely tapered conical shell, with rib angled from center of dorsal aperture to the left, extending to tip. - Species to 10 mm; shell rose colored, round in cross section, elongate conical; surface with fine transverse striations between which are minute longitudinal striations. Common; worldwide.

***Clio pyramidata*** L. (Pyramid Clio): Genus with pyramidal shell bearing strong dorsal mid rib; aperture unobstructed; protoconch acutely tapered. —Species to 21 mm; shell thin, hyaline, may have a reddish sheen; pyramidally shaped, lacking lateral spines, posterior without lateral keels or transverse grooves; dorsal rib undivided. Common, worldwide.

***Diacria trispinosa*** (Blainville) (= *Cavolinia mucronata* Quoy & Gaimard; *C. cuspidata* Delle Chiaje) (Three-spined cavoline): Genus with dorso-ventrally inflated shell, aperture with hood-like dorsal projection; edges of aperture rolled in; sides with short projecting spines.—Species to 13 mm; chestnut-brown to transparent; dorsal lip thickened to form a pad; with a long terminal spine (lost near maturity) and 2 lateral spines. Common; worldwide.

***Cavolinia gibbosa*** (Orbigny) (= *C. hargeri* (Verrill); *C. flava* (Orbigny); *C. plana* Meisenheimer; *C. gegenbauri* (Pfeffer)) (Inflated cavoline): Genus with bulbous, squat shell; aperture restricted, broad transversely; sides of shell often with spines.—Species to 10 mm, brownish; dorsal lip thin, with very short lateral spines; ventral surface with transverse keel. Common; pelagic, circumtropical.

F. **PERACLIDAE:** Euthecosomata with fragile, sinistrally coiled shell, large aperture; columella extended to form large rostrum; periostracum forms hexagonal meshwork on shell surface; operculum present. (Only 1 genus, with 5 spp. from Bda.)

***Peracle triacantha*** (Fischer) (Three-spined Peracle): To 3 mm; shell white, ribs and rostrum horny colored; nearly planospiral, spire depressed, the ultimate whorl covers the 3 preceding ones; aperture broad with 3 spine-like folds. Surface reticulated, formed by spiral lines with axial cross connections; operculum with subcentral nucleus. Uncommon.

O. **RHODOPACEA:** Opisthobranchia with vermiform shape; without shell, cephalic shield, tentacles, cerata, jaws, radula, heart and circulatory system. All possess spicules. Indications during development of an evanescent velum, foot and shell gland. No larval stage; development is directly to vermiform young. Habitat interstitial. (A single genus is known; 1 sp. from Bda.)

***Rhodope*** sp.: Genus with anus and nephridiopore on the right side at midbody.—Species elongate, worm-shaped, to 2.5 mm; spicules irregular, curved rods to 50 um long. In clean, coarse subtidal sand; rare.

O. **ONCHIDIACEA** (=SOLEOLIFERA): Opisthobranchia with posterior pulmonary sac, anus and ♀ genital pore; with 1 pair of contractile eyestalks; without shell, cephalic shield, and rhinophores. Notum broad, overlapping foot, oval in shape, bearing warts and papillae; notal margin with defensive glands. With spicules claimed to be siliceous. (A single family, Onchidiidae; with 2 spp. from Bda.)

***Onchidella floridana*** (Dall) (=*Onchidium transatlanticum* Heilprin) (Florida Onchidella): Genus with abundant minute warts and papillae; ♂ pore behind right tentacle.—Species to 25 mm; surface velvety, color slate-green, occasionally black, grayish or white; mantle margin with many minute tubercles; found intertidally, where it lives in rock cavities and crevices, emerging at low tide for short intervals to graze. Its biology, including its remarkable homing ability, has been studied by AREY (1937a, b), AREY & BARRICK (1942) and AREY & CROZIER (1918, 1921). (Color Plate 12.13.)

## Plate 149

O. **GYMNOSOMATA:** Shell-less planktonic carnivorous Opisthobranchia; foot enlarged to form natatory parapodia; mantle cavity absent; gill reduced or absent, jaws and radula usually present.

F. **CLIONIDAE:** Gymnosomata with cylindrical body tapering at posterior end; anterior with a pair of tentacles and 2 or 3 pairs of conical buccal appendages. Parapodia relatively small, oval, posterior to head; jaws and gills absent. (3 spp. from Bda.)

***Clione limacina*** (Phipps) (=*C. papillonacea* Verrill; *C. elegantissima* Dall; *C. kincaid* Agersborg) (Slug-like Clione): Genus with 1 pair of simple tentacles, 1 pair of eye tubercles; mouth with 3 pairs of conical retractile processes.— Species to 40 mm; body barrel-shaped, pointed posteriorly; fins oval; foot lobes well developed, posterior one long. Radula formula varies from 6:1:6 to 15:1:15. Food of whales. Common; worldwide in temperate seas.

***Clionina longicaudata*** (Souleyet) (Long-tailed Clione): Genus with 2 pairs of buccal cones; central radular tooth with long, thin cusp.—Species to 10 mm; small, slender, pointed posteriorly; posterior foot lobe when present is small; animal dark owing to chromatophores in the body and

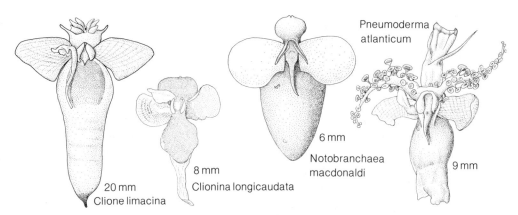

**149 GYMNOSOMATA**

wings. Jaws reduced, radula formula progressing with age, varying from 2:1:2 to 6:1:6. Uncommon; worldwide, temperate and tropical zones in limited areas.

F. **NOTOBRANCHAEIDAE:** Gymnosomata with body generally rounded, at times squat, pointed at posterior end. Anterior region set off from oval trunk by a short neck. Fins large and rounded. Proboscis with 2 pairs of conical buccal appendages without suckers. Lateral gill absent. Jaw present; radula N-1-N. (1 sp. from Bda.)

*Notobranchaea macdonaldi* Pelseneer (MacDonald's pteropod): Genus with posterior gill of 3 crests joining posteriorly; central tooth of radula with single cusp.—Species to 10 mm; brownish gray, head rounded, neck constricted; fins large; posterior foot lobe long and pointed, lateral ones triangular, 1/3 of their basis attached to the body; lateral gills simple protrusions of the skin. Uncommon; worldwide distribution in limited areas.

F. **PNEUMODERMATIDAE:** Gymnosomata with protrusible proboscis bearing suckers; gill-like structure on right side; jaws present. (1 sp. from Bda.)

*Pneumoderma atlanticum* (Oken) (=*P. violaceum* Orbigny) (Octopus pteropod): Genus with 2 sucker-bearing tentacles and fringed posterior and lateral gills.—Species to 10 mm; body cylindrical, brown to purple; posterior foot lobe short; lateral foot lobes horseshoe-shaped; sucker arms each with 40-80 short stalked and subequal suckers in maturity. Radula formula 4:1:4 or 4:0:4, median plate tricuspoid, missing in older specimens. Common; circumtropical.

### Plate 150

O. **ANASPIDEA** (=APLYSIACEA): Opisthobranchia with open sperm groove and relatively small dorsolateral parapodia that enclose a dorsal mantle cavity. Rhinophores and oral tentacles

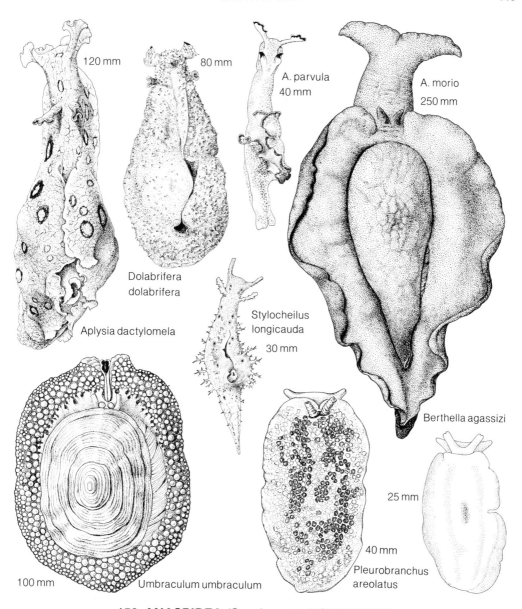

**150 ANASPIDEA (Sea hares), NOTASPIDEA**

present; cephalic shield absent; shell corneous, may be slightly calcified, reduced to a small plate within mantle cavity; gill plicate. Herbivorous, with broad multiseriate radula and foregut containing gizzard plates.

F. **APLYSIIDAE:** Anaspidea with long pleurovisceral nerve cords, visceral ganglia posterior to pedal ganglia; parapodia well separated; penis unarmed. (4 spp. from Bda.)

***Aplysia dactylomela*** Rang (=*A. aequorea* Heilprin; *A. protea* Rang) (Spotted sea hare; B Sea cat; Sea bat): Genus with smooth body, broad foot, and large dorsal natatory parapodia; internal shell plate amber colored, enclosed by mantle.—Species to 15 cm; light brown to olive-green, with black irregular, ocellated spots. When agitated will discharge a harmless, viscous, purple fluid. Eggs deposited intertidally in thin gelatinous strands. Common; in high subtidal areas on sand or rock bottoms with dense algal growth where it feeds on *Laurencia*; occasionally to 40 m. (Color Plate 12.12.)

*A. parvula* Mörch (Small sea hare): Genus as above.—Species to 5 cm; pale brown, speckled with tiny white dots; parapodial margin with clearly defined black stripe, tips of rhinophores and oral tentacles black; foot narrow, parapodia joined posteriorly. Rare; in high subtidal areas. (Color Plate 12.6.)

*A. morio* Verrill (Giant black sea hare; B Undertaker): Genus as above.—Species to 30 cm; dark brown to black; rhinophores small; tentacles large, broad; mantle foramen closed in adults. Uncommon; to 40 m. Swims readily when disturbed; sometimes found swimming under dock lights at night.

F. **NOTARCHIDAE:** Anaspidea with short pleurovisceral nerve cords, visceral ganglia dorsal to pedal ganglia; mantle often warty or papillose; parapodia much reduced, partially joined; shell usually greatly reduced or absent. Penis armed with spines. (2 spp. from Bda.)

*Dolabrifera dolabrifera* (Rang) (=*Aplysia ascifera* Rang; *D. virens* Verrill) (Warty sea cat): Genus with small, spatulate, well-calcified shell with strong apex; parapodia small, strongly fused; body flattened, with rounded posterior end.—Species to 100 mm; light olive-brown to brown, with cloudy maculations of white; mantle surface with fine, mammillose and branched tubercles. Uncommon; shallow water. (Color Plate 12.15.)

*Stylocheilus longicauda* (Quoy & Gaimard) (Long-tailed sea cat): Genus with small mantle cavity, mantle with fleshy, branched filaments (occasionally absent in small animals).—Species to 50 mm; olive-brown to khaki with numerous longitudinal, thin brown lines; iridescent blue spots ringed with orange; tentacles, rhinophores and long tapered tail with white speckles. When agitated will discharge a carmine-red fluid. Uncommon; shallow water, on open sand and in rocky rubble; feeds on *Lyngbya*. (Color Plate 12.3.)

O. **NOTASPIDEA** (=PLEUROBRANCHOMORPHA): Opisthobranchia without head shield; shell external, internal or absent; head with oral veil, oral tentacles and grooved rhinophores; mantle forming broad notum; mantle cavity with broad opening on right side, containing bipectinate gill. Jaws present, radula broad, lacking median tooth.

F. **PLEUROBRANCHIDAE:** Notaspidea with calcareous, fragile, auriform, internal shell; head partly concealed by notum; anus on right side, posterior to large, external gill. (3 spp. from Bda.)

***Berthella agassizi*** (MacFarland) (=*Pleurobranchus agassizii* MacFarland; *Bouvieria agassizii* (MacFarland) (Agassiz' pleurobranch): Genus with anus about midpoint of membrane supporting gill; shell greater than half of body length, notum relatively smooth to somewhat warty; radular teeth lacking denticles.—Species to 25 mm; notum white, orange, pink or red, smooth to slightly reticulate, extends over foot; shell large, flat, transparent, nearly rectangular, with strong growth lines and small flat spire. Gill rachis with 12-14 plumules on each side of axis. Jaw plates with 7-11 strong denticles; radular teeth shaped like detached rose thorns. Uncommon; intertidal to several meters. (Color Plate 12.9.)

***Pleurobranchus areolatus*** Mörch (=*P. atlanticus* Abbott; *P. gardineri* White) (Warty pleurobranch): Genus with anterior U-shaped notch in mantle through which 2 tube-like rhinophores protrude.—Species with dorsal surface covered with white, yellow, and brown warts, gill with about 20 pinnate leaflets per side; shell small, flat, with small spire. Uncommon; intertidal to several meters, usually under reef rubble. (Color Plate 12.4.)

F. **TYLODINIDAE:** Notaspidea with limpet-like or plate-like external shell bearing sinistral nucleus; radula broad with many teeth per row; gill on right side; soft parts much larger than shell. (2 spp. from Bda.)

***Umbraculum umbraculum*** (Lightfoot) (=*U. bermudense* (Mörch); *U. plicatulum* (von Martens)) (Atlantic umbrella shell): Genus with large, thick foot, flattened shell.—Species to 125 mm; animal yellowish brown, tuberculate; foot voluminous, sole broad, anterior deeply slit; head bearing 2 tentacles, projects slightly in front of thin-edged, fringed mantle. Gills elongate, plumose, start in front under the mantle and continue along the right side terminating free and bipinnate. The shell on top of the animal, umbrella fashion; slightly concave, strong but thin edged, ovate, calcareous, glossy with fine growth lines; area around pimple-like nucleus yellowish to light brown; periostracum thin, brownish; underside glossy, yellow, orange to brownish; attachment scar with radiating lines. Uncommon; to 80 m; will burrow beneath the sand. Several hundred eggs laid in a long, gelatinous, crenulated string. Thought to eat sponges.

### Plate 151

O. **ASCOGLOSSA** (=SACOGLOSSA): Opisthobranchia with uniseriate radula; worn teeth retained in sac (ascus); jaws absent; pharynx with muscular, pumping buccal bulb. Rhinophores absent in most primitive forms, otherwise digitiform, auriform or rolled in; oral tentacles usually absent. Cephalic shield absent except in most primitive forms. Externally shelled or shell-less. Small to moderate in size (1 mm to several centimeters; feeding as suctorial herbivores.

**SUP.F. OXYNOACEA:** Ascoglossa with external shell; ctenidium enclosed within mantle cavity; radular teeth with lateral spinules or bristles.

**F. VOLVATELLIDAE:** Oxynoacea without parapodia or tail, with reduced cephalic shield and rhinophores; shell with extended anal spout; digestive gland cladohepatic (branched), extending into dorsal mantle. (Single genus: *Volvatella*, with 1 sp. from Bda.)

*Volvatella bermudae* Clark (Bermuda Volvatella): To 7 mm; shell strong, white, translucent, smooth, appearing umbilicate, aperture large. Body white, mantle green, visible through shell; head strongly flattened, buccal mass displaced to one side. Fairly common on *Caulerpa racemosa*, especially near bay entrances; subtidal to 2 m. (Color Plate 12.5.)

**F. OXYNOIDAE:** Oxynoacea with single pair of non-natatory parapodia partially enclosing bubble-like shell; rhinophores prominent, tail long. Lacking cephalic shield; body without cerata but with simple retractile papillae. (A single genus, *Oxynoe*, with 1 sp. from Bda.)

*Oxynoe antillarum* Mörch (Antillean Oxynoe): Genus with thin, fragile, calcified shell; ultimate whorl large with capacious aperture.—Species to 20 mm; light brown to pale green; rhinophores, edges of parapodia and dorsal ridge of tail with grayish to blue-green mottling; parapodia and tail with scattered small papillae. Shell thin, vitreous; aperture large, narrow at apex end; smoothish with few growth lines. Animal not completely retractile into shell. Numerous yellow-orange eggs laid in a coiled, wide gelatinous band. Animal will secrete a whitish fluid when disturbed. Has been known to autotomize the rear portion of body. Uncommon; intertidal to 2 m, on *Caulerpa racemosa*. (Color Plate 12.8.)

**SUP.F. ELYSIACEA:** Ascoglossa lacking shell, with cerata or broad parapodia; digestive gland cladohepatic (branched), with ducts extending into parapodia or cerata. Radular teeth with or without denticles, but lacking spinules or bristles.

**F. ELYSIIDAE:** Elysiacea with parapodia, lacking cerata; rhinophores enrolled; anus antero-lateral, to right of pericardium. (At least 5 spp. of *Elysia* from Bda.)

*Elysia papillosa* Verrill (Papillose Elysia): Genus with rolled parapodia not anteriorly fused, penial stylet absent.—Species to 15 mm; pale green, with large, wide rhinophores bearing prominent brown stripe; parapodial margin light or white, often with very thin stripe at margin; pericardium about 1/4 of body length; 2 prominent veins join pericardial

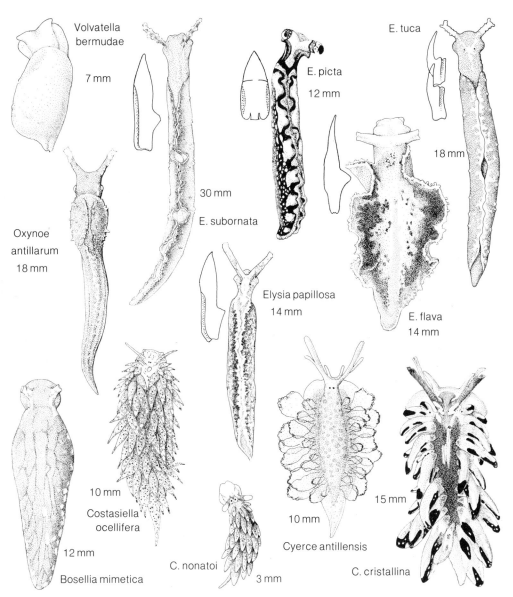

**151 ASCOGLOSSA (Sea slugs)**

hump near posterior tip. Parapodia with scattered papillae. Common; on *Udotea*, *Halimeda* and *Penicillus* to 5 m. (Color Plate 12.1.)

***E. subornata*** (Verrill) (=*E. cauze* Marcus & Marcus) (Ruffled Elysia): Genus as above.—Species to 50 mm; green to light olive; parapodial margin with broad band of variable color, white, brown, light green or pink, edged with a stripe of brown or black; small black ocelli often occur on outer parapodial surface; parapodia strongly ruffled in larger specimens. Body and narrow rhinophores with white-tipped pimple-like papillae; pericardial hump more than 1/2 of body length, with 8 pairs of veins arranged ladder-like along it. Very common; intertidal to 2 m, especially on *Caulerpa racemosa* in quiet water, but also found on other siphonalean algae. Development is direct, with extracapsular yolk. Spawns year round, but maximum activity is in late summer. (Color Plate 12.10.)

*E. tuca* Marcus & Marcus (White-rimmed Elysia): Genus as above.—Species to 15 mm; dark green with superficial black and iridescent green specks; parapodial margin with irregular patches of white; parapodia smooth, held tightly rolled, with notch at midmargin forming characteristic "spout"; rhinophores small, upper half with many white papillae. Common; to 20 m in areas of quiet water. Development lecithotrophic (short planktonic stage); egg mass similar to that of *E. subornata*. Feeds on a variety of siphonalean algae, especially *Halimeda*, *Udotea* and *Penicillus*. (Color Plate 12.7.)

*E. picta* Verrill (=*E. duis* Marcus) (Painted Elysia): Genus as above.—Species to 15 mm; deep brown to black, with iridescent bands of red, blue, green and yellow on the smooth, tightly rolled parapodia; head with a white Y-shaped mark, rhinophores with transverse red and blue bands; ventral surface with white spots. Rare; to 3 m, little is known of its biology. (Color Plate 12.14.)

*E. flava* Verrill (Yellow Elysia): Genus as above.—Species to 18 mm; head and center of back whitish, yellow or pinkish; rhinophores short; remaining portion of parapodia green with white edge; parapodia broad, minutely papillose; flanks with scattered white specks. Uncommon; high subtidal; diet unknown. (Color Plate 12.11.)

F. **BOSELLIIDAE:** Elysiacea with a broad, adherent foot and strongly denticulate teeth. (Single genus: *Bosellia*, with 1 sp. from Bda.)

*Bosellia mimetica* Trinchese (Halimeda slug): Genus with "parapodia" fused to broad foot, so that parapodia are retractile, but cannot be rolled over back; with penial spine.—Species to 15 mm; green. Eyes sessile; radular teeth short, robust. Uncommon; to 2 m, often in surf zone. Always associated with *Halimeda*. (Color Plate 13.6.)

F. **COSTASIELLIDAE:** Elysiacea with spindle-shaped cerata; rhinophores simple, rolled in or finger-like; foot rather narrowly triangular; eyes large, ad-median. (2 spp. from Bda.)

*Costasiella ocellifera* (Simroth) (=*C. lilianae* (Marcus & Marcus); *Doto ocellifera* Simroth) (Eye-bearing Costasiella): Genus with grooved, unifid rhinophores; ad-median eyes large; cerata pyriform with knobby diverticula.—Species to 13 mm; iridescent blue-green spot surrounded by orange ring on neck, orange near tip of each ceras; rhinophores long. Common; on *Avrainvillea*, to 5 m. Development is direct. (Color Plate 13.11.)

***C. nonatoi*** Marcus & Marcus (Dwarf Costasiella): Genus as above.—Species to 3 mm; black and white, with iridescence limited to small glands at tip of cerata and short rhinophores, no spot on neck. Uncommon; to 2 m. Occurs together with *C. ocellifera*, but is much less abundant, and tends to burrow within the thallus of the alga so that it is difficult to observe.

F. **CALIPHYLLIDAE:** Elysiacea with flattened cerata, enrolled or grooved bifid rhinophores; eyes ad-median; foot broad. (2 spp. in Bda.)

***Cyerce antillensis*** Engel (Antillean Cyerce): Genus with bifid rhinophores, flattened cerata without digestive diverticula; body flattened, with parapodium-like lobes.—Species to 60 mm (mostly smaller), yellow-tan; cerata nearly colorless, may be flecked with white on edges, rounded mucus glands visible within; oral tentacles digitiform, short; rhinophores brownish, flecked with white; black eyes prominent on head. Uncommon; to 3 m. On *Halimeda* and *Penicillus*. (Color Plate 13.1.)

***C. cristallina*** (Trinchese) (Crystalline Cyerce): Genus as above.—Species to 20 mm; translucent white; head, cerata and rhinophores with crimson spots and white edges; oral tentacles and rhinophores enrolled; penis terminates in a minute, curved hollow spine. Rare; to 2 m. Diet unknown. (Color Plate 13.9.)

## Plate 152

O. **NUDIBRANCHIA:** Opisthobranchia lacking shell, mantle cavity, ctenidium, plicate gill and open sperm groove. With rhinophores, often lamellate; often with oral tentacles, and neomorphic gills (cerata or circumanal gills). Exhibit strong bilateral symmetry. Carnivorous.

S.O. **DORIDACEA:** Nudibranchia with holohepatic (solid, unbranched) digestive gland and broad radula; usually with broad dorsal notum, often reinforced with calcareous spicules; anus usually on midline of notum, surrounded by tuft of gills. Rhinophores and gills often retractable into pockets in notum (cryptobranch); notum often with papillae, warts and spicules, but never with true cerata.

F. **CHROMODORIDIDAE** (=GLOSSODORIDIDAE): Doridacea with cryptobranch gills, smooth or tuberculate notum, with posterior of foot usually extending beyond notum; usually brightly colored. (9 spp. from Bda.)

***Chromodoris bistellata*** (Verrill) (Two-starred Doris): Genus with smooth notum, usually forming a broad "skirt" extending beyond foot; radula with unicuspid laterals.—Species to 20 mm; notum white with a multitude of tiny brown flecks that seem to coalesce to a solid brown with white specks; 2 white, stellate patches between the rhinophores and branchiae. Mantle edged with a fine yellow line; under portion of foot white speckled with brown. Common; under rocky rubble in protected waters to 1.5 m. (Color Plate 13.13.)

***C. clenchi*** (Russell) (= *C. neona* Marcus & Marcus) (Bill Clench's Doris): Genus as above.—Species to 10 mm; notum red with 2 irregular-shaped, opaque white blotches longitudinally placed between the branchiae and the rhinophores, and with numerous variably shaped and sized pale blue spots. Rhinophores lamellate, conical, deep violet. Branchiae opaque white with a deep violet line down the center of each. The notum margin is edged with a thin red line and an inner portion of a broad white band. Foot opaque white; the posterior portion narrow, pointed, extending beyond the mantle, is tipped with violet. Uncommon; generally found on the underside of rocky rubble in protected waters to 2 m. (Color Plate 13.2.)

***Hypselodoris zebra*** (Heilprin) (=*Chromodoris zebra*, *Glossodoris zebra*, *Felimare bayeri* Marcus & Marcus (Zebra Doris): Genus with smooth notum, slightly wider than body, tail extending well beyond notum; body with slight lateral compression, usually quite elongate, and the notum sitting high on the body; radula usually lacking median tooth, with laterals usually bicuspid.—Species to 180 mm; notum bright blue with yellow straight or wavy lines; edge portion of notum with irregular yellow markings; rhinophores lamellate, deep blue to black, branchiae blue-black edged with yellow. Undersurface of foot pale blue edged with yellow. In the young the notum is yellow with numerous irregular, elongate blue spots, a wide border of white speckled with dark spots and edged with a thin yellow line. Thought to spawn year round. Clear gelatinous egg band contains hundreds of tiny brick-red eggs; development is probably direct. Common; throughout the islands in grassy shallows to 20 m; feeds on sponges (e.g., *Dysidea etheria*). (Color Plate 13.10.)

F. **POLYCERIDAE:** Doridacea with non-retractable gill (phanerobranch); notum narrow or fused with body wall; radula usually lacking central tooth. Gills often protected by clublike extensions of notum. (3 spp. from Bda.)

***Aegires sublaevis*** Odhner (Nearly smooth Aegires): Genus with notum containing abundant spicules, hence hard in texture; notal margin very narrow.—Species to 15 mm; bright yellow with small brown spots; rhinophores protected by enlarged cephalic lobes. Found on and under rocks in exposed locations to 2 m, associated with a bright yellow sponge (*Clathrina* sp. ?). (Color Plate 13.3.)

***Gymnodoris*** sp.: Genus with oral veil bearing finger-like pro-

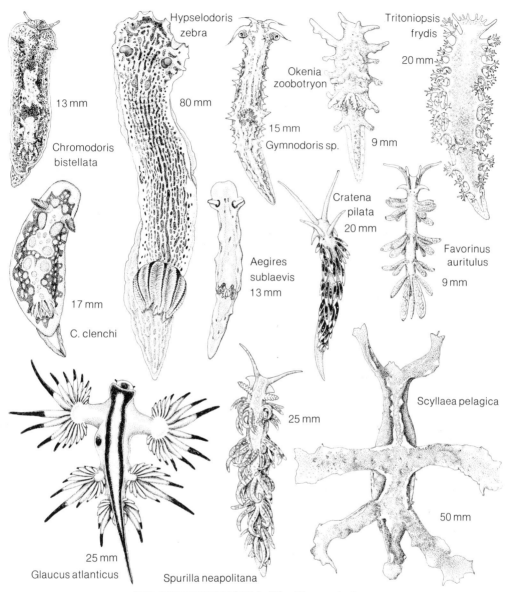

**152 NUDIBRANCHIA (Nudibranchs)**

cesses; body soft, rhinophores not retractile into cavities; notal border not distinguishable from body wall.—Species to 12 mm; white with longitudinal black streaks and many small orange spots; rhinophores white with orange spots, tipped blue-black; body with many short, blunt-pointed tubercles. Associated with bryozoans, to 3 m. (Color Plate 13.12.)

***Okenia zoobotryon*** (Smallwood) (= *O. evelinae* Marcus; *Bermudella zoobotryon*; *Polycerella zoobotryon*): Genus with stiff notum, with calcareous spicules; notal margin not clearly defined; notum bearing abundant stiff papillae with retractile, tuberculate tips; horn-like papillae anterior to rhinophores and at corner of head.—Species to 5 mm; white, mottled with light brown irregular patches that are streaked with darker brown; side and dorsum with 16-19 club-like papillae; tentacles and rhinophores elongate, cylindrical, non-retractile; tentacles 1/4 the length of the rhinophores; margin of notum and tips of papillae translucent. Will lay several cylindrical, jelly-like masses containing up to 300 eggs (see SMALLWOOD 1910, 1912). On the bryozoan *Zoobotryon,* to 3 m; common. (Color Plate 13.14.)

S.O. **DENDRONOTACEA:** Nudibranchia with cerata usually branched, tufted, or strongly tuberculate; rhinophores enclosed within sheaths; digestive gland usually at least partially cladohepatic (branched tubules); oral veil usually present.

F. **TRITONIIDAE:** Dendronotacea with body square in cross section, tufted cerata, branched rhinophore tips, and oral veil with digitiform extensions. (1 sp. from Bda.)

*Tritoniopsis frydis* Marcus (Tufted sea slug): Genus with midlateral anus, anterior genital orifice; undifferentiated lateral teeth, unicuspid median tooth.— Species to 20 mm; grayish white to orange, with brown digestive gland visible through skin; scattered white dots on notum; 2 large digitiform processes on oral veil, about 8 smaller between; about 12 cerata on each side. On base of gorgonians, to 5 m; uncommon. (Color Plate 13.8.)

F. **SCYLLAEIDAE:** Dendronotacea with flattened, leaf-like cerata and rhinophore sheaths. (With a single genus, *Scyllaea*; 1 sp. from Bda.)

*Scyllaea pelagica* L. (Sargassum nudibranch): Genus with lamellate rhinophores nearly concealed by sheaths; penis unarmed; radular teeth with strong median cusps and denticles.— Species to 50 mm; translucent, light olive-brown to orange-brown with brown and white flecks along the thinner edges of the rhinophore sheaths, body, cerata and caudal crest; cerata paired on each side; flanks with white streaks and 5 or 6 light blue spots. On pelagic *Sargassum,* occasionally stranded on shore. Feeds on hydroids and other coelenterates. (Color Plate 13.5.)

S.O. **AEOLIDACEA:** Nudibranchia with oral tentacles, unsheathed rhinophores, fusiform cerata; digestive gland always cladohepatic (branched tubules); notum absent; cerata usually contain cnidosacs in which nematocysts of prey are stored.

F. **FAVORINIDAE:** Aeolidacea with simple rhinophores; uniseriate radula; cerata in horseshoe-shaped arches; jaws triangular. (2 spp. from Bda.)

*Favorinus auritulus* Marcus & Marcus (Long-eared Favorinus): Genus with recurved propodial tentacles on anterior corners of foot; radular tooth with single elongate cusp, no denticles.—Species to 12 mm; transparent white with broad, opaque white blotch behind the tentacles, the rhinophores and along the dorsum; the area behind the rhinophores with orange center; rhinophores with 2 or 3 swellings and white tips; cerata transparent, the orange diverticula visible. Uncommon; to 5 m. Feeds on eggs of other opisthobranchs, especially sacoglossans. (Color Plate 13.15.)

*Cratena pilata* (Gould in Binney) (Hairy Cratena): Genus with short propodial corners, abundant cerata, denticulate jaw margins; radular teeth with few lateral cusps.—Species to 40 mm; body brownish to gray, cerata usually darker brown. Large individuals usually associated with *Tubularia*, small ones with thecate hydroids, spring or fall; egg mass in form of white string irregularly coiled among hydroids. (Color Plate 13.4.)

F. **AEOLIDIIDAE:** Aeolidacea with radular teeth bearing pectinate cusps; body broad, cerata in transverse rows; rhinophores lamellate or smooth; propodial tentacles present or absent. (1 sp. from Bda.)

*Spurilla neapolitana* (Delle Chiaje) (Naples Spurilla): Genus with long oral tentacles, short propodial tentacles, lamellate rhinophores.—Species to 100 mm; light olive-brown to pinkish; cerata numerous, long, recurved, tipped with white. Head, tentacles, dorsum and cerata with tiny, opaque white dots. Known to carry zooxanthellae acquired secondarily from its diet of sea anemones. Uncommon; on *Sargassum,* occasionally under rocks; to 3 m. (Color Plate 13.16.)

F. **GLAUCIDAE:** Aeolidacea with short and simple rhinophores and oral tentacles; cerata on lateral swelling of body wall. (1 sp. from Bda.)

*Glaucus atlanticus* Forster (Blue Glaucus): Genus with small head, eyeless; penis armed with hook.—Species to 50 mm; body elongate, tapering; head, tentacles and rhinophores very small; dorsum with 2 longitudinal, bright blue stripes; each side of body with 3-4 projections, terminating with a fan-like cluster of white cerata tipped with blue; underside pale grayish blue. Uncommon; occurs at sea surface where it maintains bouyancy by swallowing air. Feeds on *Physalia, Velella* and *Porpita,* transferring its prey's nematocysts into its own cerata and utilizing them for defense. (Color Plate 13.7.)

### Plate 153

S.CL. **PULMONATA:** Mainly freshwater and terrestrial Gastropoda with a shell that is typically spiral, reduced or wholly concealed by the mantle, or without a shell. Without operculum and ctenidium. Mantle cavity altered into a lung, in few cases with a secondary gill. Radula with as many as 250,000 teeth. Hermaphroditic.

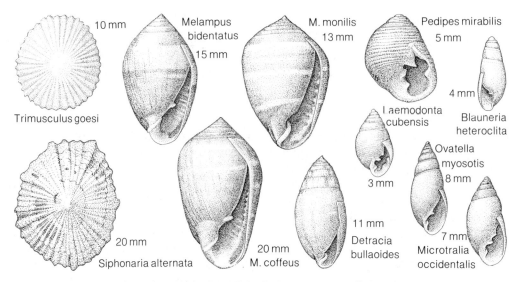

### 153 PULMONATA (Pulmonate snails)

**O. BASOMMATOPHORA:** Pulmonata with 1 pair of non-invaginable tentacles with the eyes at the base. Primarily in fresh water, with some marine and a few terrestrial species.

**SUP.F. SIPHONARIACEA:** Basommatophora with uncoiled, limpet-like shells. Pulmonary cavity sometimes with a gill.

**F. TRIMUSCULIDAE:** Siphonariacea with oval to circular, white shells. Surface with fine radiating ribs. Interior right side with weak siphonal groove. Without a secondary gill. (1 sp. from Bda.)

***Trimusculus goesi*** (Hubendick) (=*Gadinia mammillaris; Gadinia carinata* of authors) (Goes' Gadinia): Genus with circular shells.—Species to 12 mm. Shell low, conic, oval to circular, irregular in shape; surface with numerous, medium to coarse, radiating ribs. Nuclear whorls smooth, pointing posteriorly. Animal lacks tentacles and jaw; has separate ♂ and ♀ gonopores on right side of head. Uncommon; lives in the intertidal area; shells found in the beach drift.

**F. SIPHONARIIDAE:** Siphonariacea with cap-shaped shells. Surface with fine to strong radial ridges; interior highly polished. Siphonal groove on the right. Horseshoe-shaped muscle scar with opening at the right. Pulmonary cavity with gill. Air breathers; inhabit the intertidal zone of rocky shorelines. (3 spp. from Bda.)

***Siphonaria alternata*** (Say) (=*S. picta* Orbigny; *S. brunnea* Hanley; *S. alternata opalescens* Davis; *S. alternata intermedia* Davis) (Say's false limpet): Genus with flattened, patelliform shell and strong grasping foot; mantle skirt complete except for respiratory opening.—Species to 20 mm. Shell variable, high-spired to compressed; finely to coarsely ribbed. Black-brown to light tan, may be striped or mottled dark brown. Interior highly polished; brown to tan and cream; edges generally mottled dark brown. Animal lacks tentacles; common gonopore on right side of head. In tide pools and in the splash zone on rocky shoreline. Exhibits a remarkable homing ability (see COOK 1971, 1976).

SUP.F. **MELAMPIDACEA:** Basommatophora with spirally coiled, dextral shells; aperture narrow, with folds or dentition. Live under organic debris and rocks in the moist, shady, muddy areas at and slightly above the high tide debris line especially in the mangroves.

F. **MELAMPIDAE** (=ELLOBIIDAE): Melampidacea with small, round to elongate-oval shells; aperture generally toothed. (10 spp. from Bda.)

*Melampus bidentatus* Say (=*M. lineatus* Say; *M. redfieldi* Pfeiffer) (Common marsh snail): Genus with ovate-conic shells, brown to straw; may have light or dark color bands. Columella with several strong plicae.—Species to 15 mm. Shell with 5 or 6 whorls, broadest 1/3 from apex. Exterior smooth, shiny; whorls above the shoulder with spiral incised lines. Brown to light brown with 3 or 4 narrow, darker bands. Aperture narrow, widest at the base. Outer lip thin, lirate within. Columella whitish with 2 plicae, the uppermost the strongest. Common; in moist, muddy areas at and slightly above high tide line around marshes and mangroves.

*M. coffeus* (L.) (=*M. gundlachi* Pfeiffer) (Coffee bean marsh snail): Genus as above.—Species to 20 mm. Shell stocky, ovate-conic, broadest 1/4 from apex. Exterior smooth, shiny. Brown, gray-brown to tan, usually with 1-4 narrow, cream colored bands. Shoulder whorls without incised lines. Outer lip thin, 13-18 white lirae within. Columella with 2 white plicae, uppermost the stronger Common; associated with above.

*M. monilis* (Bruguière) (=*M. flavus* of authors not Gmelin; *M. flavus* var. *purpureus* Davis; *M. flavus* var. *albus* Davis) (Pear marsh snail): Genus as above.—Species to 15 mm. Shell ovate-conic, broadest 1/3 from the apex. Color dark brown to tan, may have 1-3 lighter bands. Whorls above the shoulder have a single spiral row of epidermal bristles, or short, pitted, axial scars where the bristles have worn off. Parietal wall glazed. Columella with a sharp plica well within the aperture, almost hidden, and a rounded plica at the base of the columella just below a deep spiral groove. Outer lip thin, with 8 or 9 white lirae within. Common; associated with above.

*Detracia bullaoides* (Montagu) (Bubble marsh snail): Genus with ovate-fusiform, brown shells. Larvae pelagic.—Species to 12 mm. Glossy, tan to dark brown with lighter axial flammules on upper portion of whorls; may have up to 4 whitish spiral bands of varying widths. Nucleus white, pimple-like. Aperture narrow, widest at the base. Lip thin; single white tooth at the base well within. Base of columella with 1 sharp white plica below a deep spiral notch. Common; associated with above.

***Pedipes mirabilis*** (Mühlfeld) (=*P. tridens* Pfeiffer; *P. ovalis* C. B. Adams; *P. insularis* Haas) (Admirable stepping snail): Genus with round to round-ovate shell. sculptured with many fine spiral grooves. No umbilicus. With 3 plicae on columella, 1 denticle within outer lip. Foot divided into an anterior and a posterior part; each portion moves separately allowing the animal to make short steps.—Species to 4 mm; 4 or 5 whorls. Reddish brown to straw. Many spiral, incised lines. Aperture ovate. Columella white, glossy, wide and dished with 3 sharp plicae, the uppermost the stronger. Outer lip thin-edged, thickened within by a low axial ridge and a white denticle, centrally placed. Common, associated with above.

***Laemodonta cubensis*** (Pfeiffer) (Cuban marsh snail): Genus with small, ovate, high-spired shells sculptured with many fine to coarse spiral cords. Columella with 3 plicae; outer lip with 1-3 denticles. May or may not be umbilicate.—Species to 3 mm, straw to dirty white. Apex white, enlarged, at right angles to the axis of the shell. Aperture oval; outer lip with 2 denticles within. Columella with 3 sharp plicae. Umbilicus chink-like. Periostracum hirsute, present in the juvenile, worn off in maturity. Common; associated with above.

***Ovatella myosotis*** (Draparnaud) (=*Alexia bermudensis* H. & A. Adams) (Mouse-eared marsh snail): Genus with small, brown, translucent, high-spired, ovate-conic shells. Columella with plicae. No denticles within outer lip.—Species to 8 mm, light brown to straw. Surface smooth. Aperture teardrop-shaped. Outer lip slightly thickened near the base and recurved dorsally. Parietal wall slightly glazed. Columella with 3 light colored plicae. Common; associated with above.

***Blauneria heteroclita*** (Montagu) (Blauner's sinistral marsh snail): Genus with tiny, ovate, slender, high-spired, sinistral shells. Columella with 1 plica.—Species to 3 mm. Shell ovate-conic, hyaline; spire high, blunt. Exterior smooth, shiny. Aperture elongate, teardrop-shaped. Outer lip thin, fragile; smooth within. Common; associated with above.

***Microtralia occidentalis*** (Pfeiffer) (=*Auriculastra nana* Haas) (Small western marsh snail): Genus with tiny, smooth, ovate-conic, high-spired shells. Outer lip without denticles. Columella with 2 or 3 plicae.—Species to 3 mm. Shell hyaline; apex globose, smooth. Aperture elongate, teardrop-shaped; outer lip thin, sharp in upper half, thickening in lower. Columella with 2 or 3 plicae. Common; associated with above.

R. H. Jensen & K. Clark

## Class Scaphopoda (Tusk shells)

CHARACTERISTICS: *Torted, bilaterally symmetrical MOLLUSCA with a curved, tubular, calcareous shell that is open at both ends; head and foot project from the larger, anterior opening.* Bermuda's species are from a few millimeters to 25 mm long; their shell surfaces glossy white to bluish white. Locomotion is confined to burrowing into soft substrates by means of the inflatable foot.

Of approximately 350 species, 4 have been reported from Bermuda; 3 are included here.

OCCURRENCE: Exclusively marine, Scaphopoda live partly or totally buried in sand or mud, with the anterior end downward. A small number of species are littoral, but the majority are neritic or bathyal.

Specimens may be collected by dredging or screening in sandy areas below the low tide line.

IDENTIFICATION: Present classification of the Scaphopoda is based almost exclusively on shell characters such as shape, surface sculpture and the presence of slits at the posterior opening.

On the live animal note the shape of the foot (family chacteristic!), and its function in locomotion. The head lacks eyes and tentacles, but has otocysts and captacula (long, ciliated filaments used in feeding). A well-developed radula is present. The mouth is anterior, the anus posterior. The nervous system is ganglionated and not concentrated. Auricles and gills are absent; nephridia are paired.

For anatomical studies, the specimen may be fixed in Bouin's fluid or other acid fixative so that the shell is dissolved. The mantle may then be split open and the soft parts examined. De-shelled specimens may be dropped into hot sodium or potassium hydroxide solution intact, or the buccal mass may first be dissected for preparation of the radula.

BIOLOGY: Scaphopods are dioecious, shedding gametes through the right kidney into the seawater where fertilization occurs. The animals feed by first forming a roughly ovate feeding cavity in the sand with their foot and then conveying detritus and microorganisms, such as Foraminifera and other protozoans, to the mouth by the captacula. Because gills are lacking, respiration occurs through the integument. Principal predators are mollusks, especially opisthobranch gastropods. Little is known about parasites.

### Plate 129

DEVELOPMENT: The fertilized egg undergoes spiral cleavage to form a trochophore larva that further develops into a bilaterally symmetrical veliger with bilobed mantle and shell. The larva undergoes torsion and the mantle ends fuse along the ventral margin, forming a tubular mantle and shell. The larva then sinks to the bottom and assumes an infaunal mode of life.

REFERENCES: Bermuda's species can be found in the monograph by HENDERSON (1920); observations on behavior were made by DAVIS (1968).

### Plate 154

F. **DENTALIIDAE:** Shell tapering regularly; may be sculptured or smooth. Animal has a conical foot with an encircling sheath that may be expanded laterally and interrupted dorsally. (2 spp. from Bda.)

*Dentalium semistriolatum* Guilding (Half-scratched tusk): Genus with shell that is curved and tapering and may be longitudinally sculptured, smooth or annulated. Apex commonly

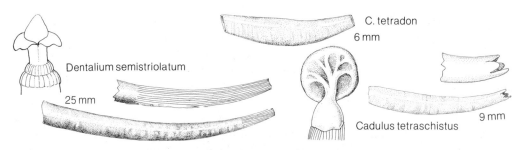

## 154 SCAPHOPODA (Tusk shells)

modified by a slit or notch.—Species with a slender, tapering shell 20-25 mm long. Surface translucent, white, and smooth. Fine longitudinal grooves extend over the posterior 1/3 of the shell; the remainder is glossy. Posterior opening is circular and unslit. Uncommon; subtidal to 200 m.

F. **SIPHONODENTALIIDAE:** Shell small and generally smooth; commonly wider in the middle than at the ends. Animal has a foot that can be expanded distally into a symmetrical disc with crenulate edge. (2 spp. from Bda.)

***Cadulus tetraschistus*** (Watson) (=*C. quadridentatus* (Dall)) (Four-toothed Cadulus): Genus with a shell that is smooth or delicately striated and may be slightly inflated about the middle.— Species with slightly arcuate shell 8-10 mm long. Shell surface bluish white, glassy smooth, with very slight growth striae. Aperture oblique, transversely oval. Four deep slits divide the posterior end into 4 teeth with beveled outside edges. The tooth on the convex side is the longest. Common; 5-100 m.

***C. tetradon*** Pilsbry & Sharp (Tetradon Cadulus): Genus as above.—Species with moderately swollen shell 5-6 mm long. Surface translucent, bluish white, glossy, with or without growth striae. Aperture oblique, circular. Four shallow slits divide the apex into 4 rounded teeth. Common; subtidal to 2,000 m.

R. H. Jensen & M. G. Harasewych

**Class Bivalvia** (=Pelecypoda, Lamellibranchia) (Clams, mussels, shipworms, etc.)

CHARACTERISTICS: *Untorted, bilaterally symmetrical MOLLUSCA with a laterally compressed, dorsally hinged, 2-valved shell that encloses the body. Head, radula and cephalic eyes lacking; a muscular foot is often present. Intestine passes through the heart.* Bivalves range in size from a few mm to over 1 m in length. The body is soft and shiny; the shell, usually hard, can be brittle in mud-dwelling or boring species. The shell surface may have striking color patterns (most often brown, yellow or red) and sculpture. All are sedentary, but range from fairly mobile, burrowing forms to byssally attached, cemented or boring species. Some (Limacea, Pectinacea) are capable of swimming short distances by clapping their valves.

Of 6 subclasses, Cryptodonta and the largely fresh water-dwelling Palaeoheterodonta are not known from Bermuda. Of approximately 7,500 species, 196 have been reported from Bermuda; 80 are included here.

OCCURRENCE: Marine and freshwater. Exclusively benthic; the majority are bottom burrowers, but others can be found on all substrates from the high tide line to hadal zones. The infaunal Bivalvia are found burrowing shallow to deep in sand, mud, peat, wood or rock; epifaunal species are attached by means of byssus (horny threads secreted by a gland in the foot) or cemented to hard substrates. Some groups are commensal, living in the burrows of, or attached to, other invertebrates.

Only a few species occur in sufficient quantities to be collected commercially in Bermuda. Collecting of living specimens for souvenirs is strongly discouraged, and several species (*Pecten ziczac, Argopecten gibbus, Macrocallista maculata* and *Pinctada imbricata*) are protected by Bermuda law.

When collecting specimens for study in intertidal areas, a diving knife is useful for prying loose cemented forms, and a small trowel and screen aid in collecting deeply burrowing and small, infaunal species. A hammer and chisel are usually required to collect species that burrow in wood, coral or rock. When diving, fanning the sand is often productive. Species from deeper water may be brought up by a dredge or bottom grab.

IDENTIFICATION: The classification system followed in this chapter is based largely on shell characters, both external and internal, and will therefore require killing the animal, preferably by alternately freezing and thawing the specimen. Prior to removal of the soft parts, however, note major features of the internal anatomy.

The bivalve body is enclosed by a mantle that secretes shell and ligament, and encloses the visceral mass and other organs. The gills are a single pair of double combs (bipectinate) in Palaeotaxodonta (the protobranch condition) where they are used for respiration only; in higher Bivalvia they are filamentous (filibranch), or the filaments are partly (pseudolamellibranch) or completely (eulamellibranch) fused to form solid sheets which, in addition, may be folded in a complex way. Filibranch and lamellibranch gills are used to filter and concentrate food particles. The mouth is often flanked by a pair of labial palps that assist in food sorting. The mantle margins may be free, and provided with tentacles and eyes (e.g., Pectinidae); or they can be fused to varying degrees, in which case the resulting inhalant and exhalant apertures are often extended to form tubular siphons that may be separate or fused, naked or covered with periostracum. The foot is variously adapted for burrowing or creeping, and may have a gland that secretes byssal threads in attached forms (e.g., Mytilidae).

The valves are held together by adductor muscles, primitively 2 (the dimyarian condition), which may be of equal (isomyarian) or unequal (anisomyarian) size; in the latter case one muscle (most often the anterior) is reduced (heteromyarian); or it may be absent altogether (monomyarian).

## Plate 131

The 2 valves, right and left, are usually of equal size (equivalve); in some genera, one is larger and overlaps the other (inequivalve). Their orientation is such that the dorsal side of the animal is where the shells are hinged, and the anterior end is usually more rounded than the often elongated or pointed posterior end. The valves are connected dorsally by an elastic, chitinous ligament that tends to spread them open when its antagonists, the adductor muscles, relax. The hinge, usually provided with teeth, represents an important systematic character. Major types are taxodont (with numerous, undifferentiated, straight or V-shaped teeth oblique to, and running along the dorsal margin); actinodont (with undifferentiated teeth radiating from the beak); isodont (with 2 equally placed, symmetrical teeth per valve resembling a ball-and-socket joint); heterodont (with 2 distinctly different types of teeth); and edentulous (without teeth). The apex of the shell is called beak or umbo; it is from here that the larval shell (prodissoconch) grows into the adult shell by accretion. The umbo is generally prosogyrate (pointing forward), more rarely opisthogyrate (pointing backward) or orthogyrate (umbones pointing towards each other). It may be flanked by auricles (wing-shaped extensions, expecially prominent in scallops). An often heart-shaped dorsal depression anterior to the umbo is called lunule; a similar depression posterior to the umbo, beneath the ligament, is called escutcheon. Although the major part of the ligament is generally located above the hinge, it may be partially or entirely inside in some groups; such an internal part, which may have a calcareous reinforcement (lithodesma), is called resilium, and inserts on a spoon-shaped depression (resilifer or chondrophore). A ligament that extends both before and behind the beak (e.g., in *Nucula*) is called amphidetic. The internal surface of a valve shows the impressions of adductor muscles (anterior and posterior muscle scars), and a fine impression caused by the edge of the mantle (pallial line). The pallial line may have a posterior, U-shaped notch (pallial sinus) in species with a retractable siphon. The external shell surface can be variously sculpted with concentric and radial structures such as lines (striae), ribs (costae), nodules or spines. The shell is made up of several layers of which the outermost (periostracum) may vary from thin and shiny to thick, coarse and stringy; the innermost (endostracum) may be chalky, or pearly-iridescent (nacreous).

Unless otherwise stated, all measurements refer to shell length.

Preserve shells dry, after removing soft parts by freezing or boiling. Shell colors should be noted, for they tend to fade. For anatomical studies fix the whole specimen in 70% ethyl alcohol or in 10% buffered formalin. Small specimens, or parts of tissue, can be preserved for histological purposes in Bouin's fixative.

BIOLOGY: The majority of bivalves are dioecious, but simultaneous, protandrous and alternating hermaphroditism have been reported in various groups. Gametes, often in enormous numbers, are generally shed into the seawater, and fertilization is external. A few species enclose their eggs in gelatinous sacs and attach them to the outside of the shell; brooding of young occurs in some groups. Most species live several years; some reach considerable age (more than 100 yr in Pacific giant clams). Filter feeding is predominant; it may be supplemented by symbiotic algae in the mantle margin (as in the giant clams), or by cellulose digestion (in wood-boring groups such as Teredinidae). Food consists mostly of small plankton and detritus particles. Boring clams and shipworms are permanently embedded in the substrate, which they excavate, with only their siphons protruding. Among the major predators are humans, fish, crustaceans, starfish, birds, and carnivorous gastropods (e.g., Naticidae, Muricidae) which drill holes through the valves. Bivalves are parasitized by ciliates, sporozoans and trematodes. Some bivalve species may be hosts to commensal decapod shrimps and crabs.

### Plate 129

DEVELOPMENT: The fertilized egg divides by spiral cleavage to form a trochophore larva, which is succeeded by a symmetrical veliger stage. The shell is first produced as a single, dorsal plate, which later becomes folded to form the 2 valves (straight-hinge larva). The foot is formed as a ventral extension (pediveliger stage). At metamorphosis, the young shed their velum and settle. Some Bivalvia brood their eggs within the gill or in the superbranchial chamber (e.g., *Teredo*) until the veliger stage.

REFERENCES: For general orientation see BOSS (1982); ABBOTT (1974) lists many of the species found in Bermuda.

Early records of Bermuda's bivalves are by DALL (1885, 1911a) and JOHNSON (1918); more recent studies include those on distribution by NICOL (1956), BOSS (1964), BOSS & MOORE (1967), WALLER (1973) and DAVIS (1973); on their contribution to reef building or erosion by MARK (1924), JAMES (1970), AURELIA (1970), LOGAN (1974) and BROMLEY (1978); and on shell formation by BEVELANDER (1951, 1952, 1963), BEVELANDER & BENZER (1948) and BEVELANDER & NAKAHARA (1966, 1967, 1969).

### Plate 155

S.CL. **PALAEOTAXODONTA:** Bivalvia characterized by a nacreous, equivalve shell with a taxodont hinge and external ligament, and by protobranch gills, which are used solely for respiration. Isomyarian. Shell margins without gape; adults non-byssate. (Only order: Nuculoida.)

SUP.F. **NUCULACEA:** Nuculoida with shells that are equivalve and truncated posteriorly. The ligament extends on both sides of the beak (amphidetic). Resilifer present or absent; pallial line simple or sinuate.

F. **NUCULIDAE:** Nuculacea with shells that are ovate to trigonal, nacreous internally, with umbos pointing posteriorly. Resilifer present, ligament internal. Pallial line simple. Animal without siphons. (2 spp. from Bda.)

*Nucula proxima* Say (Atlantic nut clam): Genus with small, trigonal shell. Lower valve margin may be finely crenulate.— Species to 7 mm. Shell greenish gray, smooth with prominent growth lines and microscopic radial sculpture. Common; dredged in sandy mud to 5 m.

S.CL. **PTERIOMORPHA:** Bivalvia with variable shell, ligament and gills. Anterior adductor muscle and foot reduced or lost in many groups. Adults are commonly byssate and modified for a sedentary, epifaunal existence, although some are secondarily free living or boring.

O. **ARCOIDA:** Pteriomorpha that are characteristically isomyarian, equivalve and have a circular to trapezoidal shell. Endostracum (inner shell layer) crossed-lamellar; gills filibranch.

SUP.F. **ARCACEA:** Arcoida with shells that are porcelaneous, and radially ornamented (ribbed). Hinge taxodont, pallial line simple. Isomyarian. Ligament elongate, external. Periostracum conspicuous. Generally byssate.

F. **ARCIDAE:** Arcacea with equivalve, trapezoidal shell in which the posterior end is elongate. Hinge with teeth in 2 straight rows meeting at the beaks. Ligament external, composed of a series of bands (duplivincular). Without ridges or shelves on inner edges of muscle scars (myophoric ridges). Commonly with a ventral byssal gape. (7 spp. from Bda.)

*Arca zebra* (Swainson) (=*A. noae* of authors; *A. occidentalis* Philippi) (Turkey wing; B Bermuda mussel): Genus with shell of moderate size, subtrapezoidal, posteriorly extended and with a large ventral byssal notch.— Species to 80 mm, elongate, irregularly ribbed, tan with brown zigzag markings. Byssally attached to rocks. Edible; hand dredged for use in the famous Bermuda mussel pie. Common; especially in inshore waters, to 15 m.

*A. imbricata* Bruguière (Mossy ark): Genus as above.—Species to 80 mm, brown, elongate, with beaded ribs. Large ventral byssal opening. Periostracum heavy, foliated. Byssally attached. Common, under rocks and rubble, intertidally to 15 m.

*Barbatia domingensis* (Lamarck) (=*Arca gradata* of authors) (White miniature ark): Genus with small, elongate to oval, equivalve, inequilateral shell.

Sculpture consists of fine radial ribs. Ligament amphidetic. Dental series slightly arched. Ventral byssal gape small.—Species to 20 mm in length, whitish, box-shaped, thick-shelled, coarsely beaded. Ligament long and narrow. Common; intertidally to 50 m, under rocks and rubble (see BRETSKY 1967).

***Anadara notabilis*** (Röding) (=*A. lienosa floridana; Arca secticostata, Arca americana, Arca jamaicensis, A. deshayesi,* all of authors) (Notable ark; B Bloody ark): Genus with rotund, heavy, equivalve shell. Strongly costate, with interlocking crenulations at the valve margins. Without byssal gape.—Species to 60 mm. Shell white, with 25-27 radial non-bifurcated ribs per valve. Concentric threads cross the ribs and are prominent in the interspaces. Interior often moss-green. Periostracum heavy, dark brown, finely hirsute. Animal bright orange-red. Common; on sandy, grassy bottoms to 15 m. Edible.

F. **NOETIIDAE:** Arcacea with subtrapezoidal to ovoid, equivalve and costate shell. Hinge taxodont; ligament vertically striate. Myophoric ridge or shelf on inner margin of both adductor muscle scars. Without ventral byssal gape. Byssus may be used to anchor animal in sand. (1 sp. from Bda.)

***Arcopsis adamsi*** (Dall) (Adam's miniature ark): Genus with small, ovoid shell. Ligament amphidetic. Short resilium between beaks.—Species to 9 mm. Shell white, rotund, with cancellate sculpture. Periostracum thin, tan. Common; intertidally to 50 m, under rocks and rubble, often found with *Barbatia domingensis*.

O. **MYTILOIDA:** Pteriomorpha with shells that are generally equivalve and very inequilateral, byssate. Anisomyarian. Anterior end of animal reduced, posterior end greatly enlarged. Shell prismato-nacreous. Gills filibranch or eulamellibranch.

SUP.F. **MYTILACEA:** Mytiloida with a thin, equivalve, inequilateral shell. Beaks near anterior end, prosogyrate (point in anterior direction). Ligament posterior to beaks, elongate. Hinge smooth, or with dysodont (with small weak teeth close to beaks) dentition. Anterior adductor muscle reduced or absent. Pallial line simple, or with shallow posterior cavity. Periostracum often hairy. Strong byssus.

F. **MYTILIDAE:** Mytilacea with medium-sized, elongate shells; umbones at pointed anterior end; posterior end rounded. Hinge dentition reduced or absent. Generally with a byssus. Some species bore into coral, coral rock or concrete with the aid of an acid secretion. (8 spp. from Bda.)

***Brachidontes domingensis*** (Lamarck) (=*B. exustus* L.) (Domingo mussel): Genus with mytiliform (triangular) shell, nearly terminal beaks, crenulated margins, dysodont teeth. Radial sculpture of bifurcating ribs.—Species to 30 mm. Inflated, arched, with 50-70 ribs at the valve margins. Shell whitish antero-ventrally, rest purplish blue.

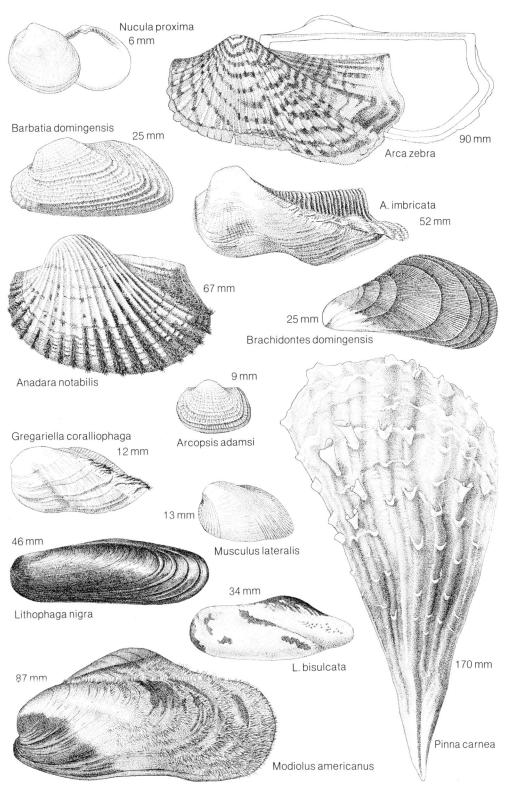

## 155 NUCULACEA, ARCOIDA (Arks), MYTILOIDA (Mussels)

Periostracum thin, golden brown. Common in clusters or singly on rocky shorelines.

***Gregariella coralliophaga*** (Gmelin) (= *Modiola opifex* (Say); *Lithophaga gossei* of authors) (Coral-boring Gregariella): Genus with thin, elongate, compressed or rotund shell. Beaks anterior. Posterior end pointed. Sculpture consists of fine radial ribs. Dysodont teeth, crenulated margins.—Species to 30 mm. Shell white, compressed or rotund, at times distorted owing to habitat. Periostracum brown, heavy posteriorly. Uncommon; burrows into coral and coral rocks, to 35 m.

***Musculus lateralis*** (Say) (Oval mussel): Genus with mytiliform shell that has radial ribs anteriorly and posteriorly, with a smooth area between. Beaks anterior. Hinge with dysodont teeth.—Species to 10 mm. Shell light brown, with a tinge of blue-green; oblong, thin. Interior iridescent; exterior with 8-10 anterior radial ribs and 18-24 posterior radial ribs. Periostracum greenish brown. Uncommon; in sandy bottoms to 10 m.

***Lithophaga nigra*** (Orbigny) (= *L. antillarum* Philippi; *L. dactylus* of authors) (Black date mussel): Genus with long, cylindrical shell that tapers posteriorly. Hinge edentulous, beaks anterior, margins smooth. Isomyarian.—Species to 50 mm. Anterior 2/3 of cylindrical shell with strong dorso-ventral ribbing. Shell white, with glossy, black-brown periostracum. Common; boring into coral and rubble; to 20 m.

***L. bisulcata*** (Orbigny) (= *Lithodomus appendiculatus* of authors) (Mahogany date mussel): Genus as above.—Species to 40 mm. Shell long, cylindrical with 2 sharp, oblique sulci. Calcareous posterior prolongation extends beyond valve margins. Shell tan with golden brown periostracum, which is generally covered by porous calcareous encrustations. Interior of valves pinkish purple. Common; boring into rubble and coral, especially *Siderastrea radians*. Found together with above.

***Modiolus americanus*** (Leach) (= *Modiola tulipa* Lamarck) (Tulip mussel; B Sand mussel; Grass scallop): Genus with inflated, smooth, anteriorly rounded shell. Hinge without teeth; ligament long.—Species to 100 mm. Shell fragile, rotund, with swollen umbones. Brown anterior; white, lavender or pink oblique streak in middle; rose to lavender posterior. Periostracum glossy brown, periostracal hairs thin. Common on grassy bottoms in inshore waters, to 7 m.

**SUP.F. PINNACEA:** Mytiloida with medium to large, equivalve, roughly triangular shells. Beaks are anterior, the hinge is long and edentulous. Long, narrow antero-ventral byssal gape, and a large posterior gape. Anisomyarian, with anterior adductor muscle greatly reduced and posterior adductor muscle large. Gills pseudolamellibranch.

F. **PINNIDAE:** Pinnacea with shells that are embedded vertically, pointed end down, in sand and attached by a byssus to rocks and shells. (1 sp. from Bda.)

*Pinna carnea* Gmelin (Amber pen shell; B Spanish oyster; Sand oyster): Genus with large, fragile, fan-shaped shell. There is a weak groove running down the middle of the interior of each valve, dividing the anterior nacreous area into two lobes.—Species to 300 mm. Shell light orange to amber, exterior smooth or with 8-12 radial rows of open spines. Common; in grassy shallows, sandy bottoms of quiet inshore waters, to 5 m.

O. **PTERIOIDA:** Pteriomorpha with inequivalve, inequilateral, nacreous or crossed-lamellar shells. Heteromyarian or monomyarian; pallial line simple. Gills filibranch or eulamellibranch. Adults byssally attached through a notch in right valve or cemented by right valve.

## Plate 156

SUP.F. **PTERIACEA:** Pterioida with right valve generally less convex than left valve. Beaks forward, hinge schizodont (median tooth of left valve broad and bifid) or edentulous (lacking teeth). Ligament external. Characteristically with byssal notch in right valve. Pallial line discontinuous anteriorly; siphons lacking.

F. **PTERIIDAE:** Pteriacea with ovate shells, usually with a short anterior and long posterior wing along a straight hinge line. Interior nacreous. Juveniles are dimyarian; anterior adductor muscle reduced or absent in the adult. (3 spp. from Bda.)

*Pteria colymbus* (Röding) (=*Avicula atlantica* Lamarck) (Atlantic winged oyster): Genus with thin, obliquely ovate shell that is longer than high and has a long posterior extension. Left valve inflated. Hinge with 1 or 2 small denticles per valve.—Species to 120 mm. Shell smooth, with concentric growth striae. Exterior deep brown with white radial rays; interior purplish to white, with non-nacreous margin. Common; byssally attached to alcyonarians offshore and on reefs; to 8 m.

*Pinctada imbricata* Röding (=*P. radiata* Leach; *P. alaperdicis* Reeve; *Meleagrina placunoides* of authors) (Atlantic pearl oyster; B Bermuda oyster; Pearl oyster): Genus with shell that is moderately thick, higher than long and lacks the pronounced posterior extension. Byssal gape in right valve. Edentulous.—Species to 80 mm. Shell thin, brittle, flat, squarish, with straight hinge line and broad ligament pit. Exterior brown to golden tan, with dark brown rays emanating from the umbos. Surface with concentric blades, fine to coarse in the young, eroded and worn in maturity. Interior with wide, non-nacreous margin. May contain pearls. Common; to 9 m, on sandy, grassy bottom byssally attached, singly or in clusters. Edible.

F. **ISOGNOMONIDAE:** Pteriacea with variably shaped inequivalve shells. Characteristically with multiple ligamental grooves and edentulous hinge in the adults. Anterior byssal gape, when present, affects both valves. Monomyarian. Interior nacreous. (3 spp. from Bda.)

*Isognomon alatus* (Gmelin) (=*Perna ephippium* of authors) (Mangrove oyster): Genus with shell that is usually higher than long, compressed, with straight hinge line. Ligamental grooves numerous, regularly arranged.— Species to 80 mm. Shell purplish to gray, roundish, thin, flat. Hinge with 8-12 ligamental grooves. Interior with non-nacreous margin. Found intertidally on mangrove roots, wharves, pilings. Once very common; decimated presumably by pollution from spilled gasoline and oil. (Color Plate 4.13.)

SUP.F. **PECTINACEA:** Pterioida with shells that are generally circular, with wing-like extensions (auricles) on both sides of the hinge. Sculpture generally radial or smooth, with fine growth lines. Hinge without teeth, but with ridges or grooves, or isodont. Ligament internal. Monomyarian. May be free, byssate or cemented to the bottom by the right valve. Byssal notch, when present, below right, anterior auricle. Interior porcelaneous.

F. **PECTINIDAE:** Pectinacea with orbicular to oval shells in which the umbones are central; anterior and posterior auricles present. May be compressed or inflated, radially ribbed or smooth. Byssal notch on or below anterior auricle of right valve. Dioecious. Mantle margin with filaments that may bear eyes. Some are capable of swimming by clapping their valves. (16 spp. from Bda.)

*Pecten ziczac* (L.) (Zigzag scallop; B Scallop; Bermuda scallop): Genus with shell in which the right (lower) valve is very convex whereas the left (upper) valve may be slightly convex, flat or concave. Auricles roughly equal in size. Shell margin crenulated. Radial ribs without scabrous surface.—Species to 130 mm. Right valve with 18-24 wide, flattened ribs; all white (rare) or reddish-brown; may be speckled with white, area near umbones white. Left valve only slightly concave, with 30-34 radiating ribs of varying width; all white to very dark brown or brown with varicolored (gray, orange, yellow, tan or white) rays. Periostracum on left valve (upper, flat) of proteinaceous/organic material, malellate, thin, most often worn off on all but the auricles, lip edging and in channels between rayed ribs. Edible; formerly very common; now rare and protected by law. On grassy, sandy bottoms, to 10 m.

*Chlamys imbricata* (Gmelin) (Little knobby scallop): Genus with shell in which both valves are convex. The anterior auricle is larger than the posterior. There is a large anterior byssal notch. Shell higher than long, with radial sculpture of major and interpolated minor scabrous ribs, producing a complexly crenulated

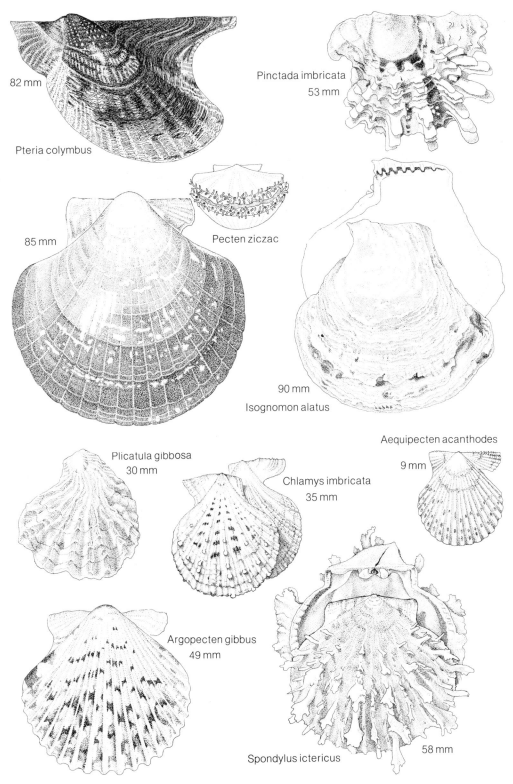

## 156 PTERIACEA, PECTINACEA (Scallops)

margin.—Species to 40 mm. Shell thin, translucent, with 8-10 major ribs (which bear prominent knobs or scales) per valve. Exterior of shell white, with red blotches below scales. Interior glossy yellow, may have a purplish stain. Common; byssally attached under rock, loose coral and rubble; inshore to reefs, to 10 m.

*Aequipecten acanthodes* (Dall) (Thorny scallop): Genus with shell in which the left valve is more convex than the right. Anterior auricle longer than posterior. Shell with circular outline and crenulated margins.—Species to 30 mm. Shell brown, red, orange, yellow or mottled; 18-20 ribs per valve. Ribs and interspaces about equal in size. Each rib bears 3 equally sized finely scabrous lines; interspaces between ribs with 2 scabrous lines. Common; on silt, sand or rubble bottom; to 50 m.

*Argopecten gibbus* (L.) (Calico scallop): Genus with shells in which the right valve is slightly more convex than the left. Auricles roughly equal in size. Radial ribs squarish in outline, with concentric scales between the ribs.—Species to 70 mm. Shell round, inflated. Upper valve usually mottled with a combination of brown, red, purple, yellow and white; ribs narrow, arched; interspaces scabrous. Lower valve white, occasionally tinted along the margin with colors of upper valve; ribs wide, squarish; interspaces narrow, scabrous. On sandy, grassy bottoms to 10 m. Once very common, now scarce and protected by law. Edible.

F. **PLICATULIDAE:** Pectinacea with small to medium-sized, subtrigonal shells; valves unequal, irregular in outline; auricles small or absent; right valve attached at umbo to substrate; surface concentrically lamellose with several broad radiating ribs. Monomyarian, adductor muscle scar small; resilium pit triangular, deeply sunk; hinge typically isodont. (1 sp. from Bda.)

*Plicatula gibbosa* Lamarck (=*P. ramosa* Lamarck) (Kitten's paw): Genus with round to trigonal shell that is attached to the substrate by the convex right valve. Surface with strong radiating ribs. One adductor muscle scar situated posteriorly.—Species to 30 mm in height. Shell grayish white with reddish brown lines on the ribs; heavy, fan-shaped, often distorted, with 5-7 prominent ribs (triangular in cross section). Shell margin crenulated. Uncommon; to 10 m.

F. **SPONDYLIDAE:** Pectinacea with round to oval, auriculate, inequivalve shells. Right valve greatly inflated, cemented to a hard substrate; lacks byssal notch. Left valve flat or slightly convex, pectiniform. Hinge isodont. Resilium triangular. Inner layers calcitic, non-nacreous; outer layer aragonitic. Radial sculpture with spines or foliaceous ornament. Monomyarian, with large muscle situated posteriorly. (1 sp. from Bda.)

*Spondylus ictericus* Reeve (=*S. americanus; S. ustulatus; S. vexillum; S. digitatus; S. ramosus; S. longitudinalis; S. ericinus; S. fimbriatus; S. spatuliferus; S. echinatus* all of authors) (Digitate thorny

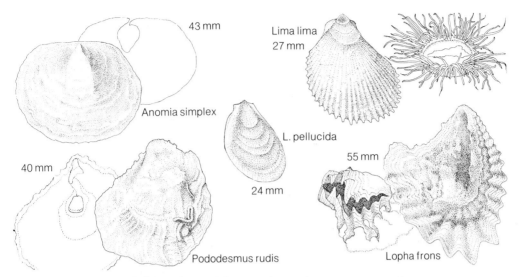

**157 ANOMIACEA, OSTREACEA (Oysters), LIMACEA**

oyster; B Rock scallop): Genus with large spinose shell that is cemented to substrate. Hinge of ball-and-socket type (isodont). Mantle may bear minute eyes.—Species to 120 mm in diameter excluding spines. Left (upper) valve with 18-25 radial rows of bifurcating or digitate spines, white, yellow, brick-red or brownish in color. Right (lower) valve inflated, generally white in color, attached to substrate. Early ornament foliaceous, brick-red, orange or yellow. At full maturity spines may be worn off. Extremely variable in color and sculpturing, hence the long synonymy. Found in shadowy areas of undercut rocky shoreline, on and under coral and Rubble, inshore to reefs. Moderately common; to 50 m. Edible.

**Plate 157**

**SUP.F. ANOMIACEA:** Pterioida with inequivalve, compressed, edentulous shells. Monomyarian, with filibranch gill and eyes along the mantle edge of the left (upper) valve. Pericardium absent.

F. **ANOMIIDAE:** Anomiacea with shells that are irregular in outline and usually attached to a hard substrate by a calcified byssus that passes through the right (lower) valve. Adductor muscle scar subcentral, with 1-3 retractor muscle scars on left valve, opposite notch in right valve. (2 spp. from Bda.)

*Anomia simplex* Orbigny (=*A. ephippium* of authors) (Common jingle shell): Genus with thin, round shell; left valve with 3 muscle scars on central area.—Species to 35 mm in diameter. Right (lower) valve conforms to substrate; left (upper) valve convex, lacks noticeable surface sculpture, but may assume sculpture of substrate. Color translucent, yellow, green or white. Uncommon; in protected inshore waters, generally on smooth substrate (bottles, shells, rocks, etc.), to 8 m.

***Pododesmus rudis*** (Broderip) (=*P. decipiens* Philippi) (Atlantic false jingle): Genus with thin, roundish shell; left valve with a single, radially striate, byssal retractor muscle scar dorsal to adductor muscle scar.—Species to 50 mm in diameter. Shell translucent white to tan; compressed, with right valve adherent to substrate and left valve slightly convex. Sculpture of numerous, fine, scabrous, radial ribs. Uncommon; in inshore waters, found singly or in clumps on rocks and rubble, to 10 m.

**SUP.F. OSTREACEA:** Pterioida with inequivalve shell irregular in outline. Edentulous, with ligament in triangular groove in each valve. Monomyarian. Pallial line, foot and byssus lacking. Cemented to substrate by left valve. Gills eulamellibranch.

**F. OSTREIDAE:** Ostreacea with very convex left valve; shell with folded and lamellar edge. Adductor muscle scar crescent-shaped, located centrally. May be incubatory or non-incubatory. Intestine does not pierce the heart. (3 spp. from Bda.)

***Lopha frons*** (L.) (=*Alectryonia limacella*; *Ostrea folium* of authors) (Leaf-like oyster): Genus with oval to subtrigonal shell. Attachment area medium to large. Valves may vary in convexity. Shell margin may be scalloped or plicate.—Species to 60 mm in height. Shell reddish brown, with 6-12 sharp, radial plicae. Interior of shell whitish; adductor muscle scar near hinge; margin finely pustulose. A variable species depending on habitat and substrate. Common; clinging to submerged branches, submerged ropes, fish pots, roots, alcyonarians, etc.; to 10 m.

**SUP.F. LIMACEA:** Pterioida with shells that are equivalve, ovate to subtrigonal, radially ribbed. Anterior and posterior auricles present, but reduced. Monomyarian. Shell slightly gaping anteriorly and posteriorly. Mantle with long tentacles, but rarely with eyes. Edentulous. Capable of swimming. Many build nests of byssal threads. (Only family Limidae, with 2 spp. from Bda.)

***Lima lima*** (L.) (=*L. scabra*; *L. multicostata*; *L. caribaea* of authors) (Spiny file clam): Genus with subtrigonal shell that is higher than long, slightly gaping at both ends. Hinge margin short.—Species to 50 mm in height. Shell white, with 26-33 scabrous, radial ribs per valve. Posterior auricle smaller than anterior. Posterior gape long and narrow. Uncommon; in offshore waters under rock and rubble, to 10 m.

***L. pellucida*** C. B. Adams (=*L. tenera*; *L. fragilis*; *L. inflata*; *L. dehiscens*; *L. hians* all of authors) (Transparent file clam): Genus as above.—Species to 30 mm in height. Shell very thin, translucent white, with 25-30 unornamented radial ribs per valve. Posterior gape large. Common; found singly or in groups under rocky rubble in quiet waters; nest builders; to 10 m.

**S.CL. HETERODONTA:** Bivalvia in which the hinge plate has differentiated into cardinal (close to the beak) and lateral (partly or completely located away from the beak) teeth. Shells may be complex crossed-lamellar or prismatic, but never nacreous. Lunule and escutcheon are usually present. The ligament is posterior to the beaks, with or without resilium. Without accessory plates or lithodesma (small calcareous plate which reinforces an internal ligament). Gills eulamellibranch. Mantle lobes usually fused. Siphons usually well developed.

**O. VENEROIDA:** Heterodonta with well-developed differentiated hinge teeth. Shell variable, but generally equivalve. Ligament posterior to beak. Mostly active burrowers in gravelly to muddy bottoms.

## Plate 158

**SUP.F. LUCINACEA:** Veneroida with equivalve shells that range from subcircular to subtrigonal. The anterior and posterior areas are delineated by low, radial folds. Beaks small, prosogyrate. Lunule asymmetric, escutcheon ill-defined. Sculpture generally concentric or smooth. Interior porcelaneous. Pallial line without sinus. Animal may have 2 siphonal apertures. Labial palps rudimentary. Generally colonize areas were oxygen and food are in low supply.

**F. LUCINIDAE:** Lucinacea with shells that are generally circular, compressed to inflated, smooth or with concentric sculpture. Radial sculpture when present more pronounced anteriorly and posteriorly. Margin smooth or crenulated. Anterior adductor muscle with characteristic ventral extension that is separate from the pallial line. Foot long, thin, cylindrical. Exhalant siphon long, eversible. Inhalant aperture papillose and open ventrally. Live completely buried in substrate. Water is drawn in anteriorly through a thin, mucus-lined tube produced by the foot. Particles are sorted by cilia on the anterior adductor muscle and passed to the mouth; the water exits posteriorly through the exhalant siphon. (11 spp. from Bda.)

***Codakia orbicularis*** (L.) (=*Lucina tigrina* of authors not L.) (Tiger Lucina): Genus with large, circular, compressed shell. Right valve with 1 anterior lateral, 3 cardinals and 1 posterior lateral teeth. Left valve with 2 anterior laterals, 2 cardinals and 2 posterior lateral teeth. Ligament partly internal. Strong radial and concentric ribs give a reticulate surface. Area within pallial line thick, chalky, pitted and/or pustulose.—Species to 90 mm in diameter. Shell white, sometimes suffused with pink dorsally; compressed, thick sculpture of fine concentric ribs and strong, dominant radial ribs; surface cancel-

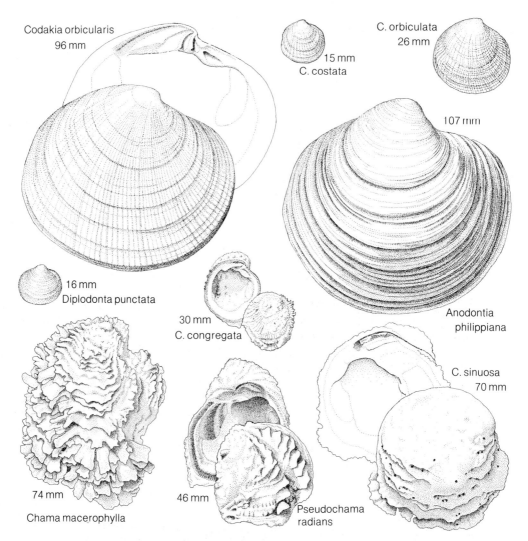

## 158 LUCINACEA (Lucines), CHAMACEA (Jewel boxes)

late. Beaks centered. Posterior margin slightly flattened. Interior white to yellow, with deep pink border dorsally or along the entire margin. Lunule small, impressed, cordate, mostly on right valve. Common; in quiet protected waters in coarse to fine sand on grassy bottoms, to 10 m.

**C. costata** (Orbigny) (Ribbed Codakia): Genus as above.— Species to 15 mm. Shell white to very pale yellow, orbicular, moderately inflated. Sculpture of concentric threads and radial ribs that become widely spaced and slightly scabrous on flattened posterior end. Surface reticulated. Lunule tiny, cordate. Common; associated with above.

**C. orbiculata** (Montagu) (=*Lucina pecten; Lucina squamosa; Lucina pectinatus* of authors) (Dwarf tiger Lucina): Genus as above.—Species to 20 mm. Shell similar to *C. orbicularis* but smaller and not as compressed; white, strong, with wide radial ribs that are continuous from beak to margin and crossed by fine concentric threads that form long concentric beads. Interior white, never colored. Common; associated with above.

***Anodontia philippiana*** (Reeve) (=*Lucina schrammi* Crosse) (Chalky buttercup): Genus with medium to large, thin, globose, slightly inequilateral shell. Surface uneven with irregular growth lines, and fine radial lines. Beaks prosogyrate. Lunule depressed. Hinge without teeth. Ligament sunken. Interior shell margin smooth.—Species to 100 mm in diameter. Shell chalky white outside and within. Interior pustulose. Anterior adductor muscle scar long and at an angle to pallial line. Uncommon; buried deep in sand or silty mud, to 30 m.

F. **UNGULINIDAE:** Lucinacea with subtrigonal, oblong, round or obliquely ovate shells. Umbones low. Surface smooth, with fine concentric threads, or punctate. Ligament and resilium marginal. Hinge with 2 cardinals, medial one bifid; laterals obscure or absent. Anterior muscle scar irregular, elongate, attached to pallial line; posterior muscle scar large. (1 sp. from Bda.)

***Diplodonta punctata*** (Say) (=*Mysia pellucida* Heilprin) (Common Atlantic diplodon): Genus with thin, white, orbicular, equivalve, inequilateral shells. Beaks prosogyrate. Surface with fine concentric threads. No lunule or escutcheon. Ligament external. Adductor muscle scars unequal in size. Pallial line broad.—Species to 12 mm. Shell white, inflated, thin, translucent, suborbicular. Surface with very fine concentric growth lines. Uncommon; to 100 m.

SUP.F. **CHAMACEA:** Veneroida with heavy, inequivalve shells that are cemented to the substrate by right or left valve. Beaks prosogyrate or opisthogyrate. Sculpture well developed, concentric and/or radial or spinose. With 2 large, subequal muscle scars. Pallial line without sinus.

F. **CHAMIDAE:** Chamacea with unequal, irregular valves. With or without lunule. Sculpture well formed, lamellose or spinose. Anisomyarian. (5 spp. from Bda.)

***Chama macerophylla*** (Gmelin) (Leafy jewel box; B Rock cockle): Genus with shell attached by the left valve. Beaks prosogyrate. Sculpture of concentric flattened spines in irregular radial rows.—Species to 75 mm. Shell white, yellow, orange, brick-red, rose, or pale to dark lavender, or in combinations. Sculpture of flattened spines or leaf-like lamellations. Lower, attached valve larger, deeper; upper valve flatter, orbiculate. Pallial line runs past the anterior adductor muscle scar. Inner margins crenulate. Teeth heterodont. Lives below the tide line to 20 m, attached to any hard object—coral, rocks, cement, wrecks, buoys, etc. Common.

***C. congregata*** Conrad (=*C. linguafelis* of authors not Reeve) (Corrugated jewel box): Genus as above.—Species to 25 mm. Shell cream with brick-red radial markings. Surface with radially waved, concentric lamellae, sometimes with short, flat spines. Lower, attached valve deep, more strongly sculpted. Interior white to red; inner margins crenulated. Uncommon; attached to a hard substrate, associated with *C. macerophylla*.

***C. sinuosa*** Broderip (=*C. bermudensis* Heilprin) (White smooth-edged jewel box): Genus as above.—Species to 75 mm. Shell white, surface with wavy, blade-like concentric lamellations, not digitate. In older specimens all sculpturing is worn off to a smooth and eroded surface. Interior may be tinted moss-green. Inner margins not crenulate. Pallial line runs to anterior adductor muscle scar and not beyond as in *C. macerophylla*. Common; inshore and offshore to reefs, to 20 m.

***Pseudochama radians*** Lamarck (=*Chama exogyra* of authors) (False jewel box): Genus with shells attached by the right valve (mirror images of *Chama*!). Beaks opisthogyrate.—Species to 70 mm. Shell whitish, thick, globular. Surface with low concentric lamellae and radiating ribs, generally eroded to a roughened, irregular surface. Interior may be stained with brown. Common; associated with *Chama congregata*.

**Plate 159**

SUP.F. **GALEOMMATACEA:** Veneroida with small, thin, equivalve, translucent shells that are covered by extensions of the middle mantle folds. Mantle with buccal, pedal and anal apertures. Hinge simple. Ligament rarely present, resilium small. Dimyarian; byssate. Inhalant current anterior, exhalant posterior. Mostly commensal on a variety of marine invertebrates. The only known parasitic bivalve belongs to this superfamily.

F. **LASAEIDAE:** Galeommatacea with small, compressed shells. Hinge slightly indented under beaks. Both valves have cardinal and lateral teeth, but in some taxa dentition may be duplicated on right valve. Ligament internal. Hermaphroditic. Brood young in mantle cavity. (3 spp. from Bda.)

***Lasaea adansoni*** (Gmelin) (=*L. rubra* Montagu; *L. bermudensis* Bush) (Adanson's red Lasaea): Genus with small, quadrate, inflated shells with prominent beaks. Hinge with large cardinal and slender lateral teeth. Anterior lateral teeth are duplicated on both valves.—Species to 3 mm, white to red with light yellow periostracum. Beaks prosogyrate. Sculpture of concentric growth lines and fine radial striae. Very common intertidally, clinging to the byssus of *Arca zebra* and in soft, silty bottoms, to 10 m.

SUP.F. **CYAMIACEA:** Veneroida with small, thick, equivalve shells. Hinge dentition consisting of an

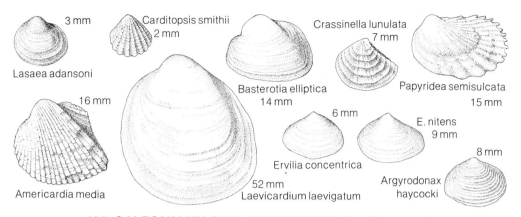

**159 GALEOMMATACEA—MACTRACEA (Cockles, a.o.)**

upper and lower series. Ligament only slightly embedded. Dimyarian. Pallial sinus lacking. Byssate. Mantle with 2 posterior apertures.

F. **SPORTELLIDAE:** Cyamiacea that have small, thick, circular shells with prominent beaks. Hinge dentition complex. Ligament external. Resilium in small pit (chondrophore) on hinge plate. Shell surface may be chalky, smooth or pustulose. (2 spp. from Bda.)

***Basterotia elliptica*** (Recluz) (= *Coralliophaga dactylus* of authors not Bruguière) (Elliptical Basterotia): Genus with subtrapezoidal, inequilateral shell. Beaks anterior. The ventral margin is slightly concave in the middle. Shell gaping posteriorly and ventrally.—Species to 9 mm. Shell white, elongate, moderately inflated; surface smooth. Hinge with a projecting cardinal tooth in each valve. Uncommon; to 10 m.

SUP.F. **CARDITACEA:** Veneroida with equivalve, inequilateral shells. Beaks anterior of center and prosogyrate. Hinge with 2 cardinal and 2 remote lateral teeth per valve. Lunule small, often depressed. Escutcheon vague. Sculpture usually of heavy radial ribs, with a crenulated, non-gaping margin. Dimyarian. Pallial lines complete, no sinus. Byssate. Viviparity common.

F. **CONDYLOCARDIIDAE:** Carditacea with minute, trigonal or ovate shells, higher than long. Radial ribs and strong concentric sculpture. Spondyliform hinge margin. Ligament internal, resilium partially hiding cardinal tooth. Laterals long. (1 sp. from Bda.)

***Carditopsis smithii*** (Dall) (= *Cardita domingensis* of authors not Orbigny) (Smith's little Cardita): Genus with trigonal, fan-shaped, subequilateral shells. Hinge plate narrow, cardinal teeth small, internal resilium rounded, laterals elongate. Prodissoconch saucer-shaped. Viviparous.—Species to 2 mm. Shell rust-brown, amber, bleaching to white. Inflated; prominent subcentral beaks. Sculpture of 10-13 rounded, beaded radial ribs, interspaces crossed by fine concentric threads. Rim between larval and adult shell prominent. Hinge line narrow and weak. Ligament mostly internal, resilium small. Common; in soft sand and mud bottoms, to 10 m (see WALLER 1973).

SUP.F. **CRASSATELLACEA:** Veneroida with thick, trigonal to circular shells that have prominent concentric sculpture. Beaks are pointed. Lunule and escutcheon are distinct. With 2 cardinal and 2 thin, elongate lateral teeth per valve. Isomyarian. Pallial line complete, no sinus.

F. **CRASSATELLIDAE:** Crassatellacea with subquadrate to trigonal shells. Anterior rounded, posterior truncated. Sculpture of heavy concentric ridges, particularly on the umbones. Periostracum dark brown. (1 sp. from Bda.)

*Crassinella lunulata* (Conrad) (=*Crassatella guadelupensis* Orbigny; *Crassatella parva* C. B. Adams) (Oval Crassinella): Genus with small triangular shell. Beaks opisthogyrate (point posteriorly). Lunule narrow. Escutcheon broad. Sculpture consists of stout concentric ribs.—Species to 9 mm. Shell white, rose, purplish brown, often rayed; compressed, roughly equilateral. Sculptured with 15-17 concentric ribs. Interior generally purplish brown. Common; on coarse sand bottoms, to 100 m.

SUP.F. **CARDIACEA:** Veneroida with trigonal, subquadrate, ovate or circular, equivalve shells, usually with prominent radial ribs. The pattern of ribs changes on the posterior slope. Hinge with 2 cardinal teeth per valve. Left valve with 1 anterior and 1 posterior lateral. Pallial line entire; brackish-water forms with pallial sinus. Anisomyarian. Mantle fused along inner and middle mantle folds to form short siphons that may bear tentacles with eyes. Animal with long, muscular foot that may be used for "jumping".

F. **CARDIIDAE:** Cardiacea with shell that is usually ovate to elliptical, inflated and cordate when viewed from either end. Shell margin crenulate. Ligament external, short and thick. (10 spp. from Bda.)

*Papyridea semisulcata* (Gray) (=*Cardium petitianum* Orbigny) (Frilled paper cockle): Genus with ovate shell that is longer than high, gaping anteriorly and posteriorly. Sculpture with radial ribs that bear imbricate scales. Hinge short.—Species to 12 mm. Shell white, moderately inflated, with 26-32 radial ribs that are indistinct anteriorly, increasing to large and scabrous posteriorly. Posterior portion of shell rose-pink, with 8-12 serrations that are extensions of the ribs. Interior radially grooved, exterior color showing through. Uncommon; dredged in sand offshore to 30 m.

*Americardia media* (L.) (Atlantic strawberry cockle): Genus with quadrate, non-gaping shell with heavy ribbing and a sharply delineated posterior slope. Hinge short.—Species to 50 mm. Shell white to yellow with reddish brown mottling; thick, inflated. Posterior slope slightly concave. Sculpture of 33-36 radial ribs. Interior porcelaneous, white. Uncommon; on sandy bottoms offshore, to 50 m.

***Laevicardium laevigatum*** (L.) (=*Cardium serratum* of authors not L. (Common egg cockle; B Toenail shell): Genus with ovate shell that is higher than long, smooth, polished, non-gaping. Radial ribs on the posterior slope weak or absent. Hinge long, arched, with prominent laterals. —Species to 50 mm; thin, obliquely ovate; bone-white, generally with lavender-rose zigzag or tent- like markings, concentric bands, or vague maculations. Color often concentrated along the posterior slope. Interior glossy, white, cream or highly colored with rose, pink, peach or lavender or a combination. Marginal edge serrated. Periostracum thin, worn, dirty gray. Common; sandy, grassy bottoms in- and offshore, to 20 m.

SUP.F. **MACTRACEA**: Veneroida with thin, equivalve, porcelaneous shells. Well-developed hinge with a single deltoid cardinal tooth in the left valve and 2 cardinal teeth in the right. External ligament small or lacking, resilium in chondrophore. Pallial sinus well developed. No byssus. Labial palps large, foot large. Mantle edge fused ventrally, siphons well developed.

F. **MESODESMATIDAE**: Mactracea with relatively heavy, compressed, inequilateral, trigonal shells. Siphons retractable, naked, nearly or completely separated. (3 spp. from Bda.)

***Ervilia concentrica*** (Holmes) (=*E. subcancellata* E. A. Smith) (Concentric Ervilia): Genus with small, thin shell. Sculpture consists of fine concentric ridges. Hinge with 1 small cardinal tooth per valve. Ligament obsolete. Resilium small. Pallial sinus large.— Species to 10 mm. Shell white, yellow or pinkish; elliptical, moderately compressed. Ends equally rounded. Umbones nearly central. Sculpture of concentric ridges may be crossed by radial threads, forming beads. Periostracum thin. Moderately common; on sandy bottoms offshore, to 100 m.

***E. nitens*** (Montagu) (Shiny Ervilia): Genus as above.— Species to 10 mm. Shell white, yellow or pinkish, thin, elliptical, compressed, flatter than *E. concentrica*. Umbones about 1/3 of the way from the anterior end. Sculpture of numerous concentric threads, strongest near the umbones. Periostracum thin. Moderately common; associated with above.

***Argyrodonax haycocki*** Dall (Haycock's silvery Donax): Genus with small, porcelaneous shell. Surface with concentric sculpture; prodissoconch minute, smooth, nearly circular. Pallial sinus large and deep. Adductor muscle scars pronounced. Isomyarian. Ligament external, weak. Resilium narrow, strong.— Species to 7 mm. Shell white, subtrigonal, thinning toward

**160 TELLINACEA (Tellins)**

the rear. Sculpture of concentric threads, coarsest and more or less irregular towards the margin. Lunule impressed, undefined. No escutcheon. Ligament short, weak. Anterior end round, longer than subtrigonal posterior end. Adductor muscle scars large, particularly the posterior one. Pallial sinus large, deep, confluent with pallial line. Shells uncommon; because no live specimens of this species are known its systematic position remains uncertain.

## Plate 160

SUP.F. **TELLINACEA:** Veneroida with shells that are mostly inequilateral, compressed, elongate to oval. Ligament external or on hinge plate. Heterodont hinge with 2 bifid cardinals and 2 well-developed laterals per valve. Dimyarian, with anterior adductor muscle larger than posterior. Pallial line and sinus distinct. With a cruciform muscle, which joins the left and right mantle lobes antero-ventral to the base of the inhalant siphon, producing 2 small, round muscle scars just below the posterior end of the pallial sinus. The burrowing animal has 2 elongate, separate siphons formed by the fusion of the inner fold of the mantle margin. It feeds on detritus sucked up from the surface of the substrate by the inhalant siphon, which acts as a vacuum cleaner.

F. **TELLINIDAE:** Tellinacea with elongate, slightly inequivalve shells which are slightly twisted to the right posteriorly. Ligament external, posterior to beaks. Labial palps slightly smaller than gills. Infaunal, lying on the left valve when buried (see BOSS 1966a, b). (15 spp. from Bda.)

*Tellina radiata* L. (=*T. radiata unimaculata* Lamarck) (Sunrise tellin): Genus with shell that is elongate to oval, with the left valve more convex than the right. Surface sculpture concentric. Color often in radial rays.—Species to 100 mm. Shell with glossy, smooth surface; cream-white (uncommon) or cream-white with radial rays of pink or yellow. Umbones red. Pallial sinus deep, almost reaching the anterior adductor muscle scar and descending to midpoint of pallial line. Interior white or suffused with yellow, pink or orange. Uncommon; offshore, deep in sand, to 30 m.

*T. laevigata* L. (=*T. laevigata* var. *stella* Davis) (Smooth tellin; B Sunset clam): Genus as above.—Species to 100 mm. Shell oval, compressed, with smooth, glossy surface and fine radial and concentric sculpturing. Cream-white, lemon-yellow to pale salmon-pink, may have a wide orange band along ventral margin; often with variably sized radial rays of salmon-pink. Beaks color of shell. Interior glossy, yellow, cream, salmon or a combination. Pallial sinus extends about 2/3 of the way to anterior adductor muscle scar, returns obliquely to pallial line and is confluent with posterior 1/3 of its length. Portion of ligament calcareous. Common; deep in sand, to 30 m.

*T. magna* Spengler (Great tellin; B China shell): Genus as above.—Species to 120 mm. Shell ovate, compressed, glossy. The left (lower) valve white; the right valve may be all white or suffused with pink. Umbones dark pink. Surface has microscopic radial and concentric sculpture. Portion of ligament thick, calcareous. Pallial sinus extends about 2/3 of the way to the anterior adductor muscle scar, returns obliquely to pallial line and is confluent for 3/4 of its length. Uncommon; associated with above. This and *T. laevigata* have hybridized and produced sterile offspring (see BOSS 1964).

*T. listeri* Röding (=*T. interrupta* Wood) (Lister's tellin): Genus as above.—Species to 90 mm. Shell white to light tan with concentric tent markings, which often occur in radial rays; elongate, solid with pronounced concentric sculpture. Interior white, suffused with yellow. Pallial sinus extends about 3/4 of the distance to the anterior adductor muscle scar, and is often connected to it by a linear scar; it descends obliquely, merging with pallial line near its midpoint. Moderately common; associated with above.

***T. gouldii*** (Gould's tellin): Genus as above.—Species to 12 mm. Shell white, solid, obliquely sub-ovate, with weak concentric sculpture that is most pronounced on the posterior end. No lateral teeth in left valve. Pallial sinus deep, extending to anterior adductor muscle scar. Its confluence with the pallial line is nearly entire. Uncommon; in sand and grassy shallows, to 20 m.

***T. mera*** Say (=*T. promera* Dall) (Pure tellin): Genus as above.—Species to 25 mm. Shell uniformly translucent white, thin, ovate, moderately inflated. Surface smooth, non-lustrous, with fine concentric striae. The pallial sinus is deep, extending 3/4 of the distance to the anterior adductor muscle scar, returning confluent with the pallial line for 1/2 its length. Uncommon; associated with above.

***T. sybaritica*** Dall (Delicate tellin): Genus as above.—Species to 11 mm. Shell white, yellow, orange, deep pink, solid or rayed; elongate, thin, moderately inflated, and strongly curved to the right posteriorly. Surface with very fine concentric sculpture. Pallial sinus nearly touching anterior adductor muscle scar, falling to pallial line and confluent with it. Uncommon; associated with above.

***T. similis*** Sowerby (=*T. decora* Say) (Candy stick tellin): Genus as above.—Species to 30 mm. Shell white, yellow or pink; may have 6-12 radial rays of pink in varying intensities; ovate, compressed, thin. Sculpture consists of concentric growth lines crossed by oblique threads. Pronounced ridge and sulcus. Pallial sinus nearly reaches anterior adductor muscle scar and descends sharply to pallial line. Interior yellow to pink. Uncommon; associated with above.

***Strigilla mirabilis*** (Philippi) (=*S. flexuosa* Say non Montagu) (Remarkable scraper shell): Genus with thin, inflated, circular to oval shell. Sculpture consists of fine concentric growth lines crossed by oblique incised lines. Lateral teeth weak in left valve.—Species to 14 mm. Shell white, shiny; exterior with 3 or 4 chevron-like rows or threads on the posterior slope. Lunule short and broad, escutcheon shallow and weak. Adductor and cruciform muscle scars weakly impressed. Pallial sinus curves into the pallial line just below the anterior adductor muscle scar. Common; in clean sand, to 30 m.

***Macoma tenta*** (Say) (=*Tellina flagellum* of authors not Dall; *M. souleyetiana* Recluz) (Elongate Macoma): Genus with oval to elongate shells that are twisted to the right posteriorly. Hinge lacks lateral teeth. Sculpture of concentric growth lines. Pallial sinus inequal in the 2 valves. Labial palps larger than gills.—Species to 20 mm. Shell white, slightly iridescent, elongate, fragile, with fine concentric threads. Posterior attenuate. Periostracum thin, golden. Pallial sinus roughly equal in both valves, extending 2/3 of the distance to the anterior adductor muscle scar and approaching the pallial line tangentially. Common; in clean sand, to 30 m.

***Psammotreta intastriata*** (Say) (= *Tellina gruneri* of authors) (Atlantic grooved Macoma): Genus with ovate to subquadrate shell, posterior twisted sharply to the right. Ligament and resilium small, sunken in a pit in hinge, almost internal. Anisomyarian.—Species to 60 mm. Shell translucent dirty white, thin; surface with very fine concentric growth lines. Twisted posterior portion of the shell with a strong, raised rib on the right valve and a corresponding sulcus on the left. Pallial sinus very large, roughly equal on both valves, almost reaching the long, narrow, anterior adductor muscle scar. Periostracum very thin, light brown. Uncommon; buried deep (to 30 cm) in sand; to 15 m.

F. **PSAMMOBIIDAE:** Tellinacea with equivalve shells that are slightly gaping, particularly at posterior end. Ligament external, posterior to beaks. Hinge with 1-3 cardinal teeth per valve. Lateral teeth weak or absent. Posterior of shell not twisted to the right. Pallial sinus present, but not deep. (4 spp. from Bda.)

***Asaphis deflorata*** (L.) (= *Capsa coccinea; Capsa spectabilis* of authors) (Gaudy Asaphis; B Purple mussel): Genus with ovate to quadrate shells, moderately inflated, numerous radial ribs. Posterior end slightly gaping.— Species to 60 mm. Shell white, yellow, salmon or purple; rounded radial ribs and fine concentric growth lines produce a slightly scabrous sculpture, particularly on the posterior end. Interior porcelaneous, color same as exterior but more brilliant; purple along the ligament ridge and heavily concentrated at the posterior end. Common; buried deep (to 15 cm) in coarse sand and gravel just above low tide line.

***Heterodonax bimaculata*** (L.) (= *Macoma eborea* Heilprin) (Twospotted false Donax): Genus with ovate shell. Hinge with 2 cardinal; and 2 indistinct lateral teeth per valve. Ligament external, posterior to beaks. Pallial sinus deep.—Species to 18 mm. Shell white, yellow, red, purplish or any combination of these; ovate, moderately inflated, rounded anteriorly and truncated posteriorly. Exterior smoothish with fine concentric growth lines. Frequently with radial rays or patches of brown color. Pallial sinus extends 2/3 of the distance to the anterior muscle scar. Common; on open protected beaches at low tide line, buried just below the surface.

F. **SEMELIDAE:** Tellinacea with shells that are generally compressed, circular to elliptical. Posterior of shell slightly twisted to the right. Ligament external, resilium internal, often with a distinct chondrophore. Heterodont hinge in which lateral teeth, when present, are close to cardinal teeth. Pallial sinus large, rounded. (3 spp. from Bda.)

***Semele proficua*** (Pulteney) f. ***radiata*** (Say) (=*S. reticulata; S. orbiculata* of authors) (Useful Semele): Genus with inequivalve shell that is circular to elliptical. Hinge with 2 cardinal teeth per valve. Lateral teeth indistinct on left valve. Chondrophore parallel to hinge line. Pallial sinus deep, not confluent with pallial line.—Species to 40 mm. Shell bone-white, yellowish or rose, margins often stained rust-brown; subcircular. Beaks prosogyrate. Lunule minute, impressed. Sculpture of moderate to strong concentric growth lines. Interior porcelaneous white, yellow or rose. Uncommon; in sand bottom, to 10 m.

***S. bellastriata*** (Conrad) (=*S. cancellata* (Orbigny)) (Cancellate Semele): Genus as above.—Species to 18 mm. Shell white, yellow, pink, rose, purple, with numerous radial ribs crossed by concentric ribs, producing a prominent cancellate surface. Resilium oblique. Interior porcelaneous; color showing through, flecked with reddish brown. Moderately common; in sand, to 10 m.

F. **SOLECURTIDAE:** Tellinacea with equivalve shells that are elongate, quadrate, widely gaping at both ends. Beaks may be central or slightly anterior. Heterodont hinge is narrow and weak. Ligament posterior to beaks. Pallial sinus present, generally shallow. (2 spp. from Bda.)

***Tagelus divisus*** (Spengler) (Divided sand clam): Genus with elongate, slightly compressed shell. Beaks nearly central. Hinge with 2 cardinal teeth per valve. Lateral teeth lacking.—Species to 40 mm. Shell elongate-quadrate, thin, fragile, smooth, with very fine concentric growth lines. Beak area white, rest of shell purplish with white radial rays. Interior same color as exterior, with a single purple, slightly raised radial rib extending from the beaks to the ventral margin. Periostracum thin, dark-straw colored. Uncommon; in silty mud and soft sandy bottoms, to 10 m.

### Plate 161

SUP.F. **ARCTICACEA:** Veneroida with medium-sized shells that are elongate to ovate, equilateral, generally equivalve; umbones well to the anterior. Exterior smooth or with some concentric growth lines. Ligament external. Hinge of each valve with 2 or 3 cardinal teeth radiating from the umbones. Pallial line generally without sinus.

F. **TRAPEZIIDAE:** Arcticacea with elongate shells. Umbones near anterior end. Hinge plate narrowed, generally with 2 cardinals in each valve; one posterior and one small anterior lateral. Pallial line mostly without sinus. (1 sp. from Bda.)

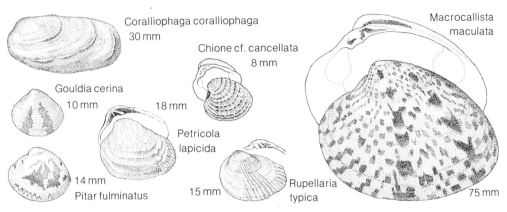

**161 ARCTICACEA, VENERACEA (Venus clams, a.o.)**

*Coralliophaga coralliophaga* (Gmelin) (=*C. carditoidea; Cypraeacardia hornbeckiana* of authors) (Coral clam): Genus with thin, irregularly shaped shells; elongate, quadrate, ovate or round. Surface with fine radial striae and concentric lamellations. Hinge reduced.—Species to 50 mm; shell bone-white, chalky, sculptured with fine radial threads and, at the posterior end, with coarse, periodic growth lamellations with wide interspaces. Umbones at the anterior end. Pallial line with sinus. Uses empty burrows of other species and is shaped accordingly, or excavates its own burrow in soft coral or coral rubble. Uncommon; to 35 m.

SUP.F. **VENERACEA:** Veneroida with ovate or ovate-quadrate, equivalve shells. Sculpture predominantly concentric; may have radial ribs, spines or lamellae, especially at the posterior portion. Umbones toward the anterior, prosogyrate (pointing forward). Ligament external, behind the umbones. Generally with 3 cardinal hinge teeth in each valve. No byssus. Mantle fused along the inner and middle mantle folds.

F. **VENERIDAE:** Veneracea with ovate or cordate (heart-shaped), variably sized, inequilateral, porcelaneous shells. Beaks forward of midline. Escutcheon (if present) and lunule well developed. Ligament external. Surface as above, but likely to have a glossy, finished look. Pallial sinus varying in size and shape, usually deep, indicative of a habit of burrowing beneath a sandy or soft substrate. No byssus. (9 spp. from Bda.)

*Chione* cf. *cancellata* (L.) (Dwarf cancellate Chione): Genus with thick, ovate-trigonal shells; surface sculpture strong, concentric, may be frilled. Lunule bounded by impressed line; escutcheon long, smooth, bounded by a small ridge. Pallial sinus small. Inner margins crenulated. Cardinal teeth smooth.—Species to 8 mm; subtriangular in shape, sculpted with numerous, rounded, radial ribs that are crossed by 13-20 thin, raised, blade-like, concentric lamellae, giving the surface a cancellate appearance. Shell bone-white, often with pale to dark brown radial rays, solid or interrupted. Lunule small, heart-shaped with fine threads that are a continuation of the concentric sculpturing; escutcheon long, right valve portion with fine radial bladed striations, left valve portion slightly larger, smoother, with 3-5 brown cross bands or maculations. Interior shiny, white or bone-white, may have faint brown markings on posterior portion. Moderately common; in soft or sandy bottoms. Full-sized (45 mm) specimens are found as fossils only. This small Bermuda species may be a different form, or just occur as recruited populations that never reach maturity.

*Gouldia cerina* C. B. Adams (=*G. bermudensis* E. A. Smith; *Transennella conradina* of authors) (Gould's wax venus): Genus with small, ovate-trigonal shells with concentric and radial sculpture, surface reticulated. Lunule long. Escutcheon lacking. Pallial sinus small.—Species to 12 mm. Shell bone-white, straw or with a tint of lavender. Brown or purplish brown patterns variable: chevrons, tent-like markings, rays or maculations, faint to dark. Sculpture of fine rounded, concentric ribs; radial ribbing strong on anterior and posterior ends. Middle section with microscopic radial lines, the concentric ribs dominant. Beaks minute, incurved. Lunule long, depressed, bounded by incised lines. No escutcheon, but area glossy, with fine incised growth lines, may have crossbars or maculations of dark brown. Pallial sinus a shallow notch. Interior porcelaneous. Common; most prevalent bivalve in some inshore waters but found also offshore, on sand bottoms, to 800 m.

*Pitar fulminatus* (Menke) (=*Chione venetiana*; *Cytherea hebraea* of authors; *Cytherea penistoni* Heilprin) (Lightning venus): Genus with ovate to subtrigonal, moderately inflated shells. Surface smooth or with very fine to coarse concentric ribbing or lamellae; may have spines or scales on posterior end. Lunule weak, escutcheon not defined. Pallial sinus deep. Right posterior cardinal split, left posterior cardinal blending with nymph (a narrow platform or reinforced area that extends posteriorly from the umbones along the dorsal margin and serves as a ligament attachment or hinge reinforcement).—Species to 18 mm, inflated, subtrigonal. Surface smooth, semi-glossy, with fine concentric growth lines. Shell all bone-white or with light brown zigzag or tent-like patterns sparsely placed or coalesced into wide patches. Beaks white to purple-brown; along the anterior side there may be a purple-brown ray that rapidly dissipates. Interior white, porcelaneous. Lunule cordate, bounded by an impressed line. The endemic form that was named *C. penistoni* by Heilprin is light brown, no patterns; umbones purplish brown, with a dark purplish ray on each side that fades rapidly.

Interior whitish lavender to purple-brown. Common; in sand or mud bottom, to 25 m.

*Macrocallista maculata* (L.) (Calico clam; B Checkerboard clam): Genus with elongate-oval to oval shells. Surface smooth or with concentric grooves. Periostracum shiny. Lunule narrow. No escutcheon. With 3 cardinal teeth in each valve. Ventral margins smooth.—Species to 80 mm. Shell light cream-coffee colored with purplish brown markings in a vague checkerboard design, rarely solid cream or purplish brown. Lunule small, vaguely defined. Interior porcelaneous with 1-3 widely spaced light lavender bands. Pallial sinus broad. Moderately common; buried about 7 mm in sandy areas, to 10 m. Easily located by the 2 gaping siphonal openings. Edible but now protected. Early reports (DALL & BARTSCH 1911; PEILE 1926; CLENCH 1942; MOORE & MOORE 1946) were all of fossil finds. Became abundant, live, in the late 1950's (see ABBOTT & JENSEN 1967).

F. **PETRICOLIDAE:** Veneracea with thin, oval to elongate shells; large siphonal and pedal gapes. Sculpture radial or obsolete. No lunule or escutcheon. With 3 cardinal teeth in left valve. Pallial sinus strong. Bore into soft limestone, coral, clay, peat, etc. Shell often distorted to conform to shape of cavity. (2 spp. from Bda.)

*Petricola lapicida* (Gmelin) (Boring Petricola): Genus with shell that is sculpted with a radial pattern in the juvenile and radially divaricate or with zigzag riblets in the mature stage.—Species to 12 mm; shell white, generally ovate, inflated; sculpture, which is not of the shell proper, is a chalky, calcareous deposit on the surface, and on the posterior end is laid down in wavy radiating ribs. Cardinal tooth bulbous and split. Siphons pale green. Bores into soft coral and limestone. Uncommon; found intertidally.

*Rupellaria typica* (Jonas) (Typical rock borer): Genus with shell that is strong, rounded anteriorly, tapering and compressed posteriorly. Sculpture chiefly radial, strongest anteriorly.—Species to 25 mm. Shell dirty white, variable in shape, anterior rounded, inflated; posterior tapered, compressed, gaping. Sculpture with numerous coarse, wavy, radial ribs; interspaces broad. Interior stained with buff or brown. Uncommon; found intertidally in bore holes in soft coral and rocks.

### Plate 162

O. **MYOIDA:** Heterodonta with thin inequilateral shells that may be equivalve or inequivalve. Long, well-developed siphons. Isomyarian or anisomyarian, with adductor muscles that can be contracted independently. One or no cardinal tooth per valve. Lunule and escutcheon absent or poorly developed. Shell

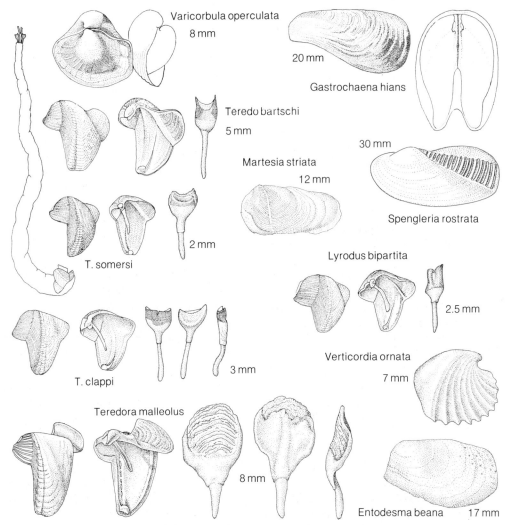

**162 MYOIDA (Shipworms, a.o.), ANOMALODESMATA**

not nacreous. Infaunal or burrowing.

**SUP.F. MYACEA:** Myoida with shells that are elongate or ovate, porcelaneous to chalky, and covered by a thin periostracum. The hinge is edentulous, with ligament that is mainly internal. With some exceptions, the valves are gaping and the pallial sinus is well developed. May be deep infaunal, commensal or burrowing.

**F. CORBULIDAE:** Myacea with sturdy, inequivalve shells. Left valve is smaller than, and fits within, the right valve. Margins are lightly appressed. Resilium in one valve. Pallial sinus small. Siphons short, united and not covered by periostracum. (1 sp. from Bda.)

***Varicorbula operculata*** (Philippi) (=*Corbula disparilis* of authors) (Oval Corbula): Genus with small, subtrigonal shell. The right valve is the larger and has concentric ribbing, whereas the smaller left valve has faint radial growth striae.—Species 5-8 mm. Shell white, glossy, longer than high. Umbones curled, pointed anteriorly. Uncommon; dredged in sand in 80-120 m.

SUP.F. **GASTROCHAENACEA:** Myoida with elongate, porcelaneous shells with a large anterior gape. Hinge edentulous; ligament external. Pallial sinus deep, siphons long, united. Gill extends into branchial siphon. Burrow into coral, rock or shell by mechanical means. May have an external protective tube to supplement their burrows. (Only family: Gastrochaenidae, with 2 spp. from Bda.)

*Gastrochaena hians* (Gmelin) (=*G. cuneiformis*; *G. cucullata*; *G. ovata* of authors; *G. mowbrayi* Davis) (Gaping boring clam): Genus with small, thin, equivalve shells. Umbones at anterior end of shell. Very large anterior gape. Produces flask-shaped burrow.—Species to 20 mm; shell white, broad posteriorly, with fine concentric sculpture. Uncommon; in coral and rubble; associated with *Lithophaga*. *G. ovata* and *G. mowbrayi* are stenomorphic (size and shape restricted by habitat) forms of *G. hians*.

SUP.F. **PHOLADACEA:** Myoida with inequilateral, edentulous, widely gaping shells usually divided by an umbonal-ventral sulcus that produces a raised rib on the interior of the valves. Chondrophore and apophysis present, ligament internal. Siphons long. Gills extend into branchial siphon. Anterior, posterior and ventral adductor muscles can be contracted independently. Burrow into peat, wood, clay, rock, coral and shell by mechanical means.

F. **PHOLADIDAE:** Pholadacea with elongate to globular shells and 1 or more accessory plates. The pedal gape may or may not be closed by one of the plates (callum) in the adult. Anterior portion of the shell with concentric ridges, commonly ribbed radially. Pallial sinus deep. Siphons united, long, enclosed in periostracal sheath. Intestine passes through heart. Lacks pallets and expanded stomach caecum. Cannot digest cellulose. (2 spp. from Bda.)

*Spengleria rostrata* (Spengler) (Truncated boring clam): Genus with thin, blade-shaped shells that are divided by an oblique furrow. The portion of the shell posterior to this furrow is crossed by pronounced vertical striae. Umbones situated about 1/3 of the way from anterior to posterior.—Species 20-30 mm in length. Shell whitish, truncated posteriorly. Sculpture consists of fine concentric lines antero-ventrally of the oblique furrow and of pronounced vertical striae postero-dorsally. In coral and rubble; moderately common.

*Martesia striata* (L.) (Striate Martesia): Genus with elongate shell truncated anteriorly. Pedal gape sealed by accessory plate (callum). Lacks dorsal anterior accessory plate (protoplax). Apophysis long and thin. Siphons capable of complete retraction. Protandrous hermaphrodites. Wood borers.—Species 20-50 mm. Shell white, pear-shaped, rotund, with pronounced umbonal-ventral sulcus. There are numerous finely denticulate riblets anterior to the sulcus, and fine concentric ridges posterior to it. Large, round accessory plate above umbone (mesoplax). Common in driftwood.

F. **TEREDINIDAE** (Shipworms): Pholadacea with greatly reduced, equivalve and auriculate shell. Apophysis, umbonal-ventral sulcus and interior rib prominent. Siphons separate or united. Pallial line situated at the valve margins. Intestine does not pass through the heart. Burrow with calcareous lining. Animal has pallets that close the burrow when the siphons are withdrawn. Burrows in wood, which it digests by means of an expanded stomach caecum. Protandrous hermaphrodites. (5 spp. from Bda.)

***Teredo bartschi*** Clapp (= *T. batilliformis* Clapp) (Bartsch's shipworm): Genus with nonsegmented, paddle-shaped pallets that have a calcareous base covered by a thin periostracum. Posterior adductor muscle. much larger than anterior. Siphons separate. Young carried in brood pouch dorsal to the gills. Species are best distinguished on characters of the pallets and not the shell.—Species with shell and pallets to 5 mm. Color white. Shell with large semicircular auricle. Pallets with long stalk and short blade that is deeply excavated at the top. Only the distal end of the blade is covered by periostracum. The calcareous portion within is hourglass-shaped.

***T. somersi*** Clapp (Somers' shipworm): Genus as above.—Species with shell and pallets to 3 mm. Color white. Auricle greatly reduced. Posterior region narrow and crescent-shaped. Pallets solid, heavy, with long stalk and short blade that extends down the stalk as a sleeve. Periostracum covering the distal portion thin, reddish brown. The calcareous portion within is crater-shaped.

***T. clappi*** Bartsch (Clapp's shipworm): Genus as above.—Species with shell and pallets to 4 mm. Auricle greatly reduced. Pallets with long stalk and short, cup-like blade that extends down the stalk as a sleeve. Periostracum covering the distal portion of the pallet is short and may have a cleft.

***Lyrodus bipartita*** (Jeffreys) (Divided shipworm): Genus with nonsegmented, paddle-shaped pallets that have a calcareous base and a large periostracal cap. Siphons short and separate. Young carried in brood pouch dorsal to gills until they reach the veliger stage.—Species with shell and pallets to 3 mm. Auricle reduced. Pallets with long stalk and triangular blade which extends down the stalk as a sleeve. Periostracum on the distal end of the pallet dark brown, with a deep longitudinal furrow on the outer face.

***Teredora malleolus*** (Turton) (=*Teredo thomsoni* Tryon) (Hammered shipworm): Genus with unsegmented oval pallets with a short stalk and a thin blade. Outer face of pallet with a thumbnail-like depression. Siphons united.—Species with shell and pallets to 10 mm. Auricle large, ovate. Pallets with short stalk and large blade. Straw-yellow periostracum on the pallets.

S.CL. **ANOMALODESMATA:** Bivalvia with prismatonacreous, short to elongate, equivalve to subequivalve shells. Isomyarian. Hinge plate and teeth weak or lacking. Primitive forms with resilium and lithodesma. Mantle margins ventrally fused. Eulamellibranchiate or septibranchiate. Most are hermaphroditic. Siphonate burrowers. (Only order: Pholadomyoida.)

SUP.F. **PANDORACEA:** Pholadomyoida with thin, inequivalve, gaping shell. Endostracum nacreous. Ligament and resilium present, sheathed with calcareous layer (lithodesma). Hinge lacks regular heterodont teeth, but may have denticles or buttresses.

F. **LYONSIIDAE:** Pandoracea with a thin shell with a slight posterior gape. Pallial sinus distinct. (1 sp. from Bda.)

***Entodesma beana*** (Orbigny) (=*Lyonsia beaui* (Orbigny); *Mytilimeria plicata* of authors) (Pearly Lyonsia): Genus with elongate, nacreous shell. Periostracum yellow. Posterior end attenuate, slightly gaping. Some species attach sand grains to the outside to the valves.—Species to 40 mm. Shell white, irregular, roughly ovate, gaping slightly at both ends. Umbones at anterior end. Sculpture of fine, concentric lines. Periostracum with brown radial rays on the posterior portion of the shell. Uncommon; within sponges in shallow water.

SUP.F. **POROMYACEA:** Pholadomyoida with round to ovate, non-gaping shells. Isomyarian. Pallial sinus shallow. Somewhat developed cardinal and lateral teeth. Gills may be reduced or absent. The mantle margins are fused.

F. **VERTICORDIIDAE:** Poromyacea with small, nacreous, radially ribbed shells. Carnivorous, capturing small animals in sticky threads secreted by pallial tentacles. Muscular stomach lined with a chitinous sheath which serves as a crushing gizzard. (1 sp. from Bda.)

***Verticordia ornata*** (Orbigny) (Ornate Verticordia): Genus with shell that is small to minute, nacreous, well ribbed externally. Pallial line simple. Ligament external.—Species to 4 mm. Shell oval, with 10-12 widely spaced, curved radial ribs on aterior 3/4 of shell. Margin strongly crenulate anteriorly. Exterior white; interior silvery nacreous. Rarely dredged, in 800 m.

R. H. Jensen & M. G. Harasewych

## Class Cephalopoda (Cuttlefishes, squids and octopuses)

CHARACTERISTICS: *Bilaterally symmetrical MOLLUSCA with well-developed head bearing 8 or 10 circumoral, mobile appendages provided with suckers and/or hooks. Pharynx with chitinous beak-like jaws and a radula. Shell mostly reduced, modified or absent, and enclosed by the mantle (except in Nautiloidea); gills as 1 (rarely 2: Nautiloidea) pair.* The size ranges from 2 cm to 20 m (!); Bermuda species are in the range of 5-100 cm. Under a shiny skin the body is surprisingly tough and muscular, and capable of considerable changes in shape in some species (e.g., *Octopus*). The coloration is variable, and rapid changes of color and patterns are an integral part of their behavior. Rapid locomotion is achieved by drawing water into the mantle cavity and then expelling it, jet-like, through the funnel; in addition, most benthic forms (many octopods and sepioids) can crawl along the bottom using arms and suckers, and pelagic forms use fins on the mantle for balance, steering and minor locomotion.

One of the 2 subclasses, the Nautiloidea, is found only in the Indo-Pacific Ocean. Of approximately 1,000 living species, about 12 are known to occur near Bermuda; the 9 included here are either regular nearshore inhabitants, or offshore species seen at the surface or stranded.

OCCURRENCE: All marine habitats of the world; benthic on coral reefs, grass flats, sand, mud, rocks; pelagic and epipelagic in bays, nearshore and open-ocean habitats. Depth range from 0 to over 5,000 m. Abundance varies depending on group, habitat and season, from isolated territorial individuals (primarily benthic octopods), to small schools with a few dozen individuals, to huge schools of oceanic species with millions of specimens.

Collecting techniques include small traps (octopods), weirs, lures and jigs (some cuttlefishes and squids), encircling seine nets (nearshore squids) and midwater and otter trawls (squids and octopods). Certain species of squids are attracted to light, then jigged or seined. Occasionally cuttlefish and octopods are caught in hand nets or are speared, but it is nearly impossible to capture free-swimming squid in this manner. Caution: the bites of cephalopods, especially octopuses, can be painful at the least, or poisonous, or get secondarily infected.

Although cephalopods are extremely important worldwide as food for human consumption, they are not caught commercially in Bermuda. Cephalopods also are important experimental animals in biomedical research. Because of the highly developed brain and sensory organs, they have a great capacity to learn and remember, rendering them valuable in behavioral and comparative neuroanatomical studies. In addition, cephalopods possess the largest single nerve axons in the animal kingdom, and are therefore used extensively in neurophysiology.

IDENTIFICATION: Major groups as well as species are readily distinguished by external characteristics.

Note the shape of the body and color patterns, which are due to innervated chromatophores (pigment-filled sacs) and iridocytes (reflective cells that give sheen). Some deep-water species may have photophores (light-producing organs). Presence and shape of fins, and the presence (cuttlefish and squids) or absence (octopuses) of a pair of long tentacles in addition to the 8 arms distinguish major taxa. Tentacles consist of a stalk devoid of suckers, and a terminal club that bears suckers and/or hooks. Arms may be connected basally by a web (Vampyromorpha and Octopoda); they usually bear 2 (1-4) rows of suckers and/or hooks that may be stalked or unstalked. In squids, the structure of the locking apparatus (ridge and groove mechanism by which mantle and funnel are locked together during expulsion of water) may be diagnostic. Mature males possess a hectocotylus, i.e., 1 or more arms modified for transferring spermatophores (sperm packets) to the female. The length of the modified tip (ligula) of the octopod hectocotylized arm, expressed as a percentage of total arm length (ligula index) may be species-distinctive. In squids, determine whether the eye is covered with a corneal membrane or open. Most species have an ink gland that opens into the rectum and releases a black fluid used to confuse predators.

Fixation for study or display specimens of squids requires an initial period (a few days) in 8-10% seawater-formalin buffered with borax or other buffering agent. Specimens should be laid out straight in deep trays so the arms and tentacles can be fixed in an untwisted state. Large, thick-mantled specimens should be thoroughly injected or cut open along the midregion of the ventral surface of the mantle with a 5-10 cm incision to ensure fixation of the viscera and reproductive organs. After fixation transfer specimens to 70% ethyl alcohol or 50% isopropyl alcohol for permanent preservation. Squids die very soon after capture and fixation should begin immediately. (Caution should be exercised if live or moribund specimens are placed in formalin in an uncovered container, for several strong pumps of the mantle may occur before death and formalin can be jetted for several feet in any direction.)

Octopuses commonly survive capture by gentle techniques, and if specimens must be preserved, they must be killed carefully to avoid strong contractions and tight coiling of the arms (which render the specimen extremely difficult to work on). Narcotize the specimen by diluting the seawater with 25% by volume of fresh water, then 50%. The process may be hastened somewhat with small amounts of ethyl alcohol (0.5-1%) or other narcotizing agent, but patience is required if coiling is to be prevented. When the specimen is completely narcotized (or dead!) dip the tips of the arms into 8-10% formalin (formulated as above), then massage the arms straight. Repeat the dipping and massaging process on increasingly longer sections of arms until entire lengths are treated. Place entire specimen in the deep tray, as with squid. Occasional arm-straightening may still be required. Fixation and preservation follow as for squid.

BIOLOGY: All cephalopods are dioecious and many, though not all, exhibit external sexual dimorphism, either in structural or size differences. Females generally are larger than males. Males of many forms possess 1 or 2 modified arms (hectocotyli) for mating. The hectocotylus may consist of modified suckers, papillae, membranes, ridges and grooves, flaps, etc., but in any case functions to transfer spermatophores from the male's mantle cavity to a locus of implantation on the female, which may occur inside the mantle cavity, around the mantle opening on the neck or head, in a pocket under the eye, around the mouth, etc. Fertilization takes place in the female as the eggs are laid. Eggs of squid generally are encased in a gelatinous matrix secreted by the nidamental glands and are laid as multi-finger-like masses (sometimes called "sea mops") attached to rocks, shells or other hard substrate on the bottom in shallow waters (inshore squids), or they are extruded as large, singular, sausage-shaped masses that drift in the open sea (oceanic squids). The fingers may contain from a few to several hundred eggs, whereas the sausages contain tens or even hundreds of thousands of eggs. The mode of reproduction and egg laying is unknown for many forms, especially oceanic and deep-sea species. Benthic octopuses lay their eggs in great, grape-like clusters and strands in lairs, under rocks and in abandoned mollusk shells, where they brood the eggs until hatching. The eggs are attached to each other, but they are not encased in a gelatinous matrix. The female of the pelagic octopus *Argonauta* constructs a thin, shell-like egg case in which she resides and lays festoons of eggs, fertilization having taken place from sperm contained in the highly modified hectocotylus that was autotomized from the male and deposited in the egg case. The life expectancy is about 1 yr in most forms, but larger species of squids and octopus must live for several years. Many species die after spawning. Many oceanic cephalopods undergo diel vertical migrations, wherein they occur at depths of about 400-800 m during the day, then ascend into the uppermost 200 m or so during the night. Whereas shallow-living cephalopods are able to conceal themselves by chromatophore-produced color patterns and chameleon-like color changes, many deep-sea forms camouflage themselves by producing bioluminescence from photophores (light-producing organs), which eliminates their silhouettes against the downwelling light in the dimly lit mid-depths. Cephalopods are active predators upon crustaceans, fishes, other cephalopods and, in the case of octopuses, bivalved mollusks. In turn, cephalopods are major food items in the diets of toothed whales, seals, pelagic birds (penguins, petrels, albatrosses, etc.) and both benthic and pelagic fishes (e.g., sea basses, tunas, billfishes). The parasites on cephalopods include dicyemid Mesozoa in the kidneys of octopods as well as digeneid Trematoda and apostomeid Ciliophora. Cestoda and Nematoda may parasitize the intestinal tracts of oceanic squids.

## Plate 129

DEVELOPMENT: Cephalopod eggs are very yolky (telolecithal) and cleavage is thus incomplete, or meroblastic, so that typical molluscan spiral cleavage is absent. Development is direct and babies hatch as miniatures of the adult (to a greater or lesser extent depending on the species). Cephalopod eggs may vary in size from about 13 mm (in some octopods) to 0.8 mm (in *Argonauta*) (ARNOLD 1971). Time of embryonic development also varies widely from a few weeks to several months, depending on the species and temperature conditions. Hatching may occur rapidly from a single clutch or be extended over a period of 2-3 weeks. Young animals often occupy different habitats from the adults. For example, the young of some species of benthic octopuses spend periods of time as planktonic organisms, and the juveniles of many deep-sea forms occur in the upper 100 m of the open ocean, then gradually move to greater depths with increasing size.

Juveniles are obtained by collecting the egg masses of nearshore squids and octopuses and letting them hatch out. Planktonic juveniles of benthic and oceanic forms can be collected by a medium mesh plankton net, but there has been little success in keeping them alive.

REFERENCES: A systematic overview of the class is given in BOSS (1982). An illustrated key to the families of Teuthoidea is by ROPER et al. (1969), Vampyromorpha were dealt with by PICKFORD (1946, 1949), and oceanic squids were reviewed by CLARKE (1966) and ROPER & YOUNG (1975; vertical distribution). Littoral Octopoda were studied by PICKFORD (1945) and VOSS & PHILLIPS (1958). Field guides to the potentially commercial species are by VOSS et al. (1973; tropical West Atlantic); ROPER & SWEENEY (1981; central Western Atlantic), and ROPER et al. (1984; worldwide). The nervous system and behavior of cephalopods are described by WELLS (1962, 1978) and YOUNG (1971).

Observations on species occurring in Bermuda were reported for *Spirula* by BRUUN (1943) and DENTON et al. (1967); for *Loligo* by LaROE (1971), ROPER (1965) and WALLER & WIKLUND (1968); for *Sepioteuthis* by ARNOLD (1965) and LaROE (1971); and for *Argonauta* by ILIFFE (1982). A review of Bermuda's inshore and oceanic species is given in VOSS (1960), and important early records include VERRILL (1881, 1901a, 1902a), HOYLE (1885a, b), HEILPRIN (1888), and VERRILL & BUSH (1900). COHEN (1976) verified the identity of *Loligo plei*.

**S.CL. COLEOIDEA** (=DIBRANCHIATA): Shell, internal, enveloped by mantle, in various degrees of reduction. With a single pair of gills, 10 (secondarily 8) cir-

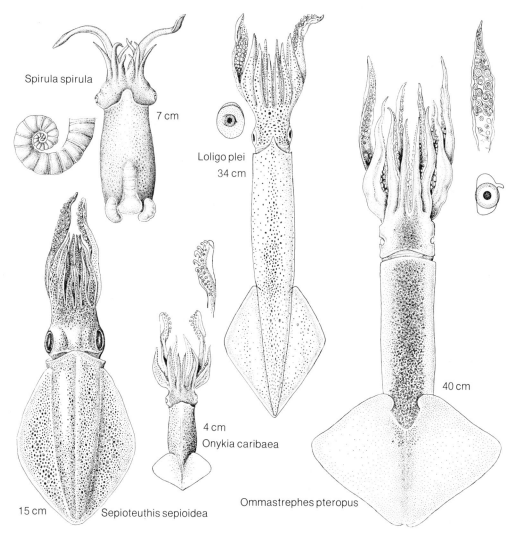

**163 SEPIOIDEA, TEUTHOIDEA (Squids)**

cumoral appendages, and a tube-like funnel (siphon). (All 4 orders represented in Bda.)

**Plate 163**

O. **SEPIOIDEA** (Cuttlefishes): Coleoidea with a calcareous or chitinous shell and 10 circumoral appendages. Tentacles (4th pair of appendages) retractile into pockets.

Posterior fin lobes free. Each tooth of radula, including rachidian, with only 1 projection. No branchial canal between afferent and efferent branchial blood vessels. Digestive gland divided or bilobed; pancreas separate. (1 sp. from around Bda.)

***Spirula spirula*** (L.) (Ram's horn shell): Internal shell open-coiled; chambers connected by a tube. Fins terminal, transversely situated. A single large photophore ("terminal disc") occurs at the blunt posterior end of the mantle. Four to 8 longitudinal rows of suckers along the arms. Whole specimens (maximum total length about 7.5 cm) are very rarely seen, for they inhabit the mid-waters off Bermuda, but the small (up to 2.5 cm diameter), white, coiled, chambered shells rise to the surface after the animals die and are commonly found washed up on the beaches after stormy weather. The species exhibits a vertical migratory behavior, occurring at 500-700 m during the day and ascending to 100-300 m at night (CLARKE 1969). The internal shell serves as a buoyancy mechanism to assist the animal in maintaining desired depths (DENTON et al. 1967).

O. **TEUTHOIDEA** (Squids): Coleoidea with a simple, rod- or feather-like, chitinous shell (pen or gladius). Ten circumoral appendages; 4th pair (tentacles) occasionally secondarily lost. Tentacles contractile, but not retractile into pockets. Posterior fin lobes fused, occasionally free. Rachidian and first lateral teeth of radula with secondary cusps. Branchial canal present between afferent and efferent branchial blood vessels. Digestive gland a single, undivided structure; pancreas separate.

S.O. **MYOPSIDA** (Inshore squids): Teuthoidea with eyes covered with corneal membrane; eye pore present. ♀ gonoducts single. Accessory nidamental glands present. Suckers present on buccal lappets. Suckers only (never hooks), on arms and tentacular clubs. (2 spp. from around Bda.)

***Loligo plei*** Blainville ( = *Doryteuthis plei*) (Arrow squid): Body long, slender especially in adult ♂. Fin length 35-58% of mantle length. Hectocotylus on left ventral arm extends to arm tip and occupies 26-50% of arm length. Maximum mantle lengths approach 35-40 cm but mature specimens may be considerably smaller. Most earlier records of this species in Bermuda were incorrectly attributed to *L. pealei* (see COHEN 1976). In all inshore waters as well as the backwaters of the reefs. COHEN (1976) described several abnormalities in Bermudian specimens (missing buccal lappets) and suggested these may be due to the small, isolated population at Bermuda. The spawning season of the Bermuda population is unknown, but mature adults in spawning condition have been recorded throughout the year across the range of *L. plei*. Eggs are laid on the bottom, attached to a hard substrate (rock, shell, coral rubble), in gelatinous strands or fingers; many strands are laid at once on the same object resulting in a large mass or "sea mop".

***Sepioteuthis sepioidea*** (Blainville) (Caribbean reef squid): Mantle stout, bluntly rounded posteriorly. Fins extend nearly entire length of mantle (except in smallest young). Hectocotylus occupies distal 26-30% of left ventral arm (4th), characterized by complete loss of suckers and greatly enlarged fleshy papillae. The species reaches a maximum mantle length of about

15 cm and a total length of about 30 cm. Around and behind coral reefs, over reef flats and turtle grass beds, surface to 20 m. The life-span is about 1 yr; sexual maturity is achieved in about 7 months. Where known, spawning occurs IX-III, possibly year round. The eggs are laid in shallow water in capsules or fingers, only 3-4 large (6×3 mm) eggs per capsule. Eggs are always laid in very cryptic spots, under flat rocks, in conch shells, well hidden from potential predators and bright light; about 130 eggs make up each "mop". After an incubation of about 36 days the hatchlings emerge at a relatively large size (10-12 mm total length), and reach sexual maturity 4-5 months after hatching.

S.O. **OEGOPSIDA** (Oceanic squids): Teuthoidea with eyes completely open, with no corneal covering or eye pore. Gonoducts of ♀ paired. Accessory nidamental glands absent. Suckers on the buccal lappets usually absent. Arms and tentacular clubs with suckers and/or hooks. (About 50 spp. from around Bda.)

*Onykia caribaea* (Lesueur): Small, compact, plump body with broad rounded fins that extend somewhat posterior to mantle tip, especially in juveniles. No light organs known. The locking apparatus on the funnel is a straight, elongate groove. The tentacular club has a distinctive rounded wrist section (carpus) consisting of about 10 each of alternating suckers and knobs and a hand section (manus) with a row of suckers along each outer margin and 2 median rows of hooks (about 10 hooks in each row). Chromatophores are dense and dark on the dorsal surface of the mantle, head and arms; in life they impart a deep bluish color characteristic of neuston species (HERRING 1967). A layer of iridophores gives a reflective sheen to the body. To about 7 cm in mantle length. An open-ocean, epipelagic species that can be caught at the surface, day or night, and is commonly associated with *Sargassum*.

***Ommastrephes pteropus*** Steenstrup (Orange-back squid): Large, streamlined, thickly muscled body with broad angular fins. Strong neck folds present. The locking apparatus on the funnel is a deeply grooved, inverted T-shape. Tentacular club with a wrist section (carpus) consisting of a single row of 3-5 alternating, small suckers and knobs; hand section (manus) with 2 longitudinal rows of small suckers along the outer margins and 2 median rows of large suckers with prominent sharp teeth. An oval patch of very tightly packed photophores lies just under the skin on the anterior end of the dorsal surface of the mantle. Similar cream-colored, individual photophores, like small grains of rice, are embedded in the muscle of the mantle ventrally. Animals are a brick-red to a deep maroon color; mantle length to 40 cm. An oceanic species living primarily in the upper few hundred meters. This powerful swimmer is often attracted in great schools to the night light of a ship, where it can be dip-netted. Little is known of its biology, but it is assumed to lay the

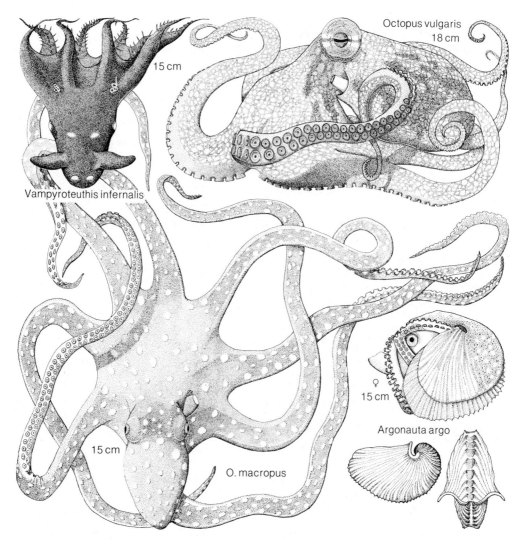

## 164 VAMPYROMORPHA, OCTOPODA (Octopuses)

*Plate 164*

O. **VAMPYROMORPHA** (Vampire squids): Coleoidea with thin, uncalcified shell, and 10 appendages of which 8 are circumoral arms with unstalked suckers and paired cirri, and 2 are retractable coiled filaments embedded in the extensive interbrachial membrane. With 1 pair of reduced, dorsal fins. Radula well developed; median rachidian with 1 denticle. (Only species of the order:)

*Vampyroteuthis infernalis* Chun (Vampire squid): Body plump, to 28 cm (including arms), dark purple-black. Eyes large, deep red. large, gelatinous, sausage-shaped egg masses, containing thousands of eggs, that have been observed on occasion offshore.

Highly differentiated photophores on and near the eyes, and near fins. Bathypelagic (300-3,000 m); rare.

O. **OCTOPODA** (Octopuses): Coleoidea with reduced, vestigial, "cartilaginous" shell and 8 circumoral appendages (arms) with unstalked suckers. Fins absent, or 1 pair of separated, paddle-shaped fins. Radula absent, or rachidian with 1 large median projection and 2 or more small lateral cusps; 1st and 2nd lateral teeth multicuspid. Branchial canal present between down-folded filaments. Digestive gland a single undivided structure with the pancreas incorporated. (5 spp. from Bda.)

*Octopus vulgaris* Cuvier (= *O. rugosus* Verrill, 1902; Peile, 1926; Robson 1929) (Common octopus; B Rock scuttle): Genus with firm, muscular body; fins on mantle and cirri on arms lacking. Hectocotylus generally external on tip of uncoiled right 3rd arm.—Species with arms not conspicuously long and slender, but stout; 1st pair always shortest; 2nd and 3rd pairs longer. Body chunky, with variably colors, commonly mottled brown, white and tan. Outer half of gills with 7-11 lamellae. Hectocotylized tip very small on the shortened 3rd right arm (ligula index less than 2.5). Maximum mantle length about 18.5 cm, total length about 1.3 m. From less than 1 m depth to 200 m; prefers a rocky, rubble, or reef habitat. Life expectancy is 1-2 yr. Males generally die after mating and females die after their eggs have hatched. Spawning generally takes place III-X. The eggs, about 3 mm long, are laid in holes in the rocks, as many as 150,000 in long strands and clusters. The mother broods the eggs, without feeding, for the 30-50 day incubation period.

*O. macropus* Risso (= *O. bermudensis* Hoyle, 1885; 1886; Verrill & Bush, 1900; Peile, 1926; *O. chromatus* Heilprin, 1888) (White-spotted octopus; B Grass scuttle): Genus as above.—Species with very long arms, 1st pair always the largest and usually the longest and the stoutest or co-equally the stoutest with the 2nd pair. Body stout; mantle lengths reach about 15 cm but the long arms give a total length of at least 1.5 m. Animal covered with large, conspicuous white spots over mantle, head and arms against a blue-green to olive background color. When disturbed, animals flush a brick-red and spots become even more prominent. Outer half of gills with 9-13 lamellae. Hectocotylized tip of 3rd right arm very long and slender (ligula index up to 14). Depth range narrow, from less than 1 m to about 17 m. The Bermudian common name implies a grass bed habitat in the flats behind the reefs and in enclosed water with grassy bottoms, but it apparently has been recorded from rocky, sandy and grassy bottoms. Relatively little is known of the biology; spawning probably occurs in winter and early spring. The eggs are extremely small and hatchlings are planktonic before settling to the bottom.

*Argonauta argo* L. (Common paper nautilus, Greater argonaut): ♀ relatively large (to 30 cm), with dorsal arms greatly expanded into glandular membranes that secrete and hold a delicate calcareous shell containing the eggs. This secondary shell is spiral, transversely wrinkled and unchambered. ♂ small (to 1.5 cm), shell-less; its relatively long (to 3 cm) hectocotylus separates at maturation and actively enters the ♀ mantle cavity at copulation. A pelagic, surface-living species that is occasionally washed ashore.

C. F. E. ROPER

# SUPERPHYLUM LOPHOPHORATA (=Tentaculata)

CHARACTERISTICS: *Solitary or colonial free-living BILATERIA with a circular or horseshoe-shaped food-catching organ (lophophore); body and coelom more or less in 3 divisions. Mostly sessile, and with some form of protective covering (tube, shell or external skeleton).*

This assemblage consists of 3 phyla (or classes according to EMIG 1982) whose relationship to each other and position within Bilateria continue to be debated (NIELSEN 1977). All 3 are represented in Bermuda: Bryozoa (p. 500), Phoronida (p. 516) and Brachiopoda (p. 518).

## Phylum Bryozoa (Sea mats, moss animals)

CHARACTERISTICS: *Small, mostly colonial, mostly sessile LOPHOPHORATA; body covered by a cuticle or permanently fastened in an exoskeleton, digestive tract U-shaped.* Although individuals (zooids) are usually less than 1 mm in size, some colonies may reach 0.5 m or more in diameter (encrusting forms) or height (bushy forms). They are colorless-transparent, or shades of yellow to red to purple. Bushy forms are soft-flaccid, flexible or rigid; encrusting forms are generally hard and brittle. The lophophore can be rapidly retracted when disturbed.

Many systematists consider Entoprocta (here treated as a subphylum) a separate phylum within the lower worms. Of some 4,000 extant bryozoan species, 87 are known from Bermuda, of which 45 are included here.

OCCURRENCE: Mostly marine and in shelf waters (rarely into oceanic depths); from tropical to polar seas. In Bermuda, they occur underneath corals, rocks, shells and larger algae, both inshore and on various reefs—outer or ledge flats, lagoon and barrier reefs—from the surface down to at least as far as examined (50 m). Many colonies add calcareous encrustations to reef frameworks or inshore rocks, but bryozoan fragments in Bermuda seldom survive abrasion to appear as recognizable grains in the loose sands accumulating across the platform.

As rapid colonizers and vigorous growers, bryozoans (together with other fouling organisms) cover ships' hulls, causing considerable loss in efficiency.

Collecting is best done by SCUBA, snorkeling or wading; some can also be obtained by dredge. Corals, rocks and shells should be turned over, and their undersides closely inspected for inconspicuous colonies.

IDENTIFICATION: Colony form is so variable that it is seldom diagnostic; consequently, attention must be concentrated upon microscopic (20-100×) inspection of the individual zooids. A wide variety of externally visible features of the zooecia suffice to distinguish Bermuda species; for many comparative studies, oriented thin sections will often be necessary as well.

Bryozoan colonies tend to be small (a few millimeters or centimeters) or inconspicuous, their form varying from encrusting thread-like networks to thin encrusting sheets, tabular or nodular masses, tuft-like clumps, and twig- or frond-like branches. They are made up of individuals (zooids). The typical, polyp-shaped feeding zooid (autozooid) is encased in an exoskeleton (zooecium) into which it can completely retract. Zooecia are box- or tube-like, chitinous or calcareous (then made of calcite or occasionally aragonite), and each bears a large, usually distal opening (aperture) through which the zooid's tentacles are extended. The bulk of the zooid, proximal to or below the tentacles, is sac-like and has a thin soft body wall lining the interior surface of the surrounding zooecium; muscle strands and digestive tract are suspended within the internal (visceral) cavity. Each zooid possesses a rudimentary nerve ganglion just below the lophophore, but generally lacks respiratory, circulatory and excretory organs. The zooecial aperture, although sometimes flush with the colony surface, is often surrounded by a raised rim (peristome) or a series of spines (apertural spines). Some peristomes are interrupted by a small notch (pseudorimule). A chitinous lid (operculum), movably hinged on 2 tiny lateral prongs (cardelles), closes off the aperture in many species. The proximal edge of the aperture frequently is deflected into a slit-like notch (sinus), or may be indented by a complex spine (lyrule).

Bryozoans, especially Cheilostomida, have elaborated tremendously upon the basic morphology of the zooecium, particularly with regard to the mechanism that protrudes the lophophore. In principle, this protrusion is effected by a raise in the coelomic fluid pressure, which in turn results from contraction of musculature. In Anascina, the exposed frontal surface contains a flexible membrane that can be bowed inward by muscles, thus increasing coelomic pressure. Various modifications of this basic plan can be seen as attempts to reduce the vulnerability of the frontal membrane. In Malacostega, the entire frontal wall is uncalcified and membranous; in some the proximal end is calcified as a narrow frontal shelf (gymnocyst) adjacent to the frontal opening (opesium), which appears as a large gaping hole after decay of the soft frontal membrane. Parallel and interior to the frontal membrane is often (e.g., in Coilostega) another shelf-like calcareous plate (cryptocyst); muscle strands extending vertically pass through round holes (opesiules) or lateral notches (opesiular indentations) puncturing the distal part of a cryptocyst. In other forms (e.g., *Beania*), the frontal membrane is over-arched by calcareous spines or ribs arising from the lateral margins, and sometimes partially fusing in the middle (e.g., Cribrimorphina) to form a perforated frontal shield (pericyst). Instead of this, some (e.g., *Scrupocellaria*) have a broad flabellate spine forming a shield-like protection (scutum) above the frontal membrane. In contrast to all the foregoing, the frontal wall in Ascophorina is well calcified, with a hydrostatic-pressure soft sac (ascus) internally below it. If smooth, imperforate, and not secondarily thickened, such a frontal wall is termed a holocyst. Where secondarily thickened, pores (pseudopores, because in life they are plugged with soft tissue) perforate the frontal wall, sometimes over the entire frontal surface (tremocyst wall), and other times only around the lateral periphery (pleurocyst wall, surrounded by marginal or areolar pores). In most species, the intrazooecial cavity is empty after death and decay of the soft parts, but in a few (e.g., Stenolaemata) is crossed transversely by calcareous plate-like partitions (diaphragms).

Many colonies are polymorphic; besides autozooids there are a number of other zooids (heterozooids) that are highly modified for various functions. Small bird's-head-like avicularia (with jaw-like "mandibles") and bristle-like vibracula protect and clean colony surfaces. Gonozooids are large hollow chambers for sheltering developing larvae (other species produce similar brood chambers, ovicells, as attachments upon otherwise unmodified autozooecia; ovicells can be endozooecial, i.e., inside the distal end of the zooecium, or hyperstomial, i.e., up on the frontal surface of the next-distal zooecium). Other heterozooids (kenozooids) take the shape of much reduced, hollow rootlets (rhizoids or radicles), encrusting threads (stolons) and space-filling chambers.

Sizes given in the illustrations usually refer to the largest dimension of a single zooid.

Most systematic work has been done with dried or bleached (in Clorox) colonies, and thus emphasizes hard-part characters; however, zooids' soft parts can sometimes be relaxed by gradual application of nembutal, and then preserved in formalin or alcohol.

BIOLOGY: Bryozoan zooids are mostly hermaphroditic, although a few species are dioecious (either as zooids or as entire colonies). Gametes generally pass into the zooid's visceral cavity, and thence out into the surrounding water via minute pores in or near the tentacles. Fertilization occurs during discharge of the ova; another zooid's spermatozoa are wafted by water currents to the first zooid's lophophore. Brooding species produce few eggs (1-5) per breeding season, whereas the number is considerably higher in non-brooding

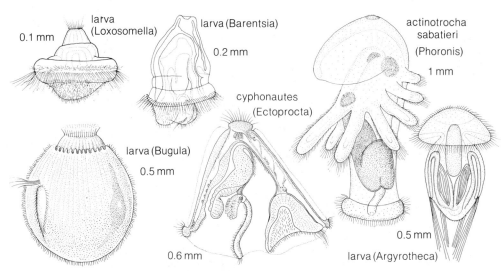

**165 LOPHOPHORATA: Development**

species. Colonies may live for several years. Bryozoans as a group tolerate moderate variations in environmental conditions; they prefer well-oxygenated, clear to moderately turbid, moderately agitated to still waters. As filter or suspension feeders, they capture small phytoplanktonic algae, bacteria, and organic detritus. In turn, bryozoans are eaten to only a minor extent by echinoids, nudibranchs and pycnogonids, and are occasionally parasitized by sporozoans and nematodes.

### Plate 165

DEVELOPMENT: After fertilization, the bryozoan zygote develops rapidly into a minute, round, ciliated, trochophore-like larva. In most species, developing larvae pass into ovicells within the parent colony where they are brooded for some time before being released (e.g., larva of *Bugula*). Many non-brooding species (e.g., *Membranipora*) have a triangular, laterally compressed feeding larva (cyphonautes). Upon release, the larva swims freely as a plankter for about a day (brooding species) to several months (cyphonautes), then settles onto a firm substrate where it metamorphoses into a typical polyp-like first zooid (ancestrula). Repeated asexual budding then produces many polyp-like zooids interconnected as a single colony. Individual zooids may undergo several degeneration-regeneration cycles, each ending in the production of "brown bodies" (excretory products?).

REFERENCES: Comprehensive systematic surveys (BASSLER 1953) and summaries (HYMAN 1951, 1959; RYLAND 1970; CUFFEY 1977) provide in-depth introductions to bryozoans. Several symposia illustrate the range among current bryozoan studies (ANNOSCIA 1968; LARWOOD 1973; POUYET 1975; WOOLLACOTT & ZIMMER 1977). Bryozoan classification followed here combines recent arrangements (BASSLER 1953; CUFFEY 1973a; RYLAND 1982) with emphasis—particularly at family level—upon local representatives.

Bermuda bryozoan ecology and systematics have been elucidated very recently (CUFFEY 1973b; RÜTZLER 1968); previous useful references include VERRILL (1900d), CANU & BASSLER (1928) and OSBURN (1940).

### Plate 166

**S.PH.  ENTOPROCTA:** Soft-bodied Bryozoa with pseudocoelomate visceral cavity; anus opening in-

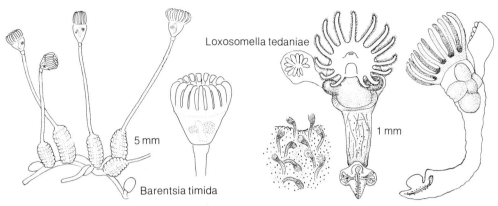

## 166 ENTOPROCTA (Entoprocts)

side the ring of tentacles. Body wall membranous, zooid consisting of a basal stalk surmounted by a cup-like calyx bearing tentacles.

F. **LOXOSOMATIDAE:** Solitary Entoprocta, with base of stalk adhering to external surface of other invertebrates. (2 spp. from Bda.)

*Loxosomella tedaniae* Rützler: Calyx higher than wide. Stomach broad, with lateral pouches. Buds arising near disto-lateral margin of stomach. Sparse to abundant at a few localities inshore; mostly on fire sponge (*Tedania ignis*).

F. **PEDICELLINIDAE:** Colonial Entoprocta with relatively isolated autozooids arising from prostrate encrusting stolons rather than from a small basal plate. (2 spp. from Bda.)

*Barentsia timida* Verrill: Basal portion of stalk greatly enlarged or widened. Very rare; at a few localities in coastal land-locked ponds (and possibly also inshore waters), on bedrock and soft-bodied invertebrates.

S.PH. **ECTOPROCTA:** Colonial Bryozoa with eucoelomate visceral cavity; anus opening outside but near the ring of tentacles. Usually with a calcareous or chitinous exoskeletal zooecium; zooid consisting of a soft-bodied trunk anchored within the zooecium and surmounted by a crown of retractable tentacles. (Includes 3 classes of which 1 (Phylactolaemata) is only represented in fresh water.)

CL. **GYMNOLAEMATA:** Ectoprocta with box-like calcareous or chitinous zooecia lacking large intrazooecial cross partitions. Colonies not regionated; zooids polymorphic. Tentacles arranged in a circular plan. (Contains 2 orders, both represented in Bda.)

### Plate 167

O. **CTENOSTOMIDA:** Gymnolaemata with inconspicuous delicate colonies; zooecia relatively

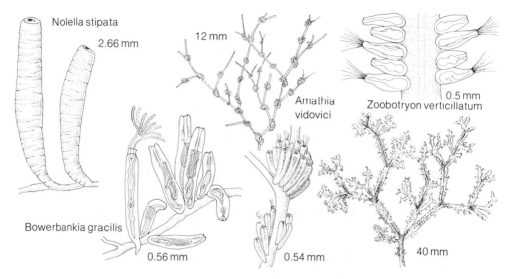

## 167 CTENOSTOMIDA (Ectoprocts 1)

isolated or loosely bundled, with thin chitinous walls; zooecial aperture same diameter as zooecium below and closed by a pleated collar rather than an operculum; no distinct ovicells. (Of 2 suborders, the Carnosina have not been recorded in Bda.)

**S.O. STOLONIFERINA:** Ctenostomida with zooids arising from stolons, which may be either delicate and encrusting (to boring), or sturdy and erect.

**F. NOLELLIDAE:** Stoloniferina with unusually long, conspicuous zooecia arising from delicate, thin, thread-like, creeping stolons. (1 sp. from Bda.)

*Nolella stipata* Gosse: Zooecia erect, elongately cylindrical, arranged singly and closely spaced along stolon, and with opaque walls covered with tiny bits of debris. Rare; at a few inshore and landlocked pond sites, on rocks and algae.

**F. VESICULARIIDAE:** Stoloniferina with small, inconspicuous, transparent zooecia attached to highly varied, transparent stolons. (3 spp. from Bda.)

*Bowerbankia gracilis* Leidy: Stolon creeping, thin, thread-like. Zooecia cylindrical, attached singly or in pairs along the stolon. Rare; in occasional barrier reefs and coastal ponds, on shells and mangrove roots.

*Amathia vidovici* (Heller): Stolon erect, thick, somewhat stiffened, dichotomously branched, bushy. Zooecia cylindrical, arranged mostly in small clusters at the branching points, with each cluster consisting of several pairs of zooecia in a double row spiralling around the branch. Uncommon; at a few inshore and landlocked pond localities, on rocks and mangrove prop roots.

*Zoobotryon verticillatum* (Delle Chiaje): Stolon erect, very thick, soft and flaccid, branching (often trifurcately), tuft-like, with an axial strand visible inside each branch. Zooecia elongate-oval, attached in an irregular row along both sides of the stolon between branching points. At some inshore localities and in coastal ponds; on submerged branches, algae and mangrove roots. Locally often common.

O. **CHEILOSTOMIDA:** Gymnolaemata with delicate to robust, non-regionated colonies of varied shapes. Zooecia simple to complex, loosely bundled to tightly packed; zooecial walls thin and chitinous to thick and calcareous; zooecial aperture noticeably smaller in diameter than zooecium below, and closed by a hinged chitinous lid (operculum). Ovicells usually developed as a conspicuous attachment on autozooids; avicularia and/or vibracula frequently present. (Includes 3 suborders all abundantly and diversely represented in Bda.)

### Plate 168

S.O. **ANASCINA:** Cheilostomida with frontal wall of zooecium membranous or chitinous, uncalcified, flexible, open to surrounding environment

INF.O. **INOVICELLATA:** Anascina with encrusting thread-like colonies, consisting of horizontal creeping stolons from which arise erect zooecia arranged in a single series. Lower (proximal) portion of zooecium tube-like, upper (distal) part flattened; zooecial walls not or only weakly calcified. No avicularia, vibracula, or permanent ovicells. (Only family: Aeteidae; with 2 spp. from Bda.)

*Aetea sica* (Couch): Encrusting delicate network. Erect zooecium usually straight. Lower tubular part of zooecium with ring-like annulations. Upper flattened portion of zooecium with a flat membranous aperture on one side; aperture about 2.5-4× longer or higher than wide. Mostly uncommon; on coral, rock and algae, at some reef (especially outer, sometimes lagoon) localities, rarely inshore or in landlocked coastal ponds.

INF.O. **MALACOSTEGA:** Anascina with encrusting or erect colonies of varied shape; most zooecial walls calcified but frontal wall membranous. Avicularia, vibracula, and ovicells sometimes present.

F. **MEMBRANIPORIDAE:** Malacostega with zooecia usually rectangular; cryptocyst (shelf-like wall, below frontal membrane) not developed to well developed; usually without spines, avicularia, or ovicells. (2 spp. from Bda.)

*Membranipora tuberculata* (Bosc): Encrusting, delicate, lacy sheet. Frontal membrane comprising most of frontal wall area; cryptocyst undeveloped. Distal corners of zooecium each with a large tubercle, so that every zooecium seems to be flanked by 4 or 6 tubercles. Abundant; on brown algae (especially floating *Sargassum*); often conspicuous in beach drift.

F. **HINCKSINIDAE:** Malacostega with zooecia rectangular to oval; cryptocyst absent to moderately developed. Spines, avicularia, and endozooecial ovicells present. (1 sp. from Bda.)

*Antropora granulifera* (Hincks): Encrusting. Cryptocyst quite broad, surrounding a large subtriangular opening (opesium). A pair of avicularia, small, pointed, directed transversely inward toward each other, placed along the distal border of the zooecium above the opesium. Sparse to abundant; at many reef (outer, lagoon and barrier) and inshore localities, on coral, rock and shells.

F. **CALLOPORIDAE:** Malacostega with shelf-like cryptocyst little developed; spines and avicularia sometimes present; ovicell hyperstomial. (4 spp. from Bda.)

*Crassimarginatella crassimarginata* (Hincks): Encrusting. Avicularium large, oval, replacing a zooecium in the budding series, and having a complete calcareous pivotal bar (so that avicularium appears theta-shaped). Ovicell small but prominent. Abundant on many outer and lagoon reefs, as well as inshore, on coral, rock and shells; especially noticeable within dark cavities in the reefs.

INF.O. **COELOSTEGA:** Anascina with encrusting or erect colonies of varied form; zooecial walls mostly calcified, but frontal wall membranous; cryptocyst well developed parallel and just interior to frontal membrane; cryptocyst perforated by small lateral openings (opesiules) or notched laterally (opesiular indentations). Avicularia, vibracula and ovicells usually present.

F. **ONYCHOCELLIDAE:** Coelostega with avicularia (onychocellaria) modified by movable mandible having membranous lateral wing-like expansions; ovicells endozooecial. (1 sp. from Bda.)

*Smittipora americana* (Canu & Bassler): Encrusting. Distal edge of cryptocyst (or proximal border of adjacent opesium opening) notched laterally by prominent opesiular indentations. Avicularia (onychocellaria) with mandible long, pointed, flanked laterally on both sides by thin membranous "wings". Usually uncommon, occasionally abundant, at some localities, mostly on outer reefs, but sometimes also on lagoon reefs and inshore; on coral rock and shells.

F. **STEGINOPORELLIDAE:** Coelostega with zooecia usually of 2 kinds (dimorphic), one being ordinary autozooecia ("A-zooecia"), the other enlarged and avicularium-like ("B-zooecia"); zooecium divided into proximal and distal chambers by a lamina descending from the cryptocyst. No true avicularia; no ovicells. (1 sp. from Bda.)

*Steginoporella magnilabris* (Busk): Encrusting. Zooecia quite large. Cryptocyst depressed

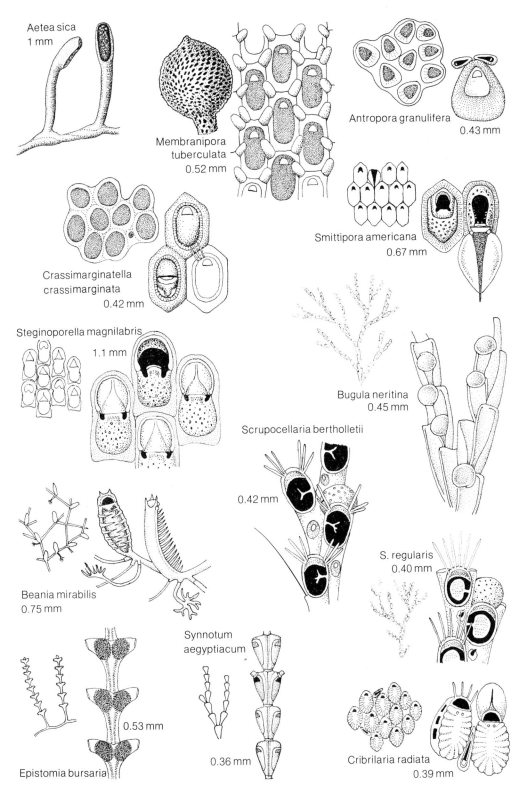

**168 ANASCINA, CRIBRIMORPHINA (Ectoprocts 2)**

below frontal membrane, perforated, with a prominent flaring lip or prong extending into the adjacent open opesium. Operculum of A-zooecium with a pair of heavy thickened chitinous ribs in an inverted-U-shaped design, that of B-zooecium similar but inverted-Y-shaped. Abundant at many reef (outer, lagoon and barrier) and inshore sites; conspicuously recognizable on corals, rocks and shells.

**INF.O. CELLULARINA:** Anascina mostly with erect tuft-like or branching colonies; zooecia usually biserially arranged; zooecial walls weakly calcified. Avicularia, vibracula and ovicells often present.

**F. BUGULIDAE:** Cellularina with erect, non-articulated (no separate joints) but flexible, chitinous zoaria; avicularia when present obviously bird's-head-like; no vibracula; no large shield-like frontal spine (scutum); ovicell hyperstomial. (3 spp. from Bda.)

*Bugula neritina* (L.): Large, bushy, flexible, red-brown tuft. Ovicell closed, white, on inner distal corner of zooecium. No avicularia. Rare to common at few localities inshore and in landlocked coastal ponds; on rocks, algae and submerged tree branches.

**F. BEANIIDAE:** Cellularina with encrusting thread-like network composed of stolon-like prolongations below erect or inclined (recumbent) zooecia; zooecia uni- or multiserially arranged, all facing in the same direction; rootlets (radicles) often present; most of frontal wall occupied by opesium. Avicularium usually borne on small stalk (pedunculate); no ovicells. (1 sp. from Bda.)

*Beania mirabilis* Johnston: Inconspicuous encrusting thread or network. Zooecia arranged uniserially. Approximately 10 large spines on each side or lateral edge of zooecium. Rare on rocks at very few inshore localities.

**F. SCRUPOCELLARIIDAE:** Cellularina with zoaria erect and articulated (with distinct flexible joints). Frontal opening (opesium) over-arched by shield-like or branched spine (scutum). (4 spp. from Bda.)

*Scrupocellaria bertholletii* (Audouin): Genus with branched colonies, branches of alternating zooids in 2 series.—Species an erect delicate tuft. Scutum large, cervicorn. Vibraculum with thin seta (bristle). Avicularium median, suboral, commonly small but may be extremely large. Sparse to abundant at many reef (outer, lagoon and barrier) and inshore localities on coral, rock, shells, algae and sponges.

*S. regularis* Osburn: Genus as above.—Species a delicate erect tuft. Scutum flat, oval, sculptured. With 2 types of avicularia: lateral avicularia with strongly hooked movable beak (mandible), and median suboral avicularia large and salient. Mostly uncommon, on coral, at some reef (outer, lagoon and barrier) localities.

F. **EPISTOMIIDAE:** Cellularina with zoaria erect tuft-like to prostrate thread-like, jointed with rootlets (rhizoids); zooecia arranged in opposite pairs (back to back) along branch; avicularia sometimes stalked (pedunculate), other times without attachment stalk (sessile). (2 spp. from Bda.)

*Epistomia bursaria* (L.): Delicate erect tuft. Oppositely paired zooecia imparting the overall appearance of a series of equilateral triangles when branch is viewed laterally. Avicularium on a long stalk, at the disto-lateral margin of the zooecium. Sparse to important, on coral and rock, at comparatively few inshore, lagoon-reef and outer-reef localities.

*Synnotum aegyptiacum* (Audouin): Delicate, tuft-like, erect to prostrate, branching only at long intervals. Zooecia long, more than 2× as long as wide, paired, resembling an acute triangle. Avicularium cylindrical, on distal edge of zooecium, alternating between right and left sides of successively adjacent pairs of zooecia; often also a stalked bulbous avicularium on frontal surface. Mostly uncommon but occasionally abundant; on many outer and lagoon reefs, as well as inshore and in landlocked coastal ponds, on coral, rock, and shells.

S.O. **CRIBRIMORPHINA:** Cheilostomida with a perforated frontal shield (pericyst; formed by partial fusion of calcareous rib-like spines) over-arching flexible frontal membrane.

F. **CRIBRILINIDAE:** Cribrimorphina with pericyst formed by 2 rows of lateral ribs (costules) fusing along median line of zooecium; shield perforated by pores or slits (lacunae); ovicell hyperstomial. (3 spp. from Bda.)

*Cribrilaria radiata* (Moll): Encrusting. Pericyst comprising most of the front of zooecium, and consisting of many laterally radiating costules (ribs) separated by small lacunae (slits) between. Abundant at many localities throughout the reef platform, outer, lagoon and barrier reefs, as well as inshore; on coral, rocks and shells. Probably the most consistently abundant and ubiquitous bryozoan species in the Bermuda reefs.

### Plate 169

S.O. **ASCOPHORINA:** Cheilostomida with frontal wall rigid and calcareous, and with flexible soft sac (ascus) located just inside or below the frontal wall.

F. **EXECHONELLIDAE:** Ascophorina with frontal shield perforated, arched, calcareous (pericyst), formed by fusion of irregular projections arising from zooecium margin. (1 sp. from Bda.)

*Exechonella antillea* (Osburn): Encrusting. Zooecia very large. Pericyst (frontal shield) perforated by large funnel-shaped pores. Sparse, on coral, at relatively few outer and lagoon reef and inshore localities.

F. **HIPPOTHOIDAE:** Ascophorina with frontal wall usually bearing transverse wrinkles of successive calcification; zooecia with no covering membrane. (2 spp. from Bda.)

*Hippothoa flagellum* Manzoni: Encrusting. Zooecia relatively isolated, connected into network by long thin tubules. Aperture prominently raised, with a distinct proximal sinus. Rare to common, on coral, rock and shells, at many inshore and reef (outer, lagoon and barrier) localities.

F. **SAVIGNYELLIDAE:** Ascophorina with zoarium erect, branched, thread-like, jointed, having only 1 zooecium per segment (between successive joints); zooecia elongated, only slightly calcified, with pores, spines and avicularia on frontal surface; ovicell hyperstomial-like ("recumbent"). (2 spp. from Bda.)

*Savignyella lafontii* (Audouin): Erect, tuft-like. Zooecium trumpet-shaped. Zooecial aperture without sinus and surrounded distally by prominent spines. Suboral avicularium conspicuously present. Uncommon; on tunicates at few inshore sites.

F. **TETRAPLARIIDAE:** Ascophorina with frontal wall a thin tremocyst; no peristome; ovicell hyperstomial; erect, tuft-like, dichotomously branched, jointed. (1 sp. from Bda.)

*Tetraplaria dichotoma* (Osburn): Zooecia in 4 rows along branch, in back-to-back pairs with each pair at right angles to the adjacent pair. Aperture semicircular, with a broad proximal sinus. Rare; on coral at few inshore, lagoon-reef, and outer-reef localities.

F. **CHEILOPORINIDAE:** Ascophorina with frontal wall a thin tremocyst; no peristome; ovicell endozooecial or absent; encrusting, sheet-like. (3 spp. from Bda.)

*Watersipora cucullata* (Busk): Zooecia large, darkly pigmented (often dark red). Frontal wall a coarsely porous tremocyst. Aperture with a broad shallow sinus and strong cardelles. Operculum brownish with paler lateral areas. Sparse; on glass bottles at few inshore sites.

F. **SMITTINIDAE:** Ascophorina with frontal wall usually perforate marginally; peristome produced and channeled; oral spines commonly present; spine or prong (lyrule) protecting entrance into compensation sac (ascus) within aperture; ovicell hyperstomial, embedded in next-distal zooecium. (8 spp. from Bda.)

*Parasmittina munita* (Hincks): Genus with well-developed lyrule and cardelles, and conspicuous areolar pores.—Species encrusting. Peristome high, vertical, with a deep proximal pseudorimule. Single or pair of falci-

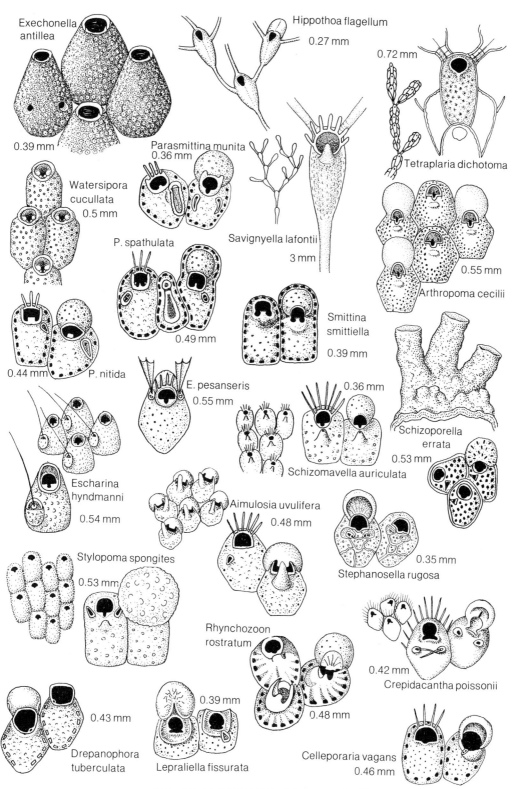

**169 ASCOPHORINA (Ectoprocts 3)**

form avicularia proximo-lateral to aperture. Sparse to abundant; on many outer and lagoon reefs as well as inshore, on coral, rock and shells.

***P. nitida*** (Verrill): Genus as above.—Species encrusting. Aperture wide, with low lateral lappet-like processes and a broad lyrule. Avicularia commonly oval and proximo-lateral to the aperture. Uncommon to important on some reefs (especially outer, but also lagoon and barrier); on coral, rock, and shells.

***P. spathulata*** (Smitt): Genus as above.—Species encrusting. Aperture elongate with a thin lyrule. Peristome forming tall lappet-like projections lateral to the aperture, and notched by a short proximal pseudorimule. Avicularia large, spatulate. Mostly sparse; on many outer and lagoon reefs, but occasionally inshore, on coral, rock, and shells.

***Smittina smittiella*** Osburn: Encrusting. Zooecium with a median suboral avicularium possessing a serrate distal border extending into the aperture; also, a wide anvil-shaped lyrule just below this extension. Mostly uncommon; on coral, rock, and shells, at many inshore and reef (outer and lagoon) localities.

F. **SCHIZOPORELLIDAE:** Ascophorina with frontal wall usually a porous tremocyst; zooecial aperture semicircular, with a distinct proximal sinus; ovicell hyperstomial. (9 spp. from Bda.)

***Arthropoma cecilii*** (Audouin): Encrusting. Zooecia large. Frontal wall evenly perforated, transparent or porcelaneous. Ovicell opaque, white. Sparse to abundant; at many outer-reef, lagoon-reef, and inshore localities, on coral, rock and shells.

***Escharina hyndmanni*** (Johnston): Genus with aperture nearly semicircular and possessing a small thin sinus on its proximal margin.—Species encrusting. A large distinct vibraculum present on a swollen chamber alongside each zooecium. Uncommon to important; on coral, rock and shell at many outer- and lagoon-reef as well as inshore sites.

***E. pesanseris*** (Smitt): Genus as above.—Species encrusting. A pair of avicularia, lateral to the aperture, which appear like a goose's foot. Ovicell small, hyperstomial-like ("recumbent"). Sparse to abundant; at many localities, mostly reef (outer and lagoon), on coral, rock and shells; occasionally inshore.

***Schizomavella auriculata*** (Hassall): Encrusting. With 6-8 long spines around the peristome. Distinct, suboral median avicularium on a prominence. Perforated, subglobose ovicell. Mostly sparse but sometimes abundant; at many, mostly reef localities (outer, lagoon and barrier), on coral, rock and shells; seldom inshore.

***Schizoporella errata*** (Waters): Encrusting sheet-like, massively nodular, or robustly tubular and branching; zoaria often quite large and intensely colored (dark red-brown or pink-purple).

Aperture with U-shaped sinus. Avicularium triangular, lateral to the aperture, on a prominence but becoming flush and then embedded with increasing calcification. Ovicell globular, porous. Common and highly conspicuous inshore, on rocks, shells, metal buoys, wooden rafts, and glass bottles, often forming massive layers; also sparse on coral in reefs (lagoon only). Many occurrences of this species have been recorded as *S. unicornis* Johnston instead.

***Stylopoma spongites*** (Pallas): Encrusting sheets, often extensive and brightly orange or pinkish colored. Aperture with a deep, V-shaped sinus. Frontal wall a coarse tremocyst. Single or paired avicularia lateral to the aperture. Huge, spherical ovicell completely covering aperture of the zooecium. Abundant and ubiquitous inshore, on shells and rocks; also important on coral in many outer, lagoon and barrier reefs.

F. **HIPPOPORINIDAE:** Ascophorina with frontal wall a smooth holocyst or a granular pleurocyst; zooecial aperture horseshoe-shaped, without a sinus; ovicell hyperstomial. (3 spp. from Bda.)

***Almulosia uvulifera*** (Osburn): Encrusting, with colony presenting a spiny appearance owing to each zooecium having a long or high-projecting umbo (central part of frontal wall, just below or proximal to aperture). Younger-stage zooecia bearing oral spines and marginal areolar pores. Ovicell broad, with a distinct proximal tongue-like prong (labellum) extending into aperture. Uncommon to important at many inshore and reef (outer, lagoon and barrier) localities; on coral, rock, shell and algae.

***Stephanosella rugosa*** Osburn: Encrusting. Zooecia small, with salient ridges on the frontal wall. One or a pair of small suboral avicularia on the median side of mammillary-shaped processes. Ovicell prominent, with radiating grooves on its flattened frontal surface. Sparse to abundant on many outer and lagoon reefs, uncommon inshore; on coral, rock and shells.

F. **CREPIDACANTHIDAE:** Ascophorina with long oral (apertural) spines, and also spines around margin of frontal wall; paired avicularia; zooecial aperture with strong cardelles; ovicell hyperstomial. (2 spp. from Bda.)

***Crepidacantha poissonii*** (Audouin): Encrusting. Aperture with a convex proximal margin. A pair of vibracula, with setae directed inwards, located on small prominences proximo-lateral to the aperture. Ovicell small, with a transverse ridge and median keel on median uncalcified portion of ovicell. Uncommon to abundant, at many inshore and reef (outer, lagoon and barrier) localities; on coral, rock and shells.

F. **SERTELLIDAE:** Ascophorina with frontal wall flanked by row of small marginal areolar pores (a pleurocyst); peristome prominent, extended, tube-like, with fissure at proximal edge; ovicell hyperstomial, deeply immersed in distal part of zooecium. Colonies most often erect lattice-like (fenestrated) fronds, but not so in Bermudian species, all of which are encrusting. (3 spp. from Bda.)

*Drepanophora tuberculata* (Osburn): Encrusting. Tooth arising from proximo-lateral margin of aperture, prominent, curving inward. Ovicell with an opening on each side at the base. Sparse, on coral, rock, shells and algae, both inshore and on some lagoon and outer reefs.

*Lepraliella fissurata* (Canu & Bassler): Encrusting. Peristome flaring, U-shaped, ending in a distal spine at each corner. Ovicell globose and with a median fissure. Sparse to abundant at many localities, both inshore and reef (outer and lagoon); on coral, rock and shells.

*Rhynchozoon rostratum* (Busk): Encrusting; sheet-like to sometimes large multilamellar gray colonies. Avicularium on a large, bulbous avicularium chamber lying proximal to the aperture, with the avicularium becoming embedded on the inner side of the peristome as calcification increases. Ovicell with a semicircular opaque frontal area. Abundant and ubiquitous inshore and on lagoon (also occasionally outer) reefs; on coral, rock, shells, algae and glass bottles.

F. **CELLEPORARIIDAE:** Ascophorina with zooecia usually heaped upon each other and unoriented except at the growing edge; avicularia of various shapes and sizes; ovicell hyperstomial or hyperstomial-like ("recumbent"). (6 spp. from Bda.)

*Celleporaria vagans* (Busk): Encrusting. Frontal wall smooth, with marginal areolar pores. Aperture large, semicircular, with a small proximal sinus. Avicularium large, on a prominence proximo-lateral to the aperture and facing medially; avicularia often large and spatulate. Ovicell hood-like. Sparse to abundant on coral, rock and shells, at relatively few outer-reef and inshore localities.

CL. **STENOLAEMATA:** Ectoprocta with tube-like zooecia, mostly calcareous-walled, often crossed internally by large calcareous partitions (diaphragms), and with colonies sometimes divisible into distinct axial and peripheral regions; zooid tentacles arranged in a circular plan.

### Plate 170

O. **CYCLOSTOMIDA** (=TUBULIPORATA, STENOSTOMATA): Stenolaemata with long, slender, tubular zooecia. Colonies delicate, of relatively isolated to tightly packed or fused zooecial tubes; colonies mostly not regionated. Zooecial walls thin, calcareous, highly porous; intrazooecial space usually empty, lacking diaphragms; no operculum; zooecial aperture of same diameter as tubular zooecium below. Ovicells devel-

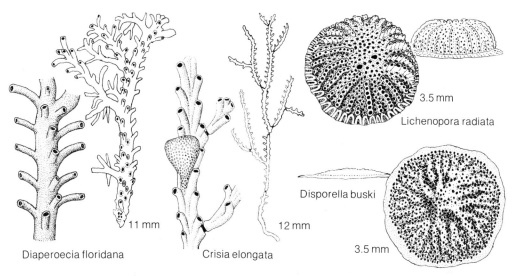

**170 CYCLOSTOMIDA (Ectoprocts 4)**

oping from modification of entire specialized zooids (gonozooids); not producing resistant resting bodies, but often displaying embryonic fission (polyembryony). (Includes several extinct and living suborders, of which 3 occur in Bda.)

S.O. **TUBULIPORINA:** Cyclostomida with encrusting or erect, thread-like or branching or sheet-like colonies; when branched, rigid and lacking flexible joints; ovicell a large, irregularly lobate, spreading gonozooid overgrowing and thus appearing pierced by adjacent autozooecia.

F. **DIAPEROECIIDAE:** Tubuliporina with thinly branching, encrusting or erect colonies; zooecial apertures commonly arranged in transverse or oblique rows across branch width and on only 1 surface of branch. (1 sp. from Bda.)

*Diaperoecia floridana* Osburn: Moderately delicate, sprawling, prostrate, or encrusting branches. Zooecial apertures opening on lateral and frontal surfaces of branch, and arranged irregularly or in vaguely transverse oblique rows (but not obviously bundled together). Peristomes long, tubular. Reverse side longitudinally striated and transversely wrinkled; sometimes held elevated above substrate by column-like props (radicles). Ovicell, when present, on front of branch, low and pierced by zooecial peristomes. Generally sparse at some localities, mostly reef (especially outer, also lagoon and barrier reefs), occasionally inshore; usually on coral, rock and shells; some previous records are as *Idmonea atlantica*.

S.O. **ARTICULINA:** Cyclostomida with erect tuft-like colonies; branches consisting of bundles of rigid calcareous zooecia separated by small flexible chitinous joints; ovicell a large swollen sac-like gonozooid inserted among the autozooecia.

F. **CRISIIDAE:** Articulina with rigid branch segments consisting of several serially arranged zooecia, rather than of pairs of zooecia placed side by side. (1 sp. from Bda.)

*Crisia elongata* Milne Edwards: Very delicate, erect tuft. Zooecial apertures opening along lateral margins of branch; 1-29 (averaging 10) zooecia between successive joints. Peristomes quite short, but bent sharply frontward; no spines. Joints clear-colored to black. Ovicell highly inflated (especially toward distal or upper end), opening through transverse slit flush with ovicell surface; ovicell on front surface of branch. Usually sparse, at some localities, mostly reef (especially outer reefs, some lagoon and barrier reefs) but occasionally inshore; generally on coral, rock and shells; frequently recorded by previous workers as *C. eburnea* or *C. denticulata*.

S.O. **RECTANGULINA:** Cyclostomida with encrusting disc- or wart-like colony; zooecia tubular, radiating outward and upward from colony center, arranged in distinct radial series (fascicles), separated from laterally adjacent series by tiny coelomic chambers (kenozooecia or "alveoli"). Ovicell formed from zooecium covered over by alveoli, often in upper center of colony. (Only family: Lichenoporidae, with 2 spp. from Bda.)

*Lichenopora radiata* (Audouin): Fairly thick (averaging about 1 mm high) disc, often light purple or lavender-colored when living. Central or top area of colony wide, occupied by tiny alveoli (and sometimes ovicell below), but not zooecia. Zooecia closely spaced, in very regular, uniserial, radial rows down lateral surface of colony. Zooecial aperture often protected partly by hood-like cover. Abundant at many localities, especially reef (outer, lagoon and barrier reefs), but also inshore; mostly on coral, rock, and shells.

*Disporella buski* (Harmer): Flattened disc, averaging under 0.5 mm high, usually white when alive. Central or top area of colony rather small; ovicell there covered by low irregular ridges rather than by well-defined alveoli. Radiating rows of widely spaced zooecia uniserial, but less regular. Upper margin of zooecial aperture often extended into 2 or 3 apertural spines. Sparse at some localities, occasionally inshore but mostly on outer and lagoon reefs; on coral, rock, shells and algae.

R. J. CUFFEY & S. S. FONDA

## Phylum Phoronida
(Horseshoe worms)

CHARACTERISTICS: *Worm-shaped, solitary, tube-dwelling LOPHOPHORATA with a horseshoe-shaped lophophore embracing the terminal mouth opening.* To 20 cm (rarely more) in length (but only a few millimeters in diameter); colorless-transparent, or with white, yellow or red pigment. Movement is

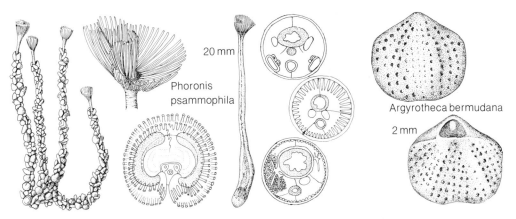

**171 PHORONIDA (Horseshoe worms), BRACHIOPODA (Lamp shells)**

limited to emergence of the anterior end from the tube.

Of only 10 species in 2 genera, 1 is known from Bermuda.

OCCURRENCE: Exclusively marine and benthic, from the intertidal to 390 m. Tubes may occur singly, vertically embedded in soft sediment, or form tangled masses; some species burrow in, or encrust, calcareous shells and rock, and 1 is associated with the tubes of Ceriantharia. Distribution is worldwide.

IDENTIFICATION: A low-power microscope is needed for live observation; species identification may require histological sectioning. To remove animals from their chitinous, rigid, sand-encrusted tube, the latter has to be broken, or split open with fine-point tweezers.

Note the shape of the lophophore and the position of mouth (inside the tentacular crown) and anus (outside between the recurving parts). The nephridia (which also act as gonoducts) open next to the anus. The posterior part of the body is swollen to form a bulb.

Fix, after anesthetization (to prevent shedding of the lophophore!), in ethanol or formaldehyde; for histology in Bouin's.

BIOLOGY: Most are dioecious, some hermaphroditic. Fertilization is internal. The life-span is probably 1 yr. All species can reproduce asexually by transverse fission or budding; autotomy of the lophophore, with subsequent regeneration, is common. Phoronids are suspension feeders, gathering small particles by way of tentacular ciliary currents. Known predators are fishes, gastropods and nematodes; gregarines and trematodes are known to live as parasites in the coelom.

*Plate 165*

DEVELOPMENT: The embryo is either expelled at an early stage, or brooded in the concavity of the lophophore. The larva, called actinotrocha, is ciliated, and has a pelagic life of 10-20 days before it settles and rapidly (within 10-30 min) metamorphoses.

Actinotrochae are not uncommonly found in plankton tows.

REFERENCES: For a general introduction see HYMAN (1959) and the little book (with an extensive bibliography) by EMIG (1979); the latter author has also given monographic accounts of systematics (1974), ecology (1973) and biology (1982).

There are no published records of species from Bermuda.

## Plate 171

***Phoronis psammophila*** Cori: Genus without epidermal collar fold below the lophophore. —Species with a horseshoe-shaped lophophore of up to 130 tentacles. Nephridia with 1 coelomic funnel; 1 (left) giant nerve fiber. To 190 mm when extended (15-40 mm contracted), flesh-colored with white spots on tentacles and base of lophophore. Dioecious; ♀ broods embryos in a single lophophoral mass. Larva is "Actinotrocha sabatieri" (with up to 12 tentacles and 3 blood masses). On sandy-muddy subtidal bottoms of inshore waters (e.g., *Thalassia* beds); locally abundant.

C. C. EMIG

## Phylum Brachiopoda (Lamp shells)

CHARACTERISTICS: *Solitary, mostly sedentary LOPHOPHORATA with body enclosed in a shell consisting of a dorsal and a ventral valve. With or without a stalk.* Shells range from 0.5 to 80 mm in size and are mostly yellowish gray; some are glassy-transparent or show bright colors.

Of 260 recent species (there are some 30,000 extinct species!) in 2 classes, the Inarticulata (Ecardines) are not known from Bermuda, and the Articulata (Testicardines) with only 1 species.

OCCURRENCE: Exclusively marine and benthic, from the shallow subtidal to the abyss. They are usually attached to hard substrates, often in clusters; few live in sediment burrows, or lie on the bottom unattached.

Collect *Argyrotheca* by examining the underside of reef corals.

IDENTIFICATION: Use a low-power microscope for observation of live animals and internal anatomy.

The bilaterally symmetrical shell is calcareous in Articulata and consists of a smaller dorsal and a larger ventral valve, the latter with an opening for the stalk (pedicle). The valves can be variously sculptured and, in *Argyrotheca*, are incompletely pierced by tubules which appear as dots on the inner surface (punctate type). In Articulata, valves are hinged together posteriorly by a tooth-and-socket arrangement. Other internal features used in classification are mostly associated with the shape of the tentacular apparatus (lophophore) and its supporting skeleton; the latter can appear as simple prongs (crura), or as a complex 3-dimensional structure.

Preserve in 75% alcohol.

BIOLOGY: Sexes are separate in most species; *Argyrotheca*, however, is probably hermaphroditic. Sex cells are shed into the coelom and discharged into the open water by way of the nephridia; in some species (probably also in *A. bermudana*) brooding up to the larval stage takes place. Brachiopods are suspension feeders, producing a ciliary current through the shell and trapping small plankton (especially diatoms) in their lophophore. They can be parasitized by gregarines and copepods, and preyed upon by fishes.

## Plate 165

DEVELOPMENT: Although the hatching larva of Inarticulata resembles an adult, that of *Argyrotheca* appears mushroom-shaped, with long larval setae. Larval development of most species is still poorly known.

REFERENCES: For an introduction to the voluminous literature see MOORE (1965).
Recent species of *Argyrotheca* are dealt with by DALL (1911a, b; 1921), and LOGAN (1975) gives observations on the ecology of Bermuda's species.

## Plate 171

**CL. ARTICULATA:** Brachiopoda with calcareous valves hinged posteriorly by a tooth-and-socket arrangement. Pedicle opening through the ventral valve. Lophophore with internal skeleton; anus absent.

**O. TEREBRATULIDA:** Articulata with punctate shell and simple lophophoral skeleton. Shell commonly smooth, teardrop-shaped.

*Argyrotheca bermudana* Dall (=*Cistella cistellula* of VERRILL 1900d): Rarely exceeds 2 mm in width. Shell cream in color, with irregular, non-divaricate, pink-red bands particularly prominent near the anterior margin; growth lines progressively crowded around the anterior margin. Valves with fairly short, subtruncate, and small deltidial plates (elements bordering the pedicle base); crura widely separate. Lophophoral support (or loop) relatively long, formed by 2 descending branches joined anteriorly on the median septum. Lophophore large, schizolophous (indented anteromedially). Pedicle well developed. On offshore and nearshore reefs and in caves, on the underside of reef-building corals (particularly *Montastrea* and *Agaricia*), often in clusters. Common but easily overlooked.

C. C. EMIG

# Phylum Chaetognatha
(Arrow-worms)

CHARACTERISTICS: *Small arrow-shaped BILATERIA with tripartite body consisting of a head with paired chitinous (!) hooks, a trunk and a tail, the 2 latter carrying rigid horizontal fins.* Length 3-30 (rarely 100) mm; colorless and often extremely transparent (especially trunk and tail).

Morphologically very uniform, Chaetognatha are one of the most isolated taxa within the animal kingdom, with possible relationships to groups as diverse as Nematoda, Annelida, Echinodermata and Chordata. Of about 65 described species, 20 have been recorded from the Bermuda region; 9 are included here.

OCCURRENCE: Exclusively marine and—with the exception of 1 benthic genus not recorded from Bermuda as yet—pelagic; in all oceans, sometimes in enormous numbers. Several species are cosmopolitan. Many species are confined to specific depth zones; some are indicators for coastal waters, whereas the majority are found in the open ocean. Diurnal vertical migrations are common.

Collect with large plankton nets.

IDENTIFICATION: Live specimens can be observed under a dissecting microscope; for species identification, preserved, unstained (!) specimens should be transferred to a flat dish or depression slide and examined under a microscope.

Note the head provided with 1 pair of eyes, and bearing curved grasping hooks and 1 or 2 rows of teeth under a retractable fold of skin. The epithelium (with a cuticle) can be thickened into a spongy "collarette". A ciliated chemoreceptive organ (corona) is situated in the neck region. The trunk (containing the ovaries and the ventral ganglion) is separated from the tail (containing the testes) by a transverse septum. Respiratory, circulatory and secretory organs are lacking. For species identification, the number and shape of lateral fins, of hooks and teeth, and the shape of testes, seminal vesicles and ovaries are important.

Fix in 5% buffered formaldehyde-seawater.

BIOLOGY: Protandric hermaphrodites (♂ germ cells ripen before ♀ germ cells); fertilization by copulation. Eggs develop in brood pouches or are shed into the ocean. Life-span may be 1 yr. Chaetognaths are passive drifters, but capable of rapid darting movements by means of a sudden stretching of the bent body. They are predators, catching Copepoda, Euphausiacea and fish larvae (of up to their own size!) in their powerful hooks. Parasites are rare, and include Rhizopoda, Nematoda, Trematoda, Cestoda and Copepoda.

DEVELOPMENT: Direct; the hatching juvenile is of near-adult shape.

REFERENCES: For general reference see ALVARINO (1965, 1969).
Bermuda's Chaetognatha have been recorded by MOORE (1949) and MORRIS (1975); for distribution in the Gulf Stream and the Sargasso Sea see GRICE & HART (1962).

## Plate 172

*Eukrohnia fowleri* Ritter-Zahony: Genus with 1 pair of lateral fins, and 1 set of teeth.—Species with 10-13 strong, gently curved hooks; with short ovaries with large seminal receptacle; seminal vesicles oval, close to end of lateral fins. Body strong, opaque; to 40 mm. A fairly rare deep-water species; 600 m and below.

*Krohnitta subtilis* (Grassi): Genus with 1 pair of lateral fins, and 1 set of teeth.—Species with 6-9 wide, transparent hooks; with short ovaries; seminal vesicles inconspicuous, touching both the lateral and caudal fin. Lateral fin extremely delicate (usually destroyed in captured specimens!). Body slender, flabby; to 16 mm. Fairly common; 0-500 m.

*Pterosagitta draco* (Krohn): The only species of this genus. With 1 pair of lateral fins, anteriorly connected with the head by means of a spongy collarette. With 2 sets of teeth, and up to 10 strong, curved hooks. Ovaries long; seminal vesicles elongated, touching the end of the lateral fins. Body firm, opaque; to 10 mm. Fairly common; 0-300 m.

*Sagitta lyra* (Krohn): Genus with 2 pairs of lateral fins and 2 sets of teeth.—Species with furcated tail fin; lateral fins faintly connected. Ovaries long; seminal vesicles oval, close to end of lateral fins. Body flaccid; to 38 mm. A common cosmopolitan; 0-600 m.

*S. minima* Grassi: Genus as above.—Species without collarette; anterior fins slightly shorter than posterior fins. Body slender, translucent to opaque, flaccid; to 10 mm. Fairly common; 0-300 m.

*S. serratodentata* Krohn: Genus as above.—Species with well-developed collarette in neck region; posterior fins large, roundish. Interior edge of hooks delicately serrated. Ovaries long; seminal vesicles horn-shaped and close to the posterior fins. Body strong, slender, opaque; to 13 mm. One of the most common species; 0-500 m.

*S. hispida* Conant: Genus as above.—Species wih well-developed collarette; posterior fins much longer than anterior fins. Ovaries long, eggs in 2 (sometimes 3) rows; seminal vesicles oval, close to end of

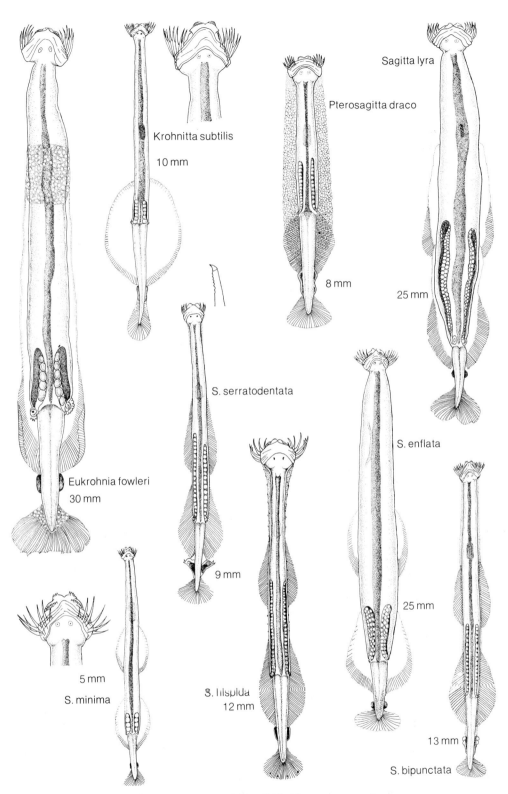

**172 CHAETOGNATHA (Arrowworms)**

lateral fins. Body strong, opaque, covered with bundles of sensory buds; to 12 mm.

***S. enflata*** Grassi: Genus as above.—Species without collarette; anterior fins short and narrow. Ovaries fairly short, eggs small and in 3 rows; seminal vesicles spherical, touching the tail fin. Body wide, flaccid, transparent; to 25 mm. Common; 0-300 m.

***S. bipunctata*** Quoy & Gaimard: Genus as above.—Species with well-developed collarette; posterior fins wider and longer than anterior fins. Ovaries long; seminal vesicles pear-shaped and close to the tail fin. Body rigid, opaque; to 18 mm. Very common; 0-200 m.

W. STERRER

## Phylum Echinodermata

CHARACTERISTICS: *Secondarily radially symmetrical (pentamerous) BILATERIA with a dermal skeleton composed of numerous ossicles of calcium carbonate in the form of calcite. Part of the coelom is converted into a system of fluid-filled vessels (water-vascular system), some of which project through the body wall as tube feet and tentacles, and assist in feeding, locomotion and respiration.*

Most echinoderms are small, a few centimeters in diameter, but some attain relatively large sizes (1 m or more). All are exclusively marine, and occur in all seas, from the intertidal to the deepest trenches. Echinoderms play an important role in the general ecology of the ocean as reworkers of sediments, scavengers, burrowers into rocky shores and predators upon commercially useful invertebrates. Their easily obtained larvae have been the subject for a great variety of embryological and biochemical studies. Gonads of echinoids and body walls of holothurians are an important food item in many parts of the world.

The 5-fold symmetry is usually conspicuous in adults, but may be often obscured in some groups, especially the holothurians, where a bilateral symmetry is usually evident. Larval stages are conspicuously bilaterally symmetrical. Three basic body forms are displayed. In one form the body is star-shaped, with arms radiating in 1 plane from a central disc. The mouth is at the center of the disc and the anus, when present, lies opposite the mouth, on the other side of the disc. The 5 arms carry the 5 main branches (radial vessels) of the water vascular system. In the 2nd form, the body is more or less spherical or cylindrical, with mouth and anus at opposite poles. There are no arms, and the 5 radial vessels lie inside the body wall. The 3rd form is similar to the 1st, except that the disc is replaced by a more or less spherical calyx, and mouth and anus lie on the upper surface of the calyx. On the basis of these differences, some recent authors refer the echinoderms to 3 subphyla. Subphylum Asterozoa includes the star-shaped classes Asteroidea and Ophiuroidea. Subphylum Echinozoa includes the spherical to cylindrical classes Echinoidea and Holothuroidea, and subphylum Crinozoa includes the class Crinoidea.

In addition to about 6,000 living species, approximately 13,000 fossil species are known. The phylum includes 5 living classes: the familiar starfish (Asteroidea, p. 523), brittle stars (Ophiuroidea, p. 527), sea urchins (Echinoidea, p. 531) and sea cucumbers (Holothuroidea, p. 537). The class Crinoidea, which includes the shallow- to deep-water feather stars and the usually deep-water sea lilies, has only once been reported from Bermuda (VERRILL 1907b, p. 285) and is not considered here.

Earlier reports on the echinoderm fauna of Bermuda (e.g., VERRILL 1900c, 1907b; CLARK 1901) were updated by CLARK (1942); a recent review of the phylum in Bermuda is by PAWSON & DEVANEY (in press).

## Class *Asteroidea* (Starfish, sea stars)

CHARACTERISTICS: *Star-shaped ECHINODERMATA with 5 or more broad arms arising from a poorly defined central disc. Digestive and reproductive systems and other organs extend into arms. Tube feet carried in distinct grooves (ambulacral furrows) on the oral surface. Madreporite (sieve plate) located on aboral surface.* With a diameter of 10-25 cm (some species to almost 1 m!), most starfish are conspicuous in the field, often brightly colored, but some live concealed under rocks or buried in sand. Movement is a usually slow gliding over the substrate by means of tube feet.

Bermuda has a surprisingly small fauna of starfish; only 7 out of approximately 1,800 living species occur in Bermuda; all are included here.

OCCURRENCE: On hard and soft bottoms to several hundred meters' depth. *Coscinasterias tenuispina* and *Linckia guildingii* are commonly found on reefs, and the former species also occurs in rocky localities and in mangroves. *Luidia clathrata,* buried in sandy areas, and *Asterina folium,* on the undersides of intertidal rocks, were abundant in VERRILL'S time but are now apparently extremely rare.

Collect in shallow water by turning over rocks; in deep water by means of a dredge.

IDENTIFICATION: Note general shape of the animal; in some the central disc is small and the arms are long, whereas in others the disc and arms are virtually indistinguishable, and the body may be essentially pentagonal. Count the arms, and determine whether or not they are of unequal size. The skeleton may be massive, especially in larger specimens, making them quite heavy and rigid, or it may be less well developed, making the body reasonably flexible.

The body is covered in a skeleton of calcareous plates (ossicles) that may assume a variety of forms. The oral surface carries a central mouth, from which radiate ambulacral furrows containing tube feet. The feet are in 2 or 4 rows in each furrow; they may carry terminal sucking discs or are simply pointed. The edge of the body is often defined by 2 series of conspicuous marginal ossicles. The upper (aboral) surface usually has a covering of ossicles that may be overlapping (imbricating), or in the form of a network (reticulate), or closely fitted together like a mosaic (tessellate); in some forms the aboral ossicles bear paxillae, little columns that carry spines or granules on their upper surfaces. Aboral plates may carry spines or short rounded projections (tubercles). Respiratory organs (papulae) project through small pores on the aboral surface, and papular pores may be scattered or grouped together in special areas. Pedicellariae, small pincer-like structures, assume a variety of forms; they may be stalked or sessile, the latter sometimes reminiscent of the valves of a clam. Pedicellariae may be present or absent on the oral and aboral surfaces.

Fix for anatomical purposes with 75% alcohol, for histology with 4% buffered formalin or Bouin's. For identification specimens can be safely dried after preservation in either alcohol or buffered formalin. If necessary, soft parts can be reconstituted by immersing specimens in detergent for several hours.

BIOLOGY: Most asteroids are dioecious, and sexes are usually indistinguishable externally. Eggs and sperms are liberated into the seawater, where fertilization occurs. Life expectancy is up to 7 yr. Regeneration of several arms is common; some species even reproduce asexually by transverse fission and subsequent regeneration. Asteroids can feed in a variety of ways. Some ingest large quantities of mud and extract organic material from it. Others are scavengers, feeding on detrital material. Some multi-armed forms are suspen-

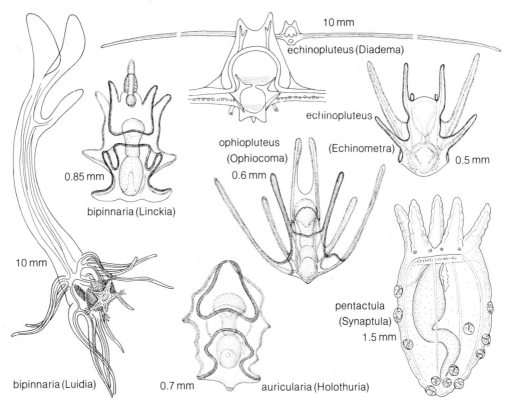

**173 ECHINODERMATA (Echinoderms): Development**

sion feeders, capturing small organisms in a net formed by their arms and tube feet. Numerous species are active predators, feeding on mollusks especially, and constituting a threat to commercial oyster and clam fisheries in many parts of the world. Asteroids are usually active, especially the predators, and display complex behavior patterns. They can often reach high population densities locally (though not in Bermuda). Predators include fishes, gastropods, and other asteroids.

The next stage is a brachiolaria, which uses its terminal adhesive discs to adhere to the substrate, while the rudimentary starfish develops at its anterior end. The starfish eventually breaks free from the brachiolaria and settles to the sea floor as a juvenile. Many species of starfish, especially in cold waters, brood young in more or less specialized ways, either near the mouth, or in pouches on the aboral surface. In these cases the larval stages are modified or omitted, and the young starfish crawls away from its mother.

*Plate 173*

DEVELOPMENT: The planktonic gastrula develops eventually into a bipinnaria larva that has a pair of sinuous ciliated bands.

REFERENCES: For general orientation see NICHOLS (1969) and JACOBSEN & EMERSON (1977); for modern classification consult SPENCER & WRIGHT (1966).

Bermudian asteroids are discussed in CLARK (1942) and PAWSON & DEVANEY (in press).

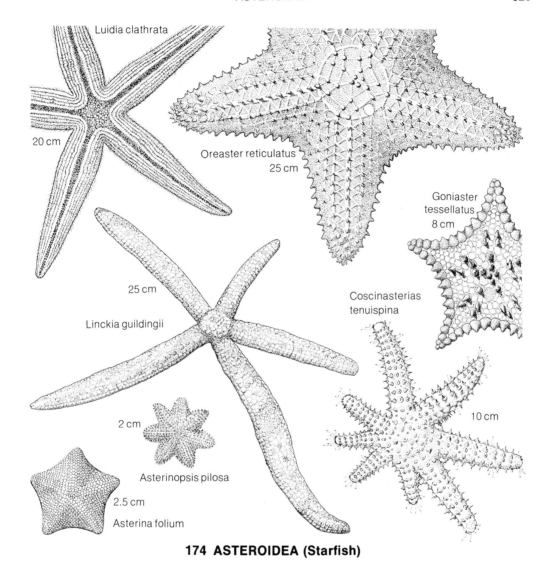

**174 ASTEROIDEA (Starfish)**

### Plate 174

**O. PLATYASTERIDA:** Asteroidea with 5 to many arms. Marginal ossicles inconspicuous. Ossicles on oral surface in regular transverse series. Tube feet without suckers. (1 living family:)

**F. LUIDIIDAE:** Platyasterida with strap-shaped arms; aboral surface covered in paxillae that are not in completely regular rows. (1 sp. from Bda.)

*Luidia clathrata* (Say): Disc and arms flattened. Upper surface covered with close-set paxillae. Color gray to pale salmon. Diameter to 200 mm. Inshore, buried in clean sand; very rare.

**O. VALVATIDA:** Asteroidea with usually 5 arms. Two rows of marginal ossicles. Upper surface of body tessellate or imbricate or paxillate. Pedicellariae valvate when present, with bases sunken into ossicles. Tube feet with suckers.

F. **OREASTERIDAE:** Valvatida with large, generally high and swollen disc with robust arms or none; marginals large; aboral skeleton reticulate, composed of stellate plates, in many forms bearing stout spines; papulae numerous in special areas. (1 sp. from Bda.)

*Oreaster reticulatus* (L.): Disc high, with numerous robust blunt tubercles; arms short and wide. Adult specimens brownish to bright orange-red. A massive species, reaching a diameter of 500 mm. In shallow water; very rare. (Color Plate 14.13.)

F. **OPHIDIASTERIDAE:** Valvatida with small disc; arms long and slender, more or less cylindrical; body normally covered by granulose membrane. Aboral skeleton tessellate. (1 sp. from Bda.)

*Linckia guildingii* Gray: Usually asymmetric, with some arms longer than others, owing to asexual reproduction by transverse fission. Diameter to 200 mm. Adults light brown. Under rocks and in crevices in reef areas; common.

F. **GONIASTERIDAE:** Pentagonal to narrowly stellate Valvatida, generally with large disc. Marginals conspicuous. Plates on oral and aboral surfaces in close contact; aboral plates flat, tabulate or paxilliform. (1 sp. from Bda.)

*Goniaster tessellatus* (Lam.): Pentagonal, with short arms. Conical spines on upper surface of fully grown specimens; spines not usually developed in specimens less than 20 mm diameter. Upper surface light red to orange; lower surface whitish. Diameter to 115 mm. In deep water (90 m), on sand. Not uncommonly brought up in dredges. (Color Plate 14.11.)

O. **SPINULOSIDA:** Asteroidea with usually 5 arms. Marginal ossicles usually absent. Aboral surface of body with reticulate or imbricating ossicles, or with no skeleton. No pedicellariae. Tube feet with suckers.

F. **ASTERINIDAE:** Spinulosida normally with minute marginal plates. Aboral skeleton composed of overlapping plates bearing grouped or single spinelets. (1 sp. from Bda.)

*Asterina folium* (Lütken): Body pentagonal, flattened, arms very short. Diameter to 25 mm. Color white to yellow or olive green or blue. In shallow water attached to undersides of rocks or pieces of coral; once very common in Bermuda, now practically absent.

*Asterinopsis pilosa* (Perrier): Body not flattened, with 6 or 7 short, broad arms. Upper and lower surfaces with clusters of minute spines. Diameter to 25 mm. Color light orange to white. On sand in deep water (90 m), occasionally in shallow water (1 m) in wave-exposed bays. (Color Plate 14.14.)

O. **FORCIPULATIDA:** Asteroidea with 5 to many arms. Marginal ossicles inconspicuous or absent. No paxillae. Upper surface of body with reticulate ossicles. Pedicellariae present. Tube feet with suckers.

F. **ASTERIIDAE:** Forcipulatida with swollen body, with small disc. Tube feet in 4 rows. (1 sp. from Bda.)

*Coscinasterias tenuispina* (Lam.): Arms elongate and spinous. Typically with 7 arms, but owing to habit of asexual reproduction by transverse fission specimens with 4, 6, 8 or 9 arms are common, and some arms are usually larger than others. Upper surface yellow-brown with darker brown or bluish to violet markings. Diameter to 220 mm; most specimens 100 mm or less. In many shallow-water habitats, on rocks, sand and mangrove roots; common. (Color Plate 14.12.)

D. L. PAWSON

## Class Ophiuroidea (Brittle stars)

CHARACTERISTICS: *Star-shaped ECHINODERMATA in which a central circular disc is sharply delimited from usually 5 long, slender arms. Digestive system, gonads and some other organ systems not extending into essentially solid arms, which are composed of ambulacral ossicles fused to form vertebrae. Tube feet not in furrows. Anus is absent. Madreporite located on oral surface of disc.* The largest Bermudian species have arms to 150 mm long. Ophiuroids are of stiff, brittle consistency, breaking readily when handled. Many are extremely colorful, although Bermudian species are mostly in shades of cream, brown, gray and black. In the field, brittle stars are usually secretive, concealing themselves in crevices or under rocks but crawling rapidly when disturbed, with pulling and "rowing" motions of their arms.

Of approximately 2,000 living species, 19 have been recorded from Bermuda; 10 are described here.

OCCURRENCE: On hard and soft bottoms to several hundred meters depth. Most species remain concealed under rocks or in crevices during the day, and emerge at night to feed. Others live buried in sand or sandy mud.

Collecting can be done intertidally or with SCUBA by overturning rocks. Species that live in interstices of corals can be obtained by immersing pieces of coral in water of lowered salinity, fresh water, or seawater containing a small quantity of formalin. Sand-dwelling forms can be collected by sieving sand collected below low-tide level.

IDENTIFICATION: Use a dissecting microscope.

Note whether or not the disc is covered by overlapping scales or by soft skin. In some forms the scales carry small spines or are overlain by closely aggregated small granules. There are usually 5 arms, although some species have 6. The arms may be capable of coiling vertically, thus rendering them prehensile (many Phrynophiurida), or, more usually, they are capable of moving only in the horizontal plane. They are either inserted laterally into the disc or give the impression of being inserted ventrally. Arms taper gradually from the base to the distal extremity, or reach their greatest width some distance from the base. The arms carry few to many short to long spines on lateral surfaces of each segment. These spines may lie on the surface of each arm segment (appressed), or they may project outwards (erect). Pores through which the tube feet emerge may be protected by 1 or more tentacle scales. The mouth is surrounded by a complex system of oral ossicles useful in classification. Five interradially placed jaws carry oral papillae around their margin; papillae are either in continuous rows or separated by gaps (diastemas). Inner edges of jaws carry teeth that run into the mouth. The base of each jaw is covered by a conspicuous oral shield. Genital slits lie on each side of arm bases; slits are usually single, sometimes double.

Fix for anatomical purposes with 75% alcohol, for histology with 4% buffered formalin or Bouin's. For identification and permanent preservation specimens can be safely dried after fixation in either alcohol or formalin.

BIOLOGY: Most ophiuroids are dioecious, but a small number of hermaphroditic species are known. Sexes are usually indistinguishable externally, except in a few sexually dimorphic species. Eggs and sperm are generally shed into the seawater, where fertilization occurs. Regeneration of lost arms is common. Life expectancy is 1-5 yr. Ophiuroids are usually selective detritus feeders; some are suspension feeders, capturing small organisms with tube feet on their upraised arms. In some areas, ophiuroids can be extremely numerous, often reaching densities of thousands of individuals per square meter. They are often found in association with other animals, living in large numbers in the cavities of sponges or crawling on the branches of sessile coelenterates. Predators include several types of fishes and starfishes. Ophiuroids can be parasitized by Protozoa, Mesozoa, Polychaeta and Gastropoda.

## Plate 173

DEVELOPMENT: The gastrula develops into a chracteristic planktonic larva (ophiopluteus), which superficially resembles the echinopluteus of echinoids except that it has fewer arms and is not laterally compressed. The rudimentary ophiuroid develops in the ophiopluteus and eventually breaks away, falling to the sea floor as a juvenile brittle star. Numerous species of ophiuroids either brood their young or liberate yolky eggs that eventually form nonfeeding vitellaria larvae.

REFERENCES: For general orientation see NICHOLS (1969) and SPENCER & WRIGHT (1966).
   Bermuda's ophiuroids have been treated briefly by CLARK (1942), and more recently by PAWSON & DEVANEY (in press).

## Plate 175

O. **PHRYNOPHIURIDA:** Ophiuroidea with disc and arms covered by skin. Upper arm plates absent or rudimentary. Arms either coiling vertically or not.

F. **OPHIOMYXIDAE:** Phrynophiurida with arms not coiling vertically. (1 sp. from Bda.)

*Ophiomyxa flaccida* (Say): Color highly variable, ranging from green to reddish to yellow, often variegated. Disc soft and slimy. Disc to 25 mm across, arms 100 mm long. Under rocks in shallow water; common. (Color Plate 14.4.)

F. **GORGONOCEPHALIDAE:** Phrynophiurida with arms simple or branching, coiling vertically, prehensile. (1 sp. from Bda.)

*Asteroporpa annulata* (Örsted & Lütken): Arms simple, long, coiling. Disc and arms with alternating white and dark brown transverse stripes. Occurs in deep water in excess of 30 m, where it coils its arms around branches of gorgonians (e.g., *Ellisella*); not uncommon. (Color Plate 14.5.)

O. **OPHIURIDA:** Ophiuroidea with disc and arms completely invested by ossicles. Arms not coiling, capable of essentially lateral movements only.

F. **OPHIURIDAE:** Ophiurida with disc provided with conspicuous naked scales on oral and aboral surfaces. No tooth-papillae at jaw apices; an unpaired infradental papilla at apex of each jaw. (1 sp. from Bda.)

*Ophiolepis paucispina* (Say): Arms inserted laterally and fused with the disc. Arm spines short, more or less appressed to arm. Disc gray,

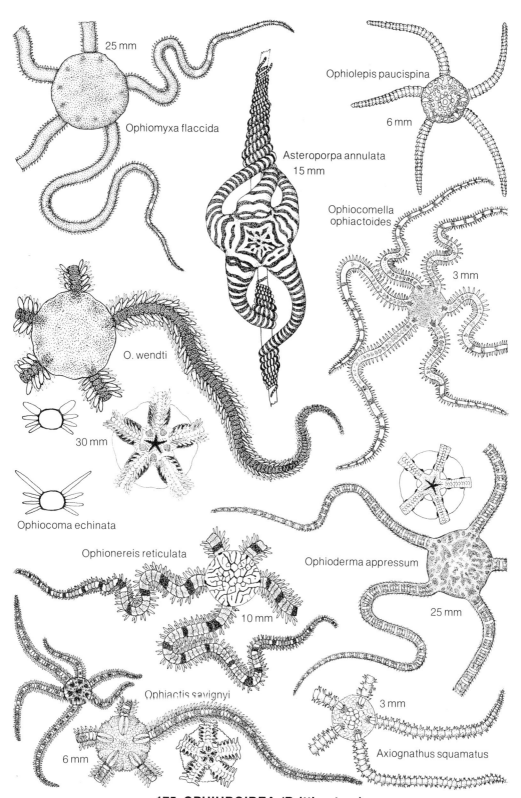

175 OPHIUROIDEA (Brittle stars)

arms often banded with brown; lower surface cream. Disc 6 mm across, arms 15 mm long. Secretive, on underside of rocks or in sand under rocks in shallow water; uncommon.

F. **OPHIOCOMIDAE:** Ophiurida with stout arms, widest at a point some distance from base. Arm spines long, erect. Spiniform tooth-papillae form cluster at apex of each jaw. (4 spp. from Bda.)

*Ophiocoma echinata* (Lam.): Genus with upper surface of disc granulate.—Species with 2 spines on each side of first arm segment, and 2 tentacle scales. Uppermost arm spine conspicuously thicker than others, widest near base of spine. Black or dark brown; no red coloration. Disc 30 mm across, arms 150 mm long. On sand under rocks in shallow water. Fertile VII-VIII. Common.

*O. wendti* (Müller & Troschel): Genus as above.—Species with 3 spines on each side of 1st arm segment, and 3 tentacle scales. Uppermost arm spine long and narrow, without bulging appearance. Black or deep brown on upper surface; lower surface rust-red. Disc 30 mm across, arms 150 mm long. On sand under rocks in shallow water. Common.

*Ophiocomella ophiactoides* (Clark): Six arms. Color green and white. Disc 3 mm across, arms 10 mm long. Under rocks and in crevices in shallow water. Common.

F. **OPHIONEREIDIDAE:** Ophiurida with naked disc scales. Free margins of jaw with continuous series of papillae. Ventral keel on midline of each ventral arm plate. (1 sp. from Bda.)

*Ophionereis reticulata* (Say): Arms inserted ventrally and not fused with disc. Arm spines moderately long, at angle to arm. Disc flat, with no granules or spines. Arms whitish or pale yellow, with conspicuous narrow bands of brown or blackish. Light-colored disc marked with characteristic network of dark lines. Disc 10 mm across, arms 75 mm long. Under rocks on sand in shallow water; common. (Color Plate 14.6.)

F. **OPHIODERMATIDAE:** Ophiuroida with disc scales of upper and lower surfaces covered with granules. An unpaired infradental papilla at apex of each jaw; no tooth-papillae at jaw apices. (2 spp. from Bda.)

*Ophioderma appressum* (Say): Arms widest at base, gently tapering, inserted laterally and firmly fused with disc. Arm spines appressed to arms, 8-10, the lowest obviously widest and longest. Four genital slits in each inter-arm area. Mottled gray or brown, arms often banded with lighter and darker shades. Disc 25 mm across, arms 125 mm long. Under rocks near low water mark; common. (Color Plate 14.3.)

F. **OPHIACTIDAE:** Ophiurida with naked scales on disc frequently carrying spines. Free margins of jaw do not carry continuous series of oral papillae; a diastema separates lateral oral papillae from infradental papillae. (2 spp. from Bda.)

*Ophiactis savignyi* (Müller & Troschel): Arms slender. Disc with scales and small spines. Color variegated green and white. Young specimens almost always with 6 arms, usually 3 larger and 3 smaller. Reproduces asexually by transverse fission. Disc 5 mm across, arms 25 mm long. Secretive, living in crevices and among rocks and coralline algae in shallow water. Common.

F. **AMPHIURIDAE:** Ophiurida with disc generally provided with conspicuous naked scales; spines on scales in some forms. No tooth-papillae at jaw apices; paired infradental papillae at apex of each jaw. (4 spp. from Bda.)

***Axiognathus squamatus*** (Delle Chiaje): Arms inserted ventrally into disc. Disc covered with fine scales. Arm spines short. Color white. Disc 3 mm across, arms 12 mm long. Secretive, in crevices and under rocks; uncommon.

D. L. PAWSON

## Class Echinoidea (Sea urchins and sand dollars)

CHARACTERISTICS: *Spherical to disc-shaped ECHINODERMATA whose skeletal plates, usually joined to form a rigid test, bear independently movable spines and other external appendages. Five radial areas (ambulacra) with plates perforated for passage of tube feet.* With a diameter of over 50 cm (spines included), *Diadema* is an impressive faunal element; most other species are in the 10-20 cm range. The regular urchins are usually dark in color (except for *Tripneustes* and some white-spined varieties of *Diadema*), whereas the irregular urchins are coffee-brown to off-white. Locomotion is slow, effected by tube feet or movement of spines. Classification is based upon combinations of characters, but for convenience in the field echinoids can be subdivided into "regular" urchins (with obvious radial symmetry, a central mouth on the underside and a central anus on the upper surface) and "irregular" forms, in which there is a more or less pronounced bilateral symmetry, with mouth central or anteriorly placed, and with anus posterior.

Of more than 800 known living species, 15 have been recorded from Bermuda; 9 are included here.

OCCURRENCE: On hard and soft bottoms to several hundred meters in depth. Regular urchins mostly on hard bottoms, or on surface of sand, irregular urchins usually buried in sand. Some species (*Lytechinus*) may occur locally in very large numbers; others (*Tripneustes*) can be found in clumps.

Collecting is best done by SCUBA; on sediment bottoms also with a dredge. Some irregular species can be collected by running fingers and toes through shallow fine sand (carefully; tests are often thin and easily crushed!).

The extremely sharp spines of *Diadema* crumble after penetration into the skin and can inflict painful wounds that should be disinfected. The gonads of some urchin species are edible but are not generally eaten in Bermuda. Sea urchin larvae are extensively used in embryology and molecular biology.

IDENTIFICATION: The external characters of live specimens are usually sufficient for determining Bermudian species.

For finer details of spine and tube feet arrangement, scrape parts of test with a knife (each tube foot is associated with 2 holes, and each spine with a single tubercle). Note the type of symmetry, as well as the presence or absence of the complex Aristotle's lantern (chewing apparatus) that controls 5 hard teeth. Identify long primary spines, usually arranged in more or less regular vertical rows on larger tubercles. Smaller secondary spines are often grouped around the primaries. The irregular urchins generally have very short spines, some of which may be modified to form flattened clavulae, which are arranged into special tracts called fascioles; the ciliated clavulae produce water currents. Further external appendages are pincer-like pedicellariae, some of which are equipped with poison glands, and the less conspicuous sphaeridia (organs of equilibrium).

In regular urchins, the test is composed of 10 vertical (meridional) rows of plates; 5 double rows of ambulacral plates (provided with spines and tube feet) alternate with 5 double rows of larger interambulacral plates (with spines only). At the anal area, each double row of interambulacral plates converges upon a single genital plate, and each double row of ambulacral plates converges upon a single ocular plate. The ring of plates (5 genitals alternating with 5 oculars) thus formed is the apical system, and encloses the periproct that carries the anus. Each genital plate carries a single genital pore; further, 1 genital plate carries the madreporite (a porous area connected to the water-vascular system). Orally, there are no specialized plates; the circumoral area (peristome) is leathery or covered with small plates and is provided with 5 pairs of bushy gills and 5 pairs of short, thick buccal podia. In irregular urchins, the test is more difficult to analyze for plates are often fused, and a pronounced (tertiary) bilateral symmetry is superimposed upon the radial symmetry. The anus is always situated outside the apical system of plates. Tube feet are absent or reduced around the circumference of the body; those in the aboral ambulacral areas are arranged in patterns resembling petals of a flower. Here the tube feet are lamellar and function as gills, whereas oral ambulacral areas have feet modified for obtaining food particles. The short spines give the body a furry texture. For detailed taxonomic study, relative proportions of the body, arrangement of certain plates and structure of pedicellariae are important.

Dimensions in figures refer to the largest diameter of the test without spines.

Fix for anatomical purposes with 75% alcohol, for histology with 4% buffered formalin or Bouin's (inject fixative also into specimen through peristome). Tests for dry storage can be treated with household bleach (the spines fall off!); in order to obtain specimens with spines in their natural position, fix them (in alcohol) before drying.

BIOLOGY: All echinoids are dioecious but the majority display no external sexual dimorphism. Sperm and eggs are shed into the seawater where fertilization takes place. Brooding occurs in some (mostly cold-water) species. Life expectancy is estimated at more than 5 yr for most species. Echinoids feed on algae, sessile animals or organic debris. *Diadema* is known to venture, at nighttime, out of the reef into the turtle grass on which it feeds. Irregular urchins ingest organic particles that they extract by means of modified podia on the oral phyllodes. Some urchins (*Lytechinus*) use their tube feet to mask themselves with shell fragments, etc., possibly as a response to light. Predators are gastropods, crabs, starfish and fishes (surgeonfish, porcupine fish, wrasses); parasites are common (Ciliata: *Biggaria, Metopus*; Turbellaria, Copepoda). Juvenile fishes sometimes take shelter between the spines of *Diadema,* and some urchins are associated with shrimps (e.g., *Gnathophylloides*).

## Plate 173

DEVELOPMENT: Shortly after fertilization the blastula becomes ciliated and free-swimming. The gastrula develops into a typical larval stage (echinopluteus) that is laterally compressed (as opposed to the otherwise similar ophiopluteus of Ophiuroidea!) and bears 4-6 pairs of arms supported by calcareous spicules. A long unpaired process is typical for the spatangoid larva. The echinopluteus is planktonic and planktotrophic (feeding on nannoplankton). After some weeks the pluteus sinks to the bottom where it rapidly (within an hour!) metamorphoses into a juvenile urchin of no more than 1 mm.

Larvae can be regularly caught in the plankton with a medium-mesh net, or obtained by rearing in the laboratory.

REFERENCES: For general orientation see NICHOLS (1969); for a systematic overview consult MORTENSEN (1928-1951).

Bermuda's echinoids have been treated by CLARK (1942) and several earlier papers, and recently by PAWSON & DEVANEY (in press). ILIFFE & PEARSE (1982) have reported on reproductive rhythms.

## Plate 176

**S.CL. PERISCHOECHINOIDEA:** Regular Echinoidea with interambulacra of 1 to many columns of plates and ambulacra of 2-20 columns. Ambulacral plates simple, not compound. (Includes all Paleozoic echinoids; 1 surviving order.)

**O. CIDAROIDA** (Pencil urchins): Perischoechinoidea with few large primary spines arising from large primary tubercles on interambulacral plates. Ambulacral plates each with a single pair of pores. Area around mouth (peristome) heavily plated. (1 sp. from Bda.)

*Eucidaris tribuloides* (Lam.) (Slate pencil urchin): Primary spines very much larger than all other spines, stout, solid, approximately as long as diameter of test. Color light to dark brown; spines sometimes banded light and dark brown. On shallow hard bottoms, often wedged into crevices; also under rocks or in "grassy" areas. Shallow water to (rarely) 450 m. Uncommon. Fertile in summer.

**S.CL. EUECHINOIDEA:** Regular Echinoidea with interambulacra each of 2 columns of plates. Ambulacral plates generally compound. (Contains the majority of all living echinoid species.)

**SUP.O. DIADEMATACEA:** Euechinoidea with rigid or flexible test. Periproct inside apical system in all living forms. Primary tubercles perforated. Teeth with no keel. Spines usually hollow.

**O. DIADEMATOIDA:** Diadematacea with fragile, depressed hemispherical test. Primary tubercles crenulate (with conspicuous indentations around base of tubercle). Primary and secondary spines extremely long, hollow, brittle. (1 sp. from Bda.)

*Diadema antillarum* (Philippi) (Spiny urchin): Primary spines approximately 3× as long as diameter of test. Color usually black, sometimes white. Spines of juveniles may be striped. In daylight in reef cavities; often emerge at night to feed. Shallow water to 400 m. Common. Spines can cause extremely painful puncture wounds. Fertile VI-VIII. Often associated with a decapod shrimp. (Color Plate 14.1.)

**SUP.O. ECHINACEA:** Euechinoidea with rigid test; periproct inside apical system, tubercles usually not perforate. Teeth with keel. Spines solid.

**O. TEMNOPLEUROIDA:** Echinacea with test usually sculptured with ridges or depressions or both; if test not sculptured, then gill slits deep and conspicuous. Anus surrounded by several plates of varying size. (2 spp. from Bda.)

*Lytechinus variegatus* (Lam.) (Purple urchin): Anus surrounded by numerous plates; peristome heavily plated. Pore-

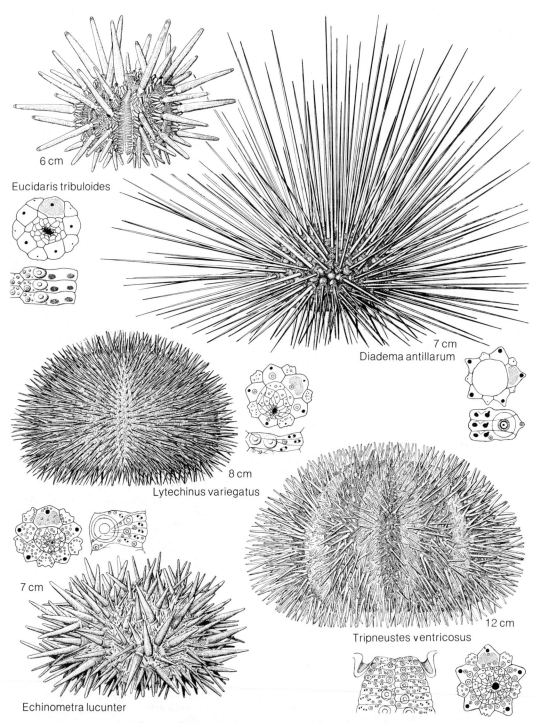

**176 CIDAROIDA—ECHINOIDA (Regular sea urchins)**

pairs in regular arcs of 3. Color deep red-violet or purple, occasionally with greenish tinge. Test diameter to 85 mm. On relatively sheltered sandy areas and among rocks in sheltered bays to depths of 100 m. Locally often abundant. Often inconspicuous owing to habit of covering upper surface of test with pieces of shell, turtle grass and algae. Fertile III-XI, probably year round.

***Tripneustes ventricosus*** (Lam.) (=*T. esculentus*) (White urchin): Anus surrounded by numerous plates; peristome not plated. Pore-pairs in 3 vertical series. Short spines white, conspicuous against dark purplish test. Test diameter to 150 mm. In shallow water with strong wave action, among rocks, or on turtle grass bottoms to depths of 30 m. Locally often abundant. Gonads ("roe") of sexually mature adults used as food in many parts of Caribbean. Fertile IV-VIII. Often associated with *Gnathophylloides mineri*.

O.   **ECHINOIDA:** Echinacea with test not sculptured; gill slits shallow, inconspicuous. Anus surrounded by several plates of varying sizes. (1 sp. from Bda.)

***Echinometra lucunter*** (L.) (Rock urchin): Test elliptical, usually conspicuously so. Spines robust, usually about 1/2 of test length. Color reddish brown. Test length to 90 mm. In the intertidal of exposed areas, embedded in special burrows in coral or sedimentary rock, to depths of 45 m; common. Fertile IV-VII.

*Plate 177*

SUP.O.   **GNATHOSTOMATA:** Euechinoidea with rigid test, periproct outside apical system. Primary tubercles usually perforate and crenulate. Spines hollow. Lantern with teeth usually present in adult. Apical system and peristome usually approximately opposite.

O.   **HOLECTYPOIDA:** Gnathostomata with hemispherical to globular or ovoid test. Ambulacra narrower than interambulacra throughout. (1 sp. from Bda.)

***Echinoneus cyclostomus*** Leske: Ambulacra not in the form of petaloids, radiating from apical system to mouth as 5 double bands. With 4 genital pores. Body elliptical. Periproct close to mouth, on same side of body as mouth. Spines very short. Test and spines light brown; tube feet dark reddish. Test length to 50 mm. Usually in sand under stones in shallow water; rare. Fertile VIII. (Color Plate 14.2.)

O.   **CLYPEASTEROIDA:** Gnathostomata with ovoid to flattened test, with ambulacra invariably as wide as or wider than interambulacra on oral surface. Ambulacra always in form of petaloids. (4 spp. from Bda.)

F.   **MELLITIDAE:** Clypeasteroida with flattened test. Five or 6 open or closed notches or holes (lunules) conspicuous in test. Four or 5 genital pores.

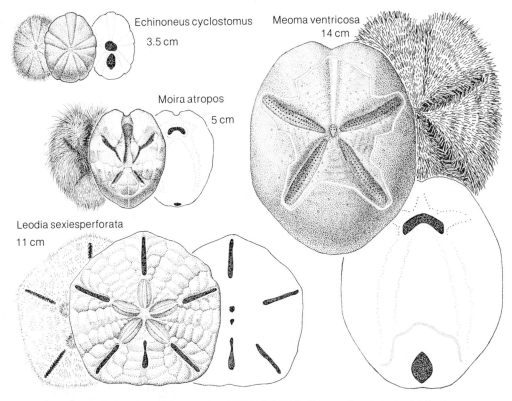

**177 HOLECTYPOIDA—SPATANGOIDA (Irregular sea urchins)**

***Leodia sexiesperforata*** (Leske) (=*Mellita sexiesperforata*) (Sand dollar): Test with 6 lunules. Four genital pores. Pale brownish. Test diameter to 130 mm. In clean sand in shallow water subject to wave action, to depths of 25 m; fairly common. Fertile IV-VII.

**SUP.O. ATELOSTOMATA:** Euechinoidea with rigid test. Periproct outside apical system. Primary tubercles usually perforate and crenulate. Primary spines hollow. Lantern absent in adult. Apical system and peristome rarely opposite.

**O. SPATANGOIDA** (Heart urchins): Atelostomata with oval test; peristome anterior, anus near posterior extremity. Petal present and well developed, often sunken into test. (4 spp. from Bda.)

**F. SCHIZASTERIDAE:** Spatangoida with peripetalous and latero-anal fasciole; no subanal fasciole. Spines generally coarse, uniform (although some have a tuft of longer spines at rear).

***Moira atropos*** (Lam.) (Heart urchin): With petals deeply sunken into upper surface of test. Test fragile. Pale brown. Test length to 55 mm. A burrower in soft mud to muddy sand, to depths of 200 m. Fairly common. Fertile summer.

F. **BRISSIDAE:** Spatangoida with peripetalous and subanal fascioles. Spines include conspicuously larger spines that are found mostly within area enclosed by fasciole.

*Meoma ventricosa* (Lam.) (Large heart urchin): Test ovoid, strong. Petals not deeply sunken. Subanal fascioles incomplete. Dark brown. Test length to 180 mm. Burrowing in coarse sand and rubble, especially in channels between patches of coral reef, to depths of 200 m. Rare. Fertile summer.

D. L. PAWSON

### Class Holothuroidea (Sea cucumbers)

CHARACTERISTICS: *Cylindrical, elongate ECHINODERMATA in which the ossicles are usually microscopic; body usually soft and leathery. Dorso-ventral axis elongated so that the animal lies on its side. Mouth terminal, surrounded by a ring of tentacles; anus at the other end of the body. Tube feet may be conspicuous, whether as 5 meridional rows or aggregated on the underside of the body (the "ventral" surface), or may be absent.* Although some species may reach 1 m in length, the majority are in the 10-40 cm range or smaller. Most Bermudian species are secretive, living under rocks or buried in sand. The predominant colors are brown, greenish or black; apodous species are usually more brightly colored. They move sluggishly; some burrow by means of podia and musculature.

There are more than 1,100 living species; 17 have been recorded from Bermuda, and 11 species are included here.

OCCURRENCE: On hard and soft bottoms from the intertidal to the deep sea; very few are pelagic. *Ocnus surinamensis* prefers a rocky substrate, where it lives in crevices. Several *Holothuria* species occur in sandy areas where they may be found on the undersides of rocks or burrowing in the sand; *Eupatinapta* also burrows in sand. The large *Isostichopus badionotus* and the small synaptid *Synaptula hydriformis* are very common and conspicuous in their preferred habitats, the latter species attaining large population densities.

Collecting is best done by hand in the intertidal, and by SCUBA in deeper water. Small burrowing species can be sieved from the sand.

IDENTIFICATION: For the common Bermudian species, external characters are usually sufficient for species identification.

Note the number and type of tentacles surrounding the mouth; these can be shield-shaped (aspidochirote), tree-like (dendrochirote) or feather-like (pinnate). Determine the arrangement of the tube feet (if any) on the body wall; they may be more or less restricted to the 5 ambulacra (radii), or concentrated in specific areas of the body (usually the ventral surface). The dorsal surface may carry more or less well-developed warts. Open the body by making a longitudinal cut through the body wall to the left or right of the mid-dorsal line. Note the conspicuous dorsal mesentery, which supports part of the intestine. Conspicuous gonads are attached to the mesentery, in the form of a single or double tuft of vesicles. The esophagus is surrounded by a firm ring of calcareous plates (calcareous ring). This structure can be relatively simple, composed of 10 or more elements, or complex, composed of a mosaic of numerous individual pieces. Near the anus, specialized caeca (Cuvierian organs) may be present in aspidochirotes.

To examine the ossicles, immerse a small piece of the body wall in liquid household bleach (sodium hypochlorite) until the tissue is dissolved. Pipette the residue into water on a microscope slide and examine under a high-power microscope. The ossicles are broadly classified into a series of simple types. "Tables" consist of a perforated disc surmounted by 2 or more erect spines or "legs". "Wheels" and "anchors" are self-explanatory. "Plates" are generally large smooth or knobbed perforated sheets of calcite. "Buttons" are essentially small plates with few perforations. "Cups" are perforated hollow hemispheres; they take a variety of shapes.

Fixation in 25% alcohol is satisfactory for general purposes. Buffered formalin is useful as a fixative for histological purposes. Any acidic fixative or preservative tends to dissolve the calcareous ossicles in the body wall, rendering positive identification extremely difficult. Specimens collected live can be relaxed with tentacles expanded in a solution of sodium sulfate (Epsom salts) or magnesium sulfate in seawater.

BIOLOGY: Most holothurians are dioecious; sexes are generally indistinguishable externally. Eggs and sperm are usually shed into the surrounding seawater where fertilization takes place. Few data are available on growth and life expectancy but it is generally estimated that holothurians may live for 5-10 yr. Holothurians are generally sluggish, remaining motionless for long periods of time. Some species feed by ingesting detritus, extracting nutrients in its passage through the intestine, and egesting waste in the form of characteristic castings. Dendrochirotids extend sticky tentacles into the water, capturing small organisms and suspended particles. Apodous forms use sticky pinnate tentacles for selective detritus feeding. Some larger holothuriids are significant reworkers of sediments. CROZIER (1918c) estimated that in Harrington Sound, Bermuda, *Isostichopus* specimens can pass between 500 and 1,000 tons of substrate through their intestines annually. They can readily regenerate lost parts, and when disturbed can eject their internal organs (autoevisceration) and grow a new set within a matter of weeks. Some species of holothuriids can eject sticky threads, Cuvierian organs, when disturbed. Holothurians have few predators; some are occasionally eaten by fishes. They have many parasites, internally (Ciliata and Tubellaria) and externally (Gastropoda, e.g., *Melanella*). The fish *Carapus* can be found hiding in the cloaca of *Actinopyga*.

**Plate 173**

DEVELOPMENT: Usually indirect, via a planktonic auricularia larva characterized by 1 sinuous ciliated locomotor band. Further development leads to a barrel-shaped doliolaria larva. Some species have direct development; a large yolky egg develops into a juvenile sea cucumber (pentactula), as in the viviparous *Synaptula*.

REFERENCES: For general orientation see NICHOLS (1969); more detailed treatment of individual species can be found in DEICHMANN (1930).
Bermuda's holothuroids are treated by CLARK (1942); CROZIER (1921a, and earlier papers) accumulated many biological observations. A recent update is by PAWSON & DEVANEY (in press).

**Plate 178**

O. **ASPIDOCHIROTIDA:** Generally large Holothuroidea (100 mm or more) with shield-shaped tentacles. Tube feet well developed, especially on lower (ventral) surface of body. Tube feet on upper surface often modified to form warts or papillae.

F. **HOLOTHURIIDAE:** Aspidochirotida with gonads as 1 tuft on left side of dorsal mesentery. Mostly tropical shallow-water forms. (8 spp. from Bda.)

*Holothuria cubana* Ludwig: Genus without teeth surrounding anus.—Species small (to 150 mm), whitish, often with 6 pairs of spots dorsally. Body wall thin and rigid. Ossicles complex tables and knobbed buttons. In sand in shallow water; uncommon.

*H. parvula* (Selenka): Genus as above.—Species small (60-70 mm), yellowish brown, with large yellow tentacles. Body wall thick, gelatinous,

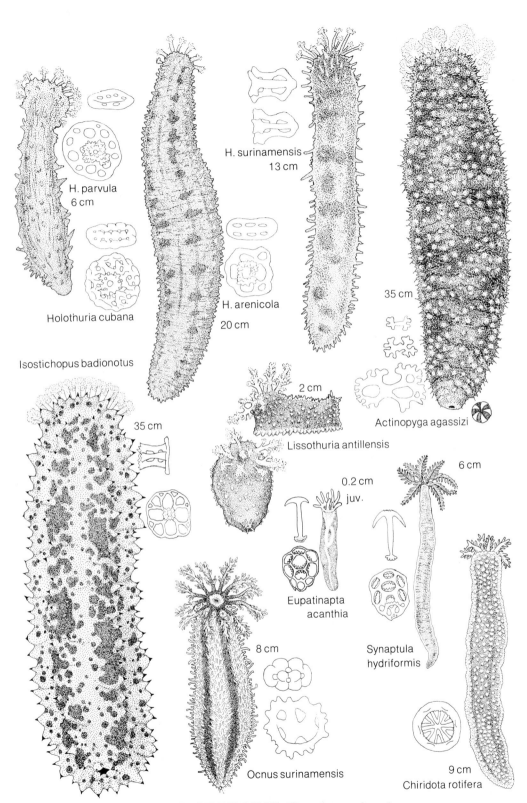

## 178 HOLOTHUROIDEA (Sea cucumbers)

containing a green pigment. Tube feet on ventral surface form a kind of sole. Ossicles tables and smooth buttons. On underside of rocks in shallow water; common. When disturbed can exude white sticky Cuvierian organs from posterior end.

***H. arenicola*** Semper: Genus as above.—Species medium-sized (250 mm), slender, with very small tentacles. Color highly variable, but often light to dark brown with double row of large darker patches dorsally. Ossicles tables, often imcomplete, and smooth buttons. An active burrower that lives deeply buried under rocks in sand or muddy sand; common.

***H. surinamensis*** Ludwig: Genus as above.—Species medium-sized (150 mm), with tube feet scattered over body surface; no pronounced distinction between dorsal and ventral surface. Skin thin, flexible, rough to touch. Color light brown to purplish, tentacles whitish. Ossicles tables with discs reduced, and large flat rods. On undersides of rocks in shallow water; often found in association with *H. parvula*; common.

***Actinopyga agassizi*** Selenka: Medium-sized to large (250 mm), with thick leathery skin. Tube feet in 3 bands ventrally, dorsally as scattered small papillae. Five hard, white teeth surround anus. Color uniform to mottled light to dark brown, sometimes yellowish. On sand and rock in shallow water on exposed coastlines; rare.

F. **STICHOPODIDAE:** Aspidochirotida with gonad in 2 tufts, 1 to each side of the dorsal mesentery. Includes some tropical and temperate forms. (1 sp. from Bermuda.)

***Isostichopus badionotus*** (Selenka) (= *Stichopus badionotus, Stichopus moebii*) (B Sea pudding): Large (400 mm), conspicuous, broad and flattened, with very thick skin. Color usually solid, ranging from light brown to almost black, with ventral surface lighter. A less common color form is light brown, blotched or spotted with dark brown to black. Ossicles tables and C-shaped bodies. A sluggish species that occurs on soft, shaded bottoms in shallow water, and shows no tendency toward concealment. Very common. (Color Plate 14.10.)

O. **DENDROCHIROTIDA:** Medium-sized or small Holothuroidea with richly branched tree-like tentacles. Tube feet well developed, either restricted to radii or scattered all over surface of body.

F. **PSOLIDAE:** Dendrochirotida with dorsal part of the body covered in large overlapping plates. The ventral surface (the sole) is soft, and well equipped with tube feet for attachment to hard substrates. (1 sp. from Bda.)

***Lissothuria antillensis*** Pawson: Small (10-20 mm), secretive, with tube feet scattered on dorsal surface of body, and with ventral tube feet in 3 distinct rows. Deep rosy red with a tinge of purple. On undersides of rocks, where the coloration renders specimens virtually invisible among surrounding coralline algae, bryozoans and *Homotrema*. Rare. (Color Plate 14.7.)

F. **CUCUMARIIDAE:** Dendrochirotida with more or less naked body that is not enclosed in large overlapping plates. Ossicles generally small and inconspicuous. (2 spp. from Bda.)

***Ocnus surinamensis*** (Semper) (=*Thyone surinamensis, Semperia bermudensis*): Medium-sized (100 mm) species, body cylindrical, tentacles richly branched, all of equal size. Tube feet in double rows on all 5 radii, with tendency to spread into interradii. Skin thick. Color brownish, tentacles dark brown. Ossicles knobbed buttons and cups. In crevices and under rocks below low tide level, with tentacles protruding. Uncommon. (Color Plate 14.9.)

O. **APODIDA:** Small to large Holothuroidea with cylindrical body and with simple pinnate tentacles. No tube feet on body.

F. **SYNAPTIDAE:** Apodida with ossicles that include anchors that project from body wall, rendering skin sticky to touch. (4 spp. from Bda.)

***Eupatinapta acanthia*** (Clark): Medium-sized (to 150 mm), translucent pinkish white. Ossicles anchors associated with anchor-plates with numerous small perforations. Burrows in sand in shallow water. Juveniles often occur interstitially in subtidal sand.

***Synaptula hydriformis*** (Lesueur): Active, medium-sized (75 mm); with 2 color forms: dark green and white, and brown and white. Ossicles anchors associated with anchor plates with few large dentate perforations. Common in shallow water among weed, such as *Ulva* and *Penicillus*; very common in landlocked saltwater ponds. Viviparous, fertile year round. (Color Plate 14.8.)

F. **CHIRIDOTIDAE:** Apodida with ossicles in the form of wheels, either aggregated into papillae or scattered in body wall. No anchors. Body wall usually smooth to touch. (1 sp. from Bda.)

***Chiridota rotifera*** (Pourtalès): Medium-sized (100 mm), pinkish red. Ossicles wheels aggregated into papillae, and curved rods. Burrows in sand, especially under stones in high-energy areas. Not uncommon. Fertile year round.

D. L. PAWSON

## Phylum Hemichordata

CHARACTERISTICS: *More or less vermiform, free-living BILATERIA with body in 3 divisions (proboscis, collar and trunk) each containing coelomic cavities (unpaired in the proboscis, paired in collar and trunk). Trunk mostly with paired gill slits. Nervous system epithelial.*

Both classes of this exclusively marine phylum are represented in Bermuda (some authors recognize a third class, Planctosphaeroidea, here treated as an appendix to Enteropneusta).

### Class Enteropneusta (Acorn worms)

CHARACTERISTICS: *Worm-shaped, free-living solitary HEMICHORDATA with numerous gill slits; proboscis and collar adapted for burrowing. Digestive tract straight.* Body length from a few centimeters to more than 2 m; color yellow to brownish, movement a slow ciliary gliding.

Of 3 familes, 2 (Harrimanidae and Spengelidae) have not been found in Bermuda yet. Of about 60 species, 4 have been reported from Bermuda; 3 are included here.

OCCURRENCE: The majority of species live in the intertidal to shallow subtidal,

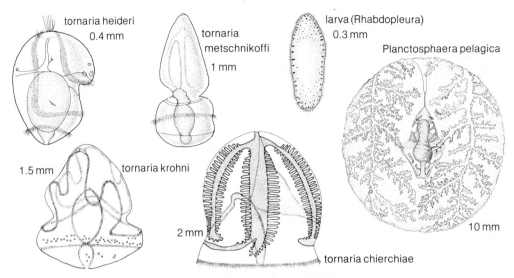

**179 HEMICHORDATA: Development**

only a few occur in deeper water. Many excavate spiral or U-shaped burrows in sediments that they line with mucus; others can be found under stones, often in small aggregations.

Collecting of intact specimens (especially of larger species) is difficult for burrows may extend 0.5 m or more below the sediment surface, and animals fragment readily when disturbed. Gentle sieving of sediment collected by SCUBA, with a spade or a dredge, may yield specimens or at least parts therof; or sediment mixed with seawater can be left to settle for a day or longer, after which specimens may move to the sediment surface.

IDENTIFICATION: Intact live animals are readily identified on the basis of external characters.

Note the thick, velvety, completely ciliated skin. Distinguish between the 3 body regions; the mouth is located ventrally between the proboscis and the collar. The anterior trunk region is dorsally pierced by 2 rows of gill slits that are often protected by lateral wing-like folds containing the gonads. Posteriorly the trunk may be undifferentiated, or the darker intestine, with bulging diverticula, may characterize a hepatic region, which is followed by a straight caudal region.

Prior to fixation for microanatomy, leave animals for a day in seawater (several changes) to allow them to empty their intestine. Anesthetize in isotonic magnesium chloride, fix in Bouin's.

BIOLOGY: Sexes are separate. Masses of eggs, embedded in mucus, are shed from the burrow and fertilized by sperm from nearby males. Asexual reproduction by fission and subsequent regeneration occurs in some species. Most acorn worms ingest sediment particles (with adhering meiobenthos) that get entangled in mucus produced by the proboscis and are then passed back to the mouth by ciliary action. Their predators are fishes and crustaceans; Protozoa, Trematoda and Copepoda are their parasites. Several species have been shown to be luminescent when disturbed at night.

*Plate 179*

DEVELOPMENT: Development may be direct via an embryo that soon stretches and

# ENTEROPNEUSTA

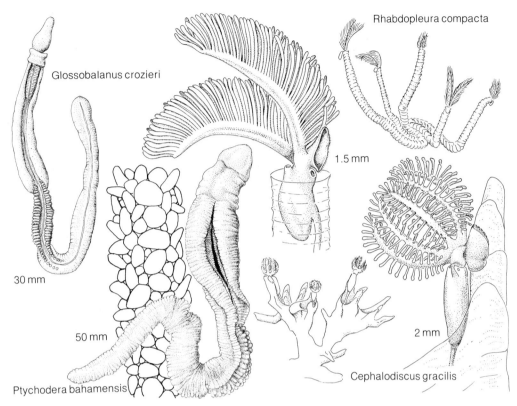

**180 HEMICHORDATA (Hemichordates)**

differentiates body regions, or indirect via a ciliated tornaria larva (several types of tornariae, named after their first describers, are known but not all have been positively assigned to adults of known species). After several weeks (to months) of planktonic existence the tornaria decreases in size and develops an equatorial restriction that represents the proboscis-collar boundary; the rather vermiform stage then settles to the bottom.

REFERENCES: For general orientation see HYMAN (1959) and VAN DER HORST (1939).
   Bermuda's species are in need of revision; the only accounts are by VAN DER HORST (1924), CROZIER (1915, 1917) and VERRILL (1901a). *Planctosphaera* was found in Bermuda by GARSTANG (1939).

*Plate 180*

F. **PTYCHODERIDAE:** Enteropneusta with pronounced regionation of the trunk (genital wings and hepatic sacculations).

***Glossobalanus crozieri*** Van der Horst: Genital folds ridge- rather than wing-shaped, not covering the gills; hepatic sacculations in 2 longitudinal rows. In sand; rare.

***Ptychodera bahamensis*** Spengel: With large gill slits and pronounced genital wings that are usually folded over the gills: lemon-yellow. In sand under stones in shallow water; fairly common.

### Plate 179

**APPENDIX:** *Planktosphaera pelagica* Spengel: Transparent, spherical larva (!) with highly sinuous ciliary bands. The adult organism is unknown. Pelagic at 200-300 m, rare.

W. STERRER

### Class Pterobranchia

CHARACTERISTICS: *Aggregated or colonial HEMICHORDATA mostly living in an externally secreted encasement (coenecium). Zooids with 1 pair or without gill slits; proboscis shield-like, collar with tentaculated arms. Digestive tract U-shaped.* Though some aggregations reach 25 cm in diameter, colonies in Bermuda form low, dark brown or whitish crusts of considerable lateral extension (20 cm and more).

Of only 3 genera with 21 species in this class, 2 are known from Bermuda; both are included here.

OCCURRENCE: Most species live in cold, deeper waters of the southern hemisphere; only a few have been reported from tropical and shallow areas although they are probably much more widespread than was believed. They form massive clumps or encrust hard substrates (rocks, shells) or sediment, sometimes in great abundance. In Bermuda, the 2 species are frequently found together.

Specimens are collected by chipping off pieces of rocks with attached colonies.

IDENTIFICATION: Small colonies can be observed live under a dissecting microscope; individuals should be removed from their tubes and placed on a depression slide.

Note the structure of the coenecium. In *Rhabdopleura*, the branching horizontal tube is partitioned and contains the black stolon (root-like structure that connects the individuals of a colony and is concerned with budding); in *Cephalodiscus* it may consist of open chambers containing clusters of zooids. Both growth forms give rise to vertical tubes from which the tentaculated arms are extended during feeding. The coenecium, secreted by the cephalic shield, is collagenous. The body of an individual zooid consists of a cephalic shield (proboscis), a collar with tentaculated arms, a sac-like trunk, and a stalk that either permanently connects to the stolon (*Rhabdopleura*) or is usually attached to the tube interior by means of a distal sucker (*Cephalodiscus*).

Anesthetize the animals in magnesium chloride and fix in Bouin's fluid.

BIOLOGY: Sexes are separate, and ripe females are usually plumper than males. Eggs, which are rich in yolk and produced only in small numbers, are released into the maternal tube where they develop. Pterobranchia are active filter feeders, creating a current by means of the cilia on their tentacles, and ingesting microalgae and small crustaceans that get caught on the sticky tentacular surface or are batted to the midline of the arm (food groove) by muscular flexing of the tentacles. Nothing is known about enemies and parasites.

### Plate 179

DEVELOPMENT: The ciliated, oval, rather unspecialized larva of *Rhabdopleura* leaves the parent's tube and, after a very short planktonic life, settles on a suitable substrate. It encapsulates itself in a dome-shaped cocoon in which it undergoes metamorphosis; it then makes a hole in the wall of the dome and builds a length of creeping tube before extending an erect tube. The colony grows by asexual budding from this primary zooid.

REFERENCES: Relatively little is known about the group beyond its anatomy. For general information consult HYMAN (1959) and VAN DER HORST (1939); more recent observations on *Rhabdopleura* have been added by STEBBING (1970) and STEBBING & DILLY (1972).

BARNES (1977) and LESTER (1985) reported the class from Bermuda.

### Plate 180

*Cephalodiscus gracilis* Harmer: Aggregated; zooids to 2 mm in length; burnt orange with blue-green flecks on trunk, red stripe on cephalic shield; 5 pairs of tentaculated arms in mature zooids. Asexual buds develop at the distal end of the stalk of the primary zooid. Stalk anchored to the coenecium by a terminal sucker. Basal portion of coenecium with a network of unpartitioned chambers giving rise to upright, vase-shaped, whitish or colorless tubes often containing more than 1 individual; openings ornamented with spines. On the undersurfaces of rocks in the shallow subtidal of agitated inshore waters; probably not uncommon.

*Rhabdopleura normani* Allman: Colonial; zooids to 1.5 mm in length; yellow-green with black pigment spots; 1 pair of tentaculated arms. Each zooid is joined to the stolon by a contractile, muscular stalk. Asexual buds develop on and remain attached to the stolon. Basal (attached) portion of coenecium can extend linearly for over 5 cm; side branches are shorter, with each branch often apparently closed off at its base by a partition giving rise to an upright, evenly annulated, erect tube. Each of these initially brownish, later transparent tubes contains a single individual. On undersurfaces of rocks and dead coral in the shallow subtidal of agitated inshore waters; probably not uncommon.

W. STERRER & S. M. LESTER

# Phylum Chordata

CHARACTERISTICS: *Variously shaped BILATERIA that possess a notochord (dorsal cartilaginous rod), a dorsal hollow nerve cord and gill clefts at least during some stage of their life cycle.*

This, the largest phylum of Deuterostomia, is composed of 3 externally very diverse subphyla: Urochordata (Tunicata; p. 545), in which the notochord when present occupies the tail only; Cephalochordata (Acrania, p. 562), in which the notochord extends into the head; and Vertebrata (Euchordata, p. 565), in which a vertebral column replaces the (embryonal) notochord.

### Subphylum Urochordata (=Tunicata)

CHARACTERISTICS: *Globose or tadpole-shaped CHORDATA with notochord, if present in the larva or the adult, in the tail only.*

The 3 major classes, all exclusively marine, are easily distinguished by external shape and habit: Ascidiacea (below) are benthic-sedentary and often rather shapeless, Thaliacea (p. 558) are planktonic and barrel-shaped, and Appendicularia (p. 561) are planktonic and tadpole-shaped. Sorberacea, a recently described class of carnivorous deep-sea tunicates (see MONNIOT & MONNIOT 1978), is not treated here although they have been found in the northern West Atlantic.

### Class Ascidiacea (Sea squirts)

CHARACTERISTICS: *Sessile UROCHORDATA enclosed in a cellulose tunic. Solitary or colonial; notochord present only in the larva.* The size range of Bermudian species is from a few millimeters to 10 cm for individuals; colonies may reach 40 cm in diameter. Many species display bright though highly variable colors and patterns. Their consistency is gelatinous to cartilaginous or leathery. Movements are restricted to contractions of body and orifices.

Of about 2,000 species known, 47 have been recorded from Bermuda; 36 are included here.

OCCURRENCE: Exclusively marine. Most species live attached to hard substrates (rocks, shells, pilings, ship bottoms and wrecks, algae, sea grasses and mangrove roots) and are particularly common in shaded areas (on overhangs, in caves and on the underside of stones). Although not recorded from Bermuda, a few very small species occur interstitially between sand grains, and others are able to live on muddy bottoms. Only a few species tolerate tidal exposure; the majority is found in the shallow subtidal; some occur at considerable depths. Many species are distributed worldwide.

Collecting is done by turning over stones or cutting off mangrove roots in shallow water, by SCUBA and with hammer and chisel, or with a dredge. Most ascidians contract immediately when touched, squirting jets of seawater through their siphons (hence the common name!).

Although a few species are used for food (e.g., around the Mediterranean, but not in Bermuda), ascidians are of commercial interest mainly as fouling organisms on marine structures.

IDENTIFICATION: The majority of species is extremely variable in shape and color and very contractile; in addition, many colonial species resorb part of their zooids during the colder season, or in response to collecting (!). Dissection of preserved (!) mature specimens, therefore, is the only means for reliable identification. Nevertheless, specimens should be given an opportunity to expand in a dish so that live features can be noted before fixation.

Differentiate between solitary (simple) and colonial (compound) species; in the latter, individuals (zooids) are connected basally by a stolon, or are partly or completely embedded in a common tunic. A product of the body wall (mantle), the tunic invests the entire body; it is composed primarily of cellulose, and often contains large numbers of calcareous spicules (e.g., in Didemnidae). Solitary species have 2 openings, an anterior buccal or branchial (intake) siphon, and a posterior atrial (exhaust) siphon. The line connecting the siphons is referred to as the dorsal side. The buccal siphon is armed internally with oral tentacles. In colonial forms, the atrial siphons of several zooids may combine, or open into a chamber that empties through a common cloaca. Within the tunic, the body can be compact (e.g., Phlebobranchiata), or divided into regions: a thorax containing the pharynx, an abdomen containing digestive and reproductive organs, and—in some forms—a postabdomen. The structure of the basket-shaped pharynx (branchial sac) with its mucus-producing ventral groove (endostyle) and its gill slits (stigmata) provide important features for identification. In some taxa (Stolidobranchia) the branchial sac may have longitudinal folds; and opposite the endostyle there may be a dorsal lamella that can be straight, or serrated to form tongue-like processes (languets). Stigmata are typically simple vertical slits in one plane (e.g., Aplousobranchia), but may be complicated by subdivision, spiralization and addition of papillae and internal longitudinal blood vessels. The gonads are typically located near the stomach (Enterogona). Most species have only 1 testis and ovary in close proximity but well delimited from each other; in some, however, gonads are doubled or even more numerous, and situated in the mantle (Pleurogona). In some families (Styelidae, Pyuridae) the internal wall of the mantle may have cushion-shaped thickenings (endocarps).

Prior to fixation and preservation (in 10% formaldehyde), specimens should be anesthetized (to prevent contraction) by adding carbon dioxide or crystals of menthol to the water.

BIOLOGY: Hermaphrodites. Male and female gonads usually alternate in gamete production, but self-fertilization may occur in some species. Solitary species usually shed eggs and sperm through the atrial siphon into the open water where fertilization takes place; some incubate their embryos in the atrial cavity. Colonial species typically brood their larvae either in the oviduct (*Ecteinascidia*) or in the atrium, which may develop special brood pouches (*Distaplia*). Asexual reproduction through budding from the stolon (e.g., *Clavelina*) or

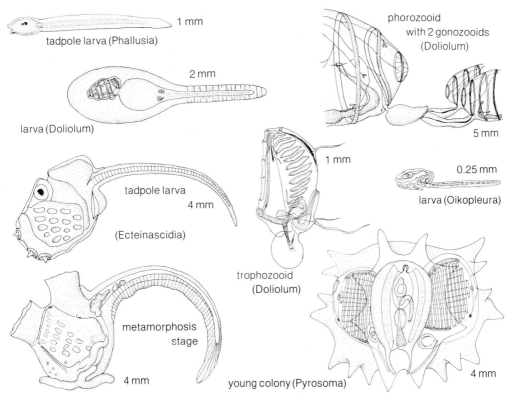

**181 UROCHORDATA (Tunicates): Development**

other parts of the body, budding at the larval stage (e.g., Didemnidae) or strobilation (multiple constrictions of the postabdomen, e.g., Polyclinidae) is common in colonial forms. The life-span is 1 to several years in solitary species, and probably longer in some colonial forms that can resorb their zooids and survive adverse environmental conditions as stolons only. All ascidians are filter feeders, drawing water into the buccal siphon by means of cilia, and removing organic particles (mostly phytoplankton) trapped in the mucus sheet that passes continuously over the inner surface of the branchial sac. Predators of ascidians are flatworms (*Thysanozoon, Pseudoceros*), mollusks (*Trivia*) and some fishes; Sporozoa, Copepoda (*Doropygus*) and Amphipoda are parasites, the latter using the host only for shelter. The tunic of some species (*Microscosmus*) can be heavily encrusted with other sedentary animals and algae.

## Plate 181

DEVELOPMENT: The larva is tadpole-shaped and does not feed. The most primitive type of tadpole larva (*Phallusia*) is small, with adhesive papillae at the anterior end, a pigmented sensory vesicle, and with a free-swimming period of 6-36 hr. More modified types include the large tadpole of *Ecteinascidia*, which has sucker-like cups around the papillae, and is free-swimming for only a few minutes to hours. In this latter type, metamorphosis is often advanced so that larval and adult organs function almost simultaneously. At metamorphosis the larva attaches itself to a suitable substrate by means of its adhesive papillae; tail and notochord degenerate, and the siphons open. In colonial forms, this primary individual (oozooid) then forms buds from which blastozooids are produced.

In many colonial species, larvae are liberated readily by gravid specimens placed in a dish with seawater.

REFERENCES: The most complete work on ascidians is by VAN NAME (1945); more recently, PLOUGH (1978) has reviewed the ascidian fauna of the U.S. east coast, including many species from Bermuda. Other useful summaries are those by BERRILL (1950), by GOODBODY (1974) on physiology and by MILLAR (1971) on biology. The studies by F. MONNIOT (1983) and C. MONNIOT (1983) are largely applicable to Bermuda.

VAN NAME's (1902) original study of Bermuda's ascidian fauna has been added to and updated by VAN NAME (1921) and BERRILL (1932); for a recent survey (which included collecting by SCUBA) see C. MONNIOT (1972a, b), F. MONNIOT (1972a) and GAILL (1972a, b). STOECKER (1978) has investigated the resistance of *Phallusia nigra* to fouling.

O. **APLOUSOBRANCHIA:** Colonial Ascidiacea with body divided into thorax, abdomen and sometimes postabdomen. Gonads situated within or under loop of intestine (Enterogona). Branchial sac without longitudinal vessels; gill slits simple.

## Plate 182

F. **POLYCLINIDAE:** Usually massive Aplousobranchia with atrial siphons opening into a common cloaca. Zooids long, with postabdomen. Asexual reproduction by strobilation of the postabdomen. (6 spp. from Bda.)

*Aplidium bermudae* van Name: Genus with elongated zooids with slender postabdomen; stomach wall with longitudinal pleats.—Species with opalescent-translucent tunic through which the long (to 3 cm) zooids can be seen. Each zooid with 4 bright red dots at the base of the buccal siphon (coloration disappears after fixation!). Cloacal tongue simple. Stomach elongated, with about 8 prominent folds. Branchial sac with 15-17 rows of stigmata. In- and offshore; very common.

*A. exile* van Name: Genus as above. —This species forms small pink to bright red cushions. Tunic translucent but not very transparent, sometimes slightly encrusted with sand. Zooids 1-1.5 cm long. Cloacal tongue simple. Stomach spherical, finely pleated (22-24 folds). Branchial sac with 12 rows of stigmata. Ovary always well separated from the intestinal coil. In sheltered subtidal locations, under stones; common. (Color Plate 15.1.)

*Polyclinum constellatum* Savigny: Yellowish cushions with a star-shaped network of yellow pigment (marking the regular arrangement of zooid siphons). Zooids with very large cloacal tongue and with stalked postabdomen. Intertidal to shallow subtidal; rare. (Color Plate 15.14.)

F. **DIDEMNIDAE:** Cushion-shaped or thin encrusting Aplousobranchia whose tunic is often white from embedded star-shaped calcareous spicules. Zooids with thorax and abdomen; gonads situated in intestinal coil. Asexual reproduction by budding from the esophagus. (6 spp. from Bda.)

*Trididemnum savignyi* (Herdman): Colonies to 15 cm diameter, thick; color gray or dark green or black, depending on the density of embedded

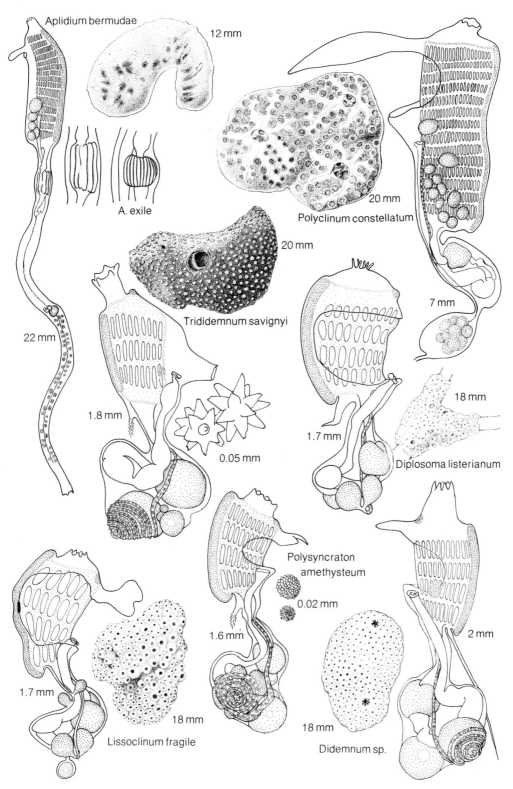

**182 POLYCLINIDAE, DIDEMNIDAE (Sea squirts 1)**

spicules. Zooids with only 3 rows of stigmata, and a tubular cloacal siphon. With 1 testis only, around which the sperm duct forms a spiral. In the first few meters below the surface, at the base of coral heads; common. (Color Plate 15.2.)

***Diplosoma listerianum*** (Milne Edwards): Very soft gelatinous crust, to 10 cm diameter; always without spicules; color from beige to chestnut. Branchial sac with 4 rows of stigmata that are bared by the enormous cloacal siphon. With 2 testes and 1 straight sperm duct. Shallow subtidal, on hard or soft (algae) substrates; moderately common. (Color Plate 15.3, 4.)

***Lissoclinum fragile*** (van Name): Very soft gelatinous crust, with or without spicules. Zooids with well-developed, bifurcated cloacal tongue. Branchial sac with 4 rows of stigmata. With 2 testes and 1 straight sperm duct. Developing oocytes usually bulge out of the intestinal coil. Shallow subtidal in sheltered locations (e.g., on mangrove roots); locally abundant. (Color Plate 15.15.)

***Polysyncraton amethysteum*** van Name: Thick crust of variable color (brown to rose or dark purple) depending on the distribution of spicules. Zooids bright red; with 4 rows of stigmata and a large cloacal siphon topped by a 2-pronged tongue. The spiral sperm duct covers the rosette- shaped, about 6-lobed testis. Spicules white, burr-like. On rocks and reefs; not uncommon.

***Didemnum*** sp.: Very white thin cushions brittle from abundant spicules. Zooids (difficult to extract from the tunic!) with 4 rows of stigmata; the large cloacal siphon may or may not carry a tongue. With 1 testis and a spiral sperm duct. This "species" (which in fact includes several species) is very common on many substrates. (Color Plate 15.5.)

### Plate 183

F. **POLYCITORIDAE:** Cushion- or grape-shaped Aplousobranchia in which the zooids are completely or incompletely covered by a common tunic. Zooids with thorax and abdomen; gonads situated in intestinal coil. Asexual reproduction by budding from stolons. (9 spp. from Bda.)

***Cystodytes dellechiajei*** (della Valle): Cushions of variable size and color; usually black, sometimes brown or pinkish, depending on the pigmentation of the zooids. Each zooid is surrounded by a capsule of Chinese-hat-shaped spicules. Zooids with 2 tubular siphons close together, and with 4 rows of stigmata. Intestinal loop elongated. Stomach wall smooth. Testis rosette-shaped, situated in the curvature of the intestine; sperm duct straight. In the shallow subtidal and on reefs; not uncommon.

***Eudistoma olivaceum*** (van Name): Genus with cushion-shaped or lobed colonies, zooids with 3 rows of stigmata, lacking a brood pouch. Without spicules; stomach smooth.—Species olive-green to yellowish, forming small cushions or lobed clusters. Zooids not very pigmented; with short, tubular siphons. Intestinal loop very long; stomach located at the end of the loop next to the gonads. Subtidal, in sheltered locations, on rock

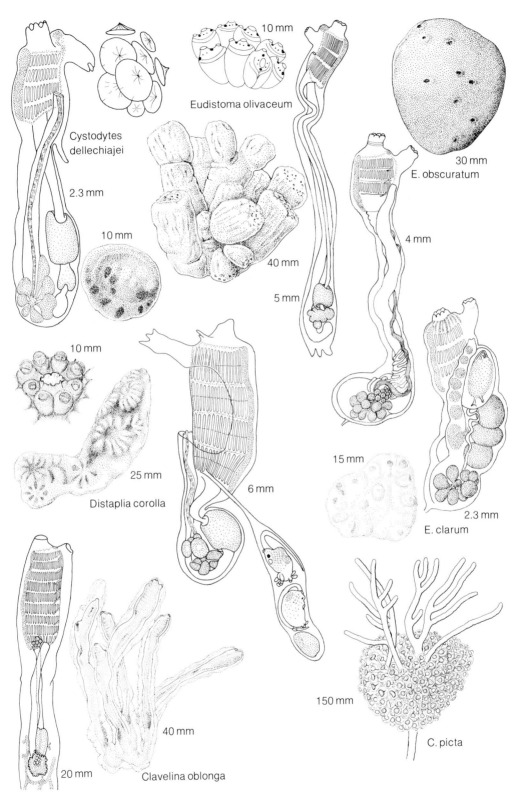

**183 POLYCITORIDAE (Sea squirts 2)**

faces and under stones; not uncommon. (Color Plate 15.8.)

***E. obscuratum*** (van Name): Genus as above.—Species with large, usually black colonies with a greenish tinge, easily mistaken for *Cystodytes dellechiajei* and *Trididemnum savignyi*. Zooids with heavily pigmented thorax. Subtidal, under stones and in reefs; common. (Color Plate 15.9.)

***E. clarum*** (van Name): Genus as above.—Species with small, transparent to white colonies. Zooids small, with strong muscle band on thorax and anterior part of abdomen. Midgut sharply constricted. Shallow subtidal, on the underside of stones; very common. (Color Plate 15.10.)

***Distaplia corolla*** Monniot: Colonies morel-shaped (capitate), of robust-cartilaginous consistency, bright red. Zooids arranged in regular rosettes around a common cloaca. With 4 rows of stigmata; each row is traversed horizontally by a blood vessel. Intestinal loop short; stomach smooth. With an external brood pouch attached via a long stalk to the mantle at the level of the 4th row of stigmata. Intertidal to 2 m; on vertical rocky surfaces only; very common. (Color Plate 15.7.)

***Clavelina oblonga*** Herdman: Genus with large zooids with numerous rows of stigmata and a stolon arising from the posterior part of the body.—Species colorless-transparent, cartilaginous; zooids connected only basally, with about 10 rows of stigmata. Stomach with 4 flat ribs. In shallow, protected areas, on mangrove roots, and on the underside of stones; common. (Color Plate 15.17.)

***C. picta*** (Verrill): Genus as above.—Species bright purple; zooids with tunic joined along the abdomen to form groups which in turn may make up grape-like clusters of 40 cm (!) diameter. Sublittoral, in clear water, on reefs, rocks, buoys, often on gorgonians; common. (Color Plate 15.13.)

O. **PHLEBOBRANCHIA:** Solitary or colonial Ascidiacea with generally undivided body. Gonads situated within loop of intestine (Enterogona). Branchial sac without folds, but generally with longitudinal vessels and papillae; gill slits (stigmata) generally straight.

*Plate 184*

F. **PEROPHORIDAE:** Colonial Phlebobranchia in which zooids are connected by thin, vine-like stolons. (5 spp. from Bda.)

***Ecteinascidia turbinata*** Herdman: Genus with oblong or clavate zooids with 15-30 rows of stigmata.—Species orange-colored, forming dense grape-like clusters 15 cm or more in diameter. Zooids to 5 cm long with very transparent tunic and up to 30 rows of stigmata; intestine forming a wide loop on left side of the body around the rosette-shaped gonads. The oviduct, in which the large transparent larvae are brooded, opens to the right side of the body. Rather ubiquitous from intertidal to the outer reefs; on exposed surfaces, often on gorgonians; very common. (Color Plate 15.12.)

***E. conklini*** Berrill: Genus as above.—Species with zooids to 10 mm long, lemon-yellow to greenish, both si-

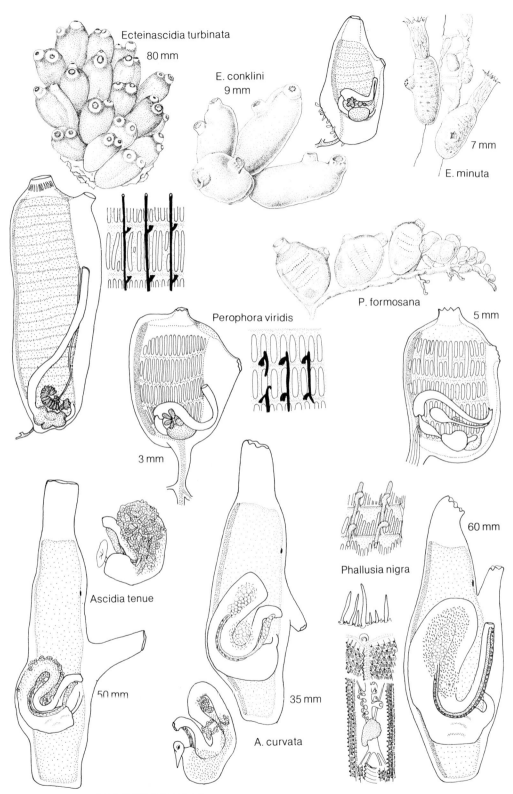

**184 PEROPHORIDAE, ASCIDIIDAE (Sea squirts 3)**

phons usually with a bright red border. With up to 15 rows of stigmata. Forms small colonies on the underside of stones; rare. (Color Plate 15.16.)

*E. minuta* (Berrill): Genus as above.—Species with zooids about 8 mm, whitish; branchial siphon long, ending in a prominent white crown. With up to 13 rows of stigmata. Forms dense carpets on the underside of stones, or muffs around mangrove roots; common.

*Perophora viridis* Verrill: Genus with small, ovate zooids with no more than 5 rows of stigmata.—Species pale greenish, 3-5 mm; with 4 rows of stigmata and incomplete longitudinal vessels. Instestine forming a tight loop around the gonads. On the underside of stones, corals, algae, to 20 m; very common.

*P. formosana* (Oka) (=*P. bermudensis*): Genus as above.—Species yellow, about 6 mm; with 5 rows of stigmata; about 1/3 of the stigmata of the 1st and 2nd rows are common to both. On the underside of stones, often together with *P. viridis*; sometimes encrusting other ascidians; common. (Color Plate 16.8.)

F. **ASCIDIIDAE:** Solitary Phlebobranchia with a thick tunic. Intestine on the left side of the branchial sac; gonads spread out over the intestinal loop and covered by short, closed renal vesicles usually containing concretions. (3 spp. from Bda.)

*Ascidia tenue* C. Monniot: Genus without acessory openings of the neural duct. - Species large (to 10 cm), laterally flattened. Tunic thin, glassy, colorless and transparent, often encrusted with sand. Intestinal tract small, compact. On the underside of stones to which it is attached with its entire left side; uncommon.

*A. curvata* (Traustedt): Genus as above.—Species to 7 cm, with a leathery, yellowish to orange-opaque tunic only rarely encrusted with sand. Intestinal tract comparatively large. On the underside of stones and on mangrove roots, often in groups; common. (Color Plate 15.11.)

*Phallusia nigra* Savigny (=*Ascidia nigra*): Tunic black, thick, shiny; never encrusted with other organisms or sand. To 10 cm. Branchial sac with numerous longitudinal vessels and papillae. Neural duct of adults usually with accessory openings (generic character!). In shallow water on rocks, moorings; very common.

O. **STOLIDOBRANCHIA:** Solitary or colonial Ascidiacea with undivided body. Gonads situated on both sides of the body (Pleurogona); intestine on the left side. Branchial sac usually with folds and with longitudinal vessels but without papillae; stigmata straight or spiral.

**Plate 185**

F. **STYELIDAE:** Solitary or colonial Stolidobranchia; branchial sac without or with 4 folds on each side. Stomach pleated; tentacles of buccal siphon simple. (12 spp. from Bda.)

*Botrylloides nigrum* Herdman: Genus colonial with 1 ovary only on each side, posterior to the testes; branchial sac without folds. With an incubatory pouch formed as

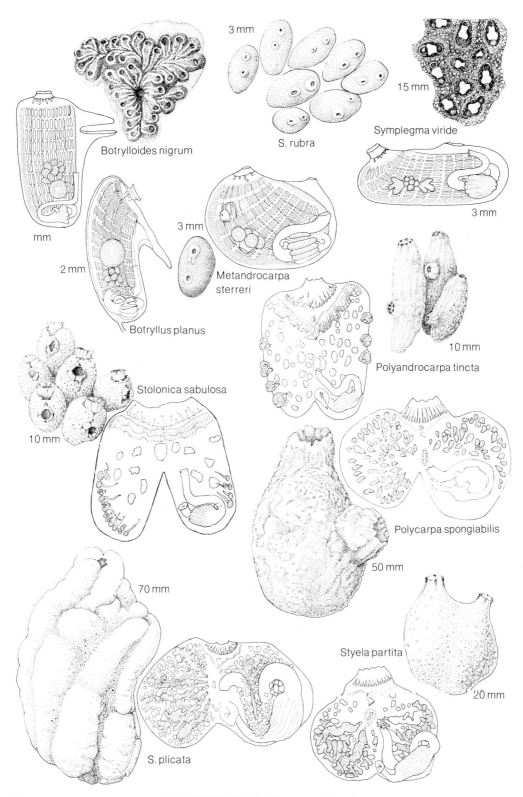

**185 STYELIDAE (Sea squirts 4)**

an outgrowth of the body wall.—Species forming fairly hard crusts, 5-10 mm thick, of highly variable color: reddish brown, gray or black, the buccal siphons often surrounded by a fine yellow or white circle. Stomach conical. On the underside of stones, on mooring chains and buoys, on sea grasses; very common, particularly in clean water. (Color Plate 16.3-6.)

***Botryllus planus*** (van Name): Genus colonial, with 1 or several ovaries on each side, anterior or dorsal to testes; branchial sac without folds. Without an incubatory pouch.—Species forming thin crusts of variable color: beige, yellow, orange, reddish or black. Stomach barrel-shaped. Under stones, on algae; with about the same distribution, and difficult to distinguish from *Botrylloides nigrum*. (Color Plate 16.1, 2.)

***Symplegma viride*** Herdman: Genus colonial, zooids in common tunic. Branchial sac without folds; 1 gonad on each side composed of an ovary and 2 testes.—Species thin, flat, encrusting; tunic reddish brown, green, purple or black, usually with an angular, bright yellow or orange patch between the siphons that disappears when the colony dies. Branchial sac with 4 longitudinal vessels on each side. ♂ and ♀ gonads mature at the same time. In shallow, protected waters, on dead coral, gorgonians, algae; common. (Color Plate 16.11, 12.)

***S. rubra*** C. Monniot: Genus as above.—Species forming small patches, with zooids bulging above common tunic. Zooids uniformly yellow, orange or carmine-red. Either ♂ or ♀ gonads mature in a colony, never both at the same time. Intertidal to shallow subtidal, on the underside of stones; uncommon.

***Metandrocarpa sterreri*** C. Monniot: Colonial, with zooids widely separated. Zooids hemispherical, 2-3 mm; bright red, with orange areas around and between siphons. Branchial sac without folds, but with 7 vessels on the left, and 8-10 on the right. Gonads either in ♂ or ♀ maturity, never both. On offshore reefs, in crevices and on the underside of corals, 5-20 m; not uncommon.

***Stolonica sabulosa*** C. Monniot: Colonial, with zooids separated, forming upright groups. Zooids to 20 mm, orange, sometimes encrusted with sand. Branchial sac on each side with 3 folds each made up of about 10 vessels. Intestine situated posteriorly. ♀ gonads on the right side of the body, ♂ gonads on both sides; both can be found in the same zooid. In areas with strong wave action, in crevices and on overhangs, 5-10 m; uncommon. (Color Plate 16.7.)

***Polyandrocarpa tincta*** (van Name): Colonial; sometimes found as isolated zooids, as small groups, or as large grape-like clusters. Zooids to 15 mm, blood-red, with a fragile tunic. Branchial sac with 4 folds on each side; gonads hermaphroditic, protruding outside the mantle. Intertidal to 10 m; in sheltered areas forming inconspicuous flat plaques on the underside of stones, or large muffs on mangrove roots; common. (Color Plate 16.10.)

***Polycarpa spongiabilis*** Traustedt: Solitary, large (to 6cm); tunic leathery, reddish-opaque, often encrusted; mantle dark brown. Branchial sac with 4 folds on each side, and with numerous vessels. Intestine massive; stomach folds internal; intestinal loop with a large endocarp. Gonads flask-shaped, hermaphroditic, numerous,

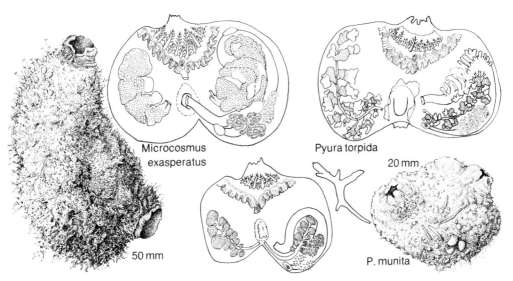

**186 PYURIDAE (Sea squirts 5)**

in both sides of the mantle. In calm, muddy inshore waters; attached to rocks, or half-buried in grassy areas; common. (Color Plate 16.9.)

*Styela partita* (Stimpson): Genus solitary, with few gonads; branchial sac with 4 folds.—Species of variable shape depending on whether the animal is attached singly or in a crowded group. To 4 cm long; tunic opaque, brownish or reddish. Siphons often striped red and white. Two gonads per side, each consisting of an elongated ovary whose blind end is surrounded by lobed testes. Mantle covered with endocarps. In shallow, protected areas; on the underside of stones, or mangrove roots where it can form massive clumps; common.

*S. plicata* (Lesueur): Genus as above.—Species large (to 8 cm); tunic thick, yellowish, with large rounded ridges that form 4 lobes each around the branchial and atrial siphons. Two gonads on the left and up to 7 on the right side. Mantle with numerous endocarps. Shallow subtidal, on rocks;

rare (probably a recent arrival in Bda).

## Plate 186

F. **PYURIDAE:** Solitary Stolidobranchia with leathery tunic; branchial sac usually with 6 or more folds on each sides. Stomach with 1 or more groups of hepatic tubules; tentacles of buccal siphon branched. (3 spp. from Bda.)

*Microcosmus exasperatus* Heller: Large (to 6 cm), irregularly egg-shaped; the yellow or red tunic often overgrown by algae, hydroids or other ascidians. Siphons red. Dorsal lamina a continuous, unserrated membrane. Intestine attached to the mantle; stomach entirely covered by the liver. One gonad on each side, consisting of a central ovary surrounded by lobed testes, all of it covered by endocarp. The left gonad lies across the intestine. In calm water, from intertidal to shallow reefs, under stones; common. (Color Plate 16.8.)

*Pyura torpida* (Sluiter): Genus with dorsal lamina serrated to form languets. Usually with 1 gonad on each side consisting of small separated ovaries and testes arranged along the genital ducts.—Species to 5 cm; the tough, opaque-red tunic often overgrown with calcareous algae, or other ascidians. Intestinal loop rather open; endocarp on gonads and intestine. In protected area; shallow subtidal, under stones and on reefs; common. (Color Plate 16.13.)

*P. munita* (van Name): Genus as above.—Species to 2 cm; the reddish tunic is usually completely encrusted with sand. The interior of the body, as seen through open siphons, is intensely blue-green. Intestinal loop rather closed; endocarp lacking. With calcareous spicules: antler-shaped ones throughout the body, and needle-shaped ones in the longitudinal vessels. In the calm intertidal, under stones; uncommon.

F. MONNIOT & C. MONNIOT

## Class Thaliacea (Salps, pyrosomids and doliolids)

CHARACTERISTICS: *Free-swimming, permanently or temporarily colonial, barrel-shaped UROCHORDATA; notochord, if present, only in the larva.* Whereas most doliolids are only a few millimeters in length, pyrosomids and salp chains may reach several meters. The body texture is soft- to rigid-cartilaginous, yet all are rather fragile. Most species are transparent to bluish; some are bright red. Pyrosomids (and to a lesser degree doliolids) luminesce by means of bacteria-filled organs. Thaliacea swim by rhythmic contraction of the circular musculature, resulting in jet propulsion.

Of about 65 recognized species, 29 are known from around Bermuda; 6 are included here.

OCCURRENCE: Pelagic. Most species live in warm oceans, usually in the upper 200 m. True mass occurrences are not uncommon at certain times of the year.

Collect large colonies by SCUBA in jars or with a hand net to avoid damaging them; smaller forms can be caught in plankton nets.

IDENTIFICATION: Live, intact colonies should be observed in the field (by snorkeling or SCUBA); smaller specimens are placed in seawater-filled dishes for study under a dissecting microscope.

Most of the barrel-shaped body is taken up by the large atrial cavity that opens to the outside through an anterior (buccal) and posterior (atrial) siphon. The atrial cavity is traversed by an S-shaped or oblique gill grid (or rod) that continues ventrally into a median, ciliated glandular groove, the endostyle. The gut loop, loose (Doliolidae) or tightly coiled (Salpidae), takes up little space in the posterior ventral part of the body. The body wall contains prominent muscle rings of which the 2 terminal ones act as sphincters. Shape, number and arrangement of muscles and gill slits (stigmata) are important systematic characters. Identification is made difficult by morphological differences in the various stages of the life cycles. In Doliolidae, for instance, the oozooid has 9 muscle bands, whereas phorozooids and gonozooids have 8. Old nurses that no longer have a trailing spur of phorozooids or trophozooids are particularly difficult to identify (see DEVELOPMENT).

Fix in buffered formaldehyde (4%); for histology in Lang's (Pyrosomida), in copper sublimate (Doliolida) and chromic-acetic acid (hard Salpida).

BIOLOGY: Hermaphrodites with a complex life cycle of alternating sexual and asexual, solitary and colonial stages. Eggs, a few at a time, are shed into the sea where fertilization occurs (Doliolida), or are incubated in the atrial wall. All Thaliacea are filter feeders, trapping phytoplankton (and small zooplankton) in a mucus net secreted by the endostyle and drawn across the gills. Hyperiid Amphipoda (e.g., *Phronima*) live in association with many Thaliacea, with some genera stealing food from the host's atrial cavity and others parasitizing or consuming the host. Medusae and fish can be important predators.

## Plate 181

DEVELOPMENT: In Pyrosomida, each zooid of the colony produces a single egg that, after fertilization within the parent atrium, grows into a small oozooid with a ring of 4 buds. After having been expelled into the open water, the oozooid degenerates, and the 4 primary buds form the colony by secondary budding.

In Salpida, 1 to a few eggs develop in the atrial wall of the mother (blastozooid). These then break loose and grow into the oozooid. The oozooid in turn asexually buds a ventral stolon with a trailing chain of individuals (blastozooids) that eventually separate to form sexually reproducing adults. Blastozooids have usually fewer muscles, and are smaller and more asymmetric than oozooids. In Bermuda waters there can be marked cyclic variations in abundance and in the proportion between the 2 generations. Some species show diel vertical migrations in excess of 500 m.

Doliolida have one of the most complex life cycles in the animal kingdom, mainly because of the several forms and functions that blastozooids assume. Eggs are shed into the atrial cavity and develop into free-swimming tadpole larvae that eventually metamorphose into the oozooid. The ventral stolon of the oozooid produces buds that migrate up along the side of the body to attach themselves, in 3 rows, to a dorsal stolon. The buds of the 2 lateral rows remain permanently attached and assume responsibility for feeding the colony (trophozooids), and the oozooid (now called "nurse") is reduced to a locomotory device. The buds of the median row grow into phorozooids ("carriers") that break loose and produce a ventral stolon with buds. These finally become gonozooids, break free, and produce eggs and sperm to start the cycle over again. In old nurses the muscle bands can be widened to the point of forming a muscle sheet, and most other organs (gills, gut) are reduced.

REFERENCES: Classical accounts are by METCALF & HOPKINS (1932) for Pyrosomida, METCALF (1932) for Salpida and NEUMANN (1935a, b) for most of the class. A useful more recent monograph is by BERRILL (1950), and much new information has been added by SOEST (1975a).

SOEST (1975b) lists the Thaliacea of the Bermuda area. MOORE (1949) gives ecological data on some Bermuda species.

## Plate 187

**O. PYROSOMIDA:** Permanently colonial Thaliacea; shaped like a hollow cone or cylinder, the colony is made up of blastozooids that have their buccal siphons open to the outer surface, their atrial siphons to the common cloacal cavity. Strongly phosphorescent. (3 spp. from Bda.)

*Pyrosoma atlanticum* Péron: Colony cartilaginous, to 50 cm. Zooids irregularly arranged in the test, with truncated spines; to 7 mm. In offshore waters to 200 m, often in large aggregations.

**O. DOLIOLIDA:** Temporarily colonial, barrel-shaped Thaliacea with few to many gill slits in 2 rows. Muscle rings complete, or reduced. Life cycle complex, with oozooid, and several generations of blastozooids. (About 6 spp. from around Bda.)

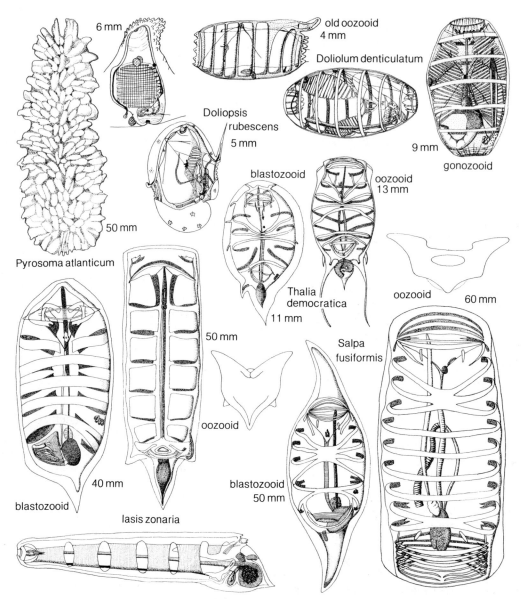

**187 THALIACEA (Salps)**

***Doliopsis rubescens*** Vogt: Gono- and phorozooid egg-shaped, with red pigment and a prominent papilla at both siphons. Muscle bands strongly reduced. Offshore; rare.

***Doliolum denticulatum*** Quoy & Gaimard: Body barrel-shaped; oozooid with 9 muscle bands, other zooids with 8. Gill slits numerous (to 45), gill deeply arched. In surface waters offshore; very common throughout the year, sometimes in large numbers. Mostly found as "old nurses".

**O. SALPIDA:** Temporarily colonial, barrel- or spindle-shaped Thaliacea with only 2, large gill slits. Muscle rings imcomplete. Life cycle an alternation of solitary (oozooid) and aggregate form (blastozooid). (About 20 spp. from around Bda.)

***Thalia democratica*** (Forskål): Solitary form with cyclindrical, smooth test carrying 2 long posterior projections. Aggregate form egg-shaped, posteriorly pentagonal and with 1 projection. Offshore; with peak occurrences 2 or 3 times per year.

***Iasis zonaria*** (Pallas): Solitary form very hard, somewhat quadrilateral when preserved. Anterior end truncated, posterior end with 3 sharp ridged processes. Body muscles broad, interrupted both dorsally and ventrally. Aggregate form irregularly cylindrical; usually carrying 5 embryos at different stages of development. Offshore; abundant throughout the year.

***Salpa fusiformis*** Cuvier: Solitary form cylindrical, firm, without projections. Aggregate form spindle-shaped, with a prominent anterior and posterior projection. Offshore; one of the most abundant salp species, particularly during the winter.

W. STERRER

## Class Appendicularia (=Larvacea)

CHARACTERISTICS: *Small, solitary, free-swimming, tadpole-like UROCHORDATA with tail and notochord persisting in adult stage.* Body length 1-2 mm (rarely to 10 mm), tail length 2-10 mm (to 60 mm). Delicate and transparent, they move by undulating tail strokes that create filtering current flows.

Of about 70 species known, more than 20 have been reported from the Sargasso Sea; 4 are included here.

OCCURRENCE: Regular and often very numerous constituents of oceanic plankton, particularly over continental shelves. Most abundant in surface layers to 100 m, they do not seem to undergo marked vertical migrations.

Most larvaceans are retained by plankton or neuston nets of medium (350 μm) mesh, but the houses they secrete fragment in nets and are lost. Specimens still in their houses have to be collected individually in jars using SCUBA.

IDENTIFICATION: Specimens (preferably live) can be placed between slide and cover slip, and observed under a low-power microscope.

Note shape and properties of the 2 body regions (the flattened tail containing the notochord and musculature; and the ovoid to cylindrical trunk) as well as position and shape of mouth and lip, gill slits, stomach (with various lobes) and anus. The surface of the trunk may show characteristic patterns of specialized glandular cells (oikoplast epithelium) that secrete the house.

Fix, after anesthesia in isotonic magnesium chloride, in neutralized formaldehyde, or in chromium-acetic acid.

BIOLOGY: Except for 1 species all are hermaphrodites. Fertilization is external, with sperm liberated through male ducts, whereas the release of eggs is by rupture of the rear of the trunk and the ovary, resulting in the death of the animal. Nothing is known yet about life expectancy. They feed on minute phytoplankton (nanoplankton) such as naked flagellates, dinoflagellates, small diatoms and Coccolithophoridae, which pass the paired coarse filters proximal in the house to be retained on the distal, fine filters. This mucous filtering device, secreted by the oikoplast epithelium, surrounds the animal completely (Oikopleuridae) or partially (Kowalevskiidae), or is carried in front of the animal (Fritillariidae). When the filters become clogged, or when the animal is disturbed (such as during collecting), it will leave the house and rebuild a new one, often within minutes. Abandoned houses, with their load of adherent nanoplankton, serve as food to other animals; Appendicularia are preyed upon by Siphonophorae, Chaetognatha and fish larvae.

### Plate 181

DEVELOPMENT: Direct. The embryo, enclosed in a mucous membrane, develops rapidly and hatches with notochord and tail muscles, whereas mouth and gills break through later. Within 24-48 hr after fertilization the young animal builds its first mucous house and starts feeding.

REFERENCES: For a general account see BERRILL (1950); the classical treatment is by LOHMANN (1933), whereas a good recent introduction is given by ALLDREDGE (1976). Modern methods of observation and collecting can be found in HAMNER et al. (1975).
  The ecology of Bermuda's neuston species is treated by MORRIS (1975).

### Plate 188

F. **OIKOPLEURIDAE:** Appendicularia with egg-shaped trunk and narrow tail. Gill slits small, round, communicating with the exterior through tubular spiracles. House large, complicated, with a pair of intake windows and a pair of filtering pipes. (About 13 spp. around Bda.)

***Oikopleura fusiformis*** Fol: Genus with tightly coiled intestine.—Species with elongated trunk and narrow tail musculature. Tail 4× trunk length. Left stomach lobe with long pointed caudal sac. With Eisen's oikoplasts, and filter grids in intake windows of house. Occasional, to 200 m.

***O. longicauda*** Vogt: Genus as above. —Species with short trunk and delicate membranous hoods; tail musculature wide. Tail 5× trunk length. Left stomach lobe with short, blunt caudal sac. Without Eisen's oikoplast, and without filter grids in intake windows of house. Very common near the surface.

F. **FRITILLARIIDAE:** Appendicularia with elongated, flattened trunk; tail short, broad, often with blunt or forked end. Gill slits small, round, opening directly to the exterior. Food trap (house) simple, protruded in front of mouth, the body being outside. (About 9 spp. around Bda.)

***Fritillaria borealis*** f. ***sargassi*** Lohmann: Gill openings small, round, well anterior to the anus. Tail musculature blunt behind, tail sharply truncated. Ovary lateral to asymmetric testis.

F. **KOWALEVSKIIDAE:** Appendicularia with cylindrical trunk and long, lanceolate tail. Heart and endostyle lacking. Gill slits wide, non-tubular; pharynx with 2 double rows of internal ciliated processes. House hemispherical, parachute-shaped, without a membranous food trap. (1 sp. around Bda.)

***Kowalevskia tenuis*** Fol: Mouth large, outer rim furnished with cilia. A rare warm-water species.

W. STERRER

### Subphylum Cephalochordata
(=Acrania) (Lancelets)

CHARACTERISTICS: *Fish-like CHORDATA with notochord extending the entire length of the body and persisting throughout life.* Small (a

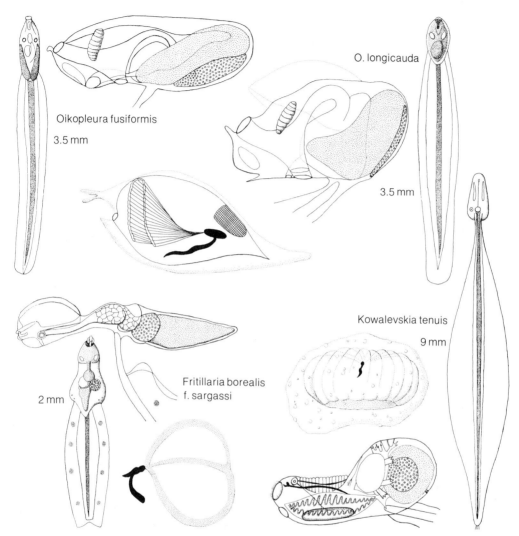

**188 APPENDICULARIA (Larvaceans)**

few centimeters), transparent to pinkish opaque. They are capable of vigorous writhing motions especially when separated from sand.

Of some 15-20 valid species in the single class Leptocardia, order Amphioxiformes, 3 have been reported from Bermuda and are included here.

OCCURRENCE: Exclusively marine; in tropical and temperate seas. They live buried superficially in medium to coarse sand, in shallow water to about 50 m, particularly at places where there is steady water movement such as between islands, at inlets and in channels.

They are collected with fine-mesh dredges; or by taking buckets of sand to the laboratory, letting it stand for a while and then treating the superficial sand layers with the magnesium chloride method. Larvae ("Amphioxides") are taken in plankton tows.

IDENTIFICATION: A dissecting microscope is sufficient for species identification. Many organ systems can easily be observed in the live animal in transmitted light.

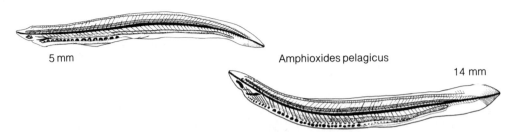

### 189 ACRANIA (Lancelets): Development

Note the absence of what one would expect of an animal of fish-like appearance and relationship: skull, jaws, teeth, backbones, scales and paired fins. The mouth, surrounded by oral tentacles (cirri) leads into the pharynx which in turn opens to the outside via an atriopore. The gut is straight and lacks a stomach. A continuous finfold runs along most of the median body contour with the exception of the ventral area between the mouth and the atriopore, expanding into a snout (rostrum) anteriorly, and a caudal fin posteriorly. The dorsal fin is segmentally divided by vertical septa into fin-ray chambers. Anterior to the ventral fin there is a pair of prominent longitudinal ridges (metapleura). Primitive visual organs are embedded throughout the length of the neural tube. The muscular system is segmented into myotomes (successive muscle blocks) whose number is a specific (though somewhat variable) character. Gonads are segmented as well.

Fix with any standard fixative, e.g., formalin.

BIOLOGY: Sexes are usually separate but similar in external appearance. Sexual products are released through the atriopore into the open water. Lancelets may reach 4-5 yr of age. Adults live buried in sand in an oblique position, ventral side up and mouth protruding. They do not seem to emerge spontaneously except at spawning time which takes place at sunset. They are filter feeders, ingesting diatoms, various protozoans, crustaceans and eggs of invertebrates.

DEVELOPMENT: Twelve hours after fertilization of the small eggs (0.1 mm diameter) the embryos hatch. Because of the "textbook" clarity of embryogenesis the lancelet has become a much-demonstrated example of chordate development. After gastrulation the embryo stretches, the mouth appears on the left side and gill openings form on the right. Subsequent larval development sees a successive increase in the number of gill openings and myotomes, and a gradual shift from the asymmetric larval features to the adult. The larva spends a pelagic life of up to 3 months before settling into the sediment. Some of the larvae, or possibly a specialized larval type, leave the littoral region for a prolonged pelagic existence; frequently caught in plankton hauls from the surface down to considerable depths and described as "Amphioxides", they grow beyond larval size and occasionally even develop gonads. These specialized larvae, which cannot be assigned with certainty to any known species, may provide a means for geographic dispersal.

REFERENCES: A good compendium on the lancelets is by BIGELOW & FARFANTE (1948); a detailed account is by DRACH (1948).
Research on species from Bermuda was carried out by DENNELL (1950a) on feeding behavior, and AREY (1915) on orientation. The systematic position of the pelagic larvae is discussed by GOLDSCHMIDT (1933).

*Plate 189*

*Plate 190*

F. **BRANCHIOSTOMIDAE:** Amphioxiformes with nearly median mouth;

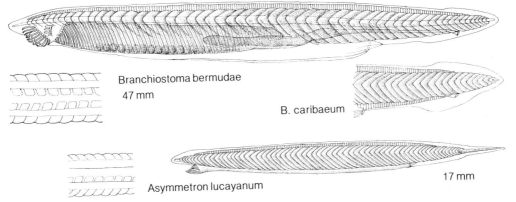

## 190 ACRANIA (Lancelets)

gonads developed on each side; both metapleura (ridges of the body wall) terminating close behind atriopore. (2 spp. from Bda.)

**Branchiostoma bermudae** Hubbs: Genus with rostrum extending only a short distance beyond preoral hood.—Species with up to 242 dorsal fin-ray chambers and 56 myotomes; upper lobe of caudal fin considerably shorter than lower lobe. With 22-28 pairs of gonads. Living specimens semitransparent and iridescent; after fixation they may become opaque. To 6 cm. Inshore in coarse sand, 1-15 m; common.

**B. caribaeum** Sundevall: Genus as above.—Species with 230-330 dorsal fin-ray chambers and 57-64 myotomes; origin of upper and lower lobe of caudal fin about opposite of each other; lobes of the same length. With 22-29 pairs of gonads. To 7 cm. Inshore and between reefs, on sandy bottom, lower tidal to almost 50 m.

F. **EPIGONICHTHYIDAE**: Amphioxiformes with nearly median mouth; gonads developed on the right side only; the right metapleuron continuous with the ventral fin, the left terminating close behind atriopare. (1 sp. from Bda.)

**Asymmetron lucayanum** Andrews: With 170-180 dorsal fin-ray chambers and 62-68 myotomes; caudal process elongated and narrow, dorsal and ventral fin narrow as well. With 26-29 gonads. To 3 cm. Inshore, in sandy areas.

M. SPINDLER

## Subphylum Vertebrata (=Euchordata)

CHARACTERISTICS: *Variously shaped CHORDATA with body typically organized into head, trunk and tail, and carrying 2 sets of paired appendages. Notochord usually present only in the embryo and replaced by vertebrae in the adult.*

Of 7 living classes, 5 have marine representatives in Bermuda: Chondrichthyes (p. 566), Osteichthyes (p. 571), Reptilia (p. 651), Aves (p. 656), and Mammalia (p. 668). The primitive, jaw-less Cyclostomata (lampreys and hagfish) have not been reported from Bermuda; and the class Amphibia does not occur in marine environments as a rule. The 3 fish-like classes (Cyclostomata, Chondrichthyes and Osteichthyes) are sometimes collectively referred to as Pisces, in contrast to the Tetrapoda (the four-limbed, mostly terrestrial other 4 classes).

## Class Chondrichthyes (Cartilaginous fishes)

CHARACTERISTICS: *Aquatic VERTEBRATA with jaws, paired fins and a skeleton consisting of cartilage that may be calcified but not ossified. Body usually covered with tooth-like placoid scales. Skull without sutures. Swim bladder absent. With a pair of nasal openings that do not open into the mouth. With 5 (more rarely 4, 6 or 7) pairs of gill slits; without a swim bladder.* Size and body shape vary considerably, from the fusiform whale shark of 12 m to flattened rays less than 0.5 m in length. Color ranges from the deep blue-gray of oceanic sharks to brownish olive in inshore sharks and rays; deep-water species are dark gray to coal-black. Sharks are capable of rapid movement by means of a well-developed tail fin that they swing from side to side in a sculling motion. The Rajiformes swim either by lateral undulations of the trunk and tail (much like sharks), or by undulating waves running along the pectoral fins from head to tail, or by bird-like vertical movements of the entire pectoral fins (Myliobatidae).

Of the 2 subclasses, Holocephali (chimaeras; with only 1 external gill opening) have not been reported from Bermuda; Elasmobranchii (=Selachii) comprise all the known sharks, rays and skates. Of about 625 known species (6 orders, 24 families), 20 have been reported from Bermuda; 7 are included here.

OCCURRENCE: Marine; only a few species venture into brackish or fresh water. The great majority of shark species inhabits warm oceans at shallow depths; rays and skates are more widely distributed both in latitude and depth. The paucity of rays and skates in Bermuda is remarkable. Most Bermudian species are pelagic, either oceanic (blue, mako and whale sharks) or inshore (spotted eagle ray). The nurse shark is demersal and can be found sleeping in reef caves during daylight.

Although the spotted eagle ray and small sharks are occasionally harpooned for "sport", sharks are regularly caught on hook and line either by sport fishing or commercial fishing using longlines. They are increasingly used for food, either fresh or smoked, in Bermuda.

The few shark attacks that have occurred at Bermuda have been on persons spearfishing, and no fatalities have been recorded. Sharks are attracted by auditory signals in the first instance, such as underwater explosions, fish struggling on the end of a spear or other low-frequency pulsating sounds. Once in the vicinity, the shark's olfactory and visual senses take over; sharks usually home in on their prey by smelling blood, and then by seeing their target. Because there is no way of knowing how hungry a shark might be at any given time, it is wise to avoid them whenever possible. Apart from the great white shark that has been implicated in numerous attacks elsewhere in the world's oceans, there would appear to be little difference in the inclination of the various pelagic carnivorous species to attack humans. Hunger of individuals seems the most important determining factor, overriding other things such as size and stimulation. Some shark species are relatively unaggressive but should be treated with respect nevertheless, be it because of their size (e.g., the whale shark), or because they bite when disturbed by divers (e.g., the nurse shark).

The 2 species of rays found in Bermuda are no threat to swimmers or boatmen unless one tries to catch or handle them, in which case the spotted eagle ray can inflict serious wounds by means of its venomous

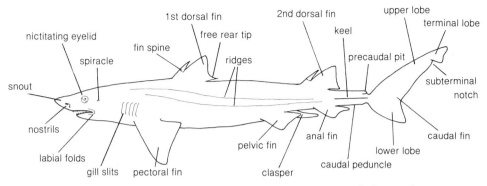

**191 CHONDRICHTHYES (Cartilaginous fishes): Schematic**

tail spines, causing shock (and possible death) of divers.

IDENTIFICATION: External features are sufficient for species identification.

### Plate 191

In contrast with Osteichthyes, Chondrichthyes generally lack countable characters. The shape and relative position of fins and the shape of head and body are the most useful general characters. The presence or absence and shape of dermal denticles, dorsal spines, caudal spines and barbels separate some groups whereas the size and shape of teeth, position of nostrils, presence and position of spiracles, presence of a nictitating membrane on the eye, labial fold, caudal keel and numbers of gill slits distinguish others. Male Chondrichthyes have lobes of pelvic fins enlarged to form claspers that function as external sex organs during copulation.

Because of their large size, scientific collection and maintenance is a problem. Whole small specimens should be fixed in 10% formalin and preserved after washing in water in 40% isopropanol or 75% ethanol; large specimens have to be mounted dry. Teeth and dermal denticles are also useful for separation of many species and can be relatively easily preserved dry.

BIOLOGY: In all elasmobranchs sexes are separate and fertilization is internal. Males copulate with females by means of claspers. In all species occurring in Bermuda mating is thought to take place in late spring or summer, although some species of Carcharhinidae have developed embryos year round. Some species are oviparous; most are viviparous (the yolk sac of the fetus forming a placenta with the maternal uterus for nutrient transfer) or ovoviviparous (without a placenta but with eggs developing within the female). Information on age and growth of Chondrichthyes is poor but larger species probably mature at 2-3 yr. Because Chondrichthyes lack a swim bladder they are negatively buoyant; pelagic sharks therefore must keep moving to stay afloat. Additionally, they have their pectoral fins locked in one plane to give their body lift and cannot use these fins for breaking or turning as most of the Osteichthyes do. All Chondrichthyes are predators; food varies, from plankton (Rhincodontidae) to fishes and mammals (Lamnidae and Carcharinidae), or bivalves, crustaceans and worms (Orectolobidae and Myliobatidae). Sharks have no predators (other than humans, and their own kind: many are known to attack each other, especially in a feeding frenzy). They do have a number of parasites such as Trematoda, Cestoda (e.g., *Crossobothrium*) and Copepoda (e.g., *Pandarus*). Sharks and rays are often accompanied by pilot fish and shark suckers (e.g., *Echeneis*).

DEVELOPMENT: In oviparous species eggs develop in rectangular or conical capsules of a horn-like material that the female deposits, preferably on sedentary benthic organisms. Freshly caught females of viviparous species often give birth to young that are quite aggressive and should be handled with care.

REFERENCES: Nomenclature used here follows ROBINS et al. (1980). The most comprehensive recent accounts are by BIGELOW & SCHROEDER (1948, 1953), who also refer extensively to material from Bermuda. The genus *Carcharhinus* has been revised by GARRICK (1982).
   Many of Bermuda's species are listed in books on tropical fish identification, e.g., RANDALL (1968), CHAPLIN & SCOTT (1972), GREENBERG (1977) and STOKES (1980). The few recent observations on sharks and rays in Bermuda are by WATERMAN (1975) and MOWBRAY (1965).

## Plate 192

O. **SQUALIFORMES** (True sharks): Elasmobranchii with 2 dorsal fins and an anal fin. With 5 gill slits; gill rakers usually absent, spiracles usually present.

F. **LAMNIDAE** (Mackerel sharks): Relatively robust fusiform Squaliformes with eyes located over mouth; lacking grooves between nostrils and mouth. With a well-developed keel on each side of caudal peduncle; caudal fin crescent-shaped, nearly symmetrical. Teeth large, sharp and relatively few. Gill slits large but not extending onto upper surface of head. (1 sp. from Bda.)

*Isurus oxyrinchus* Rafinesque (Shortfin mako): Body slender, with acutely pointed snout. Teeth strong, slender, smooth-edged, with a single cusp. First dorsal fin large, pointed, originating posterior to inner corner of pectoral fin when corner is laid back against side; 2nd dorsal fin very small. Pectoral fin long, but shorter than head length, and falcate; anal fin origin below middle of 2nd dorsal fin base. Caudal peduncle depressed; expanded laterally with a large keel on each side extending out on caudal fin. Deep blue-gray on back, gray on sides and whitish gray on belly. To 4 m. An oceanic pelagic species found near the surface. Feeds on fishes and cephalopods; a rapid swimmer that will leap out of the water when hooked. Ovoviviporous; number of young 1-6. An excellent food fish but not frequently caught.)

F. **ORECTOLOBIDAE** (Nurse sharks): Robust Squaliformes with eyes located behind mouth; with grooves connecting nostrils and mouth; mouth small, ventral, without ridges on back and without a keel on caudal peduncle. Caudal fin with poorly developed lower lobe. (1 sp. from Bda.)

*Ginglymostoma cirratum* (Bonnaterre) (Nurse shark): Head bluntly rounded; mouth short, nearly transverse; teeth small with a short central cusp and 2 cusplets on each side. The 2 posterior gill slits behind origin of pectoral fin and close to each other. Base of lst dorsal fin over base of pelvic fin; origin of 2nd dorsal fin anterior to base of anal fin. Lower lobe of caudal fin barely developed. Brown to brownish red on back, becoming tan to yellowish on belly. Young with dark spots. To over 4 m. By day found in caves in patch reefs or offshore reefs to about 40 m. Feeds at night over grass flats or sandy bottoms on invertebrates including crustaceans, mollusks and echinoderms. Occasionally caught in traps; not usually eaten.

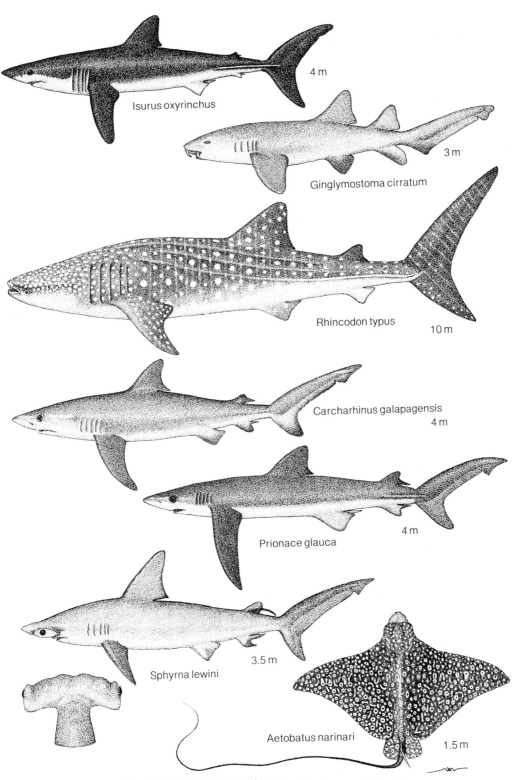

**192 CHONDRICHTHYES (Cartilaginous fishes)**

F. **RHINCODONTIDAE** (Whale sharks): Large robust Squaliformes with eyes located behind mouth; without grooves connecting nostrils and mouth. Mouth very large, transverse, terminal. Gill arches connected to form a sieve. Three ridges running lengthwise on body, the lowest ending in a strong keel on the caudal peduncle. Caudal fin crescent-shaped. (Only species in the family:)

*Rhincodon typus* Smith (Whale shark): Posterior 2 gill slits over the pectoral fin base. Snout short, truncated; mouth very wide, nostrils with short barbels and shallow grooves. Teeth numerous and very small, with hooked cusps. Upper precaudal pit present. Dark chocolate-brown above becoming yellowish below. Body and fins with white spots and bars arranged to give a "square" pattern on sides. To 12 m—the largest of fishes. Usually found singly in the open ocean near drop-offs, and on offshore banks. Feeds by filtering crustaceans, small schooling fish and cephalopods from the water. ♀ deposits egg cases more than 25 cm long. Seldom caught or used for food; harmless to divers.

F. **CARCHARHINIDAE** (Requiem sharks): Squaliformes with eyes located over mouth; with pits above and below the caudal peduncle but without a keel. Caudal fin asymmetric; lower lobe well developed but much shorter than upper lobe. Origin of 1st dorsal fin anterior to pelvic fin origin. Nictitating eyelid present; largest teeth at the sides, not the middle of jaw. (8 spp. from Bda.)

*Carcharhinus galapagensis* Snodgrass & Heller (Galapagos shark, B Dusky shark): Relatively robust, with free tip of anal fin base well separated from caudal fin origin. Origin of 1st dorsal fin posterior to pectoral fin insertions; relatively low and with rounded anterior margin. Pectoral fins relatively slender; posterior edge falcate. Second dorsal low with a nearly straight posterior margin. A low dermal ridge present on back between dorsal fins. Upper front teeth with broad cusps nearly triangular in shape. With 103-109 precaudal vertebrae. Blue-gray to lead-gray above becoming nearly white on belly. To about 3.7 m. Characteristic of oceanic islands; found inshore around shallow reefs and offshore. Viviparous, with 6-14 embryos. Feeds on fishes and large crustaceans. Caught on hook and line and on longlines; used to make shark hash and other shark products.

*Prionace glauca* (L.) (Blue shark): Slender-fusiform, with long slender snout rounded at tip. Base of 1st dorsal fin closer to pelvic fins than pectoral fins. Pectoral fins long, slender; posterior edge falcate. A weak lateral keel present on each side of caudal peduncle. No dermal ridge between dorsal fin bases. Teeth relatively slender and finely serrated. Dark blue above becoming bright blue on sides and white below. To 3.8 m. A pelagic species rarely entering shallow or lagoon water. Feeds on fishes, cephalopods and occasionally seabirds. Caught by sport fishing and on longlines.

F. **SPHYRNIDAE** (Hammerhead sharks): Relatively slender Squaliformes; the anterior portion of the head is flattened dorso-ventrally and widely expanded laterally, with the eyes at its outer edges. Nictitating membrane well developed on lower eyelid. Teeth blade-like, with single cusp. First dorsal fin anterior to origin of pelvic fin; 2nd much smaller than 1st.

***Sphyrna lewini*** (Griffith & Smith) (Scalloped hammerhead): Head hammer-shaped, its width over 3× its preoral length. Free rear tip of 1st dorsal fin well ahead of pelvic fin origin; posterior margin of anal fin deeply notched. Prenarial grooves strongly developed; teeth weakly serrated. Median indentation present on anterior margin of snout. Free rear tip of 2nd dorsal fin nearly reaching upper caudal fin origin. Anal fin base about 2× longer than base of 2nd dorsal fin. Uniform gray to grayish brown above, shading to white below. To 4.2 m. Primarily oceanic, occasionally entering shallow water. Viviparous. Feeds on fishes, cephalopods and crustaceans. Occasionally taken on hook and line and longlines.

**O. RAJIFORMES** (Rays and skates): Conspicuously dorso-ventrally flattened, disc-like Elasmobranchii with 5 pairs of gill slits on lower surface of body. Pectoral fins large, attached to side of head; eyeball attached to upper margin of orbit; eyes on top of head. Teeth in many rows, bluntly pointed or flattened. Pair of spiracles well developed, located on top of head. Tail well developed, or forming whip-like organ.

**F. MYLIOBATIDAE** (Eagle rays): Rajiformes with wing-like pectoral fins. Caudal fin indistinct; tail very slender, whip-like. Head distinctly protruding from body and slightly rounded in cross section. Eyes and spiracles more or less on side of head. Anterior portions of pectoral fins forming a snout-like subrostral lobe. A small dorsal fin located on base of tail. (1 sp. from Bda.)

***Aetobatus narinari*** (Euphrasen) (Spotted eagle ray, Whip ray): Snout pointed; teeth in a single series, flattened, pavement-like. Low dorsal fin between pelvic fins on base of tail. With 1-5 (usually 2) long, barbed, venomous tail spines close behind dorsal fin. Skin naked, smooth to touch. Upper surface chestnut-brown to olive-gray with white or yellowish spots; lower surface white. To 3.7 m length (including tail) and 2.8 m width. In harbors, bays and on reef platform, over sediment bottoms; fairly common. A rapid swimmer that will often make spectacular leaps. Feeds mostly on benthic organisms including small bivalves, crustaceans, worms, cephalopods and fish.

J. BURNETT-HERKES

## Class Osteichthyes (Bony fishes)

CHARACTERISTICS: *Aquatic VERTEBRATA with jaws, paired fins and a skeleton partly or wholly composed of true bone. Body usually covered by overlapping scales that may be modified into ossified plates, or reduced or absent. Skull with sutures. With a pair of typically double nasal openings that do not open into the mouth except in some species. With 1 pair of external gill openings and 4-5 pairs of gill arches. Swim bladder usually present.* Body size of species in Bermuda ranges from 1.5 cm (some Gobiidae) to the giant blue marlin, which reaches 4 m and 500 kg; the majority of species is in the 10-100 cm bracket. Body shape varies from fusiform, streamlined oceanic species (e.g., tunas) to compressed, deep-bodied (e.g., butterfly fishes) or depressed, flat species (e.g., flounders); from snake-like eels and morays to box-like trunkfishes and the bizarre sea horses. Colors of open-water species are generally from blue to gray above, silvery or white below; reef fishes often have bright colors and complex patterns, with nocturnal species usually in shades of red; and species occurring on sediment bottoms are in shades of gray and brown. Locomotion in most species is primarily by lateral movement of the caudal fin, with the other fins used for steering. This can result in rapid, continuous swimming (as in most oceanic species), or sudden bursts (as in many nearshore predators). The caudal fin may be assisted by the pectoral fins (as in wrasses) or by undulations of the dorsal and anal fins (as in triggerfishes) or of the entire body (as in eels and morays). Some species (e.g., wrasses) can bury themselves in the sediment, others (flying fishes) may propel themselves out of the water and glide for long distances over the waves.

The taxonomy of fishes continues to be dynamic as new information on a worldwide basis is synthesized. The class contains about 425 families, 3,880 genera and more than 18,000 species. Of about 425 species recorded from Bermuda, 180 are included here.

OCCURRENCE: Marine, brackish and fresh water, in all latitudes and all depths. Some groups are mainly pelagic and schooling (e.g., Atheriniformes, Clupeiformes) or solitary open-ocean migrants (e.g., Scombridae, Xiphidae); others are largely adapted to a benthic existence where they live more or less freely (e.g., Lophiiformes, Serranidae), temporarily buried in the substrate (e.g., Bothidae) or hidden in crevices and caves (e.g., Muraenidae, many Gobiidae and Blenniidae). Several taxa spend their entire life cycle in coral reefs (e.g., Chaetodontidae, Pomacanthidae). Most Myctophiformes and a number of Salmoniformes are meso- or bathypelagic living at depths of hundreds and thousands of meters, often ascending to the surface at night.

Fishes are collected using such traditional methods as hook and line, pots, traps and nets. Ichthyocides (fish poisons) have to be used to catch some forms and it may be necessary to crack pieces of coral or rocks to extract the more cryptic species like gobies. Bermuda Fisheries Regulations administered by the Department of Agriculture and Fisheries prohibit the use of ichthyocides and other poisons, explosive substances and fish pots for collecting except by special permit. Regulations and Fisheries Orders also protect a number of fish species.

Most Bermuda fishes are innocuous although puncture wounds from scorpion fishes can be fairly painful. Glandular material (ovaries, liver, etc.) from puffers and porcupine fishes may be toxic (tetrodotoxin) and should be avoided. Several isolated cases of ciguatera (tropical fish poisoning) during the past 2 decades have been attributed to large barracudas. The toxic fish have likely wandered to Bermuda from normally ciguatoxic areas in the Caribbean and are believed to be chance encounters. The flesh of oilfishes contains high levels of oils and waxes that cause diarrhea. Some fish (soapfishes and trunkfishes) produce secretions toxic to other fishes when kept in confined spaces. For more detail regarding poisonous and venomous fishes see HALSTEAD (1978).

Fish are important sources of protein for humans. Bermuda's commercial fishery annually yields 450,000 kg all of which is consumed locally. Groupers, snappers, porgies, grunts, parrot fish and some jacks are trapped mostly with wire arrowhead pots, whereas inshore pelagic species of jacks, little tunny, yellowtail snapper and all baitfishes are taken with modified beach seines; oceanic fishes are caught on hook and line by trolling, or by chumming (luring them by scattering bait) from anchored boats. In the past decade

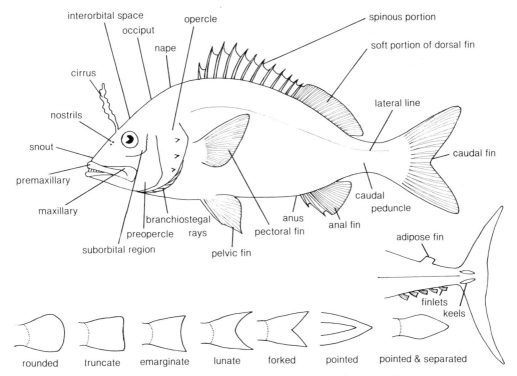

**193 OSTEICHTHYES (Bony fishes): Schematic**

**Plate 193**

catches of grouper and snapper have declined drastically, and fishermen have put more effort into catching other reef fishes and oceanic species such as tuna and wahoo.

IDENTIFICATION: Because of the morphological diversity of Osteichthyes, a wide range of characters is used to distinguish families, genera and species. External features are usually sufficient to distinguish species in the field. Although color and patterns can be useful (particularly in identifying reef fishes) colors of many species vary with habitat (shallow water vs. deep water), background color, type of substrate and activity pattern of the fish (diurnal/nocturnal). Some groups are sexually dichromatic whereas others exhibit marked ontogenetic changes in colors and patterns.

Characters of fundamental diagnostic importance are body shape (round, elongate, eel-like = anguilliform, compressed, depressed, fusiform = spindle-shaped); and the presence or absence, size, shape and position of paired and unpaired fins including the type (spinous or soft) and number of elements (rays). Relatively primitive bony fishes have ventral fins located abdominally, pectoral fins low on the sides, and a dorsal adipose fin (Salmoniformes, Myctophiformes). Finlets are located on the caudal areas of some pelagic fishes (Scombridae, Coryphaenidae, Gempylidae). The anal fin is variously modified to form an intromittent organ (gonopodium) used in copulation in some families (Poeciliidae, Idiacanthidae). The shape of the caudal fin characterizes many species; it may be continuous with or separated from the dorsal and anal fins. Spinous portions of fins are simple, unbranched, usually hard structures. Soft rays are striated and often branched. Because of the branching, counts of soft rays must be done carefully by examining the basal portion of the ray.

The skin may be naked or variously covered with simple (cycloid) or toothed (ctenoid) scales. Scales may cover virtually all surfaces of the body, head, lips and fins, or only a small portion as in some Scombridae. Ctenoid scales may have simple tooth-like projections on the exposed portions or may have species-specific elaborate projections as in Balistidae. Scales of some species are easily shed (deciduous) whereas others are fused to form plate-like structures (Syngnathidae) or solid encasements (Ostraciidae). Reinforced scales that form bony ridges or scutes are found in some families (Clupeidae, Carangidae), and their occurrence and shape are species-specific.

Patterns of sensory pores, whether forming parts of the lateral line system or variously scattered over the head and body, can be used to separate families, genera and species. The position, shape and size of the mouth and arrangement and shape of bones forming the gape (maxillary, premaxillary and dentary) are characteristic of species and higher taxa. The presence or absence and patterns of teeth in the gape, on the tongue or on the bones in the roof of the mouth (palatine and central vomerine) are important characters. Teeth may be needle-like (villiform), robust and rounded (molariform), fine and closely packed (comb-like), robust and pointed (canine), flattened and sharp-edged (incisor-like) or variously fused to form plates (beak-like). The tooth-like structures on the gill arches (gill rakers) vary with species in number and shape as do the projections in the pharynx (pharyngeal teeth).

The structure of bones on the cranium (particularly bony ridges), the shape of the opercular bones and whether these bones have spines and serrations are used to separate many species. Fleshy projections or cirri and barbels are found on some species and these may vary in size, shape and position with species. Another character of diagnostic value is the swim bladder, and whether there is a duct connecting it with the foregut (physostomous) or it is closed early in life (physoclistous). Body proportions, head length, caudal peduncle length and depth, eye diameter, body depth, relative position of anus and paired fins, location and length of vertical fins, and jaw length are examples of morphometric characters.

Specimens should be fixed in 10% formalin for 10-20 days, then washed well in water and preserved in 50% isopropyl alcohol or 80% ethyl alcohol. Specimens larger than 5 cm should be injected with formalin or have their gut cavities opened. Care must be taken to ensure solutions are not diluted by body fluids. Colors will fade rapidly and notes on color patterns must be made prior to fixation. If possible, specimens should be photographed live or immediately after death. Fins should be spread and the body flattened for ease in collecting meristic and morphometric data from the preserved specimens.

BIOLOGY: In most species sexes are separate throughout their adult life. There are, however, many groups that are hermaphroditic. Synchronous hermaphroditism is uncommon; consecutive hermaphrodites may first be male (protandrous) or female (protogynous). Protogynous hermaphrodites are found commonly among Serranidae, Labridae and Scaridae. In the latter 2 families there are species in which some males do not pass through the female stage (primary males) as well as those secondary males that have first been female. The secondary males are brightly colored but the primary males are dull or colored like the females. Sexual dichromatism and dimorphism are found among some groups (Scaridae, Labridae, Coryphaenidae, Idiacanthidae, Poeciliidae) but sexes are externally indistinguishable among most reef fishes. Eggs are usually small (1.5-3 mm in diameter), and fertilization is usually external, with eggs and sperm being released into the water simultaneously. Although pairing does take place most species engage in mass spawning, releasing floating eggs and sperm in the water column. Many species undergo migrations and form large aggregations or schools for the purpose of spawning. Such migrations and aggregation are found among the groupers (Serranidae) and common eels (Anguillidae). Reproductive strategies are varied and some species (Pomacentridae, Gobiidae) lay demersal eggs in a cluster on rocks and the males guard them. Other fishes are mouth brooders (Apogonidae), pouch brooders (Syngnathidae) or even ovoviviparous (*Gambusia*). Generally nesters, brooders and live-bearers have relatively large but few eggs.

Fishes occupy a wide variety of niches, and although some have specific food types most are fairly opportunistic. Plant material consumed ranges from minute diatoms and flagellates eaten by larval fish to macroalgae and spermatophytes such as *Sargassum* and *Thalassia*. Most herbivores have evolved some method of milling plant material. The stomach of mullet is modified into a grinding gizzard-like organ with a greatly thickened muscular wall; the pharyngeal teeth of parrot fishes form grinding plates. The comb-like teeth of chubs (Kyphosidae) and blennies (Blenniidae) and fused plate-like teeth of parrot fishes are adaptations for scraping algae from hard substrates.

Predators generally have well-developed grasping and holding teeth in both jaws, on the tongue, roof of the mouth (palatine, vomer), the pharyngeal region and on the gill arches. Predators such as barracuda (*Sphyraena*) may actively hunt their prey whereas others such as groupers (*Epinephelus, Mycteroperca*), scorpion fish (*Scorpaena*) and lizard fish (*Synodus*) are lurkers and wait for their prey to approach them, then dart out to grasp it. Frogfishes (*Antennarius*) have developed a lure on the end of their first dorsal fin ray to attract their prey, which they swallow whole. The development of extremely long snouts used for pipette feeding is found among a wide variety of fishes including coronet fish (*Fistularia*), trumpet fish (*Aulostoma*), sea horses and pipefishes (Syngnathidae) and butterfly fishes (Chaetodontidae). Billfishes (Istiophoridae) have long upper jaws and reportedly slash their prey with these bills, then return to pick up the pieces. The strong, rounded teeth found in the jaws of porgies (Sparidae), wrasses (Labridae) and triggerfishes (Balistidae) are used for crushing such prey animals as mollusks, crustaceans and sea urchins. Bottom feeders have typically developed inferior (ventral) jaws and thickened fleshy lips. Some (Labridae) have forward projecting canines for digging in the substrate; others (Mullidae) have sensory barbels used to locate buried prey. Among the most important carnivores for the energy budget of coral reefs is a group of unrelated species (members of the grouper (Serranidae), snapper (Lutjanidae), damselfish (Pomacentridae), and wrasse (Labridae) families) that forage for planktonic animals in the water column adjacent to coral reefs. All have evolved streamlined body shapes, reduced dentition and elongated gill rakers, protrusible jaws and a relative loss of skeletal robustness.

The symbiotic relationship of "cleanerfishes", which feed on ectoparasites and necrotic tissue found on the skin, gills and in mouths of other fishes, provides food for the cleaner and presumably better health for the fish being cleaned. Although obligate cleaning fishes do not occur in Bermuda, this niche is occupied by juvenile wrasses (*Bodianus, Halichoeres*) and butterfly fishes (Chaetodontidae).

The behavior of schooling among fishes has been debated and the adaptive advantage of such behavior remains unclear. Prey species such as Atherinidae, Clupeidae and Engraulidae appear to derive some benefit from school formation; generally those fishes on the perimeter of the school are the ones eaten first. Many fishes form aggregations of both sexes (Serranidae, Scaridae) for communal spawning. Mixed species schools (Lutjanidae, Pomadasyidae) are commonly found on coral reefs and beneath mangrove roots and jetties during daylight hours; at night they disperse to feed over the reef and adjacent grassy or sandy areas.

Many species, particularly nesting fishes, have well defined territories. Pomacentridae will vigorously defend territories around their nests as well as larger areas if the intruder is a potential competitor for food. The largest areas are defended against members of the same species. Territoriality is combined with hierarchical social arrangements among some groups of fishes. The highly colored "super males" of Scaridae and Labridae

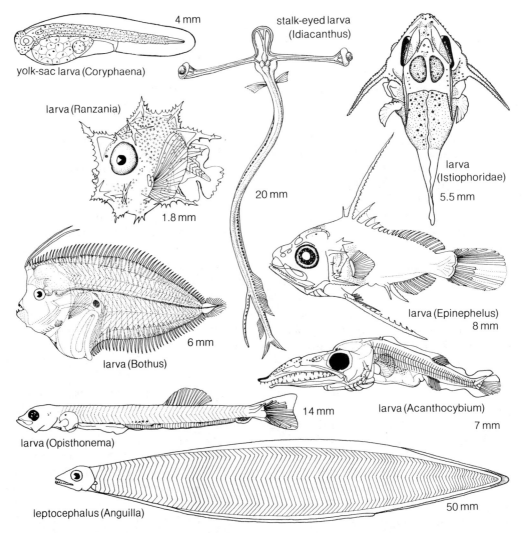

**194 OSTEICHTHYES (Bony fishes): Development**

compete for space and territorial ranges whereas non-super individuals form dense aggregations. Colors and patterns such as ocelli, bars and spots act as signals for many reef fishes, including Pomacentridae, Pomacanthidae, Chaetodontidae, Labridae and Scaridae, and change as the fish mature or become sexually active.

Fish can be hosts to a number of parasites that live externally and feed on mucus and epidermal tissue of the skin, mouth and gills, as well as internally in the gut, heart, liver, gonads, etc. Common ectoparasites include isopods (*Cymothoa*, *Alcirona*), copepods (*Caligus*), monogenetic trematodes (*Microcotyle*); endoparasites commonly found in the digestive tract are digenetic trematodes (*Alcicornis*, *Lepidapedon*), cestodes (*Acompsocephalum*), acanthocephalans (*Echinorhynchus*) and nematodes. Ongoing studies of fish diseases in Bermuda have revealed a high incidence of pathogenic protozoans, fungi and bacteria.

## Plate 194

DEVELOPMENT: Eggs hatch in as little as several hours to months after fertilization. Newly hatched fishes may resemble miniature adults (Syngnathidae) or be fragile larvae supported by a yolk sac (*Coryphaena*). Many pelagic species have sometimes bizarre flotation devices that act to increase surface area, such as the flattened, leaf-like leptocephalus of eels and the long eye stalks of *Idiacanthus*. Some devices also serve to make the larvae difficult for small predators to consume, such as well-developed opercular spines (Istiophoridae), pelvic and dorsal spines (*Epinephelus*) and dermal spines (*Ranzania*). Flat fishes (*Bothus*) have relatively normal larvae but change the entire architecture of their heads to arrange both eyes on one side of the adult. Growth rates of juvenile fishes are highly variable even among relatively long-lived species. The dolphin fish (*Coryphaena*) matures in 3-6 months whereas groupers of the same size will take 3 years; some species are annual (small gobies, flying fishes).

REFERENCES: Nomenclature used here follows ROBINS et al. (1980). Information on general ichthyology or fish biology can be found in LAGLER et al. (1977) and BOND (1979); MARSHALL (1971) provides a sketch of the evolution of fishes and their adaptations to life in the sea. The eggs and larvae of many northwest Atlantic species are described in JONES et al. (1978). Useful general treatises that include many species from Bermuda are the ongoing series FISHES OF THE WESTERN NORTH ATLANTIC (1948-1977) and JORDAN & EVERMANN (1896).

Most of Bermuda's species are included in such regional works on fish identification as B[O"]HLKE & CHAPLIN (1968), RANDALL (1968) and FISCHER (1978) and in field guides such as CHAPLIN & SCOTT (1972), STOKES (1980), GREENBERG (1977) and MOWBRAY (1965). BEEBE & TEE-VAN's (1933) field book of the shore fishes of Bermuda is fairly complete but many of the scientific names have changed. Endemism in Bermuda fishes was reviewed by COLLETTE (1962). W. BEEBE's deep-sea dives added large numbers of species (see BEEBE 1935, BEEBE & CRANE 1937 and earlier references) many of which have since changed names. SMITH-VANIZ et al. have an annotated checklist in preparation.

## Plate 195

O. **ELOPIFORMES:** Body compressed, usually slender; pelvic fins abdominal; caudal fin forked; gill openings wide; scales relatively large, cycloid. Mesocoracoid and postcleithra present. Heptocephalus with well-developed caudal fin and posterior dorsal fin.

F. **MEGALOPIDAE** (Tarpons): Elopiformes with mouth terminal or superior; pseudobranchiae absent; branchiostegal rays 23-27; dorsal rays 13-21, the last ray extended; pelvic rays 10 or 11; anal rays 22-29. Silvery fishes with large scales. (1 sp. from Bda.)

*Tarpon atlanticus* (Valenciennes) (=*Megalops atlanticus*) (Tarpon): Lower jaw projecting; body moderately deep, compressed. A plate present between 2 dentary bones of lower jaw. Pectoral fins low on body; last dorsal fin very filamentous; anal fin long, origin just behind dorsal fin base. Bluish gray on back, silvery on sides and ventrally. To 250 cm. Juveniles found in drainage canals connected to sea and in mangrove areas, larger specimens in harbors and offshore reefs to about 20 m. Feeds on schooling fishes such as sardines, herrings, mullet and anchovies, and crustaceans. Occasionally caught on hook and line but seldom used for food.

F. **ALBULIDAE** (Bonefishes): Elongate, relatively depressed, fusiform Elopiformes with lateral line; scutes on ventral surface lacking. Snout conical; mouth inferior; pectoral fins low on side of body; pelvic fins with 9 rays; anal fin well behind dorsal fin, near caudal fin peduncle. Color silvery; leptocephalus with well-

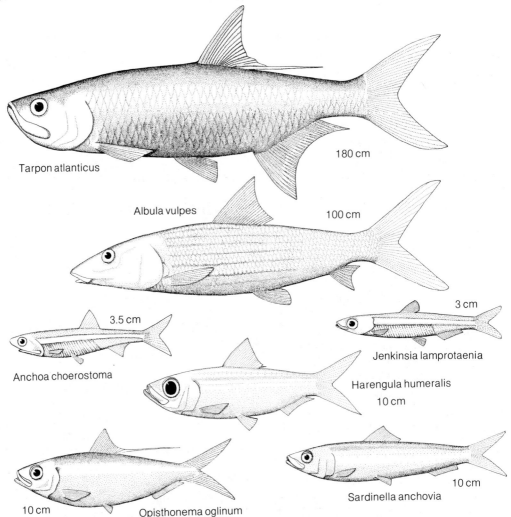

**195 ELOPIFORMES, CLUPEIFORMES (Anchovies, a.o.)**

developed caudal fin. (1 sp. from Bda.)

***Albula vulpes*** (L.) (Bonefish): Eye large, often covered with adipose tissue; mouth inferior, short, the upper jaw not reaching eye; anal fin short, with 8 or 9 rays. Color bluish-green silvery on back; silvery on sides with fine dark stripes; belly silver; a dark blotch on upper pectoral fin base. To 75 cm. Found nearshore and over lagoon on sandy and muddy areas. Feeds by grubbing in bottom for worms, crustaceans and mollusks. Caught on hook and line; an excellent game fish but seldom eaten.

**O. CLUPEIFORMES:** Body compressed; swim bladder diverticula connected to ear with 2 large vesicles within bullae of the prootic and pterotic bones. Branchiostegal rays fewer than 15; abdomen usually with scutes forming keel along midventral surface; lateral line pores absent on body; lateral line canals extending over operculum. Usually with numerous long gill rakers.

F. **ENGRAULIDAE** (Anchovies): Small silvery Clupeiformes with extremely long jaw reaching well behind eye; mouth inferior, below prominent snout. Unbranched soft rays in anterior section of dorsal and anal fins; caudal fin forked. Ventral body scutes absent; scales cycloid, deciduous. (1 sp. from Bda.)

*Anchoa choerostoma* (Goode) (B Hogmouth fry, Bermuda anchovy): Body fusiform, moderately compressed; snout relatively short and blunt. Posterior tip of maxilla pointed, nearly reaching free margin of operculum. Pseudobranch present, shorter than eye. Dorsal surface silvery bluish gray becoming translucent silvery on ventral surface; a narrow bright silver stripe on sides. To about 10 cm. Most common offshore near breakers and patch reefs but often trapped inshore by schools of little tunny, jacks and yellowtail snappers. Feeds on zooplankton. Forms dense active schools caught in cast nets and seines; highly prized as bait. An endemic species.

F. **CLUPEIDAE** (Herrings and Sardines): Small, mostly silvery Clupeiformes with fusiform or nearly round bodies; scutes mostly present along ventral surface. Lower jaw short but deep. Fins lacking spiny rays; dorsal fin single, located midway on back; caudal fin deeply forked. Lateral line absent; scales cycloid, usually deciduous. (4 spp. from Bda.)

*Harengula humeralis* (Cuvier) (Redear sardine; B Pilchard): Body fusiform, with ventral scutes; anal fin short, with less than 30 rays; pelvic fins with 8 rays. Upper jaw without central notch; hind border of gill opening with 2 fleshy knobs; hypomaxilla toothed; tooth plates extending back from tongue very slender, width less than 1/10 length; 27-28 ventral scutes. Color faintly bluish silver on back becoming silvery on sides and belly; indistinct orange spot at upper corners of operculum. To 15 cm. Inshore, pelagic, over lagoon and in harbors and bays. Feeds on zooplankton. Caught in seines and used as bait.

*Jenkinsia lamprotaenia* (Gosse) (Dwarf herring; B Blue fry): Body elongate, nearly round; snout pointed; ventral scutes absent. Premaxilla toothed; branchiostegal rays 5-6; eye relatively large; pectoral fin with 13- 14 rays. Deep bluish or greenish silver on back becoming translucent-silvery on belly. A bright midlateral silver stripe from operculum to caudal base. To 6.5 cm. Around patch reefs and rocky shorelines, often forming dense schools. Fed on by schools of little tunny and jacks. Used as bait when caught in beach seines.

*Opisthonema oglinum* (Lesueur) (Atlantic thread herring): Body compressed to fusiform, relatively deep. Belly with elongated scutes forming a distinct keel ventrally. Dorsal fin forward of center, its last ray extended, filamentous; pelvic fins with 8 rays. Back bluish green, sides and belly silver, a small black spot behind upper corner of operculum. To 25 cm. Inshore, pelagic, schooling around patch reefs and in harbors and bays. Fed on by schools of jacks and little tunny. Caught in seines and used as bait.

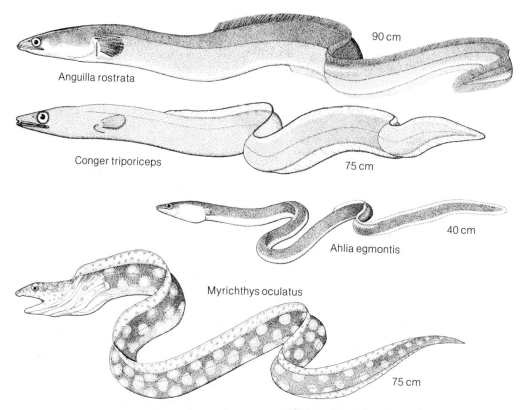

**196 ANGUILLIDAE (Eels), CONGRIDAE (Conger eels), OPHICHTHIDAE (Sand eels)**

*Sardinella anchovia* (Valenciennes) (Spanish sardine; B 'Chovy): Elongate, nearly round, with a line of ventral scutes on rounded belly (no keel). Top of head with 8-10 longitudinal striae; hypomaxilla absent; posterior margin of gill opening with 2 gently rounded fleshy knobs. Anterior rakers on lower limb of gill arches lying nearly flat; pelvic fin with 9 rays. Deep blue on back becoming silvery on sides and belly. To 30 cm. Inshore, pelagic, schooling around patch reefs offshore to breakers. Feeds on zooplankton; fed on by schooling fish (jacks, little tunny). Much sought after for bait; larger fish eaten.

**O. ANGUILLIFORMES:** Body elongate, pelvic fins absent; anal fin usually elongate; dorsal fin elongate or absent. Pectoral fins may be absent. Gill openings narrow. The 2 premaxillae, vomer and ethmoid united to form a single bone. Maxillae toothed and included in gape. Posttemporal, mesocoracoid and postcleithra absent.

### Plate 196

**F. ANGUILLIDAE** (Eels): Anguilliformes with relatively robust round body covered with small elliptical, well-embedded scales. (1 sp. from Bda.)

*Anguilla rostrata* (Lesueur) (American eel): Pelvic fins well developed;

lower jaw projects slightly; teeth in both jaws small, set in bands; small vomer teeth present; gill slits small, anterior to base of pectoral fins. Dorsal fin long, continuous with anal fin; anal fin origin just posterior to anus. Color variable, brownish gray on back, yellow on sides and silvery gray to pale yellow on belly. To 150 cm. In brackish water, particularly drainage canals with access to sea. Will move over land on wet nights and is occasionally found in wells and ponds. Feeds on small invertebrates and fishes. Spawning area believed to be Sargasso Sea southeast of Bermuda. Larvae (leptocephali) in plankton offshore.

F. **CONGRIDAE** (Conger eels): Anguilliformes with pectoral fin present; gill opening on side anterior to pectoral fin base; scales usually absent. Branchiostegal rays long, well developed but not overlapping ventrally; lower jaw not projecting; dorsal and anal fins contiguous with caudal fin; dorsal fin originates above pectoral fin. (4 spp. from Bda.)

***Conger triporiceps*** (Kanazawa) (Many-toothed conger): Dorsal fin rays segmented; a labial flange (lip-like structure) present in upper jaw as a wide band. Teeth in jaws forming 1 or 2 rows; teeth in outer row compressed forming sharp cutting edge; maxillary pores located above labial flange small. Three supratemporal pores (located dorsally across back of head) and 1 or 2 postorbital pores (located centrally behind eyes). Brownish gray on back becoming pale ventrally. Dorsal and anal fin brownish with dark bluish black margin; lateral line pores pale. To 150 cm. Mainly in sandy areas among reefs offshore, 10-60 m. Feeds on small fishes, crustaceans and other invertebrates. Occasionally caught in traps and on hook and line.

F. **OPHICHTHIDAE** (Sand eels): Long, snake-like Anguilliformes; snout pointed; nostrils widely separated, the posterior inside mouth or opening through a valve in upper lip. Gill openings small, slit-like, just anterior to pectoral fin base. Branchiostegal rays free, overlapping ventrally to form basket. Dorsal fin may be contiguous with caudal and anal fins, or discontinuous, with caudal fin rays reduced to form a sharp point. Color variable; may be banded or spotted. (11 spp. from Bda.)

***Ahlia egmontis*** (Jordan) (Key worm eel): Very slender. A fringe of fin present around tip of tail to form contiguous dorsal, anal and caudal fin. Posterior nostril on lip partly inside and partly outside mouth. Teeth on vomer, but absent on vomer shaft. Dorsal fin originating approximately above anus; pectoral fin short. Yellowish brown overall but intensity variable; may show bicolor pattern or vary from pale to dark with minute dark spots. To 40 cm. In sandy or muddy areas inshore; offshore to 20 m, in sandy pockets in reefs. Feeds on benthic invertebrates and small fishes.

***Myrichthys oculatus*** (Kaup) (Gold-spotted eel): Pectoral fins small; caudal fin forming sharp point; dorsal and anal fins not connected; dorsal fin originating on head anterior to gill openings. Teeth low, blunt and molariform; anterior nostril tubular. Body and head greenish tan to whitish, covered with paler round spots varying from gold to pale yellow surrounded by dark brownish areas. To 100 cm. In grass flats; not uncommonly seen during daylight moving on surface of sand through grass blades. Feeds mainly on crabs and other crustaceans.

### Plate 197

**F. MURAENIDAE** (Morays): Relatively robust, somewhat compressed Anguilliformes with tubular anterior nostrils at front of snout and posterior nostrils above anterior portion of eye. Teeth well developed, either depressible canines or blunt, molariform. Pelvic and pectoral fins absent; gill openings small round apertures; pharyngeal jaws well developed. Dorsal and anal fin vary from long, well developed to short and restricted to tail. Lateral line absent but few pores on head and above branchial region. (14 spp. from Bda.)

*Enchelycore nigricans* (Bonnaterre) (Viper moray): Dorsal fin originating above gill openings; teeth sharp, well developed, fang-like. Jaws arched, exposing many teeth when mouth is closed. Pectoral and pelvic fins absent. Posterior nostrils elongated into long open oval structures above anterior margin of eye. Only a few enlarged canines present, forming inner tooth series of lower jaw. Young with reticulated whitish pattern on brown; adults uniformly chestnut brown. To 100 cm. In shallow water in rocky and coral reef areas from shore to about 20 m. Feeds on fish and crustaceans. Large specimens occasionally caught on hook and line.

*Muraena miliaris* (Kaup) (Goldentail moray): Dorsal fin origin on head; posterior and anterior nostrils tubular. Teeth fang-like, without serrations. Color highly variable, may be purplish black with bright yellow spots and reticulations on caudal area; may have reverse coloration of yellowish body with brown to black freckles and reticulations, or bright golden yellow with black markings. To 50 cm. From rocky coastal areas to offshore reefs and rocky areas to more than 200 m. Occasionally caught on hook and line. Not generally used for food.

*Lycodontis funebris* (Ranzani) (Green moray): Genus with dorsal fin above branchial region or further forward above head; teeth sharp, fang-like, lacking serrations; only anterior nostril forming tube.—Species uniformly greenish; color produced by bright yellow mucus layer over bluish skin. To 190 cm. In rocky and reef areas from the shoreline to about 50 m. Feeds on fish and crustaceans. Caught in traps, larger specimens used as food.

*L. moringa* (Cuvier) (Spotted moray, B Speckled moray): Genus as above.—Species with head, fins and body dark purplish black covered with fine whitish yellow reticulations. Dorsal fin may have black margin anteriorly. Juveniles black with white lower jaw. To 120 cm. Common from shore to reefs offshore at about 50 m. Feeds on fishes and crustaceans. Caught in traps and on hook and line; occasionally used as food.

*L. vicinus* (Castelnau) (Purplemouth moray): Genus as above.—Species with small dark brown patch at corner of mouth. Two color patterns: one with distinct greenish mottlings overall, the other nearly uniform greenish brown with faint darker freckling. Dorsal and caudal fins edged with white, black submarginally. Juveniles

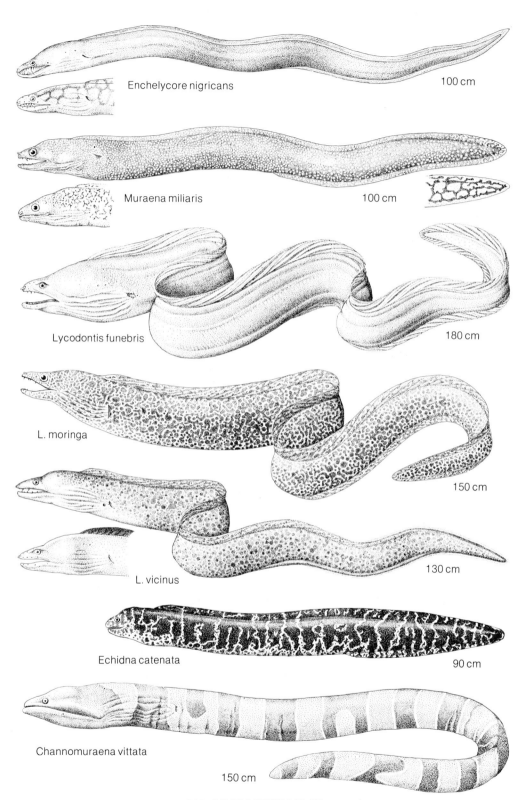

**197 MURAENIDAE (Morays)**

dark brown with white lower jaws. To 110 cm. Most common in shallow water nearshore in rocky or patch reef areas but occurs to 40 m. Feeds on fishes and crustaceans. Caught on hook and line and in traps. Occasionally used as food.

***Echidna catenata*** (Bloch) (Chain moray): Dorsal fin originating above gill openings; teeth rounded, blunt and molariform. Snout blunt. Color dark brownish black, with white to bright yellow reticulated pattern on head, body and fins. To 85 cm. In shallow water from shoreline to shallow offshore (usually less than 5 m), among loose rocks or coral rubble. Feeds on crabs and other crustaceans.

***Channomuraena vittata*** (Richardson) (Banded moray): Dorsal and anal fins confined to tip of tail; lower jaw projecting beyond upper jaw; posterior nostrils with tubes; teeth small, needle-like. Color alternating bands of reddish brown and yellowish brown or tan. To 120 cm. Only at moderate depths, 35-50 m. Feeds on crustaceans and fishes. Occasionally caught in traps by fishermen; not generally used as food.

### Plate 198

O. **MYCTOPHIFORMES:** Generally slender, usually with forked caudal fin; adipose fin present; pelvic fins abdominal, 8-12 rays; 9 principal caudal fin rays. The maxilla is excluded from the gape of the mouth (premaxilla forms gape).

F. **SYNODONTIDAE** (Lizard fishes): Elongate, cylindrical Myctophiformes; large mouth with numerous needle-like teeth; teeth also present on tongue and roof of mouth. No spines in fins. Dorsal fin single, with short base, located near middle of back. Adipose fin on back located above or slightly behind anal fin. Pelvic fins abdominal, with 8 or 9 rays. Caudal fin forked. Cycloid scales present on head and body. Lateral line present. Normally light brown fishes that lurk on the bottom, some partially bury themselves in sand or mud. (4 spp. from Bda.)

***Synodus intermedius*** (Agassiz) (Sand diver, B Snakefish): Pelvic fins with 8 rays, the inner ones much longer than the outer. Anal fin base shorter than dorsal fin base. Palatine teeth form a single pair of bands; teeth in jaws not visible when mouth is closed. Head depressed, the eyes located above midpoint of upper jaw. Scales relatively large, with 48-50 on lateral line. Color tan with dark brownish bars. Yellow to orange stripes consisting of small spots on sides. Except dorsal fin, all fins with orange highlights. To 40 cm. Common in shallow water, usually found on sandy bottom. Feeds on small fishes and sometimes takes a trolled lure.

F. **MYCTOPHIDAE** (Lantern fishes): Slender Myctophiformes with head and body compressed, eyes large, mouth terminal or inferior, jaws extending beyond posterior margin of eye. Teeth present on jaws and roof of mouth. All fins lack spines; the origin of anal fin beneath or close behind base of dorsal fin. Pelvic fins with 7 or 8 rays. Round or kidney-shaped photophores (light-emitting organs) on head and body in specific patterns along with other luminous organs. Pelagic fishes, found from surface (at night) to 2,000 m.

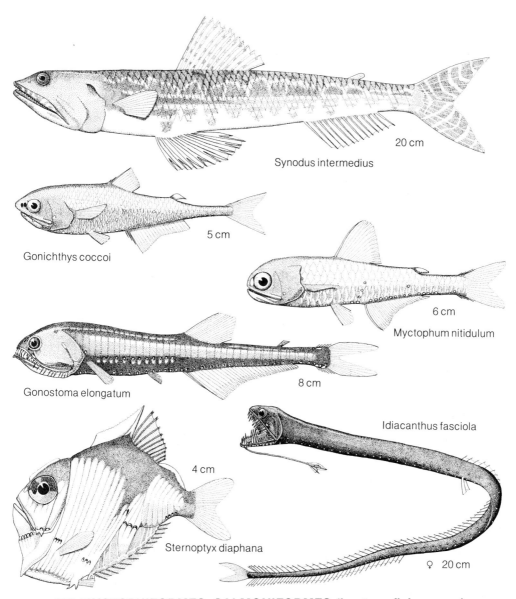

**198 MYCTOPHIFORMES, SALMONIFORMES (Lantern fishes, a.o.)**

***Gonichthys coccoi*** (Cocco) (Cocco lantern fish): Snout conical, the upper part projecting, mouth inferior. Caudal peduncle long and slender, its least depth 1/4 height of body. Origin of anal fin on or slightly in front of vertical line through end of base of dorsal fin. Gill rakers present. A single row (5-7) of photophores over base of anal fin; photophores absent on dorsal surface. Back and nape blackish; sides silvery to gold. To about 10 cm. Feeds on zooplankton. Often collected near surface in night plankton tows.

***Myctophum nitidulum*** Garman: Eye large, mouth slightly oblique, the postero-dorsal margin of the opercle sharply angular. Scales on body cycloid. A row of 3 light organs (supra-anal organs) ascending from anal region to lateral line in a slightly oblique but straight line. Four ventral light organs form a line between pelvic (ventral) fin and anal fin. The supra-anal organs are located posterior to the 3rd ventral light organ. Dorsal fin with 13 or 14 rays; anal fin with 19 or 20 rays and pectoral fin with 13 or 14 rays; origin of dorsal fin over base of ventral fin. Silvery-black on back becoming silvery on sides and ventral surfaces. ♂ with a luminous gland on upper surface of caudal peduncle consisting of 6 or 7 overlapping scale-like structures. To 10 cm. Found at or near surface during night and to 850 m during daylight. Feeds on zooplankton. Frequently caught in surface plankton tows at night.

O. **SALMONIFORMES:** Fins with soft rays only; most species with dorsal adipose fin; pectoral fin low on side of body (except in one family); pelvic fins located abdominally. Maxilla is included in the gape of the mouth.

F. **GONOSTOMATIDAE** (Bristlemouths): Elongate Salmoniformes rarely exceeding 25 cm in length. Mouth large, with teeth in jaws; well-developed gill rakers; photophores present on nearly all species, located on head, branchiostegal membrane and in 1 or more rows lengthwise on body. Color dark brown to black with silvery iridescence on sides and head. Mesopelagic to bathypelagic, but may migrate to within 50 m of surface at night. Often taken in night plankton tows.

*Gonostoma elongatum* Günther (Bristlemouth): Eyes of moderate size; series of well separated, relatively long, slender teeth on maxilla; maxilla curved convexly; premaxillae slightly concave posteriorly. Nine photophores on branchiostegal membrane; no photophores on isthmus; large luminous gland located behind the orbital photophore. Scales thin, large but deciduous. Black on back, inside of mouth and gill covers; sides may be dark silvery, bluish or with greenish iridescence. To 27 cm. Found between 700 and 1,000 m during the day and 150-200 m at night. Feeds on zooplankton (crustaceans).

F. **STERNOPTYCHIDAE** (Hatchetfishes): Salmoniformes with deep, extremely compressed body; pectoral fins low on body. Dorsal pterygiophores exposed, anterior to dorsal fin, resemble a spiny dorsal fin and may be fused into a thick plate to form a small pair of bony keels or a single elongate spine. Mouth nearly vertical. Anal fin with 11-19 rays, may be divided; dorsal fin with 8-17 rays. Photophores on head and body in species-specific patterns. Pelagic, found near the surface at night, to 1,500 m during the day. Often taken in night plankton tows near surface.

*Sternoptyx diaphana* Herman (Hatchetfish): The dorsal blade consists of a single enlarged pterygiophore; a pair of short spines at origin of anal fin; 2 pairs of short postabdominal spines in front of pelvic fin bases; anal fin single, with 13-15 rays. Abdominal photophores 10; anal 3; branchiostegal 3; isthmus 5 and preanal 3. Dark on back, silvery on sides. To 4.5 cm. Found from the surface at night to 2,000 m during daylight. Feeds on zooplankton.

F. **IDIACANTHIDAE** (Stalkeyes, Blackdragonfishes): Long, slender eel-like Salmoniformes with roundish cross section. Scales absent. Pectoral fins absent in adults, present in larvae; larvae with eyes on long stalks; ♀ with hyoid barbel and with teeth well developed, fang-like; ♂ lacking barbel, teeth and functional digestive system. Dorsal fin long, its origin anterior to midbody length, with 54-74 rays; anal fin origin under mid-dorsal fin, about 1/2 length of dorsal fin, with 29-49 rays. A pair of short, pointed bony projections anterior to and flanking each dorsal and anal ray. Luminous organs scattered over head and body. Sexual dimorphism notable. Pelagic, found in deep water to 2,000 m during the day and near the surface at night. Often taken at night in shallow plankton tows.

*Idiacanthus fasciola* Peters (Black-dragonfish, Stalkeye): ♀ black; barbel originating from hyoid; postorbital light organ much smaller than eye; 3 longitudinal rows of luminous organs on each side. ♂ dark brown or pale; sperm reservoir with duct extending to outside of body along anterior anal rays forming intromittent organ-like structure; postorbital luminous organ equal to or larger than eye. ♂ to 4.5 cm, ♀ to 30 cm. Females feed on zooplankton and small fishes; found from surface at night to 2,000 m during daylight. Occasionally taken in plankton tows near surface at night.

### Plate 199

O. **LOPHIIFORMES:** Robust, with large heads and gill opening behind head in axil of the pectoral fin. First ray of the spinous dorsal on head modified to form lure, or embedded, not visible; pelvic fins absent or anterior to pectoral fins, with 1 spine and 5 soft rays. Branchiostegal rays 5 or 6; ribs absent; swim bladder physoclistous.

F. **ANTENNARIIDAE** (Frogfishes): Short, nearly round Lophiiformes; mouth large, oblique to vertical. Gill opening small, round, located behind pectoral fin base. First dorsal fin spine free, modified to form lure; 2nd and 3rd spines free, separate and covered with skin. Pectoral fin lobe long, leg-like. Skin with short spines or naked; often with fleshy tabs or filaments. Color variable. Demersal fish that lurk on substrate for prey animals. (5 spp. from Bda.)

*Histrio histrio* (L.) (Sargassum fish): A pair of fleshy cirri present on snout anterior to 1st dorsal spine ("fishing rod"). Skin without spines but usually with fleshy tabs that may be variously branched. Pectoral fin lobe free, narrowly attached to body. First dorsal spine very short. Color variable, shades of brown, tan and yellow mottling blending with background color of floating *Sargassum* with which it is associated. To 15 cm. Feeds on fish and crustaceans found in *Sargassum* by stalking rather than attracting prey.

*Antennarius scaber* (Cuvier) (Splitlure frogfish): Relatively elongate, lacking cirri on snout in front of 1st dorsal spine. "Lure" on end of 1st dorsal spine bifid, with 2 long fleshy lobes that tend to curl. Origin of the 1st dorsal spine located in front of margin of the upper jaw. Common coloration is pale background with faint purplish tint and dark markings of blackish purple as streaks or spots or combination of the two. ♂ also exhibits a dark or black phase. Both color phases lack large round ocellar spots. Most common on rocky or coral bot-

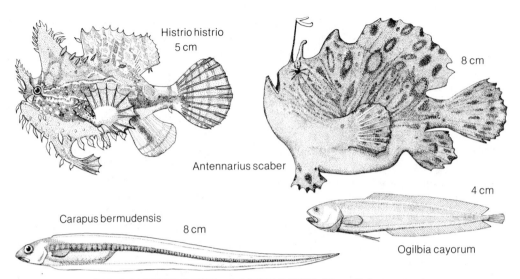

**199 LOPHIIFORMES, GADIFORMES (Frogfishes, a.o.)**

toms, 1-15 m. Large specimens occasionally caught in fish traps. Feeds on small fishes and crustaceans.

O. **GADIFORMES:** Body elongate; dorsal and anal fins usually long; pelvic fins absent or in front of pectoral fins; ectopterygoid toothless; reduction in posterior vertebrae; posterior dorsal and anal pterygiophores greater in numbers than caudal vertebrae.

F. **CARAPIDAE** (Pearlfishes): Long, tapering Gadiformes with extremely long dorsal and anal fins joined and with anus located just forward of pectoral fin origin. Pelvic fins absent; scales absent. Most species live inside bivalves, tunicates, holothurians and other echinoderms.

*Carapus bermudensis* (Jones) (Pearlfish): Long, tapering, nearly transparent, with a long low dorsal and anal fin. Silvery patches on head, gut and sometimes behind head. To 18 cm. Species is host-specific, living inside the holothurian *Actinopyga agassizi,* which is found inshore and between patch reefs from shore to 20 m. Uncommon.

F. **BROTULIDAE** (Brotulas): Relatively long, tapering Gadiformes with head broader than body, and wide gill openings. Fins lacking spines. Dorsal fin single, long, joining long anal fin in most groups. Pelvic fins jugular, below posterior part of head, close together; 1 or 2 rays in each fin. Scales small or absent. (2 spp. from Bda.)

*Ogilbia cayorum* Evermann & Kendall (Key brotula): Caudal fin distinct, separated from dorsal and anal fin by short slender peduncle. Snout rounded, mouth inferior. Color uniform greenish to brown or yellow. Color forms may represent different species; genus poorly known throughout its range. To 10 cm. Found from shallow coral rubble areas inshore to reefs offshore at 20 m, usually hiding under stones; not uncommon.

## Plate 200

**O. ATHERINIFORMES:** The opercle and preopercle with entire margins (lacking spines or serrations); branchiostegal rays 4-15; pectoral skeleton supported by Bandelot's ligament to basicranium; orbitosphenoid absent; caudal skeleton with 2-4 (usually 2) enlarged triangular hypural plates; swim bladder physoclistous. Generally pelagic, marine and freshwater.

**F. HEMIRAMPHIDAE** (Halfbeaks): Elongate Atheriniformes with lower jaw prolonged to form beak, upper jaw forming short flat triangle. Pectoral fin high on sides; pelvic fin just anterior to anal fin. Single dorsal and anal fin located posteriorly, reaching caudal peduncle. Lateral line running ventrally from pectoral fin base then posteriorly along ventral margin to caudal peduncle. Scales large, cycloid and deciduous. Caudal fin forked, lower lobe longer than upper lobe. (3 spp. from Bda.)

*Hemiramphus bermudensis* Collette (Bermuda halfbeak): Slender, with greatly prolonged lower jaw. Upper jaw scaleless. Rakers on 1st gill arch 37-45; no spines in fins; dorsal fin with 13-15 rays; anal fin with 12-14 rays. Pectoral fins short, not reaching nasal pit when folded forward. Caudal fin deeply forked. Color dark bluish above becoming silvery white below. Beak black with bright red fleshy tip; upper caudal fin lobe reddish orange in adults. To about 45 cm total length. Found only in Bermuda; an inshore pelagic form. Feeds on phytoplankton, drifting pieces of sea grass and zooplankton. Caught with beach seines and highly prized as bait. An endemic species.

**F. EXOCOETIDAE** (Flying fishes): Elongate, broadly cylindrical Atheriniformes with flattened ventral surfaces. Snout blunt; mouth small; teeth very small or absent. Dorsal and anal fins lacking spines; with short bases located just in front of caudal peduncle. Pectoral fins located high on sides; rays long, reaching beyond dorsal fin origin; rays connected by thin membrane. Pelvic fins abdominal; caudal fin deeply forked; lower lobe longer than upper. Lateral line low on body; scales large, cycloid and deciduous. Color dark blue on back becoming silvery white ventrally in adults. Juveniles to about 10 cm differ from adults in having higher dorsal fins, shorter pectoral fins, and variable, bright color patters often in reds, yellows and browns; chin barbels well developed in many species. Pelagic, oceanic; juveniles often with floating *Sargassum*. (11 spp. from Bda.)

*Cypselurus furcatus* (Mitchell) (Spotfin flying fish): Abdominal pelvic fins long, reaching well behind origin of anal fin; origin of anal fin 3 or more rays behind origin of dorsal fin. Only 1st pectoral fin ray unbranched. Snout shorter than eye diameter, blunt; palatine teeth absent. Anterior dorsal rays longest; dorsal and anal fin rays short. Pectoral fin long, reaching posterior dorsal fin. Pectoral fins relatively dark, with pale margin and pale curved stripe in central portion of fin. To 30 cm. Pelagic in oceanic waters, feeds on phytoplankton and zooplankton. Capable of leaping out of water and gliding for long distances. Occasionally netted in later summer; used primarily for bait.

*Hirundichthys affinis* (Günther) (Fourwing flying fish): Abdominal pelvic fins long, reaching well beyond origin of anal fin; origin of anal fin slightly anterior to or not more than 3 rays behind origin of dorsal fin. Dorsal and anal fins with equal or nearly equal number of rays. Only 1st ray of pectoral fin unbranched; pectoral fin long, reaching well behind origin of dorsal fin. Pectoral fins dark with pale basal triangle and a narrow white margin on posterior edge. To 25 cm. Oceanic, pelagic, feeding on zooplankton and phytoplankton. Capable of leaping out of water and gliding for long distances. Sometimes netted, used primarily for bait. Juveniles once thought to be separate species and described as *Exonautes nonsuchi* (BEEBE & TEE-VAN 1932).

F. **BELONIDAE** (Needlefishes): Long, slender Atheriniformes with both jaws prolonged into beaks containing sharp needle-like teeth. Pelvic fins thoracic, with 6 rays; dorsal and anal fins posterior; lateral line running ventrally from pectoral fin origin then posteriorly along ventral margin of body. Scales small, cycloid, deciduous. Dark blue-green on back, silvery white on belly. Pelagic, inshore and oceanic. (6 spp. from Bda.)

*Tylosaurus acus* (Lacépède) (Houndfish, Agujon, Needlefish): Body elongate, round to squarish in cross section; caudal peduncle with small black lateral keel, caudal fin deeply forked; lower lobe longer than upper. Anterior part of dorsal fin forming low lobe, dorsal fin rays 23-26; anal fin with low anterior lobe, rays 20-24. Pectoral and pelvic fins short. Dark blue above, silvery white below, with a blue stripe along sides. To about 140 cm total length. Pelagic, found mainly offshore but does enter lagoon in aggregations. Feeds mainly offshore. Caught in beach seines and on hook and line. Usually discarded by fishermen.

F. **POECILIIDAE** (Live-bearers): Elongate, nearly round Atheriniformes; eyes and mouth small; teeth small, well developed in both jaws; no spines in fins. Position of dorsal and anal fin variable; anal fin modified to form gonopodium (intromittent organ) in ♂; 3rd anal fin ray unbranched. Pectoral fins short, rounded, inserted midway between dorsum and ventrum, below lateral line. Pelvic fins thoracic. Lateral line reduced to series of separate pit organs along sides. (1 sp. from Bda.)

*Gambusia affinis* (Baird & Girard) (Mosquito fish): Dorsal fin moderately high, located midway on back; anal fin origin anterior to origin of dorsal fin. Pelvic fin short, rounded, thoracic, located midway between origin of pectoral and anal fin. Gonopodium with elbow-like projection on anterior branch of middle fin ray and 2 recurved hooks near tip. Color olive-brown on back becoming pale ventrally. Body flecked with small black spots. To 7 cm. In brackish ponds and drainage canals. Feeds on mosquito larvae, other insects and crustaceans. Introduced for the control of mosquitos in marshes, ponds, drainage ditches and individual home cisterns.

F. **CYPRINODONTIDAE** (Killifishes): Elongate, nearly round Atheriniformes; caudal peduncle deep; head flattened, mouth wide, terminal, with fine teeth on edges of jaws. Dorsal fin single, located midway on body; anal fin spineless, usually with all rays branched. Pectoral fins short, rounded and inserted low below lateral line. Pelvic fins abdominal, without spines. Caudal fin more or

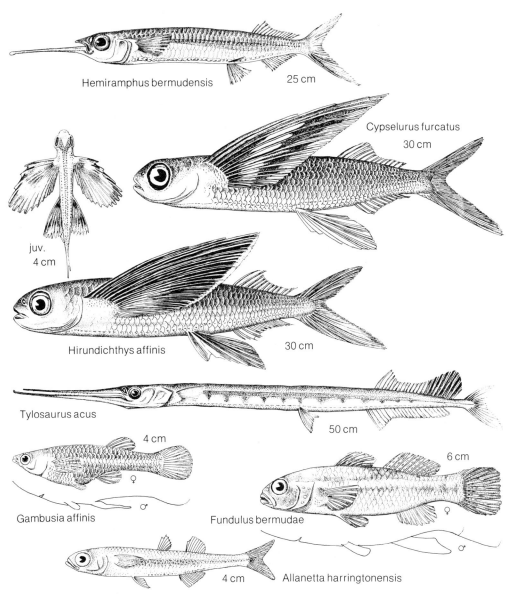

**200 ATHERINIFORMES (Flying fishes, a.o.)**

less rounded. Lateral line reduced to series of pit organs along sides. Scales large, cycloid. ♂ with elongated anal fin but no gonopodium. (1 sp. from Bda.)

***Fundulus bermudae*** Günther (Bermuda killifish): Dorsal fin low, origin anterior to origin of anal fin. Usually 30 scale rows along midlateral scale row. Sensory pores present on lower jaw. Dorsal rays 12-14, anal rays 11-12. Color variable, light brown to pale greenish yellow, darker on ventral surface. Dark indistinct greenish brown bands on sides. To 10 cm. In brackish water ponds, particularly those fringed with mangroves. Feeds on detrital matter, plants and invertebrates. An endemic species but placed with *F. heteroclitus* (L.) by some workers.

F. **ATHERINIDAE** (Silversides): Elongate Atheriniformes with moderately pointed snout; mouth terminal. First dorsal fin with 3-9 slender spines, 2nd dorsal fin well separated from 1st; anal fin preceded by 1 spine; pelvic fin abdominal, 1 spine, 5 soft rays; pectoral fins high on sides; caudal fin forked. Lateral line absent, scales large, cycloid. (1 sp. from Bda.)

*Allanetta harringtonensis* (Goode) (Reef silverside, B Rush fry): Body slender, anal fin short, with 8-13 soft rays; edge of premaxilla straight; anus well separated from origin of anal fin. Posterior half of lower jaw with a bony tooth-like projection. Body greenish blue on back, silvery to transparent on belly; a lateral broad or narrow blue stripe on sides. To 8 cm. Pelagic, inshore, rarely found beyond breakers surrounding lagoon. Feeds on zoo- and phytoplankton. Highly prized for bait.

## Plate 201

O. **BERYCIFORMES:** Relatively robust, with spinous and soft rays in dorsal and anal fins. Pelvic fins forward, more or less beneath pectoral fins; pelvic girdle attached to pectoral girdle; pelvic fin with 1 spine and 7 rays. Orbitosphenoid present on skull.

F. **HOLOCENTRIDAE** (Squirrelfishes): Elongate, somewhat compressed Beryciformes with slender caudal peduncle. Eyes large, scales ctenoid, rough to touch; bones on head serrate or with spines. Dorsal fin base long; base of spinous portion 3-4× as long as base of soft rays; pelvic fins with 1 spine, 7 soft rays; caudal fin forked; anal fin with 4 spines. (8 spp. from Bda.)

*Adioryx vexillarius* (Poey) (Dusky squirrelfish): Head small, snout short, preopercle strong, spined at angle; lobes of caudal fin nearly equal in length; rakers on lower limb of 1st gill arch 11 or 12. Spineless portion of dorsal fin separated narrowly from soft rays; dorsal fin with 13, anal fin with 9, pectoral fin with 15 soft rays. Pectoral fin with a black axil. Coloration dusky, with alternating reddish and silvery white stripes, the reddish color deepest on upper half of body; a horizontal red stripe on lower margin of eye, with parallel silvery white area below. Anterior portion of dorsal fin with dark markings, remainder of fin membranes pinkish to reddish. To 15 cm. In rocky areas nearshore, patch reefs and reefs offshore to 25 m. Forages in the water column adjacent to the reefs on zooplankton at night, hiding in caves and crevices during the day.

*Holocentrus ascensionis* (Osbeck) (Squirrelfish): Genus with short snout; preopercle with strong spine at angle; rakers on lower limb of gill arch 15-18; upper lobe of caudal fin longer than lower lobe.—Species with upper jaw long, extending to posterior portion of pupil or beyond; relatively large scales, 46-50 in lateral line series to base of caudal fin. Back and upper sides red with gold reflections, sides with alternate red and white stripes. Interspinous membrane of dorsal fin orange anteriorly, greenish posteriorly. Outer margins of caudal fin white. To 40 cm. From nearshore rocky areas to patch reefs offshore to

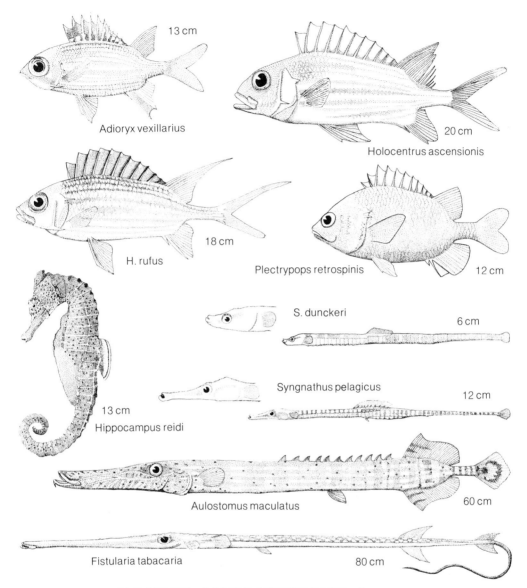

**201 BERYCIFORMES, GASTEROSTEIFORMES (Sea horses, a.o.)**

more than 90 m. Usually nocturnal, feeding on zooplankton and benthic invertebrates. Larger specimens often caught in traps or on hook and line and used as fillets.

**H. rufus** (Walbaum) (Longspine squirrelfish): Genus as above.—Species with upper jaw reaching only below center of pupil. Relatively small scales, 50-57 in lateral line series to base of caudal fin. Anterior soft dorsal rays elongated; caudal fin lobes elongated; slender. Back and head bright red becoming pink ventrally, sides with diffuse white stripes. Membranes of spinous dorsal fin with small white triangular patches on distal edge; soft dorsal, anal, caudal and pelvic fins pink to reddish. To 35 cm.

From nearshore to patch reefs and offshoee to 90 m. Inhabiting caves or crevices of reefs by day, feeding in water column on zooplankton or on benthic invertebrates at night. Larger specimens commonly caught on hook and line or in traps and used as fillets.

***Plectrypops retrospinis*** (Guichenot) (Cardinal soldierfish): Body robust, nearly round, and deep, snout short; preopercle without long spine at angle; 11 spines in anterior portion of dorsal fin; circumorbital bones with spines, those on suborbital slender and curved partly forward. Body color uniform scarlet with no black markings. Lobes of 2nd dorsal, anal and caudal fins short, rounded. To 12 cm. On patch and offshore reefs, to 25 m. Found in deep crevices and caves during day; feeds on zooplankton in water column at night.

O. **GASTEROSTEIFORMES:** Generally long, slender, with snout elongated, in most cases to extreme. Dorsal fin with spinous portion containing individual spines (vestigial in some groups). Gas bladder physoclistous.

F. **SYNGNATHIDAE** (Sea horses, pipefishes): Slender Gasterosteiformes with segmented bodies encased in numerous jointed bony rings; snout tubular; mouth small, lacking teeth. Spinous dorsal and pelvic fins absent. Eggs carried by ♂ in brood pouch. (12 spp. from Bda.)

***Hippocampus reidi*** Ginsburg (Long-snout sea horse): Body robust, head at right angles to main axis of body; caudal fin absent; caudal vertebrae forming prehensile tail; dorsal fin rays 16-19. Body covered with small round black dots on a lighter background which may itself be rather dark. Whitish bands or partial bands at regular intervals on trunk and tail. To 13 cm. Inshore in grass flats or on algae on coastal rocky areas or patch reefs. Feeds on small crustaceans, polychaete worms and other invertebrates living among algae and sea grass.

***Syngnathus dunckeri*** Metzelaar (Pugnose pipefish): Genus slender, with straight tail, caudal fin present. Ridge pattern with median trunk ridge terminating near anus, lateral tail ridge starting shortly above its end then swinging upward farther back to join superior tail ridge. Tail with 30 or more rings; median head crest low.—Species with short snout, length 2.6-2.9× the length of the head. Anal fin absent; tail rings 32-35; dorsal fin rays 24-26. Color highly variable, ♂ darker than ♀. ♀ light, without markings or with incomplete brown rings around body at regular intervals. Adult ♂ with definite narrow white rings encircling body at regular intervals, the rings margined on either side by a darker brown. To 8 cm. With floating *Sargassum* as well as in shallow-water *Thalassia* beds. Feeds on small crustaceans, polychaete worms and other invertebrates.

***S. pelagicus*** (L.) (Sargassum pipefish): Genus as above.—Species with 16-18 trunk rings and 33 or 34 tail rings. Dorsal rays 28-31; head length 5- 6.5× length of the pectoral fin. Color variable but usually with narrow white vertical lines margined with black; caudal fin dark, not distinctly marked; dorsal fin with dark stripes through middle. To 17 cm. Most often associated with pelagic *Sargassum* but takes up residence in benthic algae and sea grasses when *Sargassum* is washed ashore. Feeds on crustaceans, polychaete worms and other invertebrates.

F. **AULOSTOMIDAE** (Trumpet fishes): Gasterosteiformes with long tubular snouts and median barbel at tip of chin. Scales on body; a series of isolated dorsal spines in front of soft rayed dorsal fin. Tropical reef dwellers. (Single species in Atlantic:)

*Aulostomus maculatus* Valenciennes (Trumpet fish): Slender, slightly compressed; jaws protrusible. Dorsal spines 9-12 with separate membranes; soft rays 24- 28; caudal fin rounded, lacking a filament. Color variable, generally olivaceous brown in shallow water less than 15 m and brownish pink in deeper water. Black spots on back, sides and ventrally; irregular silvery or pale lines on head; body with indistinct stripes. To 100 cm. On reefs or near shore, often associated with gorgonian corals. Frequently rests or stalks prey in vertical orientation with head down. Feeds on small fish and crustaceans sucking them into mouth with pipette-like snout. Occasionally caught in fish traps, not used for food.

F. **FISTULARIIDAE** (Coronet fishes): Long, slender Gasterosteiformes with tubular snout. Median chin barbel missing; predorsal spines not visible; long median filament on caudal fin. (1 sp. from Bda.)

*Fistularia tabacaria* L. (Coronet fish): Body nearly round in cross section, with long, often whip-like median filament on forked caudal fin. Dorsal fin without spines and with 13-15 soft rays. Body greenish brown to gray above, shading to whitish ventrally in shallow water, or bright red above shading to pale pink ventrally in specimens from more than about 45 m. Mid-dorsal row of blue spots from head to dorsal fin and 1 or 2 rows of blue dots on sides. To 200 cm. Uncommon; juveniles found over grass flats, adults on offshore reefs to depths of 100 m or more. Feeds primarily on fishes and crustaceans. Occasionally caught on hook and line and in traps; not used for food.

O. **PERCIFORMES:** With spinous and soft rays. Pelvic fins thoracic or jugular and pelvic girdle connecting to cleithra. Pectoral fins placed high on sides, caudal fin with 17 principal rays. Branchiostegal rays 5-8 on each side, with 4 on the outer surface of the upper portion of the ceratohyal and epihyal, the remainder attached to the edge of lower section of ceratohyal. Gas bladders are physoclistous; maxillary excluded from gape by the premaxillary on each side of the upper jaw. The largest and most diverse order of fishes.

F. **SERRANIDAE** (Groupers, B Rockfishes): Bass-like Perciformes highly variable in shape and size. Caudal peduncle deep; usually 3 (occasionally 2) spines on operculum; maxilla broad, exposed at its posterior end; upper edge of opercle free. Dorsal spines 7-11; anal spines 3; pelvic fin with 1 spine and 5 soft rays, the innermost often connected to the body by a membrane. Lateral line present, not interrupted and ending at caudal fine base. All are carnivores feeding mainly on fishes and crustaceans. One of the least specialized groups of Perciformes. (27 spp. from Bda.)

## Plate 202

*Paranthias furcifer* (Valenciennes) (Creole fish, B Barber): Body robust, fusiform; caudal fin forked; snout short; dorsal fin with 9 spines, 18 or 19 soft rays. Deep reddish

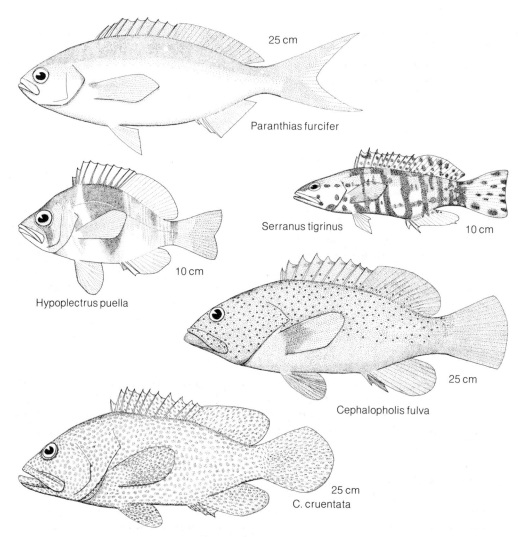

## 202 SERRANIDAE 1 (Coney, a.o.)

brown or blue on back paling to silvery pink on belly; with a series of small, widely separated white spots on back below dorsal fin; a bright red-orange spot at upper base of pectoral fin. To 35 cm. Found in aggregations on the outer reefs, 10-50 m. Feeds in water column on zooplankton from just above reef to the surface. Caught in traps and on hook and line, used for fillet.

***Hypoplectrus puella*** (Cuvier) (Butter hamlet): Relatively deep-bodied, compressed, with lower limb of preopercle with more than 3 forward pointing spines. Dorsal fin with 10 spines and 14-16 soft rays; pelvic fins long, usually reaching origin of anal fin; caudal fin lunate. Yellowish brown, with a large dark brown bar on front half of body most intense below eye; a smaller bar running from nape through pectoral fin base; 3 narrow brown bars on posterior portion of body. Pale regions between bars with pale blue lines; blue lines on head and bright blue lines on snout.

To 15 cm. In rocky areas and patch reefs from shore to 30 m. Feeds on zooplankton and in benthic communities on small crustaceans and fishes.

***Serranus tigrinus*** (Bloch) (Harlequin bass): Body elongate, slender; caudal fin truncate; branchiostegal rays 7; teeth present in jaws; dorsal fin with 10 spines, 12 soft rays; posterior edge of preoperculum with regular serrations; lower limb of opercle smooth. Greenish black on back shading to pale silvery yellow on belly; 7 irregular dark brown to black irregular bars on sides; parts of head, chest, dorsal and anal fins with small black blotches. To 10 cm. Found occasionally on shallow offshore reefs to about 30 m, feeds on zooplankton and benthic crustaceans.

***Cephalopholis fulva*** (L.) (Coney): Genus with stout body, caudal fin somewhat rounded; anal fin soft rays 7-9; dorsal fin spines 9.—Species with slightly rounded caudal fin with sharp upper and lower angles. Color highly variable, from bright scarlet to greenish brown to bright yellow, or bicolored red or brown on top, white ventrally. All color phases have 2 black spots on top of caudal peduncle and 2 on tip of lower jaw; body and soft dorsal fin membrane covered with numerous small dark edged blue spots. To 40 cm. On patch reefs and offshore reefs to 90 m. Feeds on small fishes and crustaceans. Caught in traps and on hook and line, usually sold as fillet.

***C. cruentata*** (Graysby) (=*Petrometopon cruentatum*): Genus as above.—Species with round caudal fin lacking upper and lower angles. Color reddish gray to greenish gray with numerous dense reddish brown spots on body and fin membranes. With 3 or 4 spots, either black or white, spaced along base of dorsal fin. To 35 cm. Usually found in shallow water nearshore but also to 60 m. Feeds on small fishes and crustaceans. Caught in traps and on hook and line and used as fillet.

## Plate 203

***Alphestes afer*** (Bloch) (=*Epinephelus afer*) (Mutton hamlet): Robust, with soft dorsal and anal fins covered with scales and thick skin; forward projecting spine on lower edge of preopercle; 11 dorsal fin spines, interspinous membrane notched; caudal fin rounded; anal fin with 3 spines and 8 soft rays. Color mottled brownish to yellow or orange with small scattered orange spots and dark brown blotches that may form bars, particularly at night. To 35 cm. Found mainly in shallow reefs or rocky areas and grass flats from shorelines to 50 m. Feeds on crustaceans and small fishes. Occasionally caught on hook and line or in traps, usually sold as fillet.

***Epinephelus adscensionis*** (Osbeck) (Rock hind): Genus robust, moderate to large, with soft dorsal and anal fins covered with scales and thick skin; anal fin rays 7-9, dorsal fin spines 10 or 11. Forward pointing spine on lower edge of preopercle absent.—Species with 11 dorsal spines, 16-18 dorsal rays; anal fin with 3 spines, 8 soft rays; caudal fin slightly rounded; pelvic fin origin slightly behind base of pectoral fin. Body covered with numerous reddish brown spots on a grayish background; a black saddle on caudal peduncle and 3 dark blotches on back at base of dorsal fin. To about 40 cm. In rocky and coral reef areas in shallow water and offshore to 50 m; most common at 25-35 m. Feeds on crustaceans and small fishes. Not common, caught on hook and line and in traps. Flesh excellent to eat.

***E. guttatus*** (L.) (Red hind): Genus as above.—Species with 11 dorsal spines, the interspinous membrane notched; anal fin with 3 spines, 9 soft rays; pelvic fins shorter than pectoral fins and inserted below anterior end of pectoral fin base. Body color pale to bright pink or pinkish brown covered with round small dark red spots of equal size; may exhibit indistinct bars. Edge of soft dorsal, caudal and anal fins with black margins edged with fine bluish white line. Tips of spinous dorsal fin bright yellow. To 60 cm. On patch reefs and deeper reefs offshore to 100 m. Feeds mainly on crustaceans and small fishes. Caught in traps and on hook and line; a commercially important, excellent food fish.

***E. morio*** (Valenciennes) (Red grouper, B Deer hamlet): Genus as above.—Species with 11 dorsal spines; interspinous membrane not notched; anal fin with 3 spines, 9 soft rays; caudal fin truncate to slightly lunate; pelvic fin shorter than pectorals, inserted slightly behind origin of pectoral fin. Body generally brownish red to olivaceous brown with transient pattern of whitish spots and indistinct bars. Saddle on caudal peduncle absent. Small black spots on cheeks and opercle. To 75 cm. Inshore around patch reefs and offshore to 150 m, more often on rocky outcrops than coral reefs. Feeds on crustaceans, small fishes and cephalopods. Caught in traps and on hook and line; an excellent food fish.

***E. striatus*** (Bloch) (Nassau grouper): Genus as above.—Species with 11 dorsal spines; interspinous membrane slightly notched; anal fin with 3 spines, 8 soft rays; pelvic fins shorter than pectorals, inserted below origin of pectoral fin. Caudal fin rounded to truncate. Body with an olivaceous brown color in shallow water to reddish brown in deeper water. A dark, tuning fork-shaped bar running from snout through eye to origin of dorsal fin; 5 dark vertical bars on body of which the 3rd and 4th are branched above the lateral line and joined to form a W-shaped mark. Large conspicuous black saddle on caudal peduncle. Intensity of barred pattern changeable. To more than 100 cm. Found from shallow patch reefs inshore to 100 m offshore; juveniles on grass flats and rocky outcrops inshore. Feeds on crustaceans, fishes and cephalopods. An important commercial species caught in traps and on hook and line.

***Mycteroperca bonaci*** (Poey) (Black grouper, B Black rockfish, Runner rockfish): Genus robust, somewhat elongate, with soft dorsal and anal fins covered with scales and thick skin; anal fin soft rays 11-14; dorsal fin spines 11.—Species with preopercle gently rounded and only slightly notched. Anal fin with 3 spines, 12 soft rays; caudal fin truncate. Body with light tan to brownish olive ground color often forming regular rows of rectangular blotches. Cheeks and belly with hexagonal bronze spots separated by pale gray area. Pectoral fin gray with narrow diffuse orange margin. To 170 cm. Over rocky bottom and coral reefs from shore to 90 m. Feeds on fishes, crustaceans and cephalopods. An important food fish caught in traps and on hook and line.

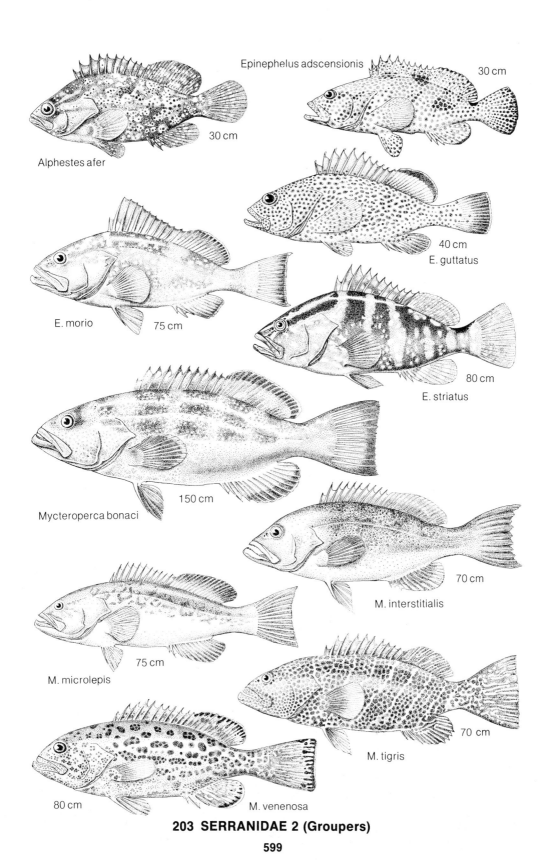

**203 SERRANIDAE 2 (Groupers)**

***M. interstitialis*** (Poey) (Yellowmouth grouper, B Monkey rockfish): Genus as above.—Species with marked notch above angle of free preopercular margin forming a serrated lobe at the angle. Rakers on 1st gill arch 24-28. Membrane on spinous dorsal fin notched. Caudal fin emarginate to lunate. May be uniformly brown overall or with brown color broken into spots with pale whitish yellow lines. Pectoral fin with yellow margin and soft dorsal, anal and caudal fins narrowly edged with yellow. Jaws, anteriorly to angle, yellow. Juveniles are bicolored, dark above, white below. To 80 cm. From shoreline to 75 m offshore, around rocky or coral reef areas. Feeds on fishes, crustaceans and cephalopods. Often caught on hook and line and in traps; an important commercial species.

***M. microlepis*** (Goode & Bean) (Gag grouper, Fine-scale rockfish): Genus as above.—Species with slight notch above angle of preopercle, the angle expanded to form smooth round lobe. Rakers on 1st gill arch 21-29. Membrane on spinous dorsal not notched. Caudal fin truncate. Scales relatively small giving the skin a smooth feel. Body generally gray overall with darker gray markings in square-like blotches. No brassy spots on sides of head; no red spots on body. To 70 cm. In shallow and grassy areas or in deeper water near rock outcroppings, rarely on offshore reefs. Feeds on fishes, crustaceans and mollusks. Caught on hook and line and in traps; a commercial species.

***M. tigris*** (Valenciennes) (Tiger grouper, B Gag rockfish): Genus as above.—Species with rounded angle of preopercle not forming lobe. Gill rakers few, 8 on lower limb of 1st gill arch. Membrane on spinous dorsal fin notched. Caudal fin truncate or slightly emarginate. Posterior nostrils much larger than the anterior ones. Body color variable, dark greenish brown-gray with 8 or 9 pale whitish bars that slope forward and down across the back. Lower sides and ventral surface pale, marbled. To 70 cm. In shallow water from surface to about 30 m; most common on shallow coral reefs in ledge flat area. Feeds on crustaceans, small fishes and cephalopods. Will swim off bottom to catch food or take a trolled bait. Caught on hook and line and in traps.

***M. venenosa*** (L.) (Yellowfin grouper, B Red rockfish, Princess rockfish): Genus as above.—Species with smooth rounded angle of preopercle without lobe. Membrane on spinous dorsal fin notched. Outer 1/4 to 1/3 of pectoral fin yellow, clearly delineated from spotted basal portion. In shallow water body color greenish olive-gray with longitudinal rows of darker blotches; soft dorsal, caudal and anal fins with irregular black margin finely edged with white. In water deeper than 30 m body is bright red with darker red markings on body and fins. To 90 cm. From shoreline to about 100 m, around reefs and rocky outcroppings. Feeds on fishes, crustaceans and mollusks. Once a commercial species, it is now only occasionally taken by hook and line and in traps.

### Plate 204

F. **GRAMMISTIDAE** (Soapfishes): Oblong, compressed, bass-like Perciformes with 2 pairs of nasal openings and the nasal organ vertically elongated; opercle with 2 or 3 spines, its entire upper edge attached to skull with membrane; upper preopercular

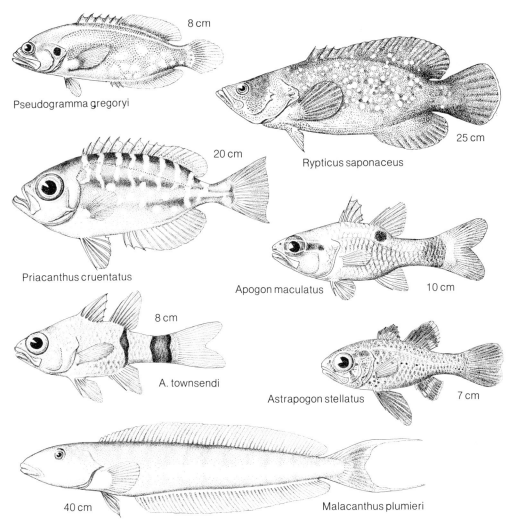

## 204 GRAMMISTIDAE—MALACANTHIDAE (Soapfishes, a.o.)

margin with 1-3 spines. Dorsal fin single, with 2-8 spines; inner pelvic fin rays attached to abdomen by a membrane; lateral line single or interrupted, pores conspicuous on preopercular margin and in area encompassed by lower lips. Color variable, from uniform to blotched, spotted and ocellated. (2 spp. from Bda.)

***Pseudogramma gregoryi*** Breder (Reef bass, Foureye basslet): With 6-8 dorsal spines and 3 anal fin spines. Lateral line incomplete. A large ocellated brown to black spot just below upper margin of operculum. Head and anterior body greenish brown to brown, the rear portion reddish brown to red uniformly colored or with paler blotches; soft dorsal, caudal and anal fins bright pink to red. To 8 cm. Cryptic, found in caves and crevices in rocks and reefs, rarely seen alive; from shallow inshore areas to reefs over 20 m deep. Feeds on crustaceans and other small invertebrates.

***Rypticus saponaceus*** (Schneider) (Soapfish): With 3 dorsal spines;

lacking anal spines and a continuous incomplete lateral line. Body color reddish brown to grayish brown covered with pale whitish spots or blotches. Caudal fin rounded with white margin. To 35 cm. Cryptic, usually lurks in caves in shallow reefs or rocky coastal areas to about 50 m. Feeds on small fishes and crustaceans. Can secrete copious quantities of mucus, making skin slimy; mucus contains a toxic protein. Infrequently caught in traps but usually discarded.

F. **PRIACANTHIDAE** (Bigeyes): Medium-sized, compressed, relatively deep-bodied Perciformes with very large eyes. Mouth large, strongly oblique, with projecting lower jaw. Dorsal fin single, with 10 spines and 10-15 soft rays without a definite notch. Pelvic fin large, origin anterior to pectoral fin; 1 spine and base joined to body by a membrane. Scales small, rough, ctenoid, completely covering head including the maxilla. Color varies from bright red to pinkish silver. Nocturnal. (3 spp. from Bda.)

*Priacanthus cruentatus* (Lacépède) (Glasseye snapper): Pelvic fins not of extreme length, shorter than head length; soft portions of dorsal and anal fins not elevated. Soft dorsal rays 13; anal soft rays 14. A well-developed spine at angle of preopercle projecting back to margin of operculum. Color highly variable, from rusty brown with silvery highlights to silvery with reddish highlights. Often shows alternating reddish and silver banded pattern on upper sides. Dorsal, anal and caudal fins with reddish spots. To about 30 cm. In caves nearshore or on reefs, 1-20 m. Seldom seen except by night diving. Feeds at night on small fishes, crustaceans, mollusks and polychaete worms, usually foraging in the water column. Seldom caught in traps, may be used as fillet.

F. **APOGONIDAE** (Cardinal fishes): Small Perciformes with large head and eyes; 2 separated dorsal fins, the 1st with 6-8 spines, the 2nd with 1 spine and 9-14 soft rays. Pelvic fin with 1 spine, 5 soft rays, origin below pectoral fin; anal fin with 2 spines and 6-18 soft rays. Color varies from red and silver to black. Nocturnal, some species are symbiotic. (11 spp. from Bda.)

*Apogon maculatus* (Poey) (Flame fish): Genus with 2 spines and 8 soft rays on anal fin; scales on body ctenoid; a series of median predorsal scales present; preopercular margin serrated; pectoral fin rays 12.— Species with body scales above and below lateral line equal in number to those in lateral line. Body color uniformly bright red with a round black spot below 2nd dorsal fin; a black saddle-like mark extending over dorsal surface of caudal peduncle; 2 sharp parallel white lines horizontally through the eye, with a dark area between. To 15 cm. Nocturnal; in caves and crevices in rocks nearshore and on offshore reefs during the day, at night feeding on zooplankton in the water column; from surface to more than 100 m.

*A. townsendi* (Breder) (Belted cardinal fish): Genus as above.— Species with a vertical bar below 2nd dorsal fin extending to base of anal fin; dark peduncular bar pronounced, with black lateral margins; body color pinkish red. Snout short. Mainly on patch reefs or offshore deeper reefs, to 30 m. Nocturnal, leaving caves or reef crevices to forage in the water column on zooplankton.

*Astrapogon stellatus* (Cope) (Conch fish): Scales on body cycloid, smooth; no scales on nape; pectoral fin rays 14-16; upper portion of free preopercular margin smooth, low portion with scalloped edge. Snout relatively long. Color brown to black with black flecks, to silvery with rows of round dots or flecks. Lives in the mantle cavity of live queen conch (*Strombus gigas*) and emerges only at night to forage in water column on zooplankton. ♂ broods eggs in its mouth.

F. **MALACANTHIDAE** (Sand tilefishes): Elongate, fusiform to robust, round-headed Perciformes with long dorsal and anal fin. The sum of dorsal and anal fin bases more than 80% of standard length. A predorsal ridge present in some species; caudal fin variable in shape, falcate, truncate or rounded. Family not well known. (2 spp. from Bda.)

*Malacanthus plumieri* (Bloch) (Sand tilefish): Body nearly round in cross section. Jaw extending to below posterior nostril in front of eye; upper lip fleshy; caudal fin deeply lunate; dorsal fin with 4 or 5 spines, 54-60 soft rays; anal fin with 1 spine, 48-55 soft rays. Body bluish green on back becoming silvery gray on belly; pale yellow bars may be present on sides. Dorsal fin with yellow margin, clear band, then yellow band, and 3 or 4 rows of bright yellow spots. Anal fin similar to dorsal; caudal fin with large areas of yellow-orange on upper and lower lobes. To 60 cm. In shallow water, 2-50 m, over open sand or coral rubble areas where it builds burrows and mounds of rubble and shell fragments. Usually enters burrows head first when frightened. Feeds on crustaceans, fishes, polychaete worms and echinoderms. Occasionally caught on hook and line.

### Plate 205

F. **ECHENEIDAE** (Remoras): Slender, somewhat depressed Perciformes with spinous dorsal fin modified to form a transversely laminated, oval sucking disc on head. Dorsal and anal fins lacking spines; pectoral fins high on body, pelvic fins originate forward beneath pectoral fins. Color variable, reddish brown to gray-black with white stripes on ventral area. Often associated with large fishes, mammals or turtles. (5 spp. from Bda.)

*Echeneis naucrates* L. (Shark sucker): Body elongate, slender, pectoral fins pointed; anal fin long, 31-41 rays; dorsal fin with 33-45 rays; disc laminae 21-27. Body generally dark tan, brown or blackish; dark lateral area bounded by a white stripe above and below extending from head to caudal peduncle. Caudal fin black in adults. Will attach to a wide variety of hosts including large sharks, rays, sea turtles and reef fishes; also lives free in reef areas. Feeds on scraps from host or small fishes. Often caught on hook and line.

F. **CARANGIDAE** (Jacks and Pompanos): Generally streamlined, silvery Perciformes, bluish or greenish above with a long anal fin always preceded by 2 stiff spines (in young specimens anal spines interconnected by membrane; in old specimens 1 or both spines may become embedded and invisible). Lateral line complete;

anteriorly arched, posteriorly straight and in most species armed with bony plates (scutes). Dorsal fins separate; ventral fins thoracic, well developed; caudal peduncle slender; caudal fin deeply forked or lunate. (22 spp. from Bda.)

***Caranx crysos*** (Mitchell) (Blue runner jack): Genus with spinous scutes on posterior part of lateral line.—Species with lower limb of 1st arch with 23-28 gill rakers. Body fusiform; caudal fin forked, dusky with blackish tips. A black spot on posterior part of operculum, level with eye. Olive to dark blue above becoming silvery ventrally. To 80 cm. Feeds on crustaceans and small fishes. In schools offshore and occasionally over shallow reefs. Caught extensively for food.

***C. ruber*** (Bloch) (Barjack; B Never Bite): Genus as above.—Species having lower limb of 1st arch with 31-35 gill rakers. Body fusiform; caudal fin slender, deeply forked, lower lobe with dark bar that crosses peduncle and continues forward along base of soft dorsal fin. On live specimens a blue stripe extends from peduncle along back to tip of snout. Bluish dorsally becoming silvery ventrally. To 65 cm. Common over reefs; feeds on zooplankton and small fishes. Caught commercially with modified beach seines for food.

***C. latus*** (Agassiz) (Horse-eye jack): Genus as above.—Species with relatively deep body with steep anterior profile. Lower limb with 1st arch with 16-18 gill rakers. Caudal fin forked and uniformly yellow. Small specimens have black tip on dorsal fin lobe. Dark blue-gray above becoming silvery on ventral surface. A black spot on posterior part of operculum, level with eye. To 100 cm. In pelagic schools, feeding on small fishes and crustaceans. A good game fish but flesh tends to be tough.

***Decapterus punctatus*** (Agassiz) (Round scad, B Round robin): Genus elongate, cigar-shaped, with detached finlet behind dorsal and anal fins; scutes covering posterior portion of lateral line.—Species with scutes covering entire straight portion of lateral line. Bluegreen on back becoming silvery ventrally, with a narrow yellow stripe on side of body at upper eye level, a black opercular spot and a row of small black spots along anterior part of lateral line. Common inshore, feeding on small clupeids, engraulids and zooplankton. Extensively used for bait.

***D. macarellus*** (Cuvier) (Mackerel scad, B Ocean robin): Genus as above.—Species with scutes only covering posterior 2/3 of straight portion of lateral line. Deep blue on back becoming silvery ventrally. A narrow golden stripe may be present on live specimens. Occurs offshore in 20-200 m. Feeds on small pelagic fishes and zooplankton. Although primarily used for bait, the flesh is firm and has good flavor.

***Elagatis bipinnulatus*** (Quoy & Gaimard) (Rainbow runner): Elongate-fusiform, with a detached finlet behind the dorsal and anal fins; scutes lacking on lateral line. Caudal fin deeply forked. Bluish green on back with 2 blue stripes on side separated by a broad yellow stripe. Yellowish below lower blue stripe, white ventrally. Pelagic but occasionally seen nearshore. To 130 cm. Feeds on small fishes and zooplankton. A recognized game fish good to eat.

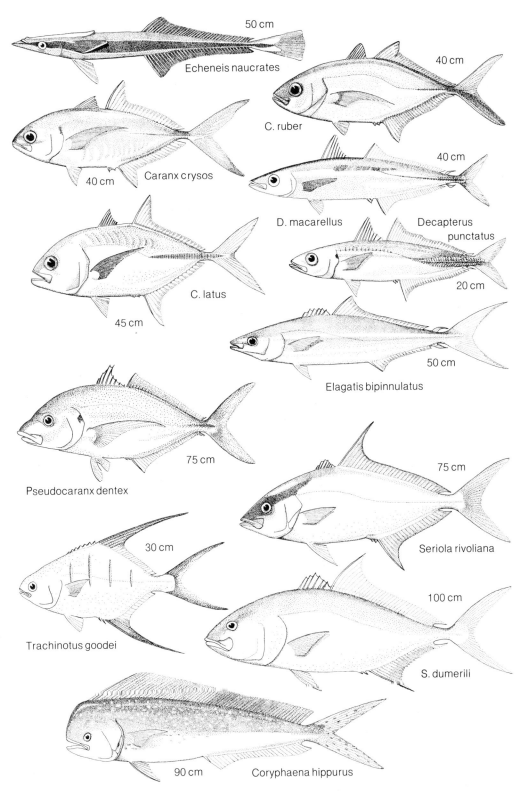

**205 ECHENEIDAE (Shark suckers), CARANGIDAE (Jacks), CORYPHAENIDAE (Dolphinfishes)**

***Pseudocaranx dentex*** (Bloch & Schneider) (Gwelly): Deep-bodied, with a relatively long snout; upper jaw not reaching anterior margin of eye. A single row of blunt conical teeth in both jaws; in large specimens lips are fleshy, teeth rounded. Spinous scutes present on posterior parts of lateral line. Caudal fin deeply forked, yellowish; pectoral fin falcate, longer than head. Dorsal fin with 8 spines, the longest spine longer than that of the 2nd dorsal fin. Lower limb of 1st arch with 23-29 gill rakers. Pale greenish blue above, silvery below, with a yellow stripe along sides and a black spot on posterior margin of operculum. Adults feed in mud and sand on mollusks, crustaceans and worms; juveniles, pelagic, feed on zooplankton and small fishes. Common over open sand or mud bottoms inshore and on reefs and banks offshore. A good food fish caught commercially.

***Trachinotus goodei*** (Jordan & Evermann) (Palometa, B Pompano): Deep-bodied, without lateral scutes or finlets. Caudal fin deeply forked, lobes of dorsal and anal fins extended, reaching beyond base of caudal fin. Dark silvery above, silvery on sides and below. Pectoral fins may be tinted yellow; dorsal and anal fin lobes are black. With 4-6 thin black bars on sides. Forms schools in surf zone of sandy beaches, feeding on small invertebrates and fishes.

***Seriola rivoliana*** (Cuvier) (Almaco-jack, B Bonito): Genus robust, elongate, with groove on caudal peduncle; lateral scutes or finlets absent; dark brownish above, silvery white below.—Species with deep body. Rakers on 1st gill arch decrease in number with growth, numbering 24-28 in fish up to 40 cm and 22-24 in larger specimens. Second dorsal fin lobe distinctly longer than 1st. Vomerine tooth patch with elliptical head and a short shaft extending posteriorly. To 120 cm and 24 kg. Feeds on small fishes. Caught on hook and line for food and sport; schools of young fish caught in beach seines.

***S. dumerili*** (Risso) (Greater amberjack): Genus as above.—Species with moderately deep body. Rakers on 1st gill arch decrease in number with growth, numbering 20-24 in fish up to 20 cm and 11-19 in fish larger than 20 cm. Second dorsal fin lobe relatively short (only slightly longer than 1st). Vomerine tooth patch with elliptic head and long shaft extending posteriorly. To 200 cm and 75 kg. A pelagic species believed to spawn offshore. Food mainly small fishes and crustaceans. An important food and sport fish.

F. **CORYPHAENIDAE** (Dolphin-fishes): Elongate, compressed Perciformes with small cycloid scales. Mouth large, with numerous fine conical teeth in bands. Dorsal and anal fins long, without spines; the dorsal fin originates on the nape, the anal fin at or before the midpoint of the body. Caudal fin deeply forked, pelvic fins well developed, fitting into grooves on body. (2 spp. from Bda.)

***Coryphaena hippurus*** (L.) (Common dolphin): Body depth tapering from head posteriorly. Forward profile of head of ♂ increases in steepness with age. Anal fin concave, extending from anus to base of caudal fin. Bands of teeth on jaws, vomer and palatines and a small oval tooth patch on tongue. Brilliant metallic blue-green with tinges of yellow and gold. Dark spots or gold blotches below dorsal fin

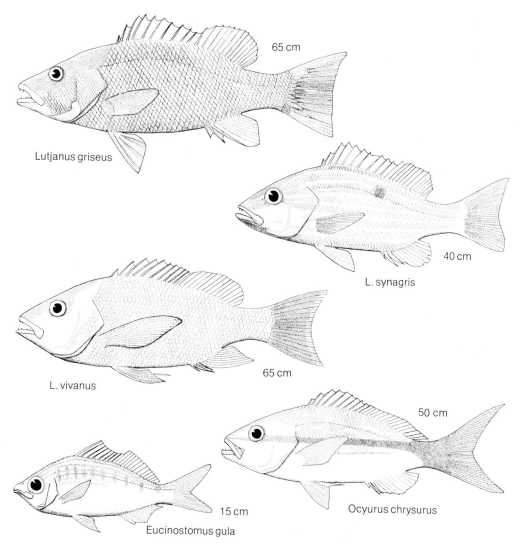

**206 LUTJANIDAE (Snappers), GERREIDAE (Mojarras)**

and scattered randomly on sides. To 200 cm and 40 kg. Pelagic, often associated with *Sargassum* and other floating objects, feeding on small fishes and crustaceans. An excellent food and game fish.

**Plate 206**

F. **LUTJANIDAE** (Snappers): Moderately compressed Perciformes with terminal mouth. Maxilla broadest posteriorly, with upper edge sliding under preorbital bone. Preopercle usually serrate, teeth present on roof of mouth. Dorsal fin single, with 10-12 spines and 9-15 soft rays. Body covered with fairly small ctenoid scales; anterior part of head scaleless. (10 spp. from Bda.)

*Lutjanus griseus* (L.) (Gray snapper): Genus robust, with interorbital region transversely convex; last rays of dorsal and anal fins not produced. Ten spines in dorsal fin; scales present on base of soft portions of dorsal and anal fins; lower limb of 1st gill arch with 16 or fewer rakers; caudal fin moderately forked to truncate.—Species with relatively slender body; color changeable, the back dark gray to greenish; the ventral surface silvery to reddish; a dark stripe often exhibited on head from angle of jaw through eye to nape. Canine teeth in upper jaw larger than in lower jaw; vomerine tooth patch with long medium posterior extension. To 65 cm; rarely exceeds 5 kg. Young are common in grass beds, mangroves and rocky areas; older specimens usually found offshore near patch reefs and to depths of 180 m, but may aggregate near wharves and jetties. Feeds primarily at night on crustaceans and small fishes. A valuable food fish.

***L. synagris*** (L.) (Lane snapper; B Whitewater snapper, Silk snapper): Genus as above.—Species with moderately deep body; color variable, usually pink to red with indistinct darker greenish vertical bars; sides and ventral surface silvery to yellowish, posterior edge of caudal fin dark red to dusky. A large diffuse dark spot on sides below anterior portion of soft dorsal fin; irregular golden stripes on head. Canines moderately developed in both jaws; vomerine tooth patch with moderate median posterior extension. Anal fin rounded. To 40 cm. Usually found over grassy areas or open sandy areas between patch reefs. Feeds mostly at night on fishes and bottom-dwelling crustaceans and worms. An excellent food fish.

***L. vivanus*** (Cuvier) (Silk snapper; B Red snapper): Genus as above.—Species with moderately deep body with bright red back becoming silvery ventrally. May exhibit fine, pale wavy yellow stripes on sides. Young specimens have dark spot on sides below soft dorsal fin. The iris is bright yellow. Vomer tooth patch with long slender median posterior extension. Pectoral fin long, reaching beyond anus; anal fin angulate. To 65 cm. Most common between 90 and 150 m near edge of island shelf. Feeds on crustaceans and small fishes, moving to shallower water at night. An excellent food fish.

***Ocyurus chrysurus*** (Bloch) (Yellowtail snapper): Body fusiform, head relatively small; caudal fin deeply forked; lower limb of 1st gill arch with 17 or more long rakers. A distinctive bright yellow stripe runs from snout to yellow caudal fin; back olive to blue with numerous large indistinct yellow spots; white on sides and ventral area below yellow stripe. To 70 cm. Young often found in grass flats, adults over coral reefs and grass flats. Adults feed in water column on small crustaceans and fishes such as herrings and sardines, young feed mainly on zooplankton. An important food fish caught with hand lines and beach seines.

F. **GERREIDAE** (Mojarras; B Shad): Small to medium-sized, compressed, silvery Perciformes. Mouth small, with small villiform teeth, the jaws highly protrusible. Most of head and body covered with conspicuous shiny scales. Scales form sheath on dorsal and anal fins. Caudal fin forked, pectoral fin long and pointed, pelvic fin with a scale-like axillary process. (4 spp. from Bda.)

***Eucinostomus gula*** (Cuvier) (Silver Jenny; B Shad): Body somewhat fusiform, preopercle entire, not serrated; color silvery or mottled but lacking vertical bars. Anal fin with 3 spines. Dorsal and anal fins dusky. Premaxillary groove running on top of snout interrupted by a transverse row of scales so that naked posterior end of groove is surrounded by scales. The 2nd of the 3 anal spines not enlarged. To 20 cm. In shallow water, especially over muddy bottoms or grass flats. Often occurring in aggregations, feeding on small bottom-dwelling invertebrates. Used as a bait fish.

### Plate 207

F. **HAEMULIDAE** (= POMADASYIDAE) (Grunts): Oblong, compressed Perciformes with strongly convex head profile. Teeth in jaws conical but not canine, lips fleshy. Opercle with 1 distinct spine; the posterior margin of suborbital not exposed. Chin with central groove and 2 pores anteriorly. Scales ctenoid, extending over head except chin, lips and snout. (7 spp. from Bda.)

*Haemulon aurolineatum* (Cuvier) (Tomtate; B White grunt): Genus with elongate body; anal fin with less or more than 11 soft rays. Preopercle finely serrate and none of the serrae directed forward. Soft sections of dorsal and anal fins densely scaled to their margins; interior of mouth red.—Species with oblong, compressed, silver-white body having a broad bronze to yellow longitudinal stripe from behind gill cover to large dark brown or black spot at base of caudal fin, and a narrower yellow stripe on back above lateral line. Dorsal fin with 13 spines and 14 or 15 soft rays; anal fin with 3 spines and 9 rays. No black blotch beneath free margin of preopercle. To 25 cm. Common inshore, aggregating near reefs or docks during the day but scattering over grass beds at night. Feeds on small bottom-dwelling invertebrates, algae and plankton.

*H. flavolineatum* Desmarest (French grunt; B Yellow grunt): Genus as above.—Species with bright yellow body with dark bronze longitudinal stripes above lateral line and oblique stripes below lateral line. Dorsal, caudal and anal fins chalky to yellow, a black blotch present beneath free margin of preopercle. Dorsal fin with 12 spines and 14 or 15 soft rays; anal fin with 3 spines and 8 soft rays. Total number of scales around caudal peduncle 22. To 25 cm. Found in aggregations around reefs and docks along shoreline to deeper (30 m) reefs offshore. Feeds mainly on small crustaceans at night over sand or grass flats. Caught with traps, seines and hook and line. Often used live as bait in fish pots.

*H. melanurum* (L.) (Cottonwick grunt): Genus as above.—Species with white, elongate body having yellow and black longitudinal stripes; back, below dorsal fin, upper half of caudal peduncle and caudal fin black. A black blotch often present below free margin of preopercle. Dorsal fin with 12 spines and 15-17 soft rays; anal fin with 3 spines and 8 soft rays. Total scales around caudal peduncle 23-25. To 35 cm. Lives in coral reef areas, to 40 m, often feeding over sand and grass flats at night on crustaceans. Caught in traps and hook and line and often sold as fillet.

*H. album* Cuvier (Margate): Genus as above.—Species with a relatively deep, silvery white body; membranes of spinous dorsal fin white; soft portion of dorsal, caudal, anal and pelvic fins dusky. Dorsal fin with 12 spines and 16 or 17 soft rays. Anal fin with 3 spines and 7 or 8 soft rays. Total number of scales around caudal peduncle 25-27. Largest species in genus, to 60 cm. A clear-water species, found over sand and grass flats near offshore reefs and breakers, occurring singly or in small aggregations. Feeds on bottom-dwelling invertebrates. Caught with traps and hand lines for food.

***H. carbonarium*** Poey (Caesar grunt): Genus as above.—Species silver-gray, with darker longitudinal stripes bronze to bright yellow. Head compressed, steel-blue with bronze stripes running from snout to behind eye; those below eye giving a blotched appearance. A black blotch present below free margin of preopercle; dorsal fin black with bronze on membrane and along base of soft portion; caudal, anal and paired fins dusky to black. Dorsal fin with 12 spines and 15 or 16 soft rays; anal fin with 3 spines and 8 soft rays. Total number of scales around caudal peduncle 22. To 35 cm. Most common on offshore reefs to 40 m, feeding mostly at night on demersal invertebrates. Caught in traps and on hand lines and often sold as fillet.

***H. sciurus*** (Shaw) (Blue-striped grunt): Genus as above.—Species bronze-yellow dorsally to cream-yellow on belly with pale blue longitudinal stripes on head and body as far as caudal fin base. Spinous dorsal fin yellow; soft rays of dorsal, caudal, and anal fins dusky. Dorsal with 12 spines and 16 or 17 soft rays; anal with 3 spines and 9 soft rays. Total number of scales around caudal peduncle 22. To 40 cm. Common inshore and on reefs to 30 m. Often found in large aggregations. Feeds mainly at night on invertebrates over grass and sand flats. Caught in traps and by hook and line and seines. Used as food or as live bait in fish pots.

### Plate 208

F. **SPARIDAE** (Porgies): Oblong, deep-bodied Perciformes with large head having scaleless snout and suborbital area. Preopercle scaled, without spines or serrations on margin. Mouth small, horizontal, the upper jaw never reaching beyond center of eye. Preorbital bone overlapping maxilla. Teeth well developed, canine-like or incisor-like in front, rounded molar-like laterally. Dorsal fin single, with 12-13 spines and 10-15 soft rays. Pectoral fins long, pointed; pelvic fins below or just behind pectoral fin bases having 1 spine and 5 soft rays; anal fin with 3 spines and 8-12 soft rays. Caudal fin emarginate or forked. (4 spp. from Bda.)

***Diplodus bermudensis*** Caldwell (Bermuda bream): Relatively small head and eyes, body silvery with bluish reflection on back. May show 8 or 9 indistinct dark vertical bars on sides; a large conspicuous black spot on upper caudal peduncle extending slightly below lateral line. With 8 well-developed flattened incisor-like front teeth and 3 rows of molar-like teeth laterally in jaws. No forward-projected spine in front of dorsal fin. With 18-21 rakers on the 1st gill arch and 62-67 pored scales on lateral line. To 40 cm. Found nearshore and on patch reefs. Omnivorous, feeding on algae, mollusks, crustaceans and other

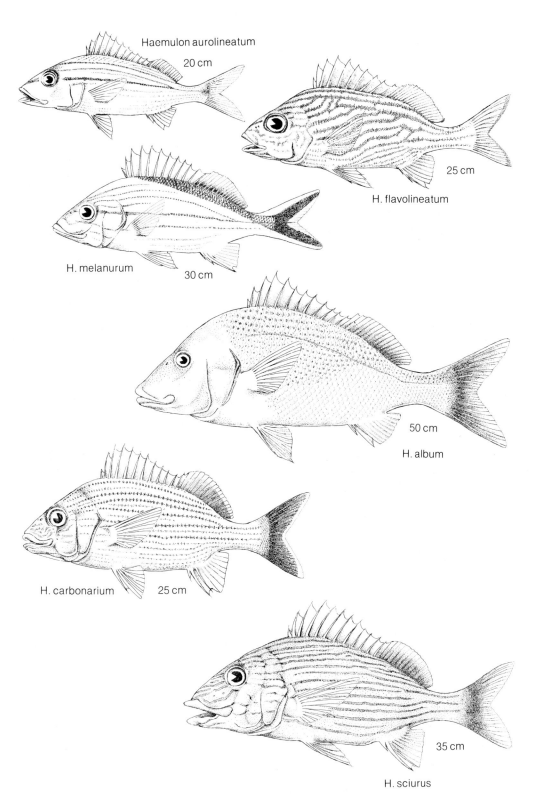

**207 HAEMULIDAE (Grunts)**

invertebrates; in harbors known to act as a scavenger. Larger specimens caught in traps and filleted as food. An endemic species.

***Lagodon rhomboides*** (L.) (Pinfish): Body silvery to olivaceous, with yellow longitudinal stripes on sides. A blackish spot near the origin of lateral line; 6 dark diffuse vertical bars on body. Pectoral, caudal and anal fins yellow, the latter with a light blue margin. Incisor-like teeth deeply notched and flattened. A forward-projected spine present in front of dorsal fin. Lateral molar-like teeth in 2 1/2 rows in each jaw; 12 dorsal and 11 anal soft rays. To 30 cm. Found in harbors and bays in shallow grassy and mangrove areas in aggregations. Feeds mainly on small crustaceans and other invertebrates. Caught in seine nets or traps and used mainly as bait or fillet.

***Calamus bajonado*** (Bloch & Schneider) (Jolthead porgy; B Bluebone porgy): Body deep, head and eyes large; slender close-set canine teeth in jaw front. The 2nd and 3rd teeth from center of upper jaw of adults enlarged but not outcurved. In upper jaw 3 rows of molar-like teeth plus an irregular series inside and toward the front. Head with steep profile and deep suborbital space; posterior nostril elongate. With 10 or 11 soft rays in anal fin; pectoral fins long and slender, with 15 rays. Color silvery, scales bluish with brassy margins. Cheek brassy without blue markings except a blue line under the eye margin. Lips and throat purplish; corners of mouth and isthmus orange. Bones are often bluish. To 70 cm. Young often found in grassy areas inshore; adults most common on offshore reefs. Feeds on sea urchins, crustaceans and mollusks. An important food fish caught mainly on hook and line and in traps.

F. **SCIAENIDAE** (Croakers): Elongate, moderately compressed Perciformes with 2 distinct dorsal fins barely connected at base. With 2 anal spines; lateral line extends to posterior edge of caudal fin. Body completely scaled except tip of rounded snout. Bony edge of opercle forked at its upper angle, forming a pair of distinct spines connected by a ridge. One or more barbels may be present on tip of lower jaw. (2 spp. from Bda.)

***Equetus acuminatus*** (Schneider) (Highhat, B Croaker): Lower jaw without barbels, its tip not projecting beyond upper jaw; mouth inferior and preopercular margin smooth. Body short, its dorsal profile strongly elevated. Less than 20 rakers on 1st gill arch. Height of anterior portion of dorsal fin elongated but not much greater than head length in adults; 40 or fewer soft rays in dorsal fin. Body with blackish horizontal stripes over silvery to pale brown sides. Leading edges of fins and trailing edges of spinous dorsal and pelvic fins white. To 25 cm. Found in rocky, reef and grassy areas in shallow waters. Feeds on small invertebrates. Occasionally caught in traps.

F. **MULLIDAE** (Goatfishes): Elongate Perciformes having convex head profile above but with ventral side of head and body flat. Chin with 2 large barbels that can be folded into grooves on throat. With 2 well separated dorsal fins; the 1st with 7 spines, the 2nd with 1 spine and 8 soft rays. Pelvic fins located below pectoral fins, with 1 spine and 5 soft rays. Caudal fin forked. (3 spp. from Bda.)

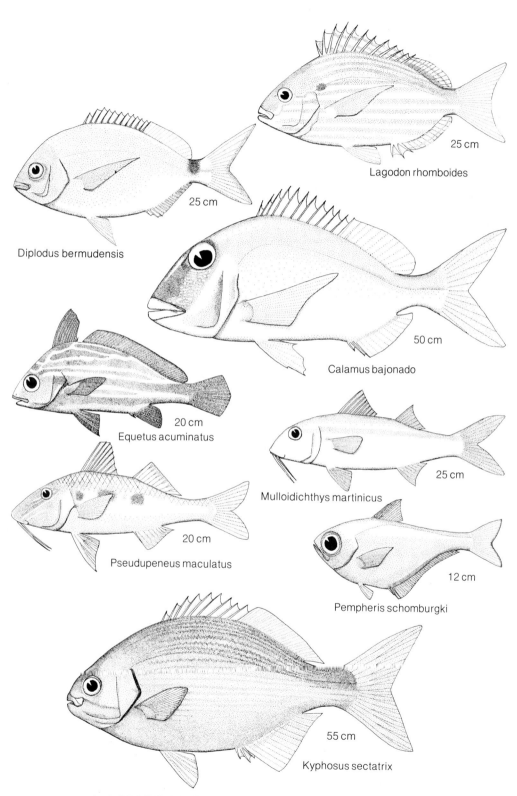

**208 SPARIDAE (Porgies)—KYPHOSIDAE (Sea chubs)**

***Pseudupeneus maculatus*** (Bloch) (Spotted goatfish): Teeth in a single series in young; in adults in 2 irregular series anteriorly, with some canine teeth in upper jaw. Snout long, with a concave profile. One spine on opercular margin; the maxilla not reaching to below anterior margin of eye. Teeth on roof of mouth lacking; 27-31 scales along lateral line. Body color sandy to olivaceous to reddish with 3 dark, square blotches on sides. Diagonal blue lines on head, the scales on back with central blue spot. To 40 cm. From shallow water to 60 m, over sand and coral rubble bottoms in reef areas. Feeds on invertebrates during daylight in sandy areas. Caught in traps and sold as fillet.

***Mulloidichthys martinicus*** (Cuvier) (Yellow goatfish): Teeth in 3 series anteriorly, all small and blunt. Snout long, maxilla not reaching below forward margin of eye. Snout profile convex. One spine on opercular margin. No teeth on roof of mouth; 34-39 scales along lateral line. Body pale on back, white on belly and sides. A bright yellow stripe from eye to caudal fin base, all fins yellow. To 45 cm. Found in shallow water inshore and over reef areas. Feeds on bottom-dwelling invertebrates in sandy areas, often in aggregations, stirring up dense clouds of fine sediment. Caught in traps and sold as fillets.

F. **PEMPHERIDAE** (Sweepers): Small, compressed, deep-bodied Perciformes with a long-based anal fin and single short-based high dorsal fin. Caudal fin forked. Snout short, maxilla reaching below center of eye. (1 sp. from Bda.)

***Pempheris schomburgki*** Müller & Troschel (Glassy sweeper, B Copper sweeper): With 5 spines and 8 or 9 soft rays in dorsal fin and 3 spines and 32-35 soft rays in anal fin. Pectoral fin short, transparent, with 17 or 18 rays; eye large. Body color coppery, with a dark band at base of anal fin. Young nearly transparent. To 15 cm. Aggregations commonly found secreted in reef caves during daytime and near reefs at night feeding on zooplankton. Swims with bobbing motion caused by using pectoral fins to maintain position.

F. **KYPHOSIDAE** (Sea chubs): Relatively deep-bodied, oval Perciformes with small head, short snout and small teeth. Each jaw with a regular row of close-set, strong, incisor-like teeth, each tooth being crescent-shaped with its base set horizontally. Pectoral fins short, dorsal fin single, caudal fin moderately forked. Scales small, covering fins except spinous portion of dorsal. Digestive tract long. (2 spp. from Bda.)

***Kyphosus sectatrix*** (L.) (Bermuda chub): With 16-18 rakers on lower limb of 1st gill arch. Dorsal fin with 11 spines and 11-13 soft rays; anal fin with 3 spines and 10-12 soft rays. Color grayish, with dull longitudinal yellow stripes on body and 2 dull yellow horizontal bands on head. Upper part of opercular membrane blackish. May display blotchy pale spots on head, body and fins. To 75 cm. Found in shallow water nearshore or over grassy areas and offshore around reefs. Feeds mainly on plants including pelagic algae, and invertebrates. Caught in traps, nets and on hook and line. Flesh dark, very firm and not always marketable. An excellent game fish and when carefully prepared highly palatable.

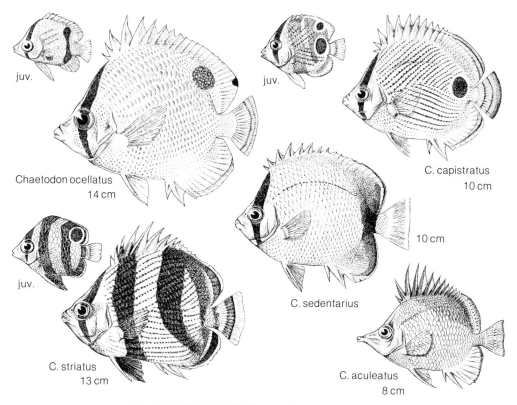

**209 CHAETODONTIDAE (Butterfly fishes)**

## Plate 209

**F. CHAETODONTIDAE** (Butterfly fishes): Perciformes with deep, strongly compressed bodies nearly round in profile. Mouth small, terminal and protractile, gape not reaching to anterior rim of orbit. Teeth arranged in brush-like bands in jaws; preopercle without strong spine at angle. Dorsal spines 12-14; the 1st not procumbent. Anal fin with 3 spines, a scaly process present at axil (base) of pelvic fin spine. Color white to silvery or yellow with various markings. An eye band usually present. All species are diurnal becoming torpid at night and usually exhibiting changes in color patterns. Often seen in pairs. (5 spp. from Bda.)

*Chaetodon ocellatus* Bloch (Spotfin butterfly fish): Genus with snout pointed in lateral aspect, may be elongated or beak-like. A dark band running from anterior base of dorsal fin through eye to base of opercle.— Species with a sharply defined black spot on posteriormost edge of dorsal fin lobe. A 2nd, larger dark spot is usually present at base of soft dorsal fin. Ground color of body white; the dorsal, anal, caudal and pelvic fins, caudal peduncle and base of pectoral fins bright yellow. Upper margin of opercle with yellow. Soft dorsal, anal and caudal fins with narrow blue-white line running parallel to edge. To 20 cm. Feeds on crustaceans and on tentacles of polychaete tube worms and zoantharians. At night may exhibit broad diffuse dark bands on sides. Found on shallow reefs in harbors and offshore. Sometimes caught in traps and used for lobster bait.

***C. capistratus*** (L.) (Foureye butterfly fish): Genus as above.—Species with a large, well-defined ocellus on side below the soft dorsal fins. Above midbody scale rows run obliquely up and back, below they run down and back; the narrow dark lines along scale rows distinctive bluish to black. A yellow-edged black band runs through eye. A yellow to dusky, black-edged band present on soft dorsal, caudal and soft anal fins. Pectoral fins not pigmented; pelvic fins pale yellow. To 15 cm. Found in reef areas and around docks or jetties nearshore. Feeds primarily on tentacles of polychaete tube worms and zoantharians and on small crustaceans.

***C. striatus*** L. (Banded butterfly fish): Genus as above.—Species with 2 broad vertical bars on sides in addition to 1 through eye and 1 across soft dorsal and anal fins and caudal peduncle. Scale rows above midbody run obliquely up and back, and below midline down and back. Except where covered with black bars, the body and fins are white or in older fish slightly tinged with yellow. To 15 cm. Found over shallow reefs and occasionally near rocky outcrops in grassy areas. Feeds primarily on tentacles of polychaete tube worms and zoantharians and on small crustaceans.

***C. sedentarius*** Poey (Reef butterfly fish): Genus as above.—Species without spots or ocelli; bands on head and across soft part of dorsal and anal fins and caudal peduncle only. Dorsal and anal fins rounded posteriorly. Edges of scales on back edged with dark yellow to give muted yellow to brown color. To 10 cm. Feeds on crustaceans, polychaete tube worms and zoantharians. Found over offshore reef areas and generally in deeper water than other members of the genus; most common at 20 m.

***C. aculeatus*** (Poey) ( = *Prognathodes aculeatus*) (Longsnout butterfly fish): Genus as above.—Species with elongated beak-like snout. A dark orange band from anterior base of dorsal fin to snout. Soft part of dorsal fin and caudal peduncle and fin whitish. Upper half of body yellow-orange shading to black at base of dorsal fin. Lower half of body white. Thin orange bands on head and across caudal peduncle. To 8 cm. Found on deep reefs (30-120 m). Feeds on tube worm tentacles, the tube feet and pedicellariae of sea urchins and small crustaceans.

**Plate 210**

F. **POMACANTHIDAE** (Angelfishes): Perciformes with deep, highly compressed body nearly round or oval in profile. Mouth small, terminal and protractile, gape not reaching to anterior rim of orbit, snout not produced. Teeth arranged in brush- like bands in jaws; preopercle with strong spine at angle. Dorsal fin with 9-15 spines, continuous, never notched; soft dorsal and anal rays may be produced to form filaments. Anal fin with 3 spines; no axillary scaly process present at base of pelvic fins. Brightly colored fishes, predominantly blue, yellow or black. All species are diurnal becoming torpid at night. (6 spp. from Bda.)

***Centropyge argi*** Woods & Kanazawa (Cherub fish): Relatively small, elongate-oval; hind margin of preorbital bone with enlarged, strong poste-

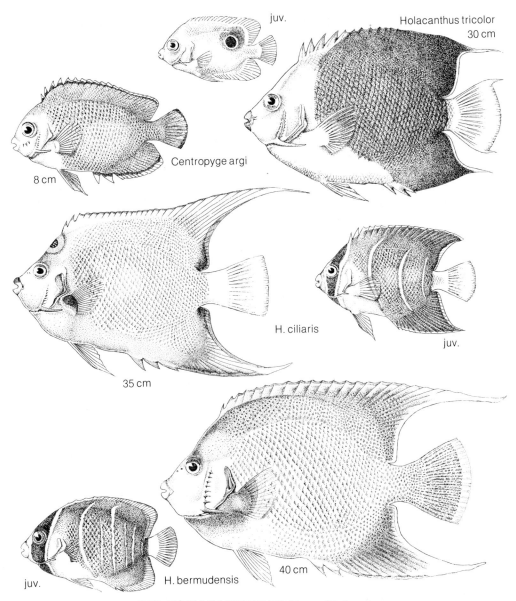

**210 POMACANTHIDAE (Angelfishes)**

riorly directed spines. Rays on soft dorsal and anal fins not projected to form filaments. Caudal fin rounded. Yellow-orange head and chest; body deep blue; a narrow blue ring around eye; blue spines on preopercular margin; pectoral fins yellowish, other fins deep blue with pale blue margins. Juveniles have same coloration as adults. To about 8 cm. Thought to feed on algae. Uncommon in less than 30 m; found over offshore reefs. Behavior resembles that of Pomacentridae with which it may be confused in the field.

***Holacanthus tricolor*** (Bloch) (Rock beauty): Genus of moderate size, round to oval; hind margin of preorbital bone lacking enlarged posteriorly directed spines. Dorsal and anal fins extended into filaments in adults. Juvenile coloration differs from that of adults.—Species having adults with short filaments posteriorly on dorsal and anal fins, the posterior contour of these fins nearly straight. Caudal fin rounded, with short filament at upper (and sometimes also lower) posterior corner. Head, chest, abdomen, anterior part of dorsal fin and pelvic fins yellow. Central and posterior portions of body and most of dorsal and anal fins black. Caudal fin and posterior edges of dorsal and anal fins bright yellow. Lips blackish. Juveniles are mostly yellow with a large black spot ocellated with blue on sides of body. With age, the spot is lost within the larger dark area that develops. To 35 cm. Found on offshore shallow reefs. Juveniles often associated with *Millepora*. Feeds on variety of invertebrates including crustaceans, sponges, coral polyps and tube worms. Larger specimens are caught in fish pots and filleted.

***H. bermudensis*** Goode (Blue angelfish): Genus as above.—Species having adults with dorsal and anal fins greatly produced into bright yellow filaments; the posterior contour of these fins concave. Caudal fin rounded, without filaments. Scales bluish brown to reddish brown with pale yellow edges. Dorsal, caudal and anal fins blue with yellow edges. Pectoral fins blue with wide yellow band and pale yellow edge. Juveniles with head and chest to origin of anal fin yellow; dark edged blue bars on body forming nearly straight lines; dark head band bordered with light blue from nape to origin of pelvic fins; pelvic and caudal fins yellow; borders of dorsal and anal fins blue. To 45 cm. Common in shallow water in rocky or reef areas from shore to deeper reefs at 40 m. Feeds on a variety of invertebrates including crustaceans, sponges, coral polyps and tube worms. Larger specimens are caught in fish pots and sold as whitefish fillet.

***H. ciliaris*** (L.) (Queen angelfish): Genus as above.—Species having dorsal and anal fins greatly produced into yellow filaments in adults, the posterior contour of these fins concave. Caudal fin rounded, without filaments. Scales bluish with bright yellow-orange edges. Head yellowish with bright blue markings on eyes, snout and opercle. Dorsal and anal fins body-colored changing to yellow near edges that are outlined with light blue. Pectoral fins yellow with black blotch spotted with blue at their base. Pelvic and caudal fins bright yellow. A large black blotch (ocellus) edged with blue at the nape. Juveniles with bluish black ground color with light blue curved bars crossing the body; head, chest, caudal and pelvic fins yellow; a dark eye band bordered with light blue extending from nape to origin of pelvic fins. To 45 cm. Uncommon; found on offshore reefs. May hybridize with *H. bermudensis* to give intermediate color patterns. Feeds on variety of invertebrates including crustaceans, sponges, coral polyps and tube worms. Larger specimens are caught in fish pots and sold as whitefish fillet.

### Plate 211

F. **POMACENTRIDAE** (Damselfishes): Small, deep-bodied, somewhat compressed Perciformes with small mouth and highly protrusible jaws. Teeth in jaws incisor-like or conical but not mo-

# OSTEICHTHYES

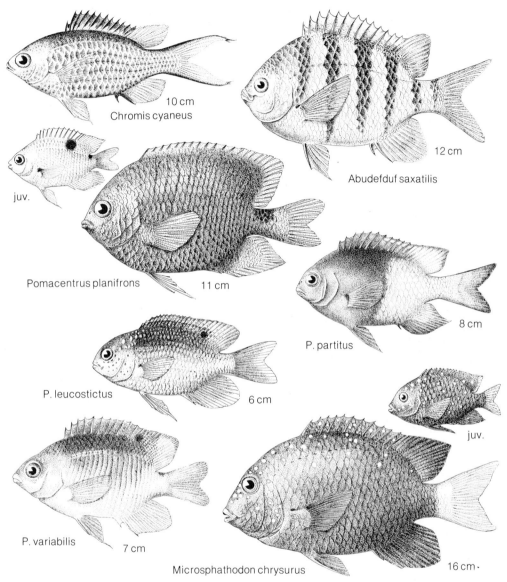

**211 POMACENTRIDAE (Damselfishes)**

lar- or canine-like. A single pair of nostrils. Dorsal fin with 10-14 spines, anal fin with 2 spines. Lower pharyngeal teeth completely fused to form a plate. Coloration may be permanent, or change with age or spawning condition. (12 spp. from Bda.)

***Chromis cyaneus*** (Poey) (Blue Chromis): Body nearly fusiform, caudal fin deeply forked, with both upper and lower lobes pointed. Teeth conical, in 2-4 rows. Upper and lower base of caudal fin with 2 or 3 projecting spines. The edge of the dorsal fin, a stripe from the head along the back to the tip of the upper caudal lobe, the outer margin of the lower caudal lobe, the leading edge of the anal fin and the eye are all black. The pectoral fins are clear and the remainder of the fish is bright blue. To 12 cm. Found on outer reefs, 10-30 m, where it feeds on plankton in the water column over the reef, often forming aggregations with the Creole wrasse (*Clepticus parrai*). Returns to shelter of reef at night or when threatened.

***Abudefduf saxatilis*** (L.) (Sergeant major): Body laterally compressed, deep; dorsal fin spines 13; preopercle entire, without spines or serrations. No notch on preorbital bone bordering jaw. Teeth incisor-like in single row. Caudal fin forked. Tips of upper and lower lobes rounded. Back and sides greenish yellow, belly bluish white. With 5 prominent vertical black bars on sides, each tapering to a point ventrally. Base of pectoral fins with a small dark spot. Upper and lower base of caudal without spines. To 20 cm. Found from shore to outer reefs, to 20 m; juveniles often in tide pools. Feeds on plankton, benthic invertebrates and algae. Schools of large fish may be aggressive and prick the skin of divers. Large specimens are sometimes caught in fish pots and used as bait.

***Pomacentrus planifrons*** Cuvier (Threespot damselfish; B Yellow damselfish): Genus moderately compressed, with single or double row of conical teeth in upper jaw; 12 dorsal spines; a serrated preopercle; and no spines on upper or lower edges of caudal fin base.—Species relatively deep-bodied, upper profile of head steep and straight. Adults brownish gray with yellowish cast having vertical dark lines following scale rows; a large black spot covering most of pectoral fin base. Vertical fins colored like body, pectorals slightly dusky, pelvics yellowish. Juveniles bright yellow, with a large black spot faintly edged in blue at base of dorsal fin at junction of spinous and soft portions; a large black spot on dorsal surface of caudal peduncle and a small one on upper pectoral base. To 12 cm. Found on shallow patch reefs offshore, to 30 m. Feeds on algae and small crustaceans; juveniles will pick parasites from other fishes.

***P. partitus*** (Poey) (Bicolor damselfish): Genus as above.—Species with relatively deep body, usually dark brown on head and anterior half of body, then abruptly orange or yellow shading posteriorly to white including caudal fin. Pectoral fins yellow except for dark base. Anterior margin of anal fin dark. To 10 cm. Found nearshore, on patch reefs and offshore to 25 m where there is a hard substrate. Feeds mainly on fine attached algae and detritus but will also eat coral polyps and other invertebrates. Nesting males are aggressive.

***P. leucostictus*** Müller & Troschel (Beau Gregory): Genus as above.—Species with relatively elongate shallow body, bright blue on the upper head and most of the back and dorsal fins; bright yellow elsewhere. Black spot on spinous blue portion of dorsal fin. No vertical dark lines on sides of body. Large males may be dark gray but with the centers of scales on nape, anterior portion of body and along back yellowish. To 10 cm. Found in shallow water on patch reefs and offshore to 10 m. Juveniles found in tide pools, grassy spots and near wharves or jetties. Feeds on algae and small crustaceans.

***P. variabilis*** (Castelnau) (Cocoa damselfish): Genus as above.—Species with relatively slender body and 19-21 pectoral fin rays. Anal fin long, extending well beyond base of caudal fin. Operculum with 3 more or less complete rows of scales. Larger fish dusky on head and back, becoming yellowish ventrally; the color transition abrupt. A dark spot always present at junction of dorsal fin and caudal peduncle. Pectoral, pelvic and anal fins bright yellow in juveniles, becoming dusky yellow in adults. Juveniles bright blue on back, yellow ventrally with a dark spot at base of dorsal fin. Found on reefs and rocky substrate from shore to depths of 20 m. Feeds on algae and small invertebrates. Nesting and courting males aggressive; may show heightened colors during this period.

***Microsphathodon chrysurus*** (Cuvier) (Yellowtail damselfish, B Jewelfish): Robust, relatively large, with teeth in upper jaw flexible, brush-like; a pronounced notch in preorbital bone bordering the jaw. Mouth small, scarcely protrusible; preopercle entire. Dorsal fin with 12 spines; caudal fin bluntly forked, upper and lower lobes rounded. Adults very dark blue to black, with brilliant iridescent blue spots scattered on dorsal and lateral surfaces. Caudal fin pale to bright yellow. Juveniles ("jewelfish") dark blue, with relatively large brilliant reflective blue spots on body and a white caudal fin. Found in shallow patch reefs offshore to 15 m. Juveniles often in tide pools. Feeds on algae and *Millepora* polyps. Juveniles will pick parasites from other species of fish. Adults are territorial and highly aggressive and will display for and attack other species of fishes and divers.

## Plate 212

F. **LABRIDAE** (Wrasses): A large family of Perciformes with wide variability in form and size. Terminal mouth varying from small to elongate and protrusible; teeth in front of jaws usually as forward-projecting canines; no teeth on roof of mouth but pharyngeal teeth well developed. Single dorsal fin; spines usually weak. Scales large, cycloid. (16 spp. from Bda.)

***Halichoeres bivittatus*** (Bloch) (Slippery Dick): Genus slender, with a continuous but abruptly curved lateral line below soft portion of dorsal fin. With 9 spines in dorsal fin. Colors highly variable with species, development and sex. Most species cover themselves with sand for shelter at night.—Species with 2 dark longitudinal stripes, one running through eye to caudal fin base, the other on lower side of body. Body color white to greenish in adult males. Yellow to orange markings on caudal fin and head, the corners of caudal fin rounded and blackish. To 20 cm. Found in a wide variety of habitats, from rocky and muddy inshore waters to reefs, sand and coral rubble offshore. Feeds mainly on crustaceans, sea urchins, polychaete worms, mollusks and brittle stars. Large specimens caught in traps and by hook and line and used as bait or discarded.

***H. garnoti*** (Cuvier & Valenciennes) (Yellowhead wrasse): Genus as above.—Species with dark lines extending diagonally upwards and back from eyes. Black dots present behind diagonal lines. Young without black dots but with median blue stripe. Young and adults bright yellow, with blue mottling on caudal and dorsal fins. Dominant males develop a vertical black bar on the side. The upper part of head and body in front of black bar bright yellow; the body behind bar and just below dorsal fin bright red to black; the remainder of body greenish. Caudal fin rounded. To 20 cm. Found on patch reefs offshore to 80 m. Feeds on crustaceans, polychaetes and other invertebrates.

***H. radiatus*** (L.) (Puddingwife; B Bluefish): Genus as above.—Species with light blue lines extending obliquely up and backwards from eye. Relatively deep-bodied; the young have large black blotches at base of dorsal fin and on caudal peduncle; 2 broad yellow stripes on body and fine pale blue bars distinct in young, becoming obscure with age. Adults yellowish olive on back shading to orange-yellow on sides. Dorsal and caudal black spots absent. Body with rows of blue spots. A small black spot at upper base of pectoral fin. Large dominant males are greenish, with dark-edged pale blue bar in middle of body. Caudal fin truncate. The largest *Halichoeres*, growing to 50 cm. Found from shallow patch reefs offshore to 40 m. Feeds on mollusks, sea urchins, crustaceans, and brittle stars. Adults frequently caught in traps and on hooks; used as fillet.

***Hemipteronotus martinicensis*** (Valenciennes) (Straight-tail razorfish): Body compressed, tapered from head. Dorsal spines 9; posterior canines absent, mouth blunt. Lateral line interrupted posteriorly, the rear portion a separate midlateral segment on caudal peduncle; caudal fin truncate or slightly rounded. ♀ light greenish gray shading to pinkish ventrally with an indistinct orange-red stripe from behind eye to caudal peduncle. A broad white area over abdomen, the ventral surface with vertical red line. Most of head and chest white, with dark area on opercle and chest. Adult males lose red, white and dark markings and develop yellow head with pale blue bands, a blue spot in axil of pectoral fin and a vertically elongate blue spot on each body scale. To 15 cm. Found over grassy and sandy areas from ledge flats to 30 m. Feeds on crustaceans, mollusks and other invertebrates. Dives headfirst into sand when disturbed.

***Thalassoma bifasciatum*** (Bloch) (Bluehead wrasse): Body elongate, cigar-shaped; dorsal fin with 8 spines, caudal fin truncate in young becoming lunate with age. Three prime color phases: the youngest fish have a black midlateral stripe continuing on head as red blotches; above stripe pale yellow (inshore) to bright yellow (on reefs); body below stripe white; a small black spot in front of dorsal fin and at upper pectoral base. On larger fish the black stripe fragments to a series of dark blotches or nearly disappers; belly pale to bright yellow. The terminal or dominant ♂ phase fish have bright blue head and green body, with 2 broad vertical black bars anteriorly separated by a pale blue interspace; pectoral fins clear, tipped with black, caudal fin falcate with black lobes. Yellow phase fish may be ♂ or ♀ and spawn in aggregations;

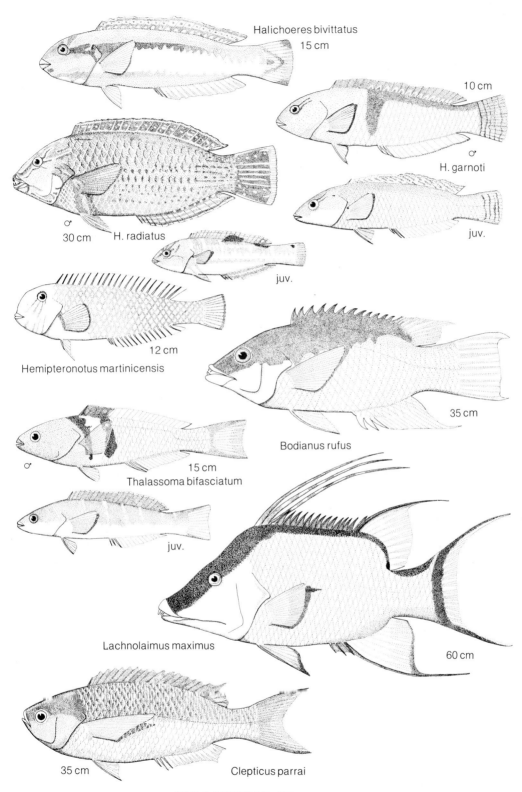

**212 LABRIDAE (Wrasses)**

blue-head ♂ phase fish mate in pairs with yellow females. To 15 cm. Found inshore on patch reefs or adjacent grass flats, and reefs offshore to 40 m. Feeds on crustaceans, mollusks and other small invertebrates. Yellow phase fish will pick parasites to supplement diet.

***Bodianus rufus*** (L.) (Spanish hogfish): Body elongate, nearly round; snout pointed, with strong teeth in front of jaw and short canines present posteriorly. Dorsal fin continuous, with 12 spines and 10 soft rays; anterior soft rays same length as spines; posterior soft rays elongate, forming a filament in dorsal and anal fins. Upper and lower lobes of caudal fin may be filamentous in adults. Color variable, generally upper 2/3 of body blue to blue-purple, the remainder bright yellow. Adults may be totally blue-black. To 40 cm. Feeds mainly on crustaceans, brittle stars, urchins and mollusks. Often taken in traps and on hand line and used for food (fillet) and lobster bait.

***Lachnolaimus maximus*** (Walbaum) (Hogfish): Largest and most deep-bodied of Labridae, with 14 spines in dorsal fin, the first 3 long and streamer-like. Interspinous membrane deeply notched; caudal fin emarginate in young, lunate in adults. Color variable, mottled brownish red with a black spot at rear base of dorsal fin; large males dark maroon on head and nape above lower edge of eye, abruptly becoming pale beneath; vertical fins blackish basally and on lobes of caudal fin. Large males have long snouts, the profile of head becoming concave. To 100 cm. Common inshore, around bases of patch reefs and offshore reefs to 30 m; most common over open bottoms where gorgonian corals are abundant. Feeds on mollusks, crustaceans, sea urchins and barnacles. Caught in traps, on hook and line, also easily speared. An excellent, highly prized food fish.

***Clepticus parrai*** (Bloch & Schneider) (Creole wrasse): Body streamlined, round, fusiform, caudal fin forked. Snout blunt; canine teeth weakly developed and relatively small in jaw front, absent posteriorly. Gill rakers well developed; upper jaw slightly protractile; mouth at oblique angle. Dorsal fin with 12 spines, 10 rays, the anterior rays the same length as spines. Color variable; blue on back shading to silver ventrally on young, to entirely purplish on adults. Found on offshore reefs from breakers to 40 m. A water column forager, feeding on zooplankton. In the field, young resemble *Chromis cyanea* but the persistent use of pectoral fins in *Clepticus* in swimming separates the 2 species. Rarely caught because they do not enter pots and seldom take a hook.

F. **SCARIDAE** (Parrot fishes): Oblong, moderately compressed Perciformes with bluntly rounded head, large scales, and teeth fused to form a pair of beak-like plates in each jaw. Pharyngeal teeth modified to form plates, the upper pharyngeals forming rows of smolariform teeth on a convex surface; the lower pharyngeal bones have molariform teeth on a concave surface. Dorsal fin slender, with 9 flexible spines and 10 soft rays. Color often bright and gaudy, varying with species, age and sex. All species are diurnal and shelter in reef crevices at night; some extrude a mucous envelope when sleeping. (14 spp. from Bda.)

### Plate 213

***Scarus coeruleus*** (Bloch) (Blue parrot fish, B Clamacore): Genus with teeth fully fused to form beak-like plates; body robust; 3 or 4 rows of scales on cheek; median predorsal scales 6 or 7; lower dental plates included within the upper when mouth is closed.—Species with 3 rows of scales below the eye; median predorsal scales 6; pectoral fin rays 16. Snout with a distinct hump in profile in adults and subadults. Small to medium fish are pale blue except upper part of head, which is yellowish; a transverse band of salmon color on chin; dental plates white. Adults deep blue; the scaled part of head bluish gray. To 70 cm. Found in rocky coastal areas or coral reefs offshore to 20 m. Feeds on benthic algae and invertebrates. Occasionally caught in traps and nets and sold as fillet.

***S. croicensis*** (Bloch) (Striped parrot fish): Genus as above.—Species with 14 pectoral fin rays; 7 predorsal scales, 3 rows of scales below the eye, and usually 6 scales in the uppermost series. Primary color phase with 3 dark brown stripes on whitish background on body, the uppermost along back, the lowermost passing through pectoral fin base. Upper part of snout yellowish in life. Terminal ♂ phase with chest and head pink below a greenish blue band at lower edge of eye, orange above; body blue-green and orange; a broad diffuse pink stripe on body anteriorly above pectoral fins. Median fins blue bordered, the central parts orange with linear blue markings. To 35 cm. Found in rocky coastal areas or coral reefs offshore to 20 m. Primary phase often forming relatively large feeding and spawning aggregations in summer. Feeds on benthic algae and invertebrates. Large specimens caught in traps and sold as fillet.

***S. guacamaia*** Cuvier (Rainbow parrot fish): Genus as above.—Species with 3 rows of scales below the eye, the lowest usually with 1 scale; median predorsal scales 6; pectoral fin rays 16. In juveniles, scales are diffuse light green in centers rimmed with light brownish orange; chest and scaled and unscaled parts of head orange-brown, with short green lines around eyes; fins dull orange with a broad streak of green extending onto membrane, margins of median fins blue. Adult fish have deeper colors, the green scales restricted to dorsal and posterior part of body; dental plates blue-green; color same in both sexes. To 100 cm. Found from rocky coastal areas offshore to reefs at 20 m. Feeds on algae and attached fauna by scraping substrate with beak. Occasionally taken in nets and traps and sold as fillet.

***S. taeniopterus*** Desmarest (Princess parrot fish): Genus as above.—Species with 14 pectoral fin rays; 7 predorsal scales; 3 rows of scales below the eye and usually 7 scales in the uppermost series. Primary color phase with 3 dark brown stripes on whitish background originating on head; the uppermost along the back, the lowermost passing through pectoral fin base. Terminal ♂ phase blue-green on head above eye, pinkish below; body blue-green and orange, with a broad pale yellow stripe anteriorly above pectoral fin; 2 narrow blue-green stripes on head, 1 through upper and 1 through lower part of eye; dorsal and anal fins blue bordered with orange centers; caudal fin blue, the upper and lower edges

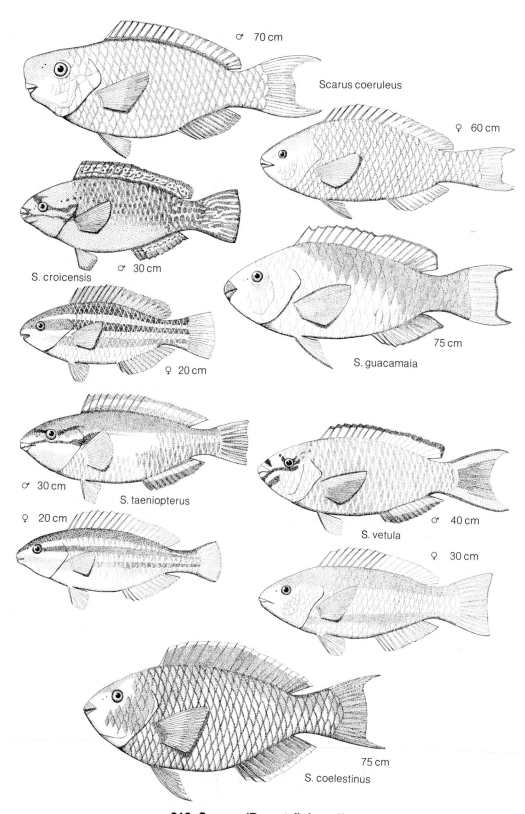

**213 *Scarus* (Parrot fishes 1)**

bright orange. To 35 cm. Found in rocky coastal areas or coral reefs offshore to 20 m. Primary phase often forming feeding and spawning aggregations during summer. Feeds on benthic algae and invertebrates. Large specimens caught in traps and sold as fillet.

**S. vetula** Bloch & Schneider (Queen parrot fish): Genus as above.—Species with 14 pectoral fin rays; 7 predorsal scales; 4 rows of scales below the eye. Primary color phase gray to dark reddish or purplish brown, with a broad whitish stripe on lower side. Terminal ♂ phase blue-green, with red or red-orange edges on scales. Head and snout blue-green, with alternating bands of orange and blue-green on lower snout and chin. Dorsal and anal fins bordered with blue and with red or orange centers; pectoral fins and caudal fins with pale or orange submarginal band, the remainder blue-green. To 55 cm. Found in rocky coastal areas or coral reefs offshore to 20 m. Feeds on benthic algae and invertebrates. Large specimens caught in traps and sold as fillet.

**S. coelestinus** Valenciennes (Midnight parrot fish, B Plum parrot fish): Genus as above.—Species with 3 rows of scales below eye, the lowest usually with 2 scales. Median predorsal scales 6; pectoral fin rays 16. Color blackish, the centers of scales diffuse bright blue. Scaled portion of head blackish, except a pale blue band across interorbital space. Unscaled parts of head bright blue; dental plates blue-green, fins dark with blue margins; color same in both sexes. To 80 cm. Found from rocky coastal areas to offshore reefs at 20 m. Feeds on algae and attached fauna by scraping rocky substrate with beak. Occasionally taken in nets or traps and sold as fillet.

### Plate 214

***Sparisoma aurofrenatum*** Valenciennes (Redband parrot fish): Genus with teeth fully fused to form beaklike plates; body robust; single row of scales below the eye; median predorsal scales 4, lower dental plates outside those of upper jaw when mouth is closed.—Species with interorbital space slightly concave or flat, membranous flap on anterior nostril longer than wide and with 4 or more lobes in adults; pectoral fin base pale. Primary color phase mottled brown to greenish brown with bluish green cast, becoming pale mottled red ventrally. A white saddle-shaped spot on caudal peduncle just behind last dorsal ray. Terminal ♂ phase with white peduncular spot; body orange to greenish blue, less mottled than primary phase; a diagonal orange stripe from corner of mouth passing below eye to upper end of gill opening; an orange spot about size of eye with dark or spotted center above pectoral fin; dorsal fin reddish; anal fin red with dark margin; caudal fin reddish with tips of lobes black. To 30 cm. Found inshore near patch reefs and offshore to 20 m. Feeds on benthic algae. Large specimens caught in traps and used as bait or filleted.

**S. crysopterum** Bloch & Schneider (Redtail parrot fish): Genus as above.—Species with interorbital space flattened. Membranous flap on anterior nostril simple or with no

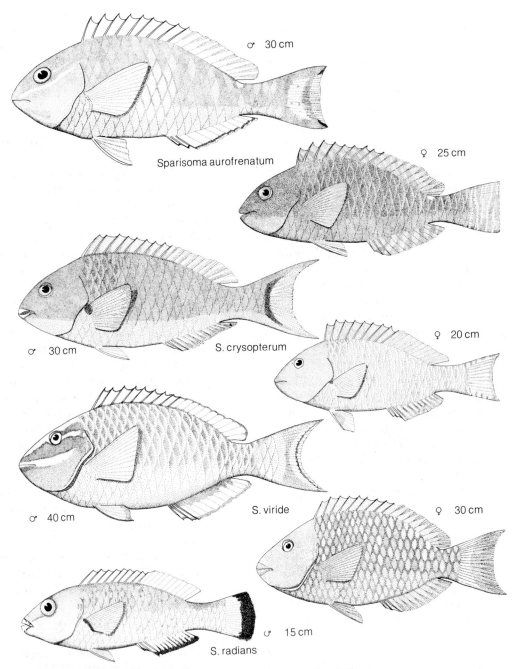

**214 *Sparisoma* (Parrot fishes 2)**

more than 6 lobate cirri; base of pectoral fin with black saddle-shaped marking at upper end. Primary phase olivaceous on back, mottled reddish on sides becoming pale ventrally. A large yellowish patch on caudal fin. Terminal ♂ phase green, the edges of scales brownish; ventral part of head and body bluish green, becoming deep blue ventrally; large patch of red in center to edge of caudal fin; dorsal, anal and pelvic fins reddish. To 35 cm. Found inshore near patch reefs and offshore on reefs to 20 m. Feeds on benthic algae and invertebrates. Large specimens caught in traps and used as bait or filleted.

***S. viride*** (Bonnaterre) (Stoplight parrot fish, B Green parrot fish): Genus as above.—Species with membranous flap of anterior nostril composite, deeply incised; pectoral fin base clear; saddle markings on peduncle lacking; caudal fin truncate, becoming lunate with age. Primary phase dark brown to reddish brown above; lower 1/3 bright red; all fins pale to bright red. Terminal ♂ phase emerald green, with 3 diagonal yellow-orange bands on head; posterior edge of gill cover bright yellow, with bright yellow spot near upper end; a large yellow patch at base of caudal fin; posterior margin of caudal fin with yellow and blue border; dorsal fin orange-yellow; anal fin bluish with yellow-orange center; pelvic fin orange-yellow bordered with blue; pectoral fin pale blue. To 70 cm. Found from rocky coastal areas to offshore reefs at 20 m. Feeds on benthic algae and invertebrates. Caught in fish traps and used as fillet.

***S. radians*** (Valenciennes) (Bucktooth parrot fish): Genus as above.—Species with somewhat rounded interorbital space; membranous flap on anterior nostril simple, ribbon-like; pectoral fin base clear or darkened for entire length. Primary phase olivaceous to gray mottled with pale dots; base and axil of pectoral fins and edges of opercle pale blue. Terminal ♂ phase greenish brown with faint pale or reddish spots; a diagonal stripe of blue and orange running from corner of jaw to rim lower edge of eye, and extending as a broken stripe towards origin of dorsal fin; pectoral fin base entirely black; a broad black band on caudal fin; anal fin dusky. The smallest species of the genus, growing to 18 cm. Found mainly in sea grass (especially *Thalassia*) beds. Feeds on epiphytes and blades of sea grass, leaving halfmoon-shaped bite marks.

**Plate 215**

**F. MUGILIDAE** (Mullets): Elongate, cylindrical Perciformes with broad, flattened head and bluntly rounded snout. Mouth small, terminal or inferior. Lateral line absent. Dorsal fins 2, the 1st with 4 slender spines. Pectoral fins set high on sides; pelvic fins about equidistant between pectoral fin base and origin of 1st dorsal fin. Color blue- green or olive on back, silvery on sides and belly; fins hyaline to dusky. (4 spp. from Bda.)

***Mugil trichodon*** Poey (Fantail mullet): Body slender; interorbital area very slightly convex; adipose tissue just reaching the pupil of eye; lips relatively thick, the lower lip with a high symphysial knob; hind end of upper jaw reaching between the posterior nostril and anterior rim of eye. Second dorsal and anal fins entirely and densely covered with small scales ex-

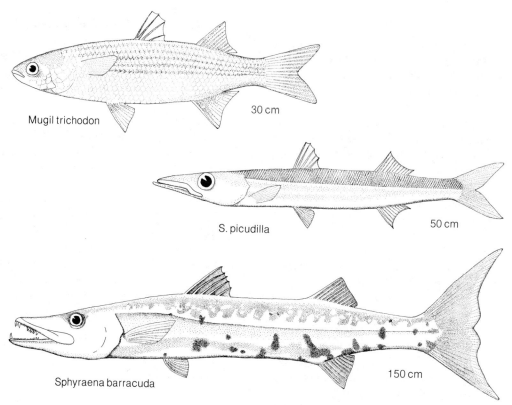

**215 MUGILIDAE (Mullets), SPHYRAENIDAE (Barracudas)**

cept along their edges. Anal fin with 3 spines and 8 soft rays, except in very small specimens where the count is 2 spines and 9 rays. Color dusky to olive with blue tints on back, silvery below; anal and pelvic fins yellowish, other fins pale with brown spots; caudal fin margin dusky; a dark bluish blotch at base of pectoral fin. To 30 cm. Found in relatively shallow water over grass flats and sandy or muddy bottoms forming small schools. Feeds on algae and other organic matter both living (bacteria, etc.) and detrital, ingested with sand and mud. Taken in cast nets and beach seines; used primarily as bait.

F. **SPHYRAENIDAE** (Barracudas): Slender, slightly compressed Perciformes with long, pointed snout, large mouth and lower jaw projecting beyond upper. Teeth canine, strong, of unequal size in jaws and on palatines. With 2 short, widely separated dorsal fins; anterior located above pelvic fins, with 5 strong spines; posterior located above anal fin. Caudal fin forked. Color gray-green or bluish black dorsally, silvery on sides and below. Body may have dark vertical bars, blotches or pale stripes. (2 spp. from Bda.)

***Sphyraena barracuda*** (Walbaum) (Great barracuda): The generic classification of the family is unsettled but a single genus in the Western Atlantic is likely.—Species with large head, the interorbital area flat to concave. Mouth large, tip of maxillary reaching beyond anterior margin of eye. Origin of 1st dorsal fin slightly behind origin of pelvic fin. Tip of pectoral fin reaches to or beyond pelvic fin origin. Color deep green or blue on back, silvery on sides becoming white ventrally. Adults with oblique dusky or dark bars on upper sides and variable numbers and shapes of scattered inky blotches on posterior portion of lower sides. Second dorsal, anal and caudal fins dusky to black with white tips. To 200 cm. Small specimens found inshore over grass flats, large specimens found singly or in aggregations over reefs or near the surface in open ocean. Large individuals believed to migrate considerable distances. Feeds mainly on fishes, also on cephalopods. Taken with seines inshore and on hook and line offshore. Often filleted and sold with skin removed. An excellent food fish, except that several large (15-20 kg) specimens have been implicated in cases of ciguatera (tropical fish poisoning).

*S. picudilla* (Poey) (Southern sennet): Genus as above.—Species with large head with pointed snout, interorbital area convex. Tip of maxillary not reaching anterior margin of eye. Origin of 1st dorsal fin above or anterior to origin of pelvic fin. Pectoral fin short, not reaching pelvic fin origin. Color dark olive above, sides and ventral surfaces silvery. With 2 faint yellow to orange longitudinal stripes on sides. To 50 cm. Found mainly in shallow waters and over lagoon areas where it usually forms large schools. Feeds mainly on small fishes and squid. Caught with seines and on hook and line. Flesh good to eat and highly prized.

## Plate 216

**F. CLINIDAE** (Clinids, Blennies): Small, mostly elongate Perciformes usually with cirri (fleshy flaps) on head and anterior nostrils, eyes and nape. Gill membranes continuous across ventral surface of head. Jaws with single row of large, canine-like teeth. Dorsal and anal fins long, dorsal fin spines flexible and more numerous than soft rays. Pelvic fin origin anterior to pectoral fin, containing 2 or 3 soft rays. All fin rays simple, unbranched. Lateral line absent in scaleless (naked) forms; when present, scales cycloid. Color variable. All forms benthic, most species found on reefs or rocky substrate. (4 spp. from Bda.)

*Labrisomus nuchipinnis* (Quoy & Gaimard) (Hairy blenny): Dorsal fin consisting of spines and soft rays. More than 60 scales, and lateral line pore present. Maxillary bone exposed when mouth is closed. Patches of small teeth behind outer row of large teeth in upper jaw. More than 2 cirri present on nostril, over eye and nape. Coloration highly variable with sex, size and substrate, from off-white with light brown bands to nearly black overall with red fins. Males tend to be dark and have red fins and belly but fewer markings. All specimens exhibit an opercular spot and may have a spot on anterior portion of dorsal fin. To 20 cm. Common inshore from shoreline to patch reefs offshore; shelters in algae, on rocks or among *Thalassia*. Feeds on small crustaceans, mollusks and other invertebrates.

*Malacoctenus gilli* (Steindachner) (Dusky blenny): Dorsal fin consisting of spines and soft rays. Scales and lateral line pores present. Maxillary bone sheathed beneath preorbital bone when mouth is closed. Lacks small teeth behind outer row of large teeth in jaws. Coloration variable, males may exhibit faint bands on upper reddish and lower greenish background, or be uniformly black with white ventral fins. Males exhibit well-defined bands and spots over body. Both sexes have longitudinal rows of small spots giving a striped effect. Opercular spot absent; a spot may be present on anterior end of base of dorsal fin and at posterior end of base of spinous portion of dorsal fin. To 8 cm. Found from shoreline to patch reefs offshore; most common nearshore and in coral rubble, 5 m. Feeds on crustaceans and other small invertebrates.

F. **BLENNIIDAE** (Combtooth blennies): Small, elongate Perciformes with broad, rounded head and cirri above the eyes. Mouth subterminal, upper jaw not protractile. Single row of closely spaced incisor-like teeth in each jaw. Dorsal and anal fins long, with fewer flexible spines than soft rays. All soft rays except in caudal fin simple, unbranched. Scales absent, lateral line pores present on entire length of body, or restricted to anterior part of body. Shallow-water, benthic. (5 spp. from Bda.)

*Entomacrodus nigricans* Gill (Pearl blenny): With 13 or 14 pectoral fin rays and 13 dorsal fin spines; the last spine small, difficult to see, resulting in deep notch between spinous and soft rays of dorsal fin, nearly separating the fin into 2 sections. Gill openings continuous ventrally. Upper lip has papillae on its outer margins and is smooth medially. Body with numerous and variable pearly-white markings on brown or black background. Dark areas may be flecked with yellow dots. To 10 cm. Found nearshore to 6 m, in tide pools, rocky slopes and in rubble or boulder-strewn areas. Feeds mainly on algae.

*Hypleurochilus bermudensis* Beebe & Tee-Van (Barred blenny): Gill openings restricted to sides of head; a long recurved canine tooth present at rear of both jaws. Dorsal fin continuous, only slightly notched. Margins of upper lip smooth. Coloration variable, ♂ generally darker than ♀; both blend coloration with substrate; 6 brownish indistinct vertical bars present on sides on a pale brown background; pale colors may have finely flecked patterns of brassy spots surrounded by fine blue margins. To 10 cm. Found in shallow water from rocky shoreline to shallow patch reefs offshore. Feeds on algae.

*Scartella cristata* (L.) (=*Blennius cristatus*) (Molly miller): Gill openings continuous ventrally. Upper top margin smooth ventrally. Dorsal fin continuous, with 12 spines. Numerous cirri present over each eye, forming ridge at top of head in line with dorsal fin. One or 2 short canine teeth posteriorly in lower jaw but none on upper jaw. Dark green to olive, with brown bars on back extending into lower portion of dorsal fin; pale spots may be present on sides and dorsal fin; head may be pinkish ventrally. To 10 cm. Found commonly in tide pools and shallow rocky areas, usually in less than 5 m. Feeds mainly on algae.

F. **GOBIIDAE** (Gobies): Small, elongate, compressed Perciformes with round or depressed head. Eyes on top of head, close together. Dorsal fin originating just behind head, either single or double, the 1st with 6-8 flexible spines, the 2nd with 1 spine and 9-25 soft rays; distance from end of 2nd dorsal fin base to base of caudal peduncle much less than basal length of 2nd dorsal fin. Pelvic fin origin below or slightly in front of pectoral fin

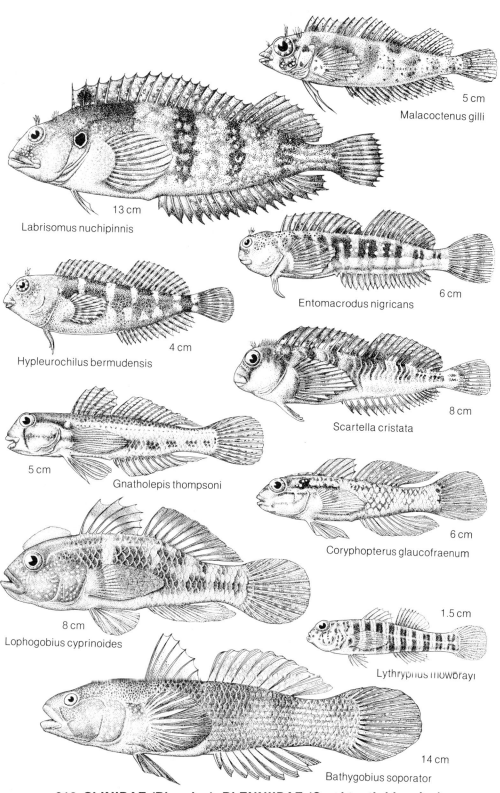

## 216 CLINIDAE (Blennies), BLENNIIDAE (Combtooth blennies), GOBIIDAE (Gobies)

origin; pelvic fins connected to form disc in some species. Lateral line absent. Color variable. (14 spp. from Bda.)

***Gnatholepis thompsoni*** Jordan (Goldspot goby): With 2 dorsal fins; pelvic fins completely united to form disc-like structure. Upper pectoral rays not free but bound by a membrane to fin. Dorsal spines 6; pores from lateral line system located on head in interorbital area; scales on body, cheeks and opercles. Mouth small, inferior. Body coloration whitish; a distinctive gold spot surrounded on sides and below by a black line located above pectoral fin base; a narrow bar passes through eye to subocular region during daylight hours. Small dots on body and fins are orange to bronze. To 7 cm. Common on sandy areas at base of rocky shore and bases of reefs offshore to 50 m. Feeds on small invertebrates.

***Coryphopterus glaucofraenum*** Gill (Bridled goby): With 2 dorsal fins; pelvic fins completely united to form disc-like struture. Upper pectoral rays not free but bound by membrane to fin. Dorsal spines 6; pores from lateral line system located on interorbit. No scales on dorsal surface of head. A distinct black spot above opercle; body may be nearly colorless or have yellow to gold brown surface markings. To 8 cm. Found on sand at base of rocky shore, patch reefs or reefs offshore to 25 m. Feeds on small invertebrates.

***Lophogobius cyprinoides*** (Pallas) (Crested goby): With 2 dorsal fins; pelvic fins completely united to form disc-like structure. Upper pectoral rays not free but bound by membrane to fin. Dorsal spines 6; pores from lateral line system located on interorbit. No scales on cheeks or opercles. A high median flap of skin on head not extending back to dorsal fin origin. Mouth terminal. Body coloration dark brown-olive, males tend to be darker than females; dorsal fin dark gray to black; posterior region paler with orange. To 10 cm. Found in mangrove and brackish areas, particularly ponds with mangrove fringe. Feeds on small invertebrates.

***Lythrypnus mowbrayi*** (Bean) (Mowbray's goby): With 2 dorsal fins; pelvic fins completely united to form disc-like structure. Upper pectoral rays not free but bound by membrane to fin. Dorsal spines 6; no pores from lateral line system on head; no scales on head, nape, cheeks, opercles or pectoral fin base. Body with dark reddish brown bands on straw-colored background. Center of pale area with thin dark band; dark spot present on lower part of pectoral fin. Subocular region with series of dark bands. To 2 cm. Lives in coral reef crevices from surface to 10 m. Found by breaking rocks apart. Probably feeds on small invertebrates.

***Bathygobius soporator*** (Valenciennes) (Frillfin goby): With 2 dorsal fins; pelvic fins completely united to form disc-like structure. Pectoral fin rays 18-21, the uppermost rays free; 37-41 lateral scales from upper end of pectoral fin base to scales overlapping base of central caudal fin rays. Coloration drab and variously blotched or marked with 4 or 5 indistinct broad bands across dorsal surface. To 15 cm. Found in shallow water, most commonly in tide pools often only 1-2 cm deep. Also on shallow sand in *Thalassia* beds to about 3 m. Feeds on small invertebrates.

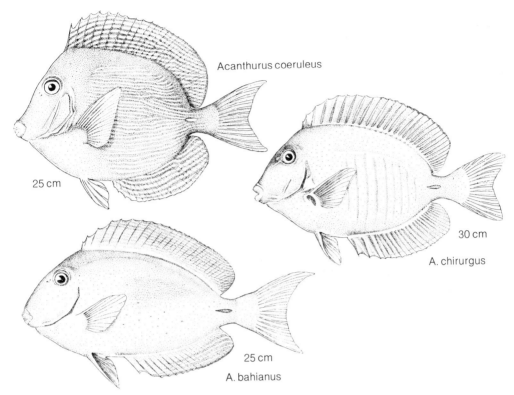

**217 ACANTHURIDAE (Surgeonfishes)**

## Plate 217

F. **ACANTHURIDAE** (Surgeonfishes): Deep-bodied, highly compressed Perciformes with lancet-like spine on side of caudal peduncle that fits into a horizontal groove. Eyes high on head; mouth terminal, low on head; head profile steep. (3 spp. from Bda.)

***Acanthurus coeruleus*** Bloch & Schneider (Blue tang): Genus with continuous dorsal fin having 9 spines; caudal fin emarginate. Teeth lobate, set close together.—Species with extremely deep body, near-round in profile. Gill rakers on 1st gill arch 13 or 14. Area around caudal spine pale yellow; body bluish, with conspicuous longitudinal lines; young bright yellow (!) lacking lines; soft anal rays 24 or 25. To 35 cm. Found on reefs and near rocky, wave-exposed shorelines. Browses on attached algae; has thin-walled stomach and does little scraping of hard substrate. Larger specimens caught in traps and sold as fillet or used as bait.

***A. chirurgus*** (Bloch) (Doctorfish): Genus as above.—Species with moderately deep profile. Rakers on 1st gill arch 16-19. Soft anal rays 22 or 23. Area around caudal spine dark, spine color whitish; body mostly light to dark brown without faint narrow longitudinal lines. Dark brown, narrow vertical bars present on body; caudal peduncle with broad white area frequently displayed. To 30 cm. Found on reefs and near rocky shores. Feeds on algae by scraping rocks. Stomach developed into gizzard-like structure. Occurs in aggregations frequently with other species of *Acanthurus*. Larger specimens caught in traps and sold as fillets or used as bait.

**A. bahianus** Castelnau (Ocean surgeonfish): Genus as above.—Species relatively elongate. Rakers on 1st gill arch 16-19; soft anal rays 22 or 23. Area around caudal spine dark; caudal spine dusky. Body mostly light to dark brown, with faint narrow longitudinal stripes; vertical bars on sides absent; caudal peduncle with a broad white area frequently displayed; outer margin of lunate caudal fin white. To 35 cm. Found on reefs and near rocky shores. Feeds on algae by scraping rocks or surface of sand. Stomach developed into gizzard-like structure. Occurs in aggregations frequently with other species of *Acanthurus*. Larger specimens caught in traps and sold as fillet or used as bait.

### Plate 218

**F. SCOMBRIDAE** (Tunas): Elongate, fusiform Perciformes with pointed snout and premaxillae forming short beak-like structure. Teeth in jaws variously developed. With 2 dorsal fins; series of finlets behind dorsal and anal fins; pectoral fins high; pelvic fins short. Caudal fin deeply forked, slender, with rays overlapping hypural plate. Caudal peduncle depressed, with at least 2 keels on each side. Lateral line simple. Scales small, usually forming a corselet behind head and around pectoral area. Color variable but generally dark blue on back, silvery on sides and below. (9 spp. from Bda.)

***Acanthocybium solandri*** (Cuvier) (Wahoo): Body very elongate-fusiform, only slightly compressed laterally. Snout about as long as rest of head. Teeth in jaws strong, triangular. Gill rakers absent. Two small keels and a large median keel between them on each side of caudal peduncle. With 7-10 dorsal and 7-10 anal finlets. Posterior portion of maxilla concealed under preorbital bone. Dark bluish black above, silvery on sides and below. Numerous irregular dark bluish vertical bars on sides extending below lateral line. To 190 cm. Found offshore, most commonly in spring and fall, and near drop-off over 75 m. Feeds on pelagic fishes and cephalopods. An excellent sport fish and food fish frequently caught using rod and reel or longline.

***Euthynnus alletteratus*** (Rafinesque) (Little tunny, B Mackerel): Body robust, fusiform; 2 dorsal fins separated by narrow space. Anterior dorsal spines much higher than others, giving fin a concave outline. Second dorsal fin lower than 1st followed by 8 finlets. Anal fin followed by 7 finlets. Scales on corselet and lateral line. Caudal peduncle slender, having central and 2 small keels at bases of fin lobes. Pectoral fins with 26 or 27 rays. Back dark blue, with complicated striped pattern extending from peduncle forward to middle of 1st dorsal fin; lower sides and belly silvery white, with several dark spots between pelvic and pectoral fins. To 120 cm. Schools of relatively small fish found inshore in bays and harbors feeding on bait fishes. Larger specimens offshore, most commonly over 75 m. Feeds on small fishes, cephalopods and crustaceans. Caught by beach seines inshore and hook and line (rod and reel) offshore. A good food fish finding a ready market locally.

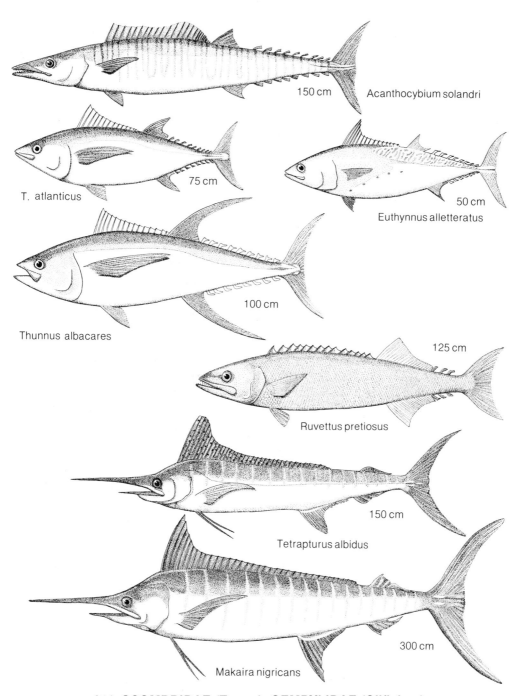

**218 SCOMBRIDAE (Tunas), GEMPYLIDAE (Oilfishes), ISTIOPHORIDAE (Billfishes)**

***Thunnus atlanticus*** (Lesson) (Blackfin tuna): Genus robust, fusiform, with 2 dorsal fins separated by a narrow space. Body lacking stripes, bars or black spots; skin covered with very small scales behind corselet. Pectoral fins with 30-36 rays. With 19-25 rakers on 1st gill arch; 2nd dorsal fin followed by 7-9 finlets. Ventral surface of liver not striated, the right lobe longest. Back dark metallic blue changing through dark gold to silvery white on sides and belly. Dorsal and anal fins and finlets dusky with silvery highlights. To 95 cm. Oceanic, most commonly found near offshore banks and drop-off over 75 m. Feeds on fishes, cephalopods and crustaceans. Caught on hook and line (rod and reel and longline). A good food fish but not as readily eaten as *T. albacares*.

***T. albacares*** (Bonnaterre) (Yellowfin tuna, B Allison tuna): Genus as above.—Species with 26-34 rakers on the 1st gill arch; 2nd dorsal fin followed by 8-10 finlets; anal fin followed by 7-10 finlets. Large specimens have very long 2nd dorsal and anal fins; pectoral fins moderately long, reaching to or beyond 2nd dorsal fin origin but not beyond end of its base. No striations on ventral surface of liver. Back metallic dark blue changing through yellow to silvery white on belly. Dorsal and anal fins and dorsal and anal finlets bright yellow; the finlets may be finely edged with black. To 195 cm. Oceanic species, most commonly found near offshore banks and drop-off over 75 m. Feeds on wide variety of fishes, cephalopods and crustaceans. Caught on hook and line (rod and reel and longline). An excellent food fish often spurned by local consumers in favor of imported tinned fish of the same species.

F.  **GEMPYLIDAE** (Snake mackerels, Oilfishes): Perciformes with elongate, somewhat fusiform bodies. Mouth large, with strong teeth in jaws. Lower jaw projecting beyond tip of upper jaw. Two dorsal fins, the 2nd shorter than 1st. Detached finlets usually present behind dorsal and anal fins. Pelvic fins small or reduced to single spine. Caudal fin moderate, forked. Keels may be present on caudal peduncle. Lateral line single or double, ending at caudal fin base. Color variable but usually brown or blue-brown to silvery on belly, body lacking distinct marks or blotches. Large, fast-swimming pelagic carnivores usually occurring at more than 150 m but often migrating to surface at night. (4 spp. from Bda.)

***Ruvettus pretiosus*** Cocco (Oilfish, B Tapioca fish): Body elongate, slightly compressed, with ventral keel between pelvic fins and anus. A single series of canine-like teeth in both jaws. Strongly pointed teeth present on vomer, palatines and pharyngeal region. First dorsal fin low, 13-15 spines; 2nd dorsal fin and anal fin higher than 1st dorsal fin, with 15-18 soft rays followed by 2 finlets. Pelvic fins well developed. Lateral line single, obscure. Scales cycloid, interspersed with rows of sharp bony tubercles. Body uniformly brown. Tips of pelvic and pectoral fins black. To 250 cm. Pelagic, most commonly found in water 75-200 m deep but rises near surface on dark nights. Feeds on fishes, cephalopods and crustaceans. Caught on longlines and vertical lines. Not generally eaten because of purgative properties of flesh.

F.  **ISTIOPHORIDAE** (Billfishes): Perciformes with elongate, compressed bodies and upper jaw prolonged into a spear that is round in cross section. Dorsal fins 2, the 1st larger; anal fins 2, the 1st larger; 1st dorsal and anal fins fold into grooves. Caudal fin large, forked, with pair of keels on depressed caudal peduncle. Body covered with embedded, narrow, pointed scales. Lateral line visible. Color of back and upper sides dark metallic blue, lower sides and belly silvery white. Spots and bars may be present on fins and body. (4 spp. from Bda.)

*Tetrapturus albidus* Poey (White marlin): Body strongly compressed, 1st dorsal fin relatively high; height of anterior spines equal to or slightly higher than body depth; anterior spines of 1st dorsal and anal fins rounded in profile. Lateral line simple. Length of pelvic fins and pectoral fins nearly equal. Anus close to origin of 1st anal fin. Scales ending in a single acute point. Body dark blue dorsally to silvery white below; 1st dorsal fin membrane blue-black dotted with numerous small black spots; remainder of body usually without markings. To about 200 cm. Oceanic, highly migratory, most commonly found in vicinity of offshore banks over water deeper than 500 m. Feeds on fishes, cephalopods and crustaceans. Taken by sport fishing and on longlines. Flesh good to eat.

*Makaira nigricans* Lacépède (Blue marlin): Body robust; 1st dorsal fin low; height of anterior part smaller than body depth; anterior spines forming distinct point on 1st dorsal and anal fins. Pelvic fins shorter than pectoral fins. Lateral line system reticulated, difficult to see in large specimens. Anus close to origin of 1st anal fin. Scales ending in 1 or 2 long acute spines. Body dark blue dorsally to silvery white on belly; about 15 indistinct vertical pale blue bars on sides. To about 400 cm. Oceanic, highly migratory, most commonly found in vicinity of offshore banks over water deeper than 1,000 m. Feeds on fishes, cephalopods and crustaceans. Taken by sport fishery and on longlines. Flesh excellent to eat.

### Plate 219

F. **NOMEIDAE** (Man-of-war fishes): Slender to deep, laterally compressed Perciformes, most species with adipose tissue well developed around large eyes; opercular and preopercular margins entire; operculum thin, with 2 weak spines. Maxillary short, covered by suborbital, rarely extending to below eye. Small conical teeth on jaws, vomer, palatines and basibranchials. With 2 dorsal fins, the 1st with about 10 spines folding into a groove; the longest spine equal to or longer than any of the soft rays. Anal fin with 1-3 spines connected by membrane to soft rays. Second dorsal and anal fins nearly equal to length. Caudal fin forked, lateral line high, not reaching caudal fin. Skin thin, subdermal mucus canal system well developed and visible in most species. Scales small, cycloid, thin and deciduous. Color variable, dark blue, silvey or black. Oceanic, epi- or mesopelagic, often associated with drifting organisms. (Probably 4 spp. from Bda.)

*Nomeus gronovii* (Gmelin) (Man-of-war fish): Small, slender, compressed, with large black pelvic fins connected to abdomen by membrane for entire length of inner ray and folding into ventral groove. Caudal

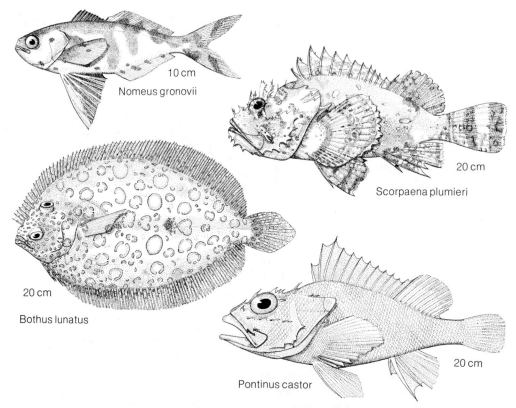

### 219 NOMEIDAE (Man-of-war fishes), SCORPAENIDAE (Scorpion fishes), BOTHIDAE (Left eye flounders)

fin deeply forked, mouth terminal. Back deep blue, either solid or in patches continued ventrally in bands or blotches; sides bright silver; a blue bar across each caudal lobe at base, and blue spots at base of anal fin extending to membrane. Dorsal and pectoral fins variably marked with blue patches. Eye blue above; iris silvery, crossed by oblique line running to angle of mouth. To 20 cm Epipelagic, oceanic, associated with Portuguese man-of-war (*Physalia*). Drifts with man-of-war until the latter is washed ashore then seeks new host. Feeds on zooplankton and tentacles of its host.

F. **SCORPAENIDAE** (Scorpion fishes): Spiny-headed Perciformes with characteristic ridge of bone (suborbital stay) extending across the cheek from the eye towards the operculum, usually with one or more spines arising from its surface. Head large, body generally robust, bass-like; mouth large, terminal; small villiform teeth on jaws and vomer in patches not rows. Pectoral fins large, usually fan-like, with 15-24 rays. Single dorsal fin with 12 or 13 spines and 9-12 soft rays. Caudal fin rounded or truncate, not forked. Anal fin with 3 spines and 5 or 6 soft rays. Fleshy skin flaps or cirri on head and body of many species. Color variable, most species blend with their background; deeper-water species therefore are scarlet. (5 spp. from Bda.)

***Scorpaena plumieri*** Bloch (Spotted scorpion fish): Head with strong bones and a pit located on top, behind the eye; dorsal fin with 12 spines; scales cycloid; preorbital bone with 3 or 4 free spinous points. Color varies with depth, from mottled brown-olive inshore to bright red with white flecks at depths of more than 20 m; a broad dark bar present posteriorly with pale area on caudal peduncle; caudal fin with 3 dark bars; a large jet-black area containing white spots in the entire axil of the pectoral fins. To 45 cm. Found in grassy areas near patch reefs inshore and on reefs offshore. Difficult to spot because of its camouflage color pattern. Preys on small fishes and crustaceans that come near. Sometimes caught in pots or hook and line but seldom kept for food. Handle with care! (Spines are covered with a protein sheath that causes a reaction like a bee sting if the skin is punctured.)

*Pontinus castor* Poey (Longsnout scorpion fish): Head profile concave, snout long, flattened dorsally; scales on body ctenoid. Pectoral fin rays 17, fin wedge- shaped; all rays on pectoral simple, unbranched. No dorsal spines protracted. Red to reddish pink overall with bright yellow flecks. Fins spotted with darker red. To 45 cm. Found in rocky areas, 75- 400 m. Feeds on fishes and invertebrates. Occasionally caught in traps or on deep set vertical lines. Not generally eaten.

O. **PLEURONECTIFORMES** (Flatfishes): Adults bilaterally asymmetric, both eyes on 1 side of head; body highly compressed, slightly rounded on eyed side, flat on blind side. Usually 6 or 7 branchiostegal rays; dorsal and anal fins long; adults with small body cavity and reduced swim bladder.

F. **BOTHIDAE** (Lefteye flounders): Pleuronectiformes with both eyes on left side of head. Preopercle exposed, its hind margin free; dorsal fin long, originating above or in front of upper eye; anal fin originating just behind pelvic fin. Single lateral line on exposed side; caudal fin separate. Mouth protractile, asymmetric, the lower jaw moderately prominent. Blind side usually pale, eyed side with various colors, usually blending with substrate. (5 spp. from Bda.)

*Bothus lunatus* (L.) (Peacock flounder): Body oval, moderately deep; lateral line with distinct curve anteriorly. Dorsal fin rays 91-99; anal fin rays 70-76. Dorsal profile of head with distinct notch in front of eyes; eyes relatively small, the upper eye not overlapping lower in adults. Numerous circles, curved spots and bright blue dots of a variable intensity on a variable background from brown-olive to creamy white. Often with 3 dark marks or spots spaced evenly on lateral line between pectoral fin base and the caudal fin. To 45 cm. Found in sandy areas interspersed with grasses, rubble or mangroves. Will partially bury itself for camouflage and feed on small fishes or crustaceans. Occasionally caught in beach seines or on hook and line.

O. **TETRAODONTIFORMES:** Without lower ribs, infraorbital bones, parietal bones and nasal bones. Posttemporal usually absent; if present simple and fused with pterotic of skull. Hyomandibular and palatine bones firmly attached to skull. Gill openings restricted to slits or round holes. Maxillae usually fused with premaxillae; scales modified as spines, shields or plates.

F. **BALISTIDAE** (Triggerfishes): Moderately to highly compressed Tetraodontiformes with thick hidelike skin having plate-like or minute rough scales. Gill opening a short vertical to oblique slit in front of pectoral fin base; branchiostegal rays hidden beneath skin. Mouth small, terminal, with 8 or fewer developed teeth in each jaw. Dorsal fin spines 1-3, separated from soft rays. Pelvic fins rudimentary, forming a series of enlarged scales encasing the end of the pelvic girdle. Bases of 2nd dorsal and anal fins nearly equal; pectoral fin short, outer margin rounded. Color variable, from drab to strikingly marked with vivid patterns. (14 spp. from Bda.)

### Plate 220

***Balistes vetula*** L. (Queen triggerfish): Genus relatively robust, with large, plate-like scales; 3 dorsal spines; teeth well developed, unequal, the central 2 on both jaws much enlarged. One or more enlarged bony scutes present behind gill opening.—Species with 2 strongly marked, curved blue stripes on head; bluish lines outlined with yellow radiating from eyes; a wide bluish band around caudal peduncle; remainder of body bluish green to yellow-brown; blue submarginal bands on dorsal, caudal and anal fins. To 45 cm. Uncommon, found mainly on offshore reefs to 45 m. Feeds on benthic invertebrates, espcially the sea urchin *Diadema*. Occasionally caught on hook and line and in traps.

***B. capriscus*** Gmelin (Gray triggerfish, B Turbot): Genus as above.—Species more or less uniformly gray overall but may exhibit 3 indistinct greenish bars across the back. Small bluish purple spots on upper body, soft dorsal and anal fins, with spots tending to form rows. To 40 cm. Found in harbors and bays, in the lagoon and on reefs offshore. Feeds on a variety of bottom-dwelling invertebrates, especially bivalves. Teeth strong, used to crush mollusks. Frequently caught on hook and line and in traps. Flesh excellent and widely used as food.

***Xanthichthys ringens*** (L.) (Sargassum triggerfish): Relatively robust, with large, plate-like scales; dorsal spines 3; teeth well developed, the central 2 on each jaw enlarged. No bony scutes behind gill opening; chin projecting; mouth supraterminal; cheek with 3 prominent, naked longitudinal grooves. Color variable; ground color silvery gray to yellow-green or pale reddish brown; cheek grooves violet to black; stripes along bases of soft dorsal and anal fins deep red, or black; spots on body deep red to violet; soft dorsal and anal fins pinkish; caudal fin variously tinted but with a central orange to red crescent shape. To 25 cm. Uncommon; juveniles associated with *Sargassum*; adults form small aggregations near drop-off, particularly around offshore banks in water more than 50 m. Feeds on *Sargassum* community and other pelagic fauna. Occasionally caught on hook and line or in traps.

***Canthidermis maculatus*** (Bloch) (Rough triggerfish): Genus robust, with large, plate-like scales; dorsal spines 3; teeth well developed, unequal, the central 2 on both jaws enlarged; no bony scutes behind gill opening; cheek closely scaled, lacking parallel grooves; lower jaw not projecting.—Species with strong keels on scales of larger specimens giving rough feel; whitish spots, sometimes

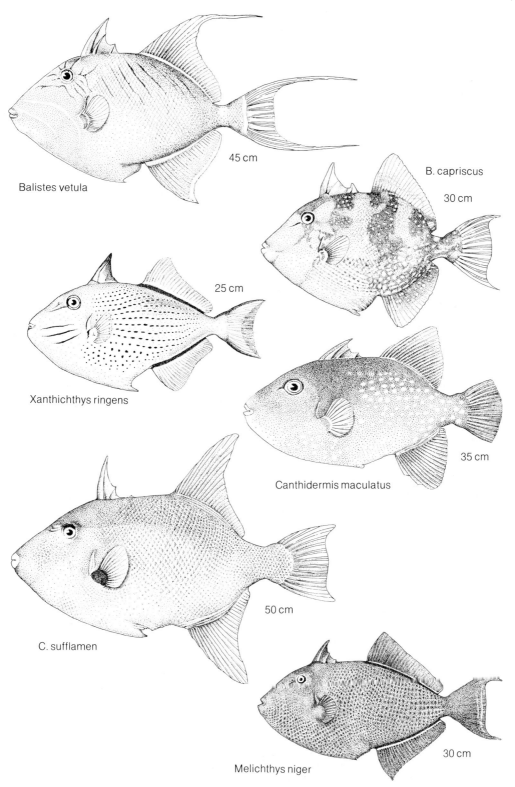

**220 BALISTIDAE 1 (Triggerfishes)**

indistinct, over body. Dorsal soft rays 24; anal rays 22. Overall color brownish gray. To 40 cm. Uncommon, pelagic, often found associated with *Sargassum*. Feeds on *Sargassum* community and forages in water column. Sometimes caught on hook and line and used as food.

***C. sufflamen*** (Mitchell) (Ocean triggerfish): Genus as above.—Species more elongate than and lacking white spots of *C. maculatus*. Dorsal rays 26 or 27; anal rays 24 or 25. Color brownish gray overall, the axils of pectoral fins black. To 65 cm. Oceanic, and on outer reefs near drop-off. Often associated with drifting *Sargassum* and flotsam. Feeds on benthic invertebrates and on *Sargassum* community. Occasionally caught on hook and line and used as food.

***Melichthys niger*** (Lacépède) (Black durgon): Robust, with large, plate-like scales; dorsal spines 3; teeth even and incisor-like; scales posteriorly with keels at center forming longitudinal ridges. Generally black with greenish yellow overtones; pale electric blue bands along bases of soft dorsal and anal fins. Edges of scales with fine orange tips giving a zigzag pattern to scale rows. Found on reefs offshore, often in large schools, most common near breakers. Grazes on benthic algae and forages in water column on zooplankton. Frequently caught in traps or on handlines in some localities, and used as fillet.

### Plate 221

***Cantherhines macrocerus*** (Hollard) (White-spotted filefish, B Hook-tail filefish): Genus highly compressed, with very small rough scales giving a velvety touch. Dorsal spines 2; 1st spine long and robust, originating above anterior portion of eye, 2nd small to vestigial; dorsal spine fits into deep groove anterior to 2nd dorsal fin. Pelvic bone with prominent external spine.—Species with 2 pairs of strong spines on each side of caudal peduncle; pectoral fin rays 14; gill rakers 29-35. Coloration variable; white spots on brownish body becoming dull to bright orange posteriorly. Soft dorsal and anal fins with pale yellow membranes, bright yellow rays. A pale to white saddle on back between 2 dorsal fins extending midway on sides. White spots may be exhibited and vary in intensity. To 45 cm. Uncommon, found mainly in clear water on offshore shallow reefs, often occurring in pairs. Feeds on sponges, gorgonians, hydroids and algae. Occasionally caught in fish traps but not used as food.

***C. pullus*** (Ranzani) (Orange-spotted filefish): Genus as above.—Species lacking caudal peduncular spines; pectoral fin rays 13; gill rakers 34-46. Generally brown to olive-brown, with indistinct pale stripes posteriorly and numerous small scattered orange spots having brownish centers. Two white spots on peduncle at bases of dorsal and anal fins. To 20 cm. Found on reefs inshore and offshore, to 50 m. Feeds on sponges, tunicates and algae. Occasionally caught in traps, may be used as fillet.

***Monacanthus ciliatus*** (Mitchell) (Fringed filefish): Genus compressed, with very small rough scales; dorsal spines 2; pelvic bone with prominent external spine. Dorsal spine originating above or behind dorsal portion of eye; no groove pres-

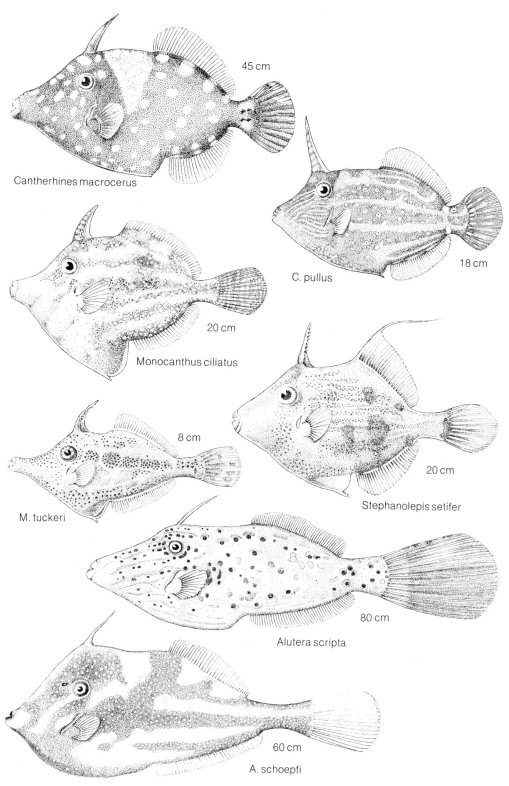

**221 BALISTIDAE 2 (Filefishes)**

ent behind dorsal spines; caudal peduncle with 2-4 pairs of enlarged spines on each side; dorsal rays not elongated and whip-like.—Species relatively deep-bodied; snout short. Coloration variable, greenish when associated with plants, brown when living among coral rubble and brown algae. May exhibit 1 or more indistinct dark stripes on sides. To 20 cm. Common in *Thalassia* beds, often seen resting or moving among grass blades with head down. Feeds on algae, detritus and small invertebrates, mainly crustaceans.

***M. tuckeri*** Bean (Slender filefish): Genus as above.—Species with much elongated snout, body only moderately deep. Body color variable, shades of brown with dark dots scattered over head and body. May exhibit an indistinct dark midlateral stripe. A network of plain lines cover head and anterior portion of body. To 8 cm. Found in grass beds, coral rubble areas and among gorgonians on shallow offshore reefs, less than about 10 m. Feeds on algae and small invertebrates.

***Stephanolepis setifer*** (Bennett) (Pygmy filefish): Compressed, deep-bodied, with very small rough scales; dorsal spines 2; pelvic bone with prominent external spine. Dorsal spine originating above posterior edge of eye; no groove present behind dorsal spines; caudal peduncle lacking spines; 2nd dorsal ray prolonged, whip-like in adult ♂. Spines on scales of fish over 2 cm are branched. Color variable, brownish to tan with irregular blotches of darker and lighter colors. To 20 cm. Common inshore and in lagoon area over grass beds and coral rubble. Feeds on algae and small invertebrates. Caught in traps and usually used for trap bait.

***Alutera scripta*** (Osbeck) (Scrawled filefish): Genus highly compressed, elongate, with only microscopic external spine on pelvic bone or without spine. Scales very small, rough and velvety to touch. Dorsal spines 2, the 2nd spine minute.—Species with 43-49 dorsal soft rays and 46-52 anal rays; pectoral rays usually 14. Caudal fin and snout both elongated to give a slender appearance. Color variable, body olive to brownish with random reticulated bright blue pattern interspersed with deep maroon to black spots. Caudal fin dusky to black. To 90 cm. Uncommon; found in grass flats and on or near coral reefs. Feeds on a variety of invertebrates including hydrozoans, gorgonians and tunicates, and algae and sea grasses. Occasionally caught on hook and line or in traps.

***A. schoepfi*** (Walbaum) (Orange filefish): Genus as above.—Species with 32-39 dorsal soft rays and 35-41 anal rays. Pectoral rays usually 12 or 13. Mottled light and dark metallic gray with small bright orange dots overall; lips sometimes black. To 60 cm. Uncommon, occasionally found over sea grass beds and sandy or muddy areas in lagoon or offshore reefs. Feeds on invertebrates, sea grasses and algae. Occasionally caught on hook and line or in traps.

### Plate 222

F. **OSTRACIIDAE** (Trunkfishes): Robust Tetraodontiformes with body nearly wholly encased in a shell of thickened scale plates forming a bony carapace, with holes for eyes, mouth, gill openings, pectoral fins, dorsal and anal fins and caudal peduncle; caudal fin triangular or

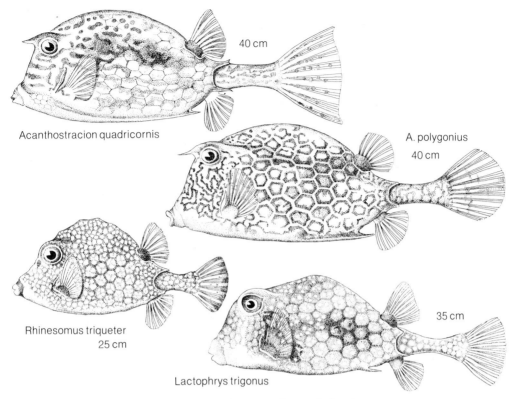

## 222 OSTRACIIDAE (Trunkfishes)

rectangular. Mouth small, terminal, with 15 or fewer teeth in each jaw; pelvic fins and spines absent; spiny dorsal fin absent; dorsal, anal and pectoral fin rays usually branched. Color variable with species, habitat and time of day. Scale plates usually hexagonal. (5 spp. from Bda.)

*Acanthostracion quadricornis* (L.) (Scrawled cowfish): Genus relatively elongate, with slender caudal peduncle and fins. A spine projecting anteriorly in front of eyes; carapace spines projecting posteriorly from corners of carapace in front of anal fin. Carapace complete around bases of soft dorsal and anal fins.—Species with 11 (rarely 12) pectoral fin rays; color pattern of irregular wavy lines on dark blotches or spots on body, and more or less horizontal parallel lines on cheek; lines and spots blackish blue to bright blue; background color grayish brown to yellowish green; mottled color pattern evident at night. To 75 cm. From shore to offshore reefs at 80 m; common over grass flats. Feeds on crustaceans, tunicates, gorgonians, anemones and other sessile invertebrates. Frequently caught in fish traps but only rarely used as food.

*A. polygonius* Poey (Honeycomb cowfish): Genus as above.—Species with 12 (rarely 11) pectoral fin rays; color pattern consisting of hexagons and near-hexagons; lines separating hexagons light, forming overall reticulated patterns. Color variable, centers of hexagons black or dark brown; pale areas blue, green, bluish green or off-white. Blotchy pattern exhibited at night. To 40 cm. Found mainly on reefs offshore to 80 m. Feeds on invertebrates including tunicates, gorgonians, sponges and crustaceans. Occasionally caught in traps; not generally eaten.

***Rhinesomus triqueter*** (L.) (=*Lactophrys triqueter*) (Smooth trunkfish): Carapace spines in front of eye lacking; carapace complete around base of dorsal fin, forming a solid bridge over caudal peduncle behind dorsal fin; spine lacking from postero-lateral edges of carapace. Generally brownish black with numerous whitish spots or patches (or the reverse), lips and bases of fins blackish. To 30 cm. Common, found in sea grass beds and dead reef or coral rubble areas inshore and on coral reefs offshore to 10 m. Feeds on variety of small benthic invertebrates including mollusks, crustaceans, tunicates, sponges and worms. Is able to secrete toxic mucus that is lethal to other fish in captivity. Caught in fish traps but not generally eaten.

***Lactophrys trigonus*** (L.) (Buffalo trunkfish): Carapace spines in front of eyes lacking; carapace incomplete and partially open behind base of dorsal fin, the notch so formed covered with skin; a large isolated scale plate present just behind carapace notch. A pair of scales on posterior corners of carapace in front of anal fin expanded to form spines. Caudal fin rounded to truncate. Color variable; usually with 2 black, poorly defined blotches, 1 behind pectoral fin, the other midlateral above carapace spine; body tan to greenish covered with small whitish spots. May also exhibit a series of irregular black markings on a bluish green background. To 45 cm. Found in sea grass beds and coral rubble areas inshore and offshore on reefs to about 50 m. Feeds on wide variety of benthic invertebrates including mollusks, crustaceans, tunicates and worms, and sea grasses. Occasionally caught in traps but not generally eaten.

**Plate 223**

F. **TETRAODONTIDAE** (Puffers): Small to moderate-sized, somewhat depressed Tetraodontiformes with a blunt head and jaws modified to form a beak of 2 heavy, powerful teeth on each jaw. Gill openings simple slits anterior to base of pectoral fins. Scales absent but numerous small spines on upper sides and belly. Pelvic fins absent. Able rapidly to inflate body by intake of water (or air). Color variable, predominantly tan or brown. (3 spp. from Bda.)

***Sphaeroides spengleri*** (Bloch) (Bandtail puffer): Dorsal rays 8; anal rays 7; nostrils small but easily visible; dorsal surface smooth, rounded, without central keel. Color on back tan with irregular dark blotches, sides and belly white; a regular row of dark spots extend from chin to caudal peduncle; caudal fin with distinct black bar. To 15 cm. Found in grass flats, patch reefs and offshore to about 30 m; common. Feeds on a variety of benthic invertebrates.

***Canthigaster rostrata*** (Bloch) (Sharpnose puffer): Dorsal rays 10, occasionally 9; anal rays 9, occasionally 8; nostrils minute, barely visible. Mid-dorsal surface posterior to eyes distinctly keeled, eyes surrounded by dark radiating lines. Color dorsally dark orange to brownish purple; lower sides and belly whitish; upper and lower portions of caudal pedun-

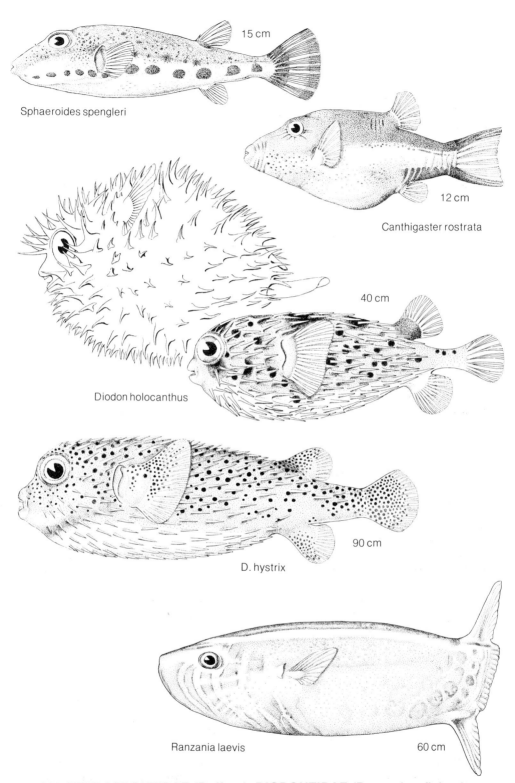

**223 TETRAODONTIDAE (Puffers), DIODONTIDAE (Porcupine fishes), MOLIDAE (Sunfishes)**

cle and fins dark (black). Vertical light blue lines on lower side of caudal peduncle. To 12 cm. Found on patch reefs inshore and offshore to about 30 m. Feeds on variety of invertebrates including crustaceans, mollusks, echinoderms, hydroids and polychaete worms, and algae and sea grasses.

F. **DIODONTIDAE** (Porcupine fishes): Small to moderate-sized, somewhat depressed Tetraodontiformes with terminal mouth; teeth fused to form parrot-like crushing beak without median suture dividing jaws into left and right sides. Pelvic fins absent; dorsal, pectoral and anal fin rays branched. Gill opening a vertical slit. Head and body with scales modified to form fixed short spines or long erectile quill-like spines. Color generally tan to whitish with dark markings. (4 spp. from Bda.)

*Diodon holocanthus* L. (Balloonfish): Genus with long, mostly double-rooted spines that fold back when fish is not inflated.—Species with head covered with small round black spots; body and head with large dark blotches. Spines on forehead longer than those just behind pectoral fin. Pale brown dorsally, shading to white ventrally. To 40 cm. Found in grass flats, patch reefs, in bays and occasionally on offshore shallow reefs. Feeds on mollusks, sea urchins and crustaceans including hermit crabs. Occasionally caught on hook and line and in beach seines.

*D. hystrix* L. (Porcupine fish): Genus as above.—Species with head, body and fins covered with small round black spots but lacking dark blotches. Spines on forehead shorter than those just behind pectoral fin. Ground color tan to olivaceous on back, shading to whitish ventrally except under surface of head which is dusky. To about 90 cm. Found on patch and shallow offshore reefs. Feeds on mollusks, sea urchins, crustaceans including hermit crabs. Occasionally caught on hook and line and in traps; not generally eaten.

F. **MOLIDAE** (Sunfishes): Highly compressed, deep-bodied Tetraodontiformes; mouth with small teeth in each jaw fused to form beak-like structure; pelvic fins absent; caudal fin lacks visible peduncular area; caudal rays modified and derived from interspinous elements; dorsal and anal fins high and short, located at posterior end of body. Gill opening a small rounded slit anterior and above pectoral fin base. (2 spp. from Bda.)

*Ranzania laevis* (Pennant) (Slender mola): Body oblong, the depth about 1/2 the height; skin smooth, divided into small hexagonal patches; caudal fin short, truncate, its base straight and slightly oblique. Dorsal and anal fins connected to the caudal fin. Pectoral fin somewhat pointed. Color on back brilliant dark purple fading to white on the belly; sides marked with pale green lines on purple areas, whitish areas with golden reflections; several irregular white spots located ventrally near caudal fin; dorsal, anal and pectoral fins pale gray, caudal fin silvery with purple highlights on fin rays. To 60 cm. Pelagic, found occasionally off Bermuda. Feeds on jellyfish, ctenophores and other planktonic animals.

J. BURNETT-HERKES

## Class Reptilia

CHARACTERISTICS: *VERTEBRATA covered with horny scales or partly enclosed in a carapace; respiration by way of lungs. Skull articulates with vertebral column by means of a single condyle; lower jaw composed of several bones and hinged to the skull via the quadrate. With 2 pairs of limbs adapted for crawling, running or swimming, or reduced. Fertilization is internal; most species lay shelled eggs, few are viviparous; larval stages are absent.*

Of 4 living orders (in 3 subclasses), Crocodylia and the primitive Rhynchocephalia do not occur in Bermuda, and Squamata (lizards and snakes) have no marine representatives on the island. Testudines (=Chelonia; turtles), the only order of the subclass Anapsida, occur with several truly marine species.

## ORDER TESTUDINES (=Chelonia) (Turtles)

CHARACTERISTICS: *REPTILIA with box-like shells made up of a dorsal carapace and a ventral plastron, both usually covered with horny epidermal scales; jaws beak-like, toothless.* Marine turtles may reach 600 kg in weight and more than 2 m (!) in length. Coloration in marine species is mostly white, yellow, green, brown and black. They are as awkward on land as they are fast and agile when diving.

Of more than 220 living species in 2 infraorders, the majority belong to Cryptodira (in which the head is vertically retractile, as opposed to Pleurodira which tuck their heads in sideways). Living sea turtles are limited to 2 families with 5 genera each represented with 1 species in Bermuda.

OCCURRENCE: Bermuda's inshore waters support populations of green turtles (*Chelonia mydas*), hawksbills (*Eretmochelys imbricata*) and loggerheads (*Caretta caretta*). The green turtle, the most abundant species, is found primarily on sea grass flats. The hawksbill, the next most common species, and the loggerhead are usually found near reefs. The leatherback (*Dermochelys coriacea*) is an occasional, transient visitor, and the Atlantic ridley (*Lepidochelys kempi*) is only known from a single specimen.

Collecting is performed with a large net or by spear. At present all species of sea turtles are protected in Bermuda. A permit from the Bermuda Department of Agriculture and Fisheries is required for collecting specimens.

There is considerable international trade in sea turtle products, especially the flesh of *C. mydas* and the shell of *E. imbricata*. Also, sea turtles and their eggs are important sources of protein for the peoples of many tropical and subtropical areas.

IDENTIFICATION: External characteristics of living or dead specimens are sufficient for determining species.

Species are identified by coloration, the pattern of scales (scutes) or lack of scutes on the dorsal portion of the shell (carapace), the presence of ridges or keels on the carapace or the plastron (ventral portion of the shell), the number of inframarginal plates that separate the carapace from the plastron, and the pattern of scales on the dorsal surface of the head. Head scales of diagnostic value include the median, unpaired, usually small frontal and the larger fronto-parietal scutes, and 1 or 2 pairs of prefrontals. The carapace is covered by 5 unpaired central scutes, 4-5 pairs of costal scutes; and 11-12 pairs of marginals which, together with the single nuchal (neck scute) and caudal (tail scute), form an outer ring.

Small sea turtles may be preserved by injecting the specimen with a 5% solution of formaldehyde and storing it in the same solution. Large specimens are usually reduced to skeletal components. The plastron is removed and the carcass eviscerated. The carcass is then skinned and as much flesh as possible scraped from the bones. If an articulated skeleton is desired the carcass is placed with a colony of dermestid beetles, which devour the remaining flesh. If a disarticulated skeleton will suffice, the carcass is placed in a tank of water, where bacterial action dissolves the tissues.

BIOLOGY: The life cycles of all species of sea turtles appear similar; that of the green turtle is best understood. Adult *C. mydas* congregate and mate in the sea near certain sand beaches. Females leave the ocean and dig nests, usually at night. All marine turtles dig nests by unvarying, alternate movements of the rear flippers, a pattern of behavior that may be unique among the Testudines. When the hatchlings leave the nest, usually at night, they quickly cross the beach and enter the sea. This sea-finding ability of hatchlings appears explicable in terms of a positive response to brightness cues provided by the seaward horizon of the beach. Once the hatchlings leave the beach they travel rapidly toward deep ocean. After reaching deep water they "disappear" for at least a year and possibly for several years; it seems certain that they spend their early life in a pelagic habitat, possibly the *Sargassum* community. Young turtles reappear in "developmental habitats", inshore feeding grounds, but it is unknown whether young turtles from a given natal beach arrive at specific developmental habitats. Upon reaching sexual maturity the turtles move to adult resident habitats, inshore areas in which mature *C. mydas* may be captured year round. Tagging studies now underway may eventually link developmental areas with particular adult habitats. Evidence obtained by tagging nesting females indicates that adult *C. mydas* make periodic long distance migrations (2,000 km and more!) between certain nesting beaches and specific adult habitats. The mechanism for their considerable navigational skill remains a mystery. Evidence for periodic migration and nest site tenacity shows that female green turtles return to nest at 2- or 3-yr intervals, nest up to 11 times a season, and typically dig their nests within 1.5 km of their last nest sites, either within a season or following remigration. Green turtles may return to their natal beach to lay their eggs, but this has not been proven. Green turtles may also remain immobile for long periods. During the winter, in areas where seawater temperatures fall below about 15°C, green turtles may lie dormant on the sea bottom for several months or more! Captive sea turtles can grow rapidly, reaching sexual maturity in 6 yr; growth in the wild is probably much slower. Bermudian green turtles may require 27 yr to attain adult size. Their apparent slow growth and the sizes they reach suggest that they may live 100 yr or more.

Apart from humans, only sharks seem to prey on adult sea turtles. Eggs and hatchlings crossing the beach are taken by lizards, crabs, birds, dogs and rats. In past centuries large numbers of sea turtles nested on Bermuda. Human exploitation and predation by rats and dogs completely destroyed these nesting colonies; no species of marine turtle nests on Bermuda today. An attempt is presently being made to reestablish the green turtle on Bermuda by importing and incubating eggs, and releasing the hatchlings into the sea.

### Plate 224

DEVELOPMENT: Sea turtle eggs incubate for about 60 days. Sex determination in green and loggerhead turtles is dependent on the temperature at which the eggs are incubated, with higher temperatures producing females and lower ones males. This may be true of all species of sea turtles.

REFERENCES: For a general introduction to the order consult CARR (1967a), ERNST & BARBOUR

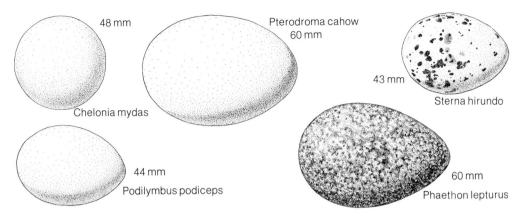

**224 TESTUDINES (Turtles), AVES (Birds): Development**

(1972), PRITCHARD (1979) and HARLESS & MORLOCK (1979). Sea turtles are treated by CARR (1967b), BUSTARD (1973), REBEL (1974) and BJÖRNDAL (1982). For sex determination in sea turtles see YNTEMA & MROSOVSKY (1980) and MORREALES et al. (1982). Winter dormancy is discussed by FELGER et al. (1976) and CARR et al. (1980).

For Bermuda's turtles see GARMAN (1884), VERRILL (1902a), BABCOCK (1937) and MOWBRAY & CALDWELL (1958). More recent papers are by BURNETT-HERKES (1974) on tagging returns, and on hatchling and juvenile behavior by FRICK (1976), IRELAND et al. (1978) and IRELAND (1979, 1980).

## Plate 225

**I.O. CRYPTODIRA:** Testudines in which the neck is vertically retractile; the cervical vertebrae can be bent in a sigmoid curve. Pelvic girdle not attached to shell. (Recent sea turtles have lost the ability to retract the head; the anatomy of their cervical vertebrae places them in this infra-order.)

**F. CHELONIIDAE:** Cryptodira with paddle-shaped forelimbs provided with claws. Carapace covered with horny scutes; retains embryonic spaces between ribs. Carapace and plastron are connected by ligaments but separated from each other by a complete row of inframarginals. Tail of adult ♂ longer than of adult ♀ (the only external sexual dimorphism). (4 spp. from Bda.)

***Chelonia mydas*** (L.) (Green turtle): Skin on dorsal surfaces of head and limbs is brown, gray or black. Ventral surfaces of the limbs are white to yellow proximally and brown to black distally. Head scales may have yellow margins. Carapace may be olive, brown, gray or near black. Color of dorsal skin tends to match color of carapace. Plastron is white to yellow. Scalation of head is symmetrical and includes a single pair of prefrontals. Carapace lacks a keel and has 5 central scutes, 4 pairs of costal scutes, 11 pairs of marginals, 1 pair of supracaudals and a single nuchal. Anterior costals do not contact the nuchal. There are 4 pairs of inframarginals. Cutting edge of jaws finely serrated. Maximum carapace length about 140 cm, maximum weight about 300 kg. The green turtle undertakes long oceanic migrations but is found primarily in coastal waters, nesting on tropical beaches where it lays about 110 eggs per nest. Adults feed almost exclusively on submerged sea grasses and algae but occasionally take mollusks, crustaceans, jellyfish and sponges. The most common turtle in Bermuda.

***Caretta caretta*** (L.) (Loggerhead turtle): Carapace and skin on dorsal surfaces of head and limbs reddish brown, ventral surfaces of limbs yellow, plastron light yellow. Pattern of head scales varies but there is always a frontal scale (occasionally divided), a fronto-parietal scale and 2 pairs of prefrontals that sometimes surround a small, extra scale. Carapace has 5 central scutes, 5 pairs of costals, 11 or 12 pairs of marginals, 1 pair of supracaudals and a single nuchal. There is a small spine or ridge at the rear of each central scute. The anterior costals are in contact with the nuchal. There are 3 pairs of inframarginals. Cutting edge of lower jaw smooth or slightly serrated. Maximum carapace length 115 cm, maximum weight about 260 kg. The loggerhead wanders widely throughout its range and may as easily be found hundreds of kilometers offshore as in coastal waters. What little is known of the travels of this species suggests that it is migratory. Although often seen in tropical areas the loggerhead mostly nests north of the Tropic of Cancer or south of the Tropic of Capricorn. Adults are omnivorous and will take a wide variety of foods including fishes, squid, jellyfish, mollusks, crustaceans, sponges, barnacles, tunicates, sea grasses and algae. Seen regularly in Bermuda waters.

***Eretmochelys imbricata*** (L.) (Hawksbill turtle): Skin on dorsal head and forelimbs dark brown or black. Scales have yellow or white borders. Ventral surfaces of the limbs are yellow to orange. Carapace usually has a "tortoise-shell" pattern and may contain streaks of black, brown, amber and olive-green. Plastron yellow to near orange. Head has 2 pairs of prefrontals. Carapace has 5 central scutes, 4 pairs of costals, 11 pairs of marginals, 1 pair of supracaudals and a single nuchal. Anterior costals do not touch the nuchal. Scutes of the carapace overlap. Vertebral keel on first 4 central scutes. There are 4 pairs of inframarginals. Cutting edge of lower jaw smooth or slightly serrated; inner surfaces of upper jaw strongly ribbed vertically. Maximum carapace length about 94 cm, maximum weight about 127 kg. The hawksbill is usually found in coastal waters. Long migrations have not been recorded for this species but there is evidence that it is migratory. Nesting is almost completely restricted to tropical areas. Adults feed on sponges, coelenterates (especially jellyfish), sea urchins, mollusks, crustaceans, fishes and occasionally algae and other vegetation. Fairly abundant in Bermuda.

***Lepidochelys kempi*** (Garman) (Atlantic ridley): Skin on dorsal head and forelimbs gray or olive-green. Ventral surfaces of limbs white to light yellow. Carapace gray or olive-green. Color of skin usually matches color of carapace. Plastron white to yellow. Head with 2 pairs of prefrontals. Carapace has a flattened appearance and has 5 central scutes, 5 pairs of costals, 12 pairs of marginals, a pair of supracaudals and a single nuchal. The anterior costals contact the nuchal. There are 4

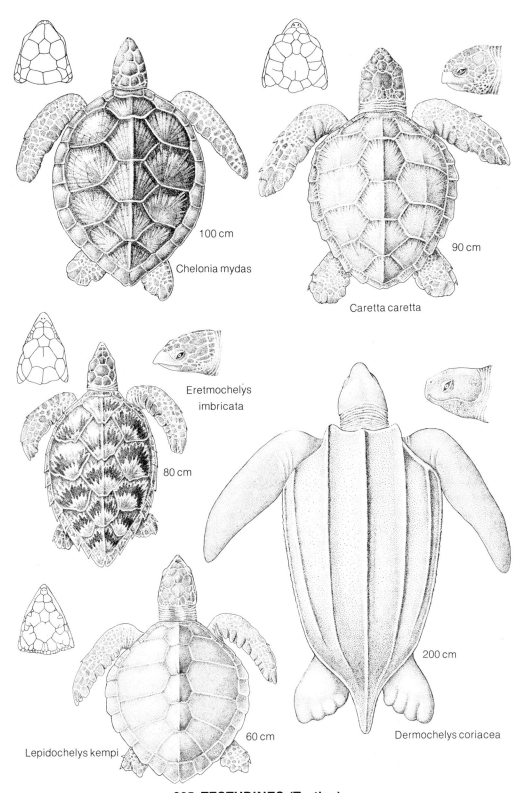

225 **TESTUDINES (Turtles)**

pairs of inframarginals. Bony surface of the upper jaw has a distinctive ridge that parallels the cutting edge. Maximum carapace length about 75 cm, maximum weight about 50 kg; the smallest of the sea turtles. Probably migratory and usually found in coastal waters, this species is known to nest in significant numbers only on a single stretch of beach near Tampico, Mexico. Adults feed almost exclusively on crustaceans, jellyfish, mollusks and fishes. There is real concern that the the species may become extinct in the near future. Very rare in Bermuda.

F. **DERMOCHELYIDAE:** Cryptodira with paddle-shaped forelimbs that lack claws. Carapace and plastron devoid of scutes and covered with a ridged, leathery skin devoid of scales. Tail of adult ♂ longer than of adult ♀. (Only species in the family:)

*Dermochelys coriacea* (L.) (Leatherback turtle): Skin on dorsal surfaces of head and limbs black, usually with white spots. Adult ♀ often with a light pink area on the crown of the head. Ventral surfaces of limbs white with black marks; carapace black, typically with white spots; plastron white to pink. Carapace has 7 longitudinal keels, plastron has 5. Upper jaw with a tooth-like cusp on each side. Maximum carapace length about 180 cm, maximum weight about 600 kg—the largest living turtle. Nesting takes place in tropical areas. Adults probably entirely pelagic, feeding almost exclusively on jellyfish. Seen occasionally off Bermuda.

J. A. FRICK & L. C. IRELAND

## Class Aves (Birds)

CHARACTERISTICS: *VERTEBRATA covered with feathers; respiration by way of lungs. Skull articulates with vertebral column by means of a single condyle; lower jaw composed of several bones and hinged to the skull via the quadrate. Jaws toothless, modified to form a horny beak; front limbs adapted for flying (or secondarily for swimming). Fertilization is internal; all lay shelled eggs; larval stages are absent.* Birds range in size from tiny hummingbirds to the 2.5 m high ostrich; most of Bermuda's seabirds are of intermediate size. Although many birds have spectacularly colorful plumage, those adapted to life on or near the ocean are mostly in shades of white, gray, brown and black. Pelagic seabirds are adapted for planing, many shorebirds for running and wading, and a number can swim and dive under water, some for long distances (e.g., cormorants).

There are approximately 28 orders and 9,000 living species in the class. Although 330 species have been recorded from Bermuda and 200 of these can be found on the island in any one year, the vast majority are just transients or vagrants during their annual migrations. The present resident and breeding avifauna of Bermuda consists of 18 species of which 3 are seabirds and 2 are marsh birds. These 5, together with the commonest migrant and wintering species, which feed mainly in the littoral and inshore waters, make up the 18 species included here.

OCCURRENCE: A very diverse class occupying every corner of the planet and adapted to live and feed in a wide range of habitats and niches, from open ocean to mountaintop and desert to rain forest. Prior to human settlement in 1612 Bermuda supported a vast population of breeding sea birds of many species, and a number of water fowl and land bird species with varying degrees of endemism. Unfortunately, the majority of these became extinct as a result of human exploitation for food, deforestation and introduced mammal predators within the first decade or so after settlement, but their bones remain common as fossils in the limestone caves and in pond sediments. In the 4 centuries since human settlement this fauna has been gradually augmented again by the introduction of various land birds from Europe, North America and the West Indies. The seabird fauna, however, has never recovered.

Stringent regulations govern the collection of birds for scientific study on Bermuda. Under the provisions of the Protection of Birds Act (1975) all resident and migratory species are fully protected with the exception of starlings, house sparrows, kiskadees and common crows. The capture of live birds for scientific study and later release requires a permit. The most efficient devices for capturing birds are Japanese mist nets, but the purchase and use of these is restricted to persons who have obtained an official bird banding permit from one of the internationally recognized bird banding centers.

During the migration periods on Bermuda many birds are found exhausted, injured or dead on roads or under utility wires. All such specimens are welcomed by the Bermuda Natural History Museum where they are prepared either as study skins or skeletons for the reference collections.

IDENTIFICATION: Birds are generally observed with the naked eye or with field glasses. Field identification is initially dependent on a sound knowledge of order and family characteristics such as the size, shape, flight behavior and habitat preferences.

Although the requirements for flight place some severe constraints on the main skeletal structures, extreme variations are found in size (eagles to hummingbirds), wing design, bill shape, length of the neck, and length and shape of the legs and feet. (1) Wing design: Shearwaters and petrels have extremely long narrow wings for planing at high speed over the open ocean; hawks and eagles have broad wings with primaries (the flight feathers of the hand), which separate like fingers at the tips for soaring in updrafts. Grouse and most doves have rounded sturdy wings for short burst of powerful flight. 2) Bill shape: This is usually the best indicator of the feeding habits of birds, which range from purely carnivorous through omnivorous to purely herbivorous. Shorebirds have extremely long bills for probing in soft mud for invertebrates. Flycatchers, swallows and goatsuckers have short flat beaks and wide mouths for catching insects; sparrows and buntings have heavy conical bills for crushing seed; and hawks, owls and many seabirds have hooked bills for holding and tearing prey. Note the position of nostrils on the beak, and whether they open in a cere (soft skin covering base of upper beak). (3) Leg shape: This provides the best indicator of the preferred habitat of a species. Seabirds and waterfowl have webbed feet, wading birds have long legs, woodpeckers have sturdy, short legs with 2 forward and 2 backward projecting toes for clinging to vertical surfaces, raptors have powerful legs with long curved talons for holding prey, and perching birds have a single powerful hind toe in opposition to the forward toes for clinging to branches. Once the family characteristics are known, identification to species is facilitated by reference to illustrated field guides and local checklists, and is achieved by closer attention to plumage pattern and color, color of bill, legs and feet, and distinctive vocalizations. It should be pointed out, though, that plumage coloration may differ strikingly between sexes, or seasonally on the same individuals as a result of moult or feather wear. In descriptions of plumages certain feather groupings are often singled out as having a particular color or pattern, e.g., primaries (flight feathers of the wrist joint), secondaries (flight feathers of the forearm), tertials (flight feathers of the upper arm), axillars (underwing coverts of the tertials) and wing coverts (rows of smaller feathers covering the flight feathers on the wing).

Birds are most commonly preserved as dried "study skins", i.e., the skin with adhering feathers is carefully removed, dried over an inserted stick and stored in a fumigated, insect-proof cabinet. Whole specimens may be injected with and stored in 75% ethanol; skeletons are preserved dry for osteological and paleontological studies.

BIOLOGY: The mean life-span of birds ranges from as little as 3 yr in most passerines (perching birds) to as much as 30 or 40 yr in some larger seabirds. Age at first breeding ranges from 1 yr in passerines to as much as 10 yr in some seabirds. All birds undergo a complete feather moult annually following the breeding season. The basic or "winter" plumage following this moult is usually unobtrusive or cryptic. In late winter or early spring the basic plumage is partly moulted (i.e., all except the flight feathers) to produce the alternate "breeding" or "summer" plumage, which is more colorful and conspicuous. During the breeding season most birds perform elaborate courtships involving the display of the colorful nuptial plumage in ritualized behavior with vocalizations by the males (although in a few species this role is performed by the female). Birds usually make a nest of grasses, twigs or pebbles, the nature and location of which vary considerably between families. Terns are ground nesters that make only a nest scrape. Tropic birds nest in cliff holes. Petrels dig soil burrows and most passerines build nests of twigs and grasses in bushes and trees. Nesting may be solitary on discrete territories defended by the males (most passerines) or colonial (most terns and some herons). Many, mainly tropical, species are sedentary, or perform only short post-breeding dispersal migrations. Others, mainly arctic and temperate breeding species, are highly migratory between arctic or temperate breeding grounds and tropical wintering grounds. All of these long distance migrants have the physiological capacity of accumulating fat reserves that serve as fuel to sustain them in flight for 100 hr or more. Such long flights are often necessary in the case of species that migrate over the open ocean between continents. Bermuda lies directly under one such major migration route which is used by many species of shorebirds, ducks, rails, herons, falcons, and some passerines in the fall. These birds leave the North American coast in the vicinity of Newfoundland or New England and do not touch land again until they reach the eastern half of the Caribbean or South America. Bermuda also lies on the route of another major migration involving pelagic birds that passes in spring. This migration is comprised of arctic breeding seabirds returning north from the south Atlantic to breed and South Atlantic breeding species migrating north around the western edge of the Sargasso Sea to spend the summer on the Newfoundland banks.

## Plate 224

DEVELOPMENT: The eggs of birds have a hard calcium shell and are variously shaped, textured, colored and patterned according to the species. Clutch size varies from 1 (petrels and tropic birds) to 3 (terns, gulls, shorebirds) and 4 or more in many passerines. The overwhelming majority of bird species incubate their eggs by sitting on them. Incubation periods vary from as little as 12 days (passerines) to as long as 53 days (the cahow). In most species both sexes share in incubation and feeding of young but there are exceptions in which either the male or the female perform all these duties. Newly hatched chicks are covered in a downy plumage and may be relatively mobile and self-sufficient (precocial), as for instance in Podicipediformes and Charadriiformes, or nearly naked or helpless for a period (altricial) as in most Passeriformes. The downy plumage of nestlings is followed by a drab or cryptic juvenile plumage that is worn for a brief period from nest departure until the pre-basic molt.

REFERENCES: For general orientation and systematic overview see AUSTIN (1961) and VAN TYNE & BERGER (1976).

The best illustrated field guides for use in Bermuda are by PETERSON (1960) and ROBBINS et al. (1966). The most recent, comprehensive checklist of Bermuda birds is by WINGATE (1973); photographs of many species have been published by SLAUGHTER (1975). Older records and checklists are by A. H. VERRILL (1901), BOWDITCH (1904), GROSS (1912), NICHOLS & MOWBRAY (1916), DWIGHT (1927), BRADLEE et al. (1931), BEEBE (1931, 1936) and BOURNE (1957). Seasonal distribution and migrations are dealt with by MOORE (1951), IRELAND & WILLIAMS (1974) and WILLIAMS et al. (1977). Reports on fossil Bermuda birds are by SHUFELDT (1916, 1922), WOOD (1923) and WETMORE (1960, 1962). The story of the "extinct", then rediscovered cahow is told by VERRILL (1901d, 1902b), BEEBE (1935b), MURPHY & MOWBRAY (1951), WINGATE (1960, 1978) and WURSTER & WINGATE (1968).

### Plate 226

O. **PODICIPEDIFORMES:** Swimming Aves with lengths 220-600 mm, sexes similar. Plumage dense and satiny; silvery white below. Bill higher than wide and sharply pointed (except in *Podilymbus*). Neck elongated. Wings short and flight whirring. Tail rudimentary. Legs set far back. Tarsi laterally compressed; toes lobe-webbed; claws extremely flattened and broad. Grebes live exclusively in water and are capable of buoyancy control like a submarine. They swim underwater with feet only. Nestlings precocial. (Only family: Podicipedidae, with 3 spp. from Bda but only 1 regular and common.)

*Podilymbus podiceps* (L.) (Pied-billed grebe; B Helldiver): Genus monotypic. Length 300-380 mm. Bill short and rounded, bearing a black band in breeding plumage. Head, neck and upper parts blackish brown; chin black in breeding plumage and white in winter and on immatures. Under parts silvery white. Wings grayish with white tips to the secondaries. Mainly migratory to Bermuda, wintering commonly in shallow enclosed bays and brackish ponds, IX-IV. Feeds on small fishes and aquatic invertebrates. Rarely resident and breeding. Builds a floating nest in cattail ponds. Eggs 2-7, pale greenish white; call: a nasal laughing sound.

O. **PROCELLARIIFORMES** (Albatrosses, shearwaters and petrels): Strictly marine Aves with long narrow wings designed for planing ove the ocean. Bill with pronounced hook at tip and covered in horny plates separated by sutures. External nostrils tubular, used to excrete brine from salt glands located in grooves of the skull above the eye orbit. Legs set far back. Hind toe small or absent, front toes long and fully webbed. Plumage dense and water repellent, typically gray or blackish above and white below. Enormous size range within order, from storm petrels (140 mm) to albatrosses (1,350 mm). (1 breeding sp. on Bda and 7 spp. commonly transient offshore.)

F. **PROCELLARIIDAE** (Shearwaters and petrels): Medium to large (length 280-914 mm) Procellariiformes with nostrils in single tube with medium septum. Outer functional primary as long as or longer than next.

*Pterodroma cahow* (Nichols & Mowbray) (Bermuda petrel, B Cahow): Length 350 mm; pigeon-sized with extremely long wings. Bill short, stout and black with pronounced hook. Webbing of toes black on outer 2/3 and pink at base. Upper parts blackish to gray with white forehead and some white on rump. Underparts white. Endemic to Bermuda. Formerly superabundant but almost exterminated by humans and introduced predators following Bermuda's colonization in 17th century. Now extremely rare but increasing under strict protection since the rediscovery of its breeding grounds in 1951. Lives on the ocean, probably ranging to the Gulf Stream and feeding on small squid, fishes and shrimp. Comes to land on Bermuda only to breed and now confined to small islets. Comes and goes from the breeding islets only at night and nests in deep burrows or crevices. Breeding season very protracted. Courtship from late X through XI, followed by a 6-week pre-laying exodus. A single white egg (57.8 mm by 43.3 mm) laid in I. Incubation 53 days. Chick covered in long gray down. Fledging period 80-100 days, with departure in late V or early VI.

O. **PELECANIFORMES** (Cormorants, pelicans, gannets, boobies, tropic birds, etc.): Fish-eating marine Aves of medium to large size. Hind toe turned forward and connected with other toes by webbing (totipalmate). Except in Phaethontidae the clavicles (wishbone) are fused to a forward jutting keel of the sternum, and there is a gular pouch.

F. **PHAETHONTIDAE** (Tropic birds): Tern-sized (405-480 mm) Pelecaniformes with 2 extremely elongate central tail feathers in adult plumage. Plumage compact and satiny, mainly white with some black markings on upper parts. Bill elongate, slightly decurved and pointed, with serrated cutting edge. Nostrils open in slit-like apertures on beak. Legs and feet extremely reduced and functional only as paddles. Flight direct, with wings beating steadily. (1 sp. breeding on Bda).

*Phaethon lepturus catesbyi* Brandt (White-tailed tropic bird, B Longtail): Length 390-400 mm excluding central tail feathers which add 500 mm. Adult: Bill orange-yellow, legs greenish yellow and webbing of feet black. Plumage satiny white with a salmon tinge especially on central tail feathers in spring. A black mark through eye. Outer web of first 5 primaries; greater wing coverts and most of tertials black. Some black barring on flanks. Range throughout the Sargasso Sea where it feeds on squid and fishes caught by diving from a height of 20-50 m. Comes to land only to breed and nests commonly in coastal cliffs throughout Bermuda, III-X. Aerial courtship display involves tail touching and a grating "tick-ek" call. Nest is a sandy depression in a cliff hole or crevice. The single

226 PODICIPEDIFORMES—FALCONIFORMES (Seabirds 1)

purplish red mottled egg (55.8 mm by 38.9 mm) is laid between IV and VIII. Incubation 43 days and fledging 65-70 days. Downy chicks pale gray. Fledglings white with black barring on upper part and lacking the elongate central tail feathers. Most fledglings depart from late VII through IX but chicks of late nesters noted as late as mid-XI.

F. **PHALACROCORACIDAE** (Cormorants): Medium to large (480-1,015 mm) Pelecaniformes with cylindrical slender bills hooked at tip. No external nostrils. Neck and body elongate. Legs short and feet large. Tail long and stiff. Plumage blackish and in most species with bare skin, usually brightly colored, on face. Flight heavy, direct and duck-like. (2 spp. from Bda.)

*Phalacrocorax auritus* (Lesson) (Double-crested cormorant): Length 750-900 mm. Adult: Head, neck and underparts glossy black. Feathers of upper parts gray-brown margined with glossy black. Tail black with 12 feathers. A tuft of black feathers on either side of head. Immature similar but sides of head, foreneck and breast whitish, grading to black on belly. Winters on Bermuda in small numbers from early X to IV or V. Rests on reef markers and buoys in harbors and sounds, feeding on fishes that it chases under water on long deep dives.

O. **CICONIIFORMES** (Herons, storks, ibises, etc.): Long-legged and long-necked wading Aves mostly dependent on shallow water for foraging. Nostrils always basal. Four toes; not webbed. Hind toe not elevated. Most species are gregarious tree nesters with eggs white-greenish or bluish. (14 North American spp. recorded commonly on Bda.)

F. **ARDEIDAE** (Herons and bitterns): Medium to large (280-1,420 mm) Ciconiiformes; bill spear-like, with some serrations distally on upper mandible and a tooth at tip. Middle toenail pectinate. Loreal area (between beak and eye) bare. Sixth cervical vertebra especially elongate, with modified articulate surface facilitating folding of neck. Patches of powder down (used in preening) concealed under main plumage. Wings broad and rounded. Plumage usually uniform gray, white or brown. Many species bear elongate plumes.

*Ardea herodias* L. (Great blue heron): Large (1,070-1,320 mm); predominantly gray with blackish gray crown, primaries and secondaries (arm feathers). Some white on head and rufous on foreneck, bend of wing and top of legs. Legs blackish. Bill olive yellow and black. The common heron of Bermuda's shoreline, harbors and islands. Non-breeding birds are present in all months of the year but common only when transient during migration in X-XI and III-IV. Feeds on small to medium-sized fishes caught in the shallows.

*Nycticorax violacea* (L.) (Yellow-crowned night heron): Length 560-700 mm. A stocky heron of medium size with a heavy rounded black bill. Adults: Uniform gray on wings and body. Head striped with black and buff and with 2 long white plumes. Legs yellow. Immature: Grayish brown speckled with white. Legs green to yellow. Non-breeding birds recorded in all months of the year but most common VIII-X when transient. A nocturnally active species that feeds almost exclusively on crustaceans. Wintering birds inhabit mangrove swamps and tidal shallows. Transients and summer residents feed mainly on land crabs (*Gecarcinus lateralis*) captured on coastal slopes. Call note: "wok". Recently reintroduced, now forming a breeding colony. Breeds III-VIII.

O. **FALCONIFORMES** (Diurnal birds of prey): Medium to large Aves with hooked bills and feet usually modified with long claws or talons. Wings broad, with primaries emarginate and separating like fingers when spread, an adaptation for soaring flight. (6 North American spp. commonly recorded on Bda.)

SUP.F. **FALCONOIDEA** (Hawks, falcons): Falconiformes with feathered head; nostrils opening in a cere, not completely piercing maxilla. Bill strongly hooked; claws long, sharp and curved. Hind toe as long or longer than shortest front toe. Hind claw as long or longer than longest front claw.

F. **PANDIONIDAE** (Ospreys): Large (540-600 mm) Falconoidea with long wings held crooked when soaring. Bill short and strongly hooked. Tarsus reticulate (with small, scale-like segments). Outer toe reversible and soles of feet studded with spines. (Only genus and species in the family:)

***Pandion haliaetus*** (L.) (Osprey, B Fish hawk): Upper parts blackish brown, under parts whitish. Head mainly white, with broad blackish brown stripe through eye. Non-breeding birds occur on Bermuda in all months of the year, but common only during migration in X-XI, whence found soaring or perching near water in the harbors and sounds. Feeds on medium-sized fishes (bream, mullet, etc.) that it catches in its talons on swooping dives.

**Plate 227**

O. **CHARADRIIFORMES** (Shorebirds, gulls, terns, auks): A diverse order of marine or wading Aves that eat animal food and share common anatomical features in the palate bones, voice box and manner of insertion of tendons connecting leg muscles to toes. Plumage dense, and muted in color with black, white, gray and brown predominating. Ground nesting, with typically 3 eggs per clutch and precocial young. Most species highly migratory.

F. **CHARADRIIDAE** (Plover, turnstones and surf birds): Chunky-bodied (150-390 mm), boldly patterned Charadriiformes with bills slightly shorter than head. Hind toe vestigial or lacking. Wings long and tertials long. (6 spp. of common occurrence on Bda.)

***Squatarola squatarola*** (L.) (Black-bellied plover): Length 270-340 mm. Bill pigeon-like and black, legs black, eyes relatively large. In breeding plumage mottled gray above with black underparts, black axillars (feathers under wings) and whitish tail. Immature and winter plumaged adults similar but underparts white except for black axillars. Non-breeding birds present on Bermuda in all months of the year but common only IX-IV when influxes of migrants arrive to spend the winter. Gregarious. Feeds mainly on insects and tideline invertebrates that it snaps up with deliberate, jerky movements. Inhabits rocky beaches and barren open areas. Call: a melancholy, whistled "wee-o-woo".

***Arenaria interpres*** (L.) (Ruddy turnstone): Length 205-240 mm. Bill black, spike-like and slightly upturned, legs bright red. Boldly patterned with chestnut, black and white in breeding plumage and brown, black and white in winter. The common shorebird of Bermuda occurring ubiquitously on rocky shores, beaches, wharves and lawns where it feeds gregariously by butting over stones and seaweed in search of small invertebrates. Non-breeding birds found in every month of the year but common only VIII-V when influxes of migrants spend the winter.

F. **SCOLOPACIDAE** (Woodcock, snipe, sandpipers): Medium to large (125-610 mm) Charadriiformes with slender, moderately to extremely long and straight or decurved bill. Neck medium to long, legs short to long and toes moderately long. Plumage usually cryptically colored; buff, gray-brown, gray or white above and white below. (27 spp. common as transients on Bda.)

***Actitus macularia*** (L.) (Spotted sandpiper): Length 180-200 mm. Adult: Bill moderately long, straight and slender. Upper parts brownish gray faintly barred with black. Outer tail feathers white with blackish bars. Underparts white spotted with black. Legs flesh-colored. Immature similar but underparts pure white and a conspicuous white spot in front of wing. A common transient from mid-VII through X when found ubiquitously on marshy mud flats, beaches and rocky coastline. Winters sparingly on the rocky shores of sheltered harbors. Feeds on marine invertebrates. Teeters continuously. Call note: "peet-weet".

***Crocethia alba*** (Pallas) (Sanderling): Length 180-210 mm. Bill and legs black. Hind toe lacking. Basic plumage pale grayish above and white below but barred with pale rufous above and washed with pale rufous below in breeding plumage. The common shorebird of surf-washed sand beaches. Feeds on interstitial organisms by running rapidly back and forth in the swash zone between waves and probing with

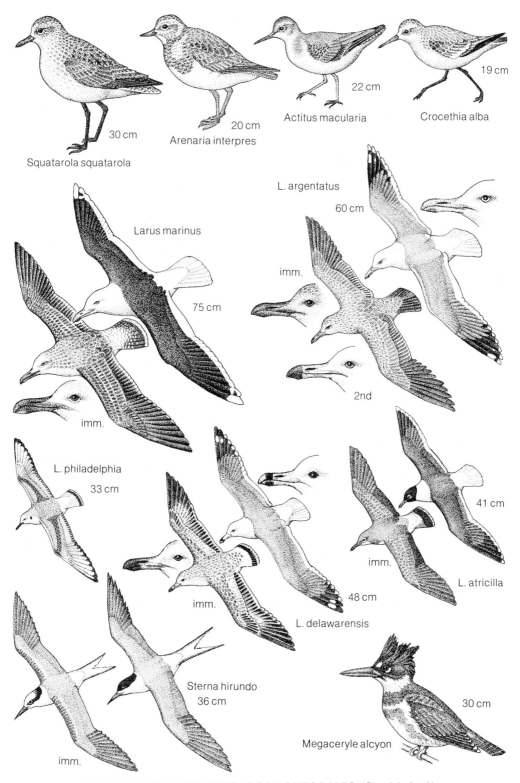

**227 CHARADRIIFORMES, CORACIIFORMES (Seabirds 2)**

the bill. Transient and wintering only (IX-IV), rare in summer.

F. **LARIDAE** (Gulls and terns): Medium to large (200-760 mm), long-winged Charadriiformes with front toes fully webbed, hind toes small and somewhat elevated. Adults typically gray on mantle with black on wing tips and head. Immatures mottled brown, gray or white.

S.F. **LARINAE** (Gulls): Mostly heavy-billed Laridae with weak hook on upper mandible and enlargement near tip of lower mandible. Tail usually square or rounded. (13 spp. recorded, 6 commonly, from Bda.)

*Larus marinus* L. (Great black-backed gull): Genus with hard, slightly curved beak, relatively broad wings and mostly rounded tail.—Species large (720-780 mm); bill heavy. Adult with distinctive black mantle, rest of plumage white. Bill yellow with red spot near tip, legs pink. Immature mottled gray-brown and white above, white below with broad blackish band on tail; bill all black. A common winter visitor between XI and IV. Feeds on carrion and fishes that it finds by soaring along the coastline or by following boats. Gregarious at favored roosting sites in barren open areas or on jetties.

*L. argentatus* Pontoppidon (Herring gull): Genus as above.—Species moderately large (590-650 mm). Adult with gray mantle and white-spotted black wing tips. Rest of plumage white. Bill yellow with red spot near tip, legs pink. Immature in 1st winter uniformly ashy brown mottled with white on upper parts, bill black. In 2nd winter mottled gray and white above and white on under parts, with broad gray band on tail; bill pink with black tip. A common winter visitor between XI and IV, feeding on carrion and fishes that are found by soaring along the coastline or following boats. Gregarious at favorite roosting sites in barren open area or on jetties.

*L. delawarensis* Ord (Ring-billed gull): Genus as above.—Species medium-sized (460-500 mm). Adult with gray mantle and white-spotted black wing tips, rest of plumage white. Bill slender, yellow with black band near tip; legs yellow. Immature mottled gray above, white below, with narrow black band on tail. Bill pink with black tip, legs pink. A common winter visitor XI-IV. Feeds on carrion and fishes that it finds by soaring along the coastline or by following boats. Gregarious at favorite roosting sites in barren open areas and on jetties.

*L. atricilla* L. (Laughing gull): Genus as above.—Species medium-sized (400-430 mm). Adult with dark gray mantle and black wing tips. Bill and feet dark red. Head black in breeding season and hooded with gray in winter. Immature brownish gray on upper parts and head, with black band on tail. Bill and legs black. Uncommon vagrant in all months of the year. Adults in breeding plumage occasional V-VIII. Seen in harbors.

***L. philadelphia*** (Ord) (Bonaparte's gull): Genus as above. —Species small (310-350 mm). Adult with gray mantle and mostly white primaries. Head black in breeding season and white with a black spot behind eye in winter. Immature similar to adult in winter but more gray and brown on mantle, tail with narrow black band at tip. Occasionally common during winter influxes between XII and IV. Seen in harbors.

S.F. **STERNINAE** (Terns): Laridae with sharp-pointed bill. Tail forked or deeply forked. Legs short. (13 spp. recorded, 6 commonly, from Bda.)

***Sterna hirundo*** L. (Common tern): Length 340-400 mm. Adult with pale gray mantle, blackish primaries and black cap, rest of plumage white. Bill red with blackish tip, legs red. Immature similar but with black cap reduced by white on forehead and some black on wing coverts (the feathers above the flight feathers). Bill and feet blackish. Folded wings extend beyond tail at rest. Summer resident, IV-XI, breeding in small numbers on tiny rocky islets in harbors and sounds. Very common as a transient in IX and X and frequent but scarce in winter. Feeds in flocks by diving and dipping on schools of fry and other bait fish in the harbors and sounds. Nest is a mere scrape on bare ground lined with leaves, twigs or shells. Eggs buff, blotched with brown and black, normally 3 per clutch and laid mainly V-VII. Incubation 25 days. Downy chicks precocial, mottled brown and black with blackish throat. Fledging takes approximately 25 days. Adults defend nest vigorously with dive-bombing attacks and harsh call notes "kee-arr, kee-arr".

O. **CORACIIFORMES** (Kingfishers and allies): A diverse assemblage of 7 well-defined families of Aves all characterized by syndactyly (front toes joined for part of their length). Most species brightly colored, and cavity- or burrow-nesting with 3-6 white or unmarked eggs. Young altricial and naked at hatching.

F. **ALCEDINIDAE** (Kingfishers): Small to medium-sized (200-460 mm) fish-eating Coraciiformes with proportionately large head and beak, short neck, stout body and small feet. Bill strong and spear-like. (1 sp. from Bda.)

***Megaceryle alcyon*** (L.) (Belted kingfisher): Length 280-350 mm. Head large and crested. Upper parts grayish blue with wide white collar around neck and white spot in front of eye. ♂ with white under parts and a broad band of gray-blue across breast. ♀ similar but with an additional band of chestnut on breast and flanks. A common wintering species VIII-V, occurring widely scattered along exposed and sheltered coastlines and on brackish ponds. Perches on trees above water and dives to catch small fish. Call note: a loud staccato rattling.

D. B. Wingate

## Class Mammalia

CHARACTERISTICS: *VERTEBRATA usually covered with hair (nearly absent in Cetacea); respiration by way of lungs. Lower jaw composed of a single pair of bones (dentaries) and hinged to the skull via the squamosal. Limbs adapted for walking, climbing, flying or swimming (hindlimbs reduced in Cetacea). Fertilization is internal; the majority (exception: the egg-laying Prototheria) give birth to live young. Young are nourished by secretions (milk) of the female's mammary glands.*

The majority of living mammals belong to the infraclass Eutheria (= Placentalia) in which developing embryos are nourished by the mother via a complex placenta; the primitive, egg-laying Prototheria and the Metatheria (marsupials) do not have marine representatives. Of about 16 living eutherian orders, 3 contain marine species: Carnivora (with seals, sea lions, walruses; and the sea otter), Sirenia (sea cows) and Cetacea; only the last occurs in Bermuda.

## ORDER CETACEA (Whales, porpoises and dolphins)

CHARACTERISTICS: *Large, nearly hairless, aquatic MAMMALIA with fusiform body; tail with lateral flukes. Pectoral limbs modified as flippers, pelvic girdle rudimentary, pelvic limbs absent.* The largest living animals (from 1 to 30 m!). Coloration is usually brown to black above, lighter to white below. All are strong, elegant, often very fast swimmers propelling themselves by vertical motions of the powerful tail fluke. Some species are able to dive to more than 2,000 m (!) and stay submerged for an hour.

Of 78 species described, 36 have been reported from the western North Atlantic; the 6 most frequently sighted off Bermuda are included here.

OCCURRENCE: Oceanic; only few species in estuaries or fresh water. Most species travel in groups (pods) of a few to more than 1,000 individuals (e.g., some dolphins). Whales occasionally strand or beach themselves, often in large numbers in the case of the toothed whales. Although not clearly understood, such mass strandings are thought to relate to a misfunction of the echolocation system, particularly in the pod leader.

The majority of whale species is now protected from commercial harvesting, and whaling, once a thriving industry worldwide, has become rather unprofitable with the depletion or near-extinction of many whale species. Bermuda had a thriving whaling industry, albeit on a small scale and mostly using the humpback whale, until the late 1800s; the last whale was killed in 1940.

IDENTIFICATION: Although higher taxa are mostly distinguished by anatomical features (such as presence of teeth or baleen, fusion of cervical vertebrae and skull structure), many species are readily identifiable in the field by body shape and size, color markings and behavior. Even individuals of certain species can now be recognized and catalogued on the basis of pigmentation (e.g., ventral fluke pattern of the humpback), callosities, dorsal fin shape and scars.

BIOLOGY: Whereas traditional knowledge of whale biology accumulated largely as a by-product of whaling, more recent data based on captive specimens as well as both tagging and direct observation of free-ranging animals from planes, from boats and by SCUBA have only begun to reveal the complex social and behavioral patterns of this order. Large species reach an age of about 50 yr, smaller ones less. Feeding is quite different in the 2 suborders, the toothed whales preying on single, sometimes large prey, and the baleen whales straining water for planktonic crustaceans, squid and small fish.

Humpbacks commonly feed by releasing air below the surface; the rising bubbles seem to concentrate prey organisms that the whales then engulf from below by distending their pleated throats. After breaking the surface they expel the water through the baleen filter. A more elaborate feeding device, the bubble net, is occasionally observed in which the humpback circles underwater while releasing air to form a corral of bubbles about the prey. The sperm whale feeds on squid that it chases on deep and long dives, achieving neutral buoyancy by changing the temperature in the oil-filled spermaceti organ in its head. Most cetaceans are gregarious; many species travel in pods of sometimes hundreds of individuals. The sperm whale lives in a complex social organization where mixed schools of mostly mature females and juveniles of both sexes are temporarily joined by single mature males for breeding. Males otherwise travel in bachelor groups that tend to become smaller as their members get older. Whales exhibit a wide range of sounds, some of which (in Odontoceti) are used for echolocation; others, such as the long, complicated and repeated "song" of male humpback whales (first recorded in the 1950's in Hawaii and Bermuda), have been related to social organization and communication.

DEVELOPMENT: After a gestation period of about a year (in large species) a single calf is born tail first (!) and nursed for several (to 18) months.

REFERENCES: A number of recent books attempt to summarize the field: SLIJPER (1976), MATTHEWS (1968), NORRIS (1966) and RIDGWAY (1972). Complete lists of living species are given by HERSHKOVITZ (1966) and RICE (1977). WINN & OLLA (1979) deal with cetacean behavior, CLARKE (1979) gives a functional interpretation of the spermaceti organ, and KANWISHER & RIDGWAY (1983) discuss physiological ecology.

Bermuda played an important role in research on the song of the humpback whale, first described in detail by PAYNE & McVAY (1971). More recent information has been added by WINN & WINN (1978) and PAYNE & PAYNE (in press), and a sound record was produced by PAYNE (1970). A catalogue of individual humpback whales, including photographs of flukes of Bermuda specimens, has been compiled by KATONA et al. (1980). A history of whaling in Bermuda is by SCHORTMAN (1969).

## Plate 228

**S.O. ODONTOCETI** (Toothed whales): Marine and freshwater Cetacea with one to many teeth. Skull asymmetric, external opening of nostrils (blowhole) single.

**F. DELPHINIDAE** (Ocean dolphins): Odontoceti with functional, conical teeth in upper and lower jaws. Head often drawn out into a beak. (2 spp. occasionally seen off Bda.)

***Delphinus delphis*** L. (Common dolphin): Maximum size 2 m, males slightly larger than females. Body color distinctive, dorsal surface almost black, sides gray and ochre, ventral surface to anus white. The curved margins of the gray and ochre areas form a tringular white patch immediately under the dorsal fin. The eye is surrounded by a black ring that extends forward and backward as a thin black line. Beak robust; dorsal fin fairly large, triangular. Seen occasionally in small family groups well offshore.

***Globicephala melaena*** Traill (Pilot whale, Pothead, Blackfish): To 6.5 m. Head bulbous, without a beak; dorsal fin larger than high and situated well within anterior half of

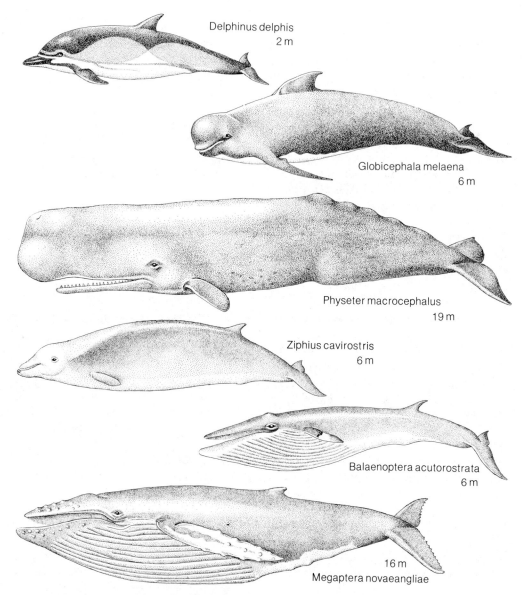

**228 CETACEA (Dolphins and whales)**

body. Flippers long and pointed. Mostly black except for a ventral white patch from chin to anus. Generally in herds of 100 or more, sometimes in smaller pods; offshore, rarely seen.

F. **PHYSETERIDAE** (Sperm whales): Odontoceti without functional teeth in the upper jaw. Skull with large antero-dorsal concavity containing the spermaceti organ. (1 sp. in Bda.)

***Physeter macrocephalus*** L. (=*P. catodon*) (Sperm whale): Males about 16 m (to 19 m), females 12.5 m (to 17 m)—the largest of toothed whales. Head massive (of males to 1/3 of body length!), blunt; blowhole at the front and slightly to the left of center, producing oblique spout. With 18-20 pairs of teeth in lower jaw, which fit into socket in the upper jaw. Dorsal fin in the form of low humps on the dorsal ridge. Flippers short, rounded. The large intestine may contain ambergris, a substance prized in perfume making. Raises tail fluke into the air prior to sounding dive. Seen occasionally, well offshore; calves sometimes stray into inshore waters.

F. **ZIPHIIDAE** (Beaked whales): Odontoceti without functional teeth in the upper jaw; lower jaw with no more than 1 or 2 pairs of functional teeth (lacking in many individuals, particularly females). (1 sp. regularly seen around Bda.)

*Ziphius cavirostris* Cuvier (Cuvier's beaked whale): To about 8 m. Adults with a white or pale gray head, back brownish, becoming lighter on underside. Lower jaw extends beyond the tip of the snout and bears, in adult males, 2 teeth (which may be badly worn in old animals). Blowhole crescent-shaped, pointing forward. Flippers short, mitten-shaped; dorsal find low, falcate. The skull is markedly asymmetrical and appears deformed. Frequently but irregularly seen off Bermuda, prone to stranding.

S.O. **MYSTICETI** (Baleen whales): Marine Cetacea without teeth but with baleen plates projecting from the palate. Skull symmetrical, external opening of nostrils (blowhole) bipartite.

F. **BALAENOPTERIDAE** (Rorquals): Mysticeti with a dorsal fin and longitudinal grooves on the throat. Baleen plates short and broad, 200 or more on each side of the upper jaw. (Of 2 genera with 6 spp., only the following 2 are fairly regularly seen off Bda.)

*Balaenoptera acutorostrata* Lacépède (Minke whale, Little piked whale): To 11 m. Easily distinguished by the wide, pure white band across the otherwise dark pectoral flipper. Upper part of body black, including rim of lower jaw; well offset against white under part. Dorsal fin falcate. Central grooves many (to 60), narrow. Spout usually indistinct. Occasionally seen, usually well offshore, I-II. One of the few whales that come close to a ship.

*Megaptera novaeangliae* Borowski (Humpback whale): Females about 16.5 m long (to 19 m), males smaller, 15 m (to 17.5 m). Easily identified by the extremely long (to 1/3 of body length!), predominantly white pectoral flippers. Dorsal fin small. Caudal edge of tail fluke with fine, irregular serrations. Ventral grooves few (to 20), wide. Head, edges of mandible and flippers with irregular protuberances. Color somewhat variable; back mostly near-black, underparts grayish to white. Spout low (4-5 m), bushy. Tail fluke raised above water prior to sounding dive. Regularly seen offshore III-V when it migrates north from its breeding and calving grounds in the Caribbean to its feeding grounds in the northern Atlantic. May come close to shore and occasionally stray inside the reef barrier.

L. MOWBRAY & W. STERRER

# REFERENCES

Abbott, R. T. 1974. *American seashells.* 2nd ed. New York: Van Nostrand Reinhold. 663 pp.

Abbott, R. T. & R. H. Jensen. 1967. Molluscan faunal changes around Bermuda. *Science* **155**: 687-688.

Abbott, R. T. & R. H. Jensen. 1968. Portuguese marine mollusks in Bermuda. *Nautilus* **81**: 86-89.

Abbott, R. T. & S. P. Dance. 1982. *Compendium of seashells.* New York: E. P. Dutton. 410 pp.

Abele, L. G. & B. E. Felgenhauer. 1982. Decapoda. In: Parker, S. P. (Ed.). *Synopsis and classification of living organisms.* Vol. 2. New York: McGraw-Hill. pp. 296-326.

Adams, C. D. 1972. *Flowering plants of Jamaica.* Kingston: University of the West Indies. 848 pp.

Ahmadjian, V. & M. E. Hale (Eds.). 1973. *The lichens.* New York: Academic. 697 pp.

Alldredge, A. 1976. Appendicularians. *Sci. Am.* **235**(1): 94-102.

Almy, Ch. C., Jr. & C. Carrion-Torres. 1963. Shallow-water stony corals of Puerto Rico. *Caribb. J. Sci.* **3**(2/3): 133-162.

Alvarino, A. 1965. Chaetognaths. *Oceanogr. Mar. Biol. Ann. Rev.* **3**: 115-194.

Alvarino, A. 1969. Los Quetognatos del Atlantico. Distribucion y notas esenciales de sistematica. *Trab. Inst. Esp. Oceanogr.* **37**: 1-290.

Anderson, D. T. 1973. *Embryology and phylogeny in amnelids and arthropods.* New York: Pergamon. 495 pp.

Anderson, O. R. 1976a. A cytoplasmic fine-structure study of two spumellarian Radiolaria and their symbionts. *Mar. Micropaleontol.* **1**: 81-99.

Anderson, O. R. 1976b. Fine structure of a collodarian radiolarian (*Sphaerozoum punctatum* Müller 1858) and cytoplasmic changes during reproduction. *Mar. Micropaleontol.* **1**: 287- 297.

Anderson, O. R. 1976c. Ultrastructure of a colonial radiolarian *Collozoum inerme* and a cytochemical determination of the role of its zooxanthellae. *Tissue & Cell* **8**: 195-208.

Anderson, O. R. 1978. Fine structure of a symbiont-bearing colonial radiolarian *Collosphaera globularis,* and 14-C isotopic evidence for assimilation of organic substances from its zooxanthellae. *J. Ultrastruct. Res.* **62**: 181-189.

Anderson, O. R. 1980. Radiolaria. In: Levandowsky, M. & S. H. Hutner (Eds.). *Biochemistry and physiology of Protozoa.* Vol. 3, 2nd ed. New York: Academic. pp. 1-42.

Anderson, O. R. 1983a. *Radiolaria.* New York: Springer-Verlag. 355 pp.

Anderson, O. R. 1983b. The radiolaria symbiosis. In: Goff, L. J. (Ed.). *Algae symbiosis. A continuum of interaction strategies.* Cambridge: Cambridge University Press. pp. 69-89.

Anderson, O. R. & A. W. H. Bé. 1976. A cytochemical fine structure study of phagotrophy in a planktonic foraminifer, *Hastigerina pelagica* (d'Orbigny). *Biol. Bull.* **151**: 437-449.

Angell, R. W. 1975. Structure of *Trichosphaerium micrum* sp. n. *J. Protozool.* **22**: 18-22.

Annoscia, E. (Ed.). 1968. Proceedings of the First International Conference on Bryozoa. *Atti Soc. Ital. Sci. Nat. Mus. Civ. Stor. Nat. Milano* **108**: 1-377.

Appy, R. G. & M. J. Dadswell. 1981. Marine and estuarine piscicolid leeches (Hirudinea) of the Bay of Fundy and adjacent waters with a key to species. *Can. J. Zool.* **59**: 183-192.

Arey, L. B. 1915. The orientation of *Amphioxus* during locomotion. *J. Exp. Zool.* **19**: 37-44.

Arey, L. B. 1937a. The physiology of the repugnatorial glands of *Onchidium*. *J. Exp. Zool.* **77**: 251-286.

Arey, L. B. 1937b. Observations on two types of respiration in *Onchidium*. *Biol. Bull.* **72**: 41-46.

Arey, L. B. & L. E. Barrick. 1942. The structure of the repugnatorial glands of *Onchidium floridanum*. *J. Morphol.* **71**: 493-521.

Arey, L. B. & W. J. Crozier. 1918. Homing habits of the pulmonate mollusk *Onchidium*. *Proc. Nat. Acad. Sci. U.S.A.* **4**: 319-321.

Arey, L. B. & W. J. Crozier. 1919. The sensory responses of *Chiton*. *J. Exp. Zool.* **29**: 157-260.

Arey, L. B. & W. J. Crozier. 1921. The natural history of *Onchidium*. *J. Exp. Zool.* **32**: 443-502.

Arnold, J. M. 1965. Observations on the mating behavior of the squid *Sepioteuthis sepioidea*. *Bull. Mar. Sci.* **15**(1): 216-222.

Arnold, J. M. 1971. Cephalopods. In: Reverberi, G. (Ed.). *Experimental embryology of marine and freshwater invertebrates*. Amsterdam: North-Holland. pp. 265-311.

Arnold, W. H. 1965. A glossary of a thousand-and-one terms used in conchology. *Veliger* **7**(suppl): 1-50.

Arnold, Z. M. 1972. Observations on the biology of the protozoan *Gromia oviformis* Dujardin. *Univ. Calif. Pub. Zool.* **100**: 1-168.

Aurelia, M. 1970. The habits of some subtidal pelecypods in Harrington Sound, Bermuda. In: Ginsburg, R. N. & S. M. Stanley (Eds.). *Reports of research: Seminar on Organism-Sediment Interrelationships, 1969*. Bermuda Biological Station Special Publication No. 6. St. George's West, Bermuda: Bermuda Biological Station. pp. 39-52.

Austin, O. L. 1961. *Birds of the world*. New York: Golden Press. 316 pp.

Ax, P. 1985. The phylogenetic position of the Gnathostomulida and Platyhelminthes within the Bilateria. In: Morris, S. C. et al. (Eds.). *The origin and relationships of lower invertebrates*. Oxford: Oxford University Press. pp. 168-180.

Babcock, H. L. 1937. The sea-turtles of the Bermuda Islands, with a survey of the state of the turtle fishing industry. *Proc. Zool. Soc. Lond.* **107**: 595-601.

Baker, E. W. & G. W. Wharton. 1959. *An introduction to acarology*. 3rd ed. New York: Macmillan. 465 pp.

Ball, G. H. 1951. Gregarines from Bermuda marine crustaceans. *Univ. Calif. Pub. Zool.* **47**: 351-368.

Barker, F. D. 1922. The parasitic worms of the animals of Bermuda. I. Trematodes. *Proc. Am. Acad. Arts Sci.* **57**(9): 215-237.

Barnard, J. L. 1958. Index to the families, genera and species of the gammaridean Amphipoda (Crustacea). *Allan Hancock Found. Occas. Pap.* **19**: 1-145.

Barnard, J. L. 1969. The families and genera of marine gammaridean Amphipoda. *U. S. Nat. Mus. Bull.* **271**: 1-535.

Barnes, R. D. 1977. New record of a pterobranch hemichordate from the Western Hemisphere. *Bull. Mar. Sci.* **27**(2): 340-343.

Barnes, R. D. 1980. *Invertebrate zoology*. 4th ed. Philadelphia: Saunders Coll./Holt, Rinehart & Winston. 1089 pp.

Bartsch, I. 1978. *Copidognathus hartwigi* n. sp. (Halacaridae, Acari), eine Meeresmilbe neu für die Fauna der Bermudas. *Acarologia* **20**: 95-96.

Bartsch I. 1983. Zur Systematik und Verbreitung der Gattung *Arhodeoporus* (Halacaridae, Acari) und Beschreibung zweier neuer Arten. *Zool. Beitr. N.F.* **28**: 1-16.

Bartsch, I. & T. M. Iliffe. 1984. The halacarid fauna (Halacaridae, Acari) of Bermuda's caves. *Stygiologia* **X**: XX-XX.

Bartsch, P. 1911. New marine molluscs from Bermuda. *Proc. U. S. Nat. Mus.* **41**: 303-306.

Bassler, R. S. 1953. Bryozoa. In: Moore, R. C. (Ed.). *Treatise on invertebrate paleontology*, Part G. Lawrence: University of Kansas Press and Geological Society of America. 253 pp.

Bayer, F. M. 1956. Octocorallia. In: Moore, R. C. (Ed.). *Treatise on invertebrate paleontology*, Part F; Coelenterata. Lawrence: University of Kansas Press and Geological Society of America. pp. 166-190, 192-231; figs. 134-138, 140-162.

Bayer, F. M. 1961. The shallow-water Octocorallia of the West Indian region. *Stud. Fauna Curacao other Caribb. Isl.* **12**(55): 1-373, 28 pls.

Bayer, F. M. & H. B. Owre. 1968. *The free-living lower invertebrates*. London: Macmillan. 229 pp.

Bayer, F. M. & A. J. Weinheimer (Eds.). 1974. *Prostaglandins from Plexaura homomalla: ecology, utilization and conservation of a major medical marine resource. A symposium.* Stud. Trop. Oceanogr. No. 12. Coral Gables: University of Miami Press. 165 pp.

Bé, A. W. H. 1959. Ecology of recent planktonic Foraminifera: Part 1—areal distribution in the western North Atlantic. *Micropaleontology* **5**: 77-100.

Bé, A. W. H. 1960. Ecology of recent planktonic Foraminifera: Part 2—bathymetric and seasonal distributions in the Sargasso Sea off Bermuda. *Micropaleontology* **6**: 373-392.

Bé, A. W. H. & O. R. Anderson. 1976.

Gametogenesis in planktonic Foraminifera. *Science* **192**: 890-892.

Bé, A. W. H. & W. H. Hamlin. 1967. Ecology of recent planktonic Foraminifera: Part 3—distribution in the North Atlantic during the summer of 1962. *Micropaleontology* **13**: 87-106.

Bé, A. W. H. et al. 1977. Laboratory and field observations of living planktonic Foraminifera. *Micropaleontology* **23**: 155-179.

Beebe, W. 1931. Notes on the birds of Nonsuch Island, Bermuda. *Aviculture* **3**(Ser. II): 86-88.

Beebe, W. 1934. *Half mile down*. New York: Duell, Sloan & Pearce. 344 pp.

Beebe, W. 1935a. Deep-sea fishes of the Bermuda Oceanographic Expeditions: Family Nessorhamphidae. *Zoologica* **20**: 25-51.

Beebe, W. 1935b. Rediscovery of the Bermuda Cahow. *Bull. N. Y. Zool. Soc.* **38**: 187-190.

Beebe, W. 1936. Recent notes on Bermuda birds. *Proc. Linn. Soc. N. Y.* **48**: 60-65.

Beebe, W. & J. Crane. 1937. Deep-sea fishes of the Bermuda Oceanographic Expeditions: Family Nemichthyidae. *Zoologica* **22**: 349-383.

Beebe, W. & J. Tee-Van. 1932. New Bermuda fish, including six new species and forty-three species hitherto unrecorded from Bermuda. *Zoologica* **8**(5): 109-120.

Beebe, W. & J. Tee-Van. 1933. *Field book of the shore fishes of Bermuda*. New York: Putnam's. (Reprinted 1970; New York: Dover Publ.) 337 pp.

Beier, M. 1963. *Ordnung Pseudoscorpionidea (Afterskorpione)*. Bestimmungsbücher zur Bodenfauna Europas 1. Berlin: Akademie-Verlag. 313 pp.

Bennitt, R. 1922. Additions to the hydroid fauna of the Bermudas. *Proc. Am. Acad. Arts Sci.* **57**: 241-259.

Bergquist, P. R. 1978. *Sponges*. Berkeley: University of California Press. 268 pp.

Bernatowicz, A. J. 1952. Seasonal aspects of the Bermuda algal flora. *Pap. Mich. Acad. Sci. Arts Lett.* **36**: 3-8.

Berrill, N. J. 1932. Ascidians of the Bermudas. *Biol. Bull.* **62**: 77-88.

Berrill, N. J. 1950. *The Tunicata, with an account of the British species*. London: Ray Society. 354 pp.

Bevelander, G. 1951. A study of calcification in molluscs with special reference to the use of $P^{32}$ and $Ca^{45}$. *N. Y. J. Dent.* **21**: 305-308.

Bevelander, G. 1952. Calcification in molluscs. III. Intake and deposition of $Ca^{45}$ and $P^{32}$ in relation to shell formation. *Biol. Bull.* **102**: 9-15.

Bevelander, G. 1963. Effect of tetracycline on crystal growth. *Nature* **198**: 1103.

Bevelander, G. & P. Benzer. 1948. Calcification in marine molluscs. *Biol. Bull.* **94**: 176-183.

Bevelander, G. & H. Nakahara. 1966. Correlation of lysosomal activity and ingestion by the mantle epithelium. *Biol. Bull.* **131**: 76-82.

Bevelander, G. & H. Nakahara. 1967. An electron microscope study of the formation of the periostracum of *Macrocallista maculata*. *Calcif. Tissue Res.* **1**: 55-67.

Bevelander, G. & H. Nakahara. 1969. An electron microscope study of the formation of the nacreous layer in the shell of certain bivalve molluscs. *Calcif. Tissue Res.* **3**: 84-92.

Biffar, T. A. 1970. The genus *Callianassa* (Crustacea, Decapoda, Thalassinidea) in south Florida, with keys to the western Atlantic species. *Bull. Mar. Sci.* **21**(3): 637-715.

Bigelow, H. B. 1918. Some Medusae and Siphonophorae from the western Atlantic. *Bull. Mus. Comp. Zool. Harv. Coll.* **62**: 365-442.

Bigelow, H. B. 1938. Plankton of the Bermuda Oceanographic Expeditions. VIII. Medusae taken during the years 1929 and 1930. *Zoologica* **23**: 99-189.

Bigelow, H. B. & I. P. Farfante. 1948. Lancelets. In: Tee-Van, J. et al. (Eds.). *Fishes of the western North Atlantic*. Part 1. New Haven: Sears Foundation for Marine Research. pp. 1-28.

Bigelow, H. B. & W. C. Schroeder. 1948. Sharks. In: Tee-Van, J. et al. (Eds.). *Fishes of the western North Atlantic*. Part 1. New Haven: Sears Foundation for Marine Research. pp. 59-576.

Bigelow, H. B. & W. C. Schroeder. 1953. Sawfishes, guitarfishes, skates and rays. In: Tee-Van, J. et al. (Eds.). *Fishes of the western North Atlantic*. Part 2. New Haven: Sears Foundation for Marine Research. pp. 1-514.

Biggar, R. B. 1932. Studies on ciliates from Bermuda sea urchins. *J. Parasitol.* **18**: 252-257.

Björnberg, T. K. S. 1972. Developmental stages in some tropical and subtropical planktonic marine copepods. *Stud. Fauna Curacao other Caribb. Isl.* **136**: 185 pp.

Björndal, K. A. 1982. *Biology and conservation of sea turtles: proceedings of the World Conference on Sea Turtles Conservation*. Washington, D.C.: Smithsonian Institution. 583 pp.

Boden, B. P., M. W. Johnson & E. Brinton. 1955. The Euphausiacea (Crustacea) of the North Pacific. *Bull. Scripps Inst. Oceanogr. Univ. Calif.* **6**: 287-400.

Bodungen, B. von et al. 1982. *The Bermuda marine environment*. Vol. III. Bermuda Biological Station Special Publication No. 18. St. George's West, Bermuda: Bermuda Biological Station. 123 pp.

Böhlke, J. E. & C. C. G. Chaplin. 1968. *Fishes of the Bahamas and adjacent tropical waters*. Wynnewood, Pa.: Livingstone. 771 pp.

Bold, H. C. & M. J. Wynne. 1978. *Introduction to the algae: structure and reproduction*. Englewood Cliffs, N.J.: Prentice-Hall. 706 pp.

Bold, W. A. van den. 1963. The ostracode genus *Orionina* and its species. *J. Paleontol.* **37**: 33-50.

Boltovskoy, E. & R. Wright. 1976. *Recent Foraminifera*. The Hague: W. Junk. 515 pp.

Bond, C. E. 1979. *Biology of fishes*. Philadelphia: W. B. Saunders. 514 pp.

Borrer, D. J. & D. DeLong. 1971. *An introduction to the study of insects*. 3rd ed. New York: Holt, Rinehart & Winston. 812 pp.

Borror, A. C. 1973. Marine flora and fauna of the northeastern United States. Protozoa: Ciliophora. *NOAA Tech. Rep. NMFS Circ.* **378**: 62 pp.

Borror, A. C. 1980. Spatial distribution of marine ciliates—micro-ecologic and biogeographic aspects of protozoan ecology. *J. Protozool.* **27**: 10-13.

Boss, K. J. 1964. Notes on a hybrid *Tellina* (Tellinidae). *Nautilus* **78**(1): 18-20.

Boss, K. J. 1966a. The subfamily Tellininae in the western Atlantic. The genus *Tellina* (Part I). *Johnsonia* **4**(45): 217-272, pls. 127-142.

Boss, K. J. 1966b. The subfamily Tellininae in the western Atlantic. The genera *Tellina* (part II) and *Tellidora*. *Johnsonia* **4**(46): 273-344, pls. 143-163.

Boss, K. J. 1982. Mollusca. In: Parker, S. P. (Ed.). *Synopsis and classification of living organisms*. New York: McGraw-Hill. pp. 945-1166.

Boss, K. J. & D. R. Moore. 1967. Notes on *Malleus* (*Parimalleus*) *candeanus* (D'Orbigny) (Mollusca: Bivalvia). *Bull. Mar. Sci.* **17**: 85-94.

Bourne, W. R. P. 1957. Breeding birds of Bermuda. *Ibis* **99**: 94-105.

Bourrelly, P. 1970. *Les Algues d'eau douce*. Tome III. Les Algues bleues et rouges, les Eugleniens, Peridiniens et Cryptomonadines. Paris: N. Boubée. 512 pp.

Bousfield, E. L. 1973. *Shallow-water Gammaridean Amphipoda of New England*. Ithaca, N.Y.: Cornell University Press (Comstock). 312 pp.

Bowditch, H. 1904. A list of Bermudian birds seen during July and August, 1903. *Am. Nat.* **38**(451/452): 555-563.

Bowman, T. E. & T. M. Iliffe. 1983. *Bermudalana aruboides*, a new genus and species of troglobitic Isopoda (Cirolanidae) from marine caves of Bermuda. *Proc. Biol. Soc. Wash.* **96**(2): 291-300.

Bowman, T. E. & T. M. Iliffe. 1985. *Mictocaris halope*, a new unusual peracaridan crustacean from marine caves on Bermuda. *J. Crust. Biol.* **5**: 58-73.

Bowman, T. E. & B. F. Morris. 1979. *Carpias* Richardson 1902, a senior synonym of *Bagatus* Nobili 1906, and the validity of *Carpias minutus* (Richardson 1902) (Isopoda; Asellota: Janiridae). *Proc. Biol. Soc. Wash.* **92**(3): 650-657.

Bowman, T. E., S. P. Garner, R. R. Hessler, T. M. Iliffe & H. L. Sanders. 1985. Mictacea, a new order of Crustacea Peracarida. *J. Crust. Biol.* **5**: 74-78.

Bradlee, T. S., L. L. Mowbray & W. F. Eaton. 1931. A list of birds recorded from the Bermudas. *Proc. Boston Soc. Nat. Hist.* **39**: 279-382.

Brady, G. S. 1880. Report on the Ostracoda dredged by the H. M. S. Challenger during the years 1873-1876. In: Thompson, Sir C. W. (Ed.). *Report on the scientific results of the voyage of the H. M. S. Challenger during the years 1873-1876*. Zoology, vol. 1, pt. 3. London: Her Majesty's Stationery Office. 184 pp. 44 pls.

Brandt, N. & C. Apstein (Eds.). 1901. *Nordisches Plankton*. Zoologischer Teil, Bd 4. Entomostraca. Kiel & Leipzig: Lipsius & Tischer. 332 pp.

Brattegard, T. 1969. Marine biological investigations in the Bahamas. 10. Mysidacea from shallow water in the Bahamas and southern Florida. Part 1. *Sarsia* **39**: 17-106.

Bretsky, S. S. 1967. Environmental factors influencing the distribution of *Barbatia domingensis* (Mollusca, Bivalvia) on the Bermuda platform. *Postilla* **108**: 1-14.

Brien, P. et al. 1973. Spongiaires. In: Grassé, P. P. (Ed.) *Traité de Zoologie*. Vol. 3, pt.1. Paris: Masson. 716 pp.

Brinkhurst, R. O. 1982. *British and other marine and estuarine oligochaetes. Keys and notes for the identification of the species*. Synopses of the British Fauna No.21. London: Cambridge University Press. 127 pp.

Brinkhurst, R. O. & D. G. Cook (Eds.). 1980. *Aquatic oligochaete biology*. New York: Plenum. 530 pp.

Brinkhurst, R. O. & B. G. M. Jamieson. 1971. *Aquatic Oligochaeta of the world*. Edinburgh: Oliver & Boyd. 860 pp.

Britton, N. L. 1918. *Flora of Bermuda*. New York: Scribner. (Reprinted 1965; New York: Hafner.) 585 pp.

Bromley, R. G. 1978. Bioerosion of Bermuda reefs. *Palaeogeogr. Palaeoclimatol. Palaeoecol.* **23**: 169-197.

Brown, W. L., Jr. 1982. Insecta. In: Parker, S. P. (Ed.). *Synopsis and classification of living organisms*. Vol. 2. New York: McGraw-Hill. pp. 326-680.

Bruun, A. F. 1943. The biology of *Spirula spirula*. *Dana Rep.* **24**: 1-46, 2 pls.

Burnett-Herkes, J. 1974. Returns of green turtles (*Chelonia mydas* Linnaeus) tagged at Bermuda. *Biol. Conserv.* **7**: 307-308.

Bustard, H. R. 1973. *Sea turtles, their natural history and conservation*. New York: Taplinger. 220 pp.

Butler, J. N. et al. 1983. *Studies of Sargassum and the Sargassum Community*. Bermuda Biological Station Special Publication No. 22. St. George's West, Bermuda: Bermuda Biological Station. 307 pp.

Cairns, S. 1976. *Guide to the commoner shallow-water gorgonians (sea whips, sea feathers and sea fans) of Florida, the Gulf of Mexico and the Caribbean region*. Sea Grant Field Guide Series No. 6. Miami: University of Miami Sea Grant Program. 74 pp.

Calman, W. T. 1912. The Crustacea of the order Cumacea in the collection of the United States National Museum. *Proc. U. S. Nat. Mus.* **41**: 603-676.

Campbell, A. S. 1964. Radiolaria. In: Moore, R. C. (Ed.). *Treatise on invertebrate paleontology*, Part D; Protista 3. Lawrence: University of Kansas Press and Geological Society of America. 195 pp.

Campbell, R. D. 1974. Cnidaria. In: Giese, A. C. & J. S. Pearse. (Eds.). *Reproduction of marine invertebrates.* Vol. 1. New York: Academic. pp. 133-199.

Cannon, H. G. 1947. On the anatomy of the pedunculate barnacle. *Lithotrya. Phil. Trans. R. Soc. Lond. B. Biol. Sci.* **233**: 89-136.

Canu, F. & R. S. Bassler. 1928. Fossil and recent Bryozoa of the Gulf of Mexico region. *Proc. U. S. Nat. Mus.* **72**(14): 1-199.

Carlgren, O. 1912. Ceriantharia. *Danish Ingolf-Exped.* **5**(3): 1-76, 5 pls.

Carlgren, O. 1923. Ceriantharia und Zoantharia. *Wiss. Ergebn. Deutschen Tiefsee-Exped. "Valdivia"* **19**(7): 241-338.

Carlgren, O. 1949. A survey of the Ptychodactaria, Corallimorpharia and Actiniaria. *K. Sven. Vetenskapsakad. Handl.* (4)**1**(1): 1-121, 4 pls.

Carr, A. 1967a. *Handbook of turtles. The turtles of the United States, Canada and Baja California.* Ithaca, N.Y.: Cornell University Press. 542 pp.

Carr, A. 1967b. *So excellent a fishe. A natural history of sea turtles.* Garden City, N.Y.: The Natural History Press. 248 pp.

Carr, A. F., L. Ogren & C. McVea. 1980. Apparent hibernation by the Atlantic loggerhead turtle *Caretta caretta* off Cape Canaveral, Florida. *Biol. Conserv.* **19**: 7-14.

Carr, N. G. & B. A. Whitton (Eds.). 1973. *The biology of blue-green algae.* Oxford: Blackwell. 676 pp.

Carr, N. G. & B. A. Whitton (Eds.). 1982. *The biology of cyanobacteria.* Berkeley: University of California Press. 700 pp.

Chace, F. A., Jr. 1942. Reports on the scientific results of the Atlantis expeditions to the West Indies, under the joint auspices of the University of Havana and Harvard University. The anomuran Crustacea. I. Galatheidea. *Torrea* **11**: 1-106.

Chace, F. A., Jr. 1972. The shrimps of the Smithsonian-Bredin Caribbean Expeditions with a summary of the West Indian shallow-water species (Crustacea: Decapoda: Natantia). *Smithson. Contrib. Zool.* **98**: 1-179.

Chace, F. A., Jr. & H. H. Hobbs, Jr. 1969. The freshwater and terrestrial decapod crustaceans of the West Indies with special reference to Dominica. *U. S. Nat. Mus. Bull.* **292**: 1-258.

Chamberlin, J. C. 1931. The arachnid order Chelonethida. *Stanford Univ. Pub. (Biol.)* **7**: 1-284.

Chamberlin, R. V. 1920. The myriopod fauna of the Bermuda Islands, with notes on variation in Scutigera. *Ann. Entomol. Soc. Am.* **13**: 271-302.

Chaplin, C. E. & P. Scott. 1972. *Fishwatchers guide to the west Atlantic coral reefs.* Wynnewood, Pa.: Livingston. 65 pp.

Chapman, V. J. 1961a. The marine algae of Jamaica: Part I. Myxophyceae and Chlorophyceae. *Bull. Inst. Jam. Sci. Ser.* **1**(1): 1-159.

Chapman, V. J. 1961b. The marine algae of Jamaica: Part II. Phaeophyceae and Rhodophyceae. *Bull. Inst. Jam. Sci. Ser.* **12**(2): 1-201.

Chen, C. 1964. Pteropod ooze from the Bermuda pedestal. *Science* **144**: 60-62.

Chen, C. & A. W. H. Bé. 1964. Seasonal distributions of euthecosomatous pteropods in the surface waters of five stations in the western North Atlantic. *Bull. Mar. Sci. Gulf Caribb.* **14**: 185-220.

Cheng, L. 1976. *Marine insects.* New York: American Elsevier. 581 pp.

Clark, H. L. 1901. Bermudian echinoderms. A report on observations and collections made in 1899. *Proc. Boston Soc. Nat. Hist.* **29**: 339-344.

Clark, H. L. 1942. The echinoderm fauna of Bermuda. *Bull. Mus. Comp. Zool. Harv. Univ.* **89**(8): 367-391.

Clark, K. B. 1982. A new *Volvatella* (Mollusca: Ascoglossa) from Bermuda, with comments on the genus. *Bull. Mar. Sci..* **32**(1): 112-120.

Clark, K. B. 1984. New records and synonymies of Bermudan opisthobranchs. *Nautilus* **98**: 85-97.

Clark, K. B. & M. Busacca. 1978. Feeding specificity and chloroplast retention in four tropical Ascoglossa, with a discussion of the extent of chloroplast symbiosis and the evolution of the order. *J. Moll. Stud.* **44**: 272-282.

Clark, K. B. & A. Goetzfried. 1978. Zoogeographic influences on development patterns of North Atlantic Ascoglossa and Nudibranchia, with a discussion of factors affecting egg size and number. *J. Moll. Stud.* **44**: 283-294.

Clarke, A. H., Jr. 1959. New abyssal molluscs from off Bermuda collected by the Lamont Geological Observatory. *Proc. Malacol. Soc. London* **33**(5): 231-238.

Clarke, G. L. 1934. The diurnal migration of copepods in St. George's Harbor, Bermuda. *Biol. Bull.* **67**: 456-460.

Clarke, M. R. 1966. A review of the systematics and ecology of oceanic squids. *Adv. Mar. Biol.* **4**: 91-300.

Clarke, M. R. 1969. Cephalopoda collected on the SOND cruise. *J. Mar. Biol. Assoc. U. K.* **49**: 961-976.

Clarke, M. R. 1979. The head of the sperm whale. *Sci. Am.* **240**: 128-141.

Clench, W. J. 1942. The genera *Dosina, Macrocallista* and *Amiantis* in the western Atlantic. *Johnsonia* **1**(3): 1-8.

Coe, W. R. 1904. The anatomy and development of the terrestrial nemertean. *Proc. Boston Soc. Nat. Hist.* **31**: 531-570.

Coe, W. R. 1936. Plankton of the Bermuda Oceanographic Expeditions. VI. Bathypelagic nemerteans taken in the years 1929, 1930 and 1931. *Zoologica* **21**: 97-113.

Coe, W. R. 1943. Biology of the nemerteans of the Atlantic Coast of North America. *Trans. Conn. Acad. Arts Sci.* **35**: 129-328.

Coe, W. R. 1944. The nemertean *Gorgonorhynchus* and the fluctuations of populations. *Am. Nat.* **78**: 94-96.

Coe, W. R. 1945. Plankton of the Bermuda Oceanographic Expeditions. XI. Bathypelagic nemerteans of the Bermuda area and other parts of the North and South Atlantic oceans, with evidence as to their means of dispersal. *Zoologica* **30**: 145-168.

Cohen, A. C. 1976. The systematics and distribution of *Loligo* (Cephalopoda, Myopsida) in the western North Atlantic, with descriptions of two new species. *Malacologia* **15**(2): 299-367.

Cole, L. J. 1904. Pycnogonida collected at Bermuda in the summer of 1903. *Proc. Boston Soc. Nat. Hist.* **31**(8): 315-328, pls. 20-22.

Collette, B. B. 1962. *Hemiramphus bermudensis*, a new halfbeak from Bermuda, with a survey of endemism in Bermudian shore fishes. *Bull. Mar. Sci. Gulf Caribb.* **12**(3): 432-449.

Collins, F. S. & A. B. Hervey. 1917. The algae of Bermuda. *Proc. Am. Acad. Arts Sci.* **53**: 1-195.

Cone, D. K. & M. Beverley-Burton. 1981. The surface topography of *Benedenia* sp. (Monogenea: Capsalidae) *Can. J. Zool.* **59**: 1941-1946.

Congdon, E. D. 1907. The hydroids of Bermuda. *Proc. Am. Acad. Arts Sci.* **42**: 463-485.

Cook, C. B. 1983. Metabolic interchange in algae—invertebrate symbiosis. *Int. Rev. Cytol.* Suppl.**4**: 177-210.

Cook, S. B. 1971. A study of the homing behavior in the limpet *Siphonaria alternata*. *Biol. Bull.* **141**: 449-475.

Cook, S. B. 1976. The role of the "home scar" in pulmonate limpets. *Bull. Am. Malacol. Union Inc.* **1976**: 34-37.

Cook, S. B. & C. B. Cook. 1981. Activity patterns in *Siphonaria* populations: heading choice and the effects of size and grazing interval. *J. Exp. Mar. Biol. Ecol.* **49**: 69-79.

Corliss, J. O. 1979. *The ciliated Protozoa: characterization, classification and guide to the literature.* 2nd ed. London: Pergamon. 455 pp.

Costello, D. P. et al. 1957. *Methods for obtaining and handling marine eggs and embryos.* Woods Hole, Mass.: Marine Biological Laboratory. 247 pp.

Coull, B. C. 1969. *Phyllopodopsyllus hermani*, a new species of harpacticoid copepod from Bermuda. *Crustaceana* **16**(2): 27-32.

Coull, B. C. 1970. Shallow water meiobenthos of the Bermuda platform. *Oecologia* **4**: 325-357.

Courtenay, W. R., Jr. 1961. Western Atlantic fishes of the genus *Haemulon* (Pomadasyidae): systematic status and juvenile pigmentation. *Bull. Mar. Sci. Gulf. Caribb.* **11**(1): 66-149.

Cowden, R. R. 1961. A cytochemical investigation of oogenesis and development to the swimming larval stage in the chiton *Chiton tuberculatum* L. *Biol. Bull.* **120**: 313-325.

Cowden, R. R. 1963. Some histochemical observations on the foot of *Chiton tuberculatum* L. *Trans. Am. Microsc. Soc.* **82**: 406-416.

Cowden, R. R. 1974. Fluorescence cytochemical studies on oocyte growth in *Chiton tuberculatum*, with special emphasis on protein end-group methods. *Differentiation* **2**: 249-255.

Cressey, R. F. 1967. Revision of the family Pandaridae (Copepoda: Caligoida). *Proc. U. S. Nat. Mus.* **121**(3570): 1-133.

Crozier, W. J. 1915. Behavior of an enteropneust. *Science* **41**: 471-472.

Crozier, W. J. 1916. On a barnacle, *Conchoderma virgatum* attached to a fish, *Diodon hystrix*. *Am. Nat.* **50**: 636-640.

Crozier, W. J. 1917. The photic sensitivity of *Balanoglossus*. *J. Exp. Zool.* **24**: 211-217.

Crozier, W. J. 1918a. Growth and duration of life of *Chiton tuberculatus*. *Proc. Nat. Acad. Sci.* **4**: 322-325.

Crozier, W. J. 1918b. Growth of *Chiton tuberculatus* in different environments. *Proc. Nat. Acad. Sci.* **4**: 325-328.

Crozier, W. J. 1918c. The amount of bottom material ingested by holothurians (*Stichopus*). *J. Exp. Zool.* **26**: 379-389.

Crozier, W. J. 1919. Coalescence of the shell-plates in *Chiton*. *Am. Nat.* **53**: 278-279.

Crozier, W. J. 1920. Sex correlated coloration in *Chiton tuberculatus*. *Am. Nat.* **54**: 84-88.

Crozier, W. J. 1921a. Notes on some problems of adaptation. 7. Regarding the pigmentation of *Stichopus moebii*. *Biol. Bull.* **41**: 112-116.

Crozier, W. J. 1921b. "Homing" behavior in *Chiton*. *Am. Nat.* **55**: 276-281.

Crozier, W. J. 1922. An observation on the "cluster formation" of *Chiton*. *Am. Nat.* **56**: 478-480.

Crozier, W. J. & L. B. Arey. 1918. On the significance of the reaction to shading in *Chiton*. *Am. J. Physiol.* **46**: 487-492.

Crozier, W. J. & L. B. Arey. 1919a. *Onchidium* and the question of adaptive coloration. *Am. Nat.* **53**: 415-430.

Crozier, W. J. & L. B. Arey. 1919b. Sensory reactions of *Chromodoris zebra*. *J. Exp. Zool.* **29**: 261-310.

Crozier, W. J. & L. B. Arey. 1920. On the ethology of *Chiton tuberculatus*. *Proc. Nat. Acad. Sci.* **5**: 496-498.

Cuffey, R. J. 1973a. An improved classification, based upon numerical-taxonomic analyses, for the higher taxa of entoproct and ectoproct bryozoans. In: Larwood, G. P. (Ed.). *Living and fossil Bryozoa.* London: Academic. pp. 549-564.

Cuffey, R. J. 1973b. Bryozoan distribution in the

modern reefs of Eniwetok Atoll and the Bermuda platform. *Pac. Geol.* **6**: 25-50.

Cuffey, R. J. 1977. Bryozoa (general characteristics), Cheilostomata, Cryptostomata, Ctenostomata, Cyclostomata (Bryozoa), Cystoporata, Ectoprocta, Entoprocta, Fenestrata, Gymnolaemata, Lophophore, Phylactolaemata, Rhabdomesonata, Stenolaemata, Trepostomata. In: Lapedes, D. N. (Ed.). *McGraw-Hill Encyclopedia of Science and Technology.* 4th ed. New York: McGraw-Hill. Vol. 2, pp. 387-389; Vol. 3, pp. 19-20, 631, 658-659, 695, 702; Vol. 4, p. 453; Vol. 5, pp. 7-8, 229; Vol. 6, p. 344; Vol. 7; p. 662; Vol. 10, p. 221; Vol. 11, p. 570; Vol. 13, p. 133; Vol. 14, pp. 89-90.

Cushman, J. A. 1918-1931. The Foraminifera of the Atlantic Ocean. *U. S. Nat. Mus. Bull.* **104**(1-8): 1-1064.

Dales, R. P. 1963. *Annelida.* London: Hutchinson. 200 pp.

Dall, W. H. 1885. List of marine mollusca from American localities between Cape Hatteras and Cape Roque, including the Bermudas. *Bull. U. S. Geol. Surv.* **24**: 1-336.

Dall, W. H. 1911a. A new genus of bivalves from Bermuda. *Nautilus* **25**: 85-86.

Dall, W. H. 1911b. A new brachiopod from Bermuda. *Nautilus* **25**(8): 86-87.

Dall, W. H. 1921. Annotated list of the recent Brachiopoda in the collection of the United States National Museum. *Proc. U. S. Nat. Mus.* **57**: 261-377.

Dall, W. H. & P. Bartsch. 1911. New species of shells from Bermuda. *Proc. U. S. Nat. Mus.* **40**: 277-288.

Darwin, C. 1851-1854. *A monograph on the sub-class Cirripedia, with figures of all the species.* London: Ray Society. 2 vols: 400 pp., 10 pls.; 684 pp., 30 pls.

Davis, J. D. 1968. A note on the behavior of the scaphopod, *Cadulus quadridentatus* (Dall) 1881. *Proc. Malacol. Soc. Lond.* **38**: 135-138.

Davis, J. D. 1973. Systematics and distribution of western Atlantic *Ervilia* (Pelecypoda: Mesodesmatidae) with notes on living *Ervilia subcancellata. Veliger* **15**: 307-313.

Davis, P. G. et al. 1978. Oceanic amoebae from the North Atlantic: culture, distribution and taxonomy. *Trans. Am. Microsc. Soc.* **97**: 73-88.

Dawes, B. 1946. *The Trematoda, with special reference to British and other European forms.* Cambridge: Cambridge University Press. 644 pp.

Dawes, C. J. 1981. *Marine botany.* New York: Wiley. 628 pp.

Day, J. H. 1967. *A monograph on the Polychaeta of South Africa.* 2 pts. London: Trustees of the British Museum (Natural History). 878 pp.

Deevey, G. B. 1968. Pelagic ostracods of the Sargasso Sea off Bermuda. *Peabody Mus. Nat. Hist. Yale Univ. Bull.* **26**: 1-125.

Deevey, G. B. 1971. The annual cycle in quantity and composition of the zooplankton of the Sargasso Sea off Bermuda. I. The upper 500m. *Limnol. Oceanog.* **16**: 219-240.

Deevey, G. B. & A. L. Brooks. 1977. Copepods of the Sargasso Sea off Bermuda: species composition, and vertical and seasonal distribution between the surface and 2000m. *Bull. Mar. Sci.* **27**(2): 256-291.

Deichmann, E. 1930. Holothurians of the western part of the Atlantic Ocean. *Bull. Mus. Comp. Zool. Harv. Univ.* **71**(3): 43-226, 24 pls.

Dennell, R. 1950a. Note on the feeding of Amphioxus (*Branchiostoma bermudae*). *Proc. Roy. Soc. Edinb. Sect. B (Biol.)* **64**: 229-234.

Dennell, R. 1950b. The occurrence of *Lysiosquilla scabricauda* (Lamarck) in Bermuda. *Proc. Linn. Soc. Lond.* **162**(1): 63-64.

Denton, E. J., J. B. Gilpin-Brown & J. V. Howarth. 1967. On the bouyancy of *Spirula spirula. J. Mar. Biol. Assoc. U. K.* **47**: 181-191.

Desikachary, T. V. 1959. *Cyanophyta.* New Delhi: Indian Council of Agricultural Research. 686 pp.

Dingle, H. 1964. A colour polymorphism in *Gonodactylus oerstedi* Hansen, 1895 (Stomatopoda). *Crustaceana* **7**(3): 236-240.

Dingle, H. 1969a. Ontogenetic changes in phototaxis and thigmokinesis in stomatopod larvae. *Crustaceana* **16**(1): 108-110.

Dingle, H. 1969b. A statistical and information analysis of aggressive communication in the mantis shrimp *Gonodactylus bredini* Manning. *Anim. Behav.* **17**(3): 561-575.

Dingle, H. & R. L. Caldwell. 1969. The aggressive and territorial behaviour of the mantis shrimp *Gonodactylus bredini* Manning (Crustacea; Stomatopoda). *Behaviour* **33**(1/2): 115-136.

Dixon, P. S. 1973. *Biology of the Rhodophyta.* Edinburgh: Oliver & Boyd. 285 pp.

Dohrn, A. 1881. Die Pantopoden des Golfes von Neapel und der angrenzenden Meeresabschnitte. *Fauna Flora Golfes Neapel* **3**: 1-252, pls. 1-17.

Drach, P. 1948. Embranchement des Cephalocordés. In: Grassé, P. P. (Ed.). *Traité de Zoologie.* Vol. 11. Paris: Masson. pp. 931-1040.

Dragesco, J. 1960. Ciliés mésopsammiques littoraux. Systématique, morphologie, écologie. *Trav. Stat. Biol. Roscoff* (N.S.) **12**: 1-356.

Dwight, J. 1927. The "new" Bermuda shearwater proves to be *Puffinus puffinus puffinus. Auk* **44**: 243.

Eales, N. B. 1960. Revision of the world species of *Aplysia* (Gastropoda, Opisthobranchia). *Bull. Brit. Mus. (Nat. Hist.) Zool.* **5**(10): 268-404.

Eason, E. H. 1964. *Centipedes of the British Isles.* London: Warne. 294 pp.

Edmunds, M. 1964. Eolid mollusca from Jamaica, with descriptions of two new genera and three new species. *Bull. Mar. Sci.* **14**: 1-32.

Ehlers, U. 1985. The phylogenetic system of the Platyhelminthes. In Morris, S. C. et al. (Eds.). *The origins and relationships of lower invertebrates*. Oxford: Oxford University Press. pp. 143-158.

Emig, C. C. 1973. Ecologie des Phoronidiens. *Bull. Ecol.* **4**(4): 339-364.

Emig, C. C. 1974. The systematics and evolution of the phylum Phoronida. *Z. Zool. Syst. Evolutionsforsch.* **12**(2): 128-151.

Emig, C. C. 1979. *British and other phoronids: keys and notes for the identification of the species*. London: Academic. 57 pp.

Emig, C. C. 1982. The biology of Phoronida. *Adv. Mar. Biol.* **19**: 1-89.

Ernst, C. H. & R. W. Barbour. 1972. *Turtles of the United States*. Lexington: University Press of Kentucky. 384 pp.

Erséus, C. 1979a. Taxonomic revision of the marine genus *Phallodrilus* Pierantoni (Oligochaeta, Tubificidae), with descriptions of thirteen new species. *Zool. Scr.* **8**: 187-208.

Erséus, C. 1979b. *Bermudrilus peniatus* n. g., n. sp. (Oligochaeta, Tubificidae) and two new species of *Adelodrilus* from the northwest Atlantic. *Trans. Am. Microsc. Soc.* **98**: 418-427.

Esterly, C. O. 1911. Calanoid Copepoda from the Bermuda Islands. *Proc. Amer. Acad. Arts Sci.* **47**(7): 219-226, 4 pls.

Fain, A. & R. Schuster. 1983. New astigmatic mites from the coastal area of the Bermuda Islands (Acari: Hyadesiidae, Saproglyphidae, Acaridae). *Entomol. Mitt. Zool. Mus. Hamburg* **7**(119): 417-434.

Farris, R. A. 1975. Systematics and ecology of Gnathostomulida from North Carolina and Bermuda. Ph.D. thesis, University of North Carolina, Chapel Hill. 246 pp.

Farris, R. A. 1977. Three new species of Gnathostomulida from the West Atlantic. *Int. Rev. gesamten Hydrobiol.* **62**: 765-796.

Fauchald, K. 1977. The polychaete worms. Definitions and keys to the orders, families and genera. *Nat. Hist. Mus. L. A. Sci. Ser.* **28**: 1-188.

Fauchald, K. & P. A. Jumars. 1979. The diet of worms: a study of polychaete feeding guilds. *Oceanogr. Mar. Biol. Ann. Rev.* **17**: 193-284.

Fauvel, P. 1923. Polychétes errantes. *Faune Fr.* **4**: 1-488.

Fauvel, P. 1927. Polychétes sédentaires. *Faune Fr.* **16**: 1-494.

Feddern, H. A. 1972. Fieldguide to the angelfishes (Pomacanthidae) in the western Atlantic. *NOAA Tech. Rept. NMFS Circ.* **369**: 1-10.

Felger, R. S., K. Cliffton & P. J. Regal. 1976. Winter dormancy in sea turtles: independent discovery and exploitation in the Gulf of California by two local cultures. *Science* **191**: 283-285.

Fewkes, J. W. 1883. On a few Medusae from the Bermudas. *Bull. Mus. Comp. Zool. Harv. Univ.* **11**: 79-90.

Fishes of the Western North Atlantic. Parts 1-7. 1948-1977. New Haven: Sears Foundation for Marine Research.

Fogg, G. et al. 1973. *The blue-green algae*. London: Academic. 459 pp.

Fraser, C. M. 1944. *Hydroids of the Atlantic coast of North America*. Toronto: University of Toronto Press. 451 pp.

Freeman, G. 1967. Studies on regeneration in the creeping ctenophore, *Vallicula multiformis*. *J. Morphol.* **123**: 71-84.

Frémy, P. 1934. Les cyanophycées des Côtes d'Europe. *Mem. Soc. Sci. Nat. Math. Cherbourg* **41**: 1-234, 66 pls.

Frick, J. A. 1976. Orientation and behavior of hatchling green turtles (*Chelonia mydas*) in the sea. *Anim. Behav.* **24**: 849-857.

Gaill, F. 1972a. Répartition du genre *Pseudodistoma* (Tuniciers): description de deux espèces nouvelles. *Cah. Biol. Mar.* **13**: 37-47.

Gaill, F. 1972b. Morphologie comparée de la glande pylorique chex quelques aplousobranches (Tuniciers). *Arch. Zool. Exp. Gén.* **113**: 295-307.

Gardiner, L. F. 1975. The systematics, postmarsupial development, and ecology of the deep-sea family Neotanaidae (Crustacea: Tanaidacea). *Smithson. Cont. Zool.* **170**: 1-265.

Garman, S. 1884. Contributions to the natural history of the Bermudas. The reptiles of Bermuda. *U. S. Nat. Mus. Bull.* **25**(5): 285-303.

Garrick, J. A. F. 1982. Sharks of the genus *Carcharhinus*. *NOAA Tech. Rep. NMFS Circ.* **445**: 1-194.

Garstang, W. 1939. Spolia Bermudiana. I. On a remarkable new type of *Auricularia* larva (*A. bermudensis*, n. sp.). *Quart. J. Microsc. Sci.* **81**: 321-345.

Gebelein, C. 1969. Distribution, morphology and accretion rate of recent subtidal algal stromatolites, Bermuda. *J. Sediment. Petrol.* **39**: 49-69.

Geitler, L. 1932. Cyanophyceae. *L. Rabenhorst's Kryptogamenflora* **14**: 1-1196.

Gemerden-Hoogeveen, G. C. H. van. 1965. Hydroids of the Caribbean: Sertulariidae, Plumulariidae and Aglaopheniidae. *Stud. Fauna Curacao other Caribb. Isl.* **22**: 87 pp.

Gerlach, S. A. 1977. Attraction to decaying organisms as a possible cause for patchy distribution of nematodes in a Bermuda beach. *Ophelia* **16**: 151-165.

Gerlach, S. A. & F. Riemann. 1973-74. The Bremerhaven checklist of aquatic nematodes. *Veröff. Inst. Meeresforsch. Bremerhaven* Suppl. **4**: 736 pp.

Gibbs, P. E. 1977. *British sipunculans*. London: Academic. 35 pp.

Gibson, R. 1972. *Nemerteans*. London: Hutchinson. 224 pp.

Giere, O. 1979. Studies on marine Oligochaeta

from Bermuda, with emphasis on new *Phallodrilus* species (Tubificidae). *Cah. Biol. Mar.* **20**: 301-314.

Giere, O. & O. Pfannkuche. 1982. The biology and ecology of marine Oligochaeta. A review. *Oceanogr. Mar. Biol. Ann. Rev.* **20**: 173-308.

Giese, A. C. & J. S. Pearse (Eds.). 1974-1979. *Reproduction of marine invertebrates*. Vols. 1-5. New York: Academic.

Gieskes, W. W. C. 1971. Ecology of the Cladocera of the North Atlantic and the North Sea. *Neth. J. Sea Res.* **5**(3): 342-376.

Giltay, L. 1934. A new pycnogonid from Bermuda. *Bull. Mus. Roy. Hist. Nat. Belgique*: **10**(42): 1-3.

Glaessner, M. F. 1969. Decapoda. In: Moore, R. C. (Ed.). *Treatise on invertebrate paleontology*, Part R; Arthropoda 4(2). Lawrence: University of Kansas and Geological Society of America. pp. 399-651.

Goldschmidt, R. 1933. A note on *Amphioxides* from Bermuda based on Dr. Beebe's collection. *Biol. Bull.* **64**: 321-325.

Golubic, S. 1976. Taxonomy of extant stromatolite-building cyanophytes. In: Walter, M. (Ed.). *Stromatolites*. Developments in Sedimentology 20. Amsterdam: Elsevier. pp. 128-140.

Golubic, S. & J. W. Focke. 1978. *Phormidium hendersonii* Howe: identity and significance of a modern stromatolite building microorganism. *J. Sediment. Petrol.* **48**: 751-764.

Goodbody, I. 1974. The physiology of ascidians. *Adv. Mar. Biol.* **12**: 2-232.

Gosner, K. L. 1971. *Guide to identification of marine and estuarine invertebrates, Cape Hatteras to the Bay of Fundy*. New York: Wiley. 693 pp.

Gould, S. J. 1968a. The molluscan fauna of an unusual Bermudian pond: a natural experiment in form and composition. *Breviora* **308**: 1-13.

Gould, S. J. 1968b. Phenotypic reversion to ancestral form and habit in a marine snail. *Nature* **220**: 804.

Grassé, P.-P. (Ed.) 1965. *Traité de Zoologie*, Tome IV; fasc. 2. Nemathelminthes (Nematodes). Paris: Masson. 731 pp.

Greenberg, I. 1977. *Guide to corals and fishes*. Miami: Seahawk Press. 64 pp.

Grell, K. G. 1973. *Protozoology*. Berlin-Heidelberg: Springer Verlag. 554 pp.

Greve, L. 1974. *Anatanais normani* (Richardson) found near Bermuda and notes on other *Anatanais* species. *Sarsia* **55**: 115-120.

Greven, H. 1980. Die Bärtierchen. In: *Die Neue Brehm-Bücherei*. Wittenberg-Lutherstadt: Ziemsen Verlag. 101 pp.

Grice, G. D. & A. D. Hart. 1962. The abundance, seasonal occurrence and distribution of the zooplankton between New York and Bermuda. *Ecol. Monogr.* **32**: 287-309.

Gross, A. O. 1912. Observations on the yellow-billed tropic-bird (*Phaeton americanus* Grant) at the Bermuda Islands. *Auk* **29**: 49-71.

Gurney, R. 1936. Notes on some decapod Crustacea of Bermuda. III. The larvae of the Palaemonidae. IV. A description of *Processa bermudensis* Rankin and its larvae. V. The first zoea of *Heteractaea ceratopus* (Stimpson). *Proc. Zool. Soc. Lond.* **1936**(37): 619-630, 7 pls.

Gurney, R. 1942. *Larvae of decapod Crustacea*. London: Ray Society. 306 pp.

Gurney, R. 1946. Notes on stomatopod larvae. *Proc. Zool. Soc. Lond.* **116**(1): 133-175.

Gygi, R. A. 1969. Wachstum und Abbau der Korallenriffe um Bermuda. *Veröff. Naturhist. Mus. Basel* **7**: 1-22.

Haas, F. 1941. Malacological notes. III. Additions to the mollusk fauna of Bermuda. *Publ. Field Mus. Nat. Hist., Zool. Ser.* **24**(17): 171-173.

Haas, F. 1943. Additions to the mollusk fauna of Bermuda. *Publ. Field Mus. Nat. Hist., Zool. Ser.* **29**(1): 19-20.

Haas, F. 1949. On some deepsea mollusks from Bermuda. *Bulleti de la Institució Catalana d'Història Nutural* **37**: 3-7.

Haas, F. 1950. On some Bermudian Ellobiidae. *Proc. Malacol. Soc. London* **28**(4): 197-199.

Haas, F. 1952. On the mollusk fauna of the land-locked waters of Bermuda. *Fieldiana: Zool.* **34**: 101-105.

Haddon, A. C. & A. M. Shackleton. 1891. A revision of the British Actiniae. Part 2. The Zoantheae. *Sci. Trans. Roy. Dublin Soc.* (2)**4**(12): 609-672, pls. 58-60.

Haeckel, E. 1887. Report on the Radiolaria collected by H. M. S. Challenger during the years 1873-1876. In: Thompson, Sir C. W. (Ed.). *Report on the scientific results of the voyage of the H.M.S. Challenger during the years 1873-1976*. Zoology, Vol. 18. London: Her Majesty's Stationery Office. 1803 pp., 140 pls.

Haig, J. 1956. The Galatheidea (Crustacea: Anomura) of the Allan Hancock Atlantic Expedition with a review of the Porcellanidae of the western North Atlantic. *Allan Hancock Atlantic Exped. Rep.* **8**: 1-44.

Hale, M. E. 1971. Morden-Smithsonian expedition to Dominica: the lichens (Parmeliaceae). *Smithson. Contrib. Bot.* **4**: 1-25.

Hale, M. E. 1979. *How to know the lichens*. 2nd ed. Dubuque, Iowa: W. C. Brown. 246 pp.

Halstead, B. W. 1978. *Poisonous and venomous marine animals of the world*. Rev. ed. Princeton, N.J.: Darwin Press. 1043 + 283 pp.

Hamner, W. M. et al. 1975. Underwater observations of gelatinous zooplankton: sampling problems, feeding biology and behavior. *Limnol. Oceanog.* **20**(6): 907-917.

Hanson, M. L. 1950. Some digenetic trematodes of marine fishes of Bermuda. *Proc. Helminthol. Soc. Wash.* **17**: 74-89.

Harasewych, M. G. & R. H. Jensen. 1979. Review of the subgenus *Pterynotus* (Gastropoda: Muricidae) in the western Atlantic. *Nemouria* **22**: 1-16.

Harless, M. & H. Morlock (Eds.). 1979. *Turtles: perspectives and research*. New York: Wiley. 695 pp.

Hart, C. W., Jr. & R. B. Manning. 1981. The cavernicolous caridean shrimps of Bermuda (Alpheidae, Hippolytidae and Atyidae). *J. Crust. Biol.* **1**(3): 441-456.

Hartmann, G. 1966-1975. Ostracoda. In: Bronn, H. G. (Ed.). *Klassen und Ordnungen des Tierreichs*, Bd. 5; Abt. I, Lief. 1-4. Leipzig & Heidelberg: C. F. Winter'sche Verlagshandlung. pp. 1-784.

Hartog, C. den. 1959. A key to the species of *Halophila* (Hydrocharitaceae) with description of the American species. *Acta Bot. Neder.* **8**: 484-489.

Hartog, C. den. 1964. The taxonomy of the seagrass genus *Halodule* Endl (Potamogetonaceae). *Blumea* **12**: 289-312.

Hartog, J. C. den. 1977. Descriptions of two new Ceriantharia from the Caribbean region . . . with a discussion of the cnidom and of the classification of Ceriantharia. *Zool. Meded. (Leiden)* **51**(14): 211-242, 6 pls.

Hartog, J. C. den. 1980. Caribbean shallow water Corallimorpharia. *Zool. Verh. Leiden* **176**: 1-83, 14 pls.

Hartwig, E. 1977. On the interstitial ciliate fauna of Bermuda. *Cah. Biol. Mar.* **18**: 113-126.

Hartwig, E. 1980a. A bibliography of the interstitial ciliates (Protozoa): 1926-1979. *Arch. Protistenk.* **123**: 422-428.

Hartwig, E. 1980b. The marine interstitial ciliates of Bermuda, with notes on their geographical distribution and habitat (1). *Cah. Biol. Mar.* **21**: 409-441.

Hazlett, B. A. & H. E. Winn. 1962a. Sound production and associated behavior of Bermuda crustaceans (*Panulirus, Gonodactylus, Alpheus* and *Synalpheus*). *Crustaceana* **4**: 2-38.

Hazlett, B. A. & H. E. Winn. 1962b. Characteristics of a sound produced by the lobster *Justitia longimanus*. *Ecology* **43**: 741-742.

Heath, H. 1911. Reports on the scientific results of the expedition to the tropical Pacific, in charge of Alexander Agassiz, by the U. S. Fish Commission steamer "Albatross", from August, 1899, to June, 1900, Commander Jefferson F. Moser, U. S. N., commanding. XIV. The Solenogastres. *Mem. Mus. Comp. Zool. Harv. Coll.* **45**(1): 1-179, 40 pls.

Heath, H. 1918. Solenogastres from the eastern coast of North America. *Mem. Mus. Comp. Zool. Harv. Univ.* **45**: 183-261, 14 pls.

Hedgpeth, J. W. 1943. Reports on the scientific results of the Atlantis expeditions to the West Indies, under the joint auspices of the University of Havana and Harvard University. Pycnogonida from the West Indies and South America collected by the Atlantis and earlier expeditions. *Proc. New England Zool. Club* **22**: 41-58, pls. 8-10.

Hedgpeth, J. W. 1948. The Pycnogonida of the western North Atlantic and the Caribbean. *Proc. U. S. Nat. Mus.* **97**: 157-342.

Heilprin, A. 1888. Contributions to the natural history of the Bermuda Islands. *Proc. Acad. Nat. Sci. Phil.* **1888**: 302-328, 3 pls.

Heilprin, A. 1889. *The Bermuda Islands: a contribution to the physical history and zoology of the Somers Archipelago. With an examination of the structure of coral reefs*. Philadelphia: Heilprin. 231 pp.

Hemleben, C. et al. 1978. "Dissolution" effects induced by shell reabsorption during gametogenesis in *Hastigerina pelagica* (d'Orbigny). *J. Foram. Res.* **9**(2): 118-124.

Hendelberg, M. 1977. *Paralinhomoeus gerlachi* (Linhomoeidae), a new marine nematode from Bermuda. *Zoon* **5**: 79-86.

Henderson, J. B. 1920. A monograph of the east American scaphopod mollusks. *U. S. Nat. Mus. Bull.* **111**: 1-177.

Henry, D. A. 1958. Intertidal barnacles of Bermuda. *J. Mar. Res.* **17**: 215-234.

Henry, D. A. & P. A. McLaughlin. 1975. The barnacles of the *Balanus amphitrite* complex (Cirripedia, Thoracica). *Zool. Verh. Leiden* **141**: 1-245.

Herman, S. S. & J. R. Beers. 1969. The ecology of inshore plankton populations in Bermuda. Part II. Seasonal abundance and composition of the zooplankton. *Bull. Mar. Sci.* **19**: 483-503.

Herring, P. J. 1967. The pigments of plankton at the sea surface. *Symp. Zool. Soc. Lond.* **19**: 215-235.

Hershkovitz, P. 1966. Catalog of living whales. *U. S. Nat. Mus. Bull.* **246**: 1-259.

Higgins, R. P. 1982a. Kinorhyncha. In: Parker, S. P. (Ed.). *Synopsis and classification of living organisms*. Vol. 1. New York: McGraw-Hill. pp. 873-877.

Higgins, R. P. 1982b. Three new species of Kinorhyncha from Bermuda. *Trans. Am. Microsc. Soc.* **101**(4): 305-316.

Hitchcock, A. S. 1971. *Manual of the grasses of the United States*. New York: Dover Publ. 2 Vols. 1051 pp.

Hoffman, R. L. 1982. Chilopoda. In: Parker, S. P. (Ed.). *Synopsis and classification of living organisms*. Vol. 2. New York: McGraw-Hill. pp. 681-688.

Höglund, N. G. & F. Rahemtulla. 1977. The isolation and characterization of acid mucopolysaccharides of *Chiton tuberculatus*. *Comp. Biochem. Physiol.* **56B**: 211-214.

Holthuis, L. B. 1958. West Indian crabs of the genus *Calappa*, with a description of three new species. *Stud. Fauna Curacao other Caribb. Isl.* **8**: 146-186.

Holthuis, L. B. & J. Zaneveld. 1958. De Kreeften de Nederlandse Antillen. *Zool. Bijdr. Leiden* **3**: 1-26.

Horst, C. J. v. d. 1924. West-Indische Enteropneusten. *Bijdragen Dierkunde Afl.* **23**: 33-60.

Horst, C. J. v. d. 1939. Hemichordata. In: Bronn, H. G. (Ed.). *Klassen und Ordnungen des Tierreichs*. Vol.

4, Abt. 4, Buch 2, Teil 2. Leipzig: Akademische Verlagsgesellschaft M. B. H. 737 pp.

Howe, M. 1918. Algae. In: Britton, N. (Ed.). *Flora of Bermuda*. New York: Scribner. pp. 489-540.

Hoyle, W. E. 1885a. Diagnosis of new species of Cephalopoda collected during the cruise of the H.M.S. Challenger, part I. The Octopoda. *Ann. Mag. Nat. Hist.* ser. 5. **15**: 222-236.

Hoyle, W. E. 1885b. Preliminary report on the Cephalopoda collected during the cruise of the H.M.S. Challenger, part I. The Octopoda. *Proc. Roy. Soc. Edinb.* **13**: 94-113.

Hughes, G. C. 1975. Studies of fungi in oceans and estuaries since 1961. I. Lignicolous, caulicolous and foliicolous species. *Oceangr. Mar. Biol. Ann. Rev.* **13**: 69-180.

Hulburt, E. M., J. H. Ryther & R. R. L. Guillard. 1960. The phytoplankton of the Sargasso Sea off Bermuda. *J. Cons. Int. Explor. Mer* **25**: 115-128.

Hulings, N. C. & J. S. Gray. 1971. A manual for the study of meiofauna. *Smithson. Contrib. Zool.* **78**: 1-83.

Humm, H. J. & S. R. Wicks. 1980. *Introduction and guide to the marine blue-green algae*. New York: Wiley. 194 pp.

Hummon, W. D. 1975. Habitat suitability and the ideal free distribution of Gastrotricha in a cyclic environment. In: Barnes, H. (Ed.). *Proceedings of the 9th European Marine Biology Symposium*. Aberdeen: Aberdeen University Press. pp. 495-525.

Hurley, D. E. 1955. Pelagic amphipods of the suborder Hyperiidea in New Zealand waters. I. Systematics. *Trans. Roy. Soc. New Zeal.* **83**: 119-194.

Hutchinson, J. 1973. *The families of flowering plants, arranged according to a new system based on their probable phylogeny*. 3rd ed. Oxford: Clarendon Press. 968 pp.

Hyman, L. H. 1939. Acoel and polyclad Turbellaria from Bermuda and the Sargassum. *Bull. Bingham Oceangr. Collect. Yale Univ.* **7**: 1-36.

Hyman, L. H. 1940. *The invertebrates: Protozoa through Ctenophora*. New York: McGraw-Hill. 726 pp.

Hyman, L. H. 1951a. *The invertebrates: Platyhelminthes and Rhynchocoela*. New York: McGraw-Hill. 550 pp.

Hyman, L. H. 1951b. *The invertebrates: Acanthocephala, Aschelminthes and Entoprocta. The pseudocoelomate Bilateria*. New York: McGraw-Hill. 572 pp.

Hyman, L. H. 1959. *The invertebrates: smaller coelomate groups*. New York: McGraw-Hill. 783 pp.

Hyman, L. H. 1967. *The invertebrates: Mollusca I*. New York: McGraw-Hill. 792 pp.

Iliffe, T. M. 1982. Argonaut—octopus in a parchment shell. *Sea Frontiers* **28**(4): 225-228.

Illg, P. L. 1958. North American copepods of the family Notodelphyidae. *Proc. U. S. Nat. Mus.* **107**(3390): 463-671.

Ingle, R. W. & P. F. Clark. 1977. A laboratory module for rearing crab larvae. *Crustaceana* **32**(2): 220-222.

Ireland, L. C. 1979. Optokinetic behavior of the hatchling green turtle (*Chelonia mydas*) soon after leaving the nest. *Herpetologica* **35**: 365-369.

Ireland, L. C. 1980. Homing behaviour of juvenile green turtles, *Chelonia mydas*. In: Amlaner, C. J. & D. W. Macdonald (Eds.). *A handbook on biotelemetry and radio tracking*. Oxford: Pergamon. pp. 761-764.

Ireland, L. C. & T. C. Williams. 1974. Radar observations of bird migration over Bermuda. In: Ganthreaux, S. A. (Ed.). *Proceedings of a conference on the biological aspects of the bird/aircraft collision problem*. Clemson, S.C.: Clemson University. pp. 383-408.

Ireland, L. C., J. A. Frick & D. B. Wingate. 1978. Nighttime orientation of hatchling green turtles, *Chelonia mydas*, in open ocean. In: Schmidt-Koenig, K. & W. T. Keeton (Eds.). *Animal migration, navigation and homing: proceedings in the life sciences*. Berlin: Springer-Verlag. pp. 420-429.

Irvine, D. E. G. & J. H. Price. 1978. *Modern approaches to the taxonomy of red and brown algae*. New York: Academic. 484 pp.

Ivanov, A. V. 1963. *Pogonophora*. London: Academic. 479 pp.

Jacobsen, M. K. & W. K. Emerson. 1977. *Wonders of starfish*. New York: Dodd, Mead. 80 pp.

James, N. P. 1970. Role of boring organisms in the coral reefs of the Bermuda platform. In: Ginsburg, R. N. & S. M. Stanley (Eds.). *Reports of research: Seminar on Organism-Sediment Interrelationships, 1969*. Bermuda Biological Station Special Publication No. 6. St. George's West, Bermuda: Bermuda Biological Station. pp. 19-28.

Jamieson, B. G. M. 1971. A review of the megascolecoid earthworm genera (Oligochaeta) of Australia. Part III. The subfamily Megascolecinae. *Mem. Queensl. Mus.* **16**: 69-102.

Jensen, K. R. 1980. A review of Sacoglossan diets, with comparative notes on radular and buccal anatomy. *Malacol. Rev.* **13**(1/2): 55-78.

Jensen, P. & S. A. Gerlach. 1976. Three new marine nematodes from Bermuda. *Veröff. Inst. Meeresforsch. Bremerhaven* **16**: 31-44.

Jensen, P. & S. A. Gerlach. 1977. Three new Nematoda-Comesomatidae from Bermuda. *Ophelia* **16**: 59-76.

Johansen, H. W. 1981. *Coralline algae, a first synthesis*. Boca Raton, Fla.: CRC Press. 239 pp.

Johnson, C. W. 1918. The *Avicula candeana* of d'Orbigny from Bermuda. *Nautilus* **32**: 37-39.

Johnson, T. W., Jr. & F. K. Sparrow, Jr. 1961. *Fungi in oceans and estuaries*. Weinheim: J. Cramer. 668 pp.

Jones, N. S. 1969. The systematics and distribution of Cumacea from depths exceeding 200 meters. *Galathea Rep.* **10**: 99-180.

Jones, N. S. 1976. British Cumaceans. *Synopses British Fauna.* New Ser. **7**: 1-62.

Jones, P. W., F. D. Martin & J. D. Hardy, Jr. 1978. *Development of fishes of the mid-Atlantic Bight.* Washington, D. C.: U. S. Dept. of the Interior; Fish and Wildlife Service. 6 Vols.

Jordan, D. S. & B. W. Evermann. 1896-1900. The fishes of North and Middle America: a descriptive catalogue of the species of fish-like vertebrates found in the waters of North America, north of the Isthmus of Panama. Parts I-IV. *U. S. Nat. Mus. Bull.* **47**: 1-3, 313, 392 pls.

Jouin, C. 1971. Status of the knowledge of the systematics and ecology of Archiannelida. *Smithson. Contr. Zool.* **76**: 47-56.

Kaestner, A. 1970. *Invertebrate zoology.* Vol. 1: Porifera, Cnidaria, Platyhelminthes, Aschelminthes, Mollusca, Annelida & related phyla. (Repr. of 1967 ed.) Melbourne, Fla.: Krieger. 597 pp.

Kaestner, A. 1979. *Invertebrate zoology.* Vol. 2: Arthopod relatives, Chelicerata, Myriapoda. (Repr. of 1968 ed.) Melbourne, Fla.: Krieger. 482 pp.

Kaestner, A. 1980. *Invertebrate zoology.* Vol. 3: Crustacea. (Repr. of 1970 ed.) Melbourne, Fla.: Krieger. 538 pp.

Kahl, A. 1930-1935. Urtiere oder Protozoa. I: Wimpertiere oder Ciliata (Infusoria), eine Bearbeitung der freilebenden und ectocommensalen Infusorien der Erde, unter Ausschluss der marinen Tintinnidae. In: Dahl, F. (Ed.). *Die Tierwelt Deutschlands.* Pts. 18, 21, 25, 30. Jena: G. Fischer. pp. 1-886.

Kanwisher, J. W. & S. H. Ridgway. 1983. The physiological ecology of whales and porpoises. *Sci. Am.* **248**(6): 102-111.

Kapraun, D. F. 1980. *An illustrated guide to the benthic marine algae of coastal North Carolina. I. Rhodophyta.* Chapel Hill: University of North Carolina Press. 206 pp.

Karling, T. G. 1978. Anatomy and systematics of marine Turbellaria from Bermuda. *Zool. Scr.* **7**: 225-248.

Katona, S. K. et al. 1980. *Humpback whales. A catalogue of individuals identified by fluke photographs.* Bar Harbour, Me.: College of the Atlantic. 169 pp.

Kensley, B. 1980. Notes on *Axiopsis (Axiopsis) serratifrons* (A. Milne Edwards) (Crustacea: Decapoda: Thalassinidea). *Proc. Biol. Soc. Wash.* **93**(4): 1253-1263.

Khan, R. A. & M. C. Meyer. 1976. Taxonomy and biology of some Newfoundland marine leeches (Rhynchobdellae: Piscicolidae). *J. Fish. Res. Board Can.* **33**: 1699-1714.

Kind, C. A. & R. A. Meigs. 1955. Sterols of marine mollusks. IV. Delata$^7$-cholestenol, the principal sterol of *Chiton tuberculatus. J. Org. Chem.* **20**: 1116-1118.

King, P. E. 1973. *Pycnogonids.* London: Hutchinson. 144 pp.

Kirsteuer, E. & K. Rützler. 1973. Additional notes on *Tubiluchus corallicola* (Priapulida), based on scanning electron microscope observations. *Mar. Biol.* **20**: 78-87.

Kofoid, C. A. & O. Swezy. 1921. *The free-living unarmored Dinoflagellata.* Berkeley: University of California Press. 562 pp.

Kohlmeyer, J. 1971. Fungi from the Sargasso Sea. *Mar. Biol.* **8**(4): 344-350.

Kohlmeyer, J. 1972. Parasitic *Haloguignardia oceanica* (Ascomycetes) and hyperparasitic *Sphaceloma cecidii* sp. n. (Deuteromycetes) in drift *Sargassum* in North Carolina. *J. Elisha Mitchell Sci. Soc.* **88**: 255-259.

Kohlmeyer, J. & E. Kohlmeyer. 1964-1969. *Icones Fungorum Maris.* Weinheim: J. Cramer. Fasc. 1-7, tabs. 1-90.

Kohlmeyer, J. & E. Kohlmeyer. 1972. Permanent microscopic mounts. *Mycologia* **64**: 666-669.

Kohlmeyer, J. & E. Kohlmeyer. 1977. Bermuda marine fungi. *Trans. Br. Mycol. Soc.* **68**: 207-219.

Kohlmeyer, J. & E. Kohlmeyer. 1979. *Marine mycology: the higher fungi.* New York: Academic. 690 pp.

Kornicker, L. S. 1981. Benthic marine Cypridinoidea from Bermuda (Ostracoda). *Smithson. Contrib. Zool.* **331**: 1-15.

Kozloff, E. N. 1969. Morphology of the orthonectid *Rhopalura ophiocomae. J. Parasitol.* **55**(1): 171-195.

Kozloff, E. N. 1971. Morphology of the orthonectid *Ciliocincta sabellariae. J. Parasitol.* **57**: 585-597.

Kramp, P. L. 1959. The hydromedusae of the Atlantic Ocean and adjacent waters. *Dana Rep.* **46**: 1-283.

Kramp, P. L. 1961. Synopsis of the medusae of the world. *J. Mar. Biol. Assoc. U. K.* **40**: 1-469.

Krantz, G. W. 1978. *A manual of acarology.* 2nd ed. Corvallis: Oregon State University Book Stores. 509 pp.

Kristensen, R. M. 1983. Loricifera, a new phylum with Aschelminthes characters from the meiobenthos. *Zeitsch. Zool. Syst. Evolutionsforsch.* **21**: 163-180.

Kristensen, R. M. & T. E. Hallas. 1980. The tidal genus *Echiniscoides* and its variability, with erection of Echiniscoididae fam. n. *Zool. Scr.* **9**: 113-127.

Kudo, R. R. 1966. *Protozoology.* 5th ed. Springfield, Ill.: Charles C. Thomas. 1188 pp.

Kunkel, B. W. 1910. The Amphipoda of Bermuda. *Trans. Conn. Acad. Arts Sci.* **16**: 1-116.

Kylin, H. 1956. *Die Gattungen der Rhodophyceen.* Lund: Gleerups Förlag. 673 pp.

Laborel, J. 1966. Contribution à l'étude des Madréporaires des Bermudes. *Bull. Mus. Natn. Hist. Nat.* (2)**38**(3): 281-300.

Lagler, K. F. et al. 1977. *Ichthyology.* New York: Wiley. 506 pp.

Land, J. van der. 1970. Systematics, zoogeography and ecology of the Priapulida. *Zool. Verh. Leiden* **112**: 1-118.

Lang, K. 1948. *Monographie der Harpacticiden.* Lund: Hakan Ohlssons Boktryckeri. 1682 pp.

LaRoe, E. T. 1971. The culture and maintenance of the loliginid squids *Sepioteuthis sepioidea* and *Doryteuthis plei. Mar. Biol.* **9**(1): 9-25.

Larwood, G. P. (Ed.). 1973. *Living and fossil Bryozoa.* London: Academic. 634 pp.

Lasserre, P. & C. Erséus. 1976. Oligochètes marins des Bermudes. Nouvelles espèces et remarques sur la distribution géographique de quelques Tubificidae et Enchytraeidae. *Cah. Biol. Mar.* **17**: 447-462.

Laubenfels, M. W. de. 1950. The Porifera of the Bermuda Archipelago. *Trans. Zool. Soc.* **27**(1): 1-154.

Lawrence, G. H. M. 1951. *Taxonomy of vascular plants.* New York: Macmillan. 823 pp.

Lebour, M. V. 1941. Notes on thalassinid and processid larvae (Crustacea Decopoda) from Bermuda. *Ann. Mag. Nat. Hist.* ser. **11**(7): 401-420.

Lebour, M. V. 1944. Larval crabs from Bermuda. *Zoologica* **29**: 113-128.

Lebour, M. V. 1945. The eggs and larvae of some prosobranchs from Bermuda. *Proc. Zool. Soc. Lond.* **114**: 462-489.

Lebour, M. V. 1949. Some euphausids from Bermuda. *Proc. Zool. Soc. Lond.* **119**(4): 823-837.

Lebour, M. V. 1950a. Notes on some larval decapods (Crustacea) from Bermuda. *Proc. Zool. Soc. Lond.* **120**(2): 369-379.

Lebour, M. V. 1950b. Notes on some larval decapods (Crustacea) from Bermuda. II. *Proc. Zool. Soc. Lond.* **120**(4): 743-747.

Lee, J. J. 1980. Nutrition and physiology of the Foraminifera. In: Levandowsky, M. & S. Hutner (Eds.). *Biochemistry and physiology of Protozoa.* Vol. 3, 2nd ed. New York: Academic. pp. 43-66.

Leedale, G. F. 1967. *Euglenoid flagellates.* Englewood Cliffs, N.J.: Prentice-Hall. 242 pp.

Lester, S. M. 1985. *Cephalodiscus* sp. (Hemichordata: Pterobranchia): observations of functional morphology, behavior and occurrence in shallow water around Bermuda. *Mar. Biol.* **85**: 263-268.

Lewin, R. A. 1981. *Prochloron* and the theory of symbiogenesis. *Ann. N. Y. Acad. Sci.* **361**: 325-329.

Lincoln, R. J. & J. G. Sheals. 1979. *Invertebrate animals. Collection and preservation.* London British Museum (Nat. Hist.) & Cambridge University Press. 150 pp.

Linton, E. 1889. Notes on Entozoa of marine fishes of New England with descriptions of several new species. *Rep. U. S. Fish. Comm. for 1886*: 453-511, 6 pls.

Linton, E. 1905. Parasites of fishes of Beaufort, North Carolina. *Bull. U. S. Fish Comm. for 1904*: 321-428, 35 pls.

Linton, E. 1907. Notes on parasites of Bermuda fishes. *Proc. U. S. Nat. Mus.* **33**: 85-126.

Linton, E. 1910. Helminth fauna of the dry Tortugas. II. Trematoda. *Carnegie Inst. Wash. Publ.* **133**: 11-98, 28 pls.

Lobban, C. S. & M. J. Wynne. 1981. *The Biology of seaweeds.* Berkeley: University of California Press. 786 pp.

Loeblich, A. R. III. 1982. Dinophyceae. In: Parker, S. P. (Ed.). *Synopsis and classification of living organisms.* Vol. 1. New York: McGraw-Hill. pp. 101-115.

Loeblich, A. R., Jr. & H. Tappan. 1964. Sarcodina, chiefly "Thecamoebians" and Foraminiferida. In: Moore, R. C. (Ed.). *Treatise on invertebrate paleontology,* Part C; Protista 2 (1&2). Lawrence: University of Kansas Press and Geological Society of America. 900 pp.

Logan, A. 1974. Morphology and life habits of the recent cementing bivalve *Spondylus americanus* Hermann from the Bermuda platform. *Bull. Mar. Sci.* **24**: 568-594.

Logan, A. 1975. Ecological observations on the recent articulate brachiopod *Argyrotheca bermudana* Dall from the Bermuda platform. *Bull. Mar. Sci.* **25**: 186-204.

Lohmann, H. 1933. Appendicularia. In: Kükenthal, W. & T. Krumbach (Eds.). *Handbuch der Zoologie.* Vol. 5, 2nd half. Berlin: Walter de Gruyter. pp. 1-202.

Lucas, M. S. 1934. Ciliates from Bermuda sea urchins. I. *Metopus. J. Roy. Micros. Soc.* **54**: 79-93.

Lucas, M. S. 1940. *Cryptochilidium (Cryptochilum) bermudense,* a ciliate from Bermuda sea urchins. *J. Morphol.* **66**: 369-390.

Lukas, K. J. 1974. Two species of the chlorophyte genus *Ostreobium* from skeletons of Atlantic and Caribbean reef corals. *J. Phycol.* **10**: 331-335.

McCain, J. C. 1968. The Caprellidae (Crustacea: Amphipoda) of the western North Atlantic. *U. S. Nat. Mus. Bull.* **278**: 1-147.

McConnaughey, B. H. 1951. The life cycle of the dicyemid Mesozoa. *Univ. Calif. Pub. Zool.* **55**: 295-336.

McConnaughey, B. H. 1963. The Mesozoa. In: Dougherty, E. C. (Ed.). *The lower Metazoa.* Berkeley: University of California Press. pp. 151-168.

McLaughlin, P. A. 1975. On the identity of *Pagurus brevidactylus* (Stimpson) (Decapoda: Paguridae), with the description of a new species of *Pagurus* from the western Atlantic. *Bull. Mar. Sci.* **25**(3): 359-376.

Mackie, G. O. 1960. Studies on *Physalia physalis* (L.). Part 2. Behaviour and histology. *Discovery Rep.* **30**: 369-407.

Maddocks, R. G. 1969. Revision of recent Bairdiidae (Ostracoda). *U. S. Nat. Mus. Bull.* **295**: 1-126.

Maddocks, R. G. 1973. Zenker's organ and new species of *Saipanetta* (Ostracoda). *Micropaleontology* **19**: 193-208.

Maddocks, R. G. 1976. Pussellinae are interstitial Bairdiidae (Ostracoda). *Micropaleontology* **22**: 194-214.

Mahnert, V. & R. Schuster. 1981. *Pachyolpium atlanticum* n. sp., ein Pseudoskorpion (Arachnida) aus der Gezeitenzone der Bermudas—Morphologie und Ökologie (Pseudoscorpiones: Olpiidae). *Rev. Suisse Zool.* **88**: 265-273.

Mahoney, R. 1973. *Laboratory techniques in zoology.* 2nd ed. London: Butterworths. 518 pp.

Manning, R. B. 1969. *Stomatopod Crustacea of the western Atlantic.* Stud. Trop. Oceanogr. No. 8. Coral Gables: University of Miami Press. 380 pp.

Manning, R. B. 1972. *Gonodactylus spinulosus* Schmitt, a West Indian stomatopod new to Bermuda. *Crustaceana* **23**(3): 315.

Manning, R. B. & A. J. Provenzano, Jr. 1963. Studies on development of stomatopod Crustacea. I. Early larval stages of *Gonodactylus oerstedii* Hansen. *Bull. Mar. Sci. Gulf Caribb.* **13**(3): 467-487.

Manton, S. M. 1977. *The Arthropoda: habits, functional morphology and evolution.* Oxford: Clarendon Press. 527 pp.

Marcus, E. 1946. *Batillipes pennaki*, a new marine tardigrade from the north and south American Atlantic coast. *Comm. Zool. Mus. Hist. Nat. Montevideo* **2**: 1-3.

Marcus, E. 1955. Opisthobranchia from Brazil. *Bol. Fac. Filos. Cienc. Let. Univ. Sao Paulo Ser. Zool.* **20**: 89-262.

Marcus, E. 1957. On Opisthobranchia from Brazil (2). *J. Linn. Soc. Lond. Zool.* **43**: 390-486.

Marcus, E. d. B.-R. 1977. An annotated checklist of the western Atlantic warm water opisthobranchs. *J. Moll. Stud. Suppl.* **4**: 1-22.

Marcus, E. d. B.-R. & E. Marcus. 1963. Opisthobranchs from the Lesser Antilles. *Stud. Fauna Curacao other Caribb. Isl.* **19**: 1-70.

Marcus, E. d. B.-R. & E. Marcus. 1967. *American opisthobranch molluscs.* Coral Gables: University of Miami Press. 256 pp.

Marcus, E. d. B.-R. & E. Marcus. 1970. Opisthobranchs from Curacao and faunistically related regions. *Stud. Fauna Curacao other Caribb. Isl.* **33**: 1-129.

Margulis, L. & K. V. Schwartz. 1982. *Five kingdoms. An illustrated guide to the phyla of life on earth.* San Francisco: W. H. Freeman. 338 pp.

Mark, E. L. 1924. Marine borers in Bermuda. *Proc. Am. Acad. Arts Sci.* **59**: 257-276.

Markert, R. E., B. J. Markert & N. J. Vertrees. 1961. Lunar periodicity in spawning and luminescence in *Odontosyllis enopla*. *Ecology* **42**: 414-415.

Markham, J. C. 1977. Preliminary note on the ecology of *Calcinus verrilli*, an endemic Bermuda hermit crab occupying attached vermetid shells. *J. Zool.* **181**: 131-136.

Markham, J. C. 1979. Epicaridean isopods of Bermuda. *Bull. Mar. Sci.* **29**(4): 522-529.

Markham, J. C. & J. J. McDermott. 1980. A tabulation of the Crustacea Decapoda of Bermuda. *Proc. Biol. Soc. Wash.* **93**(4): 1266-1276.

Marshall, N. B. 1971. *Exploration in the life of fishes.* Cambridge: Harvard University Press. 204 pp.

Marshall, S. M. & A. P. Orr. 1955. *The biology of a marine copepod.* Edinburgh: Oliver & Boyd. 188 pp.

Matthews, L. H. (Ed.) 1968. *The whale.* New York: Simon & Schuster. 287 pp.

Mauchline, J. 1971a. Euphausiacea (adults). In: *Fiches d'identification du zooplancton.* Copenhagen: Conseil International pour l'Exploration de la Mer. Sheet **134**: 1-8.

Mauchline, J. 1971b. Euphausiacea (larvae). In: *Fiches d'Identification du zooplancton.* Copenhagen: Conseil International pour l'Exploration de la Mer. Sheet **135/137**: 1-16.

Mauchline, J. 1980. The biology of mysids and euphausids. *Adv. Mar. Biol.* **18**: 1-677.

Mauchline, J. & L. R. Fisher. 1969. The biology of euphausids. *Adv. Mar. Biol.* **7**: 1-454.

Mayer, A. G. 1910a. Medusae of the world. I. The hydromedusae. *Carnegie Inst. Wash. Pub.* **109**: 1-230.

Mayer, A. G. 1910b. Medusae of the world. II. The hydromedusae. *Carnegie Inst. Wash. Pub.* **109**: 231-498.

Mayer, A. G. 1910c. Medusae of the world. III. The scyphomedusae. *Carnegie Inst. Wash. Pub.* **109**: 499-735.

Mayer, A. G. 1912. Ctenophores of the Atlantic coast of North America. *Carnegie Inst. Wash. Pub.* **162**: 1-58.

Meserve, F. G. 1934. A new genus and species of parasitic Turbellaria from a Bermuda sea cucumber. *J. Parasitology* **20**: 270-276.

Metcalf, M. M. 1932. The Salpidae: a taxonomic study. *U. S. Nat. Mus. Bull.* **100**(2): 5-193, pls. 1-14.

Metcalf, M. M. & H. S. Hopkins. 1932. *Pyrosoma*: a taxonomic study, based upon the collections of the United States Bureau of the Fisheries and the United States National Museum. *U. S. Nat. Mus. Bull.* **100**(2): 195-275, pls. 15-36.

Meyer, M. C. & R. A. Khan. 1979. Taxonomy, biology and occurrence of some marine leeches in Newfoundland waters. *Proc. Helminthol. Soc. Wash.* **46**(2): 254-264.

Millar, R. H. 1971. The biology of ascidians. *Adv. Mar. Biol.* **9**: 1-100.

Millard, N. A. H. 1975. Monograph on the Hydroida of southern Africa. *Ann. South African Mus.* **68**: 1-513.

Miller, J. D. 1984. Marine fungi in Bermuda ecosystems. *Monthly Bull., Bermuda Dept. Agriculture & Fisheries* **55**(3): 18-22.

Monniot, C. 1972a. Ascidies phlébobranches des Bermudes. *Bull. Mus. Natl. Hist. Nat. 3ᵉ sér.* **82**: 939-948.

Monniot, C. 1972b. Ascidies stolidobranches des Bermudes. *Bull. Mus. Natl. Hist. Nat.* 3ᵉ sér. **57**: 617-643.

Monniot, C. 1983. Ascidies littorales de Guadeloupe. II. Phlébobranches. *Bull. Mus. Natl. Hist. Nat.* 4ᵉ sér. **5**: 51-71.

Monniot, C. &. F. Monniot. 1978. Recent work on the deep-sea tunicates. *Oceanogr. Mar. Biol. Ann. Rev.* **16**: 181-228.

Monniot, F. 1972a. Ascidies aplousobranches des Bermudes. Polyclinidae et Polycitoridae. *Bull. Mus. Natl. Hist. Nat.* 3ᵉ sér. **82**: 959-962.

Monniot, F. 1972b. *Scaptognathides*, un noveau genre d'Acariens marins (Halacaridae) dans l'Atlantique ouest. *Acarologia* **13**: 361-366.

Monniot, F. 1983. Ascidies littorales de Guadeloupe. I. Didemnidae. *Bull. Mus. Natl. Hist. Nat.* 4ᵉ sér. **5**: 5-49.

Moore, H. B. 1949. The zooplankton of the upper waters of the Bermuda area of the North Atlantic. *Bull. Bingham Oceangr. Collect. Yale Univ.* **12**(2): 1-97.

Moore, H. B. 1951. The seasonal distribution of oceanic birds in the western North Atlantic. *Bull. Mar. Sci. Gulf Caribb.* **1**: 1-14.

Moore, H. B. & D. M. Moore. 1946. Preglacial history of Bermuda. *Geol. Soc. Am. Bull.* **57**: 207-222.

Moore, R. C. (Ed.). 1961. *Treatise on invertebrate paleontology*. Part Q; Arthropoda 3 (Crustacea, Ostracoda). Lawrence: University of Kansas Press and Geological Society of America. 442 pp.

Moore, R. C. (Ed.). 1965. *Treatise on invertebrate paleontology*. Part H; Brachiopoda. Lawrence; University of Kansas Press and Geological Society of America. 927 pp.

Morgan, C. I. 1982. Tardigrada. In: Parker, S. P. (Ed.). *Synopsis and classification of living organisms*. Vol. 2. New York: McGraw-Hill. pp. 731-739.

Morkhoven, F. P. C. M. v. 1962, 1963. *Postpalaeozoic Ostracoda*. 2 Vols. New York: Elsevier. 204 pp., 478 pp.

Morreales, S. J. et al. 1982. Temperature-dependent sex determination: current practices threaten conservation of sea turtles. *Science* **216**: 1245-1247.

Morris, B. F. 1975. The neuston of the northwest Atlantic. Ph.D. thesis. Dalhousie University. 297 pp.

Morris, B. F. & D. D. Mogelberg. 1973. *Identification manual to the pelagic Sargassum fauna*. Bermuda Biological Station Special Publication No. 11. St. George's West, Bermuda: Bermuda Biological Station. 63 pp.

Mortensen, T. 1928-51. *A monograph of the Echinoidea*. Copenhagen: C. A. Reitzel. Vols. 1-5.

Mowbray, L. S. 1965. *A guide to the reef, shore and game fish of Bermuda*. Hamilton, Bermuda: L. S. Mowbray. 64 pp.

Mowbray, L. S. & D. K. Caldwell. 1958. First record of the Ridley turtle from Bermuda, with notes on other sea turtles and the turtle fishery in the islands. *Copeia* **2**: 147-148.

Murphy, R. C. & L. S. Mowbray. 1951. New light on the Cahow, *Pterodroma cahow*. *Auk* **68**: 266-280.

Neumann, G. 1935a. Pyrosomida (Tunicata). In: Kükenthal, W. & T. Krumbach (Eds.). *Handbuch der Zoologie*. Vol. 5, Part 2. Berlin: Walter de Gruyter. pp. 226-323.

Neumann, G. 1935b. Cyclomyaria (Tunicata). In: Kükenthal, W. & T. Krumbach (Eds.). *Handbuch der Zoologie*. Vol. 5, Part 2. Berlin: Walter de Gruyter. pp. 324-400.

Newell, I. M. 1947. A systematic and ecological study of the Halacaridae of eastern North America. *Bull. Bingham Oceanogr. Collect. Yale Univ.* **10**: 1-232.

Newman, W. A. & A. Ross. 1976. Revision of the balanomorph barnacles; including a catalog of the species. *San Diego Soc. Nat. Hist. Mem.* **9**: 1-108.

Newman, W. A. &. A. Ross. 1977. A living *Tesseropora* (Cirripedia, Balanomorpha) from Bermuda and the Azores; first records from the Atlantic since the Oligocene. *San Diego Soc. Nat. Hist. Mem.* **18**(12): 207-216.

Newman, W. A., V. A. Zullo & T. H. S. Withers. 1969. Cirripedia. In: Moore, R. G. (Ed.). *Treatise on invertebrate paleontology*. Part R; Arthropoda 4(1). Lawrence: University of Kansas Press and Geological Society of America. pp. 206-295.

Nichols, D. 1969. *Echinoderms*. London: Hutchinson. 192 pp.

Nichols, J. T. & L. S. Mowbray. 1916. Two new forms of petrels from the Bermudas. *Auk* **33**: 194-195.

Nicol, D. 1956. Distribution of living glycymerids with a new species from Bermuda. *Nautilus* **70**: 48-53.

Nielsen, C. 1977. The relationships of Entoprocta, Ectoprocta and Phoronida. *Am. Zool.* **17**: 149-150.

Nielsen, C. O. & B. Christensen. 1959. The Enchytraeidae. Critical revision and taxonomy of European species. *Natura Jutlandica* **8-9**: 1-160.

Norris, K. S. (Ed.). 1966. *Whales, dolphins and porpoises*. Berkeley: University of California Press. 789 pp.

Nosek, J. 1981. A new japygid species from Bermudas, *Parajapyx schusteri* n. sp. (Insecta, Diplura). *Rev. Suisse Zool.* **88**: 327-332.

Ogilvie, L. 1928. *The insects of Bermuda*. Hamilton, Bermuda: Bermuda Department of Agriculture. 52 pp.

Olive, L. S. 1975. *The Mycetozoans*. New York: Academic. 293 pp.

Opresko, D. M. 1972. Redescriptions and reevaluations of the Antipatharians described by L. F. Pourtalés. *Bull. Mar. Sci.* **22**(4): 950-1017.

Opresko, L. et al. 1973. *Guide to the lobsters and lobster-like animals of Florida, the Gulf of Mexico and the Caribbean region.* Sea Grant Field Guide Series No. 1. Miami: University of Miami Sea Grant Program. 44 pp.

Osburn, R. C. 1940. Bryozoa of Porto Rico with a resume of the West Indian bryozoan fauna. *New York Acad. Sci., Sci. Surv. Porto Rico and Virgin Isl.* **16**(3): 321-486.

Ott, J. A. 1977. New freeliving marine nematodes from the west Atlantic. I. Four new species from Bermuda with a discussion of the genera *Cytolaimium* and *Rhabdocoma* Cobb 1920. *Zool. Anz.* **198**: 120-138.

Owre, H. B. & M. Foyo. 1967. *Copepods of the Florida current.* Fauna Caribaea No. 1, Crustacea; Part 1, Copepoda. Miami: Institute of Marine Sciences, University of Miami. 137 pp.

Parker, G. H. 1914. The locomotion of chiton. *Contrib. Bda. Biol. Stn. Res.* O.S. **3**(31): 1-2.

Parker, S. P. (Ed.). 1982. *Synopsis and classification of living organisms.* New York: McGraw-Hill. 2 Vols. 1166 pp., 1232 pp.

Pawson, D. L. & D. M. Devaney. (in press). The echinoderms of Bermuda. *Smithson. Contrib. Mar. Sci.*

Payne, K. & R. Payne (in press). Annual changes in songs of humpback whales. *Z. Tierpsychol.*

Payne, R. S. 1970. *Songs of the humpback whale.* (Sound record). Hollywood: Capitol Records.

Payne, R. S. & S. McVay. 1971. Songs of the humpback whale. *Science* **173**: 585-597.

Peile, A. J. 1926. The mollusca of Bermuda. *Proc. Malacol. Soc.* **17**: 71-98.

Peréz Farfante, I. 1969. Western Atlantic shrimps of the genus *Penaeus. Bull. Fish. Wildl. Serv.* **67**(3): 461-591.

Peréz Farfante, I. 1971. Western Atlantic shrimps of the genus *Metapenaeopsis* (Crustacea, Decapoda, Penaeidae), with descriptions of three new species. *Smithson. Contrib. Zool.* **79**: 1-37.

Pestana, H. R. 1983. Discovery of and sediment production by *Carpenteria* (Foraminifera) on the Bermuda platform. *Bull. Mar. Sci.* **33**(2): 509-512.

Peterson, R. T. 1960. *A field guide to the birds.* Boston: Houghton Mifflin. 290 pp.

Phillips, R. C. & C. P. McRoy (Eds.). 1980. *Handbook of seagrass biology: an ecosystem perspective.* New York: Garland. 353 pp.

Pickford, G. E. 1945. Le poulpe americain: a study of the littoral Octopoda of the western Atlantic. *Trans. Conn. Acad. Arts Sci.* **36**: 701-811, 14 pls.

Pickford, G. E. 1946. *Vampyroteuthis infernalis* Chun, an archaic dibranchiate cephalopod. I. Natural history and distribution. *Dana Rep.* **29**: 1-40.

Pickford, G. E. 1949. *Vampyroteuthis infernalis* Chun, an archaic dibranchiate cephalopod. II. External anatomy. *Dana Rep.* **32**: 1-132.

Pilsbry, H. A. 1907. The sessile barnacles (Cirripedia) contained in the collections of the U. S. National Museum. *U. S. Nat. Mus. Bull.* **60**: 1-122, 11 pls.

Pilsbry, H. A. 1916. The sessile barnacles (Cirripedia) contained in the collections of the U. S. National Museum; including a monograph on the American species. *U. S. Nat. Mus. Bull.* **93**: 1-366, 75 pls.

Plough, H. H. 1978. *Sea squirts of the Atlantic continental shelf from Maine to Texas.* Baltimore: Johns Hopkins University Press. 118 pp.

Pollock, L. W. 1976. Marine flora and fauna of the northeastern United States. Tardigrada. *NOAA Tech. Rep. NMFS Circ.* **394**: 1-25.

Pouyet, S. (Ed.). 1975. Bryozoa 1974. *Docum. Lab. Geol. Fac. Sci. Lyon* h.s. **3**(1/2): 1-690.

Prescott, G. W. 1968. *The algae: a review.* Boston: Houghton Mifflin. 436 pp.

Pritchard, P. C. H. 1979. *Encyclopedia of turtles.* Neptune City, N.J.: TFH Pub. 895 pp.

Provenzano, A. J. 1959. The shallow-water hermit crabs of Florida. *Bull. Mar. Sci. Gulf Caribb.* **9**(4): 349-420.

Provenzano, A. J. 1960. Notes on Bermuda hermit crabs. *Bull. Mar. Sci. Gulf Caribb.* **10**(1): 117-124.

Provenzano, A. J. & R. B. Manning. 1978. Studies on development of stomatopod Crustacea. II. The later larval stages of *Gonodactylus oerstedii* Hansen reared in the laboratory. *Bull. Mar. Sci.* **28**(2): 297-315.

Raecke, M. J. 1945. A new genus of monogenetic trematode from Bermuda. *Trans. Am. Microsc. Soc.* **64**: 300-305.

Ramazzotti, G. 1972. Il phylum Tardigrada. 2nd. ed. *Mem. Ist. Ital. Idrobiol.* **28**: 1-732.

Rammner, W. 1939. Cladocera. In: *Fiches d'identification du zooplancton.* Copenhagen: Conseil International pour l'Exploration de la Mer. Sheet 3.

Rand, T. & M. Wiles. 1985. The histopathology of *Learedius learedi* Price 1934 and *Neospirorchis schistosomatoides* Price 1934 (Digenea: Spirorchiidae) infestation in green turtles, *Chelonia mydas* L., from Bermuda. *J. Wildl. Dis.* **21**(10).

Rand, T., M. Wiles & P. Odense. (in press). The attachment of *Dermophthirius carcharhini* MacCallum, 1926 (Monogenea: Microbothriidae) to the Galapagos shark *Carcharhinus galapagensis. Trans. Am. Microsc. Soc.*

Randall, J. E. 1964. Contributions to the biology of the queen conch, *Strombus gigas. Bull. Mar. Sci.* **14**(2): 246-295.

Randall, J. E. 1968. *Caribbean reef fishes.* Neptune City, N.J.: TFH Pub. 318 pp.

Rankin, J. J. 1956. The structure and biology of

*Vallicula multiformis* gen. et sp. nov., a platyctenid ctenophore. *J. Linn. Soc. Lond. Zool.* **43**: 55-71.

Rathbun, M. J. 1918. The grapsoid crabs of America. *U. S. Nat. Mus. Bull.* **97**: 1-461.

Rathbun, M. J. 1925. The spider crabs of America. *U. S. Nat. Mus. Bull.* **129**: 1-598.

Rathbun, M. J. 1930. The cancroid crabs of America. *U. S. Nat. Mus. Bull.* **152**: 1-609.

Rathbun, M. J. 1937. The oxystomatous and allied crabs of America. *U. S. Nat. Mus. Bull.* **166**: 1-278.

Rebel, T. P. 1974. *Sea turtles and the turtle industry of the West Indies, Florida and the Gulf of Mexico.* Coral Gables: University of Miami Press. 250 pp.

Rees, G. 1969. Cestodes from Bermuda fishes and an account of *Acompsocephalum tortum* (Linton, 1905) gen. nov. from the lizard fish *Synodus intermedius* (Agassiz). *Parasitology* **59**: 519-548.

Rees, G. 1970. Some helminth parasites of fishes of Bermuda and an account of the attachment organ of *Alcicornis carangis* MacCallum, 1917 (Digenea: Bucephalidae). *Parasitology* **60**: 195-221.

Remane, A. 1929. Rotatoria. In: Grimpe, G. & E. Wagler (Eds.). *Die Tierwelt der Nord- und Ostsee.* Vol. 16, Part 7e. Leipzig: Akademische Verlagsgesellschaft Geest & Portig. pp. 1-156.

Remane, A. 1936. Gastrotricha und Kinorhyncha. In: Bronn, H. G. (Ed.). *Klassen und Ordnungen des Tierreichs.* Bd. 5. Leipzig: C. F. Winter'sche Verlagshandlung. pp. 1-242.

Remane, A., V. Storch & U. Welsch. 1980. *Systematische Zoologie.* 2nd. ed. New York: Fischer. 682 pp.

Renaud-Mornant, J. 1970. *Parastygarctus sterreri* n. sp. tardigrade marin nouveau de l'Adriatique. *Cah. Biol. Mar.* **11**: 355-360.

Renaud-Mornant, J. 1971. Tardigrades marins des Bermudes. *Bull. Mus. Natl. Hist. Nat.* **42**(6): 1268-1276.

Renaud-Mornant, J. 1982. Species diversity in marine Tardigrada. In: Nelson, D. (Ed.). *Proceedings of the Third International Symposium on Tardigrada.* Johnson City: East Tennessee State University Press. pp. 149-178.

Rice, A. L. & D. I. Williamson, 1970. Methods for rearing larval decapod Crustacea. *Helgol. Wiss. Meeresunters.* **20**: 417-434.

Rice, D. W. 1977. A list of marine mammals of the world. *NOAA Tech. Rep. NMFS SSRF* **711**: 1-15.

Rice, M. E. & M. Todorovic (Eds.). 1975, 1976. *Proceedings of the International Symposium on the Biology of the Sipuncula and Echiura.* Belgrade: Naucno Delo Press. 2 vols. 355 pp., 204 pp.

Richardson, H. 1902. The marine and terrestrial isopods of the Bermudas, with descriptions of new genera and species. *Trans. Conn. Acad. Arts Sci.* **21**: 277-310.

Richardson, H. 1905. A monograph on the isopods of North America. *U. S. Nat. Mus. Bull.* **54**: 1-727.

Riddle, L. W. 1916. The lichens of Bermuda. *Bull. Torrey Botan. Club* **43**: 145-160.

Ridgway, S. H. 1972. *Mammals of the sea. Biology and medicine.* Springfield, Ill.: Charles C. Thomas. 812 pp.

Ridley, R. K. 1969. Electron microscopic studies on dicyemid Mesozoa. II. Infusorigen and infusoriform stages. *J. Parasitol.* **55**: 779-793.

Rieger, R. M. 1977. The relationship of character variability and morphological complexity in copulatory structures of Turbellaria-Macrostomida and -Haplopharyngida. *Mikrofauna Meeresboden* **61**: 197-216.

Rieger, R. M. & W. Sterrer. 1975. New spicular skeletons in Turbellaria, and the occurrence of spicules in marine meiofauna. *Z. Zool. Syst. Evolutionsforsch.* **13**: 207-278.

Rippka, R. et al. 1979. Generic assignments, strain histories and properties of pure cultures of cyanobacteria. *J. Gen. Microbiol.* **111**: 1-61.

Robbins, C. S., B. Bruun & H. S. Zim. 1966. *Birds of North America.* New York: Golden Press. 340 pp.

Robertson, P. B. 1969. Biological investigations of the deep sea no. 48. Phyllosoma larvae of a scyllarid lobster, *Arctides guineensis*, from the western Atlantic. *Mar. Biol.* **4**(2): 143-151.

Robertson, R. 1971. Scanning electron microscopy of planktonic larval marine gastropod shells. *Veliger* **14**(1): 1-12, 9 pls.

Robins, C. R. et al. 1980. *A list of common and scientific names of fishes from the United States and Canada.* Bethesda, Md.: American Fisheries Soc. 174 pp.

Rolfe, W. D. I. 1969. Phyllocarida. In: Moore, R. C. (Ed.). *Treatise on invertebrate paleontology.* Part R; Arthropoda 4(1). Lawrence: University of Kansas Press and Geological Society of America. pp. 296-331.

Roos, P. J. 1971. The shallow-water stony corals of the Netherlands Antilles. *Stud. Fauna Curacao other Caribb. Isl.* **37**: 1-108, 53 pls.

Roper, C. F. E. 1965. A note on egg deposition by *Doryteuthis plei* (Blainville, 1823) and its comparison with other North American loliginid squids. *Bull. Mar. Sci.* **15**(3): 589-598.

Roper, C. F. E. & M. J. Sweeney. 1981. Cephalopoda. In: *FAO species identification sheets for fisheries purposes, eastern central Atlantic.* Vol. 7. Rome: FAO. pp. 1-96.

Roper, C. F. E. and R. E. Young. 1975. Vertical distribution of pelagic cephalopods. *Smithson. Contrib. Zool.* **209**: 1-51.

Roper, C. F. E., M. J. Sweeney & C. E. Nauen. 1984. *Cephalopods of the world. An illustrated and annotated species world catalogue.* FAO Species Synopsis No.125 (Vol. 3). Rome: FAO.

Roper, C. F. E., R. E. Young & G. L. Voss. 1969. An illustrated key to the families of the order Teuthoidea (Cephalopoda). *Smithson. Contrib. Zool.* **13**: 1-32.

Round, F. E. 1973. *The biology of the algae.* New York: St. Martin's Press. 278 pp.

Ruppert, E. E. 1978a. The reproductive system of Gastrotrichs. II. Insemination in *Macrodasys*: A unique mode of sperm transfer in Metazoa. *Zoomorphologie* **89**: 207-228.

Ruppert, E. E. 1978b. The reproductive system of gastrotrichs. III. Genital organs of Thanmastodermatinae subfam. n. and Diplodasyinae subfam. n. with discussion of reproduction in Macrodasyida. *Zool. Scr.* **7**: 93-114.

Russell, F. S. 1953. *The medusae of the British Isles.* Cambridge: Cambridge University Press. 530 pp.

Russell, F. S. 1970. *The medusae of the British Isles. II. Pelagic Scyphozoa, with a supplement to the first volume on Hydromedusae.* Cambridge: Cambridge University Press. 284 pp.

Rützler, K. 1968. *Loxosomella* from *Tedania ignis*, the Caribbean fire sponge. *Proc. U. S. Nat. Mus.* **124**(3650): 1-11.

Rützler, K. 1974. The burrowing sponges of Bermuda. *Smithson. Contrib. Zool.* **165**: 1-32.

Rützler, K. 1978. Sponges in coral reefs. In: Stoddart, D. R. & R. E. Johannes (Eds.). *Coral reefs: research methods.* Paris: UNESCO. 581 pp.

Ryland, J. S. 1970. *Bryozoans.* London: Hutchinson. 175 pp.

Ryland, J. S. 1982. Bryozoa. In: Parker, S. P. (Ed.). *Synopsis and classification of living organisms.* Vol. 2. New York: McGraw-Hill. pp. 743-769.

Salvini Plawen, L. v. 1978. Antarktische und subantarktische Solenogastres: Eine Monographie: 1898-1974. *Zoologica* **128**: 1-315.

Salvini Plawen, L. v. 1985. New interstitial Solenogastres (Mollusca). *Stygologia* **1**(1): 101–108.

Sargent, R. C. & G. E. Wagenbach. 1975. Cleaning behavior of the shrimp, *Periclimenes anthophilus* Holthuis and Eibl-Eibesfeldt (Crustacea: Decapoda: Natantia). *Bull. Mar. Sci.* **25**: 466-472.

Sars, G. O. 1885. Report on the Schizopoda collected by H.M.S. "Challenger" during the years 1873-1876. In: Thompson, Sir C. W. (Ed.). *Report on the scientific results of the voyage of the H.M.S. "Challenger" during the years 1873-1878.* Zoology, Vol. 13; Part 37. London: Her Majesty's Stationery Office. 228 pp., 38 pls.

Sars, G. O. 1896. *An account of the Crustacea of Norway with short descriptions and figures of all the species. II. Isopoda.* Bergen: Bergen Museum. 269 pp.

Sars, G. O. 1900. *An account of the Crustacea of Norway with short descriptions and figures of all the species. III. Cumacea.* Bergen: Bergen Museum. 115 pp.

Sawyer, T. K. 1975. Marine amoebae from surface waters of Chincoteague Bay, Virginia: Two new genera and nine new species within the families Mayorellidae, Flabellulidae and Stereomysidae. *Trans. Am. Microsc. Soc.* **94**: 71-92.

Sawyer, R. T., A. R. Lawler & R. M. Overstreet. 1975. Marine leeches of the eastern United States and the Gulf of Mexico with a key to the species. *Nat. Hist. (N. Y.)* **9**: 633-677.

Schoenberg, D. A. & R. K. Trench. 1980a. Genetic variations in *Symbiodinium* (=*Gymnodinium*) *microadriaticum* Freudenthal, and specificity in its symbiosis with marine invertebrates. I. Isoenzyme and soluble protein patterns of axenic cultures of *Symbiodinium microadriaticum*. *Proc. R. Soc. Lond. B Biol. Sci.* **207**: 405-427.

Schoenberg, D. A. & R. K. Trench. 1980b. Genetic variations in *Symbiodinium* (=*Gymnodinium*) *microadriaticum* Freudenthal, and specificity in its symbiosis with marine invertebrates. II. Morphological variation in *Symbiodinium microadriaticum*. *Proc. R. Soc. Lond. B Biol. Sci.* **207**: 429-444.

Schoenberg, D. A. & R. K. Trench. 1980c. Genetic variation in *Symbiodinium* (=*Gymnodinium*) *microadriaticum* Freudenthal, and specificity in its symbiosis with marine invertebrates. III. Specificity and infectivity of *Symbiodinium microadriaticum*. *Proc. R. Soc. Lond. B Biol. Sci* **207**: 445-460.

Schoepfer-Sterrer, C. 1969. *Chordodasys riedli* gen. nov., spec. nov., a macrodasyoid gastrotrich with a chordoid organ. *Cah. Biol. Mar.* **10**: 391-404.

Schoepfer-Sterrer, C. 1974. Five new species of *Urodasys* and remarks on the terminology of the genital organs in Macrodasyidae (Gastrotricha). *Cah. Biol. Mar.* **15**: 229-254.

Schortman, E. F. 1969. A short history of whaling in Bermuda. *Mariner's Mirror* **55**(1): 77-85.

Schroeder, P. C. & C. O. Hermans. 1975. Annelida: Polychaeta. In: Giese, A. C. & J. S. Pearse (Eds.). *Reproduction of marine invertebrates.* Vol. 3. New York: Academic. pp. 1-213.

Schultz, G. A. 1969. *How to know the terrestrial isopod crustaceans.* Dubuque, Iowa: W. Brown Co., 359 pp.

Schultz, G. A. 1972. Ecology and systematics of terrestrial isopod crustaceans from Bermuda (Oniscoidea). *Crustaceana* Suppl. **3**: 79-99.

Schultz, G. A. 1979. A new Asellota (Stenetriidae) and two, one new, Anthuridea (Anthuridae) from Bermuda (Crustacea, Isopoda). *Proc. Biol. Soc. Wash.* **91**(4): 904-911.

Schuster, R. 1979. Soil mites in the marine environment. *Recent Adv. Acarology* **1**: 593-602.

Sharp, J. 1969. Blue-green algae and carbonates—*Schizothrix calcicola* and algal stromatolites from Bermuda. *Limnol. Oceangr.* **14**: 568-578.

Sheehan, R. & F. T. Banner. 1973. *Trichosphaerium*: an extraordinary testate rhizopod from coastal waters. *Estuarine Coastal Mar. Sci.* **1**: 245-260.

Shufeldt, R. W. 1916. The bird-caves of the Bermudas and their former inhabitants. *Ibis* **10**(4): 623-635.

Shufeldt, R. W. 1922. A comparative study of some subfossil remains of birds from Bermuda, including the "Cahow". *Ann. Carnegie Mus.* **13**(3/4): 333-418, pls. 16-31.

Sieg, J. &. R. Winn. 1979. Keys to suborders and families of Tanaidacea (Crustacea). *Proc. Biol. Soc. Wash.* **91**: 840-846.

Silberhorn, G. M. 1982. *Common plants of the mid-Atlantic coast. A field guide.* Baltimore: Johns Hopkins University Press. 256 pp.

Sims, R. W. (Ed.). 1980. *Animal identification. A reference guide.* Vol. 1. Marine and brackish water animals. London: British Museum (Natural History) and Wiley. 111 pp.

Sket, B. 1979. *Atlantasellus cavernicolus* n. gen., n. sp. (Isopoda Asellota, Atlantasellidae n. fam.) from Bermuda. *Biol. Vestn. (Ljubljana)* **27**(2): 175-183.

Slaughter, R. A. 1975. *Birds of Bermuda.* Hamilton, Bermuda: Bermuda Book Store Ltd. 58 pp.

Slijper, E. J. 1976. *Whales and dolphins.* Ann Arbor: University of Michigan Press. 170 pp.

Small, E. B. & D. Lynn. 1985. Ciliophora. In: Lee, J. J., S. H. Hutner & E. C. Bovée (Eds.). *Illustrated guide to the Protozoa.* Lawrence, Kansas: Allen Press.

Smallwood, W. M. 1910. Notes on the hydroids and nudibranchs of Bermuda. *Proc. Zool. Soc. Lond.* **1910**: 137-145.

Smallwood, W. M. 1912. *Polycerella zoobotryon* (Smallwood). *Proc. Am. Acad. Arts Sci.* **47**: 607-630.

Smith, R. I. & J. T. Carlton (Eds.). 1975. *Light's manual: intertidal invertebrates of the central California coast.* Berkeley: University of California Press. 715 pp.

Smith, F. C. W. 1971. *A handbook of the common Atlantic reef and shallow-water corals of Bermuda, the Bahamas, Florida, the West Indies and Brazil.* Coral Gables: University of Miami Press. 164 pp.

Smith-Vaniz, W., P. Colin & J. Burnett-Herkes. (in prep.) *Annotated checklist of Bermuda fishes.*

Snyder, M. A. 1984. *Fusinus lightbourni* (Gastropoda: Fasciolariidae), a new species from Bermuda. *Nautilus* **98**(1): 28-31.

Soest, R. W. M. v. 1975a. Zoogeography and speciation in the Salpidae (Tunicata, Thaliacea). *Beaufortia* **23**(307): 181-215.

Soest, R. W. M. v. 1975b. Thaliacea of the Bermuda area. *Bull. Zool. Mus. Univ. Amsterdam* **5**: 7-12.

Soest, R. W. M. v. 1978. Marine sponges from Curacao and other Caribbean localities. Part I. Keratosa. *Stud. Fauna Curacao other Caribb. Isl.* **56**(179): 1-94.

Soest, R. W. M. v. 1980. Marine sponges from Curacao and other Caribbean localities. Part II. Haplosclerida. *Stud. Fauna Curacao other Caribb. Isl.* **62**(191): 1-173.

Soest, R. W. M. v. 1984. Marine sponges from Curacao and other Caribbean localities. Part III. Poecilosclerida. *Stud. Fauna Curacao other Caribb. Isl.* **66**: 1-167.

Sournia, A. (Ed.). 1978. *Phytoplankton manual.* Monographs on oceanographic methodology 6. Paris: UNESCO. 337 pp.

Southward, A. J. 1975. Intertidal and shallow-water Cirripedia of the Caribbean. *Stud. Fauna Curacao other Caribb. Isl.* **46**: 1-53.

Southward, E. C. 1968. On a new genus of Pogonophora from the western Atlantic, with descriptions of two new species. *Bull. Mar. Sci.* **18**: 182-190.

Southward, E. C. 1971a. Recent researches on the Pogonophora. *Oceanogr. Mar. Biol. Ann. Rev.* **9**: 193-220.

Southward, E. C. 1971b. Pogonophora of the northwest Atlantic: Nova Scotia to Florida. *Smithson. Contrib. Zool.* **88**: 1-29.

Southward, E. C. 1972. On some Pogonophora from the Caribbean and the Gulf of Mexico. *Bull. Mar. Sci.* **22**: 739-776.

Southwick, W. E. 1939. The "agglutination" phenomenon with spermatozoa of *Chiton tuberculatus*. *Biol. Bull.* **77**: 157-165.

Spencer, W. K. & C. W. Wright. 1966. Asterozoans. In: Moore, R. C. (Ed.). *Treatise on invertebrate paleontology.* Part U; Echinodermata 3(1). Lawrence: University of Kansas Press and Geological Society of America. pp. 4-107.

Spindler, M. et al. 1978. Light and electron microscopic observations of gametogenesis in *Hastigerina pelagica* (Foraminifera). *J. Protozool.* **24**: 427-433.

Spindler, M. et al. 1979. Lunar periodicity of reproduction in the planktonic foraminifer *Hastigerina pelagica*. *Mar. Ecol.—Prog. Ser.* **1**: 61-64.

Spoel, S. v. d. 1967. *Euthecosomata, a group with remarkable developmental stages.* Gorinchen: J. Noorduijnen Zoon N.V. 375 pp.

Spoel, S. v. d. 1976. *Pseudothecosomata, and Heteropoda (Gastropoda)*. Utrecht: Bohn, Scheltema & Holkema. 480 pp.

Stanier, R. & G. Cohen-Bazire. 1977. Phototrophic prokaryotes: the cyanobacteria. *Ann. Rev. Microbiol.* **31**: 225-274.

Stebbing, A. R. D. 1970. Aspects of the reproduction and life cycle of *Rhabdopleura compacta* (Hemichordata). *Mar. Biol.* **5**: 205-212.

Stebbing, A. R. D. & P. N. Dilly. 1972. Some observations on living *Rhabdopleura compacta* (Hemichordata). *J. Mar. Biol. Ass. U. K.* **52**: 443-448.

Steinker, D. C. 1980. Nearshore Foraminifera from Bermuda. *Compass* **57**: 129-148.

Steinker, D. C. &. W. A. Butcher. 1981. Foraminifera from mangrove shores, Bermuda. *Micron* **12**: 223-224.

Stephen, A. C. & S. J. Edmonds. 1972. *The phyla Sipuncula and Echiura*. London: British Museum (Natural History). 528 pp.

Stephenson, J. 1930. *The Oligochaeta*. Oxford: Oxford University Press. 978 pp.

Sterrer, W. 1972. Systematics and evolution within the Gnathostomulida. *Syst. Zool.* **21**: 151-173.

Sterrer, W. 1976. *Tenuignathia rikerae* nov. gen., nov. spec., a new gnathostomulid from the west Atlantic. *Int. Rev. gesamten Hydrobiol.* **61**: 249-259.

Sterrer, W. 1977. Jaw length as a tool for population analysis in Gnathostomulida. *Mikrofauna Meeresboden* **61**: 253-262.

Sterrer, W. 1982. Gnathostomulida. In: Parker, S. P. (Ed.). *Synopsis and classification of living organisms*. Vol. 1. New York: McGraw-Hill. pp. 847-851.

Sterrer, W. & R. A. Farris. 1975. *Problognathia minima* n. g., n. sp., representative of a new family of Gnathostomulida, Problognathiidae n. fam. from Bermuda. *Proc. Am. Microsc. Soc.* **94**: 357-367.

Sterrer, W. & T. M. Iliffe. 1982. *Mesonerilla prospera*, a new archiannelid from marine caves in Bermuda. *Proc. Biol. Soc. Wash.* **95**(3): 509-514.

Sterrer, W., M. Mainitz & R. M. Rieger. 1985. Gnathostomulida—enigmatic as ever. In: Morris, S. C. et al. (Eds.) *The origin and relationships of lower invertebrates*. Oxford: Oxford University Press. pp. 181-199.

Stewart, W. D. P. (Ed.). 1974. *Algal physiology and biochemistry*. Berkeley: University of California Press. 989 pp.

Stock, J. H. 1974. Pycnogonida from the continental shelf, slope and deep sea of the tropical Atlantic and east Pacific. *Bull. Mar. Sci.* **24**: 957-1092.

Stock, J. H. 1976. A new member of the crustacean suborder Ingolfiellidea from Bonaire, with a review of the entire suborder. *Stud. Fauna Curacao other Caribb. Isl.* **50**(164): 56-75.

Stoecker, D. 1978. Resistance of a tunicate to fouling. *Biol. Bull.* **155**: 615-626.

Stokes, F. J. 1980. *Handguide to the coral reef fishes of the Caribbean*. New York: Lippincott & Crowell. 160 pp.

Strenzke, K. 1955. Thalassobionte und thalassophile Collembola. In: Grimpe, G. & E. Wagler (Eds.). *Die Tierwelt der Nord- und Ostsee*. Vol. 36, Part 11f. Leipzig: Akademische Verlagsgesellschaft Geest & Portig. pp. 1-52.

Stuart, J. 1982. Hirudinoidea. In: Parker, S. P. (Ed.). *Synopsis and classification of living organisms*. Vol. 2. New York: McGraw-Hill. pp. 43-50.

Stunkard, H. W. 1954. The life history and systematic relations of the Mesozoa. *Quart. Rev. Biol.* **29**: 230-244.

Sutcliffe, W. H., Jr. 1957. Observations on the growth rate of the immature Bermuda spiny lobster *Panulirus argus*. *Ecology* **38**(3): 526-529.

Tattersall, O. S. 1955. Mysidacea. *Discovery Rep.* **28**: 1-190.

Tattersall, W. M. 1951. A review of the Mysidacea of the United States National Museum. *U. S. Nat. Mus. Bull.* **201**: 1-292.

Tattersall, W. M. & O. S. Tattersall. 1951. *The British Mysidacea*. London: Ray Society. 460 pp.

Taylor, D. L. 1974. Symbiotic marine algae: taxonomy and biological fitness. In: Vernberg, W. B. (Ed.). *Symbiosis in the sea*. Columbia: University of South Carolina Press. pp. 245-262.

Taylor, W. R. 1960. *Marine algae of the eastern tropical and subtropical coasts of the Americas*. Ann Arbor: University of Michigan Press. 870 pp.

Taylor, W. R. & A. J. Bernatowicz. 1952a. Marine species of *Vaucheria* at Bermuda. *Bull. Mar. Sci. Gulf Carib.* **2**: 405-413.

Taylor, W. R. & A. J. Bernatowicz. 1952b. Bermudian marine Vaucherias of the section Piloboloideae. *Pap. Mich. Acad. Sci. Arts Lett.* **37**: 75-85.

Taylor, W. R. & A. J. Bernatowicz. 1969. *Distribution of marine algae about Bermuda*. Bermuda Biological Station Special Publication no. 1. St. George's West, Bermuda: Bermuda Biological Station. 42 pp.

Thane-Fenchel, A. 1968. A simple key to the genera of marine and brackish-water rotifers. *Ophelia* **5**: 299-311.

Thiel, H. 1962. Untersuchungen über die Strobilisation von *Aurelia aurita* Lam. an einer Population der Kieler Förde. *Kieler Meeresforsch.* **18**(2): 198-230.

Thompson, T. E. 1976. *Biology of opisthobranch molluscs*. Vol. 1. London: Ray Society. 207 pp.

Thompson, T. E. 1977a. Jamaican opisthobranch molluscs I. *J. Mollusc. Stud.* **43**: 93-140.

Thompson, T. E. 1977b. The taxonomic status of two Bermudan opisthobranchs. *J. Mollusc. Stud.* **43**: 217-222.

Thompson, T. E. 1981. Redescription of a rare North Atlantic doridacean nudibranch, *Aegires sublaevis* Odhner. *Veliger* **23**(4): 316.

Thompson, T. E. & G. E. Brown. 1984. *Biology of opisthobranch molluscs*. Vol. 2. London: Ray Society. 229 pp.

Tomlinson, J. T. 1969. The burrowing barnacles (Cirripedia: Order Acrothoracica). *U. S. Nat. Mus. Bull.* **296**: 1-162.

Totton, A. K. 1936. Plankton of the Bermuda Oceanographic Expeditions. VII. Siphonophora taken during the year 1931. *Zoologica* **21**: 231-240.

Totton, A. K. 1960. Studies on *Physalia physalis* (L.). Part 1. Natural history and morphology. *Discovery Rep.* **30**: 301-367.

Totton, A. K. & H. E. Bargmann. 1965. *A synopsis of the Siphonophora*. London: British Museum (Natural History). 230 pp.

Treadwell, A. L. 1917. Polychaetous annelids

from Florida, Porto Rico, Bermuda and the Bahamas. *Pub. Carnegie Inst. Wash.* **251**: 255-272.

Treadwell, A. L. 1936. Polychaetous annelids from the vicinity of Nonsuch Island, Bermuda. *Zoologica* **21**: 49-68.

Treadwell, A. L. 1941. Plankton of the Bermuda Oceanographic Expeditions. Polychaetous annelids from Bermuda plankton, with eight shore species and four from Haiti. *Zoologica* **26**: 25-30.

Treece, G. 1979. Four new records of aplacophorous mollusks from the Gulf of Mexico. *Bull. Mar. Sci.* **29**: 344-364.

Tregouboff, G. 1953. Radiolaria. In: Grassé, P. P. (Ed.). *Traité de zoologie.* Vol. 1, Part 2; Protozoa. Paris: Masson. pp. 269-463.

Trench, R. K. 1974. Nutritional potentials in *Zoanthus sociatus* (Coelenterata, Anthozoa). *Helgol. Wiss. Meeresunters.* **26**: 174-216.

Trench, R. K. 1981. Cellular and molecular interactions in symbioses between dinoflagellates and marine invertebrates. *Pure Appl. Chem.* **53**: 819-835.

Turner, R. D. 1961. Pleurotomariidae in Bermuda waters. *Nautilus* **74**(4): 162-163.

Ulken, A. 1979. Phycomycetenfunde in der Sargassosee. *Veröff. Inst. Meeresforsch. Bremerhaven* **18**: 21-27.

Vachon, M. 1949. Ordre des Pseudoscorpions. In: Grassé, P. P. (Ed.). *Traité de zoologie.* Vol. 6. Paris: Masson. pp. 437-481.

Van Name, W. G. 1902. The ascidians of the Bermuda Islands. *Trans. Conn. Acad. Arts Sci.* **11**: 325-412.

Van Name, W. G. 1921. Ascidians of the West Indies. *Bull. Am. Mus. Nat. Hist.* **44**: 283-494.

Van Name, W. G. 1945. North and South American ascidians. *Bull. Am. Mus. Nat. Hist.* **84**: 1-476, 31 pls.

Van Tyne, J. & A. J. Berger. 1976.*Fundamentals of ornithology.* 2nd. ed. New York: Wiley. 808 pp.

Verrill, A. E. 1881. The cephalopods of the northeastern coast of America. Part II. The smaller cephalopods, including the "squids" and the octopi, with other allied forms. *Trans. Conn. Acad. Arts Sci.* **5**(6): 259-446, 33 pls.

Verrill, A. E. 1899. Descriptions of imperfectly known and new actinians, with critical notes on other species. *Am. J. Sci.* **7**: 41-50, 143-146, 205-218, 375-380.

Verrill, A. E. 1900a. Additions to the Anthozoa and Hydrozoa of the Bermudas. *Trans. Conn. Acad. Arts Sci.* **10**: 551-572.

Verrill, A. E. 1900b. Additions to the Crustacea and Pycnogonida of the Bermudas. *Trans. Conn. Acad. Arts Sci.* **10**: 573-582. pl. 10.

Verrill, A. E. 1900c. Additions to the echinoderms of the Bermudas. *Trans. Conn. Acad. Arts Sci.* **10**: 583-587.

Verrill, A. E. 1900d. Additions to the Tunicata and Molluscoidea of the Bermudas. *Trans. Conn. Acad. Arts Sci.* **10**: 588-594.

Verrill, A. E. 1900e. Additions to the Turbellaria, Nemertina and Annelida of the Bermudas, with revision of some New England genera and species. *Trans. Conn. Acad. Arts Sci.* **10**: 596-671.

Verrill, A. E. 1901a. Additions to the fauna of the Bermudas from the Yale Expedition of 1901, with notes on other species. *Trans. Conn. Acad. Arts Sci.* **11**: 15-62.

Verrill, A. E. 1901b. Variations and nomenclature of Bermudian, West Indian and Brazilian reef corals, with notes on various Indo-Pacific corals. *Trans. Conn. Acad. Arts Sci.* **11**: 63-168, pls. 10-35.

Verrill, A. E. 1901c. Comparisons of the Bermudian, West Indian and Brazilian coral faunae. *Trans. Conn. Acad. Arts Sci.* **11**: 169-206, pls. 10-35.

Verrill, A. E. 1901d. The story of the Cahow, the mysterious extinct bird of the Bermudas. *Popular Sci. Monthly* **60**: 22-30.

Verrill, A. E. 1902a. The Bermuda Islands; their scenery, climate, productions, physiography, natural history and geology, with sketches of their early history and changes due to man. *Trans. Conn. Acad. Arts Sci.* **11**: 413-956, 104 pls.

Verrill, A. E. 1902b. The cahow of the Bermudas, an extinct bird. *Ann. Mag. Nat. Hist.* **9**: 26-31.

Verrill, A. E. 1907a. The Bermuda Islands. Part IV. Geology and Paleontology. *Trans. Conn. Acad. Arts Sci.* **12**: 1-204.

Verrill, A. E. 1907b. The Bermuda Islands. Part V. Characteristic life of the Bermuda coral reefs. *Trans. Conn. Acad. Arts Sci.* **12**: 204-348, pls. 26-36.

Verrill, A. E. 1908a. Decapod Crustacea of Bermuda. I. Brachyura and Anomura. *Trans. Conn. Acad. Arts Sci.* **13**: 299-474, pls. 9-28.

Verrill, A. E. 1908b. Geographical distribution; origin of the Bermudian decapod fauna. *Am. Nat.* **42**: 289-296.

Verrill, A. E. 1922. Decapod Crustacea of Bermuda. II. Macrura. *Trans. Conn. Acad. Arts Sci.* **26**: 1-179, 48 pls.

Verrill, A. E. 1923. Crustacea of Bermuda. Schizopoda, Cumacea, Stomatopoda and Phyllocarida. *Trans. Conn. Acad. Arts Sci.* **26**: 181-211, pls. 49-56.

Verrill, A. E. & K. J. Bush. 1900. Additions to the marine Mollusca of the Bermudas. *Trans. Conn. Acad. Arts Sci.* **10**: 513-544, 3 pls.

Verrill, A. H. 1901. Additions to the avifauna of the Bermudas with diagnoses of two new subspecies. *Am. J. Sci.* 4th. ser. **12**: 64-65.

Vervoort, W. 1968. Report on a collection of Hydroida from the Caribbean region, including an annotated checklist of Caribbean hydroids. *Zool. Verh.* **92**: 1-124.

Volkmann, B. 1979a. *Tisbe* (Copepoda, Harpacticoida) species from Bermuda and zoogeographical considerations. *Archo. Oceanogr. Limnol.* **19** Suppl.: 1-76.

Volkmann, B. 1979b. A revision of the genus *Tisbella* (Copepoda, Harpacticoida). *Archo. Oceanogr. Limnol.* **19** Suppl.: 77-119.

Voss, G. L. 1960. Bermudan Cephalopods. *Fieldiana Zool.* **39**(40): 419-446.

Voss, G. L. & C. Phillips. 1958. A first record of *Octopus macropus* Risso from the United States with notes on its behavior, color, feeding and gonads. *Quart. J. Florida Acad. Sci.* **20**(4): 223-232.

Voss, G. L., L. Opresko & R. Thomas. 1973. *The potentially commercial species of octopus and squid of Florida, the Gulf of Mexico and the Caribbean Sea*. Sea Grant Field Guide Series 2. Miami: University of Miami Sea Grant Program. 33 pp.

Waller, R. A. & R. I. Wiklund. 1968. Observations from a research submersible: mating and spawning of the squid, *Doryteuthis plei. Bioscience* **18**(2): 110-111.

Waller, T. R. 1973. The habits and habitats of some Bermudian marine mollusks. *Nautilus* **87**: 31-52.

Wardle, R. A. & J. A. McLeod. 1952. *The zoology of tapeworms*. Minneapolis: University of Minnesota Press. 780 pp.

Waterman, S. 1975. Bermuda behemoth: the shark that wouldn't bite. *Skin Diver* **24**(4): 48-51.

Waterston, J. M. 1940. *Supplementary list of Bermuda insects*. Bermuda: Bermuda Department of Agriculture. 10 pp.

Watling, L. 1979. Marine fauna and flora of the northeastern United States. Crustacea: Cumacea. *NOAA Tech. Rep. NMFS Circ.* **423**: 1-22.

Webster, H. E. 1884. Annelida from Bermuda. *U. S. Nat. Mus. Bull.* **25**: 305-327.

Wells, J. W. 1956. Scleractinia. In: Moore, R. C. (Ed.). *Treatise on invertebrate paleontology*. Part F; Coelenterata. Lawrence: University of Kansas Press and Geological Society of America. pp. 328-444.

Wells, J. W. 1972. Some shallow water ahermatypic corals from Bermuda. *Postilla* **56**: 1-10.

Wells, M. J. 1962. *Brain and behaviour in cephalopods*. Stanford, Ca.: Stanford University Press. 171 pp.

Wells, M. J. 1978. *Octopus: physiology and behaviour of an advanced invertebrate*. London: Chapman & Hall. 471 pp.

Werner, B. 1974. *Stephanoscyphus eumedusoides* n. sp. (Scyphozoa, Coronatae), ein Höhlenpolyp mit einem neuen Entwicklungsmodus. *Helgol. Wiss. Meeresunters.* **26**: 434-463.

Werner, B. 1975. Bau und Lebensgeschichte des Polypen von *Tripedalia cystophora* (Cubozoa, class nov., Carybdeidae) und seine Bedeutung für die Evolution der Cnidaria. *Helgol. Wiss. Meeresunters.* **27**: 461-504.

Westheide, W. 1973. Zwei neue interstitielle *Microphthalmus*-Arten (Polychaeta) von den Bermudas. *Mikrofauna Meeresboden* **14**: 241-252.

Wetmore, A. 1960. Pleistocene birds in Bermuda. *Smithson. Misc. Collect.* **140**(2): 1-11.

Wetmore, A. 1962. Notes on fossil and subfossil birds. *Smithson. Misc. Collect.* **145**(2): 1-17.

Wheeler, J. F. G. 1937. Further observations on lunar periodicity. *Linn. Soc. J. Zool.* **40**: 325-345.

Wheeler, J. F. G. 1942. The discovery of the nemertean *Gorgonorhynchus* and its bearing on evolutionary theory. *Am. Nat.* **76**: 470-493.

Whittaker, R. H. 1969. New concepts of kingdoms of organisms. *Science* **163**: 150-160.

Wiedenmayer, F. 1977. *The shallow-water sponges of the Western Bahamas*. Basel: Birkhäuser. 332 pp.

Wieser, W. & F. Schiemer. 1977. The ecophysiology of some marine nematodes from Bermuda: seasonal aspects. *J. Exp. Mar. Biol. Ecol.* **26**: 97-106.

Wieser, W. et al. 1974. An ecophysiological study of some meiofauna species inhabiting a sandy beach at Bermuda. *Mar. Biol.* **26**: 235-248.

Willey, A. 1930. Harpacticoid Copepoda from Bermuda. Part 1. *Ann. Mag. Nat. Hist.* Ser. 10. **6**: 81-114.

Willey, A. 1935. Harpacticoid Copepoda from Bermuda. Part 2. *Ann. Mag. Nat. Hist.* Ser. 10. **15**: 50-100.

Williams, A. B. 1965. Marine decapod crustaceans of the Carolinas. *Fish. Bull.* **65**: 1-298.

Williams, A. B. 1974. The swimming crabs of the genus *Callinectes* (Decapoda: Portunidae). *Fish. Bull.* **72**: 685-798.

Williams, A. B., L. B. McCloskey and I. E. Gray. 1968. New records of brachyuran decapod crustaceans from the continental shelf off North Carolina, U.S.A. *Crustaceana* **15**: 41-66.

Williams, R. W. 1956a. The biting midges of the genus *Culicoides* found in the Bermuda Islands (Diptera, Heleidae). I. A description of *C. bermudensis* n. sp. with a key to the local fauna. *J. Parasitol.* **42**: 297-300.

Williams, R. W. 1956b. The biting midges of the genus *Culicoides* in the Bermuda Islands (Diptera, Heleidae). II. A study of their breeding habitats and geographical distribution. *J. Parasitol.* **42**: 300-305.

Williams, R. W. 1957. Observations on the breeding habits of some Heleidae of the Bermuda Islands (Diptera). *Proc. Entomol. Soc. Wash.* **59**: 61-66.

Williams, R. W. 1961. Parthenogenesis and autogeny in *Culicoides bermudensis* Williams. *Mosquito News* **21**: 116-117.

Williams, T. C. et al. 1977. Autumnal bird migration over the western North Atlantic ocean. *Am. Birds* **31**(3): 251-268.

Wilson, C. B. 1932. The copepods of the Woods Hole region, Massachusetts. *U. S. Nat. Mus. Bull.* **158**: 1-135. (reprinted 1972 by Junk, Lochem, Netherlands.)

Wingate, D. B. 1960. Cahow, living legend of Bermuda. *Can. Audubon* **22**(5): 145-149.

Wingate, D. B. 1973. *A checklist and guide to the birds of Bermuda*. Bermuda: D. B. Wingate. 35 pp.

Wingate, D. B. 1978. Excluding competitors from Bermuda petrel nesting burrows. In: Temple, S. E. (Ed.). *Endangered birds: management techniques for preserving threatened species*. Madison: University of Wisconsin Press. pp. 93-102.

Winn, H. E. & B. L. Olla (Eds.). 1979. *Behaviour of marine animals*. Vol. 3. Cetaceans. New York: Plenum. 438 pp.

Winn, H. E. & L. K. Winn. 1968. The song of the humpback whale *Megaptera novaeangliae* in the West Indies. *Mar. Biol.* **47**: 97-114.

Wirth, W. W. & R. W. Williams. 1957. The biting midges of the Bermuda Islands, with descriptions of five new species. *Proc. Entomol. Soc. Wash.* **59**: 5-14.

Woese, C. R. 1981. Archaebacteria. *Sci. Am.* **244**(6): 94-106.

Wood, C. A. 1923. The fossil eggs of Bermudan birds. *Ibis* 11th. ser. **5**(2): 193-207.

Wurster, C. F., Jr. & D. B. Wingate. 1968. DDT residues and declining reproduction in the Bermuda petrel. *Science* **159**(3815): 979-981.

Yamaguti, S. 1959. *Systema Helminthum*. Vol. II. The cestodes of vertebrates. New York: Interscience. 860 pp.

Yamaguti, S. 1963a. *Systema Helminthum* Vol. IV. Monogenea and Aspidocotylea. New York: Interscience. 699 pp.

Yamaguti, S. 1963b. *Systema Helminthum*. Vol. V. Acanthocephala. New York: Interscience. 423 pp.

Yamaguti, S. 1971. *Synopsis of digenetic trematodes of vertebrates*. Tokyo: Keigaku. 1074 pp., 349 pls.

Yamai, I. 1971. *Illustrations of the marine plankton of Japan*. Osaka: Hoikusha Pub. Co. 369 pp. (Text in Japanese)

Yeatman, H. C. 1957. A redescription of two parasitic copepods from Bermuda. *J. Wash. Acad. Sci.* **47**(10): 346-353.

Yntema, C. L. & N. Mrosovsky. 1980. Sexual differentiation in hatchling loggerheads (*Caretta caretta*) incubated at different controlled temperatures. *Herpetologica* **36**: 33-36.

Young, J. Z. 1971. *The anatomy of the nervous system of Octopus vulgaris*. Oxford: Oxford University Press. 690 pp.

Zullo, V. A., D. B. Bach & J. T. Carlton. 1972. New barnacle records (Cirripedia, Thoracica). *Proc. Calif. Acad. Sci.* 4th. ser. **39**(6): 65-74.

# GLOSSARY

This glossary defines the more common recurring terms, and describes some apparatus and methods for collecting, extracting and preserving organisms. Specialized terms are usually explained in the IDENTIFICATION paragraph of each chapter; for taxa consult the TAXONOMIC INDEX.

*Abdomen*  Posterior part of the body.

*Abiotic*  Non-living.

*Aboral*  Opposite the mouth (cf. oral).

*Abyssal*  Deep zone of the ocean (about 3,000-6,000 m).

*Ahermatypic*  Not reef-building (see Anthozoa).

*Antenna*  A sensory appendage (feeler), usually paired and situated on the head.

*Anterior*  At or toward the front end (in bilaterally symmetrical organisms).

*Asexual (vegetative) reproduction*  Production of new individuals by budding or release of cells that are not gametes (cf. sexual reproduction).

*Asymmetric*  Without symmetry (e.g., most Porifera) (cf. bilateral and radial symmetry).

*Autotrophic*  Type of nutrition in which inorganic molecules are synthesized into organic matter (i.e., the typical mode of plants) (cf. heterotrophic).

*Bathypelagic*  Living in deeper parts of the open ocean, below about 1,000 m.

*Benthos (adj. benthic)*  Bottom-dwelling organisms.

*Berlese apparatus*  To extract small terrestrial arthropods place sediment sample on a 2mm mesh metal sieve over a funnel. Install a gentle heat source (a light bulb) above, and a jar with 75% alcohol below the funnel. After a few hours to days most animals will have migrated out of the drying sediment and fallen into the jar.

*Bilateral symmetry*  Symmetry of an organism with equal, complementary right and left halves (e.g., a fish).

*Binomial nomenclature*  The practice of giving two names, genus and species, to all organisms. Usually followed by the name of the author (=first describer) of the species.

*Biotic*  Life-related.

*Bouin's fixative*  A good all-purpose fixative for microanatomy. Mix 15 parts saturated picric acid, 5 parts formalin (35-40%) and 1 part glacial acetic acid. Can be used warm; fixation 2-24 hr.

*Brackish*  Water with less salt than seawater.

*Buccal*  Within or near the mouth.

*Calcareous*  Made of calcium carbonate (limestone).

*Carnivorous*  Eating animals (cf. herbivorous, omnivorous).

*Caudal*  On or near the tail.

*Cell*  Smallest functional unit of an organism, consisting of a nucleus and surrounding protoplasm, and bounded by a membrane.

*Cephalothorax*  Body region combining head (cephalon) and chest (thorax).

*Chaeta (pl. chaetae)*  Bristle (of Annelida).

*Chela (pl. chelae)*  Leg that bears pincers (e.g., Crustacea).

*Chemoreceptor*  Sense organ for chemical signals, e.g., smell and taste.

*Chemosynthesis*  A process in which energy contained in inorganic molecules (such as hydrogen sulfide) is converted into energy used to synthesize organic matter —the mode of most bacteria (cf. photosynthesis).

*Chitin*  Organic compound that forms a protective surface covering mainly in Arthropoda.

*Chlorophylls*  Pigments that, in most plants, are involved in photosynthesis.

*Cilium (pl. cilia)*  Minute hair-like structure which, by beating, creates a current in the surrounding medium, and thus effects transport or locomotion. The current produced is perpendicular to the outstretched cilium (cf. flagellum).

*Cirrus*  Small, fingerlike appendage.

*Climate deterioration method*  See Oxygen depletion method.

*Coelom*  Fluid-filled cavity within the middle cell layer (mesoderm) of some animals (see Bilateria).

*Colony (adj. colonial)*  An association of incompletely separated individuals of the same species (e.g., many Cnidaria, Tentaculata, Urochordata).

*Commensalism*  A relationship in which one organism derives benefits from living with another (host), without benefitting or harming the other (cf. symbiosis, parasitism).

*Compound*  Colonial (as in many Ascidiacea); also: composed of several parts (as in setae of Polychaeta).

*Core sampler*  A sampling device that drives a hollow cylinder or box vertically into the sediment then retrieves the enclosed sample.

*Cuticle (Lat. cuticula)*  Exterior skeleton made of an organic substance such as chitin.

*Deposit feeding*  Feeding on detritus that has settled to the bottom.

*Depression slide*  A microscope slide with a central, flat depression in which small organisms, surrounded by a drop of seawater and covered with a slip, can be studied.

*Detritus*  Fragments of dead organisms; the food of many animals.

*Dilution*  Lowering of a concentration. A simple rule-of-thumb to dilute, for example, 40% formalin to 4% is to take 4 parts of formalin and add water to make 40 parts.

*Dioecious*  Having separate sexes (cf. hermaphroditic).

*Direct development*  Lacking a larva. The hatching young resembles the adult.

*Dorsal*  On or near the back of an animal.

*Dredge* A sampling device consisting of a bag-shaped net held open by a strong frame that scrapes up organisms and sediment when towed horizontally over the bottom.

*Ectoparasite* A parasite living on the outer surface of another organism (e.g., Hirudinea).

*Elutriation technique* To extract meiofauna (particularly crustaceans) from soft sediments place sediment sample in a large container and run a stream of seawater at low velocity over or through it. The runoff is passed through a fine mesh sieve in which the animals are caught.

*Embryo* An organism developing from a fertilized egg.

*Endemic* Native and restricted to a particular geographic region.

*Endoparasite* A parasite living inside another organism (e.g., Cestoda).

*Epibenthic* Upon the sea bottom surface.

*Epidermis* The outermost cell layer of a plant or animal.

*Epipelagic* Living in the surface layer of the open ocean, to about 150 m.

*Epiphytic* Growing on the outer surface of a plant (e.g., many Rhodophyta).

*Epizoic* Growing on the outer surface of an animal (e.g., some Anthozoa).

*Ethanol, ethyl alcohol* A common fixative and preservative, usually diluted to 70%.

*Euryhaline* Tolerant of a wide range of salinities (cf. stenohaline).

*Evolution* The irreversible self-organization of a system. When applied to living organisms it means the change, over successive generations, in populations as a result of the complex interplay between chance mutations in the genetic inheritance and their subsequent selection by the environment.

*Fauna* All the animals.

*Filter feeding* Type of feeding in which selected or unselected particles (detritus or plankton) are, actively or passively, filtered from the surrounding water.

*Flagellum (pl. flagella)* Minute hair-like structure which, by beating, creates a current in the surrounding medium, and thus effects transport or locomotion. The current produced is parallel to the outstretched flagellum (cf. cilium).

*Flora* All the plants.

*Formalin* A solution of formaldehyde in water (35-40%) used as a fixative; usually diluted with seawater to 4-10%. Formalin can be neutralized by keeping it (in a brown bottle) together with sodium borate (borax).

*Free-living* Living unattached to any structure including other organisms (cf. sedentary, sessile, parasitic). Often used also as the opposite of parasitic.

*Gamete* Reproductive cell, i.e., egg or sperm.

*Gametophyte* Plant generation that produces eggs and sperm; alternates with the sporophyte (see Anthophyta).

*Genus (pl. genera)* A group of closely related species; related genera are grouped into families.

*Gonopore* External opening of any reproductive system.

*Grab sampler* A sampling device consisting of 2 or more jaws that "bite" sediment out of the bottom when lowered vertically.

*Hadal* Deepest zone of the ocean (below 6,000 m).

*Herbarium specimens* Freshly collected plants are pressed between layers of paper (for details see Anthophyta). Prior to pressing, algae are spread out over a sheet of white paper in a flat, seawater-filled tray; the paper is then carefully lifted out at an oblique angle, and the water let run off.

*Herbivorous* Eating plants (cf. carnivorous, omnivorous).

*Hermaphrodite* Organism with both male and female organs and producing both eggs and sperm (cf. dioecious).

*Hermatypic* Reef-building (see Anthozoa).

*Heterotrophic* Type of nutrition in which organic compounds are obtained by consuming the bodies or products of other oganisms (i.e., the typical mode of animals) (cf. autotrophic).

*Holoplankton* Organisms that are planktonic throughout their entire life cycle (cf. meroplankton).

*Indirect development* Having a larval stage in the course of development.

*Infauna* Animals living in soft bottoms.

*Interstitial fauna* Minute animals that live in the spaces between sand grains.

*Intertidal* Zone of the seashore between high tide and low tide.

*Invertebrate* An animal without a backbone (vertebral column).

*Larva* State in the development of an animal that looks, and has a life style, different from the adult; especially common among sedentary species for which it provides a means of dispersal (cf. metamorphosis).

*Littoral* The coastal zone; pertaining to the shore.

*Locomotion* Movement from one place to another.

*Macrofauna* Benthic animals large enough to be retained by a 1-mm mesh sieve.

*Magnesium chloride (MgCl₂)* A salt useful in anesthetizing marine animals, to aid in removing them from their substrate, and to relax them prior to fixation. Dissolve enough $MgCl_2$ in fresh water (!) until the solution has the same density as seawater (check with a hydrometer). To separate interstitial fauna from sand put 0.5 l of sediment into a 2-l Erlenmeyer flask, fill up with the $MgCl_2$ solution, mix gently, and let sit for 10 min. Then shake, let the sand settle briefly, and filter the supernatant (with suspended organisms) through a plankton net (63-200 μm) mounted on a plexiglass ring. Rinse the filtrate with seawater, then place in a petri dish and cover with seawater. Organisms retained by the filter can now be sorted under a dissecting microscope; many of the smaller forms will crawl through the meshes of the filter and can be found at the bottom of the petri dish (for details see HULINGS & GRAY 1971).

*Meiofauna* Small benthic animals that will pass through a 1 mm mesh sieve but are retained by a 0.1 mm mesh.

*Meiosis* Sequence of nuclear divisions that leads to the formation of gametes.

*Meroplankton* Organisms that are part of the plankton only periodically, or only during part of their life cycle (cf. holoplankton).

*Mesopelagic* Living in intermediate depths of the open ocean, about 150-1,000 m.

*Metamorphosis* The transformation of a larva into the adult, or into another stage of development (such as a pupa in many Insecta).

*Mitosis* Nuclear division sequence that results in 2 identical daughter nuclei; the normal type of division in the growth of an organism.

*Nanoplankton* Plankton too small (5-60 μm) to be retained by a net. Can be extracted from water samples by centrifuging.

*Nekton* Actively swimming animals, able to determine the direction in which they move (cf. plankton).

*Nematocyst* Intracellular stinging capsule typical of Cnidaria.

*Neritic* Occurring nearshore, over the continental shelf (cf. oceanic).

*Neuston* The plankton living in or dependent on the surface layer of the ocean.

*Nomenclature* System or set of names used to identify organisms (cf. binomial nomenclature).

*Nucleus (pl. nuclei)* The central body of a eukaryotic cell; contains the chromosomes.

*Oceanic* Occurring offshore, beyond the continental shelf (cf. neritic).

*Omnivorous* Eating any sort of food, plant or animal (cf. herbivorous, carnivorous).

*Oral* Relating to the mouth (cf. aboral).

*Oviparous* Egg-laying (cf. ovoviviparous, viviparous).

*Ovoviviparous* Producing eggs that are fertilized and develop within, but are not nourished by, the body of the mother; the young hatch at birth (cf. viviparous, oviparous).

*Oxygen depletion method* Used to extract meiofauna from samples of sediment or algae. Buckets filled with sediment samples covered with seawater can be left for days or weeks at ambient temperature; most organisms will migrate to the surface layer which can then be treated with the Magnesium chloride or Seawater ice method. Algae are packed fairly tightly in high glass containers filled with seawater; after minutes to hours meiofauna will concentrate near the surface, particularly at the side facing the light, from where it can be pipetted off.

*Palp* Sensory appendage, feeler.

*Parasitism* Relationship in which one organism lives on or in another (its host), at the expense of the latter (cf. commensalism, symbiosis).

*Parthenogenesis* Development of an egg without fertilization by a sperm.

*Pelagic* Living in the surface water of the ocean.

*Pharynx* Anterior region, often muscular, of the alimentary canal.

*Photosynthesis* A process in which light energy is converted into chemical energy used to synthesize organic compounds from inorganic materials—the mode of most plants (cf. chemosynthesis).

*Phytoplankton* Planktonic plants (mostly minute; e.g., Dinophyta).

*Plankton* Passively drifting organisms, unable to maintain a direction (cf. nekton).

*Plankton net* A sampling device consisting of a funnel-shaped, fine mesh bag and attached short cylinder ("bucket") that collects plankton organisms filtered out by the net. Can be towed vertically or horizontally from a boat.

*Posterior* At or toward the rear end (in bilaterally symmetric organisms).

*Predator* Animal that preys on other animals.

*Primary production* The synthesis of organic matter from inorganic molecules (cf. autotrophic, photosynthesis, chemosynthesis).

*Proboscis* Trunk-like, eversible structure at the anterior end of some animals (e.g., Nemertina, Priapulida).

*Protandrous* A hermaphrodite in which the male gametes mature first (cf. protogynous).

*Protogynous* A hermaphrodite in which the female gametes mature first (cf. protandrous).

*Radial cleavage* Development pattern in which the cells divide at right angles or parallel to the vertical (polar) axis of the embryo (cf. spiral cleavage).

*Radial symmetry* Symmetry of an organism in which the body parts are equally arranged, like the spokes of a wheel, around a vertical axis that passes through the mouth (e.g., Cnidaria).

*Raptorial* Catching prey (may be used for an organ such as a claw, as well as for an organism such as an osprey).

*Rose Bengal* A solution of 0.1 g/100-200 ml of 5% formalin or 70% alcohol is useful in staining preserved meiofauna to make sorting easier.

*Rostrum* Pointed projection at the anterior end of the head (e.g., Crustacea).

*Salinity* Measure of the salt concentration of water.

*Seawater-ice extraction* Interstitial sand fauna can be separated by placing sediment in a large-diameter (5 cm) plastic tube; the bottom of the tube consists of a 125 μm mesh sieve, and the tube is suspended so that the sieve is just in contact with seawater in a petri dish below. The sediment is then covered with a cotton plug on which crushed seawater ice is placed. Animals migrate downward through the sieve and can be found on the bottom of the petri dish (for details see HULINGS & GRAY 1971).

*Sedentary* Living at the same place, usually attached, most of the time.

*Segment* One of usually many similar or identical functional units that make up the body of some phyla (particularly Annelida and Arthropoda).

*Sessile* Living permanently attached to a substrate; may also mean attached without a stem.

*Sexual reproduction* Production of new individuals by fusion of gametes (usually egg and sperm) (cf. asexual reproduction).

*Siphon* Tube leading into or out of the body of invertebrates (see Gastropoda, Ascidiacea).

*Species* A reproductively isolated group of potentially interbreeding organisms that resemble each other closely but differ from any other such group consistently in at least one clearly defined characteristic.

*Spermatophore* A packet of sperm, often elaborately shaped or wrapped, for transfer to the female (e.g., Pseudoscorpiones, Cephalopoda).

*Spicule* Minute particle of skeletal material.

*Spiral cleavage* Development pattern in which the cells divide at oblique angles to the vertical (polar) axis of the embryo (cf. radial cleavage).

*Spore* Minute reproductive cell produced by the plant sporophyte.

*Sporophyte* Plant generation that produces asexually reproductive spores; alternates with the gametophyte.

*Squeeze preparation* To study small, transparent organisms place the specimen in a drop of seawater (or magnesium chloride solution) on a microscope slide, then cover with a cover slip that has been provided with tiny "feet" of bees wax at its 4 corners. By adding or subtracting (with filter paper) fluid, and by applying pressure to the wax feet, the organism can be gently squeezed to make it more transparent and reveal details of internal organization.

*Statocyst* A gravity-sensitive organ used in orientation (e.g., Turbellaria, Hydromedusae).

*Stenohaline* Having a limited range of salinity tolerance (cf. euryhaline).

*Stylet* A dagger-like or hollow structure associated with various organs, usually in connection with penetration (e.g., Turbellaria).

*Sublittoral*  Below the coastal zone; the shallow seabed.

*Substrate*  Surface on which an organism lives.

*Supralittoral*  Above the coastal zone, but still influenced by the ocean.

*Suspension feeding*  Feedig on particles suspended in water.

*Swedmark method*  To extract meiofauna from fine sediment (silt, mud) place the sample in a large flat container, stir the surface into suspension, then filter through a fine-mesh (63 μm) sieve mounted on a plexi ring, returning the filtrate to the original sample. Examine the filter in a seawater-filled petri dish under a low power microscope. The process can be repeated as long as there are animals in the original sample.

*Symbiosis*  Close association between individuals of 2 different species (cf. commensalism, parasitism).

*Synonym*  One of 2 or more scientific names applied to the same taxon.

*Taxon*  A category or unit in the system of organisms (e.g., species, class, phylum, etc.).

*Taxonomy*  The science of classification.

*Tentacle*  Long, slender appendage.

*Thallus*  Entire body of a lower plant (e.g., Chlorophyta, Lichenes).

*Thorax*  Chest region of the body.

*Torsion (adj. torted)*  Twisting of the body (see Gastropoda).

*Trawl*  A sampling device consisting of a bag-shaped net that is spread open by a set of heavy boards at its sides, and towed behind a ship.

*Vegetative reproduction*  See Asexual reproduction.

*Ventral*  On or near the belly of an animal.

*Viviparous*  Producing embryos that develop within and are nourished by the body of the mother (cf. oviparous, ovoviviparous).

*Whole mounts*  A squeeze preparation of a microscopic organism may be turned into a whole mount by adding a fixative such as glycerol and formalin (3:1), and then sealing the cover slip on all four sides with a commercial varnish such as Eukitt.

*Zooid*  Individual of an animal colony (e.g., Bryozoa).

*Zooplankton*  Planktonic animals.

*Zooxanthellae*  Species of Dinophyta living in symbiosis with certain animals (see Anthozoa).

*Zygote*  The fertilized egg.

# TAXONOMIC INDEX

This index lists the major entries for all species, synonyms, common names and super-specific taxa, and refers to color plate numbers, but omits minor cross references and black & white plates. Composite names are alphabetized by the first word, and all entries appear essentially in the same typeface (e.g., capitals, italics) as in the text.

**A**
*Aaptos bergmanni*, 124
Abasilaria, 172
Abbreviated coral shell, 432
*Abrolophus* sp., 274
*Abudefduf saxatilis*, 620
*Abyla trigona*, 154
ABYLIDAE, 154
*Abylopsis eschscholtzi*, 155
*Acanthacarnus souriei*, 121
ACANTHARIA, 101
Acanthobdellida, 264
Acanthocephala, 223
*Acanthochitona astriger*, 397
*Acanthochitona spiculosa*, 397
ACANTHOCHITONIDAE, 397
ACANTHOCHITONINA, 397
*Acanthocybium solandri*, 636
ACANTHOLONCHIDAE, 101
*Acanthonyx petiverii*, 355
*Acanthophora spicifera*, 66
Acanthor, 224
Acanthosoma, 315
ACANTHOSTAURIDAE, 101
*Acanthostracion polygonius*, 647

*Acanthostracion quadricornis*, 647
ACANTHURIDAE, 635
*Acanthurus bahianus*, 636
*Acanthurus chirurgus*, 635
*Acanthurus coeruleus*, 635
Acari, 270
*Acartia bermudensis*, 292
*Acartia spinata*, 292
ACARTIIDAE, 292
*Acetabularia crenulata*, 41, 1.1
*Achelia gracilis*, 276
ACLIDIDAE, 416
*Acmaea punctulata*, 404
*Acmaea pustulata*, 404
ACMAEIDAE, 404
ACOCHLIDIACEA, 440
ACOELA, 200
*Acompsocephalum tortum*, 206
ACONTIARIA, 173
Acorn barnacles, 302
Acorn worms, 541
ACOTYLEA, 203
Acrania, 562
ACROTHORACICA, 304
*Actaea setigera*, 346, 11.2

*Acteocina candei,* 438
ACTEOCINIDAE, 438
ACTEONACEA, 438
*Actinactis flosculifera,* 178
*Actinia bermudensis,* 176, 6.2
*Actinia melanaster,* 176
ACTINIARIA, 172
ACTINIIDAE, 176
*Actinoides pallida,* 178
ACTINOPODEA, 101
*Actinopyga agassizi,* 540
*Actinostella flosculifera,* 178
Actinotrocha, 517
"Actinotrocha sabatieri," 518
*Actinotryx sanctithomae,* 188
Actinula, 130, 141
ACTINULIDA, 146
*Actitus macularia,* 664
Aculifera, 392
Adam's miniature ark, 464
Adams' miniature horn shell, 414
Adanson's red Lasaea, 476
Adanson's slit shell, 403
Adenophorea, 216
ADEPHAGA, 388
*Adioryx vexillarius,* 592
Admirable stepping snail, 458
*Adocia amphioxa,* 118
ADOCIIDAE, 118
*Aedes sollicitans,* 390
*Aedes taeniorhynchus,* 391
*Aegina citrea,* 148
AEGINELLIDAE, 381
AEGINIDAE, 148
*Aegires sublaevis,* 452, 13.3
AEOLIDACEA, 454
AEOLIDIIDAE, 455
*Aequipecten acanthodes,* 470
*Aequorea floridana,* 146
AEQUOREIDAE, 145
*Aesopus stearnsi,* 434
*Aetea sica,* 505
Aeteidae, 505
*Aetobatus narinari,* 571
*Agalma okeni,* 152
AGALMIDAE, 152
*Agaricia fragilis,* 180
AGARICIIDAE, 180
Agassiz' pleurobranch, 447
*Agauopsis brevipalpus,* 274
*Agauopsis ornata,* 274
*Aglaophenia latecarinata,* 139
*Aglaura hemistoma,* 148
Agujon, 590
*Ahlia egmontis,* 581
*Aimulosia uvulifera,* 513

*Aiptasia annulata,* 173
*Aiptasia pallida,* 175, 6.4
*Aiptasia tagetes,* 175
AIPTASIIDAE, 173
AIZOACEAE, 82
*Aktedrilus monospermathecus,* 260
*Alaba incerta,* 414
*Alaba melanura,* 414
*Alaba tervaricosa,* 414
Albatrosses, 659
*Albula vulpes,* 578
ALBULIDAE, 577
*Albunea paretii,* 338
ALBUNEIDAE, 338
ALCEDINIDAE, 667
*Alcicornis carangis,* 204
ALCIOPIDAE, 236
*Alcirona krebsii,* 368
ALCYONACEA, 164
ALCYONARIA, 164
*Alectryonia limacella,* 472
*Alexia bermudensis,* 458
ALICIIDAE, 173
Alima, 308
*Alima hyalina,* 310
*Allanetta harringtonensis,* 592
Allison tuna, 638
*Allodardanus bredini,* 335, 10.7
*Allogromia laticollaris,* 97
ALLOGROMIIDEA, 97
Almacojack, 606
ALPHEIDAE, 323
*Alpheopsis labis,* 323, 9.4
*Alpheopsis trigonus,* 324, 9.2
*Alphestes afer,* 597
*Alpheus armillatus,* 325, 9.5
*Alpheus cylindricus,* 326, 9.7
*Alpheus formosus,* 325
*Alpheus peasei,* 325
*Alutera schoepfi,* 646
*Alutera scripta,* 646
*Alvania auberiana,* 409
*Amathia vidovici,* 504
Amber pen shell, 467
American eel, 580
*Americardia media,* 478
AMMOTHEIDAE, 276
*Ammothella appendiculata,* 277
AMOEBIDA, 95
Amphibia, 565
Amphiblastula, 114
*Amphicaryon acaule,* 152
AMPHILOCHIDAE, 376
*Amphilochus brunneus,* 376
*Amphilonche elongata,* 101
*Amphimedon viridis,* 118, 3.9

AMPHINOMIDAE, 236
"Amphioxides," 563, 564
Amphioxiformes, 563
AMPHIPODA, 372
*Amphiroa fragilissima*, 56, 2.3, 3.8
*Amphiroa rigida*, 56
*Amphiscolops bermudensis*, 200
AMPHIURIDAE, 531
*Ampithoe rubricata*, 374
AMPITHOIDAE, 374
*Anachis catenata*, 433
*Anadara deshayesi*, 464
*Anadara lienosa floridana*, 464
*Anadara notabilis*, 464
ANADYOMENACEAE, 39
*Anadyomene stellata*, 39, 1.6
Anapsida, 651
ANASCINA, 505
Anaspidacea, 305
ANASPIDEA, 444
Ancestrula, 502
*Anchialina typica typica*, 361
*Anchistioides antiguensis*, 322
*Anchoa choerostoma*, 579
Anchovies, 579
Anemones, 159
*Anemonia antilliensis*, 176
*Anemonia elegans*, 176
*Anemonia sargassensis*, 176
Angelfishes, 616
Angiospermae, 79
*Anguilla rostrata*, 580
ANGUILLIDAE, 580
ANGUILLIFORMES, 580
Angular periwinkle, 408
ANIMALIA, 92
Animals, 92
*Anisolabis maritima*, 387
*Annalisella bermudensis*, 202
Annelida, 232
*Anodontia philippiana*, 475
ANOMALODESMATA, 491
*Anomia ephippium*, 471
*Anomia simplex*, 471
ANOMIACEA, 471
ANOMIIDAE, 471
ANOMURA, 333
*Anoplodactylus maritimus*, 277
*Anoplodactylus parvus*, 277
*Anoplodactylus petiolatus*, 277
*Anoplosolenia brasiliensis*, 71
Anostraca, 278
ANTENNARIIDAE, 587
*Antennarius scaber*, 587
ANTHOMEDUSAE, 142
ANTHOMYIIDAE, 392

Anthophyta, 79
*Anthopleura carneola*, 178
*Anthopleura catenulata*, 178
*Anthopleura varioarmata*, 178
ANTHOZOA, 159
ANTHURIDEA, 367
Antillean Cyerce, 451
Antillean Oxynoe, 448
ANTIPATHARIA, 192
*Antipathes furcata*, 193
*Antipathes hirta*, 193
*Antipathes picea*, 193
*Antipathes tanacetum*, 194
ANTIPATHIDAE, 193
Antizoea, 307
*Antropora granulifera*, 506
*Anurida maritima*, 386
AORIDAE, 376
APHYLLOPHORALES, 24
Apicomplexa, 93
Aplacophora, 392
*Aplidium bermudae*, 548
*Aplidium exile*, 548, 15.1
APLOUSOBRANCHIA, 548
APLUSTRIDAE, 438
*Aplysia aequorea*, 445
*Aplysia ascifera*, 446
*Aplysia dactylomela*, 445, 12.12
*Aplysia morio*, 446
*Aplysia parvula*, 446, 12.6
*Aplysia protea*, 445
APLYSIACEA, 444
APLYSIIDAE, 445
*Aplysilla longispina*, 117, 3.10
*Aplysilla sulfurea*, 117
APLYSILLIDAE, 117
*Aplysina fistularis*, 115
*Aplysina fistularis* f. *ansa*, 115
*Aplysina fistularis* f. *fistularis*, 115
APLYSINIDAE, 115
APODIDA, 541
*Apogon maculatus*, 602
*Apogon townsendi*, 602
APOGONIDAE, 602
Appendicularia, 561
*Apseudes propinquus*, 365
*Apseudes triangulatus*, 365
APSEUDIDAE, 365
APTERYGOTA, 385
*Arabella mutans*, 244
ARABELLIDAE, 244
Arachnactidae, 192
*Arachnanthus nocturnus*, 192, 7.10
Arachnida, 269
ARAEOLAIMIDA, 218
Araneae, 269

*Arca americana*, 464
*Arca gradata*, 463
*Arca imbricata*, 463
*Arca jamaicensis*, 464
*Arca noae*, 463
*Arca occidentalis*, 463
*Arca secticostata*, 464
*Arca zebra*, 463
ARCACEA, 463
Archaebacteria, 8
ARCHAEOBALANIDAE, 304
ARCHAEOGASTROPODA, 402
*Archaias angulatus*, 98
ARCHIANNELIDA, 257
ARCIDAE, 463
ARCOIDA, 463
*Arcopsis adamsi*, 464
ARCTICACEA, 484
*Arctides guineensis*, 332, 10.6
*Ardea herodias*, 662
ARDEIDAE, 662
*Arenaria interpres*, 664
*Arenicola cristata*, 250, 8.9
ARENICOLIDAE, 250
*Argonauta argo*, 499
*Argopecten gibbus*, 470
*Argyrodonax haycocki*, 479
*Argyrotheca bermudana*, 519
*Arhodeoporus perlucidus*, 274
*Aricidea* sp., 246
*Armadilloniscus ellipticus*, 372
*Armandia maculata*, 249
Arrow crab, 355
Arrow dwarf triton, 429
Arrow shrimp, 327
Arrow squid, 496
Arrow worms, 519
Arthropoda, 268
*Arthropoma cecilii*, 512
ARTHROTARDIGRADA, 266
ARTICULATA, 519
ARTICULINA, 515
*Articulina mucronata*, 98
*Asaphis deflorata*, 483
Aschelminthes, 197
*Ascidia curvata*, 554, 15.11
*Ascidia nigra*, 554
*Ascidia tenue*, 554
Ascidiacea, 545
ASCIDIIDAE, 554
ASCLEROCORALLIA, 188
ASCOGLOSSA, 447
ASCOMYCOTINA, 20
ASCOPHORINA, 509
Ascothoracica, 299
ASELLOTA, 370

*Aspidiophorus* sp., 216
ASPIDOCHIROTIDA, 538
*Aspidogaster ringens*, 204
ASPIDOGASTREA, 204
*Aspidosiphon elegans*, 228
ASPIDOSIPHONIDAE, 226
*Assiminea auberiana*, 409
*Assiminea modesta*, 409
*Assiminea succinea*, 409
ASSIMINEIDAE, 409
ASTACIDEA, 329
*Asteractis expansa*, 178
*Asteractis flosculifera*, 178
ASTERIIDAE, 526
*Asterina folium*, 526
ASTERINIDAE, 526
*Asterinopsis pilosa*, 526, 14.14
Asteroidea, 523
*Asteroporpa annulata*, 528, 14.5
Asterozoa, 522
*Astraea longispina*, 406
*Astraea longispinum*, 406
*Astraea phoebia*, 406
*Astrangia solitaria*, 184, 6.7
*Astrapogon stellatus*, 603
ASTROCOENIIDAE, 179
ASTROCOENIINA, 179
ASTROPHORIDA, 124
*Asymmetron lucayanum*, 565
ATELOSTOMATA, 536
ATHECANEPHRIA, 231
ATHECATA, 130, 142
ATHERINIDAE, 592
ATHERINIFORMES, 589
*Atlanta peronii*, 423
Atlantic Assiminea, 409
Atlantic carrier shell, 418
Atlantic deer cowrie, 422
Atlantic false jingle, 472
Atlantic gray cowrie, 422
Atlantic grooved Macoma, 483
Atlantic hairy triton, 428
Atlantic Modulus, 413
Atlantic nut clam, 463
Atlantic partridge tun, 427
Atlantic pearl oyster, 467
Atlantic ridley, 654
Atlantic strawberry cockle, 478
Atlantic thread herring, 579
Atlantic umbrella shell, 447
Atlantic winged oyster, 467
Atlantic wood louse, 426
Atlantic yellow cowrie, 422
ATLANTIDAE, 423
ATYIDAE, 440
Auks, 663

# INDEX

AULOSTOMIDAE, 595
*Aulostomus maculatus*, 595
*Aurelia aurita*, 158
Auricularia, 538
*Auriculastra nana*, 458
*Austrognathia microconulifera*, 212
Aves, 656
*Avicennia germinans*, 87
*Avicula atlantica*, 467
*Avrainvillea nigricans*, 32, 1.9
*Axelsonia* sp., 386
AXIIDAE, 333
AXINELLIDA, 121
AXINELLIDAE, 121
*Axiognathus squamatus*, 531
*Axiopsis serratifrons*, 333, 10.1
Axonolaimidae, 219
*Azorica pfeifferae*, 126

## B

Bacillariophyceae, 67
Bacteria, 9
BAIRDIACEA, 286
Bairdiidae, 286
*Balaenoptera acutorostrata*, 671
BALAENOPTERIDAE, 671
BALANIDAE, 304
BALANOMORPHA, 302
*Balanus amphitrite amphitrite*, 304
*Balanus eburneus*, 304
Baleen whales, 671
*Balistes capriscus*, 642
*Balistes vetula*, 642
BALISTIDAE, 641
Balloonfish, 650
Banded butterfly fish, 616
Banded coral shrimp, 328
Banded moray, 584
Bandtail puffer, 648
*Bangia atropurpurea*, 53
*Bangia fuscopurpurea*, 53
BANGIACEAE, 53
BANGIALES, 53
BANGIOPHYCIDAE, 53
Barbados keyhole limpet, 404
*Barbatia domingensis*, 463
Barber, 595
*Barentsia timida*, 503
Barjack, 604
Barnacles, 299
Barracudas, 630
Barred blenny, 632
*Bartholomea annulata*, 173, 6.5
Bartsch's shipworm, 490
*Baseodiscus delineatus*, 210
BASIDIOMYCOTINA, 23

BASOMMATOPHORA, 456
*Bassia bassensis*, 155
*Basterotia elliptica*, 477
*Bathygobius soporator*, 634
Bathynellacea, 305
*Batillaria minima*, 413
*Batillaria minima rawsoni*, 413
BATILLIPEDIDAE, 267
*Batillipes pennaki*, 268
*Batophora oerstedi*, 40
Bay bean, 83
Bdelloidea, 219
Bdellomorpha, 208
Beach hoppers, 380
Beach lobelia, 87
Beaded Florida miter, 436
Beaded periwinkle, 408
Beaked whales, 671
*Beania mirabilis*, 508
BEANIIDAE, 508
Beard worms, 230
Beau Gregory, 620
Beautiful Truncatella, 410
Beetles, 388
BELONIDAE, 590
Belted cardinal fish, 602
Belted kingfisher, 667
Bentheuphausiidae, 310
Bermuda Aclis, 417
Bermuda anchovy, 579
Bermuda bream, 610
Bermuda Caecum, 411
Bermuda chub, 614
Bermuda cone, 437
Bermuda fireworm, 238
Bermuda halfbeak, 589
Bermuda killifish, 591
Bermuda lobster, 330
Bermuda mussel, 463
Bermuda oyster, 467
Bermuda petrel, 660
Bermuda scallop, 468
Bermuda Volvatella, 448
*Bermudaclis bermudensis*, 417
*Bermudella zoobotryon*, 453
*Bermudrilus peniatus*, 262
*Beroe ovata*, 196
*Beroe punctata*, 196
BEROIDA, 196
*Berthella agassizi*, 447, 12.9
*Bertiliella* sp., 202
BERYCIFORMES, 592
*Bhawania goodei*, 236
Bicolor damselfish, 620
*Biemna microstyla*, 120
BIEMNIDAE, 120

Bigeyes, 602
*Biggaria bermudensis*, 108
*Biggaria echinometris*, 108
Bilateria, 197
Bill Clench's Doris, 452
Billfishes, 638
*Bimeria humilis*, 134
Bipinnaria, 524
Birds, 656
Bitterns, 662
Bivalvia, 460
Black Atlantic Planaxis, 413
Black corals, 192
Black date mussel, 466
Black durgon, 644
Black grouper, 598
Black mangrove, 87
Black rockfish, 598
Black salt-marsh mosquito, 391
Black-bellied plover, 664
Blackberry drupe, 432
Blackdragonfish, 587
Blackdragonfishes, 586
Blackfin tuna, 638
Blackfish, 669
Blauner's sinistral marsh snail, 458
*Blauneria heteroclita*, 458
Bleeding tooth shell, 406
Blennies, 631
BLENNIIDAE, 632
*Blennius cristatus*, 632
Bloody ark, 464
Blue angelfish, 618
Blue bleeder, 115
Blue Chromis, 619
Blue fry, 579
Blue Glaucus, 455
Blue marlin, 639
Blue parrot fish, 625
Blue runner jack, 604
Blue shark, 570
Blue tang, 635
Blue-green algae, 9
Blue-striped grunt, 610
Bluebone porgy, 612
Bluefish, 622
Bluehead wrasse, 622
*Bodianus rufus*, 624
BOLOCEROIDARIA, 173
BOLOCEROIDIDAE, 173
Bonaparte's gull, 667
Bonefish, 578
Bonefishes, 577
Bonito, 606
Bony fishes, 571
Boobies, 660
BOPYRIDEA, 370

Borage family, 86
BORAGINACEAE, 86
Boring Petricola, 487
*Borrichia arborescens*, 87
*Bosellia mimetica*, 450, 13.6
BOSELLIIDAE, 450
*Bostrychia montagnei*, 64, 2.12
BOTHIDAE, 641
*Bothus lunatus*, 641
BOTRYDIALES, 74
*Botrylloides nigrum*, 554, 11.6, 16.3-6
*Botryllus planus*, 556, 16.1&2
*Botryocladia occidentalis*, 61
*Botryocladia pyriformis*, 61
*Bougainvillia niobe*, 144
BOUGAINVILLIIDAE, 134, 144
*Bouvieria agassizii*, 447
*Bowerbankia gracilis*, 504
Box crabs, 342
Box jellies, 157
*Brachidontes domingensis*, 464
*Brachidontes exustus*, 464
Brachiolaria, 524
Brachiopoda, 518
*Brachycarpus biunguiculatus*, 322, 9.10
BRACHYCERA, 391
BRACHYRHYNCHA, 343
BRACHYSCELIDAE, 380
*Brachyscelus crusculum*, 380
BRACHYURA, 339
Brain coral, 184
Branchiobdellida, 232
*Branchiomma nigromaculata*, 255
Branchiopoda, 278
*Branchiostoma bermudae*, 565
*Branchiostoma caribaeum*, 565
BRANCHIOSTOMIDAE, 564
Branchiura, 278
BRIAREIDAE, 165
*Briareum polyanthes*, 165
Bridled goby, 634
BRISSIDAE, 537
Bristle worms, 232
Bristlemouth, 586
Bristlemouths, 586
Brittle stars, 527
Brotulas, 588
BROTULIDAE, 588
Brown algae, 41
Brown anemone, 173
Brown Sargassum snail, 414
*Bruuniella* sp., 284
BRYOPSIDACEAE, 37
*Bryopsis pennata*, 37
*Bryopsis plumosa*, 37
Bryozoa, 500
Bubble marsh snail, 457

BUCCINACEA, 433
BUCCINIDAE, 434
Bucktooth parrot fish, 629
Buffalo trunkfish, 648
Bugs, 387
*Bugula neritina*, 508
BUGULIDAE, 508
*Bulbamphiascus imus*, 296
*Bulla striata*, 439
BULLACEA, 439
BULLIDAE, 439
BULLOMORPHA, 438
Bumblebee shrimp, 323
*Bunodactis stelloides* var. *carneola*, 178
*Bunodactis stelloides* var. *catenulata*, 178
*Bunodeopsis antilliensis*, 173, 6.1
*Bunodeopsis globulifera*, 173
*Bunodosoma granuliferum*, 178, 6.12
Bur grass, 90
BURSOVAGINOIDEA, 212
Bush coral, 186
BUTOMALES, 88
Butter hamlet, 596
Butterfly fishes, 615
Buttonwood, 85
By-the-wind sailor, 131

**C**
Cabbage-head jellyfish, 158
*Cadulus quadridentatus*, 460
*Cadulus tetradon*, 460
*Cadulus tetraschistus*, 460
CAECIDAE, 410
*Caecum obesum*, 410
*Caecum plicatum*, 410
*Caecum termes*, 410
*Caecum tornatum*, 411
CAENOGASTROPODA, 407
Caesar grunt, 610
*Cafius bistriatus*, 388
Cahow, 660
*Cakile lanceolata*, 83
*Calamus bajonado*, 612
CALANOIDA, 289
*Calanopia americana*, 290
*Calappa flammea*, 342, 11.12
*Calappa gallus*, 342
*Calappa ocellata*, 342
CALAPPIDAE, 342
CALCAREA, 126
*Calcinus tibicen*, 335
*Calcinus verrilli*, 336, 10.2
*Calciosolenia murrayi*, 71
Calico clam, 487
Calico crab, 346
Calico scallop, 470
CALIGIDAE, 299

*Caligus bonito*, 299
*Caligus robustus*, 299
CALIPHYLLIDAE, 451
*Callianassa branneri*, 333, 10.10
*Callianassa longiventris*, 334
CALLIANASSIDAE, 333
*Callidactylus asper*, 342, 11.9
*Callinectes ornatus*, 345
CALLOPORIDAE, 506
*Callyspongia vaginalis*, 118
CALLYSPONGIIDAE, 118
*Caloglossa leprieurii*, 62
CALYCOPHORAE, 152
CALYCOPSIDAE, 144
Calyptopis, 311
CALYPTRAEACEA, 417
CALYPTRAEIDAE, 418
CAMPANULALES, 87
*Campanularia marginata*, 137
CAMPANULARIIDAE, 136, 146
*Canavalia rosea*, 83
Cancellate risso, 409
Cancellate Semele, 484
*Candacia ethiopica*, 292
CANDACIIDAE, 292
Candy stick tellin, 482
*Cantherhines macrocerus*, 644
*Cantherines pullus*, 644
*Canthidermis maculatus*, 642
*Canthidermis sufflamen*, 644
*Canthigaster rostrata*, 648
CAPITATA, 130, 142
*Capitella capitata*, 250
CAPITELLIDAE, 250
*Caprella danilevskii*, 381
*Caprella equilibra*, 381
CAPRELLIDAE, 381
CAPRELLIDEA, 380
*Capsa coccinea*, 483
*Capsa spectabilis*, 483
CARANGIDAE, 603
*Caranx crysos*, 604
*Caranx latus*, 604
*Caranx ruber*, 604
CARAPIDAE, 588
*Carapus bermudensis*, 588
CARCHARHINIDAE, 570
*Carcharhinus galapagensis*, 570
*Carcharodorhynchus* sp., 202
*Carcinoecetes bermudensis*, 103
CARDIACEA, 478
CARDIIDAE, 478
Cardinal fishes, 602
Cardinal soldierfish, 594
*Cardisoma guanhumi*, 354
*Cardita domingensis*, 477
CARDITACEA, 477

*Carditopsis smithii*, 477
*Cardium petitianum*, 478
*Cardium serratum*, 479
*Caretta caretta*, 654
Caribbean coral shell, 432
Caribbean reef squid, 496
Caribbean risso, 409
Caribbean Truncatella, 410
CARIDEA, 319
*Carinaria lamarcki*, 424
*Carinaria mediterranea*, 424
CARINARIACEA, 423
CARINARIIDAE, 424
Carnivora, 668
Carnosina, 504
Carpetweed family, 82
*Carpias bermudensis*, 370
*Carpilius corallinus*, 349
Cartilaginous fishes, 566
*Carybdea alata*, 158
CARYOPHYLLIIDAE, 187
CARYOPHYLLIINA, 187
CASSIDAE, 426
*Cassiopea xamachana*, 158
*Cassis madagascariensis* f. *spinella*, 426
*Cataleptodius floridanus*, 348
*Catenella repens*, 61
CATENULIDA, 199
Catesby's Risso, 409
CATOPHRAGMIDAE, 302
*Catophragmus imbricatus*, 302
Caudofoveata, 392
*Caulerpa cupressoides*, 35
*Caulerpa mexicana*, 35
*Caulerpa prolifera*, 35
*Caulerpa racemosa*, 35
*Caulerpa sertularioides*, 35, 1.8, 14.8
*Caulerpa verticillata*, 35, 1.7
CAULERPACEAE, 34
CAULERPALES, 30
*Caulleriella* sp., 249
*Cavolinia cuspidata*, 442
*Cavolinia flava*, 442
*Cavolinia gegenbauri*, 442
*Cavolinia gibbosa*, 442
*Cavolinia hargeri*, 442
*Cavolinia mucronata*, 442
*Cavolinia plana*, 442
CAVOLINIIDAE, 441
Cayenne keyhole limpet, 403
*Celleporaria vagans*, 514
CELLEPORARIIDAE, 514
CELLULARINA, 508
*Cenchrus tribuloides*, 90
Centipedes, 382
CENTRICAE, 68
*Centroceras clavulatum*, 62

*Centroderes spinosus*, 222
CENTRODERIDAE, 222
*Centropages violaceus*, 290
CENTROPAGIDAE, 290
*Centropyge argi*, 616
CEPHALASPIDEA, 438
Cephalocarida, 278
Cephalochordata, 562
*Cephalodiscus gracilis*, 545
*Cephalopholis cruentata*, 597
*Cephalopholis fulva*, 597
Cephalopoda, 492
CERAMIACEAE, 62
CERAMIALES, 62
*Ceramium byssoideum*, 62, 2.11
Ceramonematidae, 217
*Cerataulina bergonii*, 69
*Ceratium furca*, 78
*Ceratium fusus*, 79
*Ceratoconcha domingensis*, 304, 6.14
CERATOPOGONIDAE, 391
Cercariae, 204
*Cerebratulus leidyi*, 209
CERIANTHARIA, 192
*Cerianthus natans*, 192
Ceriantipatharia, 159
Cerinula, 164
CERITHIACEA, 411
CERITHIIDAE, 413
*Cerithiopsis greeni*, 414
*Cerithium bermudae*, 414
*Cerithium ferrugineum*, 414
*Cerithium litteratum*, 413
*Cerithium litteratum* f. *semiferrugineum*, 414
*Cerithium lutosum*, 414
*Cerithium variabile*, 414
CESTIDA, 195
Cestoda, 205
Cestodaria, 205
*Cestum veneris*, 195
CETACEA, 668
CHAETANGIACEAE, 54
*Chaetoceros glaudazii*, 68
*Chaetodon aculeatus*, 616
*Chaetodon capistratus*, 616
*Chaetodon ocellatus*, 615
*Chaetodon sedentarius*, 616
*Chaetodon striatus*, 616
CHAETODONTIDAE, 615
Chaetognatha, 519
*Chaetomorpha linum*, 30
*Chaetomorpha media*, 30
CHAETONOTIDA, 216
CHAETONOTIDAE, 216
CHAETOPTERIDAE, 247
Chaetosphera, 235
Chain dove shell, 433

Chain moray, 584
Chalky buttercup, 475
*Chama bermudensis*, 476
*Chama congregata*, 476
*Chama exogyra*, 476
*Chama linguafelis*, 476
*Chama macerophylla*, 475
*Chama sinuosa*, 476
CHAMACEA, 475
*Chamaesyce buxifolia*, 84
CHAMIDAE, 475
*Channomuraena vittata*, 584
CHARADRIIDAE, 663
CHARADRIIFORMES, 663
*Charonia nobilis*, 428
*Charonia tritonis*, 428
*Charonia variegata*, 428
Charophyceae, 27
Checkerboard clam, 487
CHEILOPORINIDAE, 510
CHEILOSTOMIDA, 505
Chelicerata, 268
*Chelonaplysilla erecta*, 117
Chelonethida, 269
Chelonia, 651
*Chelonia mydas*, 653
*Chelonibia testudinaria*, 304
CHELONIBIIDAE, 302
CHELONIIDAE, 653
*Chelophyes appendiculata*, 154
CHENOPODIACEAE, 81
CHENOPODIALES, 81
Cherub fish, 616
Chicken liver sponge, 125
Chilopoda, 382
Chimaeras, 566
China shell, 481
*Chione venetiana*, 486
*Chione* cf. *cancellata*, 485
*Chiridota rotifera*, 541
CHIRIDOTIDAE, 541
CHIRONOMIDAE, 391
*Chiton marmoratus*, 397
*Chiton squamosus*, 397
*Chiton tuberculatus*, 397
CHITONIDAE, 396
Chitons, 394
*Chlamys imbricata*, 468
Chlorophyta, 27
*Chondria littoralis*, 66
Chondrichthyes, 566
*Chondrilla nucula*, 125
Chondrophora, 128
*Chondrosia collectrix*, 125
CHONDROSIIDAE, 125
Chordata, 545
*Chordodasys riedli*, 215

*Chorinus heros*, 358
CHROMADORIDA, 217
*Chromis cyaneus*, 619
CHROMODORIDIDAE, 451
*Chromodoris bistellata*, 451, 13.13
*Chromodoris clenchi*, 452, 13.2
*Chromodoris neona*, 452
*Chromodoris zebra*, 452
CHRYSOMONADALES, 71
CHRYSOPETALIDAE, 236
Chrysophyceae, 70
Chrysophyta, 66
*Chrysymenia enteromorpha*, 61
CHTHAMALIDAE, 302
*Chthamalus angustitergum thompsoni*, 302
*Cicindela trifasciata*, 388
CICINDELIDAE, 388
CICONIIFORMES, 662
CIDAROIDA, 533
Ciliates, 104
Ciliophora, 104
*Cinachyra alloclada*, 126, 4.5
*Cinachyra cavernosa*, 126
*Cingulina babylonia*, 440
CIRRATULIDAE, 248
*Cirriformia punctata*, 249, 8.7
Cirripedia, 299
*Cistella cistellula*, 519
*Cittarium pica*, 405
CLADOCERA, 279
Cladocopida, 280
*Cladonema radiatum*, 132, 143
CLADONEMATIDAE, 132, 143
*Cladophora fascicularis*, 30
*Cladophora fuliginosa*, 30
*Cladophora prolifera*, 30
CLADOPHORACEAE, 30
CLADOPHORALES, 29
*Cladophoropsis membranacea*, 37
Clam shrimps, 278
Clamacore, 625
Clams, 460
Clapp's shipworm, 490
CLATHRIIDAE, 120
*Clathrina coriacea*, 126, 4.6
CLATHRINIDA, 126
Clathrinidae, 126
*Clausocalanus furcatus*, 290
*Clavelina oblonga*, 552, 15.17
*Clavelina picta*, 552, 15.13
CLAVIDAE, 134, 144
Clench's emperor helmet, 426
*Clepticus parrai*, 624
*Clibanarius anomalus*, 336
*Clibanarius antillensis*, 336
*Clibanarius tricolor*, 336
Cliff crab, 353

CLINIDAE, 631
Clinids, 631
*Clio pyramidata*, 442
*Cliona caribbaea*, 122, 4.11
*Cliona caribboea*, 122
*Cliona dioryssa*, 124, 4.10
*Cliona flavifodina*, 123
*Cliona lampa*, 124, 4.8&9
*Cliona vermifera*, 124
*Clione elegantissima*, 443
*Clione kincaid*, 443
*Clione limacina*, 443
*Clione papillonacea*, 443
CLIONIDAE, 122, 443
*Clionina longicaudata*, 443
Clitellata, 232
*Clitellio arenicolus*, 262
*Clunio* sp., 391
CLUPEIDAE, 579
CLUPEIFORMES, 578
*Clymenella somersi*, 252
CLYPEASTEROIDA, 535
*Clytia cylindrica*, 136
*Clytia fragilis*, 136
*Clytia noliformis*, 136
Cnidaria, 127
Cnidosporidia, 93
Coast spurge, 84
Coccidia, 103
Cocco lantern fish, 585
Coccogoneae, 9
COCCOLITHACEAE, 72
COCCOLITHINEAE, 71
Coccolithophorida, 70
Cocoa damselfish, 621
*Codakia costata*, 474
*Codakia orbicularis*, 473
*Codakia orbiculata*, 474
CODIACEAE, 32
*Codium decorticatum*, 32, 1.13
*Codium intertextum*, 32
*Codium taylori*, 32
Coelenterata, 111
COELODENDRIDAE, 102
*Coelodendrum ramosissimum*, 103
COELOMYCETES, 23
COELOSTEGA, 506
*Coenobita clypeatus*, 334
COENOBITIDAE, 334
COENOBITOIDEA, 334
*Coenocyathus goreaui*, 187
Coenothecalia, 164
Coffee bean marsh snail, 457
*Colangia immersa*, 184
COLEOIDEA, 494
COLEOPTERA, 388
COLLEMBOLA, 386

COLOMASTIGIDAE, 376
*Colomastix pusilla*, 376
Colonial anemones, 190
*Colopisthus parvus*, 368
Colorful Atlantic Natica, 425
*Colpomenia sinuosa*, 46
*Colubraria igniflua*, 429
*Colubraria lanceolata*, 429
*Colubraria swiftii*, 429
COLUBRARIIDAE, 429
*Columbella cribraria*, 433
*Columbella mercatoria*, 433
*Columbella somersiana*, 433
COLUMBELLIDAE, 433
Comb-jellies, 194
COMBRETACEAE, 84
Combretum family, 84
Combtooth blennies, 632
Common land crab, 354
Common spider crab, 356
Common Atlantic diplodon, 475
Common baby's ear, 424
Common dolphin, 606, 669
Common dove shell, 433
Common egg cockle, 479
Common goose barnacle, 301
Common jingle shell, 471
Common marsh snail, 457
Common mud crab, 349
Common octopus, 499
Common paper nautilus, 499
Common purple sea snail, 415
Common tern, 667
Common West Indian chiton, 397
COMPOSITAE, 87
CONACEA, 436
Concentric Ervilia, 479
Conch fish, 603
*Conchoderma virgatum*, 301
*Conchoecia spinirostris*, 284
Conchostraca, 278
*Condylactis gigantea*, 176, 9.12
*Condylactis passiflora*, 176
CONDYLOCARDIIDAE, 477
*Condylostoma arenarium*, 109
Coney, 507
Conger eels, 581
*Conger triporiceps*, 581
CONGRIDAE, 581
CONIDAE, 436
Conifers, 26
*Conocarpus erectus*, 85
CONOPHORALIA, 212
*Conus agassizi*, 437
*Conus bermudensis*, 437
*Conus mindanus*, 437
*Conus mus*, 436

CONVOLVULACEAE, 86
Copepoda, 288
Copepodite, 289
*Copidognathus floridensis*, 274
*Copidognathus pulcher*, 274
*Copilia mirabilis*, 294
Copper sweeper, 614
Coracidium, 206
CORACIIFORMES, 667
Coral anemones, 188
Coral-boring Gregariella, 466
Coral clam, 485
Coral crab, 349
CORALLIMORPHARIA, 188
CORALLIMORPHIDAE, 188
CORALLINACEAE, 56
*Coralliophaga carditoidea*, 485
*Coralliophaga coralliophaga*, 485
*Coralliophaga dactylus*, 477
*Coralliophila abbreviata*, 432
*Coralliophila caribaea*, 432
CORALLIOPHILIDAE, 432
Corals, 159
*Corbula disparilis*, 488
CORBULIDAE, 488
CORIXIDAE, 387
Cormorants, 660, 662
CORONATAE, 158
Coronet fish, 595
Coronet fishes, 595
CORONULIDAE, 302
COROPHIIDAE, 376
*Corophium acutum*, 376
Corroding worm shell, 412
Corrugated Caecum, 410
Corrugated jewel box, 476
CORYCAEIDAE, 294
*Corycaeus flaccus*, 295
*Corycaeus speciosus*, 295
*Corynactis parvula*, 188, 7.5
*Coryphaena hippurus*, 606
CORYPHAENIDAE, 606
*Coryphopterus glaucofraenum*, 634
*Coscinasterias tenuispina*, 527, 14.12
*Cossura* sp., 249
COSSURIDAE, 249
*Costasiella lilianae*, 450
*Costasiella nonatoi*, 451
*Costasiella ocellifera*, 450, 13.11
COSTASIELLIDAE, 450
COSTELLARIIDAE, 436
Cottonwick grunt, 609
COTYLEA, 203
Crabs, 312
*Crassatella guadelupensis*, 478
*Crassatella parva*, 478
CRASSATELLACEA, 478

CRASSATELLIDAE, 478
*Crassibrachia sandersi*, 231
*Crassimarginatella crassimarginata*, 506
*Crassinella lunulata*, 478
*Cratena pilata*, 455, 13.4
Creole fish, 595
Creole wrasse, 624
*Crepidacantha poissonii*, 513
CREPIDACANTHIDAE, 513
*Crepidula aculeata*, 418
*Creseis acicula*, 442
*Creseis recta*, 442
Crested goby, 634
*Cribrilaria radiata*, 509
CRIBRILINIDAE, 509
CRIBRIMORPHINA, 509
Crinoidea, 522
Crinozoa, 522
*Crisia denticulata*, 516
*Crisia eburnea*, 516
*Crisia elongata*, 516
CRISIIDAE, 516
Croaker, 612
Croakers, 612
*Crocethia alba*, 664
Crocodylia, 651
*Cronius tumidulus*, 345
Cross-barred anemone, 179
*Crossobothrium angustum*, 206
CRUCIFERAE, 83
Crustacea, 277
*Cryptarachne agardhii*, 61
CRYPTOCERATA, 387
CRYPTODIRA, 653
Cryptodonta, 460
CRYPTONEMIACEAE, 58
CRYPTONEMIALES, 56
Crystalline Cyerce, 451
Ctenophora, 194
CTENOSTOMIDA, 503
Cuban marsh snail, 458
*Cubanocuma* cf. *gutzui*, 364
CUBOMEDUSAE, 157
CUBOSPHAERIDAE, 102
Cubozoa, 127
CUCUMARIIDAE, 540
Cucumber Melanella, 417
CULICIDAE, 390
*Culicoides bermudensis*, 391
*Culicoides crepuscularis*, 391
CUMACEA, 362
CUNINIDAE, 148
Cuttlefishes, 492, 495
Cuvier's beaked whale, 671
CUVIERIDAE, 441
CYAMIACEA, 476
Cyanobacteria, 9

Cyanophyta, 9
Cyatholaimidae, 217
*Cycloes bairdii*, 342
*Cyclograpsus integer*, 350
CYCLOPOIDA, 292
CYCLORHAGAE, 221
CYCLORHAGIDA, 221
CYCLORRHAPHA, 392
Cyclostomata, 565
CYCLOSTOMIDA, 514
Cydippid, 195
CYDIPPIDA, 195
*Cyerce antillensis*, 451, 13.1
*Cyerce cristallina*, 451, 13.9
*Cylindrobulla beaui*, 440
CYLINDROBULLIDAE, 440
CYLINDROLEBERIDIDAE, 283
CYLINDROPSYLLIDAE, 297
*Cymadusa filosa*, 374
CYMATIACEA, 427
CYMATIIDAE, 428
*Cymatium chlorostomum*, 428
*Cymatium muricinum*, 428
*Cymatium nicobaricum*, 428
*Cymatium pileare*, 428
*Cymatium tuberosum*, 428
*Cymbaloporetta atlanticus*, 99
*Cymbula acicularis*, 422
*Cymodocea manatorum*, 88
CYMODOCEACEAE, 88
*Cymopolia barbata*, 41, 1.12
*Cymothoa oestrum*, 368
*Cyphoma gibbosum*, 422, 5.17
Cyphonautes, 502
*Cypraea cervus*, 422
*Cypraea cinerea*, 422
*Cypraea exanthema*, 422
*Cypraea spurca*, 422
*Cypraea spurca acicularis*, 422
*Cypraeacardia hornbeckiana*, 485
CYPRAEACEA, 421
*Cypraecassis testiculus*, 426
CYPRAEIDAE, 422
CYPRIDACEA, 285
*Cyprideis* sp., 286
CYPRIDIDAE, 286
CYPRINODONTIDAE, 590
Cypris, 301
*Cypselurus furcatus*, 589
Cyrtopia, 311
Cysticercus, 206
*Cystodytes dellechiajei*, 550
CYSTONECTAE, 151
*Cystoseira fimbriata*, 50
CYSTOSEIRACEAE, 50
CYTAEIDIDAE, 144
*Cytaeis tetrastyla*, 144

CYTHERACEA, 286
*Cytherea hebraea*, 486
*Cytherea penistoni*, 486
*Cytherella lata*, 285
CYTHERELLIDAE, 284
*Cytherelloidea irregularis*, 285
CYTHERIDEIDAE, 286
*Cyttaronema reticulatum*, 217

**D**

*Dactylospora haliotrepha*, 23
Damselfishes, 618
Dandelion coral, 165
Dandelion family, 87
*Dardanus insignis*, 335, 10.9
*Dardanus venosus*, 335
Dark-star anemone, 176
Darling beetles, 390
*Darwinella rosacea*, 117, 7.11
*Dasya baillouviana*, 63, 2.7
DASYACEAE, 63
*Dasybranchus lunulatus*, 250
DASYCLADACEAE, 39
DASYCLADALES, 39
*Dasycladus vermicularis*, 40
Dead man's fingers, 126
DECAPODA, 312
*Decapterus macarellus*, 604
*Decapterus punctatus*, 604
Decorator crab, 356
Deer hamlet, 598
Deerflies, 391
DELESSERIACEAE, 62
Delicate tellin, 482
DELPHINIDAE, 669
*Delphinus delphis*, 669
Deltoid rock shell, 431
DEMOSPONGEA, 114
*Dendrilla nux*, 117
DENDROCHIROTIDA, 540
DENDRONOTACEA, 454
Dendrophylliidae, 188
DENDROPHYLLIINA, 188
*Dendropoma annulatus*, 412
*Dendropoma corrodens*, 412
*Dendropoma irregularis*, 412
DENTALIIDAE, 459
*Dentalium semistriolatum*, 459
*Deraiophorus* sp., 272
*Derbesia osterhoutii*, 37
DERBESIACEAE, 37
DERMAPTERA, 386
DERMOCHELYIDAE, 656
*Dermochelys coriacea*, 656
*Dermomurex elizabethae*, 430
*Desmacella janiae*, 115
DESMACIDONIDAE, 120

DESMODORIDA, 217
DESMOPHYCEAE, 77
DESMOSCOLECIDA, 219
Desmoscolecidae, 219
Desor larva, 209
*Detracia bullaoides*, 457
DEUTEROMYCOTINA, 23
Deuterostomia, 197
DEUTSCHLANDIACEAE, 71
*Diacria trispinosa*, 442
*Diadema antillarum*, 533, 14.1
DIADEMATACEA, 533
DIADEMATOIDA, 533
*Diaperoecia floridana*, 515
DIAPEROECIIDAE, 515
Diatomeae, 67
Diatoms, 67
DIBRANCHIATA, 494
*Dichocoenia stokesi*, 186
*Dichotomia cannoides*, 145
DICOTYLEDONEAE, 81
*Dictyocha fibula*, 73
DICTYOCHACEAE, 73
*Dictyodendrilla nux*, 117
DICTYODENDRILLIDAE, 117
*Dictyopteris justii*, 45
*Dictyosphaeria cavernosa*, 38
*Dictyota cervicornis*, 45
*Dictyota ciliolata*, 43
*Dictyota ciliolata* var. *bermudensis*, 45
*Dictyota dentata*, 43
*Dictyota dichotoma*, 43
*Dictyota divaricata*, 45, 2.2
DICTYOTACEAE, 43
DICTYOTALES, 43
*Dicyema* sp., 208
DICYEMIDA, 207
DIDEMNIDAE, 548
*Didemnum* sp., 550, 15.5
*Didymosphaeria enalia*, 22
DIGENEA, 204
*Digenia simplex*, 64
Digitate thorny oyster, 471
DIKONOPHORA, 365
Dinoflagellata, 75
Dinoflagellates, 75
DINOPHILIDAE, 258
DINOPHYCEAE, 77
DINOPHYSIALES, 77
*Dinophysis caudata* var. *pedunculata*, 77
Dinophyta, 75
*Diodon holocanthus*, 650
*Diodon hystrix*, 650
DIODONTIDAE, 650
*Diodora cayenensis*, 403
*Diodora listeri*, 404
DIOGENIDAE, 335

DIOSACCIDAE, 296
*Diphyes bojani*, 153
DIPHYIDAE, 152
DIPLEUROSOMATIDAE, 145
*Diplodasys* sp., 215
*Diplodonta punctata*, 475
*Diplodus bermudensis*, 610
*Diploneis bombus*, 70
Diplopoda, 381
*Diploria labyrinthiformis*, 184
*Diploria strigosa*, 184
*Diplosoma listerianum*, 550, 15.3&4
DIPLURA, 385
DIPTERA, 390
DISCIADIDAE, 320
*Discias atlanticus*, 320
*Discorbis mira*, 99
*Discosoma carlgreni*, 190, 7.6&7
*Discosoma sanctithomae*, 188, 7.12&13
DISCOSOMATIDAE, 188
*Discosphaera tubifera*, 73
*Disporella buski*, 516
*Distaplia corolla*, 552, 15.7
Diurnal birds of prey, 663
*Diurodrilus* sp., 258
Divided sand clam, 484
Divided shipworm, 490
Doctorfish, 635
Doctorfly, 392
*Dodecaceria* sp., 249
*Dolabrifera dolabrifera*, 446, 12.15
*Dolabrifera virens*, 446
DOLIOLIDA, 559
Doliolids, 558
*Doliolum denticulatum*, 560
*Doliopsis rubescens*, 560
Dolphinfishes, 606
Dolphins, 668
Domingo mussel, 464
DORIDACEA, 451
*Doropygus pulex*, 299
*Dorvillea sociabilis*, 244
DORVILLEIDAE, 244
*Doryteuthis plei*, 496
*Doto ocellifera*, 450
Double-crested cormorant, 662
*Drepanophora tuberculata*, 514
*Dromia erythropus*, 340
DROMIACEA, 340
*Dromidia antillensis*, 340
DROMIIDAE, 340
Dusky blenny, 631
Dusky shark, 570
Dusky squirrelfish, 592
Dwarf Atlantic Planaxis, 413
Dwarf cancellate Chione, 485
Dwarf Costasiella, 451

Dwarf herring, 579
Dwarf horn shell, 414
Dwarf periwinkle, 407
Dwarf tiger Lucina, 474
*Dynamena crisioides*, 137
*Dynamena disticha*, 137
*Dynamena quadridentata*, 137
*Dynamenella perforata*, 368
*Dysidea crawshayi*, 121
*Dysidea etheria*, 115, 3.4&5
*Dysidea fragilis* f. *algafera*, 115
*Dysidea janiae*, 115, 3.6
DYSIDEIDAE, 115

**E**
Eagle rays, 571
Earwigs, 386
Eastern salt-marsh mosquito, 390
Ecardines, 518
ECHENEIDAE, 603
*Echeneis naucrates*, 603
*Echidna catenata*, 584
ECHINACEA, 533
ECHINISCOIDEA, 268
*Echiniscoides sigismundi*, 268
ECHINISCOIDIDAE, 268
*Echinoderes bispinosus*, 222
ECHINODERIDAE, 221
Echinodermata, 522
ECHINOIDA, 535
Echinoidea, 531
*Echinometra lucunter*, 535
*Echinoneus cyclostomus*, 535, 14.2
Echinopluteus, 532
*Echinorhynchus medius*, 224
Echinozoa, 522
Echiura, 228
Echiuridae, 229
*Ecteinascidia conklini*, 552, 15.6
*Ecteinascidia minuta*, 554
*Ecteinascidia turbinata*, 552, 15.12
*Ectinosoma dentatum*, 295
ECTINOSOMIDAE, 295
ECTOCARPACEAE, 43
ECTOCARPALES, 43
*Ectocarpus confervoides*, 43
*Ectocarpus siliculosus*, 43
*Ectopleura minerva*, 143
*Ectopleura pacifica*, 130, 143
ECTOPROCTA, 503
Eels, 580
*Ehlersia cornuta*, 240
*Elagatis bipinnulatus*, 604
Elasmobranchii, 566
*Elasmopus rapax*, 377
ELEUTHERIIDAE, 143
Elizabeth's Aspella, 430

Elliptical Basterotia, 477
*Ellisella barbadensis*, 172, 14.5
ELLISELLIDAE, 172
ELLOBIIDAE, 457
Elongate Macoma, 482
ELOPIFORMES, 577
*Elphidium sagrum*, 99
*Elysia cauze*, 449
*Elysia duis*, 450
*Elysia flava*, 450, 12.11
*Elysia papillosa*, 448, 12.1
*Elysia picta*, 450, 12.14
*Elysia subornata*, 449, 12.10
*Elysia tuca*, 450, 12.7
ELYSIACEA, 448
ELYSIIDAE, 448
Emerald nerite, 407
*Emiliana huxlei*, 73
*Encentrum* sp., 220
Enchelidiidae, 218
*Enchelycore nigricans*, 582
ENCHYTRAEIDAE, 262
*Enchytraeus albidus*, 262
ENDEIDAE, 276
*Endeis spinosa*, 276
ENDOMYARIA, 176
ENGRAULIDAE, 579
ENOPLIDA, 217
*Enoplobranchus sanguineus*, 254, 8.6
*Entemnotrochus adansonianus bermudensis*, 403
Enterogona, 548, 552
*Enteromorpha flexuosa*, 29
*Enteromorpha lingulata*, 29
Enteropneusta, 541
*Entodesma beana*, 491
*Entomacrodus nigricans*, 632
Entomostraca, 278
ENTOPROCTA, 502
*Ephelota gemmipara*, 106
EPHYDRIDAE, 392
Ephyrae, 156
*Epialtus bituberculatus*, 355
EPICARIDEA, 370
*Epicystis crucifer*, 179, 6.9&10
EPIGONICHTHYIDAE, 565
*Epinephelus adscensionis*, 597
*Epinephelus afer*, 597
*Epinephelus guttatus*, 598
*Epinephelus morio*, 598
*Epinephelus striatus*, 598
*Epistomia bursaria*, 509
EPISTOMIIDAE, 509
EPITONIACEA, 415
EPITONIIDAE, 415
*Epitonium krebsii*, 416
*Epitonium lamellosum*, 416
EPIZOANTHIDAE, 191

*Epizoanthus minutus*, 191, 7.11
*Equetus acuminatus*, 612
Erect worm shell, 412
Eretmocaris, 315
*Eretmochelys imbricata*, 654
Erichthus, 308
*Eriphia gonagra*, 348
Errantia, 232
*Ervilia concentrica*, 479
*Ervilia nitens*, 479
*Ervilia subcancellata*, 479
ERYTHRAEIDAE, 274
*Escharina hyndmanni*, 512
*Escharina pesanseris*, 512
*Eubostrichus dianae*, 217
Eucarida, 305
Eucestoda, 205
*Euchaetomera tenuis*, 361
*Euchaetomera typica*, 361
*Euchelus guttarosea*, 405
*Eucheuma isiforme*, 59
Euchordata, 565
*Eucidaris tribuloides*, 533
*Eucinostomus gula*, 608
*Euclymene coronatus*, 252, 8.13
EUDENDRIIDAE, 134
*Eudendrium carneum*, 134
*Eudendrium hargitti*, 134
*Eudendrium ramosum*, 134
*Eudistoma clarum*, 552, 15.10
*Eudistoma obscuratum*, 552, 15.9
*Eudistoma olivaceum*, 550, 15.8
*Eudoxoides mitra*, 154
*Eudoxoides spiralis*, 154
EUECHINOIDEA, 533
Euglenophyceae, 75
Euglenophyta, 74
Eugregarinida, 103
Eukaryotes, 7
*Eukrohnia fowleri*, 520
EULIMACEA, 417
Eumetazoa, 111
Eumycota, 17
*Eunice cariboea*, 242
*Eunice vittata*, 242, 8.1
*Eunicea grandis*, 169
*Eunicea calyculata*, 169
*Eunicea clavigera*, 169
*Eunicea fusca*, 167
*Eunicea tourneforti*, 167, 5.15&16
*Eunicea tourneforti* f. *atra*, 169
*Euniceopsis atra*, 167
*Euniceopsis grandis*, 169
*Euniceopsis tourneforti*, 167
EUNICIDAE, 242
*Eupatinapta acanthia*, 541
*Euphausia brevis*, 311

Euphausiacea, 310
EUPHAUSIIDAE, 311
*Euphorbia buxifolia*, 84
*Euphorbia mesembrianthemifolia*, 84
EUPHORBIACEAE, 84
*Eupolymnia crassicornis*, 254, 8.12
*Eurydice littoralis*, 368
*Eurypon clavatum*, 122
EURYPONIDAE, 121
*Eurythoe complanata*, 236, 8.8
Eutardigrada, 265
EUTHECOSOMATA, 441
Eutheria, 668
*Euthynnus alletteratus*, 636
EUTREPTIACEAE, 75
*Eutreptia* cf. *viridis*, 75
*Evadne spinifera*, 279
*Evadne tergestina*, 279
*Exechonella antillea*, 509
EXECHONELLIDAE, 509
EXOCOETIDAE, 589
*Exocorallana quadricornis*, 368
*Exogone dispar*, 240
*Exogone gemmifera*, 240
*Exogone hebes*, 240
*Exogone verrugera*, 240
*Exonautes nonsuchi*, 590
Eye-bearing Costasiella, 450

**F**
Fairy shrimps, 278
FALCONIFORMES, 663
FALCONOIDEA, 663
Falcons, 663
False corals, 188
False horn shell, 413
False jewel box, 476
False Scorpions, 269
Fantail mullet, 629
*Farranula rostrata*, 295
*Favartia alveata*, 431
*Favartia cellulosa*, 431
*Favia fragum*, 182
FAVIIDAE, 182
FAVIINA, 182
FAVORINIDAE, 454
*Favorinus auritulus*, 454, 13.15
*Felimare bayeri*, 452
Ferns, 26
*Fibulia bermuda*, 121
FILIFERA, 134, 143
*Filograna implexa*, 257
FILOSIA, 96
FILOSPERMOIDEA, 212
Fine-scale rockfish, 600
Fire coral, 134
Fireworms, 236

Fish hawk, 663
Fish leeches, 264
Fish lice, 366
*Fissurella alternata*, 403
*Fissurella antillarum*, 404
*Fissurella barbadensis*, 404
*Fissurella bermudensis*, 404
FISSURELLACEA, 403
FISSURELLIDAE, 403
*Fistularia tabacaria*, 595
FISTULARIIDAE, 595
FLABELLIFERA, 368
Flagellata, 93
*Flagellophora* sp., 200
Flame fish, 602
Flaming box crab, 342
Flamingo tongue, 422
Flat crab, 352
Flatfishes, 641
Flatworms, 197
Flies, 390
*Florarctus antillensis*, 266
Florida Onchidella, 443
Florida worm shell, 412
FLORIDEOPHYCIDAE, 53
Flowering plants, 79
Flukes, 203
Flying fishes, 589
FORAMINIFERIDA, 96
FORCIPULATIDA, 526
FORFICULINA, 387
*Fortuynia* sp., 275
FORTUYNIIDAE, 275
*Fosliella farinosa*, 56
FOSSARIDAE, 417
*Fossarus orbignyi*, 417
Foureye basslet, 601
Foureye butterfly fish, 616
"Four-eyed" swimming crab, 345
Four-spotted Trivia, 421
Four-toothed Cadulus, 460
Fourwing flying fish, 590
Free-living flatworms, 198
French grunt, 609
Frilled paper cockle, 478
Frillfin goby, 634
Frilly dwarf Murex, 431
Fringed filefish, 644
*Fritillaria borealis* f. *sargassi*, 562
FRITILLARIIDAE, 562
Frog crabs, 340
Frogfishes, 587
*Frontonia marina*, 108
FUCALES, 48
*Fucellia intermedia*, 392
*Fundulus bermudae*, 591
*Fundulus heteroclitus*, 592

FUNGI, 17
FUNGI IMPERFECTI, 23
FUNGIINA, 180
Furcilia, 311

**G**
GADIFORMES, 588
*Gadinia carinata*, 456
*Gadinia mammillaris*, 456
Gag grouper, 600
Gag rockfish, 600
Galapagos shark, 570
GALATHEIDAE, 338
GALATHEOIDEA, 338
*Galaxaura marginata*, 55
*Galaxaura obtusata*, 55, 2.9
*Galaxaura subverticillata*, 55
GALEOMMATACEA, 476
GAMASINA, 272
*Gambusia affinis*, 590
GAMMARIDEA, 374
Gannets, 660
Gaping boring clam, 489
*Gardnerula corymbosa*, 15
Garlic sponge, 120
GASTEROSTEIFORMES, 594
*Gastrochaena cucullata*, 489
*Gastrochaena cuneiformis*, 489
*Gastrochaena hians*, 489
*Gastrochaena mowbrayi*, 489
*Gastrochaena ovata*, 489
GASTROCHAENACEA, 489
Gastrochaenidae, 489
Gastropoda, 397
Gastrotricha, 213
Gaudy Asaphis, 483
GECARCINIDAE, 353
*Gecarcinus lateralis*, 354, 11.8
*Geleia nigriceps*, 106
GELIDIACEAE, 55
*Gelidiella acerosa*, 55
*Gelidium pusillum*, 55
GEMPYLIDAE, 638
*Geodia gibberosa*, 125
GEODIIDAE, 125
*Geograpsus lividus*, 350
*Geonemertes agricola*, 211
GEOPHILOMORPHA, 382
GERANIALES, 83
GERREIDAE, 608
GERRIDAE, 387
*Geryonia proboscidalis*, 148
GERYONIIDAE, 148
Ghost crab, 354
Giant black sea hare, 446
GIGARTINALES, 59
Gill-bearing anemone, 173

# INDEX

*Ginglymostoma cirratum*, 568
Glasseye snapper, 602
Glass-haired chiton, 397
Glassy sweeper, 614
GLAUCIDAE, 455
*Glaucus atlanticus*, 455, 13.7
*Globicephala melaena*, 669
*Globigerinoides ruber*, 100
*Globorotalia truncatulinoides*, 100
*Glossobalanus crozieri*, 543
GLOSSODORIDIDAE, 451
*Glossodoris zebra*, 452
*Glycera abranchiata*, 242
*Glycera oxycephala*, 242
GLYCERIDAE, 241
*Glyptobairdia coronata*, 286
Gnathiidea, 366
Gnathobdellida, 264
*Gnatholepis thompsoni*, 634
*Gnathophausia* cf. *ingens*, 359
GNATHOPHYLLIDAE, 323
*Gnathophylloides mineri*, 323
*Gnathophyllum americanum*, 323
GNATHOSTOMATA, 535
*Gnathostomula tuckeri*, 212
Gnathostomulida, 211
Goatfishes, 612
Gobies, 632
GOBIIDAE, 632
Goes' Gadinia, 456
Gold-mouthed triton, 428
Gold-spotted eel, 581
Golden-brown algae, 70
Goldentail moray, 582
Goldspot goby, 634
Golf ball coral, 182
*Golfingia elongata*, 226
GOLFINGIIDAE, 226
Gonerichthus, 308
*Goniaster tessellatus*, 526, 14.11
GONIASTERIDAE, 526
*Gonichthys coccoi*, 585
*Gonionemus suvaensis*, 146
*Goniopsis cruentata*, 352, 11.13
GONODACTYLIDAE, 308
*Gonodactylus bredini*, 308
*Gonodactylus oerstedi*, 308
*Gonodactylus spinulosus*, 308
*Gonostoma elongatum*, 586
GONOSTOMATIDAE, 586
*Gonyaulax polygramma*, 77
Goodenia family, 87
GOODENIACEAE, 87
Goose barnacles, 301
Goosefoot family, 81
GORGONACEA, 165
*Gorgonia americana*, 170

*Gorgonia citrina*, 171
*Gorgonia flabellum*, 172
*Gorgonia ventalina*, 172, 5.17
GORGONIIDAE, 170
GORGONOCEPHALIDAE, 528
*Gorgonorhynchus bermudensis*, 209
Gould's tellin, 482
Gould's wax venus, 486
*Gouldia bermudensis*, 486
*Gouldia cerina*, 486
*Gracilaria debilis*, 59
GRACILARIACEAE, 59
GRAMINALES, 89
Gramineae, 90
GRAMMISTIDAE, 600
*Grania macrochaeta*, 262
GRANTIIDAE, 126
GRANULORETICULOSIA, 96
GRAPSIDAE, 349
*Grapsus grapsus*, 353
Grass scallop, 466
Grass scuttle, 499
Grasses, 89
Gray snapper, 607
Gray triggerfish, 642
Graysby, 597
Great barracuda, 630
Great black-backed gull, 666
Great blue heron, 662
Great conch, 420
Great land crab, 354
Great star coral, 184
Great tellin, 481
Greater amberjack, 606
Greater argonaut, 499
Green algae, 27
Green lucky, 407
Green moray, 582
Green parrot fish, 629
Green sea mat, 191
Green turtle, 653
Green's miniature horn shell, 414
Greenhead, 392
*Gregariella coralliophaga*, 466
Gregarinida, 103
*Gromia oviformis*, 96
GROMIIDA, 96
Groupers, 595
*Grubia coei*, 374
Grunts, 609
*Guinardia flaccida*, 68
Guinea chick lobster, 330
Gulf weed crab, 350
Gulls, 663, 666
*Guynia annulata*, 187
GUYNIIDAE, 187
Gwelly, 606

GYMNOCERATA, 387
GYMNODINIALES, 79
*Gymnodinium microadriaticum*, 79
*Gymnodoris* sp., 452, 13.12
GYMNOLAEMATA, 503
GYMNOPLEURA, 339
GYMNOSOMATA, 443
GYMNOSTOMATA, 105
*Gyrodinium spirale*, 79

**H**

HADROMERIDA, 122
HAEMULIDAE, 609
*Haemulon album*, 609
*Haemulon aurolineatum*, 609
*Haemulon carbonarium*, 610
*Haemulon flavolineatum*, 609
*Haemulon melanurum*, 609
*Haemulon sciurus*, 610
Hagfish, 565
Hairy blenny, 631
Hairy crab, 346
Hairy Cratena, 455
HALACARIDAE, 272
*Halacarus ctenopus*, 274
*Halammohydra* sp., 146
HALAMMOHYDRIDAE, 146
HALECHINISCIDAE, 266
HALECIIDAE, 135
*Halecium bermudense*, 135
*Halecium nanum*, 136
Half-scratched tusk, 459
Halfbeaks, 589
*Halichoeres bivittatus*, 621
*Halichoeres garnoti*, 621
*Halichoeres radiatus*, 622
HALICHONDRIIDA, 121
*Haliclona molitba*, 118, 3.8
*Haliclona monticulosa*, 118, 3.2
*Haliclona permollis*, 117
*Haliclona variabilis*, 118
*Haliclona viridis*, 118
HALICLONIDAE, 117
*Halicreas minimum*, 147
HALICREATIDAE, 147
*Halicystis osterhoutii*, 37
*Halimeda incrassata*, 34, 1.5
*Halimeda monile*, 34, 1.4
Halimeda slug, 450
*Halimeda tuna*, 34
*Halisarca dujardini*, 117
HALISARCIDAE, 117
*Halistemma striata*, 152
*Halitiara formosa*, 144
*Halobates micans*, 388
*Halocordyle disticha*, 132, 143
HALOCORDYLIDAE, 132, 143

*Halocyphina villosa*, 24
HALOCYPRIDA, 284
HALOCYPRIDIDAE, 284
*Halocypris brevirostris*, 284
*Halodule bermudensis*, 88, 6.1
HALOPAPPACEAE, 72
*Halopappus adriaticus*, 72
*Halophila decipiens*, 88
*Halophiloscia couchi*, 372
*Halopteris diaphana*, 139, 2.6
*Halosarpheia fibrosa*, 21
*Halosphaeria quadricornuta*, 21
*Halosydna leucohyba*, 236
*Halymenia bermudensis*, 59
*Halymenia floresia*, 58, 2.13
*Haminoea antillarum*, 440, 12.2
Hammered shipworm, 491
Hammerhead sharks, 570
*Haplognathia* cf. *rosacea*, 212
Haplopharyngida, 198
HAPLOSCLERIDA, 117
*Haplosyllis spongicola*, 240
HAPLOTAXIDA, 260
HAPTORIDA, 106
Harbour conch, 420
*Harengula humeralis*, 579
Harlequin bass, 597
HARPACTICOIDA, 295
Harrimanidae, 541
*Hastigerina pelagica*, 99
Hat coral, 180
Hatchetfish, 586
Hatchetfishes, 586
Hawk-wing conch, 420
Hawks, 663
Hawksbill turtle, 654
Haycock's silvery Donax, 479
Heart urchin, 536
Heart urchins, 536
*Hebellopsis scandens*, 136
*Helicoprorodon gigas*, 106
Helldiver, 659
HELMINTHOCLADIACEAE, 53
*Hemiaulus hauckii*, 68
Hemichordata, 541
HEMICYTHERIDAE, 288
HEMIPTERA, 387
*Hemipteronotus martinicensis*, 622
HEMIRAMPHIDAE, 589
*Hemiramphus bermudensis*, 589
Hermit crabs, 334, 336
*Hermodice carunculata*, 236, 8.14
Herons, 662
*Herposiphonia secunda*, 64
Herring gull, 666
Herrings, 579
*Hesione picta*, 238

HESIONIDAE, 238
*Heterodonax bimaculata*, 483
HETERODONTA, 473
*Heteromysis bermudensis*, 361
HETERONEMERTINI, 209
HETEROPODA, 423
HETEROSIPHONALES, 74
*Heterosiphonia gibbesii*, 64
*Heterotanais limicola*, 366
HETEROTARDIGRADA, 266
*Heterotiara anonyma*, 144
HETEROTRICHIDA, 109
HEXACORALLIA, 172
Hexactinellida, 111
*Hexalonche amphisiphon*, 102
Hexapoda, 382
Higher crustaceans, 305
Highhat, 612
HINCKSINIDAE, 506
*Hippa cubensis*, 339, 10.8
*Hippa testudinaria*, 339
HIPPIDAE, 339
*Hippocampus reidi*, 594
HIPPOIDEA, 338
*Hippolysmata grabhami*, 328, 9.3
*Hippolyte coerulescens*, 326
*Hippolyte zostericola*, 326, 9.6
HIPPOLYTIDAE, 326
HIPPONICACEA, 417
HIPPOPODIIDAE, 152
*Hippopodius hippopus*, 152
HIPPOPORINIDAE, 513
*Hippothoa flagellum*, 510
HIPPOTHOIDAE, 510
Hirudinea, 263
*Hirundichtyhs affinis*, 589
*Histrio histrio*, 587
Hogfish, 624
Hogmouth fry, 579
*Holacanthus bermudensis*, 618
*Holacanthus ciliaris*, 618
*Holacanthus tricolor*, 617
HOLAXONIA, 165
HOLECTYPOIDA, 535
HOLOCENTRIDAE, 592
*Holocentrus ascensionis*, 592
*Holocentrus rufus*, 593
Holocephali, 566
*Holothuria arenicola*, 540
*Holothuria cubana*, 538
*Holothuria parvula*, 538
*Holothuria surinamensis*, 540
HOLOTHURIIDAE, 538
Holothuroidea, 537
HOMALORHAGIDA, 222
*Homaxinella rudis*, 121, 4.2
*Homotrema rubrum*, 99, 6.7

Honeycomb cowfish, 647
Hook-tail filefish, 644
Hoplocarida, 305
HOPLONEMERTINI, 211
HORMOGONEAE, 12
Horse-eye jack, 604
Horseflies, 391
Horseshoe crabs, 269
Horseshoe worms, 516
Horsetails, 26
Houndfish, 590
*Humicola alopallonella*, 23
Humpback whale, 671
*Hyale pontica*, 378
*Hyale prevostii*, 378
*Hyale trifolidens*, 378
HYALIDAE, 378
*Hyalina albolineata*, 436
*Hyalina avena*, 436
*Hyalina effulgens*, 436
*Hydatina physis*, 438
*Hydatina vesicaria*, 438
HYDATINIDAE, 438
HYDROCHARITACEAE, 88
*Hydroclathrus clathratus*, 46
*Hydrogamasus* sp., 272
Hydroid polyps, 128
HYDROIDA, 128
*Hydroides parvus*, 257
HYDROMEDUSAE, 140
*Hydroschendyla submarina*, 382
Hydrozoa, 127
HYMENIACIDONIDAE, 121
HYMENOMYCETES, 23
HYMENOSTOMATA, 108
HYMENOSTOMATIDA, 108
*Hyperia atlantica*, 380
*Hyperia bengalensis*, 380
HYPERIIDAE, 380
HYPERIIDEA, 380
HYPHOMYCETES, 23
*Hypleurochilus bermudensis*, 632
*Hypnea musciformis*, 61
HYPNEACEAE, 61
*Hypoplectrus puella*, 596
Hypostomata, 104
HYPOTRICHIDA, 110
*Hypselodoris zebra*, 452, 13.10
*Hypsicomus elegans*, 255
HYSTERIALES, 23

I

*Ianthella ardis*, 115
*Iasis zonaria*, 561
Ibises, 662
Icecream cone worm, 254
IDIACANTHIDAE, 586

*Idiacanthus fasciola*, 587
*Idmonea atlantica*, 515
*Idotea baltica*, 370
*Idotea metallica*, 370
*Ilyanthopsis longifilis*, 176
Imperfect fungi, 23
Inarticulata, 518
Inflated cavoline, 442
Infusoriform, 207
Infusorigen, 207
*Ingolfiella* sp., 381
INGOLFIELLIDAE, 381
INGOLFIELLIDEA, 381
INOVICELLATA, 505
Insecta, 383
Insects, 383
Inshore squids, 496
*Ipomoea pes-caprae*, 86
*Ircinia fasciculata*, 115
*Ircinia felix*, 115, 3.3
*Ircinia felix* f. *acuta*, 115
*Ircinia felix* f. *felix*, 115
*Ircinia felix* f. *fistularis*, 115
*Ircinia strobilina*, 115
Irregular worm shell, 412
*Isaurus duchassaingi*, 191, 7.2
*Ischnochiton bermudensis*, 396
ISCHNOCHITONIDAE, 396
ISCHNOCHITONINA, 396
*Isognomon alatus*, 468, 4.13
ISOGNOMONIDAE, 468
ISOPHELLIIDAE, 175
*Isophyllia dipsacea*, 187
*Isophyllia fragilis*, 187
*Isophyllia multiflora*, 187
*Isophyllia sinuosa*, 187, 6.8
*Isophyllia sinuosa* f. *multiflora*, 187
ISOPODA, 366
*Isostichopus badionotus*, 540, 14.10
ISTIOPHORIDAE, 638
*Isurus oxyrinchus*, 568
Ivory coral, 186

**J**
Jacks, 603
*Jania adherens*, 57
*Janthina communis*, 415
*Janthina janthina*, 415
*Janthina pallida*, 415
JANTHINIDAE, 415
JAPYGIDAE, 386
Jaw worms, 211
Jellyfish, 155
*Jenkinsia lamprotaenia*, 579
Jewel anemone, 188
Jewelfish, 621
*Joeropsis rathbunae*, 370

Jolthead porgy, 612
*Justitia longimanus*, 330, 10.5

**K**
*Kallymenia perforata*, 59
KALLYMENIACEAE, 59
KALYPTORHYNCHIA, 202
KARYORELICTIDA, 105
Keeled Clio, 442
*Keissleriella blepharospora*, 22
KERATOSA, 114
Keraudren's Atlanta, 423
Key brotula, 588
Key worm eel, 581
Killifishes, 591
KINETOFRAGMINOPHORA, 105
Kingfishers, 667
Kinorhyncha, 220
*Kinorhynchus fimbriatus*, 222
Kitten's paw, 470
Knobbed triton, 428
KOPHOBELEMNIDAE, 172
*Kowalevskia tenuis*, 562
KOWALEVSKIIDAE, 562
Krebs' wentletrap, 416
Krill, 310
*Krohnitta subtilis*, 520
KYPHOSIDAE, 614
*Kyphosus sectatrix*, 614
*Kytorhynchus microstylus*, 202

**L**
LABIDURIDAE, 387
LABRIDAE, 621
*Labrisomus nuchipinnis*, 631
*Lachnolaimus maximus*, 624
*Lactophrys trigonus*, 648
*Lactophrys triqueter*, 648
*Laemodonta cubensis*, 458
*Laevicardium laevigatum*, 479
LAFOEIDAE, 136
*Lagodon rhomboides*, 612
Lamarck's Carinaria, 424
Lamellibranchia, 460
Lamellose wentletrap, 416
LAMNIDAE, 568
Lampreys, 565
Lamp shells, 518
Lancelets, 562
Land crabs, 353
Land hermit crabs, 334
Lane snapper, 608
Lantern fishes, 584
*Laomedea tottoni*, 136
LAOPHONTIDAE, 297
Large heart urchin, 537
Large ivory coral, 186

Large prawn, 317
Large red spider crab, 356
Large sponge-carrying crab, 340
LARIDAE, 666
LARINAE, 666
*Larus argentatus*, 666
*Larus atricilla*, 666
*Larus delawarensis*, 666
*Larus marinus*, 666
*Larus philadelphia*, 667
Larvacea, 561
*Lasaea adansoni*, 476
*Lasaea bermudensis*, 476
*Lasaea rubra*, 476
LASAEIDAE, 476
*Latreutes fucorum*, 326
Laughing gull, 666
*Laurencia microcladia*, 66
*Laurencia obtusa*, 66, 2.10
*Laurencia papillosa*, 66, 3.8
Leaf-like oyster, 472
Leafy jewel box, 475
*Leander tenuicornis*, 320
Leatherback turtle, 656
*Lebrunia danae*, 173, 9.11
Lecithoepitheliata, 198
Leeches, 263
Lefteye flounders, 641
LEGUMINALES, 83
*Leidya bimini*, 371
LEIODERMATIIDAE, 126
*Leiodermatium pfeifferae*, 126
*Lensia subtilis*, 154
*Leodia sexiesperforata*, 536
Leopard flatworm, 203
LEPADIDAE, 301
LEPADOMORPHA, 301
*Lepas anatifera*, 301
*Lepas pectinata*, 301
*Lepidapedon trachinoti*, 204
*Lepidochelys kempi*, 654
*Lepidochitona liozonis*, 396
*Lepraliella fissurata*, 514
*Leptastacus macronyx*, 297
Leptocardia, 563
Leptocephalus, 577
*Leptochela bermudensis*, 319
*Leptochelia limicola*, 366
*Leptocylindrus danicus*, 68
*Leptodius americanus*, 348
*Leptodius floridanus*, 348
*Leptognathia longiremis*, 366
LEPTOGNATHIDAE, 366
LEPTOMEDUSAE, 144
*Leptosphaeria australiensis*, 22
*Leptosphaeria avicenniae*, 22
Leptostraca, 305

Lernaeopodoida, 288
*Lestrigonus bengalensis*, 380
Lettered horn shell, 413
*Leucandra aspera*, 127
*Leucetta floridana*, 126
*Leucetta microraphis*, 126, 4.15
LEUCETTIDA, 126
Leucettidae, 126
*Leuconia aspera*, 127
LEUCOSIIDAE, 340
*Leucosolenia canariensis*, 126
*Leucothoe spinicarpa*, 377
LEUCOTHOIDAE, 377
*Liagora ceranoides*, 53
*Liagora farinosa*, 54, 2.8
LICHENES, 24
*Lichenopora radiata*, 516
Lichenoporidae, 516
Lichens, 24
Lightbourn's Murex, 430
Lightning venus, 486
*Ligia baudiniana*, 372
*Lima caribaea*, 472
*Lima dehiscens*, 472
*Lima fragilis*, 472
*Lima hians*, 472
*Lima inflata*, 472
*Lima lima*, 472
*Lima multicostata*, 472
*Lima pellucida*, 472
*Lima scabra*, 472
*Lima tenera*, 472
LIMACEA, 472
*Limacina inflata*, 441
*Limacina trochiformis*, 441
LIMACINIDAE, 441
Limidae, 472
LIMNOMEDUSAE, 146
*Limnoria tuberculata*, 368
Limpets, 397
*Linckia guildingii*, 526
*Lindia* sp., 220
*Lindra thalassiae*, 20
*Lineus albocinctus*, 209
*Lineus albonasus*, 210
*Linuche unguiculata*, 158
*Liosina monticulosa*, 118
*Liriope tetraphylla*, 148
*Lironeca reniformis*, 370
*Lissoclinum fragile*, 550, 15.15
*Lissodendoryx isodictyalis*, 120, 3.7
*Lissothuria antillensis*, 540, 14.7
Lister's keyhole limpet, 404
Lister's tellin, 481
LITHISTIDA, 126
*Lithodomus appendiculatus*, 466
*Lithophaga antillarum*, 466

*Lithophaga bisulcata*, 466
*Lithophaga dactylus*, 466
*Lithophaga gossei*, 466
*Lithophaga nigra*, 466
*Lithophyllum intermedium*, 58
*Lithoptera tetraptera*, 101
*Lithotrya dorsalis*, 301
*Litiopa melanostoma*, 414
Little knobby scallop, 468
Little piked whale, 671
Little tunny, 636
*Littorina angulifera*, 408
*Littorina angustior*, 408
*Littorina mespillum*, 407
*Littorina ziczac*, 407
LITTORINACEA, 407
LITTORINIDAE, 407
LITUOLIDEA, 97
Live-bearers, 590
Livid Natica, 425
*Livona pica*, 405
Lizard fishes, 584
Lizards, 651
LOBATA, 195
*Lobatostoma ringens*, 204
*Lobophora variegata*, 45
*Lobopilumnus agassizii*, 348
LOBOSIA, 95
Lobsters, 312
LOCULOASCOMYCETES, 22
Locust lobsters, 332
Loggerhead turtle, 654
*Loimia medusa*, 254
*Loligo plei*, 496
Long-armed spiny lobster, 330
Long-eared Favorinus, 454
Long-horned bugs, 387
Long-snout sea horse, 594
Long-spined star shell, 406
Long-tailed Clione, 443
Long-tailed sea cat, 446
Longsnout butterfly fish, 616
Longsnout scorpion fish, 641
Longspine squirrelfish, 593
Longtail, 660
*Lopha frons*, 472
LOPHIIFORMES, 587
LOPHOGASTRIDA, 359
LOPHOGASTRIDAE, 359
*Lophogobius cyprinoides*, 634
LOPHOPHORATA, 500
Loricifera, 111, 197
*Lovenella bermudensis*, 145
LOVENELLIDAE, 145
"lower" Crustacea, 278
LOXOCONCHIDAE, 288
*Loxocorniculum* sp., 288

LOXOSOMATIDAE, 503
*Loxosomella tedaniae*, 503
*Lucifer typus*, 319
*Lucina pecten*, 474
*Lucina pectinatus*, 474
*Lucina schrammi*, 475
*Lucina squamosa*, 474
*Lucina tigrina*, 473
LUCINACEA, 473
LUCINIDAE, 473
*Luconacia incerta*, 381
*Luidia clathrata*, 525
LUIDIIDAE, 525
LUMBRICINA, 263
LUMBRINERIDAE, 244
*Lumbrineris impatiens*, 244
*Lumbrineris inflata*, 244
LUTJANIDAE, 607
*Lutjanus griseus*, 607
*Lutjanus synagris*, 608
*Lutjanus vivanus*, 608
*Lycodontis funebris*, 582
*Lycodontis moringa*, 582
*Lycodontis vicinus*, 582
*Lygdamis indicus*, 253, 8.5
*Lyngbya confervoides*, 14
*Lyngbya maiuscula*, 15
*Lyonsia beaui*, 491
LYONSIIDAE, 491
*Lyrodus bipartita*, 490
LYSARETIDAE, 244
*Lysianassa punctata*, 377
LYSIANASSIDAE, 377
*Lysidice ninetta*, 242
Lysioerichthus, 308
*Lysiosquilla scabricauda*, 310
LYSIOSQUILLIDAE, 309
*Lytechinus variegatus*, 533
*Lythrypnus mowbrayi*, 634
*Lytocarpus philippinus*, 139

**M**
MacDonald's pteropod, 444
Mackerel, 636
Mackerel scad, 604
Mackerel sharks, 568
*Macoma eborea*, 483
*Macoma souleyetiana*, 482
*Macoma tenta*, 482
*Macrocallista maculata*, 487
*Macrocoeloma subparallelum*, 355
*Macrocoeloma trispinosum nodipes*, 356, 11.6
MACROCYPRIDIDAE, 285
*Macrocyprina* sp., 285
MACRODASYIDA, 214
MACRODASYIDAE, 214
*Macrodasys* sp., 214

# INDEX 725

*Macrorhynchia clarkei,* 139
*Macrorhynchia philippina,* 139
MACROSETELLIDAE, 297
MACROSTOMIDA, 200
Macrura, 312
MACTRACEA, 479
*Madracis decactis,* 180
*Madracis mirabilis,* 180
MADREPORARIA, 179
*Maera inaequipes,* 377
*Magelona* sp., 247
MAGELONIDAE, 247
Magpie shell, 405
Mahogany date mussel, 466
MAJIDAE, 354
*Makaira nigricans,* 639
MALACANTHIDAE, 603
*Malacanthus plumieri,* 603
*Malacoctenus gilli,* 631
MALACOSTEGA, 505
Malacostraca, 305
MALDANIDAE, 252
*Mallotonia gnaphalodes,* 86
*Malmiana* sp., 264
Mammalia, 668
Man-of-war fish, 639
Man-of-war fishes, 639
Manatee grass, 88
Manca, 362, 364, 367
Mandibulata, 268
Mangrove crab, 352
Mangrove family, 86
Mangrove oyster, 468
Mantis shrimps, 306
Many-celled animals, 110
Many-toothed conger, 581
Margate, 609
MARGINELLIDAE, 435
*Marionina achaeta,* 263
*Marionina spicula,* 263
*Marionina subterranea,* 262
*Marphysa sanguinea,* 242
Marsupials, 668
*Martesia striata,* 489
Mastigomycotina, 17
Mastigophora, 93
Mastigopus, 315
*Meandra cerebrum,* 184
*Meandra labyrinthiformis,* 184
*Meandrina meandrites,* 186
MEANDRINIDAE, 186
*Megaceryle alcyon,* 667
*Megalomma lobiferum,* 255
Megalopa, 315
MEGALOPIDAE, 577
*Megalops atlanticus,* 577
*Megaptera novaeangliae,* 671

MEGASCOLECIDAE, 263
*Meioherpia atlantica,* 394
MEIOMENIIDAE, 394
*Meiosquilla lebouri,* 310
MELAMPIDACEA, 457
MELAMPIDAE, 457
*Melampus bidentatus,* 457
*Melampus coffeus,* 457
*Melampus flavus,* 457
*Melampus flavus* var. *albus,* 457
*Melampus flavus* var. *purpureus,* 457
*Melampus gundlachi,* 457
*Melampus lineatus,* 457
*Melampus monilis,* 457
*Melampus redfieldi,* 457
*Melanella intermedia,* 417
MELANELLIDAE, 417
*Meleagrina placunoides,* 467
*Melichthys niger,* 644
*Melita appendiculata,* 377
*Melita fresnelii,* 377
MELITIDAE, 377
*Mellita sexiesperforata,* 536
MELLITIDAE, 535
*Membranipora tuberculata,* 505
MEMBRANIPORIDAE, 505
*Membranobalanus declivis,* 304
*Meoma ventricosa,* 537
Mermaid's wineglass, 41
Merman's shaving brush, 34
Merostomata, 269
*Mesochaetopterus minutus,* 247
MESODESMATIDAE, 479
*Mesodinium pupula,* 106
Mesogastropoda, 402, 407
Mesomyaria, 172
*Mesonerilla prospera,* 258
*Mesophyllum mesomorphum,* 58
MESOSTIGMATA, 272
Mesotardigrada, 265
Mesozoa, 206
Metanauplius, 311
*Metandrocarpa sterreri,* 556
*Metapenaeopsis smithi,* 317
Metatheria, 668
Metatrochophore, 234
METAZOA, 110
*Metopus circumlabens,* 109
Mic Mac, 435
Microcerberidea, 366
*Microcoleus chthonoplastes,* 15
*Microcosmus exasperatus,* 557,16.8
*Microcotyle incisa,* 204
MICROHEDYLIDAE, 440
*Micromelo undatus,* 438
*Micropanope spinipes,* 346
*Microphrys bicornutus,* 355

*Microphthalmus arenarius,* 238
*Microsphathodon chrysurus,* 621
*Microtralia occidentalis,* 458
Mictacea, 361
*Mictocaris halope,* 362
Midges, 391
Midnight parrot fish, 627
MILIOLIDEA, 98
Milk conch, 420
Milky moon snail, 424
*Millepora alcicornis,* 134, 8.11
MILLEPORIDAE, 132
Minke whale, 671
*Miracia efferata,* 297
Miracidium, 204
Mites, 270
*Mithrax cornutus,* 358
*Mithrax forceps,* 356, 11.4
*Mithrax hispidus,* 356
*Mithrax pleuracanthus,* 356
*Mitra haycocki,* 437
MITRACEA, 436
Mitraria, 235
*Mitrella ocellata,* 433
*Mitrolumna biplicata,* 437
*Mnemiopsis leidyi,* 195
*Modiola opifex,* 466
*Modiola tulipa,* 466
*Modiolus americanus,* 466
MODULIDAE, 413
*Modulus modulus,* 413
*Moira atropos,* 536
Mojarras, 608
Mole crabs, 338
Mole shrimps, 333
MOLIDAE, 650
Mollusca, 392
Mollusks, 392
Molly miller, 632
*Monacanthus ciliatus,* 644
*Monacanthus tuckeri,* 646
MONERA, 9
MONHYSTERIDA, 217
Monhysteridae, 217
Monkey rockfish, 600
MONOCOTYLEDONEAE, 88
MONOGENEA, 204
MONOGONONTA, 220
MONOKONOPHORA, 365
Monoplacophora, 392
*Monostroma oxyspermum,* 29
Monstrilloida, 288
*Montastrea annularis,* 184
*Montastrea cavernosa,* 184, 6.13
*Mooreonuphis jonesi,* 242
Morays, 582
Morning-glory family, 86

*Morula nodulosa,* 432
*Morum oniscus,* 426
Mosquito fish, 590
Mosquitoes, 390
Moss animals, 500
Mosses, 26
Mossy ark, 463
Mottled shore crab, 350
Mouse cone, 436
Mouse-eared marsh snail, 458
Mowbray's goby, 634
Mud crabs, 346
Mud shrimps, 333
*Mugil trichodon,* 629
MUGILIDAE, 629
Mullets, 629
MULLIDAE, 612
*Mulloidichthys martinicus,* 614
*Munida simplex,* 338
*Muraena miliaris,* 582
MURAENIDAE, 582
*Murex nuceus,* 431
MURICACEA, 429
*Muricea atlantica,* 170, 5.13&14
*Muricea laxa,* 169
*Muricea muricata,* 169, 170
MURICIDAE, 430
*Musculus lateralis,* 466
*Mussa anectens,* 187
*Mussa dipsacea,* 187
*Mussa fragilis,* 187
*Mussa roseola,* 187
Mussel shrimps, 280
Mussels, 460
MUSSIDAE, 186
Mustard family, 83
Mutton hamlet, 597
MYACEA, 488
*Mycale microsigmatosa,* 120
MYCALIDAE, 120
Mycetozoa, 17
*Mycteroperca bonaci,* 598
*Mycteroperca interstitialis,* 600
*Mycteroperca microlepis,* 600
*Mycteroperca tigris,* 600
*Mycteroperca venenosa,* 600
MYCTOPHIDAE, 584
MYCTOPHIFORMES, 584
*Myctophum nitidulum,* 585
MYLIOBATIDAE, 571
MYODOCOPIDA, 282
MYOIDA, 487
MYOPSIDA, 496
Myriapoda, 382
*Myriastra crassispicula,* 124
*Myrichthys oculatus,* 581
*Myriochele heeri,* 252

*Myrionema amboinense,* 134
MYRTALES, 84
*Mysia pellucida,* 475
MYSIDA, 359
Mysidacea, 358
MYSIDAE, 361
*Mysidium gracile,* 361
*Mysidium integrum,* 361
Mysis, 315
Mystacocarida, 278
MYSTICETI, 671
MYTILACEA, 464
MYTILIDAE, 464
*Mytilimeria plicata,* 491
MYTILOIDA, 464
MYXILLIDAE, 120
Myxomycota, 17
Myzostomaria, 232
Myzostomida, 232
Müller's larva, 199

**N**

*Naineris laevigata,* 244
*Naineris setosa,* 245
NAJADALES, 88
NANNASTACIDAE, 363
*Nannastacus hirsutus,* 364
*Nannolaimoides decoratus,* 217
Naples Spurilla, 455
NARCOMEDUSAE, 148
Narrow periwinkle, 408
*Nassa ambiguus,* 434
NASSARIIDAE, 434
*Nassarius albus,* 434
Nassau grouper, 598
Natantia, 312
*Natica canrena,* 425
*Natica livida,* 424
*Natica marochiensis,* 424
NATICACEA, 424
Naticidae, 424
Nauplius, 281, 289, 300, 311, 315
*Nausithoë maculata,* 158
*Nausithoë punctata,* 158
Nautiloidea, 492
Nearly smooth Aegires, 452
Nectochaeta, 234
*Nectonemertes mirabilis,* 211
Needle pteropod, 442
Needlefish, 590
Needlefishes, 590
NEMALIALES, 53
NEMATOCERA, 390
Nematoda, 216
Nematogen, 207
Nematomorpha, 111, 197

*Nematonereis hebes,* 244
Nemertina, 208
NEMERTODERMATIDA, 200
Neogastropoda, 402, 407
*Neogoniolithon spectabile,* 58
NEOLORICATA, 396
*Neomeris annulata,* 39, 1.2
*Neomicrodeutopus* sp., 376
Nephropidae, 329
NEPHROPOIDEA, 329
*Nephropsis rosea,* 329
NEREIDIDAE, 240
*Nereis riisei,* 240
*Nerilla* sp., 258
NERILLIDAE, 258
*Nerita peloronta,* 406
*Nerita tessellata,* 406
*Nerita versicolor,* 406
NERITACEA, 406
NERITIDAE, 406
*Neritina virginea,* 407
*Nerocila acuminata,* 370
Netted olive, 435
Never Bite, 604
*Nibilia antilocapra,* 358
*Nicolea modesta,* 254
*Nidalia occidentalis,* 165, 5.12
NIDALIIDAE, 164
*Niphates erecta,* 118, 3.12, 7.9
NIPHATIDAE, 118
*Nitzschia closterium,* 70
No-see-ums, 391
*Nodilittorina tuberculata,* 408
NOETIIDAE, 464
*Nolella stipata,* 504
NOLELLIDAE, 504
NOMEIDAE, 639
*Nomeus gronovii,* 639
NOSTOCALES, 15
Notable ark, 464
NOTARCHIDAE, 446
NOTASPIDEA, 446
*Notobranchaea macdonaldi,* 444
NOTOBRANCHAEIDAE, 444
NOTODELPHYOIDA, 299
*Notomastus latericeus,* 250
*Notoplana* cf. *binoculata,* 203
Notostraca, 278
*Nucula proxima,* 463
NUCULACEA, 463
NUCULIDAE, 463
Nuculoida, 463
NUDA, 196
NUDIBRANCHIA, 451
Nurse shark, 568
Nurse sharks, 568
*Nycticorax violacea,* 662

## O

*Obelia dichotoma*, 136
*Obelia* spp., 146
Ocean dolphins, 669
Ocean robin, 604
Ocean skater, 388
Ocean surgeonfish, 636
Ocean triggerfish, 644
Oceanic squids, 497
Ocellated box crab, 342
*Ochetostoma baronii*, 230
*Ochetostoma erythrogrammon*, 230
*Ocinebra intermedia*, 431
*Ocnus surinamensis*, 541, 14.9
*Octactis pulchra*, 73
OCTOCORALLIA, 164
*Octolasmis forresti*, 302
OCTOPODA, 499
*Octopus bermudensis*, 499
*Octopus chromatus*, 499
*Octopus macropus*, 499
Octopus pteropod, 444
*Octopus rugosus*, 499
*Octopus vulgaris*, 499
Octopuses, 492, 499
*Oculina diffusa*, 186
*Oculina valenciennesi*, 186
*Oculina varicosa*, 186
OCULINIDAE, 186
OCULOSIDA, 102
*Ocypode quadrata*, 354
OCYPODIDAE, 354
*Ocyurus chrysurus*, 608
ODONTOCETI, 669
*Odontophora bermudensis*, 218
*Odontosyllis enopla*, 238, 8.2
OEGOPSIDA, 497
*Oenone fulgida*, 244
*Ogilbia cayorum*, 588
*Oikopleura fusiformis*, 562
*Oikopleura longicauda*, 562
OIKOPLEURIDAE, 562
Oilfish, 638
Oilfishes, 638
*Oithona nana*, 292
*Oithona plumifera*, 294
OITHONIDAE, 292
*Okenia evelinae*, 453
*Okenia zoobotryon*, 453, 13.14
OLIGOBRACHIIDAE, 231
Oligochaeta, 258
OLIGOHYMENOPHORA, 106
OLIGOTRICHIDA, 109
*Olindias phosphorica*, 146
OLINDIASIDAE, 146
*Oliva circinata*, 435
*Oliva reticularis*, 435

OLIVIDAE, 435
OLPIIDAE, 270
Olynthus, 114
*Ommastrephes pteropus*, 497
*Oncaea media*, 294
*Oncaea venusta*, 294
ONCAEIDAE, 294
*Onchidella floridana*, 443
ONCHIDIACEA, 443
Onchidiidae, 443
*Onchidium transatlanticum*, 443
Oncosphaera, 206
ONISCOIDEA, 372
ONUPHIDAE, 242
ONYCHOCELLIDAE, 506
Onychophora, 111, 197
*Onykia caribaea*, 497
*Opephora schwartzii*, 70
OPHELIIDAE, 249
OPHIACTIDAE, 530
*Ophiactis savignyi*, 530
OPHICHTHIDAE, 581
OPHIDIASTERIDAE, 526
*Ophiocoma echinata*, 530
*Ophiocoma wendti*, 530
*Ophiocomella ophiactoides*, 530
OPHIOCOMIDAE, 530
*Ophioderma appressum*, 530, 14.3
OPHIODERMATIDAE, 530
*Ophiolepis paucispina*, 528
*Ophiomyxa flaccida*, 528, 14.4
OPHIOMYXIDAE, 528
OPHIONEREIDIDAE, 530
*Ophionereis reticulata*, 530, 14.6
Ophiopluteus, 528
OPHIURIDA, 528
OPHIURIDAE, 528
Ophiuroidea, 527
OPISTHOBRANCHIA, 438
*Opisthonema oglinum*, 579
Opossum shrimps, 358
Orange filefish, 646
Orange-back squid, 497
Orange-spotted filefish, 644
*Orbicella annularis*, 184
*Orbicella cavernosa*, 184
*Orbicella hispidula*, 184
Orbigny's Fossarus, 417
ORBINIIDAE, 244
*Orbulina universa*, 100
*Orchestia* sp., 380
*Oreaster reticulatus*, 526, 14.13
OREASTERIDAE, 526
ORECTOLOBIDAE, 568
ORGANISMS, 7
ORIBATEI, 274
*Orionina bradyi*, 288

Ornate Verticordia, 491
ORTHONECTIDA, 208
*Orygmatobothrium angustum*, 206
ORZELISCIDAE, 267
*Orzeliscus belopus*, 267
*Osachila antillensis*, 342, 11.3
*Oscillatoria borneti*, 14
*Oscillatoria miniata*, 12
*Oscillatoria submembranacea*, 14
*Oscillatoria thiebautii*, 12
Oscillatoriaceae, 12
OSCILLATORIALES, 12
Osprey, 663
Ospreys, 663
Osteichthyes, 571
OSTRACIIDAE, 646
Ostracoda, 280
*Ostrea folium*, 472
OSTREACEA, 472
OSTREIDAE, 472
*Ototyphlonemertes* sp., 211
Oval Corbula, 488
Oval Crassinella, 478
Oval mussel, 466
*Ovatella myosotis*, 458
OVULIDAE, 422
OWENIIDAE, 252
*Oxygyrus keraudrenii*, 423
OXYNOACEA, 448
*Oxynoe antillarum*, 448, 12.8
OXYNOIDAE, 448
OXYRHYNCHA, 354
OXYSTOMATA, 340
*Oxytoxum tesselatum*, 77

**P**

*Pachygrapsus gracilis*, 350
*Pachygrapsus transversus*, 350
*Pachyolpium atlanticum*, 270
*Padina sanctae-crucis*, 46
*Padina vickersiae*, 45, 2.5
PAGURIDAE, 336
PAGUROIDEA, 336
*Pagurus brevidactylus*, 336
Painted Elysia, 450
Painted false Stomatella, 406
Palaeacanthocephala, 223
*Palaemon northropi*, 320, 9.8
PALAEMONIDAE, 320
Palaeoheterodonta, 460
Palaeonemertini, 208
PALAEOTAXODONTA, 462
Pale anemone, 175
*Paleonotus elegans*, 236
PALINURA, 330
*Palinurellus gundlachi*, 332
PALINURIDAE, 330

PALINUROIDEA, 330
Pallid Janthina, 415
Palometa, 606
*Palythoa caribaea*, 191, 7.8
*Palythoa grandiflora*, 190
*Palythoa mammillosa*, 190, 191
*Palythoa variabilis*, 190, 7.3&4
Pancarida, 305
PANDARIDAE, 298
*Pandarus cranchii*, 298
PANDEIDAE, 144
*Pandion haliaetus*, 663
PANDIONIDAE, 663
PANDORACEA, 491
*Panicum virgatum*, 90
*Panopeus herbstii*, 349
*Panopeus occidentalis*, 349
Pantopoda, 275
*Panulirus argus*, 330
*Panulirus guttatus*, 330, 10.11
*Panulirus laevicauda*, 330
PAPILIONACEAE, 83
Papillose Elysia, 448
*Papyridea semisulcata*, 478
PARACALANIDAE, 290
*Paracalanus parvus*, 290
*Paracerceis caudata*, 368
*Paradiscosoma carlgreni*, 190
*Parajapyx schusteri*, 386
*Paralaophonte brevirostris*, 297
*Paramonohystera wieseri*, 217
*Paramyozonaria bermudensis*, 200
*Paranebalia longipes*, 306
*Paranesidea* sp., 286
*Paranthias furcifer*, 595
*Paranthura infundibulata*, 368
PARAONIDAE, 245
*Parasmittina munita*, 510
*Parasmittina nitida*, 512
*Parasmittina spathulata*, 512
*Paraspidosiphon klunzingeri*, 228
*Parasterope muelleri*, 284
*Parastygarctus sterreri*, 266
PARATANAIDAE, 366
*Parathelges piriformis*, 372, 10.7
*Paratissa semilutea*, 392
*Paraturbanella* sp., 215
Parazoa, 111
PARAZOANTHIDAE, 192
*Parazoanthus parasiticus*, 192, 7.9
Parenchymula, 113
*Pareurystomina bissonettei*, 218
*Parhyale hawaiensis*, 378
*Pariphinotus tuckeri*, 378
*Parmelia martinicana*, 25
PARMELIACEAE, 25
Parrot fishes, 624

*Parthenope pourtalesii*, 358
PARTHENOPIDAE, 358
PASIPHAEIDAE, 319
*Paspalum vaginatum*, 90
PATELLACEA, 404
Pauropoda, 381
Pea family, 83
Peacock flounder, 641
Peanut worms, 224
Pear marsh snail, 457
Pearl blenny, 632
Pearl oyster, 467
Pearlfish, 588
Pearlfishes, 588
Pearly Lyonsia, 491
*Pecten ziczac*, 468
PECTINACEA, 468
*Pectinaria regalis*, 254, 8.3&4
PECTINARIIDAE, 253
PECTINIDAE, 468
PEDICELLINIDAE, 503
*Pedipes insularis*, 458
*Pedipes mirabilis*, 458
*Pedipes ovalis*, 458
*Pedipes tridens*, 458
Pediveliger, 462
*Pegantha clara*, 148
*Pelagia noctiluca*, 158
Pelagosphaera, 225
PELECANIFORMES, 660
Pelecypoda, 460
Pelicans, 660
PEMPHERIDAE, 614
*Pempheris schomburgki*, 614
PENAEIDAE, 317
PENAEIDEA, 317
*Penaeus duorarum*, 317
Pencil urchins, 533
*Peneroplis proteus*, 98
PENICILLARIA, 192
*Penicillus capitatus*, 34
*Penicillus dumetosus*, 34, 1.10
*Penilia avirostris*, 280
*Pennaria tiarella*, 132
PENNATAE, 69
PENNATULACEA, 172
Pentactula, 538
Pentastomida, 111, 197, 268
Peracarida, 305
*Peracle triacantha*, 442
PERACLIDAE, 442
PERCIFORMES, 595
*Percnon gibbesi*, 352, 11.5
Pericalymma, 394
*Periclimenaeus perlatus*, 323
*Periclimenes americanus*, 322
*Periclimenes anthophilus*, 322, 9.12

PERIDINIALES, 77
*Peridinium brochii*, 77
*Perinereis anderssoni*, 240
PERISCHOECHINOIDEA, 533
PERITRICHA, 108
PERITRICHIDA, 109
*Peritromus faurei*, 109
*Perna ephippium*, 468
Peron's Atlanta, 423
*Perophora bermudensis*, 554
*Perophora formosana*, 554, 15.16, 16.8
*Perophora viridis*, 554
PEROPHORIDAE, 552
*Perotrochus quoyanus*, 402
*Petaloconchus erectus*, 412
Petrels, 659
*Petricola lapicida*, 487
PETRICOLIDAE, 487
*Petrolisthes armatus*, 338, 10.3&4
*Petrometopon cruentatum*, 597
Phaeophyceae, 41
Phaeophyta, 41
*Phaethon lepturus catesbyi*, 660
PHAETHONTIDAE, 660
PHALACROCORACIDAE, 662
*Phalacrocorax auritus*, 662
*Phaleria picipes*, 390
*Phallodrilus leukodermatus*, 260
*Phallusia nigra*, 554
Pharyngobdellida, 264
*Phascolosoma antillarum*, 228
*Phascolosoma scolops*, 228
PHASCOLOSOMATIDAE, 228
Phasmidia, 216
*Phellia americana*, 175
*Phellia rufa*, 175
*Phellia simplex*, 175
PHILINACEA, 438
*Philinoglossa* sp., 439
PHILINOGLOSSIDAE, 439
PHLEBOBRANCHIA, 552
PHLIANTIDAE, 378
PHOLADACEA, 489
PHOLADIDAE, 489
Pholadomyoida, 491
PHOLIDOSKEPIA, 394
*Phormidium corium*, 14
*Phormidium hendersonii*, 14
*Phormidium penicillatum*, 14
PHORONIDA, 516
*Phoronis psammophila*, 518
*Phorus agglutinans*, 418
PHOXICHILIDIIDAE, 277
Phreatoicidea, 366
*Phronima sedentaria*, 380
PHRONIMIDAE, 380
PHRYNOPHIURIDA, 528

Phylactolaemata, 503
*Phyllactis flosculifera*, 178
Phyllocarida, 305
*Phyllopodopsyllus hermani*, 297
Phyllosoma, 315
PHYMANTHIDAE, 179
*Phymanthus crucifer*, 179
*Physalia physalis*, 151
PHYSALIIDAE, 151
*Physcia alba*, 25
PHYSCIACEAE, 25
*Physcosoma varians*, 228
*Physeter catodon*, 670
*Physeter macrocephalus*, 670
PHYSETERIDAE, 670
PHYSONECTAE, 152
Phytoflagellates, 93
Pied-billed grebe, 659
Pilchard, 579
Pilidium, 209
Pill bugs, 366
Pilot whale, 669
Pimple limpet, 404
*Pinctada alaperdicis*, 467
*Pinctada imbricata*, 467
*Pinctada radiata*, 467
Pinfish, 612
Pink conch, 420
Pink lucky, 421
*Pinna carnea*, 467
PINNACEA, 466
PINNIDAE, 467
Pipefishes, 594
*Pisania tincta*, 434
Pisces, 566
PISCICOLIDAE, 264
*Pitar fulminatus*, 486
Pitted Murex, 431
Placentalia, 668
Placozoa, 111
*Plagiostomum girardi bermudensis*, 200
*Plagusia depressa*, 352
PLANAXIDAE, 412
*Planaxis lineatus*, 413
*Planaxis nucleus*, 413
Planctosphaeroidea, 541
*Planes minutus*, 350, 11.10
*Planktosphaera pelagica*, 544
Planorbid pteropod, 441
PLANTAE, 26
Plants, 26
Planula, 130, 141, 150, 156, 164
PLATYASTERIDA, 525
PLATYCOPIDA, 284
PLATYCTENIDA, 195
Platyhelminthes, 197
*Platynereis dumerilii*, 241

*Platypodia spectabilis*, 346, 11.11
*Plectrypops retrospinis*, 594
PLEOSPORALES, 22
Plerocercoid, 206
Plerocercus, 206
*Plesiastrea goodei*, 179
*Pleurobrachia pileus*, 195
*Pleurobrachia rhododactyla*, 195
PLEUROBRANCHIDAE, 446
PLEUROBRANCHOMORPHA, 446
*Pleurobranchus agassizii*, 447
*Pleurobranchus areolatus*, 447, 12.4
*Pleurobranchus atlanticus*, 447
*Pleurobranchus gardineri*, 447
Pleurodira, 651
Pleurogona, 554
PLEURONECTIFORMES, 641
*Pleuronema coronatum*, 108
PLEUROTOMARIACEA, 402
PLEUROTOMARIIDAE, 402
*Plexaura edwardsi*, 165
*Plexaura esperi*, 167
*Plexaura flexuosa*, 165, 5.8-10
*Plexaura homomalla*, 165
*Plexaurella dichotoma*, 169
*Plexaurella nutans*, 169, 5.5-7
PLEXAURIDAE, 165
*Plexauropsis bicolor*, 167
*Plexauropsis tricolor*, 165
Plicated Caecum, 410
*Plicatula gibbosa*, 470
*Plicatula ramosa*, 470
PLICATULIDAE, 470
Ploima, 220
Plover, 663
Plum parrot fish, 627
*Plumularia setacea*, 139
PLUMULARIIDAE, 139
*Pneumoderma atlanticum*, 444
*Pneumoderma violaceum*, 444
PNEUMODERMATIDAE, 444
POCILLOPORIDAE, 179
*Pocockiella variegata*, 45
Podicipedidae, 659
PODICIPEDIFORMES, 659
*Podilymbus podiceps*, 659
PODOCERIDAE, 378
*Podocerus* sp., 378
*Podochela riisei*, 355
PODOCOPIDA, 285
*Pododesmus decipiens*, 472
*Pododesmus rudis*, 472
*Podon polyphemoides*, 279
POECILASMATIDAE, 301
POECILIIDAE, 590
POECILOCHAETIDAE, 246
*Poecilochaetus serpens*, 246

POECILOSCLERIDA, 120
Pogonophora, 230
*Polinices duplicatus*, 424
*Polinices lacteus*, 424
*Polyandrocarpa tincta*, 556, 16.10
*Polycarpa spongiabilis*, 556
*Polycerella zoobotryon*, 453
POLYCERIDAE, 452
Polychaeta, 232
*Polycirrus pennulifera*, 254
POLYCITORIDAE, 550
POLYCLADIDA, 202
POLYCLINIDAE, 548
*Polyclinum constellatum*, 548, 15.14
*Polydora* sp., 247
Polygordiidae, 257
POLYHYMENOPHORA, 109
*Polykrikos hartmanni*, 79
POLYNOIDAE, 236
*Polyophthalmus pictus*, 250
POLYPHAGA, 388
POLYPHEMIDAE, 279
Polyplacophora, 394
*Polysiphonia denudata*, 64
*Polystyliphora* sp., 200
*Polysyncraton amethysteum*, 550
*Polythrix corymbosa*, 15
POMACANTHIDAE, 616
POMACENTRIDAE, 618
*Pomacentrus leucostictus*, 620
*Pomacentrus partitus*, 620
*Pomacentrus planifrons*, 620
*Pomacentrus variabilis*, 621
POMADASYIDAE, 609
*Pomatoceros triqueter*, 257, 8.11
Pompano, 606
Pompanos, 603
*Pontella atlantica*, 290
PONTELLIDAE, 290
*Pontinus castor*, 641
PONTOCYPRIDIDAE, 285
*Pontodrilus bermudensis*, 263
Porcelain crabs, 338
PORCELLANIDAE, 338
Porcupine fish, 650
Porcupine fishes, 650
Porgies, 610
Porifera, 111
*Porites astreoides*, 182, 6.14
*Porites polymorpha*, 182
*Porites porites*, 182
PORITIDAE, 182
POROMYACEA, 491
*Porpita porpita*, 131
Porpoises, 668
Portuguese man-of-war, 151
PORTUNIDAE, 343

*Portunus anceps*, 343
*Portunus depressifrons*, 345
*Portunus ordwayi*, 343
*Portunus sayi*, 343, 11.1
*Portunus sebae*, 345
*Portunus spinimanus*, 345
PORULOSIDA, 101
POTAMIDIDAE, 413
Pothead, 669
Prasinophyceae, 27
Prawns, 317
PRAYIDAE, 152
PRIACANTHIDAE, 602
*Priacanthus cruentatus*, 602
Priapulida, 222
Priapulimorpha, 222
Prickly winkle, 408
Princess parrot fish, 625
Princess rockfish, 600
*Prionace glauca*, 570
*Prionospio cristata*, 247
*Problognathia minima*, 212
*Probopyrus latreuticola*, 371
PROCELLARIIDAE, 659
PROCELLARIIFORMES, 659
Procercoid, 206
*Processa bermudensis*, 328
PROCESSIDAE, 328
Prochlorophyta, 9
*Prognathodes aculeatus*, 616
Prokaryotes, 7
PROLECITHOPHORA, 200
*Propontocypris* sp., 285
PROROCENTRALES, 77
*Prorocentrum gracile*, 77
PROSERIATA, 200
PROSOBRANCHIA, 402
PROSTIGMATA, 272
PROSTOMATIDA, 106
*Protellopsis stebbingii*, 381
Protista, 7
PROTODRILIDAE, 258
*Protodrilus* sp., 258
Protonymphon, 276
*Protopalythoa grandis*, 190
Protostomia, 197
Prototheria, 668
PROTOZOA, 92
*Proxenetes mackfirae*, 202
PSAMMOBIIDAE, 483
*Psammotreta intastriata*, 483
*Pseudactinia melanaster*, 176, 6.3
*Pseudaxinella explicata*, 121, 4.1
*Pseudaxinella rosacea*, 121
PSEUDOCALANIDAE, 290
*Pseudocaranx dentex*, 606
*Pseudoceratina crassa*, 115, 3.1

*Pseudoceros aurolineata,* 203
*Pseudoceros crozieri,* 203, 15.12
*Pseudoceros pardalis,* 203
*Pseudochama radians,* 476
*Pseudogramma gregoryi,* 601
*Pseudominona dactylifera,* 200
PSEUDOPHYLLIDEA, 206
*Pseudoplexaura crassa,* 167
*Pseudoplexaura flagellosa,* 167
*Pseudoplexaura porosa,* 167, 5.3&4
*Pseudoplexaura wagenaari,* 167
*Pseudopterogorgia acerosa,* 170
*Pseudopterogorgia americana,* 170
Pseudoscorpiones, 269
*Pseudosquilla ciliata,* 308
*Pseudosquilla oculata,* 309
Pseudozoea, 307
*Pseudupeneus maculatus,* 614
PSOLIDAE, 540
*Pteria colymbus,* 467
PTERIACEA, 467
PTERIIDAE, 467
PTERIOIDA, 467
PTERIOMORPHA, 463
Pterobranchia, 544
*Pterodroma cahow,* 660
*Pterogorgia citrina,* 171, 5.1&2
Pteropods, 438
*Pterosagitta draco,* 520
PTERYGOTA, 386
*Pterynotus lightbourni,* 430
*Ptychodera bahamensis,* 543
PTYCHODERIDAE, 543
Puddingwife, 622
Puerto Rican red chiton, 396
Puffers, 648
Pugnose pipefish, 594
PULMONATA, 455
Pure tellin, 482
*Puriana rugipunctata,* 288
Purple mussel, 483
Purple urchin, 533
Purple-tipped anemone, 176
Purplemouth moray, 582
*Purpura patula,* 432
Pycnogonida, 275
PYCNOPHYIDAE, 222
Pygmy filefish, 646
Pyramid Clio, 442
*Pyramidella dolobrata,* 440
PYRAMIDELLIDACEA, 440
Pyramidellidae, 440
PYRENOMYCETES, 20
*Pyrgo denticulata,* 98
PYRGOMATIDAE, 304
*Pyrocystis noctiluca,* 79
*Pyrosoma atlanticum,* 559

PYROSOMIDA, 559
Pyrosomids, 558
Pyrrhophyta, 75
*Pyura munita,* 558
*Pyura torpida,* 558, 16.13
PYURIDAE, 557

**Q**
Quassia family, 84
Queen angelfish, 618
Queen conch, 420
Queen crab, 349
Queen mantis shrimp, 310
Queen parrot fish, 627
Queen triggerfish, 642
Quoy's slit shell, 402

**R**
Radiata, 111
RADIOLARIA, 101
Rainbow parrot fish, 625
Rainbow runner, 604
RAJIFORMES, 571
Ram's horn shell, 495
*Ramalina denticulata,* 25
RAMALINACEAE, 25
RANINIDAE, 340
*Ranzania laevis,* 650
*Raphanus lanceolatus,* 83
Rataria, 130
Rays, 571
RECTANGULINA, 516
Red shore crab, 353
Red algae, 50
Red anemone, 176
Red grouper, 598
Red hind, 598
Red mangrove, 86
Red rockfish, 600
Red snapper, 608
Red spider crab, 358
Red tree sponge, 121
Red-backed cleaner shrimp, 328
Red-spotted Euchelus, 405
Redband parrot fish, 627
Redear sardine, 579
Redtail parrot fish, 627
Reef bass, 601
Reef butterfly fish, 616
Reef silverside, 592
Remarkable scraper shell, 482
Remipedia, 278
Remoras, 603
*Reniera hogarthi,* 117
Reptantia, 312
Reptilia, 651
Requiem sharks, 570

Reticulated cowrie helmet, 426
*Retronectes* sp., 200
RHABDOCOELA, 200
RHABDONIACEAE, 60
*Rhabdopleura normani*, 545
*Rhabdospora avicenniae*, 23
*Rhincodon typus*, 570
RHINCODONTIDAE, 570
*Rhinesomus triqueter*, 648
RHIZANGIIDAE, 184
*Rhizoclonium hookeri*, 30
*Rhizophora mangle*, 86
RHIZOPHORACEAE, 86
RHIZOPODEA, 95
*Rhizopsammia bermudensis*, 188, 6.6
*Rhizosolenia shrubsoleii*, 68
RHIZOSTOMAE, 158
RHODACARIDAE, 272
*Rhodactis carlgreni*, 190
*Rhodactis sanctithomae*, 188
RHODOMELACEAE, 64
RHODOPACEA, 442
*Rhodope* sp., 443
Rhodophyceae, 50
Rhodophyta, 50
RHODYMENIACEAE, 61
RHODYMENIALES, 61
RHOEADALES, 83
Rhombogen, 207
*Rhombognathus* sp., 272
*Rhopalonema velatum*, 148
RHOPALONEMATIDAE, 147
*Rhopalura* sp., 208
RHYNCHOBDELLIDA, 264
Rhynchocephalia, 651
*Rhynchocinetes rigens*, 320
RHYNCHOCINETIDAE, 320
Rhynchocoela, 208
*Rhynchozoon rostratum*, 514
Ribbed Codakia, 474
Ribbon worms, 208
Ring-billed gull, 666
Ringed anemone, 173
*Rissoa minuscula*, 409
RISSOACEA, 409
RISSOIDAE, 409
*Rissoina bermudensis*, 409
*Rissoina bryerea*, 409
*Rissoina cancellata*, 409
*Rissoina catesbyana*, 409
*Rissoina chesneli*, 409
*Rissoina laevissima*, 409
*Rissoina pulchra*, 409
*Rissoina sagraiana*, 409
*Rissoina scalarella*, 409
RISSOINIDAE, 409
*Rivularia polyotis*, 15

RIVULARIACEAE, 15
Rock beauty, 617
Rock cockle, 475
Rock hind, 597
Rock scallop, 471
Rock scuttle, 499
Rock urchin, 535
Rockfishes, 595
Rooster-tail conch, 420
Rorquals, 671
Rose coral, 187
ROTALIIDEA, 99
Rotifera, 219
Rough triggerfish, 642
Round robin, 604
Round scad, 604
Roundworms, 216
Rove beetles, 388
Ruddy turnstone, 664
Ruffled Elysia, 449
*Runcina divae*, 439
RUNCINIDAE, 439
Runner rockfish, 598
*Rupellaria typica*, 487
Rush fry, 592
Rustic rock shell, 431
*Rutiderma sterreri*, 282
RUTIDERMATIDAE, 282
*Ruvettus pretiosus*, 638
*Rypticus saponaceus*, 601

S

*Sabella melanostigma*, 255, 8.10
SABELLARIIDAE, 252
SABELLIDAE, 255
SACCOCIRRIDAE, 257
*Saccocirrus* sp., 258
Sac fungi, 20
SACOGLOSSA, 447
*Sagitta bipunctata*, 522
*Sagitta enflata*, 522
*Sagitta hispida*, 520
*Sagitta lyra*, 520
*Sagitta minima*, 520
*Sagitta serratodentata*, 520
*Saipanetta brooksi*, 285
*Salicornia perennis*, 82
Sally Lightfoot, 353
SALMONIFORMES, 586
*Salpa fusiformis*, 561
SALPIDA, 560
Salps, 558
Salt grass, 91
Sand anemone, 178
Sand crabs, 338
Sand diver, 584
Sand dollar, 536

Sand dollars, 531
Sand eels, 581
Sand flies, 391
Sand mussel, 466
Sand oyster, 467
Sand tilefish, 603
Sand tilefishes, 603
Sanderling, 664
Sandpipers, 664
*Sapphirina auronitens*, 294
SAPPHIRINIDAE, 294
Sarcodina, 93
Sarcomastigophora, 93
*Sardinella anchovia*, 580
Sardines, 579
SARGASSACEAE, 48
*Sargassum bermudense*, 48, 2.4
Sargassum crab, 350
*Sargassum filipendula*, 48
Sargassum fish, 587
*Sargassum fluitans*, 50, 2.6
*Sargassum natans*, 48
Sargassum nudibranch, 454
Sargassum pipefish, 594
*Sargassum platycarpum*, 49
Sargassum swimming crab, 343
Sargassum triggerfish, 642
*Sarsiella absens*, 283
SARSIELLIDAE, 282
*Savignyella lafontii*, 510
SAVIGNYELLIDAE, 510
Say's false limpet, 456
*Scaevola plumieri*, 87
*Scala electa*, 416
*Scalaria clathrus*, 416
Scale worms, 236
Scallop, 468
Scalloped hammerhead, 571
SCALPELLIDAE, 301
Scaphopoda, 459
*Scaptognathides planus*, 274
SCARIDAE, 624
*Scartella cristata*, 632
*Scarus coelestinus*, 627
*Scarus coeruleus*, 625
*Scarus croicensis*, 625
*Scarus guacamaia*, 625
*Scarus taeniopterus*, 625
*Scarus vetula*, 627
SCHIZASTERIDAE, 536
*Schizomavella auriculata*, 512
Schizopoda, 305
*Schizoporella errata*, 512
*Schizoporella unicornis*, 513
SCHIZOPORELLIDAE, 512
Schramm's chiton, 397
SCIAENIDAE, 612

SCLERACTINIA, 179
SCLERAXONIA, 165
*Sclerobelemnon* cf. *theseus*, 172, 5.11
SCLEROPERALIA, 212
Sclerospongea, 111
*Scolelepis squamata*, 246
SCOLOPACIDAE, 664
*Scolymia* sp., 187
SCOMBRIDAE, 636
*Scorpaena plumieri*, 640
SCORPAENIDAE, 640
Scorpion fishes, 640
Scorpions, 269
Scrawled cowfish, 647
Scrawled filefish, 646
*Scrupocellaria bertholletii*, 508
*Scrupocellaria regularis*, 508
SCRUPOCELLARIIDAE, 508
Scurvy grass, 83
SCUTICOCILIATIDA, 108
*Scyllaea pelagica*, 454, 13.5
SCYLLAEIDAE, 454
SCYLLARIDAE, 332
*Scyllarides aequinoctialis*, 332
*Scyllarides nodifer*, 333
*Scypha ciliata*, 126
*Scyphosphaera apsteinii*, 71
Scyphozoa, 155
*Scytonema myochrous*, 16
*Scytonema polycystum*, 16
SCYTONEMATACEAE, 16
*Scytosiphon lomentaria*, 46
SCYTOSIPHONACEAE, 46
SCYTOSIPHONALES, 46
Sea anemones, 172
Sea bat, 445
Sea butterflies, 438
Sea cat, 445
Sea chubs, 614
Sea cows, 668
Sea cucumbers, 537
Sea fan, 172
Sea fans, 165
Sea ginger, 134
Sea holly, 48
Sea horses, 594
Sea lettuce, 29
Sea lions, 668
Sea mats, 190, 500
Sea otter, 668
Sea pens, 172
Sea plume, 170
Sea pudding, 540
Sea spiders, 275
Sea squirts, 545
Sea stars, 523
Sea urchins, 531

Sea wasps, 157
Sea whips, 165
Seals, 668
Seashore rush grass, 91
Seaside earwig, 387
Seaside lavender, 86
Seaside morning glory, 86
Seaside oxeye, 87
Seaside purslane, 83
Seaweed fly, 392
Sedentaria, 232
Segmented worms, 232
*Seila adamsi*, 414
Seisonidea, 219
Selachii, 566
SEMAEOSTOMAE, 158
*Semele bellastriata*, 484
*Semele cancellata*, 484
*Semele orbiculata*, 484
*Semele proficua* f. *radiata*, 484
*Semele reticulata*, 484
SEMELIDAE, 483
Semi-rusted horn shell, 414
*Semperia bermudensis*, 541
Semper's larva, 164
SEPIOIDEA, 495
*Sepioteuthis sepioidea*, 496
Sergeant major, 620
*Sergestes corniculum*, 319
SERGESTIDAE, 319
*Seriola dumerili*, 606
*Seriola rivoliana*, 606
SERPULIDAE, 255
SERRANIDAE, 595
*Serranus tigrinus*, 597
SERTELLIDAE, 513
*Sertularella conica*, 137
*Sertularella speciosa*, 137
*Sertularia cornicina*, 137
*Sertularia turbinata*, 139
SERTULARIIDAE, 136
*Sesarma ricordi*, 352
SESSILIFLORAE, 172
SESSILINA, 109
*Sesuvium portulacastrum*, 83
Shad, 608
Shade coral, 180
Shark sucker, 603
Sharpnose puffer, 648
Shearwaters, 659
Sheathed paspalum, 90
Shiny Atlantic auger, 437
Shiny Ervilia, 479
Shipworms, 460, 490
Shore crabs, 349
Shore flies, 392
Shorebirds, 663

Short-horned bugs, 387
Shortfin mako, 568
Shrimps, 312
SIBOGLINIDAE, 232
*Siboglinoides caribbeanus*, 232
*Sicyonia wheeleri*, 318
SICYONIIDAE, 318
*Siderastrea radians*, 182
*Siderastrea siderea*, 182
SIDERASTREIDAE, 182
SIDIDAE, 279
SIGILLIACEA, 285
Sigillidae, 285
SILICOFLAGELLINEAE, 73
Silk snapper, 608
Silver Jenny, 608
Silversides, 592
Silvery-clawed crab, 343
SIMAROUBACEAE, 84
*Simnia acicularis*, 422
Single-celled animals, 92
*Sinum perspectivum*, 424
*Siphonaria alternata*, 456
*Siphonaria alternata intermedia*, 456
*Siphonaria alternata opalescens*, 456
*Siphonaria brunnea*, 456
*Siphonaria picta*, 456
SIPHONARIACEA, 456
SIPHONARIIDAE, 456
SIPHONOCLADACEAE, 37
SIPHONOCLADALES, 37
SIPHONODENTALIIDAE, 460
SIPHONOPHORA, 149
*Siphonosoma cumanense*, 226
SIPHONOSTOMATOIDA, 298
Sipuncula, 224
SIPUNCULIDAE, 226
*Sipunculus norvegicus*, 226
Sirenia, 668
*Siriella thompsoni*, 361
Skates, 571
Skeleton shrimps, 380
Slate pencil urchin, 533
Slater, 372
Slender filefish, 646
Slender mola, 650
Slippery Dick, 621
Slug-like Clione, 443
Slugs, 397
Small-eyed star coral, 179
Small sea hare, 446
Small sponge-carrying crab, 340
Small star coral, 182, 184
Small western marsh snail, 458
*Smaragdia viridis*, 407
Smith's little Cardita, 477
*Smittina smittiella*, 512

SMITTINIDAE, 510
*Smittipora americana*, 506
Smooth risso, 409
Smooth-tailed spiny lobster, 330
Smooth tellin, 481
Smooth trunkfish, 648
Snails, 397
Snake mackerels, 638
Snakefish, 584
Snakes, 651
Snappers, 607
Snapping shrimps, 323
Snipe, 664
Soapfish, 601
Soapfishes, 600
Soft corals, 164
Soil centipedes, 382
SOLANALES, 86
SOLECURTIDAE, 484
Solenogastres, 393
SOLEOLIFERA, 443
SOLIERIACEAE, 59
SOLMARISIDAE, 148
*Solmissus incisa*, 149
Somers' shipworm, 490
Sorberacea, 545
Southern sea rocket, 83
Southern sennet, 631
Spanish hogfish, 624
Spanish oyster, 467
Spanish sardine, 580
SPARIDAE, 610
*Sparisoma aurofrenatum*, 627
*Sparisoma crysopterum*, 627
*Sparisoma radians*, 629
*Sparisoma viride*, 629
*Spartina patens*, 91
SPATANGOIDA, 536
Speckled moray, 582
Spelaeogriphacea, 305
Spengelidae, 541
*Spengleria rostrata*, 489
Sperm whale, 670
Sperm whales, 670
Spermatophyta, 26
Sphaeriales, 20
*Sphaerocoryne bedoti*, 132, 143
SPHAEROCORYNIDAE, 132, 143
*Sphaeroides spengleri*, 648
SPHAEROZOIDAE, 101
*Sphaerozoum punctatum*, 101
*Spheciospongia othella*, 122
*Sphyraena barracuda*, 630
*Sphyraena picudilla*, 631
SPHYRAENIDAE, 630
*Sphyrna lewini*, 571
SPHYRNIDAE, 570

Spider crabs, 354
Spiders, 269
SPINULOSIDA, 526
Spiny file clam, 472
Spiny-headed worms, 223
Spiny lobster, 330
Spiny lobsters, 330
Spiny slipper shell, 418
Spiny urchin, 533
*Spio pettiboneae*, 246
*Spiochaetopterus costarum oculatus*, 247
SPIONIDAE, 246
Spiral-shaped Vitrinella, 410
*Spirastrella coccinea*, 122
*Spirastrella dioryssa*, 124
*Spirastrella mollis*, 122, 4.14
SPIRASTRELLIDAE, 122
SPIROPHORIDA, 125
*Spirorbis formosus*, 257
SPIROTRICHA, 109
*Spirula spirula*, 495
Split-thumb, 308
Split-toe, 308
Splitlure frogfish, 587
SPONDYLIDAE, 470
*Spondylus americanus*, 470
*Spondylus digitatus*, 470
*Spondylus echinatus*, 470
*Spondylus ericinus*, 470
*Spondylus fimbriatus*, 470
*Spondylus ictericus*, 470
*Spondylus longitudinalis*, 470
*Spondylus ramosus*, 470
*Spondylus spatuliferus*, 470
*Spondylus ustulatus*, 470
*Spondylus vexillum*, 470
Sponge crabs, 340
Sponges, 111
SPONGIIDAE, 114
Spoon worms, 228
*Sporobolus virginicus*, 91
SPOROCHNACEAE, 46
SPOROCHNALES, 46
*Sporochnus bolleanus*, 46
Sporozoa, 103
SPORTELLIDAE, 477
Spotfin butterfly fish, 615
Spotfin flying fish, 589
Spotted eagle ray, 571
Spotted goatfish, 614
Spotted limpet, 404
Spotted moray, 582
Spotted sandpiper, 664
Spotted scorpion fish, 640
Spotted sea hare, 445
Spotted spiny lobster, 330
Spray crab, 352

Springtails, 386
Spurge family, 84
*Spurilla neapolitana*, 455, 13.16
*Spyridia aculeata*, 62
SQUALIFORMES, 568
Squamata, 651
*Squatarola squatarola*, 664
Squids, 492, 496
SQUILLIDAE, 310
Squirrelfish, 592
Squirrelfishes, 592
Staircase Truncatella, 410
Stalkeye, 587
Stalkeyes, 586
STAPHYLINIDAE, 388
Star lobster, 330
Starfish, 523
*Staurocladia vallentini*, 143
Stauromedusae, 155
Stearns' Aesopus, 434
*Stegias clibanarii*, 372
*Steginoporella magnilabris*, 506
STEGINOPORELLIDAE, 506
STELLETTIDAE, 124
*Stenetrium stebbingi*, 370
STENOLAEMATA, 514
*Stenoniscus pleonalis*, 372
*Stenoplax boogii*, 396
*Stenoplax rugulata*, 396
Stenopodidae, 328
STENOPODIDEA, 328
*Stenopus hispidus*, 328, 9.1
*Stenorhynchus seticornis*, 355, 11.7
STENOSTOMATA, 514
*Stephanocoenia michelinii*, 179
*Stephanolepis setifer*, 646
*Stephanosella rugosa*, 513
*Sterna hirundo*, 667
STERNINAE, 667
STERNOPTYCHIDAE, 586
*Sternoptyx diaphana*, 586
*Stichopathes lutkeni*, 194
STICHOPODIDAE, 540
*Stichopus badionotus*, 540
*Stichopus moebii*, 540
Stilbonematinae, 217
STOLIDOBRANCHIA, 554
*Stolonica sabulosa*, 556, 16.7
Stolonifera, 164
STOLONIFERINA, 504
STOMATELLIDAE, 406
Stomatopoda, 306
Stony corals, 179
Stoplight parrot fish, 629
Storks, 662
Straight-hinge larva, 462
Straight-tail razorfish, 622

Striate Martesia, 489
*Strigilla flexuosa*, 482
*Strigilla mirabilis*, 482
Striped parrot fish, 625
STROMBACEA, 418
STROMBIDAE, 419
*Strombus accipitrinus*, 420
*Strombus costatus*, 420
*Strombus gallus*, 420
*Strombus gigas*, 420
*Strombus raninus*, 420
*Strongylophora amphioxa*, 118
*Styela partita*, 557
*Styela plicata*, 557, 16.9
STYELIDAE, 554
STYGARCTIDAE, 266
Stygocaridacea, 305
*Styliola subula*, 442
*Stylocheilus longicauda*, 446, 12.3
*Stylocheiron carinatum*, 312
*Stylopoma spongites*, 513
*Stypopodium zonale*, 45, 2.1
SUBERITIDAE, 124
SUCTORIA, 106
Sunfishes, 650
Sunrise tellin, 481
Sunset clam, 481
Surf birds, 663
Surgeonfishes, 635
*Suriana maritima*, 84
Sweepers, 614
Swift's dwarf triton, 429
Swimming crabs, 343
Switch grass, 90
SYCETTIDA, 126
SYCETTIDAE, 126
*Sycon ciliatum*, 126, 4.7
SYLLIDAE, 238
*Syllis gracilis*, 240
*Symbiodinium microadriaticum*, 79
*Symethis variolosa*, 340
Symphyla, 381
*Symplegma rubra*, 556
*Symplegma viride*, 556, 16.11&12
*Synalpheus brevicarpus*, 324
*Synalpheus townsendi*, 325
SYNAPTIDAE, 541
*Synaptocochlea picta*, 406
*Synaptula hydriformis*, 541, 14.8
SYNAXIDAE, 332
Syncarida, 305
*Synchaeta* sp., 220
SYNGNATHIDAE, 594
*Syngnathus dunckeri*, 594
*Syngnathus pelagicus*, 594
*Synnotum aegyptiacum*, 509
SYNODONTIDAE, 584

*Synodus intermedius,* 584
*Synopia ultramarina,* 378
SYNOPIIDAE, 378
SYRACOSPHAERACEAE, 71
*Syringodium filiforme,* 88, 9.9

**T**
TABANIDAE, 391
*Tabanus atlanticus,* 392
*Tabanus nigrovittatus,* 392
Tadpole larva, 547
Tadpole shrimps, 278
*Tagelus divisus,* 484
TALITRIDAE, 380
TALITROIDEA, 378
TANAIDACEA, 364
TANAIDAE, 366
*Tanais cavolinii,* 366
*Tanais dulongii,* 366
*Tanais tomentosus,* 366
Tangerine sponge, 124
Tapeworms, 205
Tapioca fish, 638
Tardigrada, 265
Tarpon, 577
*Tarpon atlanticus,* 577
Tarpons, 577
Tassel plant, 84
*Tectarius muricatus,* 408
*Tedania ignis,* 120, 3.11, 11.6
TEDANIIDAE, 120
Telestacea, 164
*Tellina decora,* 482
*Tellina flagellum,* 482
*Tellina gouldii,* 482
*Tellina gruneri,* 483
*Tellina interrupta,* 481
*Tellina laevigata,* 481
*Tellina laevigata* var. *stella,* 481
*Tellina listeri,* 481
*Tellina magna,* 481
*Tellina mera,* 482
*Tellina promera,* 482
*Tellina radiata,* 481
*Tellina radiata unimaculata,* 481
*Tellina similis,* 482
*Tellina sybaritica,* 482
TELLINACEA, 480
TELLINIDAE, 481
*Telmatactis clavata,* 175
*Telmatactis cricoides,* 175
*Telmatactis solidago,* 175
*Telmatactis vernonia,* 175, 6.11
Temnocephalida, 198
TEMNOPLEUROIDA, 533
Ten-rayed star coral, 180
*Tenarea bermudense,* 58

TENEBRIONIDAE, 390
Tentaculata, 500
TENTACULIFERA, 195
TEREBELLIDAE, 254
*Terebellides stroemi,* 255
*Terebra hastata,* 437
TEREBRATULIDA, 519
TEREBRIDAE, 437
TEREDINIDAE, 490
*Teredo bartschi,* 490
*Teredo batilliformis,* 490
*Teredo clappi,* 490
*Teredo somersi,* 490
*Teredo thomsoni,* 491
*Teredora malleolus,* 491
Terns, 663, 666, 667
*Terpios aurantiaca,* 124, 4.13
Tessellated nerite, 406
*Tesseropora atlantica,* 302
Testicardines, 518
TESTUDINES, 651
*Tethya actinia,* 124, 4.3&4
TETHYIDAE, 124
TETILLIDAE, 126
TETRACLITIDAE, 302
Tetradon Cadulus, 460
TETRAGONICIPITIDAE, 297
*Tetranchyroderma* sp., 215
TETRAODONTIDAE, 648
TETRAODONTIFORMES, 641
TETRAPHYLLIDEA, 206
*Tetraplaria dichotoma,* 510
TETRAPLARIIDAE, 510
Tetrapoda, 566
*Tetrapturus albidus,* 639
TEUTHOIDEA, 496
THAIDIDAE, 431
*Thais deltoidea,* 431
*Thais rustica,* 431
*Thala floridana,* 436
*Thala foveata,* 436
*Thalassema baronii,* 230
*Thalassia testudinum,* 88
THALASSICOLIDAE, 102
THALASSINOIDEA, 333
*Thalassocypria* sp., 286
*Thalassolampe maxima,* 102
*Thalassoma bifasciatum,* 622
*Thalassonema nitzschoides,* 69
*Thalia democratica,* 561
Thaliacea, 558
THAUMASTODERMATIDAE, 215
Thecanephria, 230
THECATA, 135, 144
THECOSOMATA, 441
Thermosbaenacea, 305
Thomas Tower Triphora, 415

*Thor amboinensis,* 327, 9.11
THORACICA, 301
Thorny scallop, 470
Three-spined cavoline, 442
Three-spined Peracle, 442
Threespot damselfish, 620
*Thunnus albacares,* 638
*Thunnus atlanticus,* 638
*Thyone surinamensis,* 541
*Thyroscyphus intermedius,* 137
*Thyroscyphus marginatus,* 137
*Thysanopoda aequalis,* 311
*Thysanozoon nigrum,* 203
*Thysanoëssa gregaria,* 311
Ticks, 270
Tiger beetles, 388
Tiger flatworm, 203
Tiger grouper, 600
Tiger Lucina, 473
Tinted Cantharus, 434
Tintinnina, 104
*Tintinnopsis campanula,* 110
*Tisbe bermudensis,* 296
TISBIDAE, 295
Toenail shell, 479
TOMOPTERIDAE, 238
*Tomopteris helgolandica,* 238
Tomtate, 609
*Tonicia schrammi,* 397
*Tonna maculosa,* 427
TONNACEA, 426
Toothed whales, 669
Tornaria, 543
*Torrea candida,* 238
*Tozeuma carolinense,* 327, 9.9
*Tracheloraphis incaudatus,* 105
*Trachinotus goodei,* 606
TRACHYMEDUSAE, 147
*Trachypenaeus constrictus,* 318
*Transennella conradina,* 486
Transparent file clam, 472
TRAPEZIIDAE, 484
Tree coral, 186
Trematoda, 203
*Tretomphalus atlanticus,* 99
*Triangulocypris laeva,* 286
TRICHOBRANCHIDAE, 254
*Trichocorixa reticulata,* 387
*Trichodesmium thiebautii,* 12
TRICHOSIDA, 95
*Trichosphaerium micrum,* 96
Tricladida, 198
*Tricoma hopperi,* 219
*Trididemnum savignyi,* 548, 15.2
Triggerfishes, 641
*Triloculina carinata,* 98
TRIMUSCULIDAE, 456

*Trimusculus goesi,* 456
*Triphora bermudensis,* 415
*Triphora mirabile,* 415
*Triphora turristhomae,* 415
TRIPHORIDAE, 415
*Tripneustes esculentus,* 535
*Tripneustes ventricosus,* 535
TRITONIIDAE, 454
*Tritoniopsis frydis,* 454, 13.8
*Trivia quadripunctata,* 421
TRIVIACEA, 421
TRIVIIDAE, 421
TROCHACEA, 404
*Trochammina inflata,* 98
TROCHIDAE, 405
Trochiform pteropod, 441
Trochophore, 225, 229, 234, 396, 401, 462, 502
*Trona cervus peilei,* 422
Tropic birds, 660
True crabs, 339
True fungi, 17
True lobsters, 329
True sharks, 568
True spiders, 269
Trumpet fish, 595
Trumpet fishes, 595
Trumpet triton, 428
Truncated boring clam, 489
*Truncatella caribaeensis,* 410
*Truncatella clathrus,* 410
*Truncatella piratica,* 410
*Truncatella pulchella,* 410
*Truncatella pulchella* f. *bilabiata,* 410
*Truncatella scalaris,* 410
*Truncatella subcylindrica,* 410
*Truncatella succinea,* 410
TRUNCATELLIDAE, 410
Trunkfishes, 646
Tube anemones, 192
TUBIFICIDAE, 260
TUBIFICINA, 260
*Tubiluchus corallicola,* 223
TUBULARIIDAE, 130, 142
TUBULIPORATA, 514
TUBULIPORINA, 515
Tufted sea slug, 454
Tulip mussel, 466
Tunas, 636
Tunicata, 545
TURBANELLIDAE, 215
Turbellaria, 198
*Turbinaria turbinata,* 50
TURBINIDAE, 406
*Turbo pica,* 405
*Turbonilla pupoides,* 440
Turbot, 642
Turkey wing, 463

Turnstones, 663
TURRIDAE, 437
TURRITELLIDAE, 411
*Turritopsis nutricula*, 134, 144
Turtle grass, 88
Turtles, 651
Tusk shells, 459
Two-plaited miter-like turrid, 437
Two-pronged bristletails, 385
Two-spotted false Donax, 483
Two-starred Doris, 451
TYLODINIDAE, 447
*Tylos niveus*, 372
*Tylosaurus acus*, 590
TYPHLOPLANOIDA, 202
Typical rock borer, 487
*Typosyllis hyalina*, 240

**U**

*Udotea cyathiformis*, 32
*Udotea flabellum*, 34
UDOTEACEAE, 32
*Ulosa bermuda*, 121
*Ulosa ruetzleri*, 121, 4.12
*Ulva fasciata*, 29
*Ulva lactuca*, 29, 1.11
ULVACEAE, 29
ULVALES, 29
*Umbraculum bermudense*, 447
*Umbraculum plicatulum*, 447
*Umbraculum umbraculum*, 447
Undertaker, 446
*Unela* sp., 440
UNGULINIDAE, 475
Uniramia, 381
Upside-down jellyfish, 158
Urochordata, 545
*Urodasys nodostylis*, 214
*Uronychia transfuga*, 110
UROPODIDAE, 272
UROPODINA, 272
*Urostrongylum caudatum*, 110
Useful Semele, 484
*Utriculastra canaliculata*, 438

**V**

*Vallicula multiformis*, 195
*Valonia macrophysa*, 38, 1.3
*Valonia ventricosa*, 38
VALONIACEAE, 38
VALVATIDA, 525
VALVIFERA, 370
Vampire squid, 498
Vampire squids, 498
VAMPYROMORPHA, 498
*Vampyroteuthis infernalis*, 498
*Varicorbula operculata*, 488

Varicose Alaba, 414
Variegated nerite, 406
*Vaucheria bermudensis*, 74
VAUCHERIACEAE, 74
*Velella velella*, 131
VELELLIDAE, 130
Veliger, 401, 462
VENERACEA, 485
VENERIDAE, 485
VENEROIDA, 473
VERBENACEAE, 87
VERMETIDAE, 412
*Vermetus lumbricalis*, 412
*Vermicularia knorri*, 412
*Vermicularia spirata*, 412
Vermiform, 207
*Vermiliopsis bermudensis*, 257
*Verongia fistularis*, 115
Verrucomorpha, 301
Vertebrata, 565
*Verticordia ornata*, 491
VERTICORDIIDAE, 491
Vervain family, 87
VESICULARIIDAE, 504
Vestibulifera, 104
VEXILLIDAE, 436
*Vexillifera pagei*, 95
Viper moray, 582
Virginia nerite, 407
Viruses, 7
*Vitrinella helicoidea*, 410
VITRINELLIDAE, 410
VOLUTACEA, 435
*Volvaria avena* var. *southwicki*, 436
*Volvatella bermudae*, 448, 12.5
VOLVATELLIDAE, 448
*Vorticella patellina*, 109

**W**

Wahoo, 636
Walruses, 668
Warty crab, 348
Warty pleurobranch, 447
Warty sea cat, 446
Water bears, 265
Water boatmen, 387
Water striders, 387
*Watersipora cucullata*, 510
*Weltneria hessleri*, 305
West Indian Alvania, 409
West Indian Cymbula, 422
West Indian top shell, 405
West Indian worm shell, 412
Whale shark, 570
Whale sharks, 570
Whales, 668
Wharf lice, 366

Wharf louse, 372
Wheel animalcules, 219
Whip ray, 571
White basket shell, 434
White crab, 343
White grunt, 609
White marlin, 639
White miniature ark, 463
White smooth-edged jewel box, 476
White urchin, 535
White-lined Marginella, 436
White-rimmed Elysia, 450
White-speckled anemone, 175
White-spotted dove shell, 433
White-spotted filefish, 644
White-spotted octopus, 499
White-tailed tropic bird, 660
Whitewater snapper, 608
Wide-mouthed Purpura, 432
Wilk, 405
Winged insects, 386
Wingless insects, 385
Woodcock, 664
Woody glasswort, 82
Worm mollusks, 393
*Wrangelia penicillata*, 62
Wrasses, 621
Wrinkled chiton, 396
*Wurdemannia miniata*, 55
WURDEMANNIACEAE. 55

**X**

*Xanthichthys ringens*, 642
XANTHIDAE, 346
*Xanthodius denticulatus*, 348
*Xanthodius parvulus*, 348
Xanthophyceae, 73
*Xenobalanus globicipitis*, 302
*Xenophora conchyliophora*, 418
*Xenophora trochiformis*, 418

XENOPHORIDAE, 418
XESTOLEBERIDIDAE, 288
*Xestoleberis* sp., 288
*Xytopsues osburnensis*, 120

**Y**

Yellow box crab, 342
Yellow damselfish, 620
Yellow Elysia, 450
Yellow goatfish, 614
Yellow grunt, 609
Yellow-crowned night heron, 662
Yellow-green algae, 73
Yellowfin grouper, 600
Yellowfin tuna, 638
Yellowhead wrasse, 621
Yellowmouth grouper, 600
Yellowtail damselfish, 621
Yellowtail snapper, 608

**Z**

*Zanclea costata*, 132, 143
ZANCLEIDAE, 132, 143
*Zebina browniana*, 409
Zebra Doris, 452
Zigzag periwinkle, 407
Zigzag scallop, 468
ZIPHIIDAE, 671
*Ziphius cavirostris*, 671
ZOANTHARIA, 172
Zoanthella, 164
ZOANTHIDAE, 190
ZOANTHIDEA, 190
Zoanthina, 164
*Zoanthus proteus*, 191
*Zoanthus sociatus*, 191, 7.1
Zoea, 315
*Zoobotryon verticillatum*, 504
Zooflagellates, 93
Zygomycotina, 17

THE LIBRARY
ST. MARY'S COLLEGE OF MARYLAND
ST. MARY'S CITY, MARYLAND  20686

QH 110 .M37 1986

Marine fauna and flora of
   Bermuda

| JY 7 86 | | | |
|---|---|---|---|
| FE 9 87 | | | |
| MR 16 87 | | | |
| AP 13 87 | | | |
| | | | |
| | | | |
| | | | |
| | | | |
| | | | |
| | | | |

LIBRARY OF ST. MARY'S COLLEGE OF MARYLAND
ST. MARY'S CITY, MD  20686

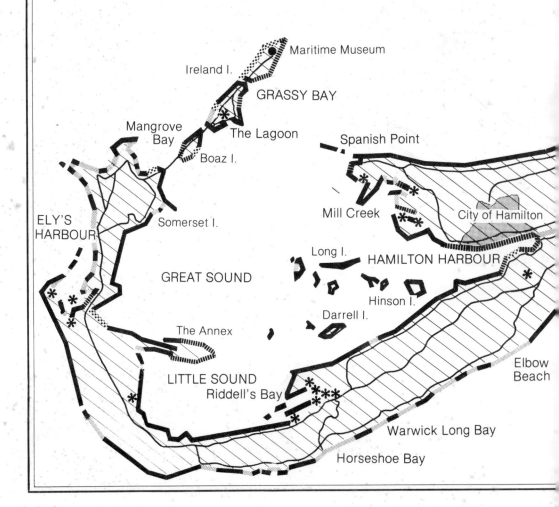